Modular Protein Domains

Edited by
G. Cesareni, M. Gimona,
M. Sudol, M. Yaffe

Further Titles of Interest

J. Buchner, T. Kiefhaber (Eds.)

Protein Folding Handbook

2005

ISBN 3-527-30784-2

R. J. Mayer, A. J. Ciechanover, M. Rechsteiner (Eds.)

Protein Degradation, Vol. 1

2005

ISBN 3-527-30837-7

K. H. Nierhaus, D. N. Wilson (Eds.)

Protein Biosynthesis and Ribosome Structure

2004

ISBN 3-527-30638-2

E. Keinan (Ed.)

Catalytic Antibodies

2004

ISBN 3-527-30688-9

S. Brakmann, A. Schwienhorst (Eds.)

Evolutionary Methods in Biotechnology

2004

ISBN 3-527-30799-0

Modular Protein Domains

Edited by
Giovanni Cesareni, Mario Gimona,
Marius Sudol, Michael Yaffe

WILEY-VCH Verlag GmbH & Co. KGaA

Editors:

Prof. Dr. Giovanni Cesareni
Department of Biology
Tor Vergata University of Rome
Via della Ricerca Scientifica
00133 Rome
Italy
cesareni@uniroma2.it

Univ.-Doz. Dr. Mario Gimona
Consorzio Mario Negri Sud
Department of Cell Biology and Oncology
Via Nazionale 8a
66030 Santa Maria Imbaro
Italy
gimona@dcbo.negrisud.it

Prof. Dr. Marius Sudol
Weis Center for Research
Geisinger Clinic
100 North Academy Avenue
Danville, PA 17822-2608
USA
msudol1@geisinger.edu

Prof. Dr. Michael Yaffe
Center for Cancer Research
Department of Biology
Massachusetts Institute of Technology
77 Massachusetts Avenue
Cambridge, MA 02139
USA
myaffe@mit.edu

Library of Congress Card No.: Applied for
British Library Cataloging-in-Publication Data: A catalogue record for this book is available from the British Library.

Bibliographic information published by Die Deutsche Bibliothek
Die Deutsche Bibliothek lists this publication in the Deutsche Nationalbibliografie; detailed bibliographic data is available in the internet at http://dnb.ddb.de.

Printed in the Federal Republic of Germany
Printed on acid-free paper

Cover Design Nicholas Navin, Cold Spring Harbor
Typesetting Manuela Treindl, Laaber
Printing betz-druck GmbH, Darmstadt
Bookbinding Litges & Dopf Buchbinderei GmbH, Heppenheim

ISBN 3-527-30813-X

Preface

Since the pioneering discovery of Watson and Crick, the scientific community has come to accept the general principle that biological information is stored in an encrypted code within the base sequence of DNA. As in other forms of classical cryptography, gene transcription and protein translation are therefore subject to rules of linguistics, since they convert information from one linear sequence (nucleic acids) into another (amino acids). In a similar, albeit more complex way, proteins can also be described by a set of linguistic 'rules' governing semantics, grammar, and syntax. However, the cryptographic complexity of proteins encodes not only sequence, but somehow explicitly specifies folding, structure, and biological function as well. How, then, can one learn to read this 'language of proteins'?

One of the most powerful approaches to 'cracking the protein code' has involved sequence comparisons between and within species, a task now greatly simplified by the ever-expanding genome sequencing efforts. Early efforts in this area led to a realization that proteins can be grouped into families and superfamilies, in which individual family members perform related, though not identical, tasks. The second remarkable insight was that unrelated proteins often share significant portions of sequence similarity and that these regions of similar sequence function as independently folded modules or domains capable of performing a specific biochemical reaction or event. Thus, a large number of functionally diverse proteins can be thought of as molecules built by combining a limited number of structurally-stable folded domains. Within these multidomain proteins, each domain is capable of autonomous function but also interacts with the other domains to allow for higher levels of regulation and control.

In this light, the human proteome becomes the equivalent of a dictionary, with the individual proteins corresponding to single entries and the modular domains akin to the words that constitute each entry. The task that lies before us, learning how to read the dictionary and decipher the rules of 'protein linguistics', then becomes a bit simpler – we can start by learning to read the words – that is, recognizing the domains and understanding their functions. Once we accomplish this, we can extend our understanding beyond simple word (domain) recognition, to understanding the biological meaning conveyed in each sentence. This book is an attempt to codify the 'words' of the signaling language spoken by proteins in living organisms: the modular protein-interaction domains, nodules, and Lego blocks described in the Prologue by Sir Tom Blundell or the sockets as Harel Weinstein defines them in his inspiring Epilogue. The compilation of articles in

Modular Protein Domains. Edited by G. Cesareni, M. Gimona, M. Sudol, M. Yaffe

ISBN: 3-527-30813-X

this book provides only a glimpse of the current knowledge of modular protein domains – as evidenced by the fact that we can recognize some 'protein sentences' as conveying large amounts of information while others seem to have no identifiable words at all. It is our sincere hope that this book will stimulate others to extend our understanding of protein modules by identifying and characterizing new domains and by discovering the grammatical rules by which domains interconnect to convey a broader biological meaning by the proteins that contain them.

Leonardo da Vinci told us: "Realize that everything connects with everything else". Indeed it does, in any integrated and efficient system that responds to the environment and has an intrinsic strategy to be flexible, grow, and adapt. Whether the system is biological, human-engineered, or a combination of both (see the Epilogue), it usually has a complex 'wiring' that interconnects its multiple primary elements to make the system an integrated whole. Modular protein-interaction domains are these primary integrating elements.

The representative repertoire of protein- and lipid-interacting domains are described here by experts in the field, including bench scientists, who originally defined them or significantly contributed to their characterization, and bioinformaticians, structural biologists, and protein chemists, who were instrumental in the rapid progress of the protein domain field. The Prologue and Epilogue in the book reflect this 'perfect marriage'.

As editors we thank all the contributors very much. When we approached the potential authors with invitations, several of them remarked that they do not write chapters for books, they publish original research reports only, but in the same breath they said that they would gladly do it. Although we still do not know what exactly changed their minds, we appreciate their decisions and efforts. The exciting 'intellectual covers' of the book, the Prologue and Epilogue, were written after most of the chapters were completed, when we sent Sir Tom Blundell and Harel Weinstein a large package of partially edited manuscripts (that at least temporarily overflowed their e-mail inboxes) to comment upon. We thank them for their insightful and clearly inspiring comments. We see their thoughts as trend-setting vectors of current and future research on modular protein-interaction domains. The editorial process reminded us that in constructing a book about our field, as in primary laboratory research, we are never experts, just permanent students. Thank you, we learned a lot. Finally, we owe a large debt of gratitude to Frank Weinreich and the entire Wiley-VCH team for invaluable assistance.

To our readers: Enjoy!

August 2004

Gianni Cesareni
Mario Gimona
Marius Sudol
Michael Yaffe

Contents

Modular Protein Domains. Edited by G. Cesareni, M. Gimona, M. Sudol, M. Yaffe

ISBN: 3-527-30813-X

List of Contributors

Rein Aasland
Department of Molecular Biology
University of Bergen
Thormolensgt. 55
5020 Bergen
Norway
aasland@mbi.uib.no

Linda J. Ball
Structural Genomics Consortium
University of Oxford
Botnar Research Centre
Oxford, OX3 7LD
United Kingdom
linda.ball@sgc.ox.ac.uk

Krzysztof Bialek
Department of Chemical Biology
GBF–German Research Centre
for Biotechnology
Mascheroder Weg 1
38124 Braunschweig
Germany

Sir Tom L. Blundell
Department of Biochemistry
University of Cambridge
80 Tennis Court Road
Cambridge, CB2 1GA
United Kingdom
tom@cryst.bioc.cam.ac.uk

Peer Bork
EMBL-Heidelberg
Biocomputing Unit
Meyerhofstrasse 1
69117 Heidelberg
Germany

Giovanni Cesareni
Department of Biology
Tor Vergata University of Rome
Via della Ricerca Scientifica
00133 Rome
Italy
cesareni@uniroma2.it

Matthew L. Cheever
Molecular Biology Program
University of Colorado Health
Sciences Center
4200 East Nith Avenue
Denver, CO 80262
USA

Stefano Confalonieri
IFOM–Istituto FIRC di Oncologia
Moleculare
Via Adamello 16
20139 Milan
Italy

Modular Protein Domains. Edited by G. Cesareni, M. Gimona, M. Sudol, M. Yaffe

ISBN: 3-527-30813-X

Arvin C. Dar
Samuel Lunenfeld Research Institute
Mount Sinai Hospital and
Department of Molecular and
Medical Genetics
University of Toronto
600 University Avenue
Toronto, Ontario M5G 1X5
Canada

Pier Paolo Di Fiore
IFOM–Istituto FIRC di Oncologia
Moleculare
Via Adamello 16
20139 Milan
Italy
difiore@ifom-firc.it

Daniel Durocher
Samuel Lunenfeld Research Institute
Mount Sinai Hospital
600 University Avenue
Toronto, Ontario M5G 1X5
Canada
durocher@mshri.on.ca

Joel C. Eissenberg
Department of Biochemistry and
Molecular Biology
St. Louis University School of Medicine
1402 South Grand Blvd.
St. Louis, MO 63104
USA
eissenjc@slu.edu

Andrew E. H. Elia
Center for Cancer Research
Massachusetts Institute of Technology
77 Massachusetts Avenue
Cambridge, MA 02139
USA

Ronald Frank
Department of Chemical Biology
GBF–German Research Centre
for Biotechnology
Mascheroder Weg 1
38124 Braunschweig
Germany
frank@gbf.de

Christian Freund
Forschungsinstitut für Molekulare
Pharmakologie
Campus Berlin-Buch
Robert-Rössle-Strasse 10
13125 Berlin
Germany
freund@fmp-berlin.de

Toby J. Gibson
EMBL-Heidelberg
Biocomputing Unit
Meyerhofstrasse 1
69117 Heidelberg
Germany

Mario Gimona
Consorzio Mario Negri Sud
Department of Cell Biology and
Oncology
Via Nazionale 8a
66030 Santa Maria Imbaro
Italy
gimona@dcbo.negrisud.it

Gerald D. Gish
Samuel Lunenfeld Research Institute
Mount Sinai Hospital
600 University Avenue
Toronto, ON M6P 3C6
Canada

James H. Hurley
Laboratory of Molecular Biology
National Institute of Diabetes and
Digestive and Kidney Diseases
National Institutes of Health, DHHS
Bethesda, MD 20892
USA
jh8e@nih.gov

Thomas Jarchau
Institut für Klinische Biochemie
und Pathobiochemie
Universität Würzburg
Versbacher Strasse 5
97078 Würzburg
Germany

Brian K. Kay
Argonne National Laboratory
Biosciences Division
9700 South Cass Avenue
Argonne, IL 60439
USA
bkay@anl.gov

David Keleti
Department of Biochemistry and
Biophysics
University of Pennsylvania School
of Medicine
422 Curie Boulevard
Philadelphia, PA 19104
USA

Sepideh Khorasanizadeh
Department of Biochemistry and
Molecular Genetics
University of Virginia Health System
1300 Jefferson Park Avenue
Charlottesville, VA 22908
USA

Laurence A. Lasky
Latterell Venture Partners
Four Embarcadero Center
Suite 2500
San Francisco, CA 94111
USA
larry@lvpcapital.com

Mark A. Lemmon
Department of Biochemistry and
Biophysics
University of Pennsylvania School
of Medicine
422 Curie Boulevard
Philadelphia, PA 19104
USA
mlemmon@mail.med.upenn.edu

Veli-Pekka Lehto
Department of Pathology
Haartman Institute
University of Helsinki
Haartmaninkatu 3
00290 Helsinki
Finland
veli-pekka.lehto@helsinki.fi

Ivica Letunic
EMBL-Heidelberg
Biocomputing Unit
Meyerhofstrasse 1
69117 Heidelberg
Germany
letunic@embl.de

Rune Linding
EMBL-Heidelberg
Biocomputing Unit
Meyerhofstrasse 1
69117 Heidelberg
Germany
linding@embl.de

Olli Lohi
Department of Pathology
Haartman Institute
University of Helsinki
Haartmaninkatu 3
00290 Helsinki
Finland

Ben Margolis
HHMI, Departments of Internal
Medicine and Biological Chemistry
University of Michigan Medical School
1150 W. Medical Center Drive
Ann Arbor, MI 48109
USA
bmargoli@umich.edu

Bruce J. Mayer
Department of Genetics and
Developmental Biology
University of Connecticut Health
Center
263 Farmington Avenue
Farmington, CT 06030-3301
USA
bmayer@neuron.uchc.edu

Piers Nash
The Ben May Institute for Cancer
Research
The University of Chicago
5481 South Maryland Avenue
Chicago, IL 60637
USA

Michael Overduin
University of Birmingham
School of Medicine
Institute for Cancer Studies
Birmingham B13 9SG
United Kingdom
m.overduin@bham.ac.uk

Vimal Parkash
Department of Pathology
Haartman Institute
University of Helsinki
Haartmaninkatu 3
00290 Helsinki
Finland

Tony Pawson
Samuel Lunenfeld Research Institute
Mount Sinai Hospital
600 University Avenue
Toronto, ON M&P 3C6
Canada
pawson@mshri.on.ca

Pål Puntervoll
Department of Molecular Biology
University of Bergen
Thormolensgt. 55
5020 Bergen
Norway

Kalle Saksela
Institute of Medical Technology
University of Tampere
Tampere University Hospital
33104 Tampere
Finland

Michael D. Scholle
Argonne National Laboratory
Biosciences Division
9700 South Cass Avenue
Argonne, IL 60439
USA

Frank Sicheri
Samuel Lunenfeld Research Institute
Mount Sinai Hospital and
Department of Molecular and Medical Genetics
University of Toronto
600 University Avenue
Toronto, Ontario M5G 1X5
Canada
sicheri@mshri.on.ca

Sachdev S. Sidhu
Department of Protein Engineering
Genentech, Inc.
1 DNA Way
South San Francisco, CA 94080
USA

Nicholas J. Skelton
Department of Protein Engineering
Genentech, Inc.
1 DNA Way
South San Francisco, CA 94080
USA

Fred J. Stevens
Argonne National Laboratory
Biosciences Division
9700 South Cass Avenue
Argonne, IL 60439
USA

Marius Sudol
Weis Center for Research
Geisinger Clinic
100 North Academy Avenue
Danville, PA 17822-2608
USA
msudol1@geisinger.edu

Andrzej Swistowski
Department of Chemical Biology
GBF–German Research Centre for Biotechnology
Mascheroder Weg 1
38124 Braunschweig
Germany

Linton M. Traub
Department of Cell Biology and Physiology
University of Pittsburgh School of Medicine
3500 Terrace Street
Pittsburgh, PA 15261
USA

Raymond C. Trievel
Department of Biological Chemistry
University of Michigan Medical School
1150 W. Medical Center Drive
Ann Arbor, MI 48109
USA

Ismo Virtanen
Department of Pathology
Haartman Institute
University of Helsinki
Haartmaninkatu 3
00290 Helsinki
Finland

Harel Weinstein
Department of Physiology and Biophysics
Institute of Computational Biomedicine
Cornell University Medical College
1300 York Avenue
New York, NY 10021
USA
haw2002@med.cornell.edu

Urs Wiedemann
Forschungsinstitut für Molekulare Pharmakologie
Campus Berlin-Buch
Robert-Rössle-Strasse 10
13125 Berlin
Germany

Steven J. Winder
University of Sheffield
Department of Biomedical Science
Western Bank
Sheffield SI0 2TN
United Kingdom

Leanne E. Wybenga-Groot
Samuel Lunenfeld Research Institute
Mount Sinai Hospital and
Department of Molecular and Medical Genetics
University of Toronto
600 University Ave.
Toronto, Ontario M5G 1X5
Canada

Michael B. Yaffe
Center for Cancer Research
Massachusetts Institute of Technology
77 Massachusetts Avenue
Cambridge, MA 02139
USA
myaffe@mit.edu

Kelley S. Yan
Department of Physiology and Biophysics
Mount Sinai School of Medicine
One Gustave Levy Place
New York, NY 10029
USA

Ming-Ming Zhou
Department of Physiology and Biophysics
Mount Sinai School of Medicine
One Gustave Levy Place
New York, NY 10029
USA
ming-ming.zhou@mssm.edu

Jürgen Zimmermann
Forschungsinstitut für Molekulare Pharmakologie
Campus Berlin-Buch
Robert-Rössle-Strasse 10
13125 Berlin
Germany

Prologue:
An Overview of Protein Modular Domains as Adaptors

Sir Tom L. Blundell

Those of us who were beginning our research careers in the 1960s were amazed by the complexity of the protein sequences and structures that were emerging. How would we be able to grasp the meaning of it all and indeed find some patterns among the apparent disorder? Two ideas gave us some hope. The first developed from the confirmation that proteins were indeed made up of helices and strands as predicted earlier by Pauling. Could there also be higher orders of organization in protein structure that would provide simplifying features and even a method of description and classification? The second idea built on the observation that many proteins appeared to have evolved divergently, not only those with similar sequences and functions, but also some with quite different functions but similar structures – I remember lactalbumin and lysozyme were all the talk at the time! Was it possible that Nature had exploited just a few folds and used them again and again for slightly different functions? Forty years later we find that both ideas have been useful. Nowhere are they better illustrated than in the modular adaptor domains found in many eukaryotic proteins involved in cell regulation, signaling, and the cytoskeleton. This fascinating book celebrates the coming of age of such domains and allows us to assess their roles in living organisms.

Protein domains represent the minimum levels of organization that provide a stable globular core; modular domains are usually constructed from several motifs or supersecondary structures, each composed of helices or strands – $\alpha\alpha$, $\beta\beta$, $\beta\alpha\beta$, and so on – and individually usually unstable, packed together to bury mainly hydrophobic residues. Thus, modular domains tend to be stable and fold on their own. Indeed, Ball et al. (Chapter 4) suggest that their stability and ability to fold is why they are often found at the N-termini of proteins. In SPRED, a stable EVH1 domain is followed by a low-complexity region; presumably, once the stable module has been synthesized on the ribosome and correctly folded, it protects the following sequence from proteolysis until it is recognized by other protein units. The stability of such domains has allowed molecular and structural biologists to 'divide and conquer': they have been able to cleave them out of larger proteins or express them as independent units. This reductionist approach has allowed their structures to be determined by NMR and X-ray crystallography, as exemplified in almost every

Modular Protein Domains. Edited by G. Cesareni, M. Gimona, M. Sudol, M. Yaffe

ISBN: 3-527-30813-X

chapter of this book. It has also allowed their specificities to be defined by screening against peptides and their partners to be identified by using the yeast two-hybrid system, phage expression libraries, or protein arrays, as described by Bialek, Swistowski, and Frank in Chapter 20. Such studies have shown that they are not only folding modules, but also functional modules.

But they are also evolutionary modules. When the first module – the SH2 in the protein kinase Src – was recognized, no one was prepared for the discovery that it occurs so many times in the human genome, mainly but not entirely involved in binding phosphotyrosyl peptides. Or that Src would contain another widely distributed domain, SH3, which binds regions of polypeptide with a high propensity for adopting polyproline helices. The 'molecular Lego' that Nature has exploited by splicing modular domains into a multitude of proteins is a hugely efficient way of developing a complex network, as Sudol notes in his thoughtful chapter.

The use of modular domains enables cellular activity to be controlled in many fascinating ways. Perhaps most importantly, it is used to colocate enzyme and substrate, so creating specific signaling cascades and pathways; for example, the PTB domain of IRS1 recognizes a phosphorylated NPxY motif of the insulin receptor so that the downstream pathway is correctly activated. There are other modules designed to recognize virtually every post-translational modification, including methyl lysine by chromo domains, implicated in gene silencing mechanisms, acetyl lysine by bromo domains, which recruit proteins for chromatin remodeling or gene transcription, and ubiquinated proteins by ubiquin-binding domains, which target to the proteasome and elsewhere. Still others link different signaling pathways, allowing cross-talk, and of course redundancy, which appears essential to cell survival.

However, there are many modular domains that recognize other features of proteins. My favorite is the PDZ domain, which recognizes a C-terminal group; it is often duplicated in a chain and contrives to line proteins tail-to-tail, as we learn from Laskey, Skelton, and Sidhu in Chapter 13. There are 400 of them in humans, so there seems to have been some evolutionary pressure against anarchy in our species! Still others, like the calponin homology domain (Chapter 16 by Gimona and Winder), bind elements of the cell cytoskeleton, so integrating the dynamic and structural aspects of the cell.

But not all modular domains are involved in protein–protein interactions. Pleckstrin homology (PH) and phox homology (PX) domains bind phospholipids, so locating key signaling and regulatory components next to the membrane (see Chapters 19 and 20). There are also many that bind carbohydrates and nucleic acids that are not included here; indeed, another lengthy book would be required if they were.

Most protein adaptor domains are relatively small, comprising 60 to 130 amino acids. This is sufficient for a shallow concave binding site that can interact with a flexible peptide, carbohydrate, or lipid. These binding sites differ from most enzyme active sites, which usually lie between two domains and are able to take two or more molecules out of the aqueous environment. The nature of the binding site of adaptor modules has a number of interesting consequences for the functions of protein modules. First, they tend to bind continuous epitopes of protein targets. This means that the epitope must be flexible on the surface of a globular protein or in a region of constitutively disordered sequence. This is partly the reason that regions of proline-rich sequence are recognized by so many different kinds of domains, including the SH3, WW, GYF, UEV, EVH1 domains and profilin proteins, representing several different folds. Proline-rich sequences tend to adopt a polyproline helix structure, as they cannot form β strands or α helices, and these often exist outside regions of globular structure or in very exposed regions on the surface of globular proteins. This is a very different story from that of many of the protein complexes involved in signaling: these have a flatter interaction surface between two preformed globular proteins, for example, that between the cyclin-dependent kinase Cdk6 and its protein inhibitor INK4.

A second consequence of the size of modular domains is that the surface area of interaction is relatively small, typically around 1000 $Å^2$. This is well below the average values of between 1500 and 2000 $Å^2$ for other protein–protein complexes, such as the 1 : 1 FGF–FGFR complex, which involves two receptor domains linked together, or the Cdk6–INK4 complex, which involves several ankyrin repeats of the inhibitor. This is partially compensated for by the greater number of hydrogen bonds formed by the small modular domains with their ligands, as observed by Marko Hyvonen and myself several years ago. Nevertheless, the core binding regions of the small adaptor domain modules often interact with a protein epitope of only five or so amino acids, with consequent low affinities, often in the micromolar region, and much more rarely in the nanomolar region. This means that the 1:1 complexes are not too stable as binary complexes, allowing the fast off-rates required of a dynamic signaling system, in which complexes are transient and disassembly is as important as assembly.

The small size of the core binding region also means that there is often insufficient interaction to allow specificity. There are just too many sequences likely to be around in the cell to allow unique interactions. This may be resolved by having additional binding regions in variable loops outside the core regions, for example, a long loop on the Csk SH3 domain is critical for recognizing its partner, proline-enriched phosphatase (PEP). Sometimes, although rarely, ligands wrap around the module. However, there is evidence in yeast, as Sudol discusses in Chapter 3, that there has been 'selection against' certain sequences as well as 'selection for', but whether this occurs in the higher eukaryotes is really not yet clear.

Another consequence of these relatively small core binding regions and low affinities is that modules are often found in clusters. This may be a further way that the system attains selectivity. For example, PLC-γ has two PH domains, an SH2 and an SH3 domain spliced in and around the lipase. In a similar way, there are an SH2 and an SH3 domain in Src in addition to the kinase and other disordered regions. It appears that there is then cooperativity between the different domain interactions, leading to a more stable multimodular complex. This is analogous to the multiprotein complexes found elsewhere in signaling pathways, where very often, binary complexes are of low affinity but lead cooperatively to stable multi-protein complexes; for example, the FGF–FGFR complex is relatively unstable, but the 2FGF–2FGFR–heparin complex is quite stable and is required for signaling. We have speculated that this minimizes incorrect signaling that might occur through opportunistic binary interactions. Similar factors are likely to be in play with modular adaptor proteins.

It would be surprising if the adaptor modules were monofunctional. So much more can be gained by interacting with more than one component of the regulatory system. Thus, we learn from Margolis and Traub (Chapter 6) that many PTB domains, which are distant relatives of PH domains, not only keep the overall structural fold characteristic of PH domains but can often bind lipids. Thus, Shc PTB probably evolved from a PH domain, retaining the ability to associate with the membrane but acquiring the ability to recognize a tyrosine-phosphorylated receptor through a different surface region. Anything that the Lego designers have invented was probably discovered by Nature long before!

It is becoming apparent that this wonderful and flexible adaptor concept has probably been one of the most powerful contributors to the evolution of eukaryotes. Indeed, it has done so while not increasing substantively – or at least as much as we thought – the number of genes. Now that we know that there are rather few genes in higher eukaryotes, it is easy to see that it might be a simpler strategy to obtain a more complex, sensitive, and responsive system by increasing the interactions between components than by increasing the number of components. In the coming years, systems biologists will no doubt occasionally stand back from their holistic vision to admire the molecular mechanisms of the cross-talk that make living systems work.

Furthermore, if we want to intervene in order to fight disease, we must understand the nature of these complex interacting systems. It is noteworthy that mutations can lead to disease, for example, the mutations in EVH1 that are characteristic of Wiscott–Aldrich syndrome occur in adaptor modules. It is also often noted that ~50% of the drug discovery activity in pharmaceutical companies is targeted at regulatory and signaling systems. There is still surprise when a tight-binding drug has little effect on the whole organism; the answer may lie in the cross-talk between pathways. But adaptor modules are also an opportunity for drug design. They might even be used in gene therapy, to disrupt signaling pathways. Furthermore, the concave surfaces are likely to be far more 'druggable' than the flat interfaces that constitute many other signaling protein complexes, although the shallow nature of most binding sites and the weak affinities still make this a challenge.

This book is thus a treasure trove for scientists of different disciplines and interests. It must be essential reading, not only for academic students of molecular, structural, cell, and systems biology, but also for practitioners in the pharmaceutical and biotechnology industries interested in therapeutic intervention.

1
The SH2 Domain: a Prototype for Protein Interaction Modules

Tony Pawson, Gerald D. Gish, and Piers Nash

1.1
The Multidomain Nature of Signaling Proteins and Identification of the SH2 Domain

The discovery of viral oncogenes in the 1970s initiated an intense effort to understand the biochemical and functional properties of their protein products. The striking effects of such transforming proteins on cellular behavior suggested that understanding their modus operandi might reveal new principles underlying the molecular basis for cellular regulation and oncogenesis. Indeed, analysis of the Src, Abl, and Fps/Fes retroviral oncoproteins led to the realization that these proteins have intrinsic protein-tyrosine kinase activity, which correlates with their ability to transform cells from a normal to a cancerous phenotype [1–4]. Furthermore, the normal progenitors of these cytoplasmic proteins, as well as the cell surface receptors for growth factors and insulin, proved to be tyrosine kinases, suggesting that tyrosine phosphorylation is a common physiological mechanism through which cells respond to mitogenic or metabolic hormones [5–8]. Such hormones frequently induce receptor clustering, which stimulates receptor kinase activity, and this effect can be mimicked by oncogenic mutations that constitutively activate the receptor in the absence of an external ligand [9–11].

A logical approach toward understanding the normal and malignant properties of cytoplasmic or receptor tyrosine kinases (RTK) was therefore to pursue the biological consequences of tyrosine phosphorylation. By focusing on the tyrosine kinases themselves, it became apparent that these enzymes are activated by intermolecular autophosphorylation within their kinase domains, leading to a sustained increase in kinase activity [12–14]. Subsequent analysis has shown that such autophosphorylation causes a structural reorganization of the catalytic domain, primarily through movement of the activation loop, which otherwise occludes the kinase active site and interferes with ATP and substrate binding [15, 16]. These observations were consistent with a straightforward model in which tyrosine kinases might regulate the enzymatic activity of exogenous substrates through phosphorylation near their active sites. Curiously, however, receptor tyrosine kinases (RTK) themselves were often found to be the most abundant phosphotyrosine (pTyr)-

Modular Protein Domains. Edited by G. Cesareni, M. Gimona, M. Sudol, M. Yaffe

ISBN: 3-527-30813-X

containing proteins in growth factor-stimulated cells, raising the possibility that tyrosine phosphorylation might have unexpected biochemical functions. Furthermore, cytoplasmic protein-tyrosine kinases appeared to have extended polypeptide sequences in addition to the catalytic domain, suggesting that kinases such as Fps, Src, and Abl might have novel, noncatalytic properties. We therefore set out to test the notion that a cytoplasmic tyrosine kinase such as v-Fps might contain multiple functional domains.

Our group had used site-directed mutagenesis, a technique that was then in its infancy, to establish that tyrosine phosphorylation within the C-terminal catalytic domain of the v-Fps cytoplasmic oncoprotein stimulates its kinase activity and transforming potential [13]. We devised a variation of this approach to scan the 130-kDa v-Fps retroviral protein for folded domains that participate in its transforming activity. To this end, we inserted a dipeptide (or oligopeptide) motif at any one of numerous sites throughout the v-Fps protein, with the idea that such an insertion might disrupt the folded structure, and consequently the function, of a globular domain into which it was incorporated, but leave distinct domains intact [17]. At the time we knew that v-Fps had an N-terminal region encoded by the retroviral Gag gene fused to sequences derived from the cellular Fps tyrosine kinase, with the kinase domain located at the extreme C terminus [18]. The insertion mutagenesis approach identified three v-Fps domains that appeared to collaborate to induce cellular transformation. Not surprisingly, insertions in the C-terminal region of v-Fps corresponding to the kinase domain led to a loss of kinase and transforming activities. Insertions in an N-terminal v-Fps sequence caused a significant loss of transforming activity without an obvious effect on catalytic efficiency. This region has more recently been termed the FCH (Fes/Cip4 homology) domain and has been implicated in cytoskeletal localization [19, 20]. Consistent with the rationale for this mutagenesis approach, insertions in the FCH domain destroyed recognition of the v-Fps protein by two anti-Fps monoclonal antibodies that recognize conformation-dependent epitopes [21]. In contrast, insertion in the Gag sequence or in the central region of v-Fps had little effect.

Most interestingly, insertions (and subsequently deletions) in the sequence immediately N-terminal to the kinase domain influenced both kinase activity and substrate recognition and inhibited cellular transformation [17, 22]. Although this region was not necessary for kinase activity per se, it formed a protease-resistant structure that also contained the kinase domain, was important for full kinase activity in vivo, and promoted the tyrosine phosphorylation of selected cellular proteins [22–24]. Inspection of the amino acid sequences of Src and Abl showed that this region of ~100 amino acids N-terminal to the v-Fps kinase domain was relatively conserved in these other cytoplasmic tyrosine kinases and similarly located adjacent to the kinase domain, and we therefore termed it the Src homology 2 (SH2) domain (with the kinase domain implicitly being the SH1 region) [22] (Figure 1.1).

Taken together, these data indicated that v-Fps, and likely other cytoplasmic tyrosine kinases, were composed of multiple folded domains. The SH2 sequence was identified as a conserved noncatalytic domain, but one with the potential to

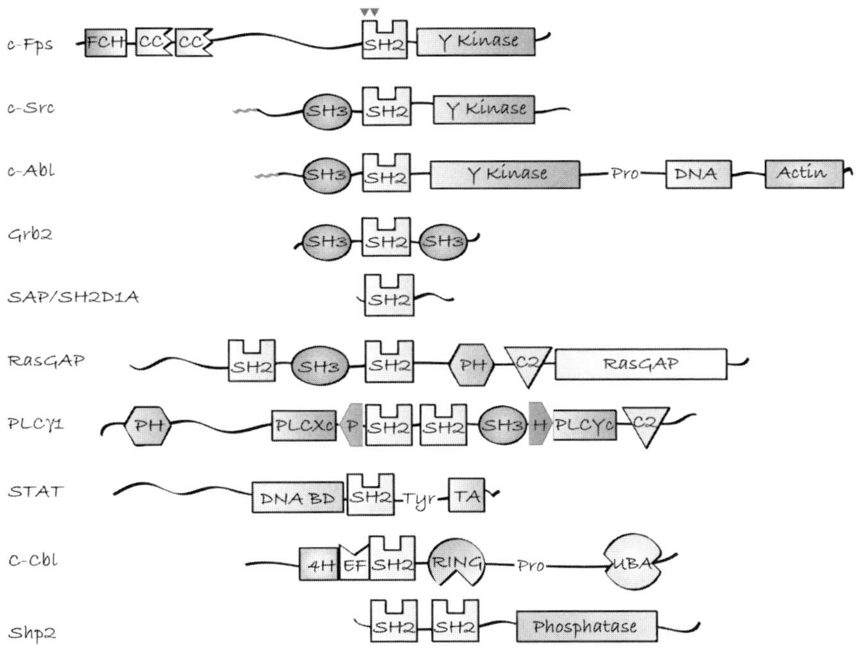

Figure 1.1 The domain organization of several SH2-containing proteins. Arrowheads mark the positions of insertions that led to identification of the SH2 domain in v-Fps.
Y kinase = tyrosine kinase domain; CC = coiled coil; Pro = SH3 binding motifs; BD = binding domain; actin = actin binding domain; PLC X/Yc = split PLC catalytic domain; Tyr = tyrosine phosphorylation site; 4H = 4-helix bundle; EF = EF hard-like domain; TA = transactivation domain; PH = pleckstrin homology; SH = Src homology; DNA = DNA-binding domain; Actin = actin-binding domain; UBA = ubiquitin-associated domain.

regulate both tyrosine kinase activity and substrate recognition and to promote malignant transformation. Such results suggested that the ability of cytoplasmic tyrosine kinases to recognize their substrates is not an exclusive property of the catalytic domain, but also depends on modular noncatalytic domains that target the kinase to specific substrates for phosphorylation.

Support for these ideas came from cloning of the viral Crk oncogene by Hidesaburo Hanafusa [25] and of cDNAs for mammalian phospholipase C (PLC) γ1 and Ras GTPase-activating protein (GAP) [26–28]. Intriguingly, the v-Crk oncoprotein comprised only an SH2 domain and a sequence present in Src, Abl, PLC-γ1, and RasGAP (but absent from Fps), which Hanafusa termed SH3 and which was subsequently shown by the Baltimore lab to bind proline-rich motifs [29]. Despite this lack of intrinsic catalytic activity, the v-Crk oncoprotein stimulated endogenous tyrosine kinase activity in transformed cells and enhanced the phosphorylation of a 130-kDa protein ($p130^{cas}$) at tyrosine, consistent with the view that the SH2 and SH3 domains might nucleate the formation of multiprotein complexes involved in tyrosine kinase signaling. Sequence analysis indicated that both PLCγ1 and RasGAP contain two SH2 domains and an SH3 domain. Taken with the earlier v-Fps data, these results suggested that SH2 domains are a common element of otherwise disparate enzymes and adaptors that function downstream of tyrosine kinases to control intracellular pathways such as those involved in phospholipid metabolism and Ras GTPase signaling. In support of this idea, activated RTKs were found to induce PLCγ1 and RasGAP phosphorylation at tyrosine, and we also observed that RasGAP associates with two other pTyr-containing proteins in cells with activated or oncogenic tyrosine kinases [30–34]. The variable domain organization of proteins such as Fps, Src, Crk, PLCγ1, and RasGAP indicated that the SH2 and SH3 domains might function in a cassette-like, combinatorial fashion and have broad functions in controlling cellular behavior in response to extracellular signals [35].

The precise means through which SH2 domains might link tyrosine kinases to specific biochemical pathways therefore became a central issue. In this regard, a number of groups found that activated receptor tyrosine kinases (RTK) physically associate with targets such as PLCγ, phosphatidylinositol (PI) 3′ kinase, and RasGAP [30, 36]. Furthermore, autophosphorylation of the PDGF β receptor at a specific tyrosine site (Tyr751) in the noncatalytic kinase insert region proved to specifically recruit PI 3′ kinase to the activated receptor [37]. Extrapolating from these data, we identified SH2 domains as the unifying element in coupling phosphorylated RTKs to their diverse intracellular targets. In these experiments, we found that isolated SH2 domains of different cytoplasmic signaling proteins, including PLCγ1, RasGAP, and Src, share a common ability to bind selectively to activated, autophosphorylated RTKs, as well as to tyrosine-phosphorylated docking proteins (e.g., $p62^{Dok\text{-}1}$) [38, 39]. Of interest, different SH2 domains bound distinct sets of pTyr-containing proteins from growth factor-stimulated cells, suggesting that SH2 domains impose an element of specificity in tyrosine kinase signaling through the selective recognition of tyrosine-phosphorylated proteins. These results indicated that endogenous SH2 domains are independently folding modules that bind selectively to activated RTKs and cytoplasmic proteins in a pTyr-dependent fashion. In a similar vein, recombinant v-Crk protein was found to associate with both tyrosine kinase activity and a series of pTyr-containing proteins from transformed cells [40, 41], consistent with the idea that aberrant binding of SH2 domains to pTyr-containing proteins stimulates malignant transformation. SH2 domain–pTyr interactions were thereby revealed as important in physiological signaling from RTKs and in the induction of malignant transformation by deregulated tyrosine kinase activity.

Subsequent data confirmed that SH2 domains directly recognize short pTyr-containing sequences. The binding sites for SH2 domains on the activated EGF receptor were mapped to the noncatalytic C-terminal tail of the receptor, which contains multiple autophosphorylation sites [42], and SH2 domains were found to bind to denatured pTyr-containing proteins and specific phosphopeptides corresponding to RTK autophosporylation sites [43–46].

1.2
SH2 Domains as a Prototype for Interaction Domains

SH2 domains can be viewed as the forerunner of the large family of interaction domains that control many aspects of signal transduction and cellular regulation, through their ability to bind variously to protein, phospholipid, nucleic acid, or small molecule ligands [47, 48] (Figure 1.2). Such domains have a modular design and mediate simple binary interactions, frequently by binding short peptide motifs. Through their ability to recognize a broad range of post-translational modifications and to discriminate between related binding sequences, interaction domains play a central role in the dynamic and specific cellular responses to extracellular and internal signals. Indeed, the scheme for RTK signaling, in which cytoplasmic sequences of the activated receptor recruit the interaction domains of target proteins, is broadly applicable to a wide range of cell surface receptors and intracellular regulators.

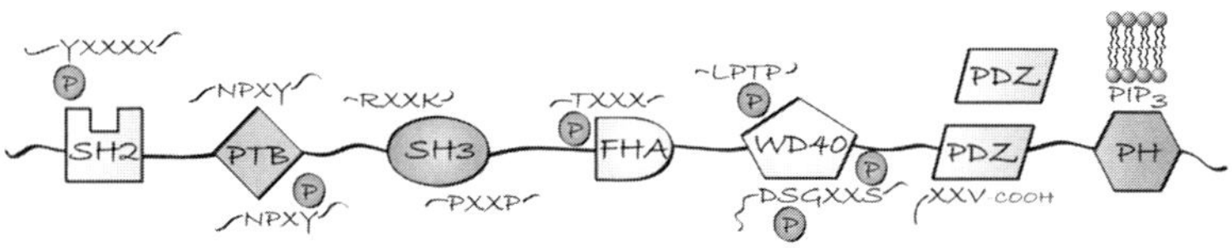

Figure 1.2 Summary of the binding properties of several interaction domains.

Although rather simple at first glance, interaction domains are remarkably versatile. Individual domains can have multiple ligands, and when covalently linked they can function as adaptors to join otherwise distinct polypeptides into a common signaling pathway. The reiterated use of interaction domains can also generate multiprotein assemblies with complex properties. Furthermore, interaction domains can provide allosteric regulation and switch-like behavior to associated catalytic domains and target enzymatic domains to their physiological substrates. In addition, interaction domains can serve as the organizing principle for extended protein networks, which underlie the systems properties of eukaryotic cells [49].

1.3
Structure and Binding Properties of SH2 Domains

The majority of SH2 domains bind short phosphopeptide motifs, which are themselves components of larger polypeptides, such as activated RTKs [50, 51]. A critical feature of phosphopeptide recognition involves association of the SH2 domain with the phosphorylated tyrosine; this interaction alone provides about half the binding energy and therefore results in a ~1000-fold increase in affinity as compared with an unphosphorylated peptide [52–54]. As a consequence, the physiological association of SH2 domains with their binding partners is usually phosphorylation-dependent. An analysis of SH2 domain binding sites on cellular phosphoproteins, as well as the in-vitro binding preferences of SH2 domains for phosphorylated peptides, indicated that SH2 domains also bind selectively to residues immediately C-terminal to the pTyr, in a fashion that varies from one SH2 domain to another [52, 55]. For example, the SH2 domains of the p85 subunit of PI 3′ kinase typically bind sites on activated RTKs that contain a Met that is three residues C-terminal to the pTyr (the +3 position), and in vitro they bind preferentially to phosphopeptides having the consensus motif pTyr-x-x-Met (where x is any amino acid). The generality of this concept was revealed by using SH2 domains to probe a phosphopeptide library in which a central pTyr is flanked by degenerate positions. This approach showed that different SH2 domains preferentially recognize distinct amino acids at the C-terminal +1 to +3 positions [56, 57]. This technique identified, not only residues that are favored by a particular SH2 domain, but also amino acids that are inhibitory to binding. This is relevant, because the specificity of a particular protein–protein interaction in vivo depends on both permissive forces that direct an interaction domain to its intended target as well as inhibitory effects that block binding to nonphysiological partners.

The dominance of pTyr recognition is necessary for SH2-mediated interactions to be phospho-dependent, but also dictates that the affinity of SH2 domains for the C-terminal specificity residues is not too high, or complex formation could no longer be regulated by phosphorylation. Indeed, the dissociation constants of SH2 domains for their optimal phosphopeptide ligands are typically in the range of 500 nM–1 μM [58]. This relatively modest affinity may have additional biological consequences. Since signaling from cell surface receptors is highly dynamic, the protein–protein interactions that control intracellular pathways must not be too strong, or signaling events would become irreversible. A case in point is provided by the intramolecular interaction in the Src tyrosine kinase between the SH2 domain and a pTyr site (Tyr527) in the C terminus that induces an autoinhibited state (see below). The phosphorylated Tyr527 motif is a poor ligand for the Src SH2 domain, but the interaction proceeds because the two interacting partners are tethered within the same molecule. Replacing the Tyr527 site with an optimal binding motif for the Src SH2 domain yields a Src tyrosine kinase that is locked in the autoinhibited state and cannot be activated, because the inhibitory interaction is too strong [59]. In this example, an SH2-mediated interaction is designed to be weak, so that it can be readily reversed.

Similarly, although specificity in protein–protein interactions is biologically important, there is often significant flexibility in the binding of interaction domains to their targets. A given SH2 domain may interact physiologically with phosphorylated sites on multiple different proteins, which may allow for cell-specific functions or the formation of more complex signaling networks within a single cell [60]. Conversely, a single phosphorylated motif on an RTK may potentially provide a docking site for several distinct SH2-containing targets [61].

Structural analysis of a number of SH2 domains in association with phosphopeptide ligands has revealed the molecular basis for these biochemical observations (Figure 1.3) [50, 51]. SH2 domains typically have an N-terminal region that forms an α helix (αA) and a central antiparallel β sheet containing strands βA, βB, βC,

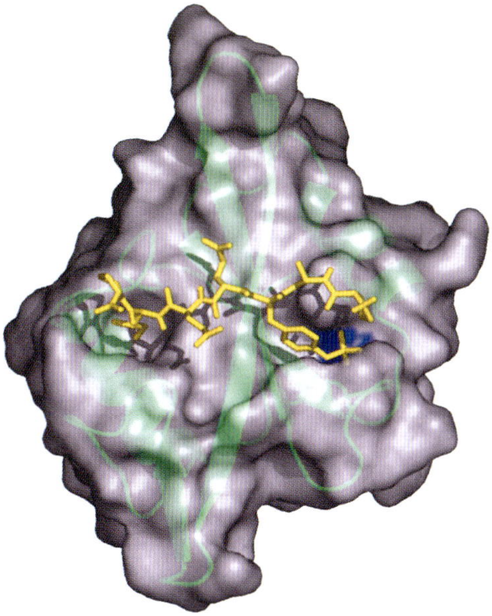

Figure 1.3 Structure of the Src SH2 domain bound to a pTyr-Glu-Glu-Ile peptide (PDB: 1SPS) [67]. The surface of the SH2 domain is in pink, and the secondary structural elements of the SH2 domain in green. The αA helix is to the right and αB helix to the left. The Arg βB5 residue critical for pTyr binding is in blue. The N-terminal pTyr of the peptide (yellow) occupies the pTyr binding pocket. The peptide runs over the central β sheet of the SH2 domain, the +1 and +2 glutamates contact the surface of the domain, and the sidechain of the +3 Ile (to the left) fits in a hydrophobic pocket.

and βD. The C-terminal region has a smaller β sheet (with strands βD′, βE, and βF), followed by another α helix (αB), and a short β strand (βG) that associates with the core β sheet. This creates a bipartite structure with two discrete binding pockets located on either side of the core β sheet [62–64]. Residues in the most highly conserved N-terminal region of the SH2 domain form a positively charged pocket that makes numerous contacts with the pTyr residue. In terms of binding free energy, the most prominent interaction involves a bidentate ionic association between an invariant buried Arg (ArgβB5) in the SH2 domain and two oxygens of the pTyr phosphate (Figure 1.4); substitution of this Arg effectively eliminates phosphopeptide binding [65, 66]. The second binding surface on the SH2 domain is much more variable between different SH2 domains and is responsible for the selective recognition of C-terminal residues. In the Src (or Lck) SH2 domain bound to a pYEEI peptide, the peptide runs orthogonally over the β sheet and adopts an extended conformation so that residues C-terminal to the pTyr contact the more variable specificity pocket. Here, the +1 and +2 glutamates make long-range contacts with the surface of the SH2 domain, while the sidechain of the +3 Ile is buried in a hydrophobic cavity [67, 68] (Figure 1.3).

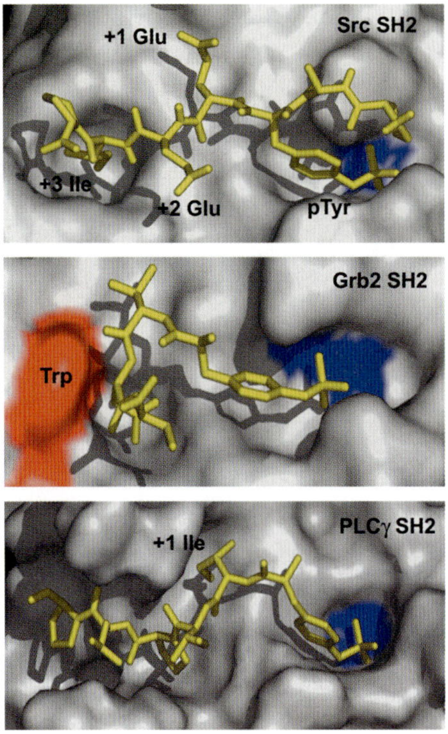

Figure 1.4 Different modes of SH2 domain–phosphopeptide recognition shown for the Src, Grb2, and PLC-γ1 (C terminal) SH2 domains (PDP: 1SPS; PDB: 1TZE; PDB: 2PLD) [67, 70, 72]. In each panel, the surface of the SH2 domain is in gray, and the Arg βB5 required for pTyr recognition in blue. The pTyr binding pocket is to the right, and the more variable surface involved in selective recognition of C-terminal residues is to the left. The phosphopeptides are in yellow, with the N-terminal pTyr to the right and the more C-terminal residues to the left. The +3 Ile of the *po*YEEI peptide occupies a hydrophobic pocket on the Src SH2 domain, as also shown in Figure 1.3. The *po*YVNV peptide bound to the Grb2 SH2 domain is forced into a β turn by the Trp residue at the EF1 position (shown in red). The C-terminal residues of the *po*YIIPLPD peptide bound to the PLC-γ1 SH2 domain occupy an extended hydrophobic cleft.

1.4 Different Modes of SH2 Domain–Phosphopeptide Recognition

The N-terminal SH2 domain of the Shp-2 tyrosine phosphatase and the C-terminal SH2 domain of phospholipase C-γ1 also bind phosphopeptides, such as those from the platelet-derived growth factor receptor (PDGFR), in an extended conformation [69, 70]. However, in contrast to Src, the specificity region of these SH2 domains has a long hydrophobic groove that accommodates at least five residues C-terminal to the pTyr, primarily through hydrophobic interactions (Figure 1.4). In PLC-γ1, the +1 Ile of the Tyr1021 motif from the βPDGFR projects into the start of this hydrophobic cleft, facilitated by a cysteine at the βD5 position of the SH2 domain, which provides space for the Ile sidechain. In the Src SH2 domain this space is filled by a tyrosine at βD5, which therefore selects against Ile at the +1 position [71]. The SH2 domain of the Grb2 adaptor has yet another binding mode, conferred by a Trp at the EF1 position (the first residue in the loop between the βE and βF strands) (Figure 1.4). In the Src SH2 domain the corresponding EF1 residue is a Thr that forms part of the +3 hydrophobic binding pocket. In Grb2, the EF1 Trp fills this pocket and projects from the surface of the SH2 domain, forcing the phosphopeptide into a β turn that is accommodated by an Asn at the +2 position of the peptide, which makes hydrogen bonds with backbone atoms of the SH2 domain [72–74]. In the SH2 domain of the Grb7 adaptor, an insertion in the EF loop also occludes the +3-binding pocket and imposes a β-turn conformation on the bound phosphopeptide, which favors a +2 Asn in a similar fashion to Grb2 [75]. These data show how subtle variations in the sequences and structures of SH2 domains can influence their binding preferences. The biological relevance of this selectivity is suggested by the fact that physiological binding sites for SH2-containing proteins often conform to those identified by in-vitro peptide-binding studies and structural analyses.

To test whether the ability of SH2 domains to select specific phosphopeptide motifs is biologically significant, we examined the effects of manipulating their binding selectivity [76]. Examination of multiple SH2 domain sequences revealed the presence of a Trp at the EF1 position that was unique to the Grb2 SH2 domain and which we thought might account for its propensity to bind Asn (as subsequently validated by structural analysis). We found that replacing the EF1 Thr of the Src SH2 domain with Trp resulted in strong selection for peptides having an Asn at the +2 position; structural analysis subsequently revealed that this mutant form of the Src SH2 domain recognizes a physiological Grb2 binding site (pTyr-Val-Asn-Val) in a virtually identical fashion to Grb2 itself [77]. To explore the biological effects of this specificity switch, we made use of the fact that inactivation of the Grb2 gene in *Caenorhabditis elegans* (called *sem-5*) results in multiple defects, including abnormal development of the vulva, which can be rescued by reexpression of wild-type Grb2. We replaced the SH2 domain of Grb2 with that from Src and found that the chimeric Src-Grb2 adaptor, in which the Src SH2 domain is flanked by the Grb2 SH3 domains, was inefficient in rescuing the defect caused by loss of the endogenous gene. However, the same chimeric adaptor, but with Trp replacing Thr at the EF1 position of the Src SH2 domain, restored biological activity to nearly wild-type levels. This experiment showed that SH2 domain selectivity can be attributed to specific residues and demonstrated that the biological function of an SH2 domain correlates with its binding selectivity.

The properties of SH2 domains discussed above are reflected in a number of interaction domains. The ability of post-translational modifications to regulate modular protein–protein interactions is shared by a variety of domains that selectively recognize peptide motifs after phosphorylation at tyrosine (PTB domain) or serine/threonine (FHA, WD40, MH2, Polo Box, BRCT, FF, and WW domains and 14-3-3 proteins, see Chapter 8) [78, 79], acetylation of lysine residues (bromodomains), arginine methylation (chromodomains) [80], proline hydroxylation (VHL) [81, 82], or monoubiquitination (UIM, CUE, UBA domains) [83]. Many of these domains show a similar structural organization to SH2 domains, in the sense that they have a cassette-like design and typically have distinct binding pockets for the modified amino acid and flanking residues, with the latter providing specificity among different members of the same domain class.

1.5 Signaling Pathways and Networks

SH2 domains are embedded in numerous proteins having widely different biochemical functions, which are nonetheless all directly involved in tyrosine kinase signaling owing to the pTyr-binding properties of their SH2 domains. This observation suggests an evolutionary rationale for the pervasive use of interaction domains in cell signaling. According to this scheme, new signaling pathways and cellular functions could be generated through the simple device of inserting a novel interaction domain into an already-existing polypeptide or by organizing interaction domains into new combinations. This would provide the relevant protein with innovative physical connections and would thereby link previously separate cellular processes into a common complex or pathway. This approach has several potential advantages. Incorporating an interaction domain into a specific protein is simpler than evolving a complex mechanism of allosteric regulation. In particular, the same

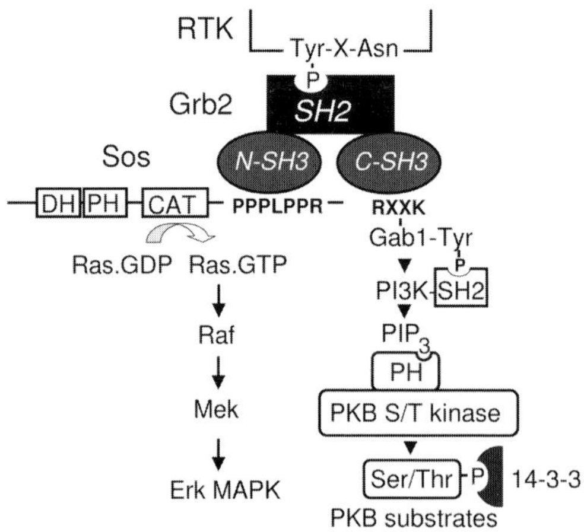

Figure 1.5 A series of modular protein–protein and protein–phospholipid interactions mediate signaling from pTyr-X-Asn motifs on activated RTKs. The N-terminal and C-terminal SH3 domains of Grb2 bind proline-rich and basic motifs, respectively.

interaction domain can be readily coupled to many different types of protein and thereby endow them with a common function, such as pTyr recognition. In addition, interaction domains can themselves provide autoregulatory and switch-like behavior when linked to catalytic domains, through a combination of intra- and intermolecular interactions [84]. In contrast, a regulatory device that depends entirely on conformational change (e.g., induced by phosphorylation) may not be easily transferred from one polypeptide to another.

The various types of SH2 domain proteins can be organized into a number of different classes:

- **Adaptors:** Proteins of the Grb2, Crk, and Nck families are composed exclusively of SH2 and SH3 domains and link specific pTyr motifs recognized by the adaptors' SH2 domains to cellular pathways controlled by proteins that bind selectively to their SH3 domains. Each adaptor recruits a distinct set of SH3-binding proteins and thereby controls a different facet of cellular organization. Grb2, for example, binds to pTyr-X-Asn motifs through its SH2 domain (as discussed above), to the Sos guanine nucleotide exchange factors (GEF) of the Ras GTPase through its N-terminal SH3 domain [85–90], and to the Gab-1 or Gab-2 scaffolding proteins through its C-terminal SH3 domain [91, 92] (Figure 1.5).

  Neither Grb2 nor Sos are notably phosphorylated at tyrosine, and thus activation of Ras by exchange of GDP for GTP may depend primarily on physical recruitment of the Grb2–Sos complex to sites of RTK activation at the plasma membrane, where Ras is also localized. GTP-bound Ras has a number of targets, most significantly the Raf protein-serine/threonine kinase that stimulates the Erk MAP kinase pathway.

  In contrast, Grb2-associated Gab proteins become tyrosine-phosphorylated at sites that bind additional SH2 domain proteins [93], such as PI 3′ kinase, which consequently generates PI-3,4,5-P_3 in the plasma membrane. PIP_3, in turn, is recognized by the PH domain of the protein-serine/threonine kinase PKB/Akt [94], which is activated after recruitment to the membrane and phosphorylates a number of targets, such as Bad, Forkhead transcription factors, and Tuberin involved in apoptosis, cell cycle control, and cell growth [95]. Interestingly, the serine/threonine sites phosphorylated by PKB are often themselves directly recognized by 14-3-3 proteins, which regulate the activity of their associated phosphoproteins [96].

  Because it can bind multiple targets through its SH3 domains, the Grb2 adaptor couples a single pTyr site to a series of interconnected signaling pathways, which act in a cooperative fashion to control the growth, proliferation, survival, and differentiation of growth factor-stimulated cells. These signaling events are assembled from simple, modular protein–protein and protein–phospholipid interactions, which act in concert to form a network with a broad influence on cell behavior.

  The SH2 and SH3 domains of Nck have different binding specificities compared with Grb2. Most strikingly, the Nck SH3 domains bind a series of proteins that regulate the actin cytoskeleton [97], such as N-WASP and the Pak-serine/threonine kinase [98–100]. This correlates with the ability of Nck to couple tyrosine kinases to cytoskeletal reorganization in cultured cells and intact organisms [101, 102]. Crk, in contrast, links specific pTyr sites to proteins involved in membrane dynamics and cell adhesion, including activators of the Rap and Rac GTPases [103–108]. SH2/SH3 adaptors therefore provide a simple tool through which tyrosine kinases influence a range of cellular activities. Many other types of cell surface receptors, such as TNF receptors, Toll-like receptors, guidance receptors, adhesion molecules and integrins, TGFβ receptors, and Wnt receptors, similarly associate with their cytoplasmic targets through modular interaction domains, often in the form of adaptor proteins [109, 110].

- **Scaffolds:** Scaffold proteins bind multiple signaling components with complementary functions. This applies to SH2 domain proteins such as Blnk (SLP-65) and SLP-76 [111, 112], which act in B cells and T cells as hubs to organize a number of key signaling proteins (including Tec family tyrosine kinases, PLCγ, Nck, and Vav) [113–115] and therefore play a critical role in signaling from antigen receptors [116–118].

- **Enzymes:** A number of RTK targets are enzymes with intrinsic SH2 domains. These include proteins that regulate phospholipid metabolism, such as phospholipase Cγ1 and γ2 and the SHIP 5′-inositol phosphatase [119]. PI 3′ kinase has a noncatalytic p85 adaptor subunit that contains SH2 domains, separated by a coiled-coil region that associates with the p110 catalytic subunit [95, 120]; however, in contrast to other adaptors, p85 is a dedicated component of the PI 3-kinase complex. Other SH2-containing enzymes regulate Ras, Rho, and Rab family GTPases, as well as tyrosine phosphorylation [121–128]. These enzymes are commonly activated by the binding of their SH2 domains to pTyr sites, as discussed in more detail below, and are also frequently substrates for tyrosine phosphorylation.

- **Regulators:** A substantial family of SH2-containing proteins control the duration and localization of tyrosine kinase signals. c-Cbl is an E3 protein–ubiquitin ligase that promotes RTK monoubiquitination and internalization [129–131], and SOCS proteins inhibit cytokine receptor signaling both by directly binding and inhibiting the activity of receptor-associated JAK-tyrosine kinases and by inducing their ubiquitination [132]. The APS protein, in contrast, binds and stabilizes the activated form of the insulin receptor (see below).

- **Transcription factors:** The Stat proteins are SH2-containing transcriptional regulators that provide a direct link between activated cytokine receptors at the plasma membrane, which they bind through their SH2 domain, and gene expression in the nucleus, where they relocalize after phosphorylation and dimerization, as discussed below.

1.6
Plasticity of SH2 Domains

The human genome encodes some 115 SH2 domains. An analysis of existing exon/intron boundaries suggests that these SH2 domains likely evolved from a common ancestor, which may then have been selected for its versatility [133]. Consistent with this possibility, SH2 domains show significant flexibility in their binding properties, beyond their conventional capacity to discriminate between distinct pTyr-containing peptide motifs.

The SAP/SH2D1A protein, for example, is composed almost entirely of a single SH2 domain [134], which binds to a tyrosine-based motif in the T-cell coreceptor SLAM [135] (Figure 1.6). Homotypic engagement of the SLAM extracellular region leads to tyrosine phosphorylation of its cytoplasmic tail, which regulates signaling pathways in stimulated T cells that control the pattern of cytokine production and proliferation of CD8+ T cells. This in turn ensures a limited lymphoid response to viral infection. Mutations in the human SAP gene (see below) have revealed that its product is essential for SLAM signaling, and this has uncovered unusual properties of the SAP SH2 domain. First, it binds preferentially to an extended motif having the consensus sequence Thr-Ile-(p)Tyr-x-x-Val/Ile, which conforms to its physiological binding site on SLAM (Tyr281) [136, 137]. Surprisingly, the SAP SH2 domain binds an unphosphorylated Tyr281 SLAM peptide with a K_d of ~300 nM, and there is only about a 3-fold increase in affinity upon tyrosine phosphorylation [136–138]. As a consequence, SAP can bind to both unphosphorylated and phosphorylated forms of SLAM in T cells. Biochemical and structural analysis has shown that the SAP SH2 domain effectively has three binding pockets, one for the central (p)Tyr, another for the +3 Val, and a third that accommodates the Thr at the −2 position of the peptide ligand [136, 137] (Figure 1.7). It is this unusual interaction with an N-terminal peptide residue that enhances pTyr-independent binding to the SAP SH2 domain. Presumably, the great majority of SH2 domains do not exploit this potential to bind unmodified peptides, because a

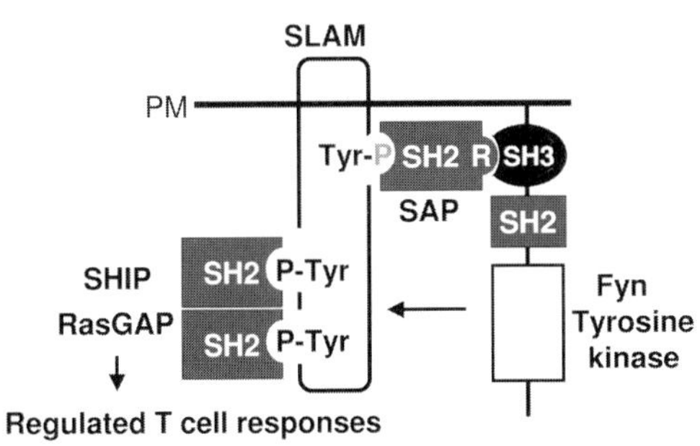

Figure 1.6 The SAP SH2 domain acts as an adaptor to link the SLAM receptor in T cells to the Fyn tyrosine kinase, which consequently phosphorylates SLAM at sites that engage additional SH2-containing proteins. Loss-of-function mutations in the SAP SH2 domain cause X-linked lymphoproliferative syndrome. R = Arg78; PM = plasma membrane.

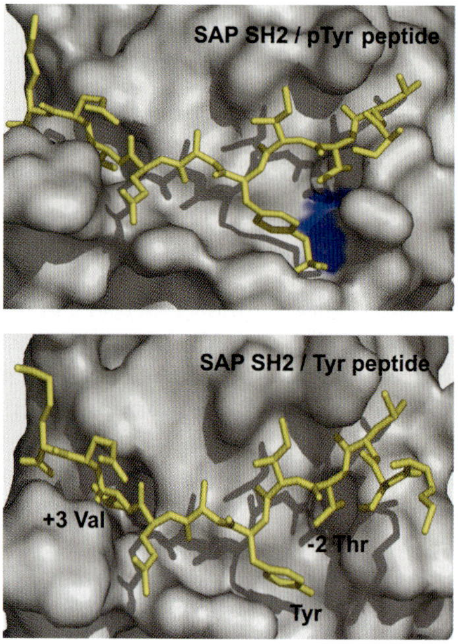

Figure 1.7 Recognition by the SAP SH2 domain of a Tyr-based motif in SLAM in both its unphosphorylated and phosphorylated states. The peptide binding surface of the SAP SH2 domain is in grey, and the Arg βB5 residue in blue. The SH2 domain is complexed with either the phosphorylated SLAM peptide (TIpYAQV) (top) or its unphosphorylated counterpart (bottom). Note the three binding pockets for the –2 Thr, (p)Tyr, and the +3 Val, respectively.

relatively high affinity for the unphosphorylated ligand would nullify the switch-like properties of tyrosine phosphorylation in inducing SH2 domain recognition.

The SAP SH2 domain also has an adaptor function, due to a second binding surface that associates with the SH3 domain of a Src-like cytoplasmic tyrosine kinase, Fyn [139]. Although SH3 domains usually recognize proline-rich motifs, some bind to basic sites [140]; in the Fyn–SAP complex, the Fyn SH3 domain recognizes such a basic surface on the SAP SH2 domain, focused on Arg78 [141, 142]. This surface of the SAP SH2 domain does not overlap the conventional peptide binding site, and the SAP SH2 domain can therefore bind simultaneously to SLAM and the Fyn tyrosine kinase. The SAP-binding site on the Fyn SH3 domain, however, overlaps the regions involved in PxxP recognition and the autoinhibitory intramolecular interaction that represses Fyn activity. SAP therefore bridges SLAM to active Fyn, which consequently phosphorylates SLAM at additional tyrosine sites for the recruitment of SH2-containing effectors such as the SHIP 5′-inositol phosphatase and RasGAP (Figure 1.6).

A number of more general points can be taken from this scheme. First, the reiterated use of rather simple binary interactions can be employed to assemble more sophisticated multiprotein complexes. Second, interaction domains such as

SH2 and SH3 can potentially employ their primary binding surface to engage distinct types of peptide ligand (i.e., phosphorylated or unphosphorylated peptides for SH2 domains, proline-rich or basic motifs for SH3 domains). Third, a single domain can use distinct surfaces to bind distinct types of ligand, as shown by the ability of the SAP SH2 domain to engage both SLAM and Fyn. These are themes that are frequently encountered in the analysis of interaction domains. For example, PTB domains can bind both tyrosine phosphorylated and unphosphorylated peptide ligands [143–145], as well as phospholipids [146]; indeed, a number of interaction domains with functions ranging from phosphoinositide recognition to the binding of proline-rich sequences (such as PH and EVH1 domains, see Chapters 4 and 17) have the same fold as PTB domains [147].

1.7 SH2 Domain Dimerization

The versatility with which SH2 domains bind pTyr-containing sequences is exemplified by the SH2 domain of the APS adaptor, which recognizes the activation loop of the insulin receptor kinase (IRK) catalytic domain. Upon insulin stimulation, the activation loop of IRK is autophosphorylated on three tyrosine residues and is consequently displaced from an inhibitory conformation that occludes the active site. One of these phosphorylated tyrosines (pTyr1163) interacts with residues in the kinase domain to lock the activation segment in place, but the other two (pTyr1158 and pTyr1162) are exposed [148]. The APS SH2 domain binds to these two pTyr residues in an unusual fashion [149]. The C-terminal region of the APS SH2 domain forms an unusually long αB helix, and this allows two SH2 domain monomers to dimerize through an extensive hydrophobic interface. pTyr1158 of IRK occupies the normal pTyr-binding pocket of the SH2 domain, but the subsequent residues of the IRK activation loop are sterically blocked from following a typical path perpendicular to the central β sheet, due to the C-terminal helix of the other monomer in the APS SH2 dimer. Rather, the activation loop binds in the plane of the β sheet, facilitated by numerous specific interactions with the IRK kinase domain. Notably, there is a second pTyr-binding site in the APS SH2 domain, which is occupied by pTyr1162 of the IRK activation loop. Thus, two IRK chains are bridged by two (dimeric) APS SH2 domains. APS may therefore enhance and prolong the activated state of IRK by protecting the phosphorylated activation loop from phosphatases and stabilizing the active form of the kinase domain (Figure 1.8).

These results emphasize several points made above (i.e., the ability of SH2 domains to bind multiple ligands – here, the phosphorylated IRK and another APS SH2 domain – and the plasticity of SH2 domain pTyr-binding sites), but also demonstrate that SH2 domains can form homodimers. The Grb10 SH2 domain also forms a dimer through an interface that involves the C-terminal α helix (αB) [150]. In addition the SH2 domains of the STAT transcription factors dimerize in response to cytokine stimulation (see below) [151]. In this case, however, dimerization is induced by phosphorylation of a tyrosine just C-terminal to the SH2 domain,

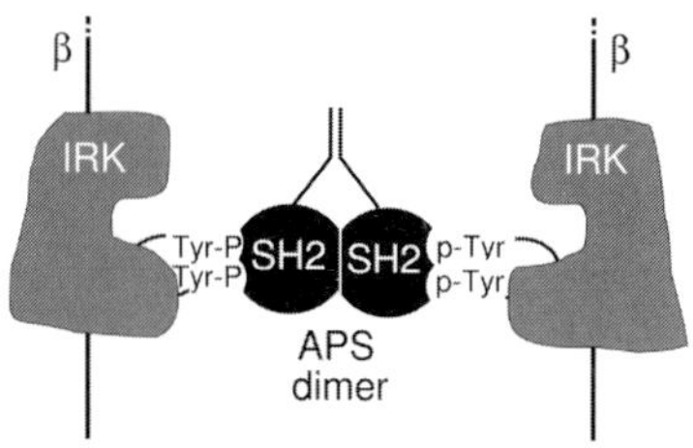

Figure 1.8 The APS SH2 domain homodimerizes and binds the phosphorylated activation loop of the insulin receptor kinase (IRK). IRK is itself a heterotetramer; only the cytoplasmic regions of its β subunits are shown.

which leads to reciprocal intermolecular SH2 domain–pTyr interactions between two STAT proteins. Oligomerization is a common theme among interaction domains, which frequently self-associate into homo- or heterodimers.

1.8
Tandem SH2 Domains

A number of proteins, including the Syk and ZAP-70 cytoplasmic tyrosine kinases, PLC-γ, RasGAP, and the Shp-1 and Shp-2 tyrosine phosphatases, have two SH2 domains arranged in tandem (see Figure 1.1) and bind preferentially to doubly phosphorylated sites in which the two pTyr residues are closely spaced. Tandem SH2 domains appear to bind such bisphosphorylated motifs with significantly increased affinity and specificity, as compared to the interaction of a single SH2 domain with a monophosphorylated peptide [152]. This principle is demonstrated by the Syk and ZAP-70 tyrosine kinases, which mediate signaling by immunoreceptors, notably the B-cell and T-cell antigen receptors (BCR/TCR), and Fc receptors [153]. Upon activation of antigen receptors, Src family tyrosine kinases phosphorylate BCR or TCR signaling subunits on ITAM sequences (for immunoreceptor tyrosine-based activation motif). ITAMs have the consensus YxxI/L-x_{6-8}-YxxI/L, and their efficient binding to the Syk or ZAP-70 tandem SH2 domains requires that both tyrosines be phosphorylated [154]. Each of the two pTyr motifs within an ITAM binds in a conventional orientation to a single SH2 domain, such that the N-terminal pTyr associates with the C-terminal SH2 domain, and the C-terminal pTyr with the N-terminal SH2 domain. Strikingly, the pTyr binding pocket of the N-terminal SH2 domain of ZAP-70 is formed in part by residues from the C-terminal SH2 domain [155]. As a consequence, binding of the tandem ZAP–70 SH2 domains to doubly phosphorylated ITAM motifs in the TCR complex is strongly dependent on the spacing of the two phosphorylated tyrosines. The Syk SH2 domains are autonomous and more flexible and can bind a wider range of ITAM motifs with variable spacing between the pTyr residues [156, 157]. These data make the point that multiple interaction domains can be joined to create more-specific protein–protein interactions than those displayed by any one domain alone.

Indeed, many signaling proteins contain several distinct interaction domains, often with quite different binding properties, which can potentially bind cooperatively to their targets.

1.9
Composite and Complex Interaction Domains

The c-Cbl protein is an E3 protein–ubiquitin ligase with an N-terminal SH2 domain, which binds specific pTyr motifs on autophosphorylated RTKs, and a central ring domain that recruits an E2 ubiquitin ligase and thus induces the monoubiquitination of receptors associated with the SH2 domain. The c-Cbl SH2 domain is unusual, in the sense that it is embedded in a larger folded structure that also contains a four-helix bundle and an EF hand [158]. Although phosphopeptides bind the SH2 domain in a conventional fashion, the associated subdomains may participate in target recognition.

The SH2 domain of STAT transcription factors is also part of a larger structural unit, since, unlike most SH2 domains, those from STAT proteins are not functional when expressed in isolation from their surrounding sequences. The core region of STATs contains a helical coiled-coil sequence, a DNA binding region, and a linker region followed by the SH2 domain; C-terminal to this core lies a transactivation sequence that regulates transcription. Activated cytokine receptors at the plasma membrane induce phosphorylation of a STAT tyrosine residue located immediately C-terminal to the SH2 domain, and this results in STAT dimerization through reciprocal SH2–pTyr interactions. STAT dimers are recruited to the nucleus, where they bind specific promoters to regulate gene expression. Structural analysis of the dimeric core region shows that the four domains (coiled-coil, DNA binding, linker, SH2) are interlocked through a common hydrophobic core and extensive interfaces between the domains [159, 160]. Of interest, the phosphate binding loop of the SH2 domain interacts directly with the linker domain, which in turn associates with the DNA binding domain; thus, there is potential for direct coupling between pTyr binding to the SH2 domain and DNA binding. The sequence following the SH2 domain of one monomer is organized so that, upon phosphorylation, it can engage only the SH2 domain of a partner STAT, rather than undergoing intramolecular interaction. STAT proteins therefore provide an elegant example of a functionally and structurally complex polypeptide built from smaller protein–protein and protein–nucleic acid interaction domains.

1.10
Allosteric Regulation

In enzymes such as the Src and Abl cytoplasmic tyrosine kinases or the Shp-2 protein-tyrosine phosphatase (PTP), reversible intramolecular interactions between the SH2 and catalytic domains provide the basis for autoinhibition of enzymatic activity and for switch-like activation after these repressive contacts are broken. The Shp-2 tyrosine phosphatase has two tandem SH2 domains, followed by a phosphatase domain and a C-terminal tail. In the autoinhibited state, the N-terminal (N-) SH2 domain undergoes an extensive intramolecular interaction with the active site of the phosphatase domain [161]. The D′E loop of the SH2 domain buries into the PTP catalytic cleft, directly engages the catalytic Cys nucleophile and the phosphate binding cradle of the active site, and locks the PTP domain in an open, inactive conformation. This interaction also distorts the conformation of the N-SH2 domain by imposing movements of the αB helix and the EF loop that occlude the phosphopeptide binding cleft; as a consequence, although the phosphopeptide binding site of the N-SH2 domain is exposed to solvent in the autoinhibited state, it is not properly configured for ligand recognition. The C-SH2 domain, in contrast, makes minimal contact with the N-SH2 or PTP domains, but could promote Shp-2 activation by binding one pTyr site on a bisphosphorylated peptide. This would increase the local concentration of the doubly phosphorylated motif and promote binding of the second pTyr site to the N-SH2 domain. The resulting reorganization of the N-SH2 domain would then break its interaction with the PTP domain, resulting in activation of the phosphatase. In this scheme, the N-SH2 domain has two distinct ligand binding sites (for a phosphopeptide or the PTP domain), but, in contrast to the examples discussed above, these interactions display negative cooperativity and are mutually incompatible. In Shp-2, the specific recognition of a bisphosphorylated motif by the tandem SH2 domains is thus precisely coupled to activation of the phosphatase. This device ensures that Shp-2 phosphatase activity is turned on only as the SH2 domains bind an appropriate target.

A related mechanism is exploited by the Src and Abl tyrosine kinases, although the details are very different. In both Src and Abl, an SH3 domain is followed by an SH2 domain, which precedes the kinase domain. In Src family kinases, autoinhibition involves phosphorylation of a Tyr (Tyr527) in the extreme C-terminal tail by another tyrosine kinase, Csk [124]. This induces an intramolecular interaction with the SH2 domain and positions the SH3 domain to bind the linker region between the SH2 and kinase domains (Figure 1.9) [162, 163]. In this autoinhibited conformation, the conventional binding clefts of both the SH2 and SH3 domains engage internal ligands and are therefore protected from adventitious interactions with other polypeptides. At the same time, these interactions inactivate the kinase domain, although in contrast to Shp-2, this is achieved indirectly. Dynamic simulations and mutagenesis experiments have suggested that the linker between the SH2 and SH3 domains forms a rigid clamp in the autoinhibited state [164]. Since the SH2 domain interacts with the large lobe of the kinase and the SH3 domain with the small lobe, this clamp prevents flexibility at the intervening active site required for the exchange of ATP and ADP. Also, in this autoinhibited conformation, the regulatory tyrosine in the activation loop is sequestered and therefore less prone to autophosphorylation and resulting kinase activation, and the C helix in the small lobe undergoes a rotation that removes a key glutamate from the active site.

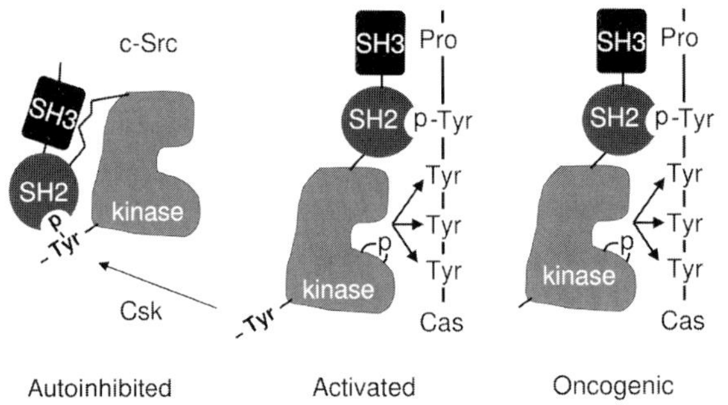

Figure 1.9 The SH2 and SH3 domains of the c-Src tyrosine kinase regulate its activity and substrate specificity. Activated c-Src can be targeted to its substrates through SH2/SH3-mediated interaction, which promotes the progressive phosphorylation of proteins such as Cas. Loss of the regulatory C terminus converts Src into a constitutively active tyrosine kinase.

This autoinhibited state can be broken in several ways. Most simply, dephosphorylation of Tyr527 leads both to dissociation of the SH2 and SH3 domains, which can then bind exogenous proteins with appropriate motifs, and to activation of the kinase domain, which can phosphorylate targets engaged by the adjacent interaction domains. Alternatively, an exogenous protein with high-affinity binding motifs for the Src SH3 or SH2 domains can compete for the inhibitory intramolecular interactions and thereby activate and localize the kinase domain. Therefore, as with Shp-2, the autoinhibited state both sequesters the interaction domains and inactivates the catalytic domain, and activation involves coordinate stimulation of the kinase domain and tethering of the SH2/SH3 domains to potential substrates. In the v-Src oncoprotein, deletion of the C-terminal tail removes the autoinhibitory Tyr527 site, and the tyrosine kinase becomes constitutively active (Figure 1.9).

The Abl kinase lacks an inhibitory phosphorylation site C-terminal to the kinase domain, but undergoes a very similar mode of autoinhibition as Src. In the Abl1b isoform, an N-terminal myristate group is inserted into the large helical lobe of the kinase, inducing a conformational distortion that creates a direct interface between the large lobe of the kinase domain and the SH2 domain [165]. This interaction, which involves the αA helix of the SH2 domain, is incompatible with phosphopeptide binding and positions the SH3 domain for interaction with the SH2-kinase linker, in a similar fashion to Src. The precise mechanism through which this inactivates the kinase domain is somewhat different for Abl than for Src, but the principles are the same. Interestingly, the specificity of the kinase inhibitor Gleevec (STI-571/Imanitib) for Abl as compared with Src is due to the distinct autoinhibited conformation imposed on the Abl kinase by its associated SH2 domain [165]. Gleevec binds exclusively to Abl in the autoinhibited conformation and therefore disturbs the equilibrium between inactive and active kinases, with beneficial effects in the treatment of chronic myelogenous leukemia. Under physiological conditions, Abl could be activated through loss of the N-terminal myristate group, which would abolish the conformation of the large lobe required for SH2 binding, or through competition by exogenous proteins for interactions with the SH2 and SH3 domains [166].

The regulatory device in which interaction domains undergo an intramolecular interaction with a catalytic domain or another interaction surface, resulting in reversible autoinhibition, is emerging as a common theme in intracellular regulatory proteins. Similarly, signaling enzymes are typically localized within the cell and docked onto their substrates through noncatalytic protein–protein and protein–phospholipid interactions.

1.11
SH2 Domains and Disease

Mutations in either SH2 domains or their binding motifs are associated with a number of human disorders. Missense mutations in the gene for human Shp-2 (*PTPN11*) cause Noonan's syndrome (NS), a developmental disorder associated with cardiac defects, facial dysmorphism, and skeletal abnormalities [167]. The substitutions that cause NS map to the autoinhibitory interface between the N-SH2 and PTP domains. As a consequence, these mutations abrogate the inhibitory interaction between the SH2 and phosphatase domains without disrupting pTyr binding or catalytic activity. NS therefore results from activating mutations that selectively block the ability of the N-SH2 domain to inhibit the PTP domain. Somatic mutations that activate Shp-2 are also common in juvenile myelomonocytic leukemia (JMML) and are present in other myeloid leukemias [168, 169]. The Shp-2 mutations in JMML are almost all in the N-SH2 domain and block autoinhibition, resulting in aberrant phosphatase activation. Although it is counterintuitive that a tyrosine phosphatase would promote malignant cell proliferation, Shp-2 activates signaling through growth stimulatory pathways such as Erk MAP kinase by dephosphorylating inhibitory pTyr sites.

Loss-of-function mutations in the Btk cytoplasmic tyrosine kinase cause a B-cell immunodeficiency termed X-linked agammaglobulinemia (XLA), which is associated with a deficiency in mature B cells and susceptibility to bacterial infection. Btk has an N-terminal PH domain, followed by SH3, SH2, and kinase domains, which adopt an extended conformation with little interaction between the domains [170]. Btk is activated by the binding of its PH domain to PI-3,4,5-P_3 at the membrane [171] and by tyrosine phosphorylation, and associates with the phosphorylated Blnk/SLP-65 scaffold through its SH2 domain [172]. Btk in turn stimulates a number of downstream pathways involved in B cell activation, for example, through the phosphorylation of PLC-γ2 and consequent calcium release [173]. XLA missense mutations can affect either the PH, SH3, SH2, or kinase domains and result in a loss of ligand binding, kinase activity, or protein stability [174, 175]. A distinct inherited immune defect, X-linked lymphoproliferative syndrome (XLP; also called Purtilo's syndrome or Duncan's disease), results from mutations in the *SAP/SH2D1A* gene [135, 176, 177]. Young boys with XLP are unable to control infections of Epstein–Barr virus (EBV) and typically develop a fulminant infectious mononucleosis, which is usually fatal as a result of severe tissue necrosis and organ failure. Surviving patients commonly develop malignant lymphomas. Many SAP mutations cause substitutions in the SH2 domain, resulting in either a loss of binding to SLAM or protein destabilization (or both) (Figures 1.6 and 1.7) [178, 179]. In the absence of SAP, the Fyn tyrosine kinase cannot be recruited to SLAM or related receptors, and this aspect of lymphoid regulation is lost.

SH2 domains are also indirectly influenced by a wide range of disease-causing mutations. Notably, oncogenic mutations in RTKs or cytoplasmic tyrosine kinases activate ectopic pTyr–SH2 domain interactions that promote intracellular signaling and lead to a malignant state. An example is the Tyr177 site located in the BCR region of the chimeric Bcr–Abl oncoprotein that causes chronic myelogenous leukemia. In the context of Bcr–Abl, Tyr177 is phosphorylated, and because it lies in a Tyr-x-Asn motif, recruits the SH2 domain of the Grb2 adaptor [180, 181], which in turn interacts through its SH3 domains with Sos and Gab-2 (Figure 1.10). These latter proteins stimulate the Ras-MAP kinase and PI 3′ kinase pathways, in a fashion that correlates with malignancy [182].

Similarly, pathogenic proteins encoded by human viruses or bacteria can form aberrant interactions with the SH2 domains of host cell proteins and thus rewire cellular functions, to the advantage of the pathogen. Examples include the latent membrane protein (LMP) 2A of EBV, which spans the membrane of infected B cells 12 times and has an N-terminal cytoplasmic region with three sites of tyrosine phosphorylation; these pTyr motifs bind to Src family tyrosine kinases and to the tandem SH2 domains of Syk. By recruiting these B-cell tyrosine kinases, LMP2A appears to interfere with normal signaling through the BCR and at the same time provide a survival signal [183, 184]. This helps to maintain viral latency by simultaneously blocking activation of the infected B cell, which would induce a lytic state, and keeping the cell alive [185]. In a related vein, the CagA protein of *Helicobacter pylori* is phosphorylated by Src family kinases at multiple tyrosine motifs

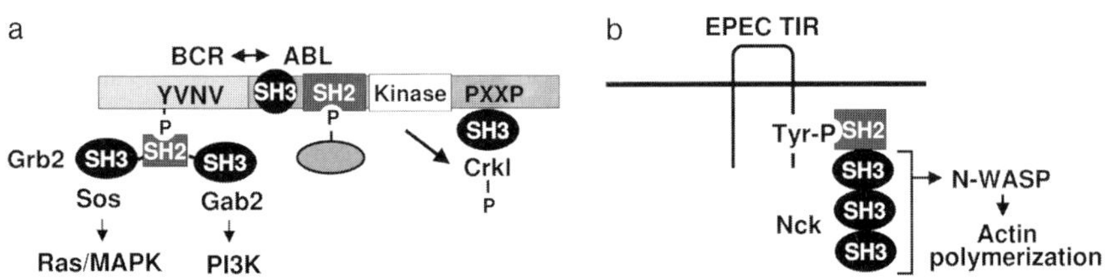

Figure 1.10 Aberrant SH2–pTyr interactions in cancer and bacterial infection. (a) Schematic of the Bcr–Abl tyrosine kinase. Tyr177 in the Bcr-encoded region is phosphorylated and binds Grb2. The Crkl adaptor binds a proline-rich motif in the C terminus and is itself a substrate for phosphorylation. (b) The TIR protein encoded by enteropathogenic *Escherichia coli* (EPEC) is phosphorylated at a tyrosine motif in its C-terminal cytoplasmic tail that binds the SH2 domain of the Nck adaptor, which recruits regulators of the actin cytoskeleton through its SH3 domains.

that bind the SH2 domains of Shp-2, which is important for morphological changes and increased motility of infected epithelial cells [186]. As a further example, the Tir protein of enteropathogenic *E. coli* is inserted into the host plasma membrane and is phosphorylated at a cytoplasmic Tyr site that recruits the Nck SH2/SH3 adaptor and its binding partner N-WASP to stimulate actin polymerization [187]. As a consequence, the bacterium induces abnormal actin polymerization (Figure 1.10), resulting in massive surface protrusions to which the bacterium adheres. In a striking example of convergent evolution, the A36R protein of vaccinia virus employs a similar motif to recruit Nck and induce actin polymerization, which drives the virus particle through the cell [188].

These observations regarding SH2 domains establish a precedent for the role of protein–protein interactions in disease states, and indeed there are now many examples of mutations that cause human disease by interfering with the functions of interaction domains. Another consideration is that naturally occurring non-synonomous polymorphisms might be a predisposing factor in disease by influencing protein–protein interactions. For example, the SKG strain of mice spontaneously develop chronic arthritis, caused by a polymorphism that results in a Trp-to-Cys substitution at the start of the C-SH2 domain of ZAP-70 [189]. This impairs the association of ZAP-70 with TCR and leads to attenuated T-cell signaling. This in turn favors the survival of autoreactive T cells that would otherwise be eliminated through negative selection and apoptosis.

Finally, it is likely that small molecules that modify protein–protein interactions will become more prominent in the treatment of disease. Such compounds could potentially block disease-associated protein interactions, stabilize existing complexes, or drive the interactions of proteins that are not normally associated. Indeed, immunosuppressant drugs such as cyclosporin and rapamycin function in the latter mode, by bridging an interaction between immunophilins and the protein phosphatase calcinuerin or the Tor protein kinase, respectively, which leads to inhibition of these enzymes and blockage of T-cell activation. Although protein–protein interactions are generally considered to be problematic drug targets, recent data have suggested the feasibility of inhibiting specific protein–protein interactions in therapeutically useful ways. For example, a small molecule that blocks the binding of p53 to Mdm2 stimulates the p53 pathway and has antitumor activity against cancer cells that retain wild-type p53 [190].

1.12 Summary

SH2 domains have a very simple function, namely, to recognize specific pTyr-containing sequences and thereby to couple tyrosine kinases to intracellular pathways that control cell growth, proliferation, metabolism, and differentiation. However, they also exhibit complex properties, through their ability to regulate both the substrate specificity and catalytic activity of enzymes in which they are embedded. In addition, by acting cooperatively with other interaction domains and by providing the basis for networks of interacting proteins (as in activated lymphocytes), they can exert a broad effect on cellular behavior. Perhaps most importantly, these functional attributes of SH2 domains have yielded a framework for the discovery and analysis of a large family of interaction domains, which has crystallized a new way of thinking about the lexicon of cellular organization.

References

1 Eckhart, W., Hutchinson, M. A., Hunter, T., An activity phosphorylating tyrosine in polyoma T antigen immunoprecipitates. *Cell* **1979**, *18*, 925–933.

2 Sefton, B. M., Hunter, T., Beemon, K., Eckhart, W., Evidence that the phosphorylation of tyrosine is essential for cellular transformation by Rous sarcoma virus. *Cell* **1980**, *20*, 807–816.

3 Witte, O. N., Dasgupta, A., Baltimore, D., Abelson murine leukaemia virus protein is phosphorylated in vitro to form phosphotyrosine. *Nature* **1980**, *283*, 826–831.

4 Pawson, T., Guyden, J., Kung, T. H., Radke, K., Gilmore, T., Martin, G. S., A strain of Fujinami sarcoma virus which is temperature-sensitive in protein phosphorylation and cellular transformation. *Cell* **1980**, *22*, 767–775.

5 Ushiro, H., Cohen, S., Identification of phosphotyrosine as a product of epidermal growth factor-activated protein kinase in A-431 cell membranes. *J. Biol. Chem.* **1980**, *255*, 8363–8365.

6 Ek, B., Westermark, B., Wasteson, A., Heldin, C. H., Stimulation of tyrosine-specific phosphorylation by platelet-derived growth factor. *Nature* **1982**, *295*, 419–420.

7 Petruzzelli, L. M., Ganguly, S., Smith, C. J., Cobb, M. H., Rubin, C. S., Rosen, O. M., Insulin activates a tyrosine-specific protein kinase in extracts of 3T3-L1 adipocytes and human placenta. *Proc. Natl. Acad. Sci. USA* **1982**, *79*, 6792–6796.

8 Hunter, T., Cooper, J. A., Epidermal growth factor induces rapid tyrosine phosphorylation of proteins in A431 human tumor cells. *Cell* **1981**, *24*, 741–752.

9 Schreiber, A. B., Libermann, T. A., Lax, I., Yarden, Y., Schlessinger, J., Biological role of epidermal growth factor-receptor clustering: investigation with monoclonal anti-receptor antibodies. *J. Biol. Chem.* **1983**, *258*, 846–853.

10 Downward, J., Yarden, Y., Scrace, G., Totty, N., Stockwell, P., Ullrich, A., Schlessinger, J., Waterfield, M. D., Close similarity of epidermal growth factor receptor and v-erb-B oncogene protein sequences. *Nature* **1984**, *307*, 521–527.

11 Stern, D. F., Heffernan, P. A., Weinberg, R. A., p185, a product of the *neu* proto-oncogene, is a receptor-like protein associated with tyrosine kinase activity. *Mol. Cell Biol.* **1986**, *6*, 1729–1740.

12 Rosen, O. M., Herrera, R., Olowe, Y., Petruzzelli, L. M., Cobb, M. H., Phosphorylation activates the insulin receptor tyrosine protein kinase. *Proc. Natl. Acad. Sci. USA* **1983**, *80*, 3237–3240.

13 Weinmaster, G., Zoller, M. J., Smith, M., Hinze, E., Pawson, T., Mutagenesis of Fujinami sarcoma virus: evidence that tyrosine phosphorylation of P130gag-fps modulates its biological activity. *Cell* **1984**, *37*, 559–568.

14 Hubbard, S. R., Till, J. H., Protein tyrosine kinase structure and function. *Annu. Rev. Biochem.* **2000**, *69*, 373–398.

15 Hubbard, S. R., Wei, L., Ellis, L., Hendrickson, W. A., Crystal structure of the tyrosine kinase domain of the human insulin receptor. *Nature* **1994**, *372*, 746–754.

16 Mohammadi, M., Schlessinger, J., Hubbard, S. R., Structure of the FGF receptor tyrosine kinase domain reveals a novel autoinhibitory mechanism. *Cell* **1996**, *86*, 577–587.

17 Stone, J. C., Atkinson, T., Smith, M. E., Pawson, T., Identification of functional regions in the transforming protein of Fujinami sarcoma virus by in-phase

insertion mutagenesis. *Cell* **1984**, *37*, 548–558.

18 Shibuya, M., Hanafusa, H., Nucleotide sequence of Fujinami sarcoma virus: evolutionary relationship of its transforming gene with transforming genes of other sarcoma viruses. *Cell* **1982**, *30*, 787–795.

19 Aspenstrom, P., A cdc42 target protein with homology to the non-kinase domain of FER has a potential role in regulating the actin cytoskeleton. *Curr. Biol.* **1997**, *7*, 479–487.

20 Takahashi, S., Inatome, R., Hotta, A., Qin, Q., Hackenmiller, R., Simon, M. C., Yamamura, H., Yanagai, S., Role for Fes/Fps tyrosine kinase in microtubule nucleation through is Fes/CIP4 homology domain. *J. Biol. Chem.* **2004**, *278*, 49129–49133.

21 Stone, J. C., Pawson, T., Correspondence between immunological and functional domains in the transforming protein of Fujinami sarcoma virus. *J. Virol.* **1985**, *55*, 721–727.

22 Sadowski, I., Stone, J. C., Pawson, T., A non-catalytic domain conserved among cytoplasmic protein-tyrosine kinases modifies the kinase function and transforming activity of Fujinami sarcoma virus $P130^{gag\text{-}fps}$. *Mol. Cell Biol.* **1986**, 4396–4408.

23 DeClue, J. E., Sadowski, G. S., Martin, G. S., Pawson, T., A conserved domain regulates interactions of the *v-fps* protein-tyrosine kinase with the host cell. *Proc. Natl. Acad. Sci. USA* **1987**, *84*, 9064–9068.

24 Koch, A., Moran, M., Sadowski, I., Pawson, T., The common *src* homology region 2 domain of cytoplasmic signaling proteins is a positive effector of v-*fps* tyrosine kinase function. *Mol. Cell Biol.* **1989**, *9*, 4131–4140.

25 Mayer, B. J., Hamaguchi, M., Hanafusa, H., A novel viral oncogene with structural similarity to phospholipase C. *Nature* **1988**, *332*, 272–275.

26 Stahl, M. L., Ferenz, C. R., Kelleher, K. L., Kriz, R. W., Knopf, J., Sequence similarity of phospholipase C with the non-catalytic region of *src*. *Nature* **1988**, *332*, 269–272.

27 Trahey, M., et al., Molecular cloning of two types of GAP complementary DNA from human placenta. *Science* **1988**, *242*, 1697–1700.

28 Vogel, U. S., Dixon, R. A. R., Schaber, M. D., Diehl, R. E., Marshall, M. S., Scolnick, E. M., Sigal, I. S., Gibbs, J. A., Cloning of bovine GAP and its interaction with oncogenic *ras* p21. *Nature* **1988**, *335*, 90–93.

29 Ren, R., Mayer, B. J., Cicchetti, P., Baltimore, D., Identification of a ten-amino acid proline-rich SH3 binding site. *Science* **1993**, *259*, 1157–1161.

30 Margolis, B., et al., EGF induces tyrosine phosphorylation of phospholipase C-II: a potential mechanism for EGF signalling. *Cell* **1989**, *57*, 1101–1107.

31 Meisenhelder, J., Suh, P.-G., Rhee, S. G., Hunter, T., Phospholipase C is a substrate for the PDGF and EGF receptor protein-tyrosine kinases in vivo and in vitro. *Cell* **1989**, *57*, 1109–1122.

32 Wahl, M. I., Nishibe, S., Suh, P.-G., Rhee, S. G., Carpenter, G., Epidermal growth factor stimulates tyrosine phosphorylation of phospholipase C (II) independently of receptor internalization and extracellular calcium. *Proc. Natl. Acad. Sci. USA* **1989**, *86*, 1568–1572.

33 Molloy, C. J., Bottaro, D. P., Fleming, T. P., Marshall, M. S., Gibbs, J. B., Aaronson, S. A., PDGF induction of tyrosine phosphorylation of GTPase activating protein. *Nature* **1989**, *342*, 711–714.

34 Ellis, C., Moran, M., McCormick, F., Pawson, T., Phosphorylation of GAP and GAP-associated proteins by transforming and mitogenic tyrosine kinases. *Nature* **1990**, *343*, 377–381.

35 Pawson, T., Non-catalytic domains of cytoplasmic protein-tyrosine kinases: regulatory elements in signal transduction. *Oncogene* **1988**, *3*, 491–495.

36 Kumjian, D. A., Wahl, M. I., Rhee, S. G., Daniel, T. O., Platelet-derived growth factor (PDGF) binding promotes physical association of PDGF receptor with phospholipase C. *Proc. Natl. Acad. Sci. USA* **1989**, *86*, 8232–8236.

37 Kazlauskas, A., Cooper, J. A., Autophosphorylation of the PDGF receptor in the kinase insert region regulates interactions with cell proteins. *Cell* **1989**, *58*, 1121–1133.

38 Anderson, D., Koch, C. A., Grey, L., Ellis, C., Moran, M. F., Pawson, T., Binding of SH2 domains of phospholipase Cγ1, GAP, and Src to activated growth factor receptors. *Science* **1990**, *250*, 979–982.

39 Moran, M. F., Koch, C. A., Anderson, D., Ellis, C., England, L., Martin, G. S., Pawson, T., Src homology region 2 domains direct protein–protein interactions in signal transduction. *Proc. Natl. Acad. Sci. USA* **1990**, *87*, 8622–8626.

40 Mayer, B. J., Hanafusa, H., Association of the v-*crk* oncogene product with phosphotyrosine-containing proteins and protein kinase activity. *Proc. Natl. Acad. Sci. USA* **1990**, *87*, 2638–2642.

41 Matsuda, M., Mayer, B. J., Fukui, Y., Hanafusa, H., Binding of oncoprotein, p47gag-crk, to a broad range of phosphotyrosine-containing proteins. *Science* **1990**, *248*, 1537–1539.

42 Margolis, B., Li, N., Koch, A., Mohammadi, M., Hurwitz, D. R., Zilberstein, A., Ullrich, A., Pawson, T., Schlessinger, J., The tyrosine phosphorylated carboxyterminus of the EGF receptor is a binding site for GAP and PLC-gamma. *EMBO J.* **1990**, *9*, 4375–4380.

4.3 Mayer, B. J., Jackson, P. K., Baltimore, D., The noncatalytic src homology region 2 segment of abl tyrosine kinase binds to tyrosine-phosphorylated cellular proteins with high affinity. *Proc. Natl. Acad. Sci. USA* **1991**, *88*, 627–631.

44 Escobedo, J. A., Kaplan, D. R., Kavanaugh, W. M., Turck, C. W., Williams, L. T., A phosphatidylinositol-3′ kinase binds to platelet-derived growth factor receptors through a specific receptor sequence containing phosphotyrosine. *Mol. Cell Biol.* **1991**, *11*, 1125–1132.

45 Koch, C. A., Moran, M. F., Anderson, D., Liu, X., Mbamalu, G., Pawson, T., Multiple SH2-mediated interactions in v-*src*-transformed cells. *Mol. Cell Biol.* **1992**, *12*, 1366–1374.

46 Reedijk, M., Liu, X., van der Geer, P., Letwin, K., Waterfield, M. D., Hunter, T., Pawson, T., Tyr721 regulates specific binding of the CSF-1receptor kinase insert to PI 3′-kinase SH2 domains: a model for SH2-mediated receptor–target interactions. *EMBO J.* **1992**, *11*, 1365–1372.

47 Koch, C. A., Anderson, D., Moran, M. F., Ellis, C., Pawson, T., SH2 and SH3 domains: Elements that control interactions of cytoplasmic signaling proteins. *Science* **1991**, *252*, 668–674.

48 Pawson, T., Nash, P., Assembly of cell regulatory systems through protein interaction domains. *Science* **2003**, *300*, 445–452.

49 Tong, A. H., et al., A combined experimental and computational strategy to define protein interaction networks for peptide recognition modules. *Science* **2002**, *295*, 321–324.

50 Kuriyan, J., Cowburn, D., Modular peptide recognition domains in eukaryotic signaling. *Annu. Rev. Biophys. Biomol. Struct.* **1997**, *26*, 259–288.

51 Bradshaw, J. M., Waksman, G., Molecular recognition by SH2 domains. *Adv. Protein Chem.* **2002**, *61*, 161–210.

52 Piccione, E., Case, R. D., Domchek, S. M., Hu, P., Chaudhuri, M., Backer, J. M., Schlessinger, J., Shoelson, S. E., Phosphatidylinositol 3-kinase p85 SH2 domain specificity defined by direct phosphopeptide/SH2 domain binding. *Biochemistry* **1993**, *32*, 3197–3202.

53 Lemmon, M. A., Ladbury, J. E., Thermodynamic studies of tyrosyl-phosphopeptide binding to the SH2 domain of p56lck. *Biochemistry* **1994**, *33*, 5070–5076.

54 Bradshaw, J. M., Mitaxov, V., Waksman, G., Mutational investigation of the specificity determining region of the Src SH2 domain. *J. Mol. Biol.* **2000**, *299*, 521–535.

55 Domchek, S. M., Auger, K. R., Chatterjee, S., Burke, T. R. Jr., Shoelson, S. E., Inhibition of SH2 domain/phosphoprotein association by a nonhydrolyzable phosphonopeptide. *Biochemistry* **1992**, *31*, 9865–9870.

56 Songyang, Z., et al., Identification of phosphotyrosine peptide motifs which bind to SH2 domains. *Cell* **1993**, *72*, 767–778.

57 Songyang, Z., et al., Specific motifs recognized by the SH2 domains of Csk, 3BP2, fes/fps, Grb2, SHPTP1, SHC, Syk and vav. *Mol. Cell Biol.* **1994**, *14*, 2777–2785.

58 Ladbury, J. E., Lemmon, M. A., Zhou, M., Green, J., Botfield, M. C.,

Schlessinger, J., Measurement of the binding of tyrosyl phosphopeptides to SH2 domains: a reappraisal. *Proc. Natl. Acad. Sci. USA* **1995**, *92*, 3199–3203.

59 Porter, M., Schindler, T., Kuriyan, J., Miller, W. T., Reciprocal regulation of Hck activity by phosphorylation of Tyr(527) and Tyr(416). Effect of introducing a high affinity intramolecular SH2 ligand. *J. Biol. Chem.* **2000**, *275*, 2721–2726.

60 Pawson, T., Protein modules and signalling networks. *Nature* **1995**, *373*, 573–580.

61 Ponzetto, C., et al., A multifunctional docking site mediates signaling and transformation by the hepatocyte growth factor/scatter factor receptor family. *Cell* **1994**, *77*, 261–271.

62 Waksman, G., et al., Crystal structure of the phosphotyrosine recognition domain (SH2) of the v-*src* tyrosine kinase complexed with tyrosine phosphorylated peptides. *Nature* **1992**, *358*, 646–653.

63 Overduin, M., Rios, C. B., Mayer, B. J., Baltimore, D., Cowburn, D., Three-dimensional solution structure of the src homology 2 domain of c-abl. *Cell* **1992**, *70*, 697–704.

64 Booker, G. W., Breeze, A. L., Downing, A. K., Panayotou, G., Gout, I., Waterfield, M. D., Campbell, I. D., Structure of an SH2 domain of the p85α subunit of phosphatidylinositol-3-OH kinase. *Nature* **1992**, *358*, 684–687.

65 Marengere, L. E. M., Pawson, T., Identification of residues in GAP SH2 domains that control binding to tyrosine phosphorylated growth factor receptors and p62. *J. Biol. Chem.* **1992**, *267*, 22779–22786.

66 Mayer, B. J., Jackson, P. K., Van Ettern R. A., Baltimore, D., Point mutations in the abl SH2 domain coordinately impair phosphotyrosine binding in vitro and transforming activity in vivo. *Mol. Cell Biol.* **1992**, *12*, 609–618.

67 Waksman, G., Shoelson, S., Pant, N., Cowburn, D., Kuriyan, J., Binding of a high affinity phosphotyrosyl peptide in the src SH2 domain: crystal structures of the complexed and peptide-free forms. *Cell* **1993**, *72*, 779–790.

68 Eck, M. I., Shoelson, S. E., Harrison, S. C., Recognition of a high affinity phosphotyrosyl peptide by the Src homology 2 domain of p56^{lck}. *Nature* **1993**, *362*, 87–91.

69 Lee, C.-H., Kominos, D., Jacques, S., Margolis, B., Schlessinger, J., Shoelson, S. E., Kuriyan, J., Crystal structures of peptide complexes of the amino-terminal SH2 domain of the Syp tyrosine phosphatase. *Structure* **1994**, *2*, 423–438.

70 Pascal, S. M., Singer, A. U., Gish, G., Yamazaki, T., Shoelson, S. E., Pawson, T., Kay, L. E., Forman-Kay, J. D., Nuclear magnetic resonance structure of an SH2 domain of phospholipase C-gamma1 complexed with a high affinity binding peptide. *Cell* **1994**, *77*, 461–472.

71 Songyang, Z., Gish, G., Mbamalu, G., Pawson, T., Cantley, L. C., A single point mutation switches the specificity of group III Src homology (SH) 2 domains to that of group I SH2 domains. *J. Biol. Chem.* **1995**, *270*, 26029–26032.

72 Rahuel, J., et al., Structural basis for specificity of GRB2-SH2 revealed by a novel ligand binding mode. *Nat. Struct. Biol.* **1996**, *3*, 586–589.

73 McNemar, C., et al., Thermodynamic and structural analysis of phosphotyrosine polypeptide binding to Grb2-SH2. *Biochemistry* **1997**, *36*, 10006–10014.

74 Ogura, K., et al., Conformation of an Shc-derived phosphotyrosine-containing peptide complexed with the Grb2 SH2 domain. *J. Biomol. NMR* **1997**, *10*, 273–278.

75 Ivancic, M., Daly, R. J., Lyons, B. A., Solution structure of the human Grb7-SH2 domain/erbB2 peptide complex and structural basis for Grb7 binding to ErbB2. *J. Biomol. NMR* **2003**, *27*, 205–219.

76 Marengere, L. E. M., Songyang, Z., Gish, G. D., Schaller, M. D., Parsons, T., Stern, M. J., Cantley, L. C., Pawson, T., SH2 domain specificity and activity modified by a single residue. *Nature* **1994**, *369*, 502–550.

77 Kimber, M. S., Nachman, J., Gish, G., Pawson, T., Pai, E., Structural basis for specificity switching by the Src SH2 domain. *Molecular Cell* **2000**, *5*, 1043–1049.

78 Yaffe, M. B., Phosphotyrosine-binding domains in signal transduction. *Nat. Rev. Mol. Cell Biol.* **2002**, *3*, 177–186.

79 YAFFE, M. B., ELIA, A. E., Phosphoserine/threonine-binding domains. *Curr. Opin. Cell Biol.* **2001**, *13*, 131–138.

80 MARMORSTEIN, E., Protein modules that manipulate histone tails for chromatin regulation. *Nat. Rev. Mol. Cell Biol.* **2001**, *2*, 422–432.

81 JAAKKOLA, P., et al., Targeting of HIF-alpha to the von Hippel–Lindau ubiquitylation complex by O_2-regulated prolyl hydroxylation. *Science* **2001**, *292*, 468–472.

82 IVAN, M., et al., HIFalpha targeted for VHL-mediated destruction by proline hydroxylation: implications for O_2 sensing. *Science* **2001**, *292*, 464–468.

83 HICKE, L., DUNN, R., Regulation of membrane protein transport by ubiquitin and ubiquitin-binding proteins. *Ann. Rev. Cell Dev. Biol.* **2003**, *19*, 141–172.

84 DUEBER, J. E., YEH, B. J., CHAK, K., LIM, W. A., Reprogramming control of an allosteric signaling switch through modular recombination. *Science* **2003**, *301*, 1904–1908.

85 OLIVIER, J. P., et al., A Drosophila SH2–SH3 adaptor protein implicated in coupling the sevenless tyrosine kinase to an activator of Ras guanine nucleotide exchange, Sos. *Cell* **1993**, *73*, 179–191.

86 SIMON, M. A., DODSON, G. S., RUBIN, G. M., SH3–SH2–SH3 protein is required for $p21^{ras}$ activation and binds to sevenless and Sos proteins in vitro. *Cell* **1993**, *73*, 169–177.

87 ROZAKIS-ADCOCK, M., FERNLEY, R., WADE, J., PAWSON, T., BOWTELL, D., The SH2 and SH3 domains of mammalian Grb2 couple the EGF-receptor to mSos1, an activator of Ras. *Nature* **1993**, *363*, 83–85.

88 LI, N., BATZER, A., DALY, R., SKOLNIK, E., CHARDIN, P., BAR-SAGI, D., MARGOLIS, B., SCHLESSINGER, J., Guanine nucleotide releasing factor hSos1 binds to Grb2 and links receptor tyrosine kinases to Ras signaling. *Nature* **1993**, *363*, 85–88.

89 RAABE, T., OLIVIER, J. P., DICKSON, B., LIU, X., GISH, G. D., PAWSON, T., HAFEN, E., Biochemical and genetic analysis of Drk SH2/SH3 adaptor protein of *Drosophila*. *EMBO J.* **1995**, *14*, 2509–2518.

90 CHENG, A. M., et al., Mammalian Grb2 regulates multiple steps in embryonic development and malignant transformation. *Cell* **1998**, *95*, 793–803.

91 LOCK, L. S., ROYAL, I., NAUJOKAS, M. A., PARK, M., Identification of an atypical Grb2 carboxyl-terminal SH3 domain binding site in Gab docking proteins reveals Grb2-dependent and –independent recruitment of Gab1 to receptor tyrosine kinases. *J. Biol. Chem.* **2000**, *275*, 31536–31545.

92 SCHAEPER, U., FUCHS, K. P., SACHS, M., KEMPKES, B., BIRCHMEIER, W., Coupling of Gab1 to c-Met, Grb2, and Shp2 mediates biological responses. *J. Cell Biol.* **2000**, *149*, 1419–1432.

93 GU, H., NEEL, B. G., The "Gab" in signal transduction. *Trends Cell Biol.* **2003**, *13*, 122–130.

94 THOMAS, C. C., DEAK, M., ALESSI, D. R., VAN AALTEN, D. M., High-resolution structure of the pleckstrin homology domain of protein kinase B/akt bound to phosphatidylinositol (3, 4, 5)-trisphosphate. *Curr. Biol.* **2002**, *12*, 1256–1262.

95 CANTLEY, L. C., The phosphoinositide 3-kinase pathway. *Science* **2002**, *296*, 1655–1657.

96 YAFFE, M. B., How do 14-3-3 proteins work? Gatekeeper phosphorylation and the molecular anvil hypothesis. *FEBS Lett.* **2002**, *513*, 53–57.

97 RIVERA, G. M., BRICENO, C. A., TAKESHIMA, F., SNAPPER, S. B., MAYER, B. J., Inducible clustering of membrane-targeted SH3 domains of the adaptor protein Nck triggers localized actin polymerization. *Curr. Biol.* **2004**, *14*, 11–22.

98 BUDAY, L., WUNDERLICH, L., TAMAS, P., The Nck family of adapter proteins: regulators of actin cytoskeleton. *Cell Signal* **2002**, *14*, 723–731.

99 ZHAO, Z. S., MANSER, E., Interaction between PAK and Nck: a template for Nck targets and role of PAK autophosphorylation. *Mol. Cell Biol.* **2000**, *20*, 3906–3917.

100 ROHATGI, R., NOLLAU, P., HO, H. Y., KIRSCHNER, M. W., MAYER, B. J., Nck and phosphatidylinositol 4, 5-bisphosphate synergistically activate actin polymerization through the N-WASP-Arp2/3 pathway. *J. Biol. Chem.* **2001**, *276*, 26448–26452.

101 GARRITY, P. A., RAO, Y., SALECKER, I., MCGLADE, J., PAWSON, T., ZIPURSKY, S. L., *Drosophila* photoreceptor axon guidance and targeting requires the Dreadlocks

SH2/SH3 adaptor protein. *Cell* **1996**, *85*, 639–650.

102 Bladt, F., Aippersbach, E., Gelkop, S., Strasser, G. A., Nash, P., Tafuri, A., Gertler, F. B., Pawson, T., The murine Nck SH2/SH3 adaptors are important for the development of mesoderm-derived embryonic structures and for regulating the cellular actin network. *Mol. Cell Biol.* **2003**, *23*, 4586–4597.

103 Tanaka, S., et al., C3G, a guanine nucleotide-releasing protein expressed ubiquitously, binds to the Src homology 3 domains of CRK and GRB2/ASH proteins. *Proc. Natl. Acad. Sci. USA* **1994**, *91*, 3443–3447.

104 Ohba, Y., et al., Requirement for C3G-dependent Rap1 activation for cell adhesion and embryogenesis. *EMBO J.* **2001**, *20*, 3333–3341.

105 Kiyokawa, E., Hashimoto, Y., Kobayashi, S., Sugimura, H., Kurata, T., Matsuda, M., Activation of Rac1 by a Crk SH3-binding protein, DOCK180. *Genes. Dev.* **1998**, *12*, 3331–3336.

106 Ichiba, T., Kuraishi, Y., Sakai, O., Nagata, S., Groffen, J., Kurata, T., Hattori, S., Matsuda, M., Enhancement of guanine-nucleotide exchange activity of C3G for Rap1 by the expression of Crk, CrkL, and Grb2. *J. Biol. Chem.* **1997**, *272*, 22215–22220.

107 Reddien, P. W., Horvitz, H. R., CED-2/CrkII and CED-10/Rac control phagocytosis and cell migration in *Caenorhabditis elegans. Nat. Cell Biol.* **2000**, *2*, 131–136.

108 Feller, S. M., Crk family adaptors-signalling complex formation and biological roles. *Oncogene* **2001**, *20*, 6348–6371.

109 Pawson, T., Nash, P., Protein–protein interactions define specificity in signal transduction. *Genes. Dev.* **2000**, *14*, 1027–1047.

110 Pawson, T., Specificity in signal transduction: From phosphotyrosine–SH2 domain interactions to complex cellular systems. *Cell* **2004**, *116*, 191–203.

111 Jackman, J. K., Motto, D. G., Sun, Q., Tanamoto, M., Turck, C. W., Peltz, G. A., Korezky, G. A., Findell, P. R., Molecular cloning of SLP-76, a 76-kDa tyrosine phosphorylation associated with Grb2 in T cells. *J. Biol. Chem.* **1995**, *270*, 7029–7032.

112 Fu, C., Turck, C. W., Kurosaki, T., Chan, A. C., BLNK: a central linker protein in B cell activation. *Immunity* **1998**, *9*, 93–103.

113 Bubeck Wardenburg, J., Pappu, R., Bu, J. Y., Mayer, B., Chernoff, J., Straus, D., Chan, A. C., Regulation of PAK activation and the T cell cytoskeleton by the linker protein SLP-76. *Immunity* **1998**, *9*, 607–616.

114 Su, Y. W., Zhang, Y., Schweikert, J., Koretzky, G. A., Reth, M., Wienands, J., Interaction of SLP adaptors with the SH2 domain of Tec family kinases. *Eur. J. Immunol.* **1999**, *29*, 3702–3711.

115 Chiu, C. W., Dalton, M., Ishiai, M., Kurosaki, T., Chan, A. C., BLNK: molecular scaffolding through 'cis'-mediated organization of signaling proteins. *EMBO J.* **2002**, *21*, 6461–6472.

116 Pivniouk, V., Tsitsikov, E., Swinton, P., Rathbun, G., Alt, F. W., Geha, R. S., Impaired viability and profound block in thymocyte development in mice lacking the adaptor protein SLP-76. *Cell* **1998**, *94*, 229–238.

117 Clements, J. L., Yang, B., Ross-Barta, S. E., Eliason, S. L., Hrstka, R. F., Williamson, R. A., Koretzky, G. A., Requirement for the leukocyte-specific adapter protein SLP-76 for normal T cell development. *Science* **1998**, *281*, 416–419.

118 Minegishi, Y., Rohrer, J., Coustan-Smith, E., Lederman, H. M., Pappu, R., Campana, D., Chan, A. C., Conley, M. E., An essential role for BLNK in human B cell development. *Science* **1999**, *286*, 1954–1957.

119 Kalesnikoff, J., Sly, L. M., Hughes, M. R., Buchse, T., Rauh, M. J., Cao, L. P., Lam, V., Mui, A., The role of SHIP in cytokine-induced signaling. *Rev. Physiol. Biochem. Pharmacol.* **2003**, *149*, 87–103.

120 Fu, Z., Aronoff-Spencer, E., Backer, J. M., Gerfen, G. J., The structure of the inter-SH2 domain of class IA phosphoinositide 3-kinase determined by site-directed spin labeling EPR and homology modeling. *Proc. Natl. Acad. Sci. USA* **2003**, *100*, 3275–3280.

121 Hall, C., et al., Alpha 2-chimerin, an SH2-containing GTPase-activating protein for the ras-related protein p21rac derived by alternate splicing of the human n-chimerin gene, is selectively expressed in brain regions and testes. *Mol. Cell Biol.* **1993**, *13*, 4986–4998.

122 Crespo, P., Schuebel, K. E., Ostrom, A. A., Gutkind, J. S., Bustelo, X. R., Phosphotyrosine-dependent activation of Rac-1 GDP/GTP exchange by the *vav* proto-oncogene product. *Nature* **1997**, *385*, 169–172.

123 Barbieri, M. A., Kong, C., Chen, P. I., Horazdovsky, B. F., Stahl, P. D., The SRC homology 2 domain of Rin1 mediates its binding to the epidermal growth factor receptor and regulates receptor endocytosis. *J. Biol. Chem.* **2003**, *278*, 32027–32036.

124 Brown, M. T., Cooper, J. A., Regulation, substrates and functions of Src. *Biochem. Biophys. Acta* **1996**, *1287*, 121–149.

125 Pendergast, A. M., The Abl family kinases: mechanisms of regulation and signaling. *Adv. Cancer Res.* **2002**, *85*, 51–100.

126 Greer, P., Closing in on the biological functions of Fps/Fes and Fer. *Nat. Rev. Mol. Cell Biol.* **2002**, *3*, 278–289.

127 Smith, C. I., Islam, T. C., Mattsson, P. T., Mohamed, A. J., Nore, B. F., Vihinen, M., The Tec family of cytoplasmic tyrosine kinases: mammalian Btk, Bmx, Itk, Tec, Txk and homologs in other species. *BioEssays* **2001**, *23*, 436–446.

128 Neel, B. G., Gu, H., Pao, L., The 'Shp'ing news: SH2 domain-containing tyrosine phosphatases in cell signaling. *Trends Biochem. Sci.* **2003**, *28*, 284–293.

129 Joazeiro, C. A., Wing, S. S., Huang, H., Leverson, J. D., Hunter, T., Liu, Y. C., The tyrosine kinase negative regulator c-Cbl as a RING-type, E2-dependent ubiquitin–protein ligase. *Science* **1999**, *286*, 309–312.

130 Haglund, K., Sigismund, S., Polo, S., Szymkiewicz, I., Di Fiore, P. P., Dikic, I., Multiple monoubiquitination of RTKs is sufficient for their endocytosis and degradation. *Nat. Cell Biol.* **2003**, *5*, 461–466.

131 Mosesson, Y., Shtiegman, K., Katz, M., Zwang, Y., Vereb, G., Szollosi, J., Yarden, Y., Endocytosis of receptor tyrosine kinases is driven by monoubiquitylation, not polyubiquitylation. *J. Biol. Chem.* **2003**, *78*, 21323–21326.

132 Wormald, S., Hilton, D. J., Inhibitors of cytokine signal transduction. *J. Biol. Chem.* **2004**, *279*, 821–824.

133 Manning, C. M., Mathews, W. R., Fico, L. P., Thackeray, J. R., Phospholipase C-gamma contains introns shared by src homology 2 domains in many unrelated proteins. *Genetics* **2003**, *164*, 433–442.

134 Latour, S., Veillette, A., Molecular and immunological basis of X-lined lymphoproliferative disease. *Immunol. Rev.* **2003**, *192*, 212–224.

135 Sayos, J., et al., The X-linked lymphoproliferative-disease gene product SAP regulates signals induced through the co-receptor SLAM. *Nature* **1998**, *395*, 462–469.

136 Poy, F., Yaffe, M. B., Sayos, J., Saxena, K., Morra, M., Sumegi, J., Eck, M., Crystal structures of the XLP protein SAP reveal a class of SH2 domains with extended, phosphotyrosine-independent sequence recognition. *Molecular Cell* **1999**, *4*, 555–561.

137 Hwang, P. M., et al., A "three-pronged" binding mechanism for the SAP/SH2D1A SH2 domain: structural basis and relevance to the XLP syndrome. *EMBO J.* **2002**, *21*, 314–323.

138 Li, S.-C., Gish, G., Yang, D., Coffey, A. J., Forman-Kay, J. D., Ernberg, I., Kay, L. E., Pawson, T., Novel mode of ligand binding by the SH2 domain of the human XLP disease gene product SAP/SH2D1A. *Curr. Biol.* **1999**, *9*, 1355–1362.

139 Latour, S., Gish, G., Helgason, C. D., Humphries, R. K., Pawson, T., Veillette, A., Regulation of SLAM-mediated signal transduction by SAP, the X-linked lymphoproliferative gene product. *Nat. Immunol.* **2001**, *2*, 681–690.

140 Berry, D. M., Nash, P., Liu, S. K., Pawson, T., A high-affinity Arg-X-X-Lys SH3 binding motif confers specificity for the interaction between Gads and SLP-76 in T cell signaling. *Curr. Biol.* **2002**, *12*, 1336–1341.

141 Latour, S., Roncagalli, R., Chen, R., Bakinowski, M., Shi, X., Schwartzberg, P., Davidson, D., Veillette, A., Binding of SAP SH2 domain to FynT

SH3 domain reveals a novel mechanism of receptor signalling in immune regulation. *Nat. Cell Biol.* **2003**, *5*, 149–154.

142 Chan, B., et al., SAP couples Fyn to SLAM immune receptors. *Nat. Cell Biol.* **2003**, *5*, 155–160.

143. Borg, J. P., Ooi, J., Levy, E., Margolis, B., The phosphotyrosine interaction domains of X11 and FE65 bind to distinct sites on the YENPTY motif of amyloid precursor protein. *Mol. Cell Biol.* **1998**, *16*, 6229–6241.

144 Dhalluin, C., et al., Structural basis of SNT PTB domain interactions with distinct neurotrophic receptors. *Mol. Cell* **2000**, *6*, 921–929.

145 Zwahlen, C., Li, S.-C., Kay, L. E., Pawson, T., Forman-Kay, J. D., Multiple modes of peptide recognition by the PTB domain of the cell fate determinant Numb. *EMBO J.* **2000**, *19*, 1505–1515.

146 Ravichandran, K. S., Zhou, M. M., Pratt, J. C., Harlan, J. E., Walk, S. F., Fesik, S. W., Burakoff, S. J., Evidence for a requirement for both phospholipid and phosphotyrosine binding via the Shc phosphotyrosine-binding domain in vivo. *Mol. Cell Biol.* **1997**, *17*, 5540–5549.

147 Blomberg, N., Baraldi, E., Nilges, M., Saraste, M., The PH superfold: a structural scaffold for multiple functions. *Trends Biochem. Sci.* **1999**, *24*, 441–445.

148 Hubbard, S. R., Crystal structure of the activated insulin receptor tyrosine kinase in complex with peptide substrate and ATP analog. *EMBO J.* **1997**, *16*, 5572–5581.

149 Hu, J., Liu, J., Ghirlando, R., Saltiel, A. R., Hubbard, S. R., Structural basis for recruitment of the adaptor protein APS to the activated insulin receptor. *Mol. Cell* **2003**, *12*, 1379–1389.

150 Stein, E. G., Ghirlando, R., Hubbard, S. R., Structural basis for dimerization of the Grb10 Src homology 2 domain: implications for ligand specificity. *J. Biol. Chem.* **2003**, *278*, 13257–13264.

151 Levy, D. E., Darnell, J. E., Stats: transcriptional control and biological impact. *Nat. Rev. Mol. Cell Biol.* **2002**, *3*, 651–662.

152 Ottinger, E. A., Botfield, M. C., Shoelson, S. E., Tandem SH2 domains confer high specificity in tyrosine kinase signaling. *J. Biol. Chem.* **1998**, *273*, 729–735.

153 Chan, A. C., Shaw, A. S., Regulation of antigen receptor signal transduction by protein tyrosine kinases. *Curr. Opin. Immunol.* **1996**, *8*, 394–401.

154 Grucza, R. A., Bradshaw, J. M., Mitaxov, V., Waksman, G., Role for electrostatic interactions in SH2 domain recognition: salt-dependence of tyrosyl-phosphorylated peptide binding to the tandem SH2 domain of the Syk kinase and the single SH2 domain of the Src kinase. *Biochemistry* **2000**, *39*, 10072–10081.

155 Hatada, M. H., et al., Molecular basis for interaction of the protein tyrosine kinase ZAP-70 with the T-cell receptor. *Nature* **1995**, *377*, 32–38.

156 Futterer, K., Wong, J., Grucza, R. A., Chan, A. C., Waksman, G., Structural basis for Syk tyrosine kinase ubiquity in signal transduction pathways revealed by the crystal structure of its regulatory SH2 domains bound to a dually phosphorylated ITAM peptide. *J. Mol. Biol.* **1998**, *281*, 523–537.

157 Kumaran, S., Grucza, R. A., Waksman, G., The tandem Src homology 2 domain of the Syk kinase: a molecular device that adapts to interphosphotyrosine distances. *Proc. Natl. Acad. Sci. USA* **2003**, *100*, 14828–14833.

158 Meng, W., Sawasdikosol, S., Burakoff, S. J., Eck, M. J., Structure of the amino-terminal domain of Cbl complexed to its binding site on ZAP-70 kinase. *Nature* **1999**, *398*, 22–23.

159 Chen, X., Vinkemeier, U., Zhao, Y., Jeruzalmei, D., Darnell, J. E., Kuriyan, J., Crystal structure of tyrosine phosphorylated STAT-1 dimer bound to DNA. *Cell* **1998**, *93*, 827–839.

160 Becker, S., Groner, B., Muller, C. W., Three-dimensional structure of the Stat3beta homodimer bound to DNA. *Nature* **1998**, *394*, 145–151.

161 Hof, P., Pluskey, S., Dhe-Paganon, S., Eck, M. J., Shoelson, S. E., Crystal structure of the tyrosine phosphatase SHP-2. *Cell* **1998**, *92*, 441–450.

162 Sicheri, F., Moarefi, I., Kuriyan, J., Crystal structure of the Src family tyrosine kinase Hck. *Nature* **1997**, *385*, 582–585.

163 Xu, W., Harrison, S. C., Eck, M. J., Three-dimensional structure of the

tyrosine kinase c-Src. *Nature* **1997**, *385*, 595–602.

164 Young, M. A., Gonfloni, S., Superti-Furga, G., Roux, B., Kuriyan, J., Dynamic coupling between the SH2 and SH3 domains of c-Src and Hck underlies their inactivation by C-terminal tyrosine phosphorylation. *Cell* **2001**, *105*, 115–126.

165 Nagar, B., et al., Structural basis for the autoinhibition of c-Abl tyrosine kinase. *Cell* **2003**, *112*, 859–872.

166 Hantschel, O., Nagar, B., Guettler, S., Kretzschmar, J., Dorey, K., Kuriyan, J., Superti-Furga, G., A myristoyl/phosphotyrosine switch regulates c-Abl. *Cell* **2003**, *112*, 845–857.

167 Tartaglia, M., et al., Mutations in PTPN11, encoding the protein tyrosine phosphatase SHP-2, cause Noonan syndrome. *Nat. Genet.* **2001**, *29*, 465–468.

168 Tartaglia, M., et al., Somatic mutations in PTPN11 in juvenile myelomonocytic leukemia, myelodysplastic syndromes and acute myeloid leukemia. *Nat. Genet.* **2003**, *34*, 148–150.

169 Loh, M. L., et al., Somatic mutations in PTPN11 implicate the protein tyrosine phosphatase SHP-2 in leukemogenesis. *Blood* **2004**, *103*, 2325–2331.

170 Marquez, J. A., et al., Conformation of full-length Bruton tyrosine kinase (Btk) from synchrotron X-ray solution scattering. *EMBO J.* **2003**, *22*, 4616–4624.

171 Salim, K., et al., Distinct specificity in the recognition of phosphoinositides by the pleckstrin homology domain of dynamin and the Bruton's tyrosine kinase. *EMBO J.* **1996**, *15*, 6241–6250.

172 Hashimoto, S., et al., Identification of the SH2 domain binding protein of Bruton's tyrosine kinase as BLNK: functional significance of Btk-SH2 domain in B-cell antigen receptor-coupled calcium signaling. *Blood* **1999**, *94*, 2357–2364.

173 Watanabe, D., Hashimoto, S., Ishiai, M., Matsushita, M., Baba, Y., Kishimoto, T., Kurosaki, T., Tsukada, S., Four tyrosine residues in phospholipase C-gamma 2, identified as Btk-dependent phosphorylation sites, are required for B cell antigen receptor-coupled calcium signaling. *J. Biol. Chem.* **2001**, *276*, 38595–38601.

174 Vertie, D., et al., The gene involved in X-linked agammaglobulinaemia is a member of the *src* family of protein-tyrosine kinases. *Nature* **1993**, *361*, 226–226.

175 Vihinen, M., Mattsson, P. T., Smith, C. I., Bruton tyrosine kinase (BTK) in X-linked agammaglobulinemia (XLA). *Front Biosci.* **2000**, *5*, D917–D928.

176 Morra, M., Howie, D., Grande, M. S., Sayos, J., Wang, N., Wu, C., Engel, P., Terhorst, C., X-linked lymphoproliferative disease: a progressive immunodeficiency. *Annu. Rev. Immunol.* **2001**, *19*, 657–682.

177 Coffey, A. J., et al., Host response to EBV infection in X-linked lymphoproliferative disease results from mutations in an SH2-domain encoding gene. *Nature Gen* **1998**, *20*, 129–135.

178 Morra, M., et al., Characterization of SH2D1A missense mutation identified in X-linked lymphoproliferative disease patients. *J. Biol. Chem.* **2001**, *276*, 36809–36816.

179 Li, C., Iosef, C., Jia, C. Y., Gkourasas, T., Han, V. K., Shun-Cheng Li, S., Disease-causing SAP mutants are defective in ligand binding and protein folding. *Biochemistry* **2003**, *42*, 14885–14892.

180 Puil, L., Liu, X., Gish, G., Mbamalu, G., Bowtell, D., Pelicci, P. G., Arlinghaus, R., Pawson, T., BCR-ABL oncoproteins bind directly to activators of the ras signalling pathway. *EMBO J.* **1994**, *13*, 764–773.

181 Pendergast, A. M., et al., BCR-ABL-induced oncogenesis is mediated by direct interaction with the SH2 domain of the GRB-2 adaptor protein. *Cell* **1993**, *75*, 175–185.

182 Sattler, M., et al., Critical role for Gab2 in transformation by BCR/ABL. *Cancer Cell* **2002**, *1*, 479–492.

183 Alber, G., Kim, K.-M., Weiser, P., Riesterer, C., Carsetti, R., Reth, M., Molecular mimicry of the antigen receptor signalling motif by transmembrane proteins of the Epstein–Barr virus and the bovine leukaemia virus. *Curr. Biol.* **1993**, *3*, 333–339.

184 Longnecker, R., Miller, C. L., Regulation of Epstein–Barr virus latency by latent membrane protein 2. *Trends Microbiol.* **1996**, *4*, 38–42.

185 Merchant, M., Caldwell, R. G., Longnecker, R., The LMP2A ITAM is essential for providing B cells with development and survival signals in vivo. *J. Virol.* **2000**, *74*, 9115–9124.
186 Higashi, H., et al., *Helicobacter pylori* CagA induces Ras-independent morphogenetic response through SHP-2 recruitment and activation. *J. Biol. Chem.* **2004**, *279*, 17205–17216.
187 Gruenheid, S., DeVinney, R., Bladt, F., Goosney, D., Gelkop, S., Gish, G. D., Pawson, T., Finlay, B. B., Enteropathogenic *E. coli* Tir binds Nck to initiate actin pedestal formation in host cells. *Nat. Cell Biol.* **2001**, *3*, 856–859.
188 Frischknecht, F., Moreau, V., Rottger, S., Gonfloni, S., Reckmann, I., Superti-Furga, G., Way, M., Actin-based motility of vaccinia virus mimics receptor tyrosine kinase signaling. *Nature* **1999**, *401*, 926–929.
189 Sakaguchi, N., et al., Altered thymic T-cell selection due to a mutation of the ZAP-70 gene causes autoimmune arthritis in mice. *Nature* **2003**, *426*, 454–460.
190 Vassilev, L. T., et al., In vivo activation of the p53 pathway by small-molecule antagonists of MDM2. *Science* **2004**, *303*, 844–848.

2
SH3 Domains

Bruce J. Mayer and Kalle Saksela

2.1
Brief Overview

In many ways the Src homology 3 (SH3) domain serves as the archetype for the ever-growing family of modular protein binding domains. Much of what we know (or think we know) about the numerous other domains described in this volume can be traced back to concepts and experimental approaches first tested on the SH3 domain. It is one of the most commonly found modular domains in all eukaryotic genomes, attesting to its usefulness to the cell and its adaptability to a variety of specific circumstances. As befits a domain with such a long and rich history, a number of excellent reviews are available describing its structure, binding properties, and biological activities in great detail [1–3]. In this chapter we first discuss the historical context of our understanding of these domains and then focus on a few outstanding questions on which rapid progress is being made.

SH3 domains are relatively short (~60 residues) protein modules whose primary activity is to bind to proline-rich peptides. The domain is found in a large number of intracellular proteins, many of which are involved in signaling and regulation. Although many such examples contain a single SH3 domain, proteins with multiple copies are common, and as many as five SH3 domains can be found in a single protein. The overall structure of the domain is defined by a compact β sandwich consisting of five major strands, the first three of which are connected by variable loops that for historical reasons are generally referred to as the RT loop, the N-Src loop, and the distal loop; the fourth and fifth strands are separated by a 3_{10} helix (Figure 2.1). A hydrophobic groove on the surface is defined by a series of highly conserved hydrophobic residues and is adapted to bind specifically to extended peptides that adopt the left-handed polyproline-2 (PPII) helical conformation.

Proline is a defining feature of most SH3 binding sites, not only because its presence favors the PPII helix, but also because its unique N-substituted structure allows it to make specific contacts not possible for the other 19 amino acids. Proline is also an excellent choice to mediate specific protein–protein interactions from a thermodynamic perspective – although quite hydrophobic, it is generally found on

Modular Protein Domains. Edited by G. Cesareni, M. Gimona, M. Sudol, M. Yaffe

ISBN: 3-527-30813-X

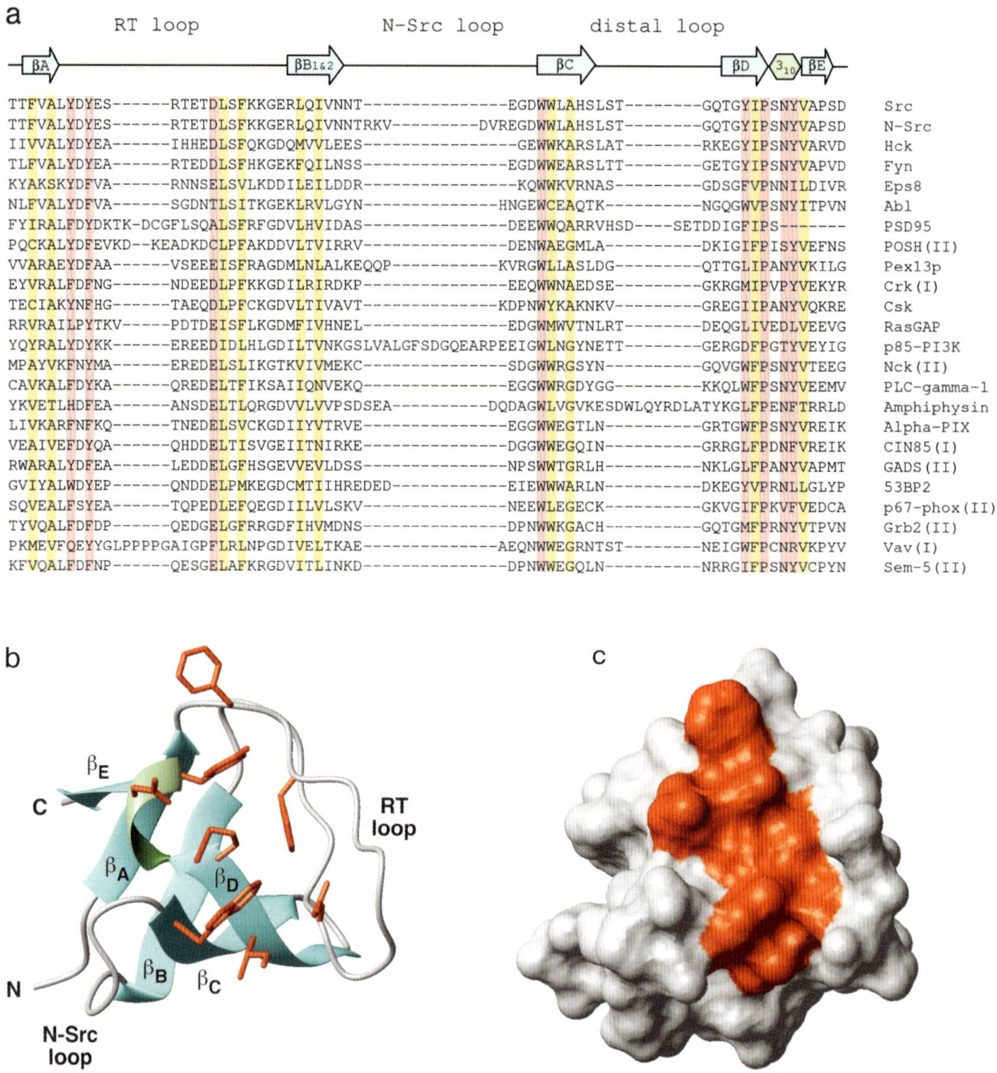

Figure 2.1 Structure of the SH3 domain. (a) Sequences of most SH3 domains discussed in the text are aligned. At top, positions of β strands and the 3_{10} helix are indicated, along with the positions of the RT, N-Src, and distal loops. Dashes indicate gaps imposed to maximize alignment. Yellow shading indicates hydrophobic core residues, and orange indicates conserved residues defining the core ligand binding site. All sequences are for human proteins, except for Sem-5 (*Caenorhabditis elegans*). A roman numeral in parentheses after a name indicates which of multiple SH3 domains present in that protein is shown (e.g., II indicates the second SH3 domain). (b) Ribbon diagram of N-terminal SH3 domain of Sem-5 (PDB number: 2SEM). Sidechains of the core ligand binding-site residues shaded orange in (a) are also depicted. (c) Surface rendering of Sem-5 SH3 in the same orientation as in (b). Again, core peptide binding-site residues are shaded in orange. In this orientation, the two X-P pockets are located at the top and middle of the domain, and the specificity pocket at the bottom. (b) and (c) were generated with the program MOLMOL.

the surface of proteins, thus driving interactions with relatively hydrophobic binding partners; furthermore, its restricted rotational freedom means the entropic cost of binding is relatively low [4]. It is therefore no surprise that proline-rich regions are the targets of a number of other protein binding domains in addition to SH3 (see Chapters 3 and 4 on the WW and EVH1 domains).

The specificity of the domain for particular short proline-rich peptides is generally modest, with affinities usually in the low micromolar range. Although it is clear that individual SH3 domains favor binding to particular sites and classes of sites, in general these differences are not black and white, but more akin to varying shades of gray. Specificity is addressed in more detail later in this chapter, but a few general themes are outlined here. First, the limited specificity of peptide–SH3 binding means that SH3-mediated interactions can be highly dependent on their environment. As detailed below, additional surfaces on the peptide or SH3 domain, or other domains on the two potential binding partners or even on other members of a multiprotein complex, can all confer much greater overall specificity to an SH3–peptide interaction. Second, moderate affinities also imply that the interactions they mediate are highly dynamic, because off-rates are necessarily very rapid. This means that SH3-mediated interactions have the potential to quickly remodel, depending on subcellular localization and available binding partners.

2.2 Historical Perspective

SH3 domains owe some of their prominence to the fact that they are found in several 'glamorous' and therefore well-studied proteins. They were first noted as a region of sequence similarity between the Src and Abl families of nonreceptor tyrosine kinases, although what we now recognize as the SH3 was not at first defined as an independent domain, but rather as part of a poorly understood N-terminal modulatory domain of these oncogenic kinases. This region also contained what Pawson and colleagues in 1986 called the Src Homology 2, or SH2, domain [5] (the conserved kinase catalytic domain was termed SH1). This view changed in 1988, with the simultaneous cloning of the Crk oncogene and a gene for a phosphatidylinositol-specific phospholipase C, PLC-γ [6, 7]. In both instances, sequence analysis revealed the presence not only of SH2 domains, but also of a clearly distinct and independent region of similarity to the N terminus of Src, which soon was called the SH3 domain.

Several factors focused immediate attention on the SH3 domain. Its presence in three completely different classes of proteins (tyrosine kinase, phospholipase, and an oncogene product with no apparent catalytic domain) implied that the domain was modular and therefore presumably conferred some independent function. Whatever this function was, it evidently worked in a variety of different protein contexts. And this function clearly had the potential to regulate the activity of enzymes in which it was found, since mutations in this region of Src could have both positive and negative effects on the ability of Src variants to transform cells.

Furthermore, it was likely that the domain played a role in signal transduction, because all three classes of proteins in which it was found figure prominently in cellular signaling pathways (many other SH3-containing examples were quickly noted, including Ras-GAP, the p85 phosphatidylinositol 3-kinase (PI3K) regulatory subunit, and several cytoskeleton-associated proteins).

Hard as it may be to imagine now in the age of Prosite, SMART, and other such resources, the concept of modular proteins was novel at the time, and it was not immediately obvious that such modules might confer specific protein binding activity. In 1990, however, several groups found that the SH2 domain (which, although clearly distinct from SH3, was found in many of the same proteins) binds specifically to tyrosine-phosphorylated peptide sequences. Importantly, SH2 binds to denatured proteins on filters, implying that linear peptide determinants mediated binding [8]. This had practical implications, because in the era before yeast two-hybrid screens, it meant that there was a straightforward way to clone the binding partners of such domains. In 1991 Cicchetti et al. [9] used a biotinylated, GST-tagged SH3 domain from Abl to probe a bacteriophage cDNA expression library. In this screen, they found two plaques that bound avidly to the GST-SH3 probe but not to a GST control. Rough deletion mapping of the corresponding clones, termed 3BP-1 and 3BP-2, rapidly focused attention on a proline-rich segment of 3BP-1 as the putative SH3 binding site.

At the time this was a somewhat perplexing and troubling result, because proline-rich regions of proteins were not thought to be terribly interesting; certainly they were incapable of folding into globular domains and thus were relegated to relatively unstructured regions about which little was known. Subsequent mapping by alanine scanning of the 3BP-1 and 3BP-2 binding sites revealed that specific residues in the proline-rich region were required for binding [10] and the core PxxP motif (where x denotes any amino acid) identified in that study defines the great majority of SH3 binding sites. However, the biological significance of these SH3–ligand interactions remained dubious at best. At this point genetics and structural biology each independently made major contributions to buttress these initial biochemical studies, revealing the importance of SH3 domains and their role in signaling as well as the molecular details of the interactions that they mediated.

2.2.1 Genetics to the Rescue

As it turned out, the observation that SH3 domains could bind proline-rich peptides provided the final piece of a longstanding puzzle involving activation of Ras. Members of the Ras family of small GTP-binding proteins were originally isolated as oncogene products, and it was known that Ras activation (by GTP binding) is a critical switch in activating downstream signaling pathways, including those promoting proliferation. But nothing was known about what actually triggered Ras activation in the cell. In the early 1990s, evidence from three genetic model organisms provided vital clues. In *Caenorhabditis elegans*, a so-called adaptor protein, Sem-5, was shown by Horvitz's group [11] to lie upstream of Ras and downstream of the Let-23 receptor tyrosine kinase in the vulval differentiation pathway; sequencing Sem-5 revealed that it consisted entirely of two SH3 domains and an SH2 domain. At roughly the same time, Rubin's group [12] showed that a *Drosophila melanogaster* protein called 'Son of Sevenless', or Sos, was upstream of Ras and downstream of the Sevenless receptor tyrosine kinase in the R7 photoreceptor differentiation pathway. Intriguingly, Sos showed sequence similarity to Cdc25 and related proteins from yeast, which had been shown to act as guanine nucleotide exchange factors (GEF) for Ras-like GTPases [13]. The sequence of Sos revealed an extensive proline-rich segment at its C terminus.

With the insight that SH3 domains might bind proline-rich peptides, a plausible model for Ras activation was immediately apparent. Sem-5 (or Grb2 as it was called in vertebrate cells) would likely be recruited to bind via its SH2 domain to tyrosine-phosphorylated receptors on the membrane. The SH3 domains in turn might be expected to bind Sos via its proline-rich tail and bring it along for the ride to the membrane. Because Ras is lipid-modified and therefore constitutively associated with membranes, recruitment of Sos to the membrane would lead to guanine nucleotide exchange on Ras and its activation via local concentration effects. Thus tyrosine phosphorylation (activation of receptor by ligand) could be coupled directly to Ras activation. This model was almost immediately validated by several groups [14] and now serves as the paradigm for changes in the subcellular localization and binding interactions of proteins as a driving force in signal transduction. The critical role of the Grb2–Sos interaction was conclusively demonstrated more recently by Pawson's group [15], who showed that defects in mouse embryonic cells lacking Grb2 could largely be rescued by expression of a fusion protein in which the Grb2 SH2 domain was appended to Sos.

2.2.2 Structure and Specificity

The structural biologists quickly moved to characterize the SH3 domain, even before it had a genetically validated role in life-or-death decisions of the cell. In 1992 the groups of Saraste [16] and Schreiber [17] solved the first SH3 domain structures by X-ray crystallography and nuclear magnetic resonance (NMR), respectively, and the NMR studies provided the first clues about the binding site for proline-rich peptides. Within two years high-resolution structures of a number of SH3 domains bound to specific proline-rich peptide ligands were available. Along with SH2 domain structures that were solved around the same time, these studies exemplified a radical new approach to structural biology. Because of rapid advances in molecular biology, the emergence of enticing new protein targets, and the new insight that many proteins were assembled from modular functional domains, researchers quickly realized that much could be learned (and journal covers could be gained) by rapidly solving the structures of individual modular pieces of important and 'hot' proteins.

These structural studies also helped establish a new understanding of protein–protein interactions, which could be mediated by families of closely related modules,

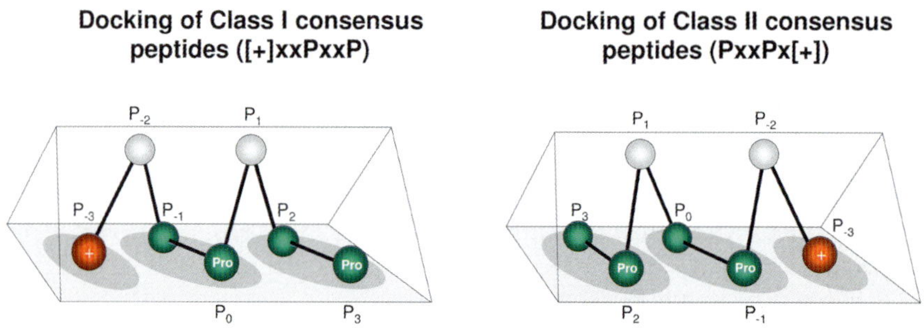

P_{-3}	P_{-2}	P_{-1}	P_0	P_1	P_2	P_3
R/K	x	Φ	P	x	x	P

P_3	P_2	P_1	P_0	P_{-1}	P_{-2}	P_{-3}
x	P	x	Φ	P	x	R/K

Atypical SH3 binding motifs with similarity to class I consensus peptides: RKxxYxxY, PxxRxxKP

Atypical SH3 binding motifs with similarity to class II consensus peptides: PxxDY, Px(P/A)xxR, ΦxRPx

Figure 2.2 Binding of Class I and Class II ligands to SH3 domains. Top, ball-and-stick diagrams of the structure of PPII helical ligands. Ligand residues are numbered according to the nomenclature of Lim et al. [24]. Five key residues of the core ligand peptide contact the surface of the SH3 domain, two in each of the two x-P pockets (green), and one in the specificity pocket (red). Below, positions of consensus residues in Class I and Class II binding sites are indicated. Depending on the binding orientation of the peptide, the PxxP-defining proline residues may occupy positions P_0 and P_3 (Class I) or P_{-1} and P_2 (Class II). The other residue of the central x-P dipeptide (position P_{-1} in Class I and P_0 in Class II) is generally hydrophobic (Φ). In both cases residues at positions P_{-2} and P_1 face away from the SH3 domain and do not participate in binding. Proline residues are common in these positions, however, and they help to stabilize the PPII helical conformation of the peptide. Ionic interactions of the positively charged residue (R or K) in position P_{-3} with acidic residues in the specificity pocket determine the orientation of binding. For Class I sites, this basic residue is at the N terminus of the peptide; for Class II sites, it is at the C terminus. Other types of specificity-pocket interactions also exist, such as in the Class I peptides that bind Abl SH3, which have a hydrophobic residue at position P_{-3}. Atypical binding motifs that bind in a recognizably similar manner to Class I and Class II sites are also indicated.

each binding to stereotypical, short, linear, peptide ligands [18]. This was a far cry from earlier prevailing notions that protein–protein interactions were mediated by extensive complementary surfaces requiring the folded tertiary structure of each partner. This insight had important ramifications, because it implied that we could learn a great deal about a protein and its potential binding partners merely by inspecting its primary sequence – if your favorite protein had an SH3 domain, you could be confident that it would bind to proteins having proline-rich PxxP motifs.

SH3 domains also served as a crucial proving ground for new approaches to identifying specific ligands. This hunt was motivated both by basic questions about SH3-mediated signaling pathways and by the hope that specific high-affinity small-molecule ligands for these domains might serve as lead compounds for drug development. The relatively small size of the SH3 domain and the ability to purify large amounts of the active domain from bacteria, along with its evident involvement in key biochemical pathways relevant to human health, made them ideal targets. Early on, Schreiber's [19] group demonstrated that combinatorial libraries of beads, each coupled to a unique peptide sequence, could be used to select preferred ligands for the Src SH3 domain. Screening of phage display libraries (where each phage expressed on its surface a unique peptide sequence) also proved a very powerful approach, by which preferred binding partners could be enriched by multiple rounds of binding and amplification [20–22].

Both approaches revealed that a PxxP core binding motif was almost always present in preferred binding sites, generally flanked by a basic residue. One remarkable discovery that arose from these studies was that the conserved basic residue could be either N-terminal or C-terminal to PxxP: for example, RxxPxxP and PxxPxR could both be selected by the Src SH3 [23]. Structural analysis of SH3 domains bound to both classes of ligands (termed Class I and Class II ligands, respectively) showed that the domain could indeed bind ligands in both orientations at the same binding site (Figure 2.2). More specifically, for most SH3 domains the peptide-binding groove consisted of two hydrophobic slots, each normally occupied by an x-P dipeptide (where x was generally hydrophobic), and a third, negatively charged, 'specificity' pocket that engaged the basic residue of the ligand [23, 24]. Recent studies have suggested that the sidechain orientation of a highly conserved tryptophan residue in the SH3 domain dictates whether it can bind Class I, Class II, or both classes of ligands [25]. The fact that peptides could adopt either an N-to-C or C-to-N orientation was unprecedented and is a consequence of the pseudo-symmetry of the left-handed PPII helical conformation. Similar behavior has since been observed for other domains that bind PPII helical ligands (WW, EVH1 domains).

The central conundrum raised by many of the structural studies of SH3 domains and by the results of various screens for preferred binding partners was the realization that there really isn't much room for specificity in the core peptide-binding groove. Only five core ligand residues directly contact the SH3 domain; two of these are the invariant prolines and one the basic residue (which is almost always arginine) (Figure 2.2). As will be more fully discussed in a later section, additional contacts between variable loops of the SH3 domain and ligand residues N-terminal or C-terminal to the core can greatly enhance specificity, and in a few examples, unconventional binding sites (such as PxxRxxKP motifs) can specifically bind to the core ligand-binding groove. But it is important to emphasize that, within the core binding site, the potential for specificity is highly limited, because only proline can be accommodated at two positions. Despite these constraints, however, a high degree of selectivity is achievable in vivo. A recent study by Lim's group [26] showed that a yeast SH3 ligand peptide was absolutely selective for its known biologically relevant binding partner, not binding detectably to any of the other 26 SH3 domains of *Saccharomyces cerevisiae*. On the other hand, this peptide bound robustly to a number of vertebrate SH3 domains, suggesting that high selectivity arose in the course of evolution through both positive and negative selection. The generality of this concept to other yeast SH3–ligand interactions or to the much more numerous potential interactions in metazoan proteomes remains to be seen.

More recent structural studies have revealed yet another way in which the versatile SH3 domain can be used to assemble protein complexes in the cell. MAGUK proteins, which contain a PDZ domain and an SH3 domain followed by a guanylate kinase-like (GK) domain, serve as homo- or heteropolymeric scaffolds for assembly of large complexes on the membrane, for example, in organizing the structure of the neuromuscular junction. It was found that the SH3 domains of MAGUK proteins could bind in cis or in trans to the GK domain, but the mechanism was unclear, because there was no obvious PxxP binding motif on the GK domain. The subsequent X-ray crystal structures of the SH3-GK segment of PSD-95 revealed not only that the SH3 and GK domains associated tightly with each other, but more remarkably, that a sixth β strand, located C-terminal to the GK domain, contributed directly to the folded SH3 domain structure (the majority of which is N-terminal to the GK domain); in addition, a long hinge region was inserted between the fourth and fifth strands of SH3 [27, 28]. This, along with additional genetic and biochemical evidence, strongly suggests that the first four strands of the SH3 domain form a folding subunit that can bind in either cis or trans orientation to a second subunit composed of the fifth SH3 strand, the GK domain, and the sixth SH3 strand, by a process of 3D domain swapping [28]. Such a cis–trans isomerization is likely to underlie both the specificity and stability of the polymeric networks of assembled MAGUK proteins.

2.3 Predicting Binding Partners

The SH3 domain is the most common of the modular protein binding domains: the human proteome is predicted to encode almost 300 different SH3 domains (current best estimate is 284, unpublished data of Wang and Saksela). The number of SH3-encoding genes is somewhat lower, since many proteins contain multiple SH3 domains, and in some, such as Src and N-Src, alternative RNA splicing contributes to SH3 domain diversity by introducing sequence variation into the SH3 loop regions. The number of potential SH3 target proteins is more difficult to predict, but appears to be considerably higher than the number of SH3 domains, creating an enormous number of theoretically possible pairs of SH3-mediated protein interactions. Knowing which ones of these interactions actually occur and are biologically meaningful would go a long way toward understanding the wiring of signaling networks that regulate fundamental aspects of cellular physiology and are deregulated in many important diseases.

As already discussed, the inherent specificity in protein recognition built into the ~60 amino acids that constitute individual SH3 domains is only one of the many factors that determine the time, place, and partnerships of SH3-mediated protein interactions. Nevertheless, it is reasonable to assume that SH3–ligand pairs that have evolved to possess the highest binding affinity and selectivity are likely to be physiologically relevant. Therefore, in addition to screening for preferred SH3 binding partners from various expression libraries, much effort has also been invested in developing tools for predicting potential SH3–ligand pairs from sequence databases.

Significant progress in this area was recently reported by Cesareni and colleagues [29], who combined experimental and computational approaches to predict SH3 interaction networks in yeast. This analysis was based on extensive phage-display screening of a random nonapeptide library to establish minimal consensus binding sequences for most (20 of the total predicted 27) SH3 domains of *S. cerevisiae*. These SH3 domains were also used in a large-scale yeast two-hybrid screening for preferred binding proteins from cDNA libraries. Of the total of 394 interactions predicted to be most likely based on computer analysis of potential target proteins, 59 were also found among the 233 SH3–ligand pairs obtained experimentally in the two-hybrid screens.

In another study, this group generated a large library of Abl SH3 variants in which 12 residues that participate in ligand recognition (based on close examination of Abl SH3–peptide complex structures) were replaced with random combinations of amino acid that occur in the corresponding positions of other SH3 domains [30]. From this degenerate library, individual SH3 domains could be selected that bound well to a peptide (APTYPPPLPP) preferred by native Abl SH3 but not to another peptide (LSSRPLPTLPSP) known to be a good ligand for Src SH3. Conversely, SH3 domains with the opposite specificity, i.e., that bind to the Src peptide but not to the Abl peptide, could also be selected. The lessons learned from this exercise were subsequently applied to predicting the relative preference of the collection of *S. cerevisiae* SH3 domains for these two peptides. Gratifyingly, the three yeast SH3 domains that bind to Abl peptide in an ELISA assay were also predicted as the three most likely Abl peptide binders, whereas the top six predicted candidates for Src peptide binding included each of the five yeast SH3 domains that actually showed good binding to this peptide.

A related computational approach was successfully developed by Brannetti et al. [31]. By combining a large amount of data from SH3–ligand structures and SH3 target peptide selection experiments, they generated an algorithm (SH3-SPOT) for searching databases to find optimal decapeptide sequences for any SH3 domain of interest. The validity of this approach was supported by the observation that the hallmarks of preferred target peptides for a given SH3 domain, such as Src, could be correctly predicted even if experimental data regarding that particular SH3 domain were excluded from the training dataset for the algorithm. We should note, however, that if the actual binding data for the entire Src family of SH3 domains were omitted from the training dataset, the accuracy for predicting peptides preferred by the Src SH3 domain was greatly reduced. Nevertheless, when this algorithm was used to examine protein sequence databases for potential partners for the SH3 domains of Abl, Hck, and Grb2, the top ten scoring decapeptides did include PxxP-containing regions of proteins, such as 3BP2 (Abl SH3), HIV-1 Nef and Cbl (Hck SH3), and Sos2 (Grb2 SH3), which had been previously reported as relevant partners of the corresponding SH3-containing proteins.

2.3.1
Core Peptide Docking vs. Extended Interactions

Together, the studies described above clearly establish that the amino acid sequence of an SH3 domain can be used to help predict its target peptide preference, which in turn has predictive value for identifying relevant partner proteins for that SH3 domain. So, does this remarkable progress mean that we can now predict the partnerships among the SH3 domains and their ligands and start using this information to model cellular signaling networks? The answer to this question appears to be 'yes and no'. Although such computational tools can be of great value in guiding research to identify the relevant SH3-mediated protein interactions underlying various normal and pathological processes in the cell, these efforts must ultimately rely on traditional (although perhaps increasingly systematic and high-throughput) experimental approaches.

A major limitation of the computational approach is that it fails to take into account molecular contacts that are unique to individual SH3–ligand interactions, and which are therefore, by definition, not predictable by homology modeling. The overall contribution of such 'atypical' contacts in ligand recognition among natural SH3-mediated interactions as a whole is presently not known. However, in most instances for which the structural basis of unusually strong or selective SH3–ligand binding has been determined, it has been found to rely on molecular contacts outside the conserved core peptide binding pockets.

Within the SH3 fold these additional specificity-determining contacts typically involve residues in the nonconserved regions of the RT and N-Src loops. This makes sense intuitively, since these are the only areas in which aligned primary structures of different SH3 domains completely diverge in sequence composition and also in length (Figure 2.1). When discussing the RT loop that connects the first and second β strands in the SH3 fold, however, we should note that this is a large structure (typically consisting of 18 amino acids, but varying between 16 and 26 in length in human SH3 domains), composed of a central highly variable loop flanked by constant stretches of four (N-terminal) and eight (C-terminal) relatively conserved amino acids with β strand character. The conserved RT loop residues provide key structural determinants for the third specificity pocket on the SH3 ligand-binding surface, which is in agreement with the relatively limited ligand selectivity provided by this surface. By contrast, the nonconserved residues in the central region of the RT loop have a greater capacity to engage in ligand-specific contacts (see below).

An interesting example how nonconserved SH3 loop residues can enhance the affinity and specificity of ligand binding was reported by Cowburn's group [32], who studied the complex of the Csk SH3 domain bound to a high-affinity (K_d 0.8 μM) target peptide derived from Csk's natural partner and regulator, PEP (proline-enriched phosphatase). This peptide contains a Class II core PxxP motif, but its strong and selective binding to the Csk SH3 also requires two hydrophobic amino acids located six residues after the PxxP motif (PPLPERTPESFIV). In fact, the consensus Class II PxxP motif of PEP is, although necessary, insufficient to mediate binding to the Csk SH3. The NMR structure revealed that after the PxxP motif, the

PEP peptide contained a 3_{10} helix, which helped to position the critical isoleucine and valine residues in a hydrophobic specificity pocket in Csk SH3. A key feature of this pocket is a 'clamp' formed by a lysine residue in the N-Src loop. Thus, the strong and functionally relevant SH3-mediated interaction between Csk and PEP has evolved to rely mainly on contacts involving the N-Src loop of the SH3 domain and determinants in the ligand that fall outside of the canonical PxxP motif, which by itself has insignificant affinity towards Csk SH3.

Although not equally well understood in structural terms, a related story can be seen in the strong dependence of the second SH3 domain of the adapter protein Nck on a critical serine residue located C-terminal to the consensus Class II motifs in its target proteins such as PAK1 (PAPPMRNTS) [33]. Interestingly, this serine residue is also a major autophosphorylation site in activated PAK kinases, and Nck SH3 fails to bind significantly to peptides containing a phosphoserine in this position. Since membrane recruitment by Nck leads to PAK activation [34], modulation of affinity of the PAK binding site for Nck SH3 by autophosphorylation may represent an important negative feedback mechanism for PAK activation. Phosphorylation sites can be found in the vicinity of many other core SH3 binding motifs, and one should remember that regulation of SH3 binding affinity via modulation of noncanonical contacts involved in ligand recognition may be more common than is currently appreciated.

Another informative study was reported by Kami et al. [35], who examined the basis of the unusually tight (K_d 24 nM) association of the SH3 domain of $p67^{phox}$ with the proline-rich region of $p47^{phox}$, another cytosolic component of NADPH oxidase. This binding involves a Class II PxxP motif in $p47^{phox}$. However, the tight association requires 20 additional amino acids immediately C-terminal to the $p47^{phox}$ PxxP motif, whereas a 10-residue peptide consisting of only this PxxP motif binds to the $p67^{phox}$ SH3 domain with modest affinity (K_d 20 μM). Structural analyses revealed that the 20 amino acids after the PxxP consensus sequence had a helix-turn-helix (HTH) structure, which cooperated with binding of the PPII-helical PxxP peptide by making additional contacts with the area of $p67^{phox}$ SH3 bordered by the RT and N-Src loops. Even when separated from the PxxP sequence, the HTH peptide was shown to bind independently to $p67^{phox}$ SH3 with a K_d of 10 μM. This led the authors to propose this interaction as an example of PxxP-independent SH3 binding. We should note, however, that interaction of the isolated HTH peptide with $p67^{phox}$ SH3 depends critically on the arginine residue that overlaps the Class II PxxP consensus motif of $p47^{phox}$ (PAVPPR). In the NMR structure, this arginine residue is seen to engage in canonical ionic interactions with the conserved negatively charged residues of the $p67^{phox}$ SH3 to coordinate binding of the $p47^{phox}$ PxxP-HTH peptide. Thus, this interaction could perhaps better be considered a novel example of a strategy to increase the modest affinity and target specificity provided by the core PxxP motif.

A somewhat different example of complex SH3–ligand recognition that is useful to consider here relates to the avid binding (K_d 0.25 μM) of the SH3 domain of the Src family tyrosine kinase Hck to the HIV-1 pathogenicity factor Nef. Similar to the studies discussed above, Lee et al. [36] concluded that the Class II consensus

PxxP motif in Nef is insufficient to mediate this strong binding, because a 12-mer peptide spanning this site (PVRPQVPLRPMT) has only a modest affinity (K_d 91 μM) for Hck SH3. Interestingly, the closely related SH3 domain of Fyn showed similar modest binding to the 12-mer Nef PxxP peptide, but unlike Hck SH3, fails to bind to full-length Nef protein with high affinity (K_d > 20 μM). Most of the striking difference in binding affinity between these two homologous SH3 domains could be assigned to a single residue in the variable region of the RT loop. When Arg96 in Fyn SH3 (numbering according to full-length Fyn) was replaced with an isoleucine, which is found in the corresponding position in the RT loop of Hck SH3, the resulting R96I mutant Fyn SH3 bound Nef with an affinity (K_d 0.38 μM) almost as high as that of Hck SH3. The molecular basis of the binding selectivity and affinity provided by this Hck-specific isoleucine residue was revealed by a crystal structure of Fyn-R96I SH3 complexed to the core domain of Nef [37]. The sidechain of this isoleucine residue is inserted into a hydrophobic pocket between two α helices in Nef, where it is packed against sidechains of multiple nonpolar Nef residues, in particular, Leu87, Phe90, Trp113, and Ile114. Notably, and in contrast to the two examples of complex SH3–ligand recognition discussed above, here the specificity-determining residues are noncontiguous and not close to the PxxP consensus sequence in the polypeptide chain of the ligand.

The experimental approach in the overwhelming majority of published studies on identification of SH3 recognition sites has been to narrow down the interaction to the shortest ligand peptide that can support efficient SH3 binding. Also, although coordinates for more than 150 structures involving an SH3 domain alone or in complex with a target peptide have been deposited in databanks, Nef remains the only native PxxP-containing intermolecular partner protein whose structure in complex with an SH3 domain has been solved. Thus, it is impossible to know whether stabilizing interactions between RT and/or N-Src loops of various SH3 domains and surfaces in their target proteins formed by contiguous or non-contiguous residues located at a distance from the core PxxP motif are widespread or if the mode of SH3 domain recognition by Nef is unusual.

There are, however, several notable examples of 3D structures of an SH3 domain complexed with a native target protein in the context of an *intra*molecular interaction. These are the autoinhibited forms of the tyrosine kinases Hck, Src, and Abl [38–40]. In all three, binding of the SH3 domain to a target site present in the linker region connecting the SH2 and kinase domains plays a key role in locking the latter into a catalytically inactive state. Competition of this intramolecular interaction by high-affinity ligands for the SH3 domain, such as Nef for Hck [41], therefore promotes activation of these kinases. Although these structures provide us important insights into the regulation of tyrosine kinases, they are less informative for understanding how optimal SH3 binding specificity and affinity is achieved. Presumably because intermolecular competition for these intramolecular SH3 interactions must be allowed, the molecular contacts involved in recognition of the internal SH3 binding sites in these kinases are clearly suboptimal. In fact, these interactions were not appreciated until the Hck and Src structures were solved, partly because the linker-region peptides of Hck, Src, and Abl that interact with their own SH3 domains substantially deviate from known SH3 binding consensus sequences, and only in Hck are two proline residues even present. This serves as a reminder that even relatively weak SH3-mediated interactions can play a vital role in regulating signaling pathways.

2.3.2 Atypical SH3 Docking Motifs

It is now increasingly evident that lack of a conventional PxxP motif in a ligand protein by no means rules out high-affinity SH3 binding. Indeed, another type of complication in predicting relevant target proteins of SH3 domains based on consensus binding motifs is the steadily increasing number of SH3 interactions that do not involve a bona fide Class I or Class II PxxP motif. For discussion's sake, we classify these interactions into three somewhat arbitrary categories: (1) interactions that involve binding of the core PxxP-motif docking surface of the SH3 to a peptide that does not conform to a classic PxxP consensus; (2) interactions that are mediated by a defined peptide element in the ligand but do not involve the PxxP-motif docking surface of the SH3 domain; and (3) complex interactions, which may or may not involve residues contributing to the PxxP docking surface, but differ fundamentally from typical SH3 target peptide recognition. The last category includes the MAGUK protein SH3–GK interaction described above, Eps8 dimerization mediated via β strand exchange between two Eps8 SH3 domains [42], the integral role of the SH3 domain of p53 binding protein-2 (53BP2) in the 53BP2–p53 complex [43], and binding of the adaptor protein Grb2 and the guanine nucleotide exchange factor Vav via heterodimerization of their SH3 domains [44]. These interactions, while fascinating, are not further discussed here.

Category 2 now consists of one well-characterized example, but, considering the evolutionary adaptability characteristic of modular protein interaction domains, it would not be surprising if additional examples came to light. Distel's group [45] studied interactions between peroxins, which are proteins involved in biogenesis and translocation of peroxisomes, and noticed that Pex5p binds to the SH3 domain of Pex13p. Surprisingly, however, they found that a short (25 amino acids) α-helical element in Pex5p is not only necessary, but also sufficient, to mediate this interaction. Furthermore, this α-helical peptide does not compete with binding of yet another peroxin, Pex14p, which recognizes Pex13p SH3 via a canonical PxxP interaction. A recent determination of the NMR structure of Pex13p, together with chemical shift experiments, identified the binding site for Pex5p on the opposite face of the Pex13 SH3 domain, away from its PxxP binding surface [46], providing a structural explanation for the earlier observations.

Reported SH3 interactions belonging to category 1 are steadily increasing in number. It appears that in some of them, evolution has resulted in an SH3 domain having a ligand-binding surface that is specialized for recognizing atypical PxxP-like peptide motifs. This seems to occur in the Eps8 family SH3 domains, which (when in monomeric form) recognize ligands, such as Abi-1and RN-tre, via a motif conforming to the consensus PxxDY [47]. Instead of Class I or Class II consensus

PxxP motifs, Eps8 SH3 also selects peptides containing the PxxDY consensus from random peptide libraries.

The situation is similar with the SH3 domains of the CIN85/CMS family of adaptor proteins, which play important roles in endocytosis and signal transduction. Using a novel peptide screening method, Kurakin et al. [48] determined that the three SH3 domains of CIN85 specifically select the consensus sequence Px(P/A)xxR. They also mapped a region conforming to this consensus as the CIN85 binding site in its established partner Cbl-b and noted that the same consensus can be found in other known CIN85/CMS-interacting proteins, including c-Cbl, BLNK, AIP1, SB1, and CD2. Moreover, their database analyses found the Px(P/A)xxR motif in additional proteins, such as PAK2, ZO-2, and $TAF_{II}70$, which were subsequently confirmed to have the capacity to interact with CIN85 SH3 domains in a GST pulldown assay.

This group also pointed out interesting similarities between this novel CIN85/CMS binding motif and another atypical proline-rich SH3 target sequence present in PAK2 (PPPVIAPRP), which serves as the binding site for the SH3 domains of α- and β-PIX (PAK interacting exchange factor) proteins [49], suggesting that PIX and CIN85/CMS SH3 domains might use similar strategies in target peptide recognition. The strength of binding of Px(P/A)xxR peptides to CIN85/CMS SH3 domains is not remarkable (K_d values close to 10 μM), whereas binding of a 22 amino-acid PAK peptide to αPIX SH3 was reported to be of unusually high affinity (K_d 24 nM [49]). This difference, however, could be explained by a differential contribution of residues outside the PxxP-like motifs of the peptides used for measuring these affinities.

Another apparent example of specialization for recognition of a nonconsensus PxxP motif is the SH3 domain of amphiphysin (an adaptor protein involved in endocytosis). When used in screening peptide libraries, this SH3 domain selected the consensus sequence ΦxRPxR, in which a hydrophobic (Φ) amino acid, and not proline, was preferentially selected in the first position [50, 51]. Although a proline residue is present in this position of the amphiphysin SH3 binding site in its partner protein dynamin (PSRPNR), the latter sequence still deviates from a classic Class II PxxP peptide, in which a hydrophobic residue generally precedes the second proline (PxΦPx[+], where + is R or K).

In contrast to the atypical interactions discussed above, binding of SH3 domains to the proline-free RKxxYxxY motif found in the immune cell adaptor protein SKAP55 and its relative SKAP55-R [52] looks like an example of an opposite adaptation. SKAP55-derived peptides spanning this motif can bind to a number of different SH3 domains, such as those of Fyn and Fyb that prefer to select consensus Class I PxxP ligands in random peptide screening experiments. The affinity of RKxxYxxY peptide-mediated binding is relatively low (K_d 20–60 μM), although well in the range typical of SH3 binding to canonical PxxP peptides. Thus, the RKxxYxxY motif appears to have evolved as an alternative, proline-independent, means for connecting SKAP55 to cellular signaling proteins containing SH3 domains that otherwise recognize ligands with a Class I PxxP consensus. NMR chemical shifts indicate that the SKAP55 peptide makes contact with many but not all of the same residues involved in typical PxxP peptide binding. The spacing of the lysine and tyrosine residues in the SKAP55 peptide, and the failure of SH3 domains that preferentially bind Class II PxxP consensus sequences to bind this peptide, also suggest structural similarities with Class I PxxP peptide binding.

Yet another interesting example of SH3 binding via an unorthodox binding site involves the consensus sequence PxxRxxKP, which was originally identified in the T-cell receptor-associated adaptor protein SLP-76 as the binding site for the C-terminal SH3 domain of Grb2 [53] and its relative GADS (Grb2-related adaptor downstream of Shc, also known as Mona, Grap2, GrpL, and Grf40) [54], but is present also in other signaling proteins, such as the deubiquitinating enzyme UBPY [55] and the GADS binding protein ASMH [56]. The structural basis of recognition of GADS SH3 by this motif was recently elucidated by two groups, leading to essentially identical conclusions. Feller's [57] group used a 13-mer human SLP-76 peptide PAPSIDRSTKPPL, which binds the GADS SH3 domain with high affinity (K_d 0.18 μM). The complex solved by Li's group [58] included the 11-mer peptide APSIDRSTKPA also known to bind tightly to GADS SH3 (K_d 0.24 μM; [54]). These studies revealed that the RSTK region of the SLP-76 peptide adopts a 3_{10}-helix conformation, which contributes to the binding in itself, but also positions the critical residues of the PxxRxxKP consensus into the three binding pockets of the PxxP-binding surface of the GADS SH3. The Ala-Pro residues of the SLP-76 peptide are accommodated by the first hydrophobic pocket, much like the first x-P dipeptide of a canonical Class II PxxP peptide (xPxΦPx[+]). Moreover, the role of the Ile residue in the SLP-76 peptide is analogous to that of the hydrophobic Φ residue of the second canonical Φ-P dipeptide. Remarkably, however, the second hydrophobic pocket of GADS SH3, where this Ile residue was found to sit, is too small to accommodate a canonical Φ-P dipeptide, thus explaining why GADS SH3 interacts poorly with typical Class II PxxP peptides. After the 3_{10} helix region in the SLP-76 peptide, however, the analog to the consensus PxxP motif recognition again appears, and the ionic interactions involving the lysine residue of the PxxRxxKP motif and the third specificity pocket of GADS SH3 were found to be similar to those seen in typical SH3–ligand complexes.

Although some of the strategies of atypical PxxP-like peptides for contacting the PxxP binding surface of SH3 domains might inherently give rise to a stronger interaction than corresponding contacts of canonical PxxP peptides, the discussion above regarding additional stabilizing interactions that involve less-conserved SH3 regions applies equally well to these non-PxxP interactions. A good example of this seems to be the GADS SH3–PxxRxxKP interaction discussed above. The same SLP-76 peptide also binds to the C-terminal SH3 domain of Grb2 [53], albeit with a much lower affinity. Feller and colleagues [57] noted that this is probably not due to major differences in docking of the PxxRxxKP motif residues to the similar PxxP binding surfaces of GADS and Grb2 SH3 domains. Instead, based on comparison of their structure with NMR cross-saturation data on Grb2 SH3–SLP-76 binding [35], they attribute the higher affinity of the GADS SH3 domain to the capacity of its RT and N-Src loops to engage in much more extensive interactions with the SLP-76 peptide.

2.4
Experimental Exploitation of SH3 Specificity

Insights gained from studies on natural SH3 binding specificity have already been successfully used to engineer molecules for the experimental and therapeutic manipulation of SH3-mediated interactions. For example, peptides with high affinity for individual SH3 domains (identified by screening random peptide libraries or by rational design) have been introduced into living cells to serve as competitors for natural ligands (see below). The core PxxP motif of the C3G protein binds strongly to the first SH3 domain of the adapter Crk (and CrkL), due an unusually tight coordination of a lysine residue in the C3G peptide by three acidic residues in Crk SH3 [59]. Kardinal et al. [60] used an optimized 28-amino-acid C3G peptide that binds to CrkLSH3 with very high affinity (35 nM) and fused it with a 17-residue amphipathic antennapedia-derived 'shuttle' peptide, which provided this fusion peptide with the capacity for receptor-independent entry into cells. When added to cultures of chronic myeloid leukemia (CML) cells, these cell-penetrating Crk/CrkL SH3 ligands disrupted Bcr–Abl/CrkL complexes and strongly reduced proliferation of both primary CML blast cells and cell lines established from Bcr-Abl–positive patients, suggesting that interaction of Crk SH3 with its effectors is important for proliferative signaling.

The affinity of the C3G-derived peptide developed by Feller and colleagues towards an unrelated control SH3 domain (Grb2 SH3) was two orders of magnitude weaker (K_d 4.27 μM) than that of Crk–CrkL. However, a concern remains that such peptides might also unintentionally target other SH3-mediated interactions that should be left untouched. Lim and colleagues [61] have reported an interesting and potentially useful solution to this problem. They noted that the entire shape of the sidechains of PxxP-defining proline residues was less important for SH3 recognition than the N-substitution on their backbone amide nitrogens. Although proline is the only N-substituted amino acid available to the cell, the organic chemist has a much richer palette from which to select. Accordingly, each of the consensus proline residues of an SH3 binding peptide from Sos (YEVPPPVPPRRR) was replaced by various nonnatural N-substituted residues (peptoids). Interestingly, when binding of these hybrid ligands to SH3 domains of Src, Grb2, Crk, and Sem-5 was examined, many of these peptoids were able to substitute for proline and support SH3 binding. Moreover, some of these substitutions led to remarkable changes in the relative affinities of these ligands for the different SH3 domains tested. One proline-to-peptoid substitution caused a greater than 100-fold increase in binding to Grb2 SH3 (to a K_d of 40 nM), without any effect on binding to Src SH3. Other examples of significantly improved binding to only one or a subset of the tested SH3 domains were also seen. Equally important when considering the development of specific inhibitory ligands was that some proline-to-peptoid substitutions led to a substantial decrease in binding to one or two of the SH3 domains without affecting interaction with the others.

These studies demonstrated that the pockets accommodating the conserved proline residues of SH3-ligand peptides present a major untapped source of additional selectivity that can be exploited in development of SH3-specific inhibitors by using peptoids as proline mimetics. In a follow-up study on peptide-peptoid ligands, Nguyen et al. [62] focused specifically on the issue of optimal selectivity, which they assert is more critical than overall affinity when developing therapeutically useful SH3 inhibitors. Combinatorial peptoid substitutions of multiple PxxP-defining as well as other proline residues were successfully introduced into the Sos peptide used in their previous study, and peptoids were also tested in other peptides that already contained flanking sequences optimized for binding to Src SH3 or Crk SH3. One of these ligands bound Crk SH3 with a remarkable affinity of K_d 8 nM, while showing over 1000-fold weaker binding to both Src SH3 and Grb2 SH3 (K_d 45.6 and 13.7 μM, respectively). Additional modifications of this ligand gave rise to variants that had lost part of their affinity for Crk SH3 (K_d values of 0.5 μM and 1.98 μM), but more importantly, showed virtually no binding to the two other SH3 domains (K_d values at or above 1 mM). Thus, within their test set of SH3 domains, Lim and colleagues succeeded in developing SH3 ligands with truly orthogonal binding selectivity.

The success described above resulted partly from the higher selectivity of peptoids as compared to proline residues in recognizing the hydrophobic binding pockets of SH3 domains, but also relied on the unusually tight accommodation by the Crk SH3 domain of a Class II 'consensus' lysine residue presented in the context of an optimal peptide. Examples of similar specificity determinants close to the PxxP motifs of other SH3 ligands were discussed above. Nevertheless, in general, such selective PxxP-peptide flanking motifs are in short supply, and targeting of a peptide to bind to a specific SH3 domain remains a major challenge. This is particularly true if one would like to selectively target one SH3 domain but spare closely related domains. Indeed, it is hardly a coincidence that studies on enhancing the specificity of SH3–ligand recognition, be it via PxxP flanking sequence optimization or peptoid substitutions, have usually focused on SH3 domains representing members from widely divergent SH3 subfamilies.

An innovative approach reported by Schreiber and colleagues [63] for generating SH3 ligands that could have unnaturally high capacities to discriminate different SH3 domains is to append to the N terminus of a short PxxP peptide (PLPPLP) a nonpeptide moiety generated by combinatorial synthesis from a diverse set of organic monomers. From a library of such compounds, several molecules were identified that bind to Src SH3 with affinities similar to those of optimized ligands selected by Src SH3 from random libraries of longer peptides. One of these hybrid compounds bound to Src SH3 with a K_d of 3.4 μM and showed over 50-fold lower affinity for PI3K SH3 (K_d 170 μM). The mode of recognition of Src SH3 by these peptide/nonpeptide compounds was examined by NMR spectroscopy [64]. Similar to amino acid residues N-terminal to the conserved proline residues of Class II SH3 ligands, the nonpeptide moieties of these compounds contact the SH3 specificity pocket, but via binding interactions very different from those seen in peptide–SH3 complexes. Although the binding affinities reported by Combs et al. [63] are relatively modest compared to some of the engineered SH3 ligand peptides discussed earlier, this study provides a promising proof-of-concept. Combinatorial synthesis of more complex nonpeptide moieties and/or using this strategy in combination with longer peptides already optimized by other means could be a powerful way of generating SH3 ligands with desired binding properties.

An approach for engineering of SH3 binding specificity that is directly opposite to the studies described above was pursued by Hiipakka et al. [65]. Prompted by the critical role of the variable region of the RT loop of the Hck SH3 domain in binding to HIV-1 Nef (see above) and the sequence dissimilarity of this region in different SH3 domains, they generated a large phage library displaying Hck SH3 variants carrying six randomized residues in place of the variable region of the RT loop (residues EAIHHE; see Figure 2.1). From this library several individual SH3 domains (named RRT-SH3, for random RT loop) were selected that bind to Nef 30- to 40-fold better than wild-type Hck, thus pushing the affinity of this interaction into the low nanomolar range. When expressed in cells, these RRT-SH3 domains could potently inhibit intracellular functions of Nef [66]. Of note, the majority of cellular proteins that were found to coprecipitate with wild-type Hck SH3 failed to associate with the modified, Nef-targeted, RRT-SH3 domains. This suggests that the same variable region of the RT loop that plays a role in Nef binding is also involved in recognition of normal cellular partners of Hck SH3. Thus, the improved affinity toward one partner protein of Hck SH3, such as Nef, achieved via RT loop manipulation is also accompanied by enhanced binding selectivity in consequence of disrupted RT loop interactions with other potential partner proteins. This is an encouraging concept when considering the use of ligand-targeted RRT SH3 domains as competitive inhibitors for intracellular SH3-mediated processes.

Among natural interactions, a critical role for SH3-specific determinants in the RT loop in ligand binding was conclusively demonstrated only for the Nef/Hck SH3 complex. However, recent unpublished work by Hiipakka and coworkers has established RRT-SH3 engineering as a feasible approach for generating tightly binding SH3 domains targeted against divergent SH3 binding proteins, such as Sos, PI3K, PAK2, and CD28, which are normally not high-affinity partners of Hck SH3. Thus, modifications involving six nonconserved residues in the RT loop are sufficient to target Hck SH3 to proteins having canonical PxxP motifs that normally have no particular preference for Hck SH3 or might even be poorly suited for binding. This highlights the potential of the RT loop in providing affinity and specificity to SH3–ligand recognition and suggests that the RRT-SH3 domain approach might be a generally applicable strategy for modulating SH3 interactions for research and perhaps even for therapeutic purposes.

2.5 Conclusion

SH3 domains have long been a testing ground for new ideas and new methods in protein analysis, and there is every indication that they will continue to play a prominent role. The power of newly available genomic, proteomic, and computational approaches, coupled with the detailed structural and biochemical information that is already available, are providing unprecedented insights into the normal biological roles of these domains. There can be little doubt that these insights will ultimately give rise to specific and effective strategies for rationally manipulating their binding activities in vivo, both in the laboratory and in the clinic.

Acknowledgments

BJM is grateful to M. Schiller for help in preparing Figure 2.1 and for helpful discussions. Research in BJM's laboratory was partially supported by a grant from the National Institutes of Health (CA82258). KS would like to thank Marita Hiipakka and Jinghuan Wang for help and discussions, and his laboratory was supported by the Academy of Finland and Medical Research Fund of Tampere University Hospital.

References

1 Larson, S. M., Davidson, A. R., The identification of conserved interactions within the SH3 domain by alignment of sequences and structures. *Protein Sci.* **2000**, *9*, 2170–2180.

2 Mayer, B. J., SH3 domains: complexity in moderation. *J. Cell Sci.* **2001**, *114*, 1253–1263.

3 Zarrinpar, A., Bhattacharyya, R. P., Lim, W. A., The structure and function of proline recognition domains. *Sci. STKE* **2003**, *2003(179)*, RE8.

4 Kay, B. K., Williamson, M. P., Sudol, M., The importance of being proline: the interaction of proline-rich motifs in signaling proteins with their cognate domains. *FASEB J.* **2000**, *14*, 231–241.

5 Sadowski, I., Stone, J. C., Pawson, T., A noncatalytic domain conserved among cytoplasmic protein-tyrosine kinases modifies the kinase function and transforming activity of fujinami sarcoma virus $P130^{gag\text{-}fps}$. *Mol. Cell. Biol.* **1986**, *6*, 4396–4408.

6 Stahl, M. L., Ferenz, C. R., Kelleher, K. L., Kriz, R. W., Knopf, J. L., Sequence similarity of phospholipase C with the non-catalytic region of src. *Nature* **1988**, *332*, 269–272.

7 Mayer, B. J., Hamaguchi, M., Hanafusa, H., A novel viral oncogene with structural similarity to phospholipase C. *Nature* **1988**, *332*, 272–275.

8 Mayer, B. J., Jackson, P. K., Baltimore, D., The noncatalytic src homology region 2 segment of abl tyrosine kinase binds to tyrosine-phosphorylated cellular proteins with high affinity. *Proc. Natl. Acad. Sci. USA.* **1991**, *88*, 627–631.

9 Cicchetti, P., Mayer, B. J., Thiel, G., Baltimore, D., Identification of a protein that binds to the SH3 region of Abl and is similar to Bcr and GAP-rho. *Science* **1992**, *257*, 803–806.

10 Ren, R., Mayer, B. J., Cicchetti, P., Baltimore, D., Identification of a 10-amino acid proline-rich SH3 binding site. *Science* **1993**, *259*, 1157–1161.

11 Clark, S. G., Stern, M. J., Horvitz, H. R., C. elegans cell-signalling gene sem-5 encodes a protein with SH2 and SH3 domains. *Nature* **1992**, *356*, 340–344.

12 Simon, M. A., Bowtell, D. D. L., Dodson, G. S., Laverty, T. R., Rubin, G. M., Ras1 and a putative guanine nucleotide exchange factor perform crucial steps in signaling by the sevenless protein tyrosine kinase. *Cell* **1991**, *67*, 701–716.

13 Crechet, J. B., Poullet, P., Mistou, M. Y., Parmeggiani, A., Camonis, J., Boy-Marcotte, E., Damak, F., Jacquet, M., Enhancement of the GDP–GTP exchange of RAS proteins by the carboxyl-terminal domain of SCD25. *Science* **1990**, *248*, 866–868.

14 McCormick, F., How receptors turn ras on. *Nature* **1993**, *363*, 15–16.
15 Cheng, A. M., et al., Mammalian Grb2 regulates multiple steps in embryonic development and malignant transformation. *Cell* **1998**, *95*, 793–803.
16 Musacchio, A., Noble, M., Pauptit, R., Wierenga, R., Saraste, M., Crystal structure of a Src-homology 3 (SH3) domain. *Nature* **1992**, *359*, 851–855.
17 Yu, H., Rosen, M. K., Shin, T. B., Seidel-Dugan, C., Brugge, J. S., Schreiber, S. L., Solution structure of the SH3 domain of Src and identification of its ligand-binding site. *Science* **1992**, *258*, 1665–1668.
18 Harrison, S. C., Peptide–surface association: the case of PDZ and PTB domains. *Cell* **1996**, *86*, 341–343.
19 Yu, H., Chen, J. K., Feng, S., Dalgarno, D. C., Brauer, A. W., Schreiber, S. L., Structural basis for the binding of proline-rich peptides to SH3 domains. *Cell* **1994**, *76*, 933–945.
20 Sparks, A. B., Quilliam, L. A., Thorn, J. M., Der, C. J., Kay, B. K., Identification and characterization of Src SH3 ligands from phage-displayed random peptide libraries. *J. Biol. Chem.* **1994**, *269*, 23853–23856.
21 Feng, S., Kasahara, C., Rickles, R. J., Schreiber, S. L., Specific interactions outside the proline-rich core of two classes of Src homology 3 ligands. *Proc. Natl. Acad. Sci. USA* **1995**, *92*, 12408–12415.
22 Rickles, R. J., Botfield, M. C., Weng, Z., Taylor, J. A., Green, O. M., Brugge, J. S., Zoller, M. J., Identification of Src, Fyn, Lyn, PI3K and Abl SH3 domain ligands using phage display libraries. *EMBO J.* **1994**, *13*, 5598–5604.
23 Feng, S., Chen, J. K., Yu, H., Simon, J. A., Schreiber, S. L., Two binding orientations for peptides to the Src SH3 domain: development of a general model for SH3–ligand interactions. *Science* **1994**, *266*, 1241–1247.
24 Lim, W. A., Richards, F. M., Fox, R. O., Structural determinants of peptide-binding orientation and of sequence specificity in SH3 domains. *Nature* **1994**, *372*, 375–379.
25 Fernandez-Ballester, G., Blanes-Mira, C., Serrano, L., The tryptophan switch: changing ligand-binding specificity from type I to type II in SH3 domains. *J. Mol. Biol.* **2004**, *335*, 619–629.
26 Zarrinpar, A., Park, S. H., Lim, W. A., Optimization of specificity in a cellular protein interaction network by negative selection. *Nature* **2003**, *426*, 676–680.
27 Tavares, G. A., Panepucci, E. H., Brunger, A. T., Structural characterization of the intramolecular interaction between the SH3 and guanylate kinase domains of PSD-95. Mol. *Cell* **2001**, *8*, 1313–1325.
28 McGee, A. W., Dakoji, S. R., Olsen, O., Bredt, D. S., Lim, W. A., Prehoda, K. E., Structure of the SH3-guanylate kinase module from PSD-95 suggests a mechanism for regulated assembly of MAGUK scaffolding proteins. *Mol. Cell* **2001**, *8*, 1291–1301.
29 Tong, A. H., et al., A combined experimental and computational strategy to define protein interaction networks for peptide recognition modules. *Science* **2002**, *295*, 321–324.
30 Panni, S., Dente, L., Cesareni, G., In vitro evolution of recognition specificity mediated by SH3 domains reveals target recognition rules. *J. Biol. Chem.* **2002**, *277*, 21666–21674.
31 Brannetti, B., Via, A., Cestra, G., Cesareni, G., Citterich, M. H., SH3-SPOT: an algorithm to predict preferred ligand to different members of the SH3 gene family. *J. Mol. Biol.* **2000**, *298*, 313–328.
32 Ghose, R., Shekhtman, A., Goger, M. J., Ji, H., Cowburn, D., A novel, specific interaction involving the Csk SH3 domain and its natural ligand. *Nat. Struct. Biol.* **2001**, *8*, 998–1004.
33 Zhao, Z. S., Manser, E., Lim, L., Interaction between PAK and Nck: a template for Nck targets and role of PAK autophosphorylation. *Mol. Cell Biol.* **2000**, *20*, 3906–3917.
34 Lu, W., Katz, S., Gupta, R., Mayer, B. J., Activation of Pak by membrane localization mediated by an SH3 domain from the adaptor protein Nck. *Curr. Biol.* **1997**, *7*, 85–94.

35 Kami, K., Takeya, R., Sumimoto, H., Kohda, D., Diverse recognition of non-PxxP peptide ligands by the SH3 domains from p67(phox), Grb2 and Pex13p. *EMBO J.* **2002**, *21*, 4268–4276.
36 Lee, C.-H., Leung, B., Lemmon, M. A., Sheng, J., Cowburn, D., Kuriyan, J., Saksela, K., A single amino acid in the SH3 domain of Hck determines its high affinity and specificity in binding to HIV Nef protein. *EMBO J.* **1995**, *14*, 5006–5015.
37 Lee, C.-H., Saksela, K., Mirza, U. A., Chait, B. T., Kuriyan, J., Crystal structure of the conserved core of HIV-1 Nef complexed with a Src family SH3 domain. *Cell* **1996**, *85*, 931–942.
38 Sicheri, F., Moarefi, I., Kuriyan, J., Crystal structure of the Src family tyrosine kinase Hck. *Nature* **1997**, *385*, 602–609.
39 Xu, W., Harrison, S. C., Eck, M. J., Three-dimensional structure of the tyrosine kinase c-Src. *Nature* **1997**, *385*, 595–602.
40 Nagar, B., et al., Structural basis for the autoinhibition of c-Abl tyrosine kinase. *Cell* **2003**, *112*, 859–871.
41 Moarefi, I., LeFevre-Bernt, M., Sicheri, F., Huse, M., Lee, C.-H., Kuriyan, J., Miller, W. T., Activation of the Src-family tyrosine kinase Hck by SH3 domain displacement. *Nature* **1997**, *385*, 650–653.
42 Kishan, K. V., Scita, G., Wong, W. T., Di Fiore, P. P., Newcomer, M. E., The SH3 domain of Eps8 exists as a novel intertwined dimer. *Nat. Struct. Biol.* **1997**, *4*, 739–743.
43 Gorina, S., Pavletich, N. P., Structure of the p53 tumor suppressor bound to the ankyrin and SH3 domains of 53BP2. *Science* **1996**, *274*, 1001–1005.
44 Nishida, M., Nagata, K., Hachimori, Y., Horiuchi, M., Ogura, K., Mandiyan, V., Schlessinger, J., Inagaki, F., Novel recognition mode between Vav and Grb2 SH3 domains. *EMBO J.* **2001**, *20*, 2995–3007.
45 Barnett, P., Bottger, G., Klein, A. T., Tabak, H. F., Distel, B., The peroxisomal membrane protein Pex13p shows a novel mode of SH3 interaction. *EMBO J.* **2000**, *19*, 6382–6391.
46 Pires, J. R., Hong, X., Brockmann, C., Volkmer-Engert, R., Schneider-Mergener, J., Oschkinat, H., Erdmann, R., The ScPex13p SH3 domain exposes two distinct binding sites for Pex5p and Pex14p. *J. Mol. Biol.* **2003**, *326*, 1427–1435.
47 Mongiovi, A. M., Romano, P. R., Panni, S., Mendoza, M., Wong, W. T., Musacchio, A., Cesareni, G., Di Fiore, P. P., A novel peptide-SH3 interaction. *EMBO J.* **1999**, *18*, 5300–5309.
48 Kurakin, A. V., Wu, S., Bredesen, D. E., Atypical recognition consensus of CIN85/SETA/Ruk SH3 domains revealed by target-assisted iterative screening. *J. Biol. Chem.* **2003**, *278*, 34102–34109.
49 Manser, E., et al., PAK kinases are directly coupled to the PIX family of nucleotide exchange factors. *Mol. Cell* **1998**, *1*, 183–192.
50 Grabs, D., Slepnev, V. I., Songyang, Z., David, C., Lynch, M., Cantley, L. C., De Camilli, P., The SH3 domain of amphiphysin binds the proline-rich domain of dynamin at a single site that defines a new SH3 binding consensus sequence. *J. Biol. Chem.* **1997**, *272*, 13419–13425.
51 Cestra, G., et al., The SH3 domains of endophilin and amphiphysin bind to the proline-rich region of synaptojanin 1 at distinct sites that display an unconventional binding specificity. *J. Biol. Chem.* **1999**, *274*, 32001–32007.
52 Kang, H., Freund, C., Duke-Cohan, J. S., Musacchio, A., Wagner, G., Rudd, C. E., SH3 domain recognition of a proline-independent tyrosine-based RKxxYxxY motif in immune cell adaptor SKAP55. *EMBO J.* **2000**, *19*, 2889–2899.
53 Lewitzky, M., et al., The C-terminal SH3 domain of the adapter protein Grb2 binds with high affinity to sequences in Gab1 and SLP-76 which lack the SH3-typical P-x-x-P core motif. *Oncogene* **2001**, *20*, 1052–1062.
54 Berry, D. M., Nash, P., Liu, S. K., Pawson, T., McGlade, C. J., A high-affinity Arg-X-X-Lys SH3 binding motif confers specificity for the interaction between Gads and SLP-76 in T cell signaling. *Curr. Biol.* **2002**, *12*, 1336–1341.

55 Kato, M., Miyazawa, K., Kitamura, N., A De-ubiquitinating enzyme UBPY interacts with the SH3 domain of Hrs binding protein via a novel binding motif Px(V/I)(D/N)RxxKP. *J. Biol. Chem.* **2000**, *275*, 37481–37487.

56 Asada, H., et al., Grf40, a novel Grb2 family member, is involved in T cell signaling through interaction with SLP-76 and LAT. *J. Exp. Med.* **1999**, *189*, 1383–1390.

57 Harkiolaki, M., et al., Structural basis for SH3 domain-mediated high-affinity binding between Mona/Gads and SLP-76. *EMBO J.* **2003**, *22*, 2571–2582.

58 Liu, Q., Berry, D., Nash, P., Pawson, T., McGlade, C. J., Li, S. S., Structural basis for specific binding of the Gads SH3 domain to an RxxK motif-containing SLP-76 peptide: a novel mode of peptide recognition. *Mol. Cell* **2003**, *11*, 471–481.

59 Wu, X., Knudsen, B., Feller, S. M., Zheng, J., Sali, A., Cowburn, D., Hanafusa, H., Kuriyan, J., Structural basis for the specific interaction of lysine-containing proline-rich peptides with the amino-terminal SH3 domain of c-Crk. *Structure* **1995**, *3*, 215–226.

60 Kardinal, C., et al., Cell-penetrating SH3 domain blocker peptides inhibit proliferation of primary blast cells from CML patients. *FASEB J.* **2000**, *14*, 1529–1538.

61 Nguyen, J. T., Turck, C. W., Cohen, F. E., Zuckermann, R. N., Lim, W. A., Exploiting the basis of proline recognition by SH3 and WW domains: design of N-substituted inhibitors. *Science* **1998**, *282*, 2088–2092.

62 Nguyen, J. T., Porter, M., Amoui, M., Miller, W. T., Zuckerman, R. N., Lim, W. A., Improving SH3 domain ligand selectivity using a non-natural scaffold. *Chem. Biol.* **2000**, *7*, 463–473.

63 Combs, A. P., Kapoor, T. M., Feng, S., Chen, J. K., Daude-Snow, L. F., Schreiber, S. L., Protein structure-based combinatorial chemistry: discovery of non-peptide binding elements to Src SH3 domains. *J. Am. Chem. Soc.* **1996**, *118*, 287–288.

64 Feng, S., Kapoor, T. M., Shirai, F., Combs, A. P., Schreiber, S. L., Molecular basis for the binding of SH3 ligands with non-peptide elements identified by combinatorial synthesis. *Chem. Biol.* **1996**, *3*, 661–670.

65 Hiipakka, M., Poikonen, K., Saksela, K., SH3 domains with high affinity and engineered ligand specificities targeted to HIV-1 Nef. *J. Mol. Biol.* **1999**, *293*, 1097–1106.

66 Hiipakka, M., Huotari, P., Manninen, A., Renkema, G. H., Saksela, K., Inhibition of cellular functions of HIV-1 Nef by artificial SH3 domains. *Virology* **2001**, *286*, 152–159.

3
The WW Domain

Marius Sudol

3.1
Introduction and Brief History of Module Discovery

As with the SH3 (Src homology 3) domain, the WW (tryptophan–tryptophan) domain also emerged from work on oncogenic viruses that was initiated at the Rockefeller University in New York in a virology laboratory focused on neoplastic transformation of cells by Rous sarcoma virus [1–3]. Biological analyses of host-dependent mutants of Rous sarcoma virus and site-directed mutation of the *Src* gene itself provided insights into the molecular mechanism of Src signaling [2, 4, 5]. Both approaches revealed a propensity of the N-terminal region of Src kinases to form discrete, apparently specific, protein–protein complexes [2]. The accumulated data prompted a hypothesis proposing that these complexes represent an important facet of Src function by regulating the enzymatic activity of Src [4, 5]. The identification of two protein–protein interaction modules, namely the SH2 and SH3 domains, within the very N-terminal region of Src kinases [6] reinforced the initial insight and prompted many laboratories to search for protein partners of SH domains. The WW domain was identified as an imperfect repeat of 38 amino acids in a differently spliced form of the Yes kinase-associated protein, YAP [7]. Surprisingly and uniquely, the 38 amino acids were added by a splicing event to one of the YAP isoforms that already had a single copy of the related sequence upstream [8]. The repeated sequence was distinct and easily noticed by visual inspection [8]. The Yes kinase is the closest relative of Src, and YAP was isolated as one of the SH3 domain-binding partners of Yes and Src kinases [9].

Examination of the consensus sequence in the first alignment of WW domains revealed the presence of two highly-conserved tryptophan (W) residues, spaced 20–22 amino acids apart, and led to its being named the WW domain [7]. Although intuitively we already considered this repeat a domain, at that point there were no data indicating that the short repeat of 38 amino acids mediated protein–protein complexing or that it had a defined structure. Both results came swiftly to confirm our predictions. Using functional screens with a labeled WW domain of YAP as a probe, two partial cDNAs were isolated which had different sequences except for

Modular Protein Domains. Edited by G. Cesareni, M. Gimona, M. Sudol, M. Yaffe

ISBN: 3-527-30813-X

one short common region with a PPPPY motif, which was subsequently shown to be required for binding [10]. By sequentially replacing each of these five positions with alanine, a consensus sequence of PPxY was established [10]. Identification of the ligand prompted structural analysis of the domain–peptide complex and nuclear magnetic resonance analysis revealed a compact three-stranded β-sheet structure of the YAP WW domain [11]. At that point the sequence repeat with two conserved tryptophans became a 'bona fide' WW domain. It was clear to us that the domain was related to the SH3 domain by its ability to recognize proline-rich peptides, yet in many aspects it was distinct from it.

3.2
Structure of the WW Domain–Ligand Complex

High-resolution crystal structures and nuclear magnetic resonance structures of several WW domains and their complexes with ligands were solved [12–14]. Each WW domain folds as a compact triple-stranded β sheet [15] (Figures 3.1 and 3.2). The fold is stable in the absence of cofactors, ligands, or disulfide bridges. Based on the recognition of proline-rich ligands, the WW domains were classified into four groups (Table 3.1). Group I WW domains recognize a PPxY binding motif, Group II WW domains interact with PPLP motifs, and Group III WW domains recognize polyprolines flanked or interrupted by R or K residues. Group IV is represented only by several WW domains and recognize a motif containing a proline residue preceded by phosphoserine or phosphothreonine, *po*(S/T)P [16]. For several

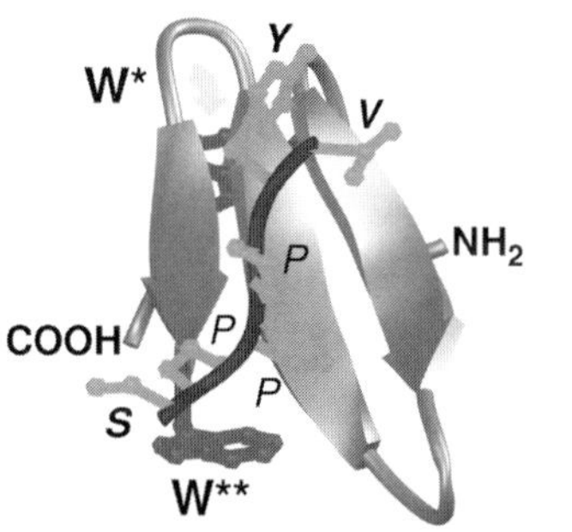

Figure 3.1 Structure of the WW domain of dystrophin in complex with SPPPYV peptide derived from β-dystroglycan. N and C termini of the domain are indicated. The sidechains of the first and second conserved tryptophan in the domain are indicated (W* and W**, respectively). The first conserved tryptophan is hidden behind the surface involved in ligand binding. Prolines and the tyrosine of the ligand are labeled in italics (*P* and *Y*). Note how the aromatic ring of the second tryptophan stacks with the nearby proline (for more details see [12, 15]).

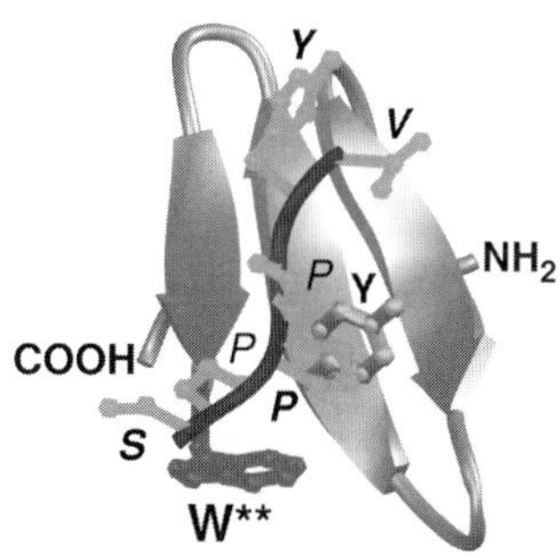

Figure 3.2 Structure of the WW domain of dystrophin in complex with SPPPYV peptide derived from β-dystroglycan. Note how the sidechain tyrosine (Y) in the second β strand of the domain stacks with one of the prolines of the ligand. The tyrosine–proline and the second tryptophan–proline stacks constitute an aromatic cradle, a common molecular arrangement found in SH3, EVH1, and profilin complexes [12].

Table 3.1 Major classes of WW domains categorized according to whether they recognize proline-rich or proline-containing ligands (for more details, see [16] and the text).

Class of WW (Example)	*Consensus*	*Ligand Protein*
Class I (YAP)	PPxY	ErbB4
Class II (FE65)	PPLP	Mena
Class III (FBP)	RPPP(R)	SmB
Class IV (Pin1)	*po*(S/T)P	RNA Pol II

members of Group I WW domains, the ligand phosphorylation on the tyrosine of the PPxY core negatively regulates complex formation in vitro and in cell culture (reviewed in [16]).

The two signature tryptophans of the WW domain play different roles; the first one seems to be involved in maintaining the stability of the domain [15]; and the second conserved tryptophan participates in ligand binding [10, 15] (Figure 3.1). One interesting feature of the WW domain was revealed by a high-resolution crystal structure of the dystrophin WW domain in complex with the β-dystroglycan peptide [12]. It involves the second conserved tryptophan. The rings of aromatic sidechains of the WW domain form hydrophobic stacks with proline rings of the ligand (Figure 3.2). This arrangement is reinforced by a dominant hydrogen bond formed between a carbonyl oxygen of proline and the indole nitrogen in the tryptophan [12]. This structural arrangement, with minor variations, is a common feature found also among complexes of SH3 and EVH1 domains and profilin proteins. It was called an 'aromatic cradle' [12].

Using computer-aided screens of publicly accessible sequence databanks, we identified several hundred proteins or open reading frames that harbor WW domains. Examples of well defined and intensively studied WW domain-containing proteins are shown in Figure 3.3.

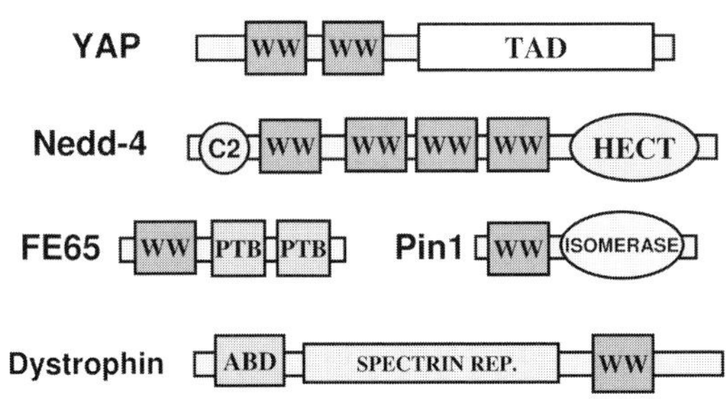

Figure 3.3 Modular structures of proteins that harbor WW domains. As with other protein domains such as the SH2 and SH3 domains, the WW domain occurs in a variety of proteins that carry out diverse functions. YAP is a transcriptional coactivator. Nedd-4 is one of the E3-ubiquitin ligases. FE65 is an adaptor protein. Pin1 is the enzyme isoprolyl isomerase. Dystrophin is a cytoskeletal protein. TAD: transcription activation domain; C2: membrane-binding domain; HECT: E3-ubiquitin ligase domain; PTB: phosphotyrosine binding domain; ABD: actin binding domain composed of calponin homology 1 (CH1) and CH2 domains; SEPCTRIN REP: spectrin repeats.

3.3
WW Domains and Human Diseases

3.3.1
From Liddle's Syndrome to Liddle's Disease

In 1963, Liddle and colleagues described a rare syndrome of a salt-sensitive form of hypertension with clinical characteristics similar to a primary hyperaldosteronism [17]. The affected patients suffered from hypertension and metabolic alkalosis, both symptoms also observed in individuals with hyperproduction of aldosterone by the adrenal glands. However, Liddle demonstrated that his patients had normal levels of mineralocorticoids. Moreover, using an amiloride analog to selectively block a sodium channel, Liddle was able to decrease the blood pressure in patients ingesting a low-salt diet. At that point he proposed the term 'pseudoaldosteronism' – today it is called the Liddle syndrome.

In 1994 Shimkets et al. [18] were the first to demonstrate the genetic basis of Liddle's syndrome. A mutation truncating the C-terminal region of the β subunit of ENaC (amiloride-sensitive sodium channel) was the first genetic lesion shown to correlate with Liddle disease. A subsequent study mapped other mutations, mostly deletions that encompass in common a 12-residue sequence containing the PPPNY motif [19]. At that time, the PPxY consensus motif was already recognized as a WW domain ligand core, and work in the laboratories of Rossier [20] and Rotin [21] revealed a WW domain-containing ubiquitin ligase, NEDD4, as a functional partner of ENaC. A simple scheme of the molecular basis of Liddle disease emerged and was confirmed by numerous biochemical and physiological studies [22, 23]. In the center of the scheme, the ENaC subunits (β and γ) interact through their PPxY regions with WW domains of Nedd4 ubiquitin ligase (Figure 3.4). The complex is required for fast turnover of the ENaC channel and its degradation by a proteasome pathway. When the complex is compromised because of mutations in the PPxY cores, the high concentration of ENaC results in excessive activity of the channel,

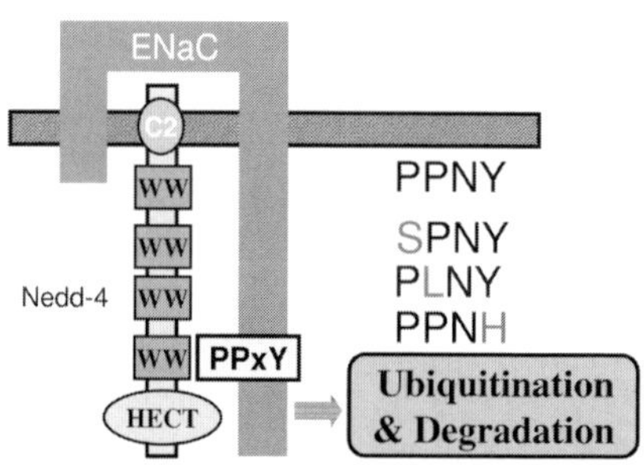

Figure 3.4 A simple scheme of the complex between the β subunit of amiloride-sensitive sodium channel ENaC and Nedd-4 ubiquitin ligase. HECT is the ubiquitin ligase domain, C2 is a membrane-binding domain. The short sequences on the right indicate the wild-type sequence of the PPxY motif that binds WW domains of Nedd4 followed by three point mutants described in individual patients with Liddle syndrome (see text for more details).

giving rise to sodium imbalance and hypertension. Interestingly, the systematic mutagenesis performed on the first proline-rich ligand of a WW domain, which defined the PPxY consensus [10], was reconfirmed in Nature. Missense single-point mutations in the gene encoding the β subunit of ENaC, which were reported to cause Liddle syndrome, mapped precisely to the core residues of the PPxY motif: P615S, P616L, and Y618H [24–26]. Independent patients, each harboring one of the three single-point mutations within the PPxY core showed full symptoms of Liddle disease.

It is most likely that the ENaC–Nedd4 pair is not the only functional complex that is formed to regulate the channel. Other WW domains are known to interact with ENaC [27], and it would be interesting to understand how the channel is regulated by non-Nedd4-like proteins.

3.3.2
Amyloid Precursor Protein: APP and FE65

FE65 was isolated by Esposito et al. [28] in a screen designed to clone brain-enriched or brain-specific cDNAs. The structure of the FE65 protein predicted by the DNA sequence turned out to represent a typical adaptor protein composed of a WW domain and two PTB domains [29]. FE65 became a recognized entry on the signal transduction map when Russo et al. [30], and later, several other groups, showed a specific complex between APP and the second PTB of FE65. Recently, the APP–FE65 complex was given a functional dimension. A report by Cao and Sudhof [31] elegantly documented that the γ-secretase–generated C-terminal fragment (CTF) of APP was able to translocate to the cell nucleus to form a multicomponent complex including FE65, the histone acetylase Tip60, and most likely, several other proteins (Figure 3.5). This complex was shown to activate transcription via Gal4 or LexA reporters only when the FE65 WW domain was intact, suggesting a requirement for the WW domain ligand in the process. Indeed, the list of FE65 WW domain

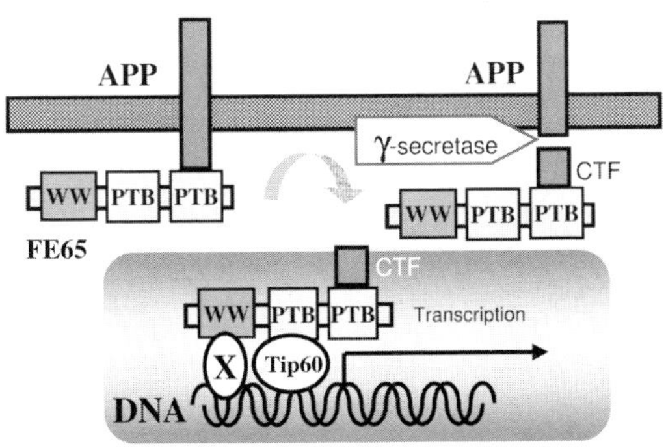

Figure 3.5 Signaling by amyloid precursor protein (APP) complex. The carboxy-terminal fragment (CTF) of APP that is cleaved by γ-secretase forms a complex with FE65 adaptor protein. The complex translocates to the cell nucleus and, together with other components such as Tip60 histone acetylase and FE65WW domain ligands (X-unknown), it regulates transcription (see text for a more detailed explanation).

ligands derived from the mapping data of human WW domains by AxCell Biosciences Corporation [27] contains mostly nuclear proteins that are able to regulate or affect transcription. It should be important to test the FE65WW domain ligands revealed by the WW domain mapping [27] as potentially critical components of the multicomponent APP–FE65 complex in regulating transcription. With the general availability of inexpensive advanced genomic chips, one should be able to decipher the differential profiles of gene expression regulated by APP–FE65 or APP–FE65 with mutated WW domains in neuronal cells. Finally, it would be critical to determine if any of the gene expression patterns that are regulated by the APP–FE65 complex are also subject to changes in neuronal cells derived from brains affected by Alzheimer's disease.

3.3.3 Dystrophin WW Domain and Muscular Dystrophy

The muscular dystrophies are a group of diseases that affect skeletal muscle and are characterized by progressive muscle wasting [32]. Duchenne and Becker muscular dystrophies are X-linked recessive diseases caused by mutations in the dystrophin gene [33]. This gene encodes several isoforms, among which is the 427-kD α isoform called dystrophin, which is expressed in muscle and brain. Dystrophin consists of three distinct regions: the N-terminal actin binding domain, the long spectrin-like rod region, and the cysteine-rich C-terminal region (Figure 3.3). Analysis of the dystrophin gene from affected patients has yielded a patho-functional map of dystrophin. According to the map, the most severe forms of Duchenne or Becker muscular dystrophies were observed when deletions affected either the N- or the C-terminal region of dystrophin.

The identification of the WW domain in the C-terminal region of dystrophin immediately prompted a search for potential partner proteins within the known repertoire of proteins constituting the multicomponent dystrophin–glycoprotein complex [34, 35]. The search was facilitated by the extensive biochemical data on the dystrophin-assembled complexes and the presumed presence of a PPxY core in the potential partners [34, 35]. β-Dystroglycan was instantly selected as the primary candidate ligand for the dystrophin WW domain [34]. At the C-terminal tail, which was shown to be required for interaction with dystrophin [35], the β-dystroglycan protein was found to harbor two PPxY cores. Subsequent biochemical and structural studies documented a direct contact between the most terminal PPxY core of β-dystroglycan and the dystrophin WW domain [12, 36].

The study of the dystrophin–β-dystroglycan complex instructed us about two important and more general issues. We realized that some of the WW domains require longer flanking sequences to maintain their binding activity [12, 36]. We showed that a short N-terminal flank could cooperate with a very long C-terminal flanking sequence. Later, through the work of Chung and Campanelli [37], we realized that the very same WW domain could also retain binding activity when a much shorter C-terminal flanking sequence was 'helped' by somewhat longer N-terminal flank.

The biology of the dystrophin WW domain–β dystroglycan complex is being investigated now in vivo. We are generating a knock-in mouse having a mutated WW domain of dystrophin. Based on limited patient data having deletions of the exons encoding the human WW domain, the expected phenotype of mutated mouse is that of muscular dystrophy of the Duchenne type (which in mice is much milder than in humans).

3.4 Emerging Directions and Recent Developments

3.4.1 AxCell's Map

Recently a protein–protein interaction map for the human WW domain family was generated [27]: 57 human WW domains were expressed as GST fusions and probed with 1930 proline-rich peptides containing cores recognized by major classes of WW domains [16], with each peptide corresponding to the known protein or ORF in the human proteome. Colorimetric assay based on the alkaline phosphatase–streptavidin–biotin system was used to measure the apparent binding constants between GST-tagged WW domains and biotinylated peptides. A network of more than 69 000 interactions was deciphered and now serves as a blueprint for identification of new signaling pathways that utilize WW domains.

This comprehensive effort [27] represents the first protein–protein interaction map of a domain in the human proteome.

Apart from its value for projecting potential signaling routes and networks, the WW domain mapping provides a unique resource for understanding the structure–function relationship of the WW domain–ligand interface. More directly, the data inform us about structural determinants of the WW domain, i.e., residues within the domain that are responsible for recognition of specific ligands.

3.4.2 ErbB4 Receptor Protein–Tyrosine Kinase and its WW Domain-containing Adaptor, YAP

We focused our research on ErbB4 when Carpenter and colleagues [38, 39] showed that the ErbB4 receptor protein-tyrosine kinase is proteolytically processed by membrane proteases in response to the cognate ligand, and the resulting C-terminal fragment (CTF) was found in the cell nucleus to regulate transcription. We were attracted by the relative simplicity of the Notch-type signaling and its relative novelty for the family of ErbB receptors. The decision to investigate ErbB4 signaling also stemmed from several observations, all pointing to the cognate complex between ErbB4 and YAP. Screening of molecular repertoires of proline-rich peptides on SPOT membranes, together with a screen of phage-displayed peptide libraries for 'super binders' to the first WW domain of YAP, resulted in a ΨPPPPYP/R consensus sequence [40], where Ψ (psi) designates aliphatic amino acids. This consensus matches the sequence of the most C-terminal PPxY motif (LPPPPYR) of ErbB4. We noted that ErbB4 contained three perfect PPxY motifs as potential binding sites of WW domains [41]. The Carpenter group [38] showed that CTF of ErbB4 has the potential to regulate transcription, but it had to be mutated to reveal its transcriptional regulation potential. Previously, YAP was shown to interact with PPxY containing sequences of nuclear proteins to regulate transcription [42]. Finally, the comprehensive mapping of human WW domains confirmed that the most-terminal PPxY of ErbB4 binds YAP with relatively high affinity [27]. We realized that we had in our hands a formerly missing link. With all the suggestive evidence and the available reagents, we documented that YAP was able to associate physically with the full-length ErbB4 receptor and functionally with the ErbB4 cytoplasmic fragment in the nucleus [41]. As expected, the YAP–ErbB4 complex was mediated by the first WW domain of YAP and the most C-terminal PPxY motif of ErbB4 [41]. The wild-type CTF of ErbB4 did not show transcriptional activity by itself. However, in the presence of YAP its transcriptional activity was revealed (Figure 3.6) [41]. Using immunofluorescent staining, we showed that the CTF of ErbB4 and full-length YAP partly colocalize in the cell nucleus, further supporting the validity of the functional complex.

To understand the ErbB4 signaling pathway, one should identify transcripts that are regulated by the ErbB4–YAP complex in response to the ligand (neuregulin) or TPA (tumor promoter) treatment – the latter is also known to initiate the proteolytic processing of ErbB4. Since the levels of ErbB4 and ErbB2 expression plus the resulting stoichiometry of ErbB4–ErbB2 heterodimers correlate with breast and cerebellar cancers [43], the precise description of the transcriptional program affected by the ErbB4–YAP complex should provide a molecular tool for a better understanding of these two types of cancers.

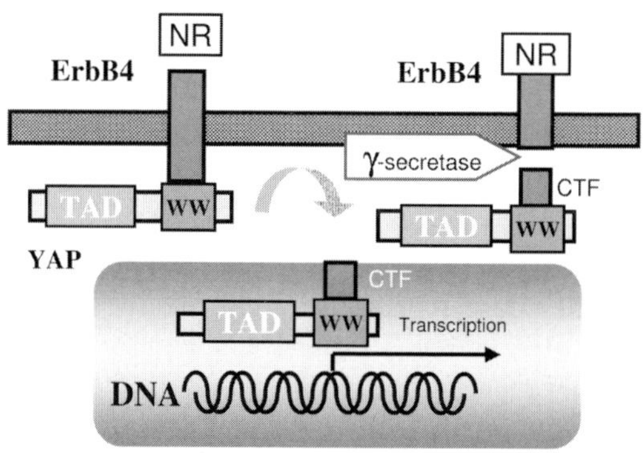

Figure 3.6 Signaling by ErbB4 receptor. Upon stimulation with ligand (NR-neuregulin), the ErbB4 receptor undergoes proteolytic processing that involves γ-secretase. The released CTF fragment, which contains PPxY motifs, forms a specific complex with the WW domain of YAP, a known cotranscriptional regulator. Both proteins translocate to the cell nucleus to regulate gene expression [41]. TAD: transcriptional activation domain.

3.4.3
Membrane Proteins with PPxYs Implicated in Cancer

The examples of ENaC, β-dystroglycan, and especially the ErbB4 receptor as membrane proteins with distinct PPxY motifs within their cytoplasmic tails, which represent sites for formation of WW domain-mediated signaling complexes, turned out to be molecular scenarios amenable to fruitful experimental study. While studying these three complexes we kept searching for other *membrane* proteins that would contain PPxY motifs and could be implicated in human diseases. The most attractive candidate found in that search was the *STAG1/PMEPA1* gene [44, 45]. The gene transcript and protein are robustly induced by androgen and found as over-expressed products in prostate and kidney tumors. The protein product is < 40 kD, has a short extracellular domain, and is followed by a transmembrane region and a relatively long cytoplasmic tail with two perfect PPxY motifs. Little is known about WW domain partners of the STAG1/PMEPA1 protein and nothing about the extracellular ligand(s). It is also not known if the protein can undergo ligand-induced dimerization or be phosphorylated on the tyrosines of the PPxY motifs. If indeed the up-regulation of the *STAG1/PMEPA1* gene accelerates androgen pathways that promote neoplastic changes in prostate and kidney, then interfering with the downstream WW domain-mediated signals may have direct ramifications for cancer therapy.

3.5
Concluding Remarks

The study of the WW domain has raised many important questions that are relevant to the entire field of modular protein domains. Several of these questions are mentioned here and, where possible, partial answers and comments follow.

Are all WW domains and other protein domains well demarcated, independently folded, autonomous units that maintain their binding function even when removed from their original parent proteins? The data seem to indicate that these signature features of protein domains hold true for majority of them [e.g., 19, 27]. However, at one extreme there are domains that are truly exemplary 'Lego blocks' and at the opposite extreme there are a notable number of domains that require significantly longer flanking sequences to achieve full binding function – a Gaussian distribution it seems.

Can a 'gold standard' of interaction for each domain–ligand complex be generated using techniques of alanine scan mutagenesis or molecular repertoires combined with high-throughput binding screens? More precisely, can we generate sequential deletions of proteins harboring the domains and their cognate ligands and create a comprehensive interaction matrix? Such data would identify minimal sequences required for domain–ligand binding and the overall binding profile of a specific complex.

Is the specificity of domain–ligand pairs revealed only in the full protein and cell compartment contexts in which the interaction pair is presented? A lucid and elegant study from the laboratory of Lim [46] suggests that some protein domains do have unique cognate ligands and achieve fidelity in complex formation. The SH3 domain of Sho1, a yeast protein regulating the high-osmolarity stress-response pathway, and the proline-rich ligand Pbs2 seem to interact with perfect specificity. The remaining 26 SH3 domains present in the yeast proteome do not recognize Pbs2 in vitro or in vivo [46]. Perhaps, upon careful analysis of all known protein domains, we will find more examples of optimized specificity of domain–ligand binding within interaction networks, including those sets that are composed of overlapping recognition elements such as SH3 and WW domains, for example [47]. One could go even further by speculating that some, but not all, domain–ligand pairs have achieved fidelity of binding through the processes of negative and positive selection. Since promiscuous interactions may lead to evolutionarily significant disadvantages [47], one may suggest again that complex protein–protein interaction systems are under evolutionary pressure to achieve the highest degree of binding specificity.

Does the apparent plasticity of protein modules, as revealed by engineered molecular repertoires of domains followed by successful selection toward new ligands [48–50], point to them as attractive molecular 'capsules' for tailored gene therapy? A successful interference with the Rous sarcoma virus budding process, by cis expression of the YAP WW domain that targeted PPxY-containing viral Gag, shows that an isolated WW domain can be used as a reagent to modulate in vivo processes [51]. I anticipate that using wild-type or mutated protein domains as modulators of signaling pathways will be fruitful.

Can the comprehensive data of the WW domain protein–protein interaction map [27] be analyzed to decipher other rules of the 'protein recognition code' [52–54], given the data describing the fidelity in complex formation among some domain–ligand pairs and the apparently strong selection pressures on protein–protein networks to achieve a high level of specificity? The protein–protein interaction map provides data that allow better understanding of specific residues within the WW domain binding surface, which are responsible for recognition of different classes of proline-rich ligands. However, truly general rules of a protein recognition code will probably emerge from concerted approaches that include structural, genetic, and biochemical analyses of biologically meaningful domain–ligand complexes.

A decade ago my laboratory entered the field of modular protein domains by serendipity. The analogy between modular protein domains and Lego blocks, which implicitly stipulates reiterated and combinatorial use of domains by Nature to create repertoires of functional complexes [6, 55], is an attractive, widely accepted concept. Only recently did I learn the origin of the name Lego – the word is an abbreviated fusion of two Danish words 'leg godt', meaning 'have fun playing'. Indeed, designing and executing experiments with protein modules has been an engaging activity at times reminiscent of play. I look forward to new challenges and to further exploration of possibilities of tailoring the modular domains into powerful probes for profiling signaling events and for identifying leads for the development of therapeutic interventions.

Acknowledgements

I would like to thank all the members of my laboratory who contributed to the work on the WW domain. Special thanks are due to Aaron Einbond, Henry Chen, Xavier Espanel, and Akihiko Komuro for their outstanding dedication. Our work has been supported by grants from the NIH (CA45757; AR45626; DK62345), Human Frontier Science Program Organization (RG234), and Alzheimer's and Muscular Dystrophy Associations. The help of Amjad Farooq with the WW structure figures is much appreciated. Comments of Drs. David Foster, Richard Jove, and Roman Osman significantly improved the manuscript.

I apologize that many references are not included in this review. More than 300 papers have been published on the WW domain, and complete coverage of the subject was not possible.

References

1 Sudol, M., The WW module competes with the SH3 domain? *Trends Biochem. Sci.* **1996**, *21*, 162–163.

2 Jove, R., Hanafusa, H., Cell transformation by the viral *src* oncogene. *Annu. Rev. Cell Biol.* **1987**, *3*, 31–56.

3 Mayer, B. J., SH3 domains: complexity in moderation. *J. Cell Sci.* **2001**, *114*, 1253–1263.

4 Hirai, H., Varmus, H. E., Site-directed mutagenesis of the SH2- and SH3-coding domains of c-src produces varied phenotypes, including oncogenic activation of c-src. *Mol. Cell. Biol.* **1990**, *10*, 1307–1318.

5 Hirai, H., Varmus, H. E., Mutations in src homology regions 2 and 3 of activated chicken c-src that result in preferential transformation of mouse or chicken cells. *Proc. Natl. Acad. Sci. USA* **1990**, *87*, 8592–8596.

6 Pawson, T., Nash, P., Assembly of cell regulatory systems through protein interaction domains. *Science* **2003**, *300*, 445–452.

7 Bork, P., Sudol, M., The WW domain: a new signaling site in dystrophin? *Trends Biochem. Sci.* **1994**, *19*, 531–533.

8 Sudol, M., Bork, P., Einbond, A., Kastury, K., Druck, T., Negrini, M., Huebner, K., Lehman, D., Characterization of the mammalian *YAP* (yes-associated protein) gene and its role in defining a novel protein module, the WW domain. *J. Biol. Chem.* **1995**, *270*, 14733–14741.

9 Sudol, M., Yes-associated protein (YAP65) is a proline-rich phosphoprotein that binds to the SH3 domain of the *Yes* proto-oncogene product. *Oncogene* **1994**, *9*, 2145–2152.

10 Chen H. I., Sudol, M., The WW domain of Yes-associated protein binds a novel proline-rich ligand that differs from the consensus established for SH3-binding modules. *Proc. Natl. Acad. Sci. USA* **1995**, *92*, 7819–7823.

11 Macias, M. J., Hyvonen, M., Baraldi, E., Schultz, J., Sudol, M., Saraste, M., Oschkinat, H., The structure of the WW domain in complex with a proline-rich peptide. *Nature* **1996**, *382*, 646–649.

12 Huang, X., Roy, F., Zhang, R., Joachimiak, A., Sudol, M., Eck, M. J., Recognition of a proline motif in beta-dystroglycan by an 'embedded' WW domain in human dystrophin. *Nat. Struct. Biol.* **2000**, *7*, 634–638.

13 Kanelis, V., Rotin, D., Forman-Kay, J. D., Solution structure of a Nedd4 WW domain–ENaC peptide complex. *Nat. Struct. Biol.* **2001**, *8*, 407–412.

14 Macias, M. J., Gervais, V., Civera, C., Oschkinat, H., Structural analysis of WW domains and design of a WW prototype. *Nat. Struct. Biol.* **2000**, *7*, 375–379.

15 Macias, M. J., Wiesner, S., Sudol, M., WW and SH3 domains, two different scaffolds to recognize proline-rich ligands. *FEBS Lett.* **2002**, *513*, 30–37.

16 Sudol, M., Hunter, T., NeW Wrinkles for an Old Domain. *Cell* **2000**, *103*, 1001–1004.

17 Liddle, G. W., Bledsoe, T., Coppage, W. J. S., A familial renal disorder simulating primary aldosteronism but with negligible aldosterone secretion. *Trans. Assoc. Am. Physicians* **1963**, *76*, 199–213.

18 Shimkets, R. A., et al., Liddle's syndrome: heritable human hypertension caused by mutations in the beta-subunit of the epithelial Na channel. *Cell* **1994**, *79*, 407–414.

19 Sudol, M., Structure and function of the WW domain. *Prog. Biophys. Mol. Biol.* **1996**, *65*, 113–132.

20 Schild, L., Lu, Y., Gautschi, I., Schneeberger, E., Lifton, R. P., Rossier, B. C., Identification of a PY motif in the epithelial Na channel subunits as target sequence for mutations causing channel activation found in Liddle syndrome. *EMBO J.* **1996**, *15*, 2381–2387.

21 Staub, O., Dho, S., Henry, Correa, P. C., J., Ishikawa, T., McGlade, J., Rotin, D., WW domains of Nedd4 bind to the proline-rich PY motifs in the epithelial Na channel deleted in Liddle's syndrome. *EMBO J.* **1996**, *15*, 2371–2380.

22 Schild, L., Canessa, C. M., Shimkets, R. A., Gautschi, I., Lifton, R. P., Rossier, B. C., A mutation in the epithelial sodium channel causing Liddle disease increases channel activity in the *Xenopus laevis* oocyte expression system. *Proc. Natl. Acad. Sci. USA* **1995**, *92*, 5699–5703.

23 Snyder, P. M. Price, M. P., McDonald, F. J., Adams, C. M., Volk, K. A., Zeiher, B. G., Stokes, J. B., Welsh, M. P., Mechanism by which Liddle's syndrome mutations increase activity of a human epithelial Na channel. *Cell* **1995**, *83*, 969–978.

24 Hansson, J. H., Schild, L., Lu, Y., Wilson, T. A., Gautschi, I., Shimkets, R., Nelson-Williams, C., Rossier, B. C., Lifton, R. P., A de novo missense mutation of the beta subunit of the epithelial sodium channel causes hypertension and Liddle syndrome, identifying a proline-rich segment critical for regulation of channel activity. *Proc. Natl. Acad. Sci. USA* **1995**, *92*, 11495–11499.

25 Tamura, H., Schild, L., Enomoto, N., Matsui, N., Marumo, F., Rossier, B. C., Sasaki, S., Liddle disease caused by a missense mutation of the beta subunit of the epithelial sodium channel gene. *J. Clin. Invest.* **1996**, *97*, 1780–1784.

26 Inoue, J., Iwaoka, T., Tokunaga, H., Takamune, K., Naomi, S., Araki, M., Takahama, K., Yamaguchi, K., Tomita, K., A family with Liddle's syndrome caused by a new missense mutation in the beta subunit of the epithelial sodium channel. *J. Clin. Endocrinol. Metab.* **1998**, *83*, 2210–2213.

27 Hu, H., et al., A map of WW domain family interactions. *Proteomics* **2004**, *4*, 643–655.

28 Esposito, F., Ammendola, R., Duilio, A., Costanzo, F., Giordano, M., Zambrano, N., D'Agostino, P., Russo, T., Cimino, F., Isolation of cDNA fragments hybridizing to rat brain-specific mRNA's. *Dev. Neurosci.* **1990**, *12*, 373–381.

29 Sudol, M., Sliwa, K., Russo, T., Functions of WW domain in nucleus. *FEBS Lett.* **2001**, *490*, 190–195.

30 Russo, T., Faraonio, R., Minopoli, G., De Candia, P., De Renzis, S., Zambrano, N., FE65 and protein network centered around the cytosolic domain of the Alzheimer's beta-amyloid precursor protein. *FEBS Lett.* **1998**, *434*, 1–7.

31 Cao, X., Sudhof, T. C., A transcriptionally active complex of APP with Fe65 and histone acetyltransferase Tip60. *Science* **2001**, *293*, 115–120.

32 O'Brien, K. F., Kunkel, L. M., Dystrophin and muscular dystrophy, past, present and future. *Mol. Genet. Metab.* **2001**, *74*, 635–644.

33 Spence, H. J., Chen, Y. J., Winder, S. J., Muscular dystrophies, the cytoskeleton and cell adhesion. *BioEssays* **2002**, *24*, 542–552.

34 Einbond, A., Sudol, M., Towards prediction of cognate complexes between the WW domain and proline-rich ligands. *FEBS Lett.* **1996**, *84*, 1–8.

35 Jung, D., Yang, B., Meyer, J., Chamberlain, J. S., Campbell, K. P., Identification and characterization of the dystrophin anchoring site on beta-dystroglycan. *J. Biol. Chem.* **1995**, *270*, 27305–27310.
36 Rentschler, S., Linn, H., Deininger, K., Bedford, M. T., Espanel, X., Sudol, M., The WW domain of dystrophin requires EF-hands region to interact with beta-dystroglycan. *Biol. Chem.* **1999**, *380*, 431–442.
37 Chung, W., Campanelli, J. T., WW and EF hand domains of dystrophin-family proteins mediate dystroglycan binding. *Mol. Cell. Bio. Res. Comm.* **1999**, *2*, 162–171.
38 Ni, C. Y., Murphy, M. P., Golde, T. E., Carpenter, G., Gamma-secretase cleavage and nuclear localization of ErbB-4 receptor tyrosine kinase. *Science* **2001**, *294*, 2179–2181.
39 Ni, C. Y., Yuan, H., Carpenter, G., Role of ErbB-4 carboxyl-terminus in gamma secretase cleavage. *J. Biol. Chem.* **2003**, *278*, 4561–4565.
40 Linn, H., Ermekova, K., Rentschler, S., Sparks, A., Kay, B., Sudol, M., Using molecular repertoires to identify high-affinity peptide ligands of the WW domain of human and mouse YAP. *Biol. Chem.* **1997**, *378*, 531–537.
41 Komuro, A., Nagai, M. Navin, N., Sudol, M., WW domain-containing protein YAP associates with ErbB-4 and acts as a co-transcriptional activator for the carboxyl-terminal fragment of ErbB-4 that translocates to the nucleus. *J. Biol. Chem.* **2003**, *278*, 33334–33341.
42 Yagi, R., Chen, L. F., Shigesada, K., Murakami, Y., Ito, Y., A WW domain-containing yes-associated protein (YAP) is a novel transcriptional co-activator. *EMBO J.* **1999**, *18*, 2551–2562.
43 Carpenter, G., ErbB-4: mechanism of action and biology. *Exp. Cell. Res.* **2003**, *284*, 66–77.
44 Xu, L. L., Shanmugam, N., Segawa, T., Sesterhenn, I. A., McLeod, D. G., Moul, J. W., Srivastava, S., A novel androgen-regulated gene, *PMEPA1*, located on chromosome 20q13 exhibits high level expression in prostate. *Genomics* **2000**, *66*, 257–263.
45 Rae, F. K., Hooper, J. D., Nicol, D. L., Clements, J. A., Characterization of a novel gene, *STAG1/PMEPA1*, up-regulated in renal cell carcinoma and other solid tumors. *Mol. Carcinogenesis* **2001**, *32*, 44–53.
46 Zarrinpar, A., Park, S.-H., Lim, W., Optimization of specificity in cellular protein interaction network by negative selection. *Nature* **2003**, *426*, 676–680.
47 Bedford, M. T., Chan, D. C., Leder, P., FBP WW domains and the Abl SH3 domain bind to a specific class of proline-rich ligands. *EMBO J.* **1997**, *16*, 2376–2383.
48 Espanel, X., Sudol, M., A single point mutation in a group I WW domain shifts its specificity to that of group II WW domains. *J. Biol. Chem.* **1999**, *274*, 17284–17289.
49 Toepert, F., Knaute, T., Guffler, S., Pires, J. R., Matzdorf, T., Oschkinat, H., Schneider-Mergener, J., Combining SPOT synthesis and native peptide ligation to create large arrays of WW protein domains. *Angew. Chem., Int. Ed.* **2003**, *42*, 1136–1140.
50 Kasanov, J., Pirozzi, G., Uveges, A. J., Kay, B. K., Characterizing Class I WW domains defines key specificity determinants and generates mutant domains with novel specificities. *Chem. Biol.* **2001**, *8*, 231–241.
51 Patnaik, A., Wills, J. W., In vivo interference of Rous sarcoma virus budding by cis expression of a WW domain. *J. Virol.* **2002**, *76*, 2789–2795.
52 Sudol, M., From Src Homology modules to other signaling domains: proposal of the 'protein recognition code'. *Oncogene* **1998**, *17*, 1469–1474.
53 Tong, A. H. Y., et al., A combined experimental and computational strategy to define protein interaction networks for peptide recognition. *Science* **2002**, *295*, 321–324.
54 Pannini, S., Dente, L., Cesareni, G., In vitro evolution of recognition specificity mediated by SH3 domains reveals target recognition rules. *J. Biol. Chem.* **2002**, *277*, 21666–21674.
55 Das, S., Smith, T. F., Identifying nature's protein Lego set. *Adv. Protein Chem.* **2000**, *54*, 159–183.

Web Resources Relating to the WW Domain

WW domain page: http://www.bork.embl-heidelberg.de/Modules/ww_refs.html – this site was created in 1994 and is regularly updated by the Sudol and Bork laboratories.

SMART database: http://smart.embl-heidelberg.de/

Pfam database: http://www.sanger.ac.uk/Software/Pfam/

4
EVH1/WH1 Domains

Linda J. Ball, Urs Wiedemann, Jürgen Zimmermann, and Thomas Jarchau

4.1
Introduction

Drosophila enabled (Ena)/vasodilator-stimulated phosphoprotein (VASP) homology 1 (EVH1) domains, sometimes referred to as Wiskott–Aldrich syndrome protein homology 1 (WH1) domains, are a family of small (~115 residues; ~13 kDa), noncatalytic, protein–protein interaction modules essential for linking their host proteins to various signaling pathways. Here, we take a close look at the sequence signatures and structural features that define EVH1 domains and distinguish them from related domains. Using information derived from multiple sequence alignments and phylogenetic trees, together with the available biological data and high-resolution structures of several representative domains and their complexes, we now suggest a classification scheme for EVH1 domains. Mechanisms of peptide–ligand recognition, modulation of binding specificity, and grouping of the domains in light of their evolutionary history are also discussed.

To date, there are four main families of proteins that contain EVH1 domains: the actin regulatory Ena/VASP proteins, the synaptic scaffolding (Homer/Vesl) proteins, the Wiscott–Aldrich syndrome protein (WASP), the closely related N-WASP (also involved in modulating the actin cytoskeleton), and the Sprouty-related EVH1 domain-containing (Spred) proteins which regulate the Ras–MAP kinase pathway. The Ran binding domains (RanBDs) of the Ran binding proteins and nucleoporins possess a striking similarity to EVH1 domains in both sequence and structure. A high degree of structural similarity to the pleckstrin homology (PH) domains and phosphotyrosine binding (PTB) domains, found in a widespread number of eukaryotic signaling proteins, is also observed, despite a sequence identity of less than 10%. This has led to suggestions that EVH1 domains may share a common ancestry with PH and PTB domains and may comprise one of six subfamilies belonging to the PH domain superfamily [1].

Modular Protein Domains. Edited by G. Cesareni, M. Gimona, M. Sudol, M. Yaffe

ISBN: 3-527-30813-X

4.2
Occurrence and Distribution of EVH1 Domains

Extensive searches of the SMART [2, 3], Pfam [4], SwissProt/SPTrembl [5], GenPept, and PIR databases [6] revealed a total of 519 EVH1 domain and RanBD sequences. After purging, 206 nonredundant sequences remained, 120 of which could be assigned to protein families. These were aligned based on the known 3D structures of representative members from each family, using the algorithm DALI [7]. Figure 4.1 shows the alignment of a selection of these sequences, which were chosen to represent each of the main protein families and species in which EVH1 domains are found. A more thorough breakdown of the species distribution of these proteins is provided in Table 4.1. The sequences were then used to calculate a phylogenetic tree (Figure 4.2) which shows the relationships between EVH1 domains from the different host proteins [8–10]. To avoid crowding, only a reduced selection of 45 domains, representing all branches of the tree, is shown in the figure.

Figure 4.1 Sequence alignment of EVH1 domains and RanBD domains. Sequences were obtained from an extensive search of the SMART database [114] using EVH1 (called WH1 in SMART) and RanBD sequence sets, and a search of the SwissProt/SPTrembl, GenPept, and PIR databases using a hidden Markov model (HMM) [115], based on a sequence profile containing well known representatives of these domains. Eight sequences were added manually, either because they were not readily available from the databases or were not found by the three HMMs used. The VASP_DD protein sequence, for example, is not yet available in the databases and was obtained from the original publication [15] and cross-checked against the *Dictyostelium discoideum* genome sequence. The resulting set of 519 domain sequences was edited manually to remove redundant sequences, yielding 206 nonredundant domain sequences, of which 120 were well annotated. From this set, representative sequences were selected for the alignment shown here, based on pairwise structural alignments using DALI [7] with manual refinement. Residues are colored as follows: red: acidic; blue: basic; green: hydrophobic; yellow: aromatic; purple: polar residues; dark green: Gly; and light brown: Pro. Sequences are grouped according to protein family, and secondary structure elements were obtained from the available structures using DSSP [116]. Asterisks denote sequences of domains shown in Figures 4.5 and 4.6. Sequence numbers and assignments of secondary structural elements above the alignment are according to the EVH1 domain of Mena (SP: Q03173, PDB: 1evh). The secondary structure assignment below the alignment represents the consensus conservation observed across the EVH1, RanBD, PTB, and PH families. The characteristic aromatic triad (Y16, W23, and F77; Mena numbering) of the Ena/VASP-family EVH1 domains are highlighted by boxes for comparison of the different groups. Numbers on the right represent the region of the sequence displayed. SwissProt accession codes are given in the figure.
Abbreviations of the organism names are appended to each protein name as follows: AG = *Anopheles gambiae*, AT = *Arabidopsis thaliana*, BD = *Babesia divergens*, BR = *Brachydanio rerio*, BB = *Babesia rodhaini*, BT = *Bos taurus*, CA = *Candida albicans*, CE = *Caenorhabditis elegans*, CF = *Canis familiaris*, DD = *Dictyostelium discoideum*, DM = *Drosophila melanogaster*, EN = *Encephalitozoon cuniculi*, EG = *Eremothecium gossypii*, FR = *Fugu rubripes*, GG = *Gallus gallus*, GL = *Giardia lamblia*, HM = *Hirudo medicinalis*, HS = *Homo sapiens*, LE = *Lycopersicon esculentum*, LM = *Leishmania major*, MM = *Mus musculus*, NC = *Neurospora crassa*, OA = *Ovis aries*, PF = *Plasmodium falciparum*, PY = *Patinopecten yessoensis*, RN = *Rattus norvegicus*, SC = *Saccharomyces cerevisiae*, SP = *Schizosaccharomyces pombe*, SS = *Sus scrofa domestica*, XL = *Xenopus laevis*.

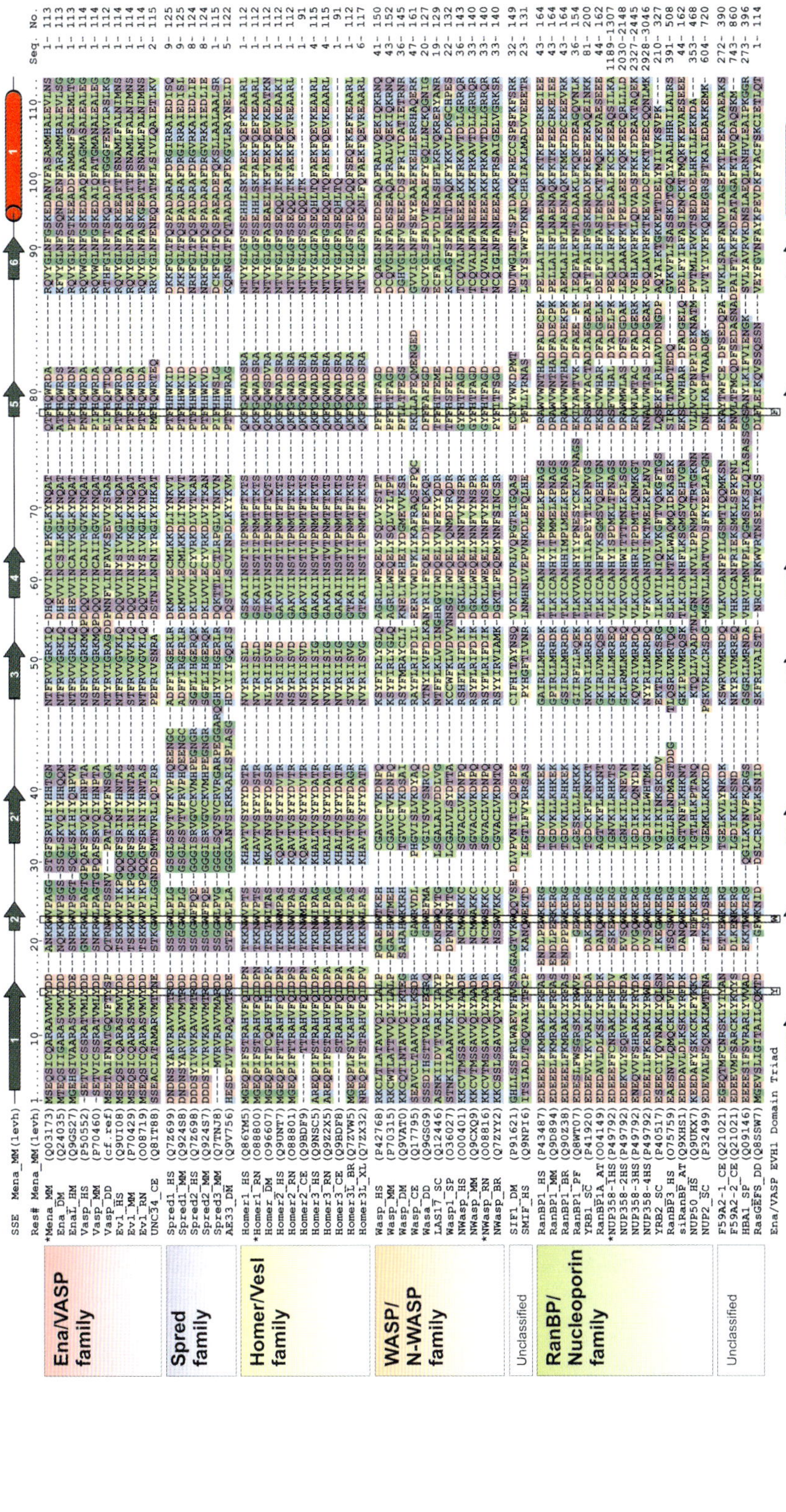

Figure 4.1 (Legend see p. 74)

Table 4.1 Number and origin of EVH1 and RanBD domains. The organism name abbreviations are according to Figure 4.1.

Protein family	*Number of domains*	*Organisms in which domain has been found to date*
Ena/VASP	19	HS MM RN CF GG DM CE HM DD
Spred	7	HS MM DM
Homer	20	HS MM RN OA BR XL DM CE
WASP/N-WASP	16	HS MM RN BT BR DM CE DD SC SP EG
RanBP/Nucleoporin	58	HS MM BT SS BR XL DM EN LM PF SC SP CA AT LE BD
Unclassified	86	HS MM RN BR FR XL AG DM CE PY GL PF DD NC SP AT BB
Total	206	

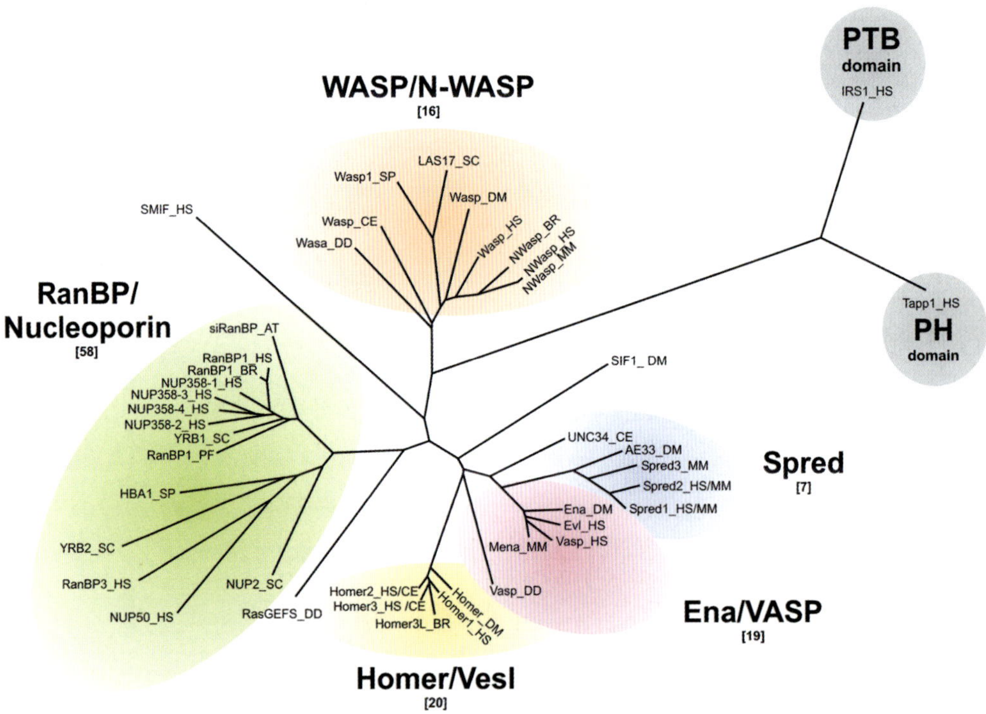

Figure 4.2 Phylogentic tree of EVH1, RanBD, PH, and PTB domain sequences. The phylogenetic tree is based on the alignment shown in Figure 4.1 with known loop regions removed. The number of domain sequences known to date are given in brackets below each group. Representative sequences of the PH and PTB domains have been incorporated into the alignment based on a structural alignment using DALI [7]. The sequence relatedness was calculated with PROTDIST using the Jones–Taylor–Thornton mutation model. This was then used by FITCH to calculate an unrooted tree in which branch lengths represent sequence similarity. All programs were part of the Phylip package version 3.6.a2 [10]. The tree was created using the TreeView application [8].

The classification of the EVH1 domains of these proteins alone gives a biologically meaningful grouping of protein host families in terms of molecular function. The total number of sequences occurring in each major branch is given in brackets. Such a representation helps us to immediately see the groupings obtained based on sequence and structural alignments. Each of these different groups of host proteins is discussed below.

4.2.1 Proteins Containing EVH1 Domains

The EVH1 domains were first identified in the Ena/VASP protein family [11, 12], which includes *Drosophila* Ena, its mammalian orthologs (VASP, Mena, and EVL (Ena/VASP-like protein)), the *Caenorhabditis elegans* ortholog Unc-34, and the *Dictyostelium* DdVASP, among others (for reviews see [13–15]). The members of this family localize to highly dynamic areas of actin reorganization, such as the leading edge of lamellipodia, the tips of filopodia, adherens-type cell–matrix and cell–cell junctions, and other dynamic membrane regions (for reviews see [16, 17]), where they are involved in regulating the rate of actin polymerization. They are essential players in a wide range of physiological and developmental processes, including axon guidance [18–21], platelet aggregation [22, 23], T-cell activation, phagocytosis, and migration of neutrophils, fibroblasts, and neurons [24–29]. The Ena/VASP proteins are also engaged in actin-based intracellular motility and cell-to-cell spreading of the pathogenic bacterium *Listeria monocytogenes* [30]. The proteins show both positive and negative regulatory activities toward actin-based processes [13, 16].

The Homer/Vesl proteins are a family of synaptic scaffolding proteins that are constitutively expressed in brain and enriched at excitatory synapses [31, 32]. They function as molecular adaptors, binding to and targeting neurotransmitter receptors and other scaffolding proteins to the post-synaptic density (PSD), a specialized protein complex at the synaptic junction [33, 34]. Homer/Vesl proteins have generated interest as potential mediators of synaptic plasticity having important implications for learning and memory formation [35]. Their N-terminal EVH1 domains are responsible for interactions with various receptors in the signaling pathways of these proteins.

Like Ena/VASP, the WASP/N-WASP proteins are involved in spatial and temporal regulation of the actin cytoskeleton. These proteins were originally discovered in various cells of the hematopoietic system, where they play roles in promoting actin polymerization in response to upstream intracellular signals from the Cdc42 and phosphatidylinositol 4,5-bisphosphate (PIP_2) signaling pathways [36–39]. Missense mutants in their N-terminal EVH1 domains (historically called WH1 domains) result in the X-linked recessive disorder Wiscott–Aldrich syndrome (WAS), characterized by immunodeficiency, eczema, and thrombocytopenia [36, 40–44]. These symptoms are consistent with cytoskeletal defects in hematopoetic cells and with roles for WASP in multiple actin-based motility processes [36]. A knockout mutation in the more ubiquitously expressed gene for N-WASP was lethal to developing embryos and resulted in defects in many actin-based processes. WASP/N-WASP proteins are also recruited by intracellular pathogens, such as the *Shigella* bacterium and the vaccinia virus, to support their own actin-based motility [45, 46]. The interaction occurs via the EVH1 domain as previously observed in the Ena/VASP proteins.

The Spred family is the most recently discovered family of proteins found to contain EVH1 domains. So far, three paralogs of the Spred protein have been identified in mouse and humans, with Spred2 being the most ubiquitously expressed isoform [47]. Spred proteins suppress the signaling pathway of Ras-, Raf-, and mitogen-activated protein (MAP) kinase, which regulates differentiation in neuronal cells and myocytes [48]. Mechanistically, Spred inhibits early steps in the activation of MAP kinases, resulting in down-regulation of the Ras–MAP kinase signaling pathway [48]. The *Drosophila* ortholog of Spred is the AE33 protein, which is believed to be involved in regulating photoreceptor development [49].

The closest known relatives of the EVH1 domains are the Ran-binding domains (RanBDs) found in the Ran-binding protein/nucleoporin family of nuclear transport proteins. These proteins are essential for the trafficking of cytoplasmic proteins through the nuclear pore complex (NPC), the directionality of transport being tightly controlled by their interaction with the small GTP-binding protein Ran [50–52]. The RanBDs form a large family having a very high sequence homology to EVH1 domains [53–55] (Figure 4.1). The structurally related phosphoinositide-binding PH domains and the PTB domains, which bind primarily to peptides containing phosphorylated tyrosines, are not included in Figure 4.1, because of their low sequence identity to the EVH1 and RanBD families.

Figure 4.3 Domain organization in proteins containing EVH1 domains, RanBDs, PTB, and PH domains. Domain abbreviations: EVH1 = Ena/VASP-homology 1; EVH2 = Ena/VASP-homology 2; GBD = GTP-binding domain; VPH = verprolin homology domain; CA = cofilin homology and acidic domain; KBD = kinase binding domain; SPR = sprouty-related domain; RanBD = ran binding domain; ZnF-RBZ = zinc finger domains; PPIase = proline isomerase; TPR (Pfam) = tetratrico-peptide repeat; GRIP (Pfam) = golgin-97, RanBP2alpha, Imh1p, and P230/golgin-245; PTB = phosphotyrosine binding; PDZ = postsynaptic density/disc-large/ZO-1; SH2 = Src homology 2; SH3 = Src homology 3; PH = pleckstrin homology; RhoGAP = GTPase-activator protein for Rho-like GTPases; DYNc = dynamin GTPase c; GED = dynamin GTPase effector domain. SwissProt accession codes for the specific examples shown in the figure are: VASP/Evl = P50552 (human); Ena/Mena = Q03173 (mouse); WASP = P42768 (human); N-WASP = O00401 (human); Homer1 = Q9Z216 (mouse); Homer2 = Q9UNT7 (human); Spred = Q7Z698 (human); nucleoporin 50 = O08587 (rat); RanBP2/Nup358 like = P49782 (human); RanBP2L/BS63 like = Q99666 (human); amyloid β binding protein = P98084 (mouse); tensin = Q9UPS7 (human); Fe65 = O00213 (human); Rho GTPase-activating protein = Q7ZWQ2 (*Xenopus*); syntrophin = Q8BNW6 (mouse); dynamin-1 = Q05193 (human).

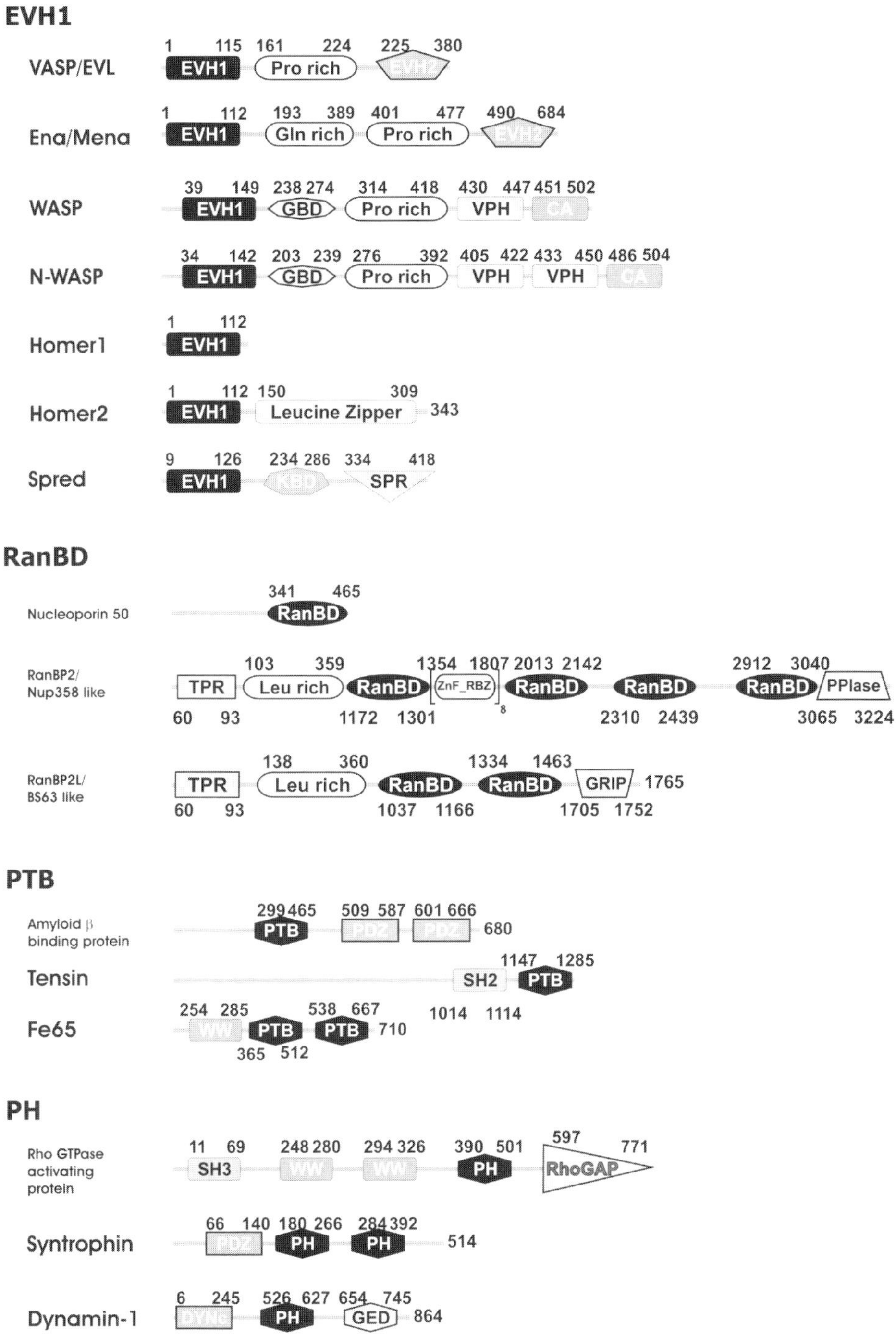

Figure 4.3 (Legend see p. 78)

4.2.2
Modular Architecture of EVH1-containing Proteins: Domain Location, Domain Combinations, and Copy Number

Proteins containing EVH1, RanBD, PTB, and PH domains generally possess highly modular multidomain structures. Different domains within the same parent protein can therefore bind simultaneously to distinct interaction partners or substrates in response to specific signals. The combination of domain types that occur together within a host protein and the overall domain architecture are crucial factors in determining function. Figure 4.3 summarizes the domain structures for representative examples of the EVH1 domain-, RanBD-, PTB domain-, and PH domain-containing proteins.

The Ena/VASP proteins all share an overall tripartite domain structure. The N-terminal EVH1 domain binds FPPPP-containing motifs exposed on the surfaces of receptors or focal adhesion proteins such as zyxin [56], vinculin [57], and the *Listeria* surface protein ActA [30]. A central, low complexity, proline-rich region contains profilin and both SH3 and WW domain binding sites [58]. In the larger *Drosophila* Ena and mouse Mena sequences, an additional Gln-rich region precedes this (Figure 4.3). All share a conserved C-terminal EVH2 domain required for tetramerization and for F-actin and G-actin binding [59, 60]. The C-terminal region is responsible for making direct contacts with the actin cytoskeleton.

The domain architecture of the Homer/Vesl family is more variable. Homer1 contains only an N-terminal EVH1 domain followed by a low complexity region. In contrast, its close relative Homer 2 consists of an N-terminal EVH1 domain, a low complexity linker region, and a leucine zipper motif at the C terminus responsible for clustering [33]. The EVH1 domain interacts with neurotransmitter receptors, binding selectively to PPxxF-containing motifs found in the C termini of group I metabotropic glutamate receptors (mGluRs), inositol-1,4,5-trisphosphate receptors (IP3Rs), ryanodine receptors (RyRs), and Shank family proteins [31, 61, 62].

The WASP/N-WASP family contain a more complex domain structure, comprising an N-terminal EVH1 domain, a short basic motif, a GTPase binding domain (GBD), a proline-rich region, and a C-terminal region containing either a verprolin-homology (VPH) and cofilin-acidic (CA) domain (WASP) or two tandem VPH domains followed by a CA domain (N-WASP). The latter region is often called, as a whole, the verprolin-cofilin-acidic (VCA) domain (Figure 4.3). The WASP EVH1 domain targets a 25-residue proline-rich peptide in the WASP interacting protein (WIP), which binds by wrapping itself around the entire WASP EVH1 domain [36]. This is in contrast to the interactions of Ena/VASP and Homer/Vesl EVH1 domains, which bind to much shorter (6–12 amino acids) target peptides. The C-terminal verprolin-cofilin-acidic (VCA) region interacts directly with and activates the actin-related protein (Arp)2/3 actin-nucleating complex to promote actin polymerization [63, 64]. The basic motif, GBD, and the proline-rich region have been shown to be involved in autoinhibitory interactions that block VCA activity. These interactions are relieved by binding of the upstream activators, phosphatidylinositol 4,5-bisphosphate (PIP_2), the GTPase Cdc42, or SH3 domains, respectively [36].

The Spred proteins possess a domain structure comprising an N-terminal EVH1 domain, a central c-Kit binding domain (KBD), and a C-terminal cysteine-rich Sprouty-related (SPR) domain [48] (Figure 4.3). To date, no binding partner has been identified for the EVH1 domain, although the fairly limited tissue distribution of some Spred isoforms (expressed in glandular epithelium, intestinal lymphoid tissues, tonsils, and, intriguingly, the invasive trophoblast of the developing embryo) suggests that their binding partners may be significantly different from those of the other EVH1 domains. Differences in sequence at specific positions known to be important for ligand recognition in the Ena/VASP and Homer/Vesl EVH1 domains also suggest that the Spred EVH1 domain may reveal a novel binding mode. The C-terminal SPR domain is essential for localization of Spred to the plasma membrane [48]. Both the EVH1 and SPR domains are required for suppression of neuronal cell differentiation [48].

In all the protein families discussed above, EVH1 domains occur in single copies located exclusively at the N termini of their host proteins. Only one hypothetical protein from *C. elegans* (YKC2; SwissProt: p41993) was found that contains two putative EVH1 domains, separated by about 450 amino acids. This is in contrast to the RanBDs, PTB domains, and PH domains, which show a much broader range of location and occurrence [2, 65]. Many of the proteins containing RanBD, PTB, and PH domains are much larger than the EVH1-containing proteins and possess correspondingly more varied domain architectures. Selected examples are shown in Figure 4.3. The smallest member of the RanBP/nucleoporin proteins, nucleoporin 50, contains just one RanBD located at its C terminus. However, very little is understood about the structure and nature of the N terminus of these proteins. The RanBP2 proteins are up to an order of magnitude larger than the EVH1-containing proteins, with the largest composed of more than 3000 residues. The RanBDs in these large proteins are found at a variety of positions in the sequence and often occur several times within the same protein, separated by stretches of 150–450 amino acids.

In light of the variable location and copy numbers of other domains, the conserved N-terminal location and singular occurrence of EVH1 domains raises interesting questions as to why this is so and whether it may be one of the defining features of EVH1 domains in general. One explanation may lie in the folding pathways of the EVH1 domain-containing proteins. Protein synthesis occurs from the N to the C terminus, and folding in eukaryotes is believed to occur mainly in a cotranslational manner. It is therefore favorable for autonomously folding, highly stable domains to be synthesized and leave the ribosome ahead of the low-complexity regions and oligomerization regions, which are also present in the EVH1 domain-containing proteins. Such regions are largely unstructured and, in the absence of their binding partners, are more prone to proteolysis and nonspecific aggregation, once released into the cytoplasm. In contrast, the domains that coexist in the host proteins of RanBDs, PTB, and PH domains are generally highly stable modules capable of independent folding.

EVHI domains bind specifically to proline-rich motifs (PRMs) in peptides exposed on the surfaces of their binding partners.

The N-terminal location of EVH1 domains is likely to facilitate their access to PRMs on bulky target molecules and may contribute to a segmental polarity of the EVH1 host protein that separates the adaptor module (EVH1) from its various effector domains. Certainly, given this strong preference for an N-terminal location, it is clear that EVH1 domains can occur only once in any given protein. The use of domain repeats, either in a cis configuration on the same polypeptide chain or in a trans configuration on identical chains in a quarternary structure, can achieve synergistic functions for the host protein, such as allowing the clustering of multiple binding partners or increasing the binding affinity for a single ligand. Since the Ena/VASP proteins tetramerize via their C-terminal EVH2 domains [59, 60], this brings together four EVH1 domains in each tetramer, thereby providing a mechanism for clustering of Ena/VASP proteins. Often, the target PRMs of EVH1 domains occur in close tandem repeats in the sequence of the binding partner. This could provide an additional mechanism for increasing binding affinity. Similarly, the Homer2 protein contains a leucine zipper motif at its C terminus, which is also involved in clustering [32, 66]. The EVH1 domains of each polypeptide chain in the oligomerized protein must therefore be located far away from the C-terminal oligomerization regions in the final structure, to allow unhindered access to their target peptides. This is clearly easier to achieve when these domains are well separated in the sequence.

In summary, a conserved, unique N-terminal location appears to be a characteristic feature of EVH1 domains, setting them clearly apart from structurally related domains such as RanBDs, PTB, and PH domains. The EVH1 domains may confer a segmental polarity to their hosts that is required for functional or biogenetic reasons, resulting in the topological separation of this exposed terminal adaptor domain from the different types of genetically fused effector domains with which they coexist.

4.2.3
Classification of EVH1 Domains

To date, there is considerable discrepancy in the nomenclature and classification of EVH1 domains in the various databases. The name 'EVH1 domain' is frequently used interchangeably with the name 'WH1 domain', although the EVH1 nomenclature is generally the most widely used in the fields of biochemistry and molecular cell biology, owing to the early functional connotations resulting from the identification of the first ligand [67]. Further confusion arises when different databases (for example, SMART, Pfam, InterPro [68]) refer to only one of these different names and classify them in conflicting schemes. Here, we refer to EVH1/WH1 domains as EVH1 domains and classify them based on their sequence conservation, domain co-occurrence, structural similarities, and ligand binding preferences, where this information is available.

From the sequence alignment and phylogenetic analysis (Figures 4.1 and 4.2), it is clear that the EVH1 domains cluster into four main groups, which we have named after their primary host proteins: the Ena/VASP class, the Homer/Vesl class, the WASP/N-WASP class, and the Spred class. The Ena/VASP and Homer/Vesl classes have already been referred to as Class 1 and Class 2 EVH1 domains, based on their selectivities for peptides containing either FPPPP or PPxxF motifs, respectively [69]. We suggest keeping this nomenclature and adding the more distantly related WASP/N-WASP EVH1 domains (Figure 4.2), which bind a LPPPEPY-containing peptide in WIP, as Class 3, and the more recently described family of Spred EVH1 domains, which contain distinct putative ligand binding residues but for which no binding partner is yet known, as Class 4. Several EVH1 domains fall between these main classes, such as those from the *Drosophila* Still-life type 1 protein [70], the *C. elegans* hypothetical YKC2 protein, and the *Dictyostelium* RasGEFS. These could either provide putative evolutionary links between the different classes already known or be singular representatives of new, so-far-undiscovered families. Further detailed information on their structures, functions, and ligand-binding characteristics is needed before solid conclusions can be drawn about these proteins.

The RanBDs cluster to form a large class of their own, which includes domains from the Ran-binding proteins and nucleoporins, as well as the HBA1 protein from the yeast *Schizosaccharomyces pombe*. Both the alignment in Figure 4.1 and the tree in Figure 4.2 show clearly that the RanBDs are most closely related to the EVH1 domains of the Homer/Vesl family, as well as to the RasGEFS and *Drosophila* Still-life type 1 protein. However, despite this relationship, it would be inaccurate to include RanBDs as a subclass of EVH1 domains based on the criteria used here. It is possible that the RanBDs and EVH1 domains may instead stem from a common ancestor, which at some distant time may have descended from the same origin as the other PH and PH-like domains.

The specific binding of the Homer EVH1 domain to C-terminal sequences has previously led to the suggestion that this EVH1 domain may be a divergent PDZ domain [31, 66]. Inspection of Homer EVH1 and PDZ domain sequences, however, reveals almost no similarity apart from a common four-amino-acid 'typical' motif constituted by turn-promoting residues, an invariant Phe, and a hydrophobic residue (GLGF in Homer). This motif is involved in C-terminal COOH recognition in PDZ domains [71, 72]. However, the different locations of this motif in the PDZ and Homer EVH1 structures rule out any functional relationship and make it unlikely that there is a meaningful evolutionary link between the two types of domain [53].

4.3 Structures of EVH1 Domains and Their Complexes

4.3.1 High-resolution Structures of EVH1 and Related Domains

High resolution 3D structures of seven EVH1 domains representing three of the four classes of EVH1 domain, from the Ena/VASP, WASP/N-WASP, and Homer/Vesl families, have now been solved, several in complex with their target proline-

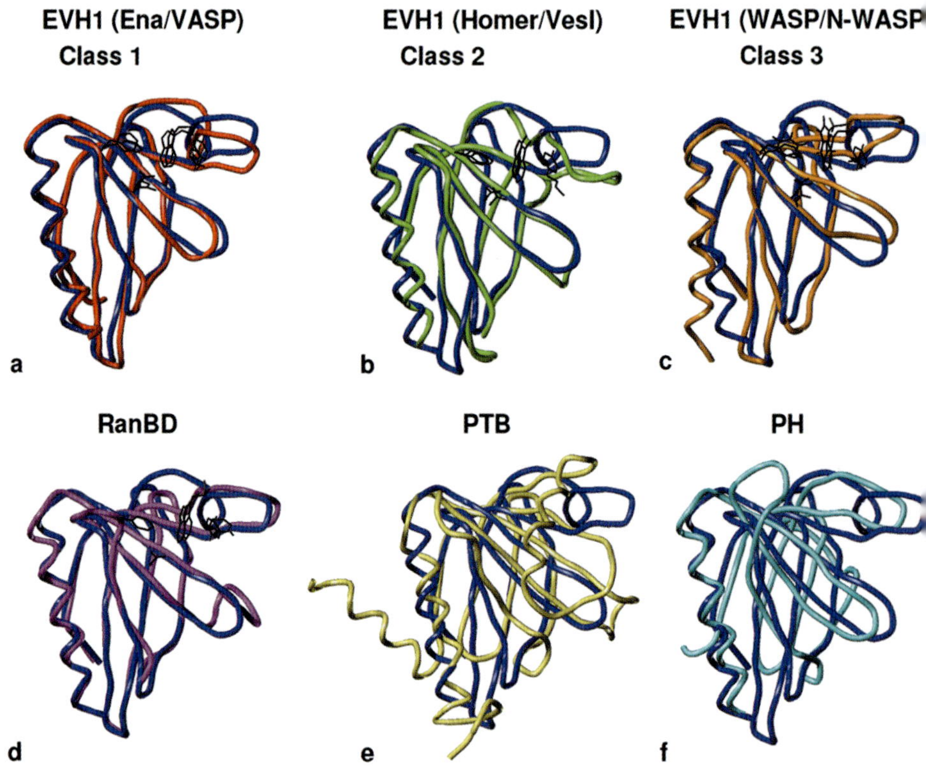

Figure 4.4 Comparison of 3D structures. Superpositions of the backbone (N, Cα, and C′) atoms of the EVH1 domain of Mena with representative members of each of the related groups shown in the phylogenetic tree. The Mena EVH1 domain (blue) is overlaid with (a) the Class 1 EVH1 domain from VASP (red); (b) the Class 2 Homer EVH1 domain (green); (c) the Class 3 N-WASP EVH1 domain (orange); (d) the first RanBD (RanBD1) of Nup358 (magenta); (e) the PTB domain of Numb (yellow); and (f) the PH domain of DAPP1/PHISH (cyan). Structures were aligned by using DALI [7]. The locations of the exposed aromatic clusters are shown in black. PDB (and SwissProt) accession codes are Mena. EVH1 = 1evh (Q03173); VASP EVH1 = 1egx (P50552); Homer EVH1 = 1ddv (O88800); N-WASP EVH1 = 1mke (O08816); RanBD = 1rrp (P49792); PTB = 2nmb (P16554); PH = 1fao (Q9UHF2).

rich ligand [33, 36, 73–77]. Structures of the RanBDs from both the large RanBP2 and the smaller, nucleoporin-like RanBP1 proteins are also known: the former in complex with Ran-GTP [54] and the latter in a ternary complex with Ran-GTP bound to RanGAP [78]. The structures of 5 PTB and 16 PH domains from a wide range of different host proteins are also available (for reviews see [79, 80] and Chapters 6 and 17).

Superposition of the Cα traces of representatives from each of the three classes of EVH1 domain for which structures are now known and the related RanBDs, PTB domains, and PH domains with the Class 1 EVH1 domain from murine Ena (Mena) are shown in Figure 4.4. The overall folds are essentially the same, forming a compact, parallel β sandwich capped along one side by a long α helix. Mutagenesis studies have highlighted several specific core residues as being important for stabilization of the EVH1 fold [81]. The main differences between the different families occur, not surprisingly, in the loop regions, where sequence variability is also highest (Figure 4.1). The structures of the different classes of EVH1 domains show very little difference (mean backbone rmsd values of VASP, Homer, and WASP EVH1 domains relative to the EVH1 domain of Mena are 1.39, 2.82, and 2.0 Å, respectively). The similarity to the RanBD fold is quite remarkable (rmsd 1.6 Å). Despite low sequence identity (~10%), the PTB and PH domains show high structural homology to both EVH1 domains and RanBDs. The agreement is closest in the most highly structured regions (average rmsd values relative to Mena EVH1 are 2.7 and 2.4 Å, respectively) but much less in the loop regions, where long insertions and deletions occur. Additional elements of secondary structure are also present in several members of these more distantly related families (for example, the PTB domain of the human SHC protein shown in the figure).

A distinguishing feature of EVH1 domains is the highly conserved triad of surface-exposed aromatic sidechains, Y16, W23, and F77 (outlined by boxes in the alignments in Figure 4.1; numbering relative to Mena). These come together in the 3D structure to form an aromatic cluster (shown in black in Figure 4.4), which provides a hydrophobic docking site for the proline-rich peptide ligands targeted by EVH1 domains. From the degree of conservation alone, it is clear that W23 is a highly important residue. It is completely conserved in all families of EVH1 domains and in almost all of the RanBDs. From inspection of the 3D structures, one can see that a large area of this sidechain makes important hydrophobic contacts with the core, thus involving it in fold stabilization. The mutation W23L in human VASP results in an insoluble, aggregated protein [75], in agreement with similar inactivating mutations analyzed in a yeast two-hybrid system [82]. Nevertheless, the indole proton of W23 (Hε1) is oriented toward the surface and is available as a H-bond–donating group in ligand interactions. Thus, in EVH1 domains and RanBDs, W23 fulfils a crucial role in domain stabilization while simultaneously exposing one functional group for ligand recognition. The other two residues of the triad are less strictly conserved. Variation in these allow corresponding variations in the geometry and properties of the binding site, enabling the different families of EVH1 domains to bind specifically to distinct consensus ligands.

The conservation in fold and of important functional residues relates the above domains very closely. Furthermore, as this fold has been found only in signaling proteins and only in eukaryotes, this suggests that EVH1, Ran-binding, and PTB domains all comprise subfamilies of the PH superfamily, which may have arisen from a common distant ancestor [65]. Over time, these subfamilies have diverged, but have retained the stable PH fold as a scaffold upon which to build distinct specialized ligand-recognition sites, leading to the rich degree of functional diversity observed today [65].

4.3.2
Structures of EVH1 Complexes and Determinants of Ligand Specificity

The EVH1 domains of Ena/VASP and Homer/Vesl families bind peptides that are 6–12 amino acids long and contain proline-rich motifs (PRMs) 4–6 amino acids long. The binding affinities of the core motifs in isolation are extremely low (K_d values in the millimolar range), but are increased to biologically significant levels by the presence of additional core-flanking epitopes, which make additional contacts with the domain surface. The Ena/VASP EVH1 domains bind specifically to FPPPP motifs found in focal adhesion-associated proteins like zyxin [56], vinculin [57], and the *Listeria* ActA [30], whereas the Homer/Vesl EVH1 domains bind specifically to PPxxF motifs from the group I mGluRs [31], IP3R receptors [61], ryanodine receptors, and Shank proteins [62, 83, 84]. In contrast, the N-WASP EVH1 domain binds a much longer proline-rich peptide, having a minimum length of 25 residues (residues 461–485) and does not bind a 10-residue ligand of the Mena EVH1 domain from ActA, which contains a PRM (DFPPPPT) very similar to that found in the WIP peptide (DLPPPEP) [67]. It should be noted that the peptide–N-WASP complex was expressed from a single construct in which the 25-residue WIP peptide was fused by a 5-amino-acid linker to the N terminus of N-WASP [36]. This clearly favors binding for entropic reasons. It is not known whether binding occurs to an equivalent independent peptide, because isolated N-WASP EVH1 domains were found by these authors to be insoluble.

The structures of representative complexes from each of the three families of EVH1 domains for which structures are now available are shown in Figure 4.5 [36, 73, 76]. No structure is yet available for the Spred EVH1 domains. The overall construction of the ligand-recognition sites is generally similar in all classes of EVH1 domains. The exposed Trp sidechain (W23; Mena numbering) is usually located at the centre of the aromatic triad (Figures 4.4 to 4.6) and is oriented in a plane almost perpendicular to the domain surface. On one or both sides, at approximately 90° to this plane and almost parallel to the domain surface, lie the flat rings of either Tyr or Phe sidechains. This perpendicular arrangement of aromatic rings results in rectangular hydrophobic pockets on each side of the Trp, well suited to the recognition of peptide ligands that adopt structures close to that of the left-handed PPII helix structure, characterized by backbone angles $\phi = -78°$ and $\varphi = +146°$ [85–87]. The FPPPP-containing peptides bound by the Ena/VASP EVH1 domains and the LPPPEP region of the WIP peptide bound by the N-WASP EVH1 domain are good examples of this. The indole proton of the central Trp forms a hydrogen bond to a backbone carbonyl oxygen in the peptide, which anchors the ligand into place. The sidechains of the peptide residues surrounding this carbonyl (usually prolines) then pack closely into the rectangular hydrophobic pockets on either side of the W23 sidechain (Figure 4.6). In the Ena/VASP EVH1 domains, Y16 and F77 (Mena numbering) comprise the Trp-flanking aromatic residues, and the peptide P(2) and P(5) of the Class 1 EVH1-binding motif FPPPP binds to either side of W23 (underlined residues are those whose sidechains make the closest contacts with the domain; yellow in Figure 4.6). The indole proton of

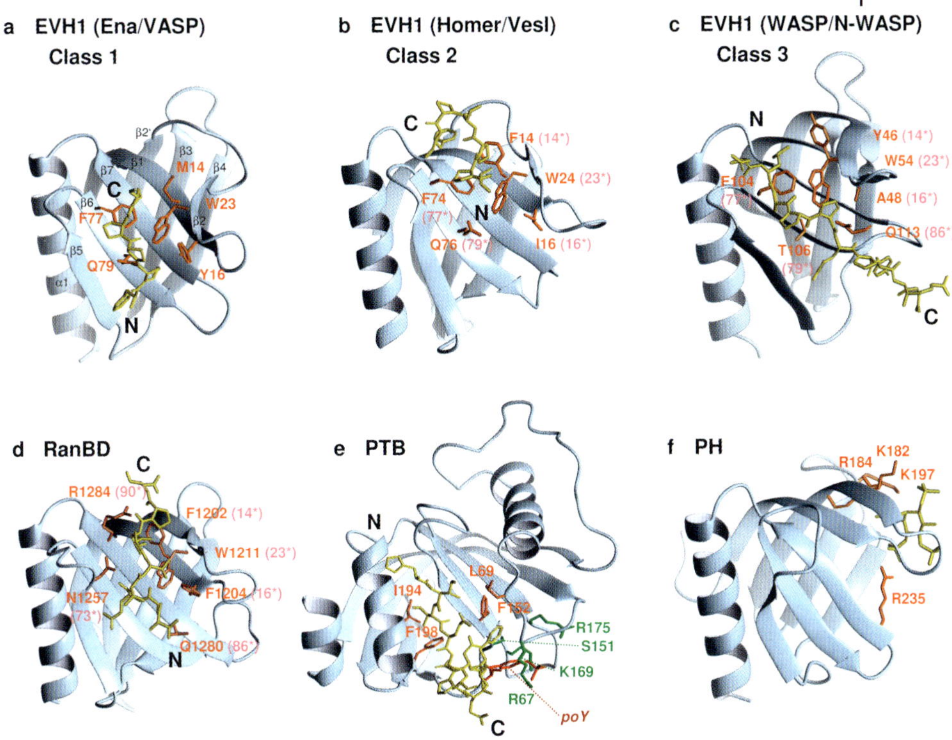

Figure 4.5 Structures of complexes of EVH1 and related domains with their respective ligands. (a) Mena EVH1 domain with FPPPP peptide. (b) Homer EVH1 domain with the TPPSPF peptide. (c) N-WASP EVH1 domain with the minimal 25-residue peptide from WIP fused to its N terminus. Only the N-terminal 11 amino acids of this peptide (DLPPPEPYNQT) that bind in the 'PRM-binding groove' are shown. (d) The first RanBD (RanBD1) from Nup358 with the Ran protein. Only the C-terminal 10 amino acids (EVAQTTALPD) of Ran relevant to this discussion are shown. (e) The PTB domain of the SHC protein with the 12-residue peptide HIIENPQ*po*YFSDA phosphorylated at tyrosine. (f) The PH domain from DAPP1/PHISH with inositol-1,3,4,5-tetrakisphosphate. All classes of EVH1 domains and the RanBD share similar exposed clusters of aromatic sidechains in their peptide binding grooves, as described in the text. Numbers with asterisks refer to the equivalent positions in the sequence of Mena. PDB accession codes are as in Figure 4.4, except for the PTB domain complex (PDB: 1shc; SwissProt: P29353).

W23 makes a hydrogen bond to the carbonyl of P(3). An additional hydrogen bond between the domain's Q79 sidechain and the ligand P(2) supports the interaction. In each class of EVH1 domain, an H-bond–donating residue is almost always located at this position (Figure 4.1). The N-terminal F(1) of this sequence makes a further close hydrophobic contact with the domain, which is important for anchoring the peptide and for determining the orientation of the otherwise highly symmetric ligand (Figure 4.6).

Variations in the geometries of the PRM binding sites give rise to the observed differences in ligand preference between the different domain families. The complex of the Homer EVH1 domain bound to the Class 2 EVH1-binding motif TPPxxF

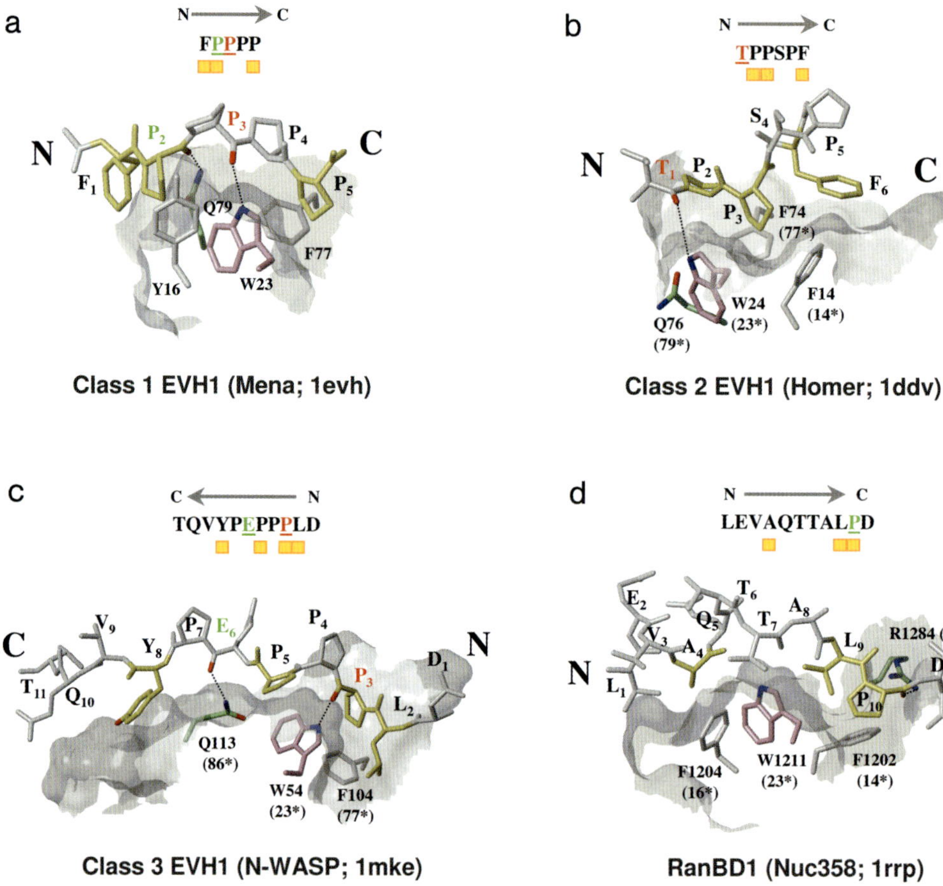

Figure 4.6 Comparison of peptide binding interfaces for the different classes of EVH1 domains (a–c) and RanBDs (d). Hydrogen bonds in the interface are shown by dotted lines where distances are less than 2.6 Å. The conserved W23 (Mena numbering) is colored pink. In all EVH1 domains the conserved W23 forms an important hydrogen bond via its indole proton to a carbonyl oxygen in the peptide backbone. The second conserved H-bond–donating residue Q79 (Mena) is shown in green. The location of this residue is conserved in Ena/VASP and Homer/Vesl EVH1 domains, but differs in the EVH1 domains of N-WASP. (Asterisks indicate numbering when aligned to Mena). Ligand sidechains that make close hydrophobic contacts with the domain are colored yellow. The peptide sequence is shown above each complex. H-bond–acceptor residues are underlined and colored as follows: red = hydrogen bond to Trp; green = hydrogen bond to Gln (EVH1) or Arg (RanBD).

shows a very different binding mode from that of Ena/VASP EVH1 domains bound to their FPPPP motifs (Figure 4.6). In the Homer/Vesl proteins, I16 replaces Y16 (Mena), thereby altering the shape of the binding pocket that accommodates the sidechain of P(2) in the Ena/VASP EVH1 complexes [69]. Simultaneously, the hydrophobic residue M14 of Ena/VASP domain is replaced by the aromatic F14 in Homer/Vesl (Figures 4.1 and 4.5) providing a new binding pocket for the sidechain of P(3) in the Homer peptide. Hence, the loss of an aromatic sidechain at position

16 and the gain of a new aromatic sidechain at position 14 change the location and geometry of the PRM-recognition triad. The two consecutive prolines and terminal Phe of the TPPxxF motif make the closest hydrophobic contacts to the Homer EVH1 domain surface, with the exposed indole proton of Homer W24 forming a hydrogen bond to the carbonyl oxygen of the N-terminal peptide Thr (Figure 4.6).

The complex of the N-WASP EVH1 domain with the N-terminal 11 amino acids of the 25-residue WIP minimal binding sequence $_1$DLPPPEPYNQT$_{11}$ is shown in Figures 4.5 and 4.6. The remainder of the peptide wraps around the back of the EVH1 domain [36] and is omitted from the figures for clarity. The PRM-binding triad of N-WASP is most closely related to that of Homer (also apparent from the sequence alignment in Figure 4.1). As in Homer, the conserved Y16 residue of the Ena/VASP EVH1 domains is lost (replaced by A48; equivalent to I16 in Homer) and thus no longer forms part of the peptide binding site. Instead, residues Y46 (14*), W54 (23*), and F104 (77*) in the N-WASP EVH1 domain are equivalent to F14, W24, and F74 in Homer (numbers with asterisks refer to the equivalent positions in Mena), forming almost identical aromatic clusters in the two proteins. However, the WIP peptide does not bind N-WASP in the same manner as observed for the Homer complex. Instead of making close contacts with Y46 (14*), the WIP peptide contacts a groove comprising W54 (23*), F104 (77*), T106 (79*), and Q113 (86*) on the N-WASP surface. The peptide P(3) and P(5) flank the exposed W54 sidechain on either side of the hydrogen bond between this residue and the carbonyl of P(3). An additional hydrogen bond from Q113 (86*) to the peptide E(6) helps to anchor the ligand into position (Figure 4.6). The most striking difference between the N-WASP–WIP complex and the Ena/VASP and Homer/Vesl EVH1–peptide complexes is that the WIP peptide binds in the reverse orientation to that observed in the other complexes. One reason for this may be the Y(8) sidechain of the WIP peptide, which makes several close contacts with the domain surface, just as the F(1) of the FPPPP motif contacts the Class 1 EVH1 domains. Y(8) is also oriented so that a hydrogen bond may be possible from its hydroxyl proton to the backbone carbonyl of N-WASP G109 (82*). It is therefore likely that Y(8) plays a role in orienting the WIP ligand, and it will be interesting to find out whether other WASP ligands exist in which this pattern is conserved.

Binding studies to date have shown that the Spred EVH1 domain does not bind the FPPPP motif recognized by the Ena/VASP EVH1 domains (Zimmermann et al., personal communication). The replacement of the conserved Class 1 EVH1 domain Y16 with an Arg in Spred is almost certainly an important factor in determining the ligand preference of this family (Figures 4.1, 4.4 and 4.5). Work in progress will reveal more details on the binding mode of this class of EVH1 domains in the future.

All the substitutions seen in the exposed PRM-recognition sites are very conservative and have no noticeable effect on the overall stability of the fold. Nevertheless, they are critically placed and provide sufficient external variation to tailor the different families of EVH1 domains to recognize highly specific consensus target sequences. This allows EVH1 domains to be used as molecular adaptors by diverse families of host proteins to mediate their localization to very different signaling proteins. Interestingly, the PRM-binding interfaces of the EVH1 domains described above and in Figure 4.6 have many features in common with those of other protein interaction domains that bind specifically to proline-rich sequences. Examples include the well known families of the SH3, WW, GYF, and UEV domains, as well as the small, actin-binding, profilin protein. This mechanism for proline-rich peptide recognition is therefore not specific to EVH1 domains, but is rather widely used in many different signaling pathways (Ball et al., 2004).

4.3.3
Comparisons with RanBDs, PTB Domains, and PH Domains

There are four sites of contact between Ran and the RanBD. One of these involves a groove on the RanBD surface, which binds the C-terminal fragment of Ran. The interaction surface is analogous to the peptide interaction surface of the EVH1 domains. Figure 4.5 shows the first RanBD (RanBD1) of the nuclear pore complex protein Nup358 in complex with the Ran protein [54]. For clarity, only the 11 C-terminal amino acids $_1$LEVAQTTALPD$_{11}$ of Ran, which bind the groove relevant to this discussion, are shown in the figure. The RanBDs are the closest relatives to the EVH1 domains (as shown by the phylogenetic tree in Figure 4.2) and employ very similar mechanisms of ligand recognition. Both families of domains utilize clusters of exposed aromatic sidechains to recognize their peptide ligands. The completely conserved W23 of the EVH1 domain (Mena numbering) is also conserved in the majority of RanBDs, where it is also necessary for fold stabilization. As in the EVH1 domains, this important residue is flanked in RanBDs by two additional aromatics to form hydrophobic binding pockets that accommodate specific peptide sidechains (yellow in Figure 4.6). The interaction is supported by one or more additional H-bond–donating residues exposed at the ligand binding interface. The W1211 (W23*) sidechain of the Nup358 RanBD1 is positioned so that it can form a hydrogen bond with the backbone carbonyl of the peptide T(7) in the same way as observed for the EVH1 domains, but this distance in the NMR structure of the N-WASP–WIP complex is longer than usually expected for hydrogen bonds and the geometry of the interaction is not ideal. The interaction of the ligand with the domain is stabilized by an additional hydrogen bond between the domain R1284 (R90*) sidechain and the ligand P(10). The Ran protein binds RanBD in the same orientation as the Ena/VASP and Homer EVH1 domains. However, since the peptide fragment shown here is only a small fragment of a much larger ligand, there are likely to be other factors that also contribute to its binding orientation and affinity, which shall not be discussed here.

For comparison, the PTB domain of the human SHC protein with a phosphotyrosine peptide from a TRKA receptor [88] and the PH domain of DAPP1/PHISH with inositol 1,3,4,5-tetrakisphosphate [89] are also shown in Figure 4.6. Although their folds are visibly very similar, it is clear that the binding sites of these domains have little if anything in common with those of the EVH1 domains or RanBDs. Thus, the aromatic peptide recognition clusters clearly evolved long after the above domains had diverged from the PH family.

4.4 Biological Function and Signaling Pathways Involving EVH1 Domains

The ability of proteins to target their binding partners in a highly specific manner is the basis for the assembly of multiprotein complexes required for signaling in all living cells. EVH1 domains mediate protein–protein interactions in a diverse range of signaling cascades, depending on their host protein and site of action.

4.4.1 Ena/VASP Interactions

The Ena/VASP proteins are involved in the generation and maintenance of cell polarization processes by localized actin polymerization and reorganization in a variety of specialized cytoskeletal substructures [14, 16]. During epithelial sheet formation, they bind FPPPP motifs within zyxin and vinculin via their EVH1 domains. This localizes the Ena/VASP proteins to premature epithelial contact sites called adhesion zippers, which are involved in sealing epithelial layers [29, 90]. Ena/VASP proteins are also involved in barrier formation in the endothelium [91, 92] and in regulating integrin-mediated platelet adhesion [22, 23].

Interactions between activated T-cells and antigen-presenting cells lead to the formation of a contact structure called the immunological synapse. Ena/VASP proteins are recruited by FPPPP motifs within the EVH1-binding protein Fyb/SLAP to a signaling complex which, together with WASP, supports localized actin-filament polymerization at these synapses [28]. Additionally, phagocytosis by macrophages involves assembly of phagocytic cups, to which Ena/VASP proteins are again recruited by the Fyb/Slap protein. This leads to the Fc cell-surface-receptor–mediated remodeling of the actin cytoskeleton that accompanies particle internalization [26].

The Ena/VASP proteins also have important roles in migration of neurons, neutrophils, and fibroblasts [14, 16]. During the establishment of synaptic contacts in developing nervous systems, growth cones of axons are navigated by receptor-mediated recognition of guidance proteins [93]. FPxxP binding sites for Class 1 EVH1 domains are found in the intracellular portion of the *Drosophila* and *C. elegans* axon guidance receptor Robo/Sax-3 [20, 94]. In the human and murine transmembrane semaphorin Sema6A-1 guidance protein [95], the corresponding EVH1 binding motifs are VPPKP. Furthermore, the *C. elegans* Ena/VASP ortholog Unc-34 is required for appropriate response of the axon guidance receptors Unc-5 and Unc-40/DCC to the guidance protein netrin [94, 96].

The locomotion of fibroblasts is a complex process requiring coordination of forward protrusion, attachment, contraction, and rear detachment, which translocates the cell body. Ena/VASP proteins were found to be negative regulators of cell migration [25]. At the subcellular level, they enhance F-actin polymerization at the leading edge of highly dynamic, transiently formed lamellipodia, to which they are localized by EVH1-mediated interactions [97]. Molecular understanding of these processes has greatly benefited from studies of the intracellular pathogenic bacterium *Listeria monocytogenes* [98, 99]. The EVH1 domains of the Ena/VASP proteins are recruited to the bacterial cell surface by tandem FPPPP motifs in the *Listeria* surface protein ActA [67]. There, they are involved in actin polymerization, resulting in the assembly of dynamic 'comet' tails, which generate the motile force necessary for bacterial propulsion through the cytoplasm [98, 99].

In summary, Ena/VASP proteins form a key link between signaling pathways and cytoskeletal dynamics by acting as regulators of actin filament assembly. These proteins are components of the actin polymerization machinery and are believed to function mechanistically by delivery of polymerization-competent actin monomers, exclusion of polymerization-terminating proteins, detachment of F-actin filaments, and/or inhibition of 'Y' branch formation in actin filament arrays, depending on the cellular context in which they are found [13, 100].

4.4.2 Homer/Vesl Interactions

Homer/Vesl proteins are adaptor proteins involved in clustering, anchoring, and modulation of neurotransmitter receptor proteins in excitatory glutamatergic synapses of the central nervous system [84], which are among the most elaborate junctions existing between cells. The apical postsynaptic plasma membrane region of dendritic spines differentiates into an electron-dense subcellular compartment called the postsynaptic density (PSD). The PSD contains, together with various multidomain PSD scaffold proteins, the NMDA (*N*-methyl-D-aspartate) and metabotropic glutamate receptors (mGluRs) [34]. The Class 2 EVH1 domains of Homer/Vesl proteins bind a PPxxF motif in the multidomain scaffold protein Shank/ProSAP [83], which is engaged in further interaction with NMDA receptors [62]. They also bind the Class 2 EVH1-binding motif (TPPSPF) in the C termini of group 1 mGluRs [31]. Thus, the clustering of Homer/Vesl proteins [59, 60] provides a mechanism for indirectly linking two glutamate receptor types. This may be relevant to synaptogenesis or to specific types of receptor crosstalk. The Homer/Vesl EVH1 domains are also known to bind the PPKKF motif within inositol trisphosphate receptors (IPRs) and the C termini of group 1 mGluRs [61]. Here, they are involved in the glutamate-induced release of Ca^{2+} from intracellular stores. The monomeric Homer1 isoform, which lacks a self-association domain [32, 66], is up-regulated by neural activity [35]. This protein is expected to competitively disrupt the signaling complexes assembled by the oligomeric Homer/Vesl isoforms [61]. The Homer/Vesl proteins thus regulate the composition and stability of multiprotein complexes comprising different synaptic receptor proteins. This implies a role in the modulation of synaptic plasticity, which is clearly important with respect to learning and memory formation.

4.4.3 WASP/N-WASP Interactions

The founding member of the WASP/N-WASP proteins, the Wiskott–Aldrich syndrome (WAS) protein, was originally discovered during a search for the genetic defect responsible for WAS, a rare, X-linked, recessive immunodeficiency disease with altered functions of many types of hematopoetic cells [101]. Among other roles, WASP is expected to be involved in biogenesis of the immunological synapse and in the early stages of T-cell cytoskeletal polarization (see above) [101]. The protein–protein interactions involved in these processes are not yet well understood. Understanding of the molecular function of WASP has been gained from analysis of cellular model systems [101], which identified WASP/N-WASP proteins as highly regulated effectors of Rho GTPases. Following release from an autoinhibitory state by Rho GTPases, WASP activates the actin-nucleating Arp2/3 complex via its C-terminal VCA effector domain. The EVH1 domain of N-WASP binds a consensus LPPPEPY motif in the C-terminal region of WIP and its homologs, CR16 and verprolin [36], to regulate N-WASP–mediated actin polymerization and filopodium formation. Various viral and bacterial pathogens have evolved strategies to hijack the actin-nucleating activity of Arp2/3, using pathogen-encoded virulence factors that are direct or indirect upstream activators of N-WASP [98]. Whereas the IcsA protein of *Shigella flexneri* directly mimics the host cell Rho GTPase Cdc42, to activate N-WASP [102], the viral membrane protein A36R of vaccinia virus [103] and the bacterial protein Tir of extracellular enteropathic *Escherichia coli* (EPEC) [104, 105] both recruit the host-encoded adaptor protein Nck, together with WIP and N-WASP, to initiate localized actin polymerization, which then supports either intracellular motility (*Shigella*, vaccinia) or contact formation to the host cell (EPEC). Pathogen-triggered binding of the EVH1 domain of N-WASP to PRMs of host proteins therefore functionally resembles the interaction of Ena/VASP EVH1 domains with the PRMs of the *Listeria monocytogenes* virulence factor ActA, discussed above. However, unlike the other virulence factors, ActA is able to activate the Arp2/3 complex directly.

4.4.4 Spred Interactions

Spred proteins were originally isolated in a screen for new tyrosine-kinase–binding proteins. The proteins are involved in regulation of differentiation in neuronal cells and myocytes [48]. The proteins were recently described as negative regulators of the signaling pathways of Ras, Raf, and mitogen-activated protein (MAP) kinase [47, 48]. Mechanistically, Spred inhibits early steps in the activation of MAP kinases by association with Ras, thereby suppressing phosphorylation and activation of Raf. This results in down-regulation of the Ras–MAP kinase signaling pathway [48]. The specific binding partner of the Spred EVH1 domain is currently unknown, although Spreds' inhibitory activity is lost when their EVH1 domain is substituted for a Class 3 EVH1 domain of WASP, indicating a highly specific interaction. To date, only one other Spred family member, the *Drosophila* AE33 [49], has been identified. It is known to be involved in photoreceptor cell specification during *Drosophila* eye development. However, its role in these developmental pathways has not yet been elucidated.

4.5
Emerging Research Directions and Recent Developments

One of the most important reasons for studying sequence similarities and obtaining detailed high-resolution structures of proteins and their complexes is that comparisons may then be made with proteins for which little functional data are available. This information also facilitates the annotation and classification of hypothetical proteins revealed by genome sequencing projects and provides a rational basis for the prediction of protein structures and the identification of candidate ligands for novel proteins.

4.5.1
Use of Sequence and Structural Data in Prediction of Binding Partners

Comparisons of sequence and fold reveal important relationships between different domains and provide insights into their evolutionary origins. Since the overall fold of a domain is determined by a small number of key residues in the hydrophobic core, structural alignments performed with algorithms such as DALI [106] and FSSP [107, 108] can provide important clues about function and ancestry even where sequence identity is very low. However, a fold similarity on its own is insufficient for the prediction of binding partners, as demonstrated in detail for the EVH1 domains discussed in this chapter. For this, close inspection of the conserved residues exposed on the surface of the domain can be very revealing. Often, groups of surface-exposed residues, although separated by many amino acids in the primary sequence, come together in the 3D structure to form a ligand binding site. These residues are usually conserved within a given domain family, and sometimes across many families that share common mechanisms of ligand recognition. Good examples are the PRM-binding domains, which include the SH3, WW, EVH1, GYF, and UEV domains and the single-domain actin-binding protein profilin. All six families of PRM-binding domains contain exposed groups of aromatic residues termed 'aromatic clusters' or 'aromatic cradles', which are responsible for binding proline-rich motifs [53, 109]. The presence of this cluster is now considered a signature for PRM recognition. Thus, if the structure of a new domain is solved, detailed comparisons with similar structures for which the function is better understood can narrow the search for binding partners significantly. A thorough understanding of the common features that define different types of domain–ligand interfaces is ultimately needed to enable us to derive rules for the prediction of binding partners from sequence information alone.

4.5.2
Use of Structural Data from Complexes to Guide the Rational Design of New Ligands

Knowledge of the molecular determinants of binding affinity and specificity is necessary for rational structure-based design of inhibitors to hinder interactions between specific proteins and to modulate cellular processes. To understand the subtle factors that modify specificity and affinity, it is necessary to characterize, structurally and biochemically, the most important features of the interactions we wish to inhibit. Only then can this knowledge be used to guide the design of novel target peptides or nonpeptide molecules having new activities. SPOT analyses [110] have helped enormously in understanding which residues of a known binding sequence are critical for binding and which may be replaced with little or no consequence. This allows the derivation of consensus ligand sequences, as demonstrated for the FPPPP motifs recognized by Class 1 EVH1 domains. Here, the consensus peptide sequence was found to be FPxϕP (where ϕ is a hydrophobic residue and x is any residue), meaning that the central two residues can be replaced in searches for higher-affinity partners [69, 75]. Knowledge of variable and conserved residues is a crucial first step toward the prediction, modification, and design of peptide binding partners, as successfully demonstrated experimentally for EVH1 domains [111].

The design of inhibitors for EVH1 domains is currently under way. Such inhibitors should be useful in several different ways: (1) to study the effects of modulating EVH1-mediated signaling cascades; (2) as molecular tags to monitor the formation and dissociation of EVH1-mediated interactions within the cell; and (3) as lead molecules for the development of future generations of novel therapeutics. The dose-dependent modulation of EVH1 domain binding activity would mean that it should be possible to develop treatments of diseases for which partial inhibition of EVH1-mediated events would be desirable (for example, pathologically altered adhesion and motility in inflammatory diseases and metastatic states, or even the spreading of intracellular pathogens).

4.6 Concluding Remarks

In this chapter we have taken a close look at the sequence signatures, domain occurrence/co-occurrence, and structural features that define the EVH1 domain. Subtle differences in the patterns and nature of exposed residues have been shown to lead to different ligand specificities. Combining sequence similarity, phylogeny, structural information, and ligand preference, we have classified the known EVH1 domains into four main categories. The stability of any domain fold is clearly an important factor in its evolution, and EVH1 domains are no exception. This is demonstrated by the high conservation of residues that comprise the hydrophobic core. The specific functionality then built upon the stable scaffold depends on subtle variations in the solvent-exposed residues, particularly those located within and around ligand binding sites. These can alter specificities dramatically, depending on their charge, hydrophobicity, size, and H-bond forming abilities. More distant relatives show greater differences in their patterns of surface residues, leading to widely different functions although they utilize a common global fold. Even seemingly minor variations at critical sequence locations can alter ligand binding specificities considerably. This provides an economical evolutionary mechanism for gradual functional diversification starting from a stable fold, as has been observed repeatedly during studies of the protein universe [112, 113]. Detailed studies of these variations provide clues about the domain's ancestry and phylogeny. Furthermore, knowing which residues can be varied without destabilizing the domain fold, and which contextual combinations of domain folds are tolerated in different host protein repertoires, is a crucial first step in the understanding of protein evolution and protein–protein interactions in general. Such knowledge is essential for the rational design of inhibitors, interaction interfaces, and novel proteins in the future.

References

1 Murzin, A. G., Brenner, S. E., Hubbard, T., Chothia, C., Scop: a structural classification of proteins database for the investigation of sequences and structures. *J. Mol. Biol.* **1995**, *247*, 536–540.

2 Schultz, J., Copley, R. R., Doerks, T., Ponting, C. P., Bork, P., Smart: a web-based tool for the study of genetically mobile domains. *Nucleic Acids Res.* **2000**, *28*, 231–234.

3 Letunic, I., Goodstadt, L., Dickens, N. J., Doerks, T., Schultz, J., Mott, R., Ciccarelli, F., Copley, R. R., Ponting, C. P., Bork, P., Recent improvements to the Smart domain-based sequence annotation resource. *Nucleic Acids Res.* **2002**, *30*, 242–244.

4 Bateman, A., Birney, E., Cerruti, L., Durbin, R., Etwiller, L., Eddy, S. R., Griffiths-Jones, S., Howe, K. L., Marshall, M., Sonnhammer, E. L., The Pfam protein families database. *Nucleic Acids Res.* **2002**, *30*, 276–280.

5 O'Donovan, C., Martin, M. J., Gattiker, A., Gasteiger, E., Bairoch, A., Apweiler, R., High-quality protein knowledge resource: Swiss-Prot and Trembl. *Brief Bioinform.* **2002**, *3*, 275–284.

6 Wu, C. H., et al., The protein information resource: an integrated public resource of functional annotation of proteins. *Nucleic Acids Res.* **2002**, *30*, 35–37.

7 Holm, L., Sander, C., Protein structure comparison by alignment of distance matrices. *J. Mol. Biol.* **1993**, *233*, 123–138.

8 Page, R. D., Treeview: an application to display phylogenetic trees on personal computers. *Comput. Appl. Biosci.* **1996**, *12*, 357–358.

9 Felsenstein, J., An alternating least squares approach to inferring phylogenies from pairwise distances. *Syst. Biol.* **1997**, *46*, 101–111.

10 Felsenstein, J., PHYLIP: Phylogeny inference package (version 3.2). *Cladistics* **1989**, *5*, 164–166.

11 Haffner, C., Jarchau, T., Reinhard, M., Hoppe, J., Lohmann, S. M., Walter, U., Molecular cloning, structural analysis and functional expression of the proline-rich focal adhesion and microfilament-associated protein VASP. *EMBO J.* **1995**, *14*, 19–27.

12 Gertler, F. B., Niebuhr, K., Reinhard, M., Wehland, J., Sonario, P., Mena, a relative of VASP and *Drosophila* enabled, is implicated in the control of microfilament dynamics. *Cell* **1996**, *87*, 227–239.

13 Krause, M., Dent, E. W., Bear, J. E., Loureiro, J. J., Gertler, F. B., Ena/VASP proteins: regulators of the actin cytoskeleton and cell migration. *Annu. Rev. Cell Dev. Biol.* **2003**, *19*, 541–564.

14 Kwiatkowski, A. V., Gertler, F. B., Loureiro, J. J., Function and regulation of Ena/VASP proteins. *Trends Cell Biol.* **2003**, *13*, 386–392.

15 Han, Y. H., Chung, C. Y., Wessels, D., Stephens, S., Titus, M. A., Soll, D. R., Firtel, R. A., Requirement of a vasodilator-stimulated phosphoprotein family member for cell adhesion, the formation of filopodia, and chemotaxis in dictyostelium. *J. Biol. Chem.* **2002**, *277*, 49877–49887.

16 Reinhard, M., Jarchau, T., Walter, U., Actin-based motility: stop and go with Ena/VASP proteins. *Trends Biochem. Sci.* **2001**, *26*, 243–249.

17 Krause, M., Bear, J. E., Loureiro, J. J., Gertler, F. B., The Ena/VASP enigma. *J. Cell Sci.* **2002**, *115*, 4721–4726.

18 Gertler, F. B., Doctor, J. S., Hoffmann, F. M., Genetic suppression of mutations in the *Drosophila* Abl proto-oncogene homolog. *Science* **1990**, *248*, 857–860.

19 Gertler, F. B., Comer, A. R., Juang, J.-L., Ahern, S. M., Clark, M. J., Liebl, E. C., Hoffmann, F. M., Enabled, a dosage-sensitive suppressor of mutations in the *Drosophila* abl tyrosine kinase, encodes an abl substrate with SH3 domain-binding properties. *Genes. Dev.* **1995**, *9*, 521–533.

20 Bashaw, G. J., Kidd, T., Murray, D., Pawson, T., Goodman, C. S., Repulsive axon guidance: Abelson and enabled play opposing roles downstream of the roundabout receptor. *Cell* **2000**, *101*, 703–715.

21 Dickson, B. J., Rho GTPases in growth cone guidance. *Curr. Opin. Neurobiol.* **2001**, *11*, 103–110.

22 Aszodi, A., Pfeifer, A., Ahmad, M., Glauner, M., Zhou, X. H., Ny, L., Andersson, K. E., Kehrel, B., Offermanns, S., Fassler, R., The vasodilator-stimulated phosphoprotein (VASP) is involved in cGMp- and cAMP-mediated inhibition of agonist-induced platelet aggregation, but is dispensable for smooth muscle function. *EMBO J.* **1999**, *18*, 37–48.

23 Hauser, W., Knobeloch, K.-P., Eigenthaler, M., Gambaryan, S., Krenns, V., Geiger, J., Glazova, M., Rhode, E., Walter, U., Zimmer, M., Megakaryocyte hyperplasia and enhanced agonist induced platelet activation in VASP knockout mice. *Proc. Nat. Acad. Sci. USA* **1999**, *96*, 8120–8125.

24 Anderson, S. I., Behrendt, B., Machesky, L. M., Insall, R. H., Nash, G. B., Linked regulation of motility and integrin function in activated migrating neutrophils revealed by interference in remodelling of the cytoskeleton. *Cell Motil. Cytoskeleton* **2003**, *54*, 135–146.

25 Bear, J. E., Loureiro, J. J., Libova, I., Fässler, R., Wehland, J., Gertler, F. B., Negative regulation of fibroblast motility by Ena/VASP proteins. *Cell* **2000**, *101*, 717–728.

26 Coppolino, M. G., Krause, M., Hagendorff, P., Monner, D. A., Trimble, W., Grinstein, S., Wehland, J., Sechi, A. S., Evidence for a molecular complex consisting of Fyb/Slap, Slp-76, Nck, VASP and WASP that links the actin cytoskeleton to fcgamma receptor signalling during phagocytosis. *J. Cell. Sci.* **2001**, *114*, 4307–4318.

27 Goh, K. L., Cai, L., Cepko, C. L., Gertler, F. B., Ena/VASP proteins regulate cortical neuronal positioning. *Curr. Biol.* **2002**, *12*, 565–569.

28 Krause, M., Sechi, A. S., Konradt, M., Monner, D., Gertler, F. B., Wehland, J., Fyn-binding protein (Fyb)/Slp76-associated protein (Slap), VASP proteins and the Arp2/3 complex link T cell receptor (TCR) signaling to the actin cytoskeleton. *J. Cell Biol.* **2000**, *149*, 181–194.

29 Vasioukhin, V., Bauer, C., Yin, M., Fuchs, E., Directed actin polymerization is the driving force for epithelial cell–cell adhesion. *Cell* **2000**, *100*, 209–219.

30 Chakraborty, T., et al., A focal adhesion factor directly linking intracellularly motile *Listeria monocytogenes* and *Listeria ivanovii* to the actin-based cytoskeleton of mammalian cells. *EMBO J.* **1995**, *14*, 1314–1321.

31 Brakeman, P. R., Lanahan, A. A., O'Brien, R., Roche, K., Barnes, C. A., Huganir, R. L., Worley, P. F., Homer: a protein that selectively binds metabotropic glutamate receptors. *Nature* **1997**, *386*, 284–288.

32 Xiao, B., Tu, J. C., Petralia, R. S., Yuan, J. P., Doan, A., Breder, C. D., Ruggiero, A., Lanahan, A. A., Wenthold, R. J., Worley, P. F., Homer regulates the association of group 1 metabotropic glutamate receptors with multivalent complexes of Homer-related, synaptic proteins. *Neuron* **1998**, *21*, 707–716.

33 Irie, K., Nakatsu, T., Mitsuoka, K., Miyazawa, A., Sobue, K., Hiroaki, Y., Doi, T., Fujiyoshi, Y., Kato, H., Crystal

structure of the Homer1 family conserved region reveals the interaction between the EVH1 domain and own proline-rich motif. *J. Mol. Biol.* **2002**, *318*, 1117–1126.

34 Garner, C. C., Nash, J., Huganir, R. L., PDZ domains in synapse assembly and signalling. *Trends Cell Biol.* **2000**, *10*, 274–280.

35 Kato, A., Ozawa, F., Saitoh, Y., Hirai, K., Inokuchi, K., *Vesl*, a gene encoding VASP/Ena family related protein, is upregulated during seizure, long-term potentiation and synaptogenesis. *FEBS Lett.* **1997**, *412*, 183–189.

36 Volkman, B. F., Prehoda, K. E., Scott, J. A., Peterson, F. C., Lim, W. A., Structure of the N-WASP EVH1 domain–WIP complex: insight into the molecular basis of Wiskott–Aldrich syndrome. *Cell* **2002**, *111*, 565–576.

37 Carlier, M. F., Ducruix, A., Pantaloni, D., Signalling to actin: the cdc42-N-WASP-Arp2/3 connection. *Chem. Biol.* **1999**, *6*, R235–240.

38 Pollard, T. D., Blanchoin, L., Mullins, R. D., Molecular mechanisms controlling actin filament dynamics in nonmuscle cells. *Annu. Rev. Biophys. Biomol. Struct.* **2000**, *29*, 545–576.

39 Higgs, H. N., Pollard, T. D., Activation by cdc42 and pip(2) of Wiskott–Aldrich syndrome protein (WASP) stimulates actin nucleation by Arp2/3 complex. *J. Cell Biol.* **2000**, *150*, 1311–1320.

40 Derry, J. M., Ochs, H. D., Francke, U., Isolation of a novel gene mutated in Wiskott–Aldrich syndrome. *Cell* **1994**, *78*, 635–644.

41 Kolluri, R., Shehabeldin, A., Peacocke, M., Lamhonwah, A. M., Teichert-Kuliszewska, K., Weissman, S. M., Siminovitch, K. A., Identification of WASP mutations in patients with Wiskott–Aldrich syndrome and isolated thrombocytopenia reveals allelic heterogeneity at the *was* locus. *Hum. Mol. Genet.* **1995**, *4*, 1119–1126.

42 Villa, A., et al., X-linked thrombocytopenia and Wiskott–Aldrich syndrome are allelic diseases with mutations in the *wasp* gene. *Nat. Genet.* **1995**, *9*, 414–417.

43 Greer, W. L., Shehabeldin, A., Schulman, J., Junker, A., Siminovitch, K. A., Identification of *wasp* mutations, mutation hotspots and genotype–phenotype disparities in 24 patients with the Wiskott–Aldrich syndrome. *Hum. Genet.* **1996**, *98*, 685–690.

44 Zhu, Q., Watanabe, C., Liu, T., Hollenbaugh, D., Blaese, R. M., Kanner, S. B., Aruffo, A., Ochs, H. D., Wiskott–Aldrich syndrome/X-linked thrombocytopenia: *Wasp* gene mutations, protein expression, and phenotype. *Blood* **1997**, *90*, 2680–2689.

45 Suzuki, T., Miki, H., Takenawa, T., Sasakawa, C., Neural Wiskott–Aldrich syndrome protein is implicated in the actin-based motility of *Shigella flexneri*. *EMBO J.* **1998**, *17*, 2767–2776.

46 Moreau, V., Frischknecht, F., Reckmann, I., Vincentelli, R., Rabut, G., Stewart, D., Way, M., A complex of N-WASP and WIP integrates signalling cascades that lead to actin polymerization. *Nat. Cell Biol.* **2000**, *2*, 441–448.

47 Kato, R., Nonami, A., Taketomi, T., Wakioka, T., Kuroiwa, A., Matsuda, Y., Yoshimura, A., Molecular cloning of mammalian Spred-3 which suppresses tyrosine kinase-mediated ERK activation. *Biochem. Biophys. Res. Commun.* **2003**, *302*, 767–772.

48 Wakioka, T., Sasaki, A., Kato, R., Shouda, T., Matsumoto, A., Miyoshi, K., Tsuneoka, M., Komiya, S., Baron, R., Yoshimura, A., Spred is a Sprouty-related suppressor of RAS signalling. *Nature* **2001**, *412*, 647–651.

49 DeMille, M. M., Kimmel, B. E., Rubin, G. M., A *Drosophila* gene regulated by rough and glass shows similarity to Ena and VAS*Gene*, P. **1996**, *183*, 103–108.

50 Gorlich, D., Transport into and out of the cell nucleus. *EMBO J.* **1998**, *17*, 2721–2727.

51 Dingwall, C., Kandels-Lewis, S., Seraphin, B., A family of Ran binding proteins that includes nucleoporins. *Proc. Natl. Acad. Sci. USA* **1995**, *92*, 7525–7529.

52 Beddow, A. L., Richards, S. A., Orem, N. R., Macara, I. G., The Ran/TC4 GTPase-binding domain: identification by expression cloning and characterization of a conserved sequence motif. *Proc. Natl. Acad. Sci. USA* **1995**, *92*, 3328–3332.

53 Callebaut, I., Cossart, P., Dehoux, P., Evh1/WH1 domains of VASP and WASP proteins belong to a large family including Ran-binding domains of the RanBP1 family. *FEBS Lett.* **1998**, *441*, 181–185.

54 Vetter, I. R., Nowak, C., Nishimoto, T., Kuhlmann, J., Wittinghofer, A., Structure of a Ran-binding domain complexed with Ran bound to a GTP analogue: implications for nuclear transport. *Nature* **1999**, *398*, 39–46.

55 Stewart, M., Baker, R. P., 1.9 Å resolution crystal structure of the *Saccharomyces cerevisiae* Ran-binding protein mog1p. *J. Mol. Biol.* **2000**, *299*, 213–223.

56 Reinhard, M., Jouvenal, K., Triquier, D., Walter, U., Identification, purification and characterisation of a zyxin-related protein that binds the focal adhesion and microfilament protein VASP (vasodilator stimulated phosphoprotein). *Proc. Natl. Acad. Sci. USA* **1995**, *92*, 7956–7960.

57 Reinhard, M., Rüdiger, M., Jokusch, B. M., Walter, U., VASP interaction with vinculin: a recurring theme of interactions with proline rich motifs. *FEBS Lett.* **1996**, *399*, 103–107.

58 Reinhard, M., Giehl, K., Abel, K., Haffner, C., Jarchau, T., Hoppe, V., Jockusch, B. M., Walter, U., The proline-rich focal adhesion and microfilament protein VASP is a ligand for profilins. *EMBO J.* **1995**, *14*, 1583–1589.

59 Zimmermann, J., Labudde, D., Jarchau, T., Walter, U., Oschkinat, H., Ball, L. J., Relaxation, equilibrium oligomerization, and molecular symmetry of the VASP (336–380) EVH2 tetramer. *Biochemistry* **2002**, *41*, 11143–11151.

60 Bachmann, C., Fischer, L., Walter, U., Reinhard, M., The EVH2 domain of the vasodilator stimulated phosphoprotein mediates tetramerization, F-actin binding and actin bundle formation. *J. Biol. Chem.* **1999**, *274*, 23549–23557.

61 Tu, J. C., Xiao, B., Yuan, J. P., Lanahan, A. A., Leoffert, K., Li, M., Linden, D. J., Worley, P. F., Homer binds a novel proline rich motif and links group 1 metabotropic glutamate receptors with IP3 receptors. *Neuron* **1998**, *21*, 717–726.

62 Naisbitt, S., Kim, E., Tu, J. C., Xiao, B., Sala, C., Valtschanoff, J., Weinberg, R. J., Worley, P. F., Sheng, M., Shank, a novel family of postsynaptic density proteins that binds to the nmda receptor/psd-95/gkap complex and cortactin. *Neuron* **1999**, *23*, 569–582.

63 Machesky, L. M., Insall, R. H., Signaling to actin dynamics. *J. Cell Biol.* **1999**, *146*, 267–272.

64 Rohatgi, R., Ma, L., Miki, H., Lopez, M., Kirchhausen, T., Takenawa, T., Kirschner, M. W., The interaction between N-WASP and the Arp2/3 complex links cdc42-dependent signals to actin assembly. *Cell* **1999**, *97*, 221–231.

65 Ponting, C. P., Schultz, J., Copley, R. R., Andrade, M. A., Bork, P., Evolution of domain families. *Adv. Protein Chem.* **2000**, *54*, 185–244.

66 Kato, A., Ozawa, F., Saitoh, Y., Fukazawa, Y., Sugiyama, H., Inokuchi, K., Novel members of the VESL/Homer family of PDZ proteins that bind metabotropic glutamate receptors. *J. Biol. Chem.* **1998**, *273*, 23969–23975.

67 Niebuhr, K., Ebel, F., Frank, R., Reinhard, M., Domann, E., Carl, U. D., Walter, U., Gertler, F. B., Wehland, J., Chakraborty, T., A novel proline-rich motif present in acta of *Listeria monocytogenes* and cytoskeletal proteins is the ligand for the Evh1 domain, a protein module present in the Ena/VASP family. *EMBO J.* **1997**, *16*, 5433–5444.

68 Mulder, N. J., et al., Interpro: an integrated documentation resource for protein families, domains and functional sites. *Brief Bioinform.* **2002**, *3*, 225–235.

69 Ball, L. J., Jarchau, T., Oschkinat, H., Walter, U., Evh1 domains: structure, function and interactions. *FEBS Lett.* **2002**, *513*, 45–52.

70 Sone, M., Hoshino, M., Suzuki, E., Kuroda, S., Kaibuchi, K., Nakagoshi, H., Saigo, K., Nabeshima, Y., Hama, C., Still life, a protein in synaptic terminals of *Drosophila* homologous to GDP–GTP exchangers. *Science* **1997**, *275*, 543–547.

71 Nourry, C., Grant, S. G., Borg, J. P., PDZ domain proteins: plug and play! *Sci. STKE* **2003**, *2003*, RE7.

72 van Ham, M., Hendriks, W., PDZ domains: glue and guide. *Mol. Biol. Rep.* **2003**, *30*, 69–82.

73 Prehoda, K. E., Lee, D. J., Lim, W. A., Structure of the enabled/VASP homology 1 domain–peptide complex: a key component in the spatial control of actin assembly. *Cell* **1999**, *97*, 471–480.

74 Fedorov, A. A., Fedorov, E., Gertler, F., Almo, S. C., Structure of Evh1, a novel proline rich ligand-binding module involved in cytoskeletal dynamics and neural function. *Nat. Struct. Biol.* **1999**, *6*, 661–665.

75 Ballet, L. J., et al., Dual epitope recognition by the VASP Evh1 domain modulates polyproline ligand specificity and binding affinity. *EMBO J.* **2000**, *19*, 4903–4914.

76 Beneken, J., Tu, J. C., Xiao, B., Nuriya, M., Yuan, J. P., Worley, P. F., Leahy, D. J., Structure of the Homer Evh1 domain–peptide complex reveals a new twist in polyproline recognition. *Neuron* **2000**, *26*, 143–154.

77 Barzik, M., Carl, U. D., Schubert, W. D., Frank, R., Wehland, J., Heinz, D. W., The N-terminal domain of Homer/VESL is a new class II Evh1 domain. *J. Mol. Biol.* **2001**, *309*, 155–169.

78 Seewald, M. J., Korner, C., Wittinghofer, A., Vetter, I. R., Rangap mediates GTP hydrolysis without an arginine finger. *Nature* **2002**, *415*, 662–666.

79 Yan, K. S., Kuti, M., Zhou, M. M., PTB or not PTB: that is the question. *FEBS Lett.* **2002**, *513*, 67–70.

80 Lemmon, M. A., Ferguson, K. M., Abrams, C. S., Pleckstrin homology domains and the cytoskeleton. *FEBS Lett.* **2002**, *513*, 71–76.

81 Ahern-Djamali, S. M., Comer, A. R., Bachmann, C., Kastenmeier, A. S., Reddy, S. K., Hua, P., Beckerle, M. C., Walter, U., Hoffmann, F. M., Mutations in *Drosophila* enabled and rescue by human VASP indicate important functional roles for Evh1 and Evh2 domains. *Mol. Biol. Cell* **1998**, *9*, 2157–2171.

82 Carl, U. D., Pollmann, M., Orr, E., Gertler, F. B., Chakraborty, T., Wehland, J., Aromatic and basic residues within the Evh1 domain of VASP specify its interaction with proline rich ligands. *Curr. Biol.* **1999**, *9*, 715–718.

83 Tu, J. C., et al., Coupling of mglur/Homer and psd-95 complexes by the Shank family of postsynaptic density proteins. *Neuron* **1999**, *23*, 583–592.

84 Xiao, B., Tu, J. C., Worley, P. F., Homer: a link between neural activity and glutamate receptor function. *Curr. Opin. Neurobiol.* **2000**, *10*, 370–374.

85 Cowan, P. M., McGavin, S., Structure of poly-L-proline. *Nature* **1955**, *176*, 501–503.

86 Adzubei, A. A., Sternberg, M. J. E., Left-handed polyproline II helices commonly occur in globular proteins. *J. Mol. Biol.* **1993**, *229*, 472–493.

87 Williamson, M. P., The structure and function of proline rich regions in proteins. *Biochemical J.* **1994**, *297*, 249–260.

88 Zhou, M. M., et al., Structure and ligand recognition of the phosphotyrosine binding domain of SH*Nature,* C. **1995**, *378*, 584–592.

89 Ferguson, K. M., Kavran, J. M., Sankaran, V. G., Fournier, E., Isakoff, S. J., Skolnik, E. Y., Lemmon, M. A., Structural basis for discrimination of 3-phosphoinositides by pleckstrin homology domains. *Mol. Cell.* **2000**, *6*, 373–384.

90 Vasioukhin, V., Fuchs, E., Actin dynamics and cell–cell adhesion in epithelia. *Curr. Opin. Cell Biol.* **2001**, *13*, 76–84.

91 Lawrence, D. W., Comerford, K. M., Colgan, S. P., Role of VASP in reestablishment of epithelial tight junction assembly after Ca^{2+} switch. *Am. J. Physiol. Cell Physiol.* **2002**, *282*, C1235–1245.

92 Comerford, K. M., Lawrence, D. W., Synnestvedt, K., Levi, B. P., Colgan, S. P., Role of vasodilator-stimulated phosphoprotein in PKA-induced changes in endothelial junctional permeability. *FASEB J.* **2002**, *16*, 583–585.

93 Song, H., Poo, M., The cell biology of neuronal navigation. *Nat. Cell Biol.* **2001**, *3*, E81–88.

94 Yu, T. W., Hao, J. C., Lim, W., Tessier-Lavigne, M., Bargmann, C. I., Shared receptors in axon guidance: Sax-3/robo signals via UNC-34/enabled and a netrin-independent UNC-40/dcc function. *Nat. Neurosci.* **2002**, *5*, 1147–1154.

95 Klostermann, A., Lutz, B., Gertler, F., Behl, C., The orthologous human and murine semaphorin 6a-1 proteins (sema6a-1/sema6a-1) bind to the enabled/vasodilator stimulated phosphoprotein-like protein (EVL) via a novel carboxyl-terminal zyxin like domain. *J. Biol. Chem.* **2000**, *275*, 39647–39653.
96 Colavita, A., Culotti, J. G., Suppressors of ectopic UNC-5 growth cone steering identify eight genes involved in axon guidance in *Caenorhabditis elegans*. *Dev. Biol.* **1998**, *194*, 72–85.
97 Bear, J. E., et al., Antagonism between Ena/VASP proteins and actin filament capping regulates fibroblast motility. *Cell* **2002**, *109*, 509–521.
98 Goldberg, M. B., Actin-based motility of intracellular microbial pathogens. *Microbiol. Mol. Biol. Rev.* **2001**, *65*, 595–626.
99 Cameron, L. A., Giardini, P. A., Soo, F. S., Theriot, J. A., Secrets of actin-based motility revealed by a bacterial pathogen. *Nat. Rev. Mol. Cell Biol.* **2000**, *1*, 110–119.
100 Samarin, S., Romero, S., Kocks, C., Didry, D., Pantaloni, D., Carlier, M. F., How VASP enhances actin-based motility. *J. Cell Biol.* **2003**, *163*, 131–142.
101 Thrasher, A. J., WASP in immune-system organization and function. *Nat. Rev. Immunol.* **2002**, *2*, 635–646.
102 Suzuki, T., Sasakawa, C., N-WASP is an important protein for the actin-based motility of *Shigella flexneri* in the infected epithelial cells. *Jpn. J. Med. Sci. Biol.* **1998**, *51 Suppl.*, S63–68.
103 Frischknecht, F., Moreau, V., Rottger, S., Gonfloni, S., Reckmann, I., Superti-Furga, G., Way, M., Actin-based motility of vaccinia virus mimics receptor tyrosine kinase signalling. *Nature* **1999**, *401*, 926–929.
104 Kalman, D., Weiner, O. D., Goosney, D. L., Sedat, J. W., Finlay, B. B., Abo, A., Bishop, J. M., Enteropathogenic *E. coli* acts through WASP and arp2/3 complex to form actin pedestals. *Nat. Cell Biol.* **1999**, *1*, 389–391.
105 Gruenheid, S., DeVinney, R., Bladt, F., Goosney, D., Gelkop, S., Gish, G. D., Pawson, T., Finlay, B. B., Enteropathogenic *E. coli* Tir binds Nck to initiate actin pedestal formation in host cells. *Nat. Cell Biol.* **2001**, *3*, 856–859.
106 Holm, L., Sander, C., Dali: a network tool for protein structure comparison. *Trends Biochem. Sci.* **1995**, *20*, 478–480.
107 Holm, L., Sander, C., The fssp database of structurally aligned protein fold families. *Nucleic Acids Res.* **1994**, *22*, 3600–3609.
108 Holm, L., Sander, C., Dictionary of recurrent domains in protein structures. *Proteins* **1998**, *33*, 88–96.
109 Zarrinpar, A., Bhattacharyya, R. P., Lim, W. A., The structure and function of proline recognition domains. *Sci. STKE* **2003**, *2003*, RE8.
110 Frank, R., Overwin, H., *Methods in Molecular Biology*, Humana Press, Totowa, NJ **1996**.
111 Zimmermann, J., Kühne, R., Volkmer-Engert, R., Jarchau, T., Walter, U., Oschkinat, H., Ball, L. J., Design of N-substituted peptomer ligands for Evh1 domains. *J. Biol. Chem.* **2003**, *278*, 36810–36818.
112 Koonin, E. V., Wolf, Y. I., Karev, G. P., The structure of the protein universe and genome evolution. *Nature* **2002**, *420*, 218–223.
113 Wolf, Y. I., Karev, G., Koonin, E. V., Scale-free networks in biology: new insights into the fundamentals of evolution? *Bioessays* **2002**, *24*, 105–109.
114 Schultz, J., Milpetz, F., Bork, P., Ponting, C. P., Smart, a simple modular architecture research tool: identification of signaling domains. *Proc. Natl. Acad. Sci. USA* **1998**, *95*, 5857–5864.
115 Eddy, S. R., Profile hidden Markov models. *Bioinformatics* **1998**, *14*, 755–763.
116 Kabsch, W., Sander, C., Dictionary of protein secondary structure: pattern recognition of hydrogen-bonded and geometrical features. *Biopolymers* **1983**, *22*, 2577–2637.

5
The GYF Domain

Christian Freund

5.1
Introduction

The GYF domain was discovered as a C-terminal fragment of the protein CD2BP2, which confers binding specificity to the cytoplasmic domain of the T-cell adhesion molecule CD2 in a yeast two-hybrid screen [1]. Subsequent investigation of this fragment showed that the C-terminal 62 amino acids of CD2BP2 form an independent folding unit that retains full binding capacity [2]. The structure of the isolated domain, in conjunction with mutational data, revealed the GYF (Gly-Tyr-Phe) sequence of the domain as part of a bulge-helix-bulge motif that is essential for the binding of the CD2 ligand. The target sequence for the GYF domain within the CD2 tail comprises two proline-rich motifs that are highly conserved among species and that have been shown to mediate CD2-dependent signal transduction [3]. This initial data placed the GYF domain in the superfamily of recognition domains for proline-rich sequences that so far comprises seven folds, namely profilin [4], SH3 [5, 6], WW [7], EVH1 [8], GYF [1, 2], UEV [9, 10], and probably the substrate binding domain of prolyl 4-hydroxylase [11]. Sequence data comparisons soon identified a set of proteins that contain GYF domains; Figure 5.1 shows the alignment of GYF domain sequences of diverse origin. The analysis revealed the GYF domain to be present in most if not all eukaryotic genomes, but the number of GYF domain-containing proteins is small in a given species. Interestingly, *Arabidopsis thaliana* has the largest number of GYF domain-containing proteins, and only in plants is the GYF domain found with other known protein domains within the same protein (see for example the SMART database [12]). Specifically, SET, SwiB, and C_2H_2 domains flank GYF domains in some of the plant proteins, indicating a role of these proteins in histone modification, chromosome segregation, and RNA binding, respectively. In the animal kingdom, the lack of other annotated protein interaction domains within GYF domain-containing proteins does not allow straightforward biochemical characterization of these proteins within a known functional context. However, structural and functional characterization of the CD2BP2-GYF–CD2 interaction allows us to draw certain conclusions about GYF

Modular Protein Domains. Edited by G. Cesareni, M. Gimona, M. Sudol, M. Yaffe

ISBN: 3-527-30813-X

Figure 5.1 Sequence alignment of GYF domains of various origins. Residues conserved to greater than 75% are shown as white letters on a dark grey background, and amino acids conserved in more than 60% of the aligned GYF domains appear as black letters on a light grey background. Residue 25 (CD2BP2 GYF domain numbering) is also highlighted in white type on a light grey background, since a hydrophobic amino acid (I, M, L or V) is always found at this position. Multiple sequence alignments and sequence relationships (as indicated by vertical distance in the dendrogram) were obtained with the pileup routine of the GCG package.
Individual GYF domains are labeled with the DDBJ/EMBL/GenBank accession numbers, and the DNA sequences can be found at http://www.ensembl.org/ and http://portal.tmri.org/Rice.
The abbreviations SDG2, SDG25, SINFRU, and NP0470 were used for the CAB10279, BAB10481, SINFRUP00000160281, and NP_047030 GYF domains, respectively.
In SDG2 and SDG25, two GYF domains are predicted to be present in each of the proteins (labeled as SDG2_1, SDG2_2, SDG25_1, and SDG25_2). The species containing the respective GYF domain is indicated after the accession number. The final two sequences (SDG25_1 and Q9VKV2) are included in the alignment despite the absence of certain hallmark residues of the fold. Domain borders are well defined only for the N-terminal sequences, and the C terminus of GYF domains varies significantly. For the alignment sequence lengths of approximately 60 amino acids were assumed, according to the size of the CD2BP2-GYF domain.

domain function in general [13]. The observed competition in ligand binding of the CD2BP2-GYF domain and the Fyn-SH3 domain [13] emphasizes the need to investigate the binding specificities of proline-rich-sequence binding domains across domain borders. Furthermore, results obtained for the localization of CD2BP2-GYF and Fyn-SH3 show that inducible subcellular compartmentalization of proline-rich sequences within target proteins is likely to select for the appropriate binding partner within a living cell.

5.2 Structure of the CD2BP2-GYF Domain and Its Interaction with the CD2 Signaling Peptide SHRPPPPGHRV

The GYF domain displays a ββαββ topology in which the extremely short β strands of the sheet are antiparallel (Figure 5.2, upper left panel). The conserved amino acids of the domain constitute part of the hydrophobic core (W4, Y6, M25, F34), are important for the structural display of the two bulges flanking the helix (G18, P19, G32), or are primarily involved in ligand binding (F20, W28, Y33).

Interestingly, an array of aromatic sidechains occupy the space between the tilted α helix and the sheet, thereby creating a single lipophilic center that interacts with the core of the peptide ligand (Figure 5.2, upper right panel). The peptide adopts an extended conformation and forms a polyproline typeII helix involving residues Pro4–Pro7 (Figure 5.2). The major binding surface of the GYF domain accommodates Pro6 and Pro7 of the ligand and is defined by the aromatic residues Tyr6, Trp8, Tyr17, Tyr20, Trp28, Tyr33, and Phe33 of the GYF domain (Figure 5.2, lower left panel). The floor of the binding pocket is defined by the hydrophobic core residues Tyr6 and Phe34, and Phe20, Trp28, and Tyr34 define the walls on three sides. The backbone carbonyl group of Pro4 in the ligand is H-bonded to the sidechain NH of Trp28 of the protein, and Pro6 of the ligand is oriented almost parallel to the aromatic ring of Trp28. Since Trp28 is the most highly conserved amino acid of the GYF domain family, a similar role for this residue can be assumed for GYF domains in general. Trp8 and Tyr17 are tilted away from the almost parallel main axes of the conserved aromatic residues and thereby open the fourth side of the binding pocket just enough to place Gly8 of the ligand at the edge of the major binding pocket. This glycine allows a kink in the backbone of the ligand to occur, thereby preventing collision of sidechain atoms with the protein. The dihedral angles for Gly8 ($\phi = 76^{\circ}$, $\psi = -80^{\circ}$) are in the disallowed region of the Ramachandran plot and support the important role of glycine at this position of the ligand. Gly8 of the ligand also terminates the proline-helix part; this kink within the ligand conformation is probably stabilized by a hydrogen bond between the Pro7 carbonyl group and the backbone amide group of His9. Pro4 and Pro5 do not contribute significantly to the interaction surface but are likely to stabilize the PPII helical conformation of residues Pro4–Pro7. Arg3 and Arg10 of the ligand are close to negatively charged sidechains of the protein, and the sidechain of Arg10 allows hydrophobic contacts to be made between the aliphatic groups of its sidechain and the Trp8 aromatic ring of the GYF domain.

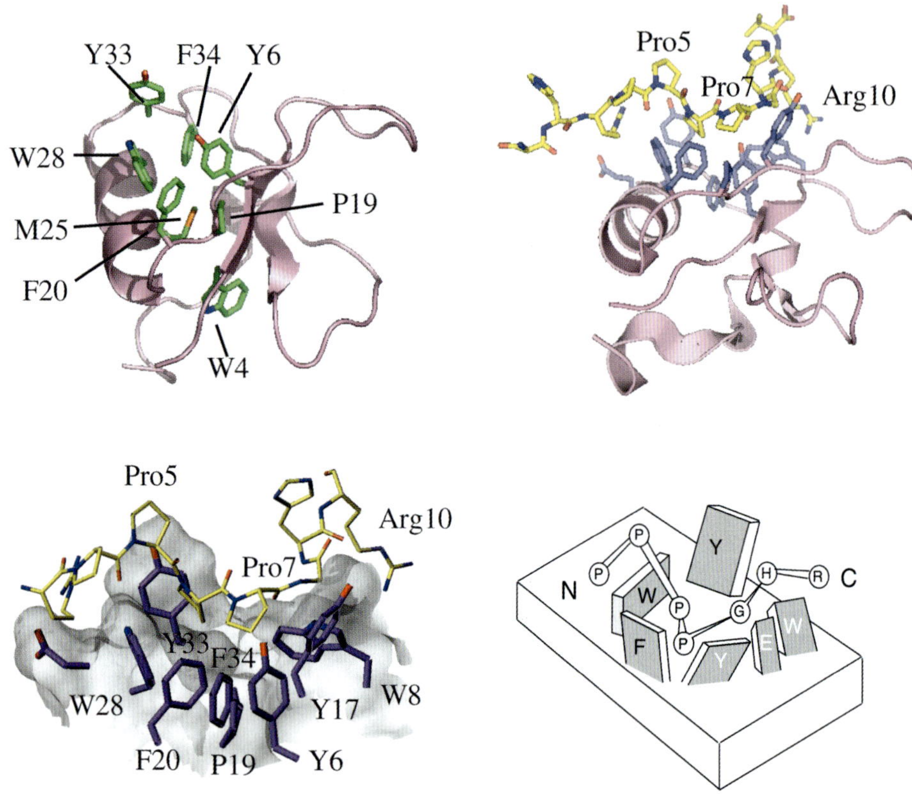

Figure 5.2 Structure of the GYF domain and the GYF domain–ligand complex. The structure of the isolated GYF domain is shown in the upper left panel (PDB code: 1GYF). The sidechains of residues that are highly conserved within the GYF domain family are highlighted in green and marked according to the numbering in Figure 5.1. The upper right and the lower left panels show the structure of the complex of the CD2BP2-GYF domain and the CD2 peptide SHRPPPPGHRV (PDB code: 1L2Z). Residues comprising the binding site of the domain are shown in blue, and the peptide is presented as a ball-and-stick structure in yellow. The amino acids of the peptide are displayed in three-letter code, and the protein residues involved in binding in single-letter code. An enlargement of the GYF domain binding site is additionally shown as a translucent surface in the lower left panel to emphasize the central pocket that accommodates the core of the peptide. Shown schematically in the lower right panel are the conserved residues of the GYF domain family, which contribute largely to the interaction surface (black letters), and the less-conserved binding-site residues of the CD2BP2-GYF domain (white letters).

Structural analysis and amino acid sequence comparison allows GYF domains to be classified into several tentative subclasses (Figure 5.1). The CD2BP2 subclass contains tryptophan instead of aspartate at position 8, and the loop connecting β strands 1 and 2 is two to three amino acids longer than in the other subclasses. The W8-to-D8 substitution certainly changes the lipophilic potential of the binding site and suggests that a different spectrum of ligands can be bound by this type of GYF domain. Proteins containing the CD2BP2-type GYF domain share large regions of sequence homology N-terminal to the GYF domain, and in this class the GYF domain is always present at the C terminus of all the respective proteins. The CD2BP2 subclass is present in evolutionarily distant species ranging from humans to flies to yeast. This broad sequence conservation implies that the CD2BP2 subclass separated early in evolution. Another class of GYF domains is present only in plants. In this class the signature F/M/L-K/R/E-I/V-W sequence is preserved C-terminal to the GYF sequence motif, and other subclasses may be defined based on amino acid conservation within the C terminus of the domain (Figure 5.1). Since the sequences C-terminal to the GYF motif are more diverse and thought to be structurally rather than functionally important, it is unlikely that these sequence-based classifications translate into separate functional classes with distinct ligand binding specificities.

5.3
Molecular and Signaling Function of GYF Domains

5.3.1
Sequence Specificity of the CD2BP2 GYF Domain

Substitution analysis of the CD2 peptide spotted onto a cellulose membrane reveals the contribution of individual residues in the peptide SHRPPPPGHRV to binding to the CD2BP2 GYF domain (Figure 5.3) [14]. Here, the wild-type sequence is synthesized at all positions of the first column and each wild-type residue is replaced by all naturally occurring amino acids in the following matrix of 11 $\times$ 20 peptides.

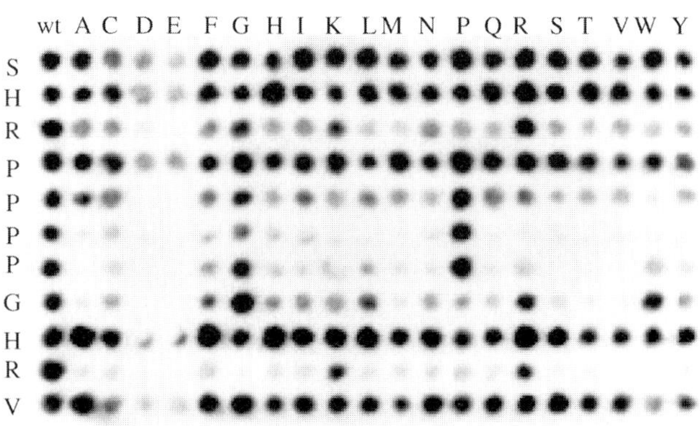

Figure 5.3 Substitution analysis of the CD2 peptide SHRPPPPGHRV and binding to the CD2BP2-GYF domain. Peptides were synthesized on a cellulose membrane and incubated with a 40 μM solution of GST-GYF. After 3 washing steps with TBS, bound GST-GYF protein was detected with an anti-rabbit antibody/anti-rabbit HPR-conjugated antibody pair. Pierce Pico chemiluminescence substrate was added and, after a 1-min incubation, emitted light was detected. The wild-type sequence SHRPPPPGHRV is present in all spots of the first column, and each residue of the sequence was replaced by all other naturally occurring amino acids according to the scheme presented in the first row.

There is a strong requirement for proline at positions 6 and 7, but prolines at positions 4 and 5 can be replaced by most of the other amino acids. This finding correlates well with the numerous van der Waals interactions observed between prolines 6 and 7 and the aromatic sidechains of the domain in the structure of the complex (Figure 5.2). For position 8, glycine results in the strongest signal; however, several other amino acids are allowed at this position. Surprisingly, tryptophan is tolerated at position 8, indicating that alternative modes of ligand binding to the GYF domain might exist. Positively charged amino acids are preferred at positions 3 and 10, but most amino acids are tolerated at the other positions. In general, the substitution analysis can be rationalized by the structure of the GYF domain in complex with the CD2 peptide (Figure 5.2), with a few notable exceptions that call for structural investigations. Examples of the latter are the R3 to G3 and the G8 to L8/R8/W8 substitutions, all of which show comparable spot intensities to the wild-type sequence. Although the backbone conformation of GYF-domain residues is not expected to change upon binding of these ligands, rotation of aromatic sidechains of the GYF domain and flexibility within the ligand most likely account for the observed binding behavior.

5.3.2
Spliceosomal Proteins Contain Binding Motifs for CD2BP2-GYF

Based on the results of the substitution analysis of the CD2 peptide (Figure 5.3), the human genome was searched for sequences that are putative binding motifs of CD2BP2-GYF. Several proteins that contain sequences similar to the peptide signature RxxPPGxR were identified as components of the splicing machinery. Specifically, the SmB/B′ nucleoprotein contains several C-terminal sequence stretches that closely resemble the CD2 binding sequence. An SmB/B′-derived peptide of the sequence GRGTPMGMPPPGMRPPPPGM-RGLL was further investigated by peptide spot analysis and NMR spectroscopy [14]. This peptide bound with an affinity comparable to the affinity of the PPPPGHRSQAPSHRPPPPGHRV peptide derived from CD2 [9]. GST pulldown experiments using a GST-GYF construct to evaluate binding under physiological conditions identified the SmB/B′ protein as an interactor with the CD2BP2-GYF domain [14]. This result is in agreement with the observation that CD2BP2 is a component of the pre-spliceosome [15] and reveals a function of CD2BP2 in addition to mediating CD2-triggered signal transduction.

5.3.3
Phage Display of CD2BP2-GYF

Although substitution analysis can be used to explore the local minimum of the free energy of binding of a given sequence, combinatorial methods assist in deriving more general requirements for sequence recognition by GYF domains. Phage display is the most widely used method for exploring the sequence space of proline-rich recognition modules. GeneVIII-based vector systems (e.g., [16]) that allow the multivalent display of peptides are particularly well suited for screening low-affinity binding modules. A biased library of the topology xxPPPxxx was used for the CD2BP2-GYF domain, and the results compared well with those of an X_9 peptide library screen (Kofler et al., unpublished results). The results after three rounds of panning with the biased library and after six rounds of panning with the X_9 library were very similar. Both libraries selected the sequence P-P-G-Φ (Φ = hydrophobic amino acid) as a core motif in most of the sequenced clones. Several natural proteins contain this sequence motif, and further experiments will show whether these proteins interact with the CD2BP2-GYF domain in vivo. In conclusion, a combined approach that uses the substitution analysis of known physiological peptide ligands and phage display-based searches for novel ligands are most likely to reveal functionally important interactions of CD2BP2 and other GYF domain-containing proteins.

5.3.4
Sequence Repetition in GYF Domain-mediated Interactions

Sequence repetition is an efficient means of enhancing the overall affinity of the GYF domain–CD2 interaction [13], and it is conceivable that local concentration enhancement is a general strategy by which physiological binding partners attract GYF domain-containing proteins. The CD2BP2-GYF domain–CD2 interaction was analyzed by systematically shortening the length of the peptide HPPPPPGHRSQAP-SHRPPPPGHRV (Figure 5.4). Although this analysis corroborates the central role of the PPPGHR motif for binding, it also shows that the repetitive longer peptide sequence binds with higher overall affinity than the short single motif. A minimum of three to four amino acids for the linker peptide connecting the two binding motifs seems to be sufficient for higher-affinity binding (right panel in Figure 5.4).

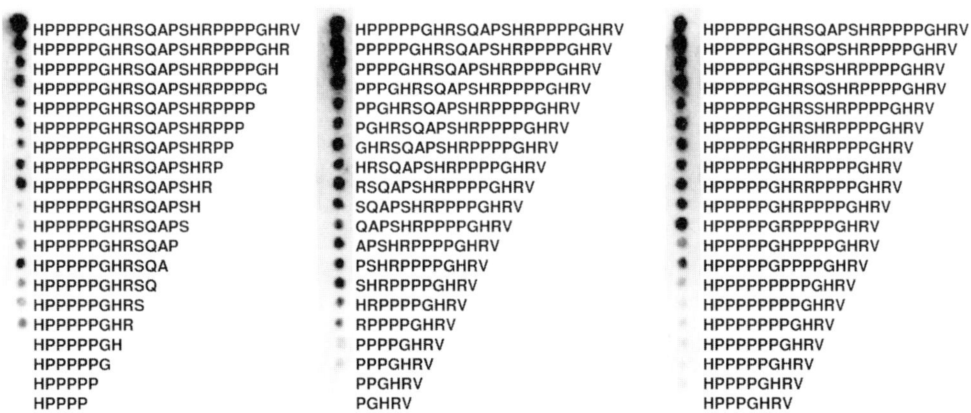

Figure 5.4 Peptide spot analysis of the CD2 sequence HPPPPPGHRSQAPSHRPPPPGHRV and truncation derivatives of this sequence. A GST fusion protein of the CD2BP2-GYF domain was used in this analysis, and experimental conditions were as described for Figure 5.3.

Shorter linker sequences most likely interfere with the proposed binding of two GYF domains to a double motif-containing peptide sequence [13]. In this respect, the GYF domain-containing proteins often contain coiled-coil regions within the same protein. Since coiled-coil regions tend to dimerize or oligomerize, it will be interesting to see whether these regions lead to greater avidity enhancement of GYF domain-mediated interactions. For database search profiles, the importance of motif repetition implies that suboptimal sequences present in multiple copies within a given protein might be more efficient in recruiting GYF domain-containing proteins than a protein with a single, more optimal binding sequence.

5.3.5
Functional Relevance of the CD2BP2-GYF Domain Interaction with CD2

5.3.5.1 Competitive Binding of CD2BP2-GYF and Fyn-SH3 to the CD2 Tail in Vitro

The CD2BP2 protein was identified as one of the interaction partners of the cell adhesion molecule CD2 in T cells. CD2 contains five proline-rich stretches in its cytoplasmic domain, which are responsible for the interaction with various intracellular proteins. The protein CD2AP [17] plays an important role in T-cell polarization and links CD2 to the actin cytoskeleton, and the CD2BP1 protein has been implicated in adhesion [18]. The CD2BP2 protein and the src kinase Fyn were both independently characterized as proteins that modulate the interleukin-2 response of CD2-stimulated T cells [1, 19]. Previous experiments, which used a truncated CD2 tail variant, identified the tandem PPPPGHR-x_7-PPPPGHR amino acid sequence in the CD2 cytoplasmic domain, which is well conserved among species, as the element responsible for the CD2-mediated production of interleukin-2 [2]. Thus, both CD2BP2 and Fyn might interact with PPPPGHR motifs. Indeed, in NMR titration experiments, where either an ^{15}N-labeled CD2BP2-GYF or an ^{15}N-labeled Fyn-SH3 domain were mixed with increasing amounts of a PPPPGHR-x_7-PPPPGHR containing CD2 peptide, the interaction site could be mapped. CD2BP2-GYF and Fyn-SH3 bind to the single PPPPGHR motif with K_d values of 190 μM and ~300 μM, respectively. The apparent affinity for the tandem motif PPPPGHR-x_7-PPPPGHR is 5–10 fold higher, which highlights the importance of sequence repetition for enhancement of the overall affinity of proline-rich sequence recognition [13].

NMR competition experiments further increased our understanding of whether the CD2BP2-GYF and Fyn-SH3 domains can compete for the same binding sequence in the context of the entire CD2 cytoplasmic domain. First, a ^{15}N–^{1}H correlation spectrum of isolated ^{15}N-labeled GYF domain (at 0.2 mM) was recorded; a region of this spectrum is shown in Figure 5.5 (black resonances). Subsequently, a substoichiometric amount of unlabeled CD2 cytoplasmic domain (0.08 mM) was added. The ^{15}N–^{1}H correlation spectrum shows a change in chemical shift for the resonances of the binding site residues (blue, Figure 5.5). In a final step, unlabeled Fyn-SH3 domain (0.4 mM) was added in two-fold excess over the protein. The corresponding spectrum (red, Figure 5.5) demonstrates that the N–H resonances of binding site residues shift in the direction of the unbound GYF domain.

The fraction of CD2-bound GYF domain is significantly decreased in the presence of the Fyn-SH3 domain. Thus, under limiting CD2 concentrations, there is competition between the two domains for the same binding site within CD2 [13].

5.3.5.2 In Vivo Compartmentalization of CD2 Binding Proteins

The functional relevance of the binding of CD2BP2 and Fyn to CD2 was elucidated by coimmunoprecipitation experiments [13]. A fraction of CD2 is inducibly recruited to lipid rafts upon CD2 cross-linking [20], and CD2BP2 is exclusively located in the detergent-soluble cellular fraction, whereas Fyn resides exclusively in the raft fraction [21, 22]. Neither CD2BP2 nor Fyn protein localization is affected by CD2 cross-linking, revealing that CD2BP2 and Fyn are localized in distinct cellular compartments in vivo, whereas CD2 shuttles inducibly from non-raft to the raft membrane compartment. Moreover, anti-Fyn monoclonal antibodies specifically precipitate CD2 in the raft fractions of CD2-stimulated cells, but there is little if any CD2 coprecipitated from the raft fraction of unstimulated T cells. However, CD2BP2 remains in the detergent-soluble fraction, where it associates with CD2 in T cells [1]. As such, CD2BP2 might help prevent the translocation of CD2 into lipid rafts before CD58 ligation of the CD2 ectodomains. Augmentation of IL-2 production (150%–200%) in Jurkat T cells transfected with a CD2BP2 GYF domain-only construct is in agreement with the following hypothesis: cellular expression of the isolated GYF domain is likely to release CD2 from CD2BP2 and CD2BP2-associated proteins, thereby fostering raft translocation of CD2 and subsequent signal-transduction events triggered by the activation of Fyn kinase [23].

5.3.6 Other GYF Domain–Containing Proteins

Recently, three other GYF domain-containing proteins have been described. In the yeast *Saccharomyces cerevisiae*, proteins interacting with Lin1p were identified in a yeast two-hybrid screen [24]. Lin1p contains a C-terminal GYF domain that is proposed to interact with the essential splicing factor Prp8. Prp8 is part of the U5 snRNP spliceosome component and interacts closely with the GU dinucleotide at the 5′ splice site. The N terminus of Prp8 contains a number of PPG(F/Y/L) motifs that closely resemble the consensus motif found in the phage-display screen of CD2BP2-GYF. Sequence comparison shows that Lin1p belongs to the CD2BP2-GYF domain subclass, which contains a tryptophan at position 8 (Figure 5.1). It is therefore plausible to assume that the Lin1p GYF domain binds to sequences with a similar profile as the CD2BP2-GYF domain. Since Lin1p also interacts with chromosomal remodeling factors, it constitutes a possible link between mRNA splicing and chromosome segregation.

In mouse, the proteins GIGYF1 and GIGYF2 (for Grb10-interacting GYF protein 1 and 2) have recently been identified [25]. Both proteins interact with a proline-rich stretch within the Grb10 protein. Grb10 is an adapter protein that binds to the intracellular domains of activated tyrosine kinase receptors, including insulin-like

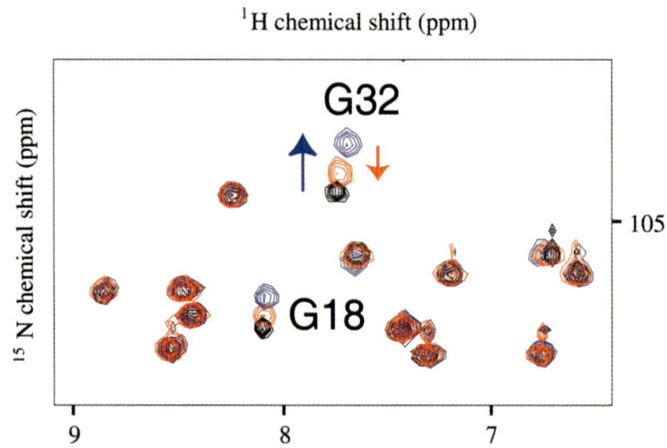

Figure 5.5 $^{15}N–^{1}H$ NMR correlation spectra of ^{15}N-labeled CD2BP2 GYF domain and its binding to the CD2 cytoplasmic domain in the absence and presence of the SH3 domain of Fyn tyrosine kinase. The resonances of the isolated GYF domain are shown in black, the resonances of the GYF domain in the presence of 0.1 mM CD2 cytoplasmic tail are blue, and the GYF domain signals after addition of 0.4 mM Fyn-SH3 domain to the GYF-peptide mixture are red. The resonances of G18 and G32 are shifted upon addition of the peptide, and this shift is reversed upon addition of the Fyn-SH3 domain.

growth factor 1 receptor (IGF-I receptor). In mouse fibroblasts expressing the IGF-I receptor, the GIGYF1 protein binds transiently to the receptor tail via Grb10. Overexpression of a GYF domain-containing fragment of GIGYF1 in fibroblast cells significantly increases IGF-I–stimulated receptor tyrosine phosphorylation. Three proline-rich sequence motifs within the Grb10 protein were identified as putative GYF-domain binding sites. Deletion analysis showed that excision of any two of the three proline motifs is necessary and sufficient to result in complete loss of binding [25].

The yeast protein Smy2 contains a GYF domain and has been cloned as a suppressor of MyoD, a cytoskeletal motor protein [26]. In a yeast two-hybrid screen, Smy2 interacts with MUD2 and MSL5, two proteins involved in pre-mRNA splicing [27]. In addition to an RNA binding domain, MSL5 contains a number of proline-rich sequences that are potential GYF-domain interaction sites. A third GYF domain-containing protein in *S. cerevisiae*, YPL105C, is sequence-related to Smy2, and its highly homologous GYF domain is most likely to bind to very similar target sequences as the Smy2 GYF domain. The functional consequence of this redundancy is not known, but these findings point to a role of GYF domain-containing proteins in splicing or in processes that are functionally coupled to splicing, such as pre-mRNA retention or transcription [28, 29].

5.4 Emerging Research Directions and Recent Developments

Compared with our understanding of other proline-rich recognition domains, little is known about the GYF domain. Although the requirements for ligand binding were established for SH3, WW, and EVH1 domains in the past 10 years (reviewed in [30]), deciphering the recognition code for GYF domains remains an important task. Extensive knowledge of ligand binding requirements is available only for the CD2BP2-GYF domain, and investigating the sequence space of ligands for other GYF domains will pave the way for the identification of cognate interaction partners of GYF domain-containing proteins. I expect that a single GYF domain will prove to be able to bind to a variety of naturally occurring target sequences in vitro, and it will therefore be essential to establish functionally relevant interactions in vivo. Since results obtained so far suggest that sequence repetition is an effective means of enhancing the overall affinity of an interaction, it is likely that multiple binding motifs within natural targets will be preferred. A further important question relates to the finding that the repertoires of sequences recognized by the different proline-rich recognition domains overlap. It has been demonstrated that SH3 domains and WW domains can compete for the same sequences in vitro [31], and similarly, GYF and SH3 domains bind to the same sequence motif in CD2 [13]. However, the results for CD2 show that the localization of the potential binding partners is critical for the interaction to occur. Future experiments that allow the identification of interacting proteins under more physiological conditions are thus required to show whether cooperativity and competition among proline-rich sequence recognition modules occur within the living cell and whether they play a role in biological processes. Since most of these recognition domains dissociate from their respective binding partners with high off-rates, it is probably more appropriate to describe the various scenarios of proline-rich sequence recognition within a statistical framework rather than assuming a simple switch function that alters binding. However, the latter mechanism may occur if regulatory mechanisms such as phosphorylation or compartmentalization alter the accessibility of a binding domain's surface. There is increasing evidence that autoinhibition may be a major mechanism for restricting a protein's binding potential unless a proper stimulus is obtained [32]. Studying the function of GYF and other adapter domains within the context of the entire protein will therefore be necessary in order to discern whether the binding behavior of the isolated domain varies from the binding properties of the domain within the full-length protein.

5.5 Concluding Remarks

In contrast to the findings regarding SH3 and WW domains, the occurrence of GYF domains in eukaryotic proteins has remained relatively constant during evolution, suggesting a conserved role of GYF domain-containing proteins in a yet-to-be-identified biological process. The reason for the variability of the SH3 and WW domains in comparison to the GYF domain might reside in the properties of the scaffold. SH3 and WW domains are β-sheet structures with extended loop regions of variable sequence that are responsible for the specificity of the respective domain. The GYF domain ligand binding site converges on a single hydrophobic hot spot defined by the relatively rigid packing of aromatic residues from a bulge-helix-bulge motif and a small β sheet. The contribution of specificity-determining sites to the free energy of binding might therefore be small. In addition, the affinity of an interaction defined by a single surface pocket of the domain may restrict the strength of possible interactions in comparison with SH3 domains, which use several surface pockets for ligand binding. However, since structural and functional data of GYF domains are scarce, it is well possible that the GYF domain scaffold is more ubiquitously present than suggested by sequence-based databank annotations. Many GYF domains do not strictly obey the conserved signature defined in the initial studies, and it is conceivable that more remotely related sequences exist that do not bind to consensus ligands but yet are built on the conserved features of the scaffold. Certainly, biophysical investigations of the structure and stability of the domain, biochemical studies of specificity and intramolecular regulation, and cell biological experiments defining the location and physiological binding partners have to complement each other to elucidate the exact role of GYF domain-containing proteins in eukaryotic species.

Acknowledgements

I am indebted to Ellis Reinherz and Gerhard Wagner for their continuous interest and support in the elucidation of the structure of the GYF domain of CD2BP2. I am grateful to Michael Kofler in my laboratory for stimulating discussions and his ongoing enthusiasm for GYF domain structure and function throughout his PhD thesis work, and to Katja Heuer for carefully reading the manuscript. I also wish to thank Rudolf Volkmer-Engert and Angelika Ehrlich for membrane spot synthesis. Part of the work presented in this chapter was made possible by grants from the BMBF (0311879), the Deutsche Forschungsgemeinschaft (DFG: FR1325/2-1) and the Volkswagen Stiftung (1/77 955).

References

1 Nishizawa, K., Freund, C., Li, J., Wagner, G., Reinherz, E. L., Identification of a proline-binding motif regulating CD2-triggered T lymphocyte activation. *Proc. Natl. Acad. Sci. USA* **1998**, *95*, 14897–14902.

2 Freund, C., Dötsch, V., Nishizawa, K., Reinherz, E. L., Wagner, G., The GYF domain is a novel structural fold that is involved in lymphoid signaling through proline-rich sequences. *Nat. Struct. Biol.* **1999**, *6*, 656–660.

3 Chang, H. C., Moingeon, P., Pedersen, R., Lucich, J., Stebbins, C., Reinherz, E. L., Involvement of the PPPGHR motif in T cell activation via CD2. *J. Exp. Med.* **1990**, *172*, 351–355.

4 Carlsson, L., Nystrom, L. E., Sundkvist, I., Markey, F., Lindberg, U., Actin polymerizability is influenced by profilin, a low molecular weight protein in non-muscle cells. *J. Mol. Biol.* **1977**, *115*, 465–483.

5 Mayer, B. J., Hamaguchi, M., Hanafusa, H., A novel viral oncogene with structural similarity to phospholipase C. *Nature* **1988**, *332*, 272–275.

6 Stahl, M. L., Ferenz, C. R., Kelleher, K. L., Kriz, R. W., Knopf, J. L., Sequence similarity of phospholipase C with the non-catalytic region of src. *Nature* **1988**, *332*, 269–272.

7 Bork, P., Sudol, M., The WW domain: a signalling site in dystrophin? *Trends Biochem. Sci.* **1994**, *19*, 531–533.

8 Niebuhr, K., et al., A novel proline-rich motif present in ActA of *Listeria monocytogenes* and cytoskeletal proteins is the ligand for the EVH1 domain, a protein module present in the Ena/VASP family. *EMBO J.* **1997**, *16*, 5433–5444.

9 Sancho, E., et al., Role of UEV-1, an inactive variant of the E2 ubiquitin-conjugating enzymes, in in vitro differentiation and cell cycle behavior of HT-29-M6 intestinal mucosecretory cells. *Mol. Cell Biol.* **1998**, *18*, 576–589.

10 Pornillos, O., Alam, S. L., Davis, D. R., Sundquist, W. I., Structure of the Tsg101 UEV domain in complex with the PTAP motif of the HIV-1 p6 protein. *Nat. Struct. Biol.* **2002**, *9*, 812–817.

11 Myllyharju, J., Kivirikko, K. I., Identification of a novel proline-rich peptide-binding domain in prolyl 4-hydroxylase. *EMBO J.* **1999**, *18*, 306–312.

12 Schultz, J., Milpetz, F., Bork, P., Ponting, C. P., SMART, a simple modular architecture research tool: identification of signaling domains. *Proc. Natl. Acad. Sci. USA* **1998**, *95*, 5857–5864.

13 Freund, C., Kühne, R., Yang, H., Park, S., Reinherz, E. L., Wagner, G., Dynamic interaction of CD2 with the GYF and the SH3 domain of compartmentalized effector molecules. *EMBO J.* **2002**, *21*, 5985–5995.

14 Kofler, M. M., Heuer, K., Zech, T., Freund, C., Recognition sequences for the GYF domain reveal a possible spliceo somal function for CD2BP2. *J. Biol. Chem.* **2004**, *279*, 28292–28297.

15 Hartmuth, K., et al., Protein composition of human prespliceosomes isolated by a tobramycin affinity-selection method. *Proc. Natl. Acad. Sci. USA* **2002**, *99*, 16719–16724.

16 Felici, F., Castagnoli, L., Musacchio, A., Jappelli, R., Cesareni, G., Selection of antibody ligands from a large library of oligopeptides expressed on a multivalent exposition vector. *J. Mol. Biol.* **1991**, *222*, 301–310.

17 Dustin, M. L., et al., A novel adaptor protein orchestrates receptor patterning and cytoskeletal polarity in T-cell contacts. *Cell* **1998**, *94*, 667–677.

18 Li, J., et al., A cdc15-like adaptor protein (CD2BP1) interacts with the CD2 cytoplasmic domain and regulates CD2-triggered adhesion. *EMBO J.* **1998**, *17*, 7320–7336.

19 Lin, H., Hutchcroft, J. E., Andoniou, C. E., Kamoun, M., Band, H., Bierer, B. E., Association of p59(fyn) with the T lymphocyte costimulatory receptor CD2: binding of the Fyn Src homology (SH) 3 domain is regulated by the Fyn SH2 domain. *J. Biol. Chem.* **1998**, *273*, 19914–19921.

20 Yang, H., Reinherz, E. L., Dynamic recruitment of human CD2 into lipid rafts: linkage to T cell signal trans-

duction. *J. Biol. Chem.* **2001**, *276*, 18775–18785.

21 Alland, L., Peseckis, S. M., Atherton, R. E., Berthiaume, L., Resh, M. D., Dual myristylation and palmitylation of Src family member p59fyn affects subcellular localization. *J. Biol. Chem.* **1994**, *269*, 16701–16705.

22 Shenoy-Scaria, A. M., Dietzen, D. J., Kwong, J., Link, D. C., Lublin, D. M., Cysteine3 of Src family protein tyrosine kinase determines palmitoylation and localization in caveolae. *J. Cell Biol.* **1994**, *126*, 353–363.

23 Fukai, I., Hussey, R. E., Sunder-Plassmann, R., Reinherz, E. L., A critical role for p59(fyn) in CD2-based signal transduction. *Eur. J. Immunol.* **2000**, *30*, 3507–3515.

24 Bialkowska, A., Kurlandzka, A., Proteins interacting with Lin 1p, a putative link between chromosome segregation, mRNA splicing and DNA replication in *Saccharomyces cerevisiae*. *Yeast* **2002**, *19*, 1323–1333.

25 Giovannone, B., Lee, E., Laviola, L., Giorgino, F., Cleveland, K. A., Smith, R. J., Two novel proteins that are linked to insulin-like growth factor (IGF-I) receptors by the Grb10 adapter and modulate IGF-I signaling. *J. Biol. Chem.* **2003**, *278*, 31564–31573.

26 Lillie, S. H., Brown, S. S., Suppression of a myosin defect by a kinesin-related gene. *Nature* **1992**, *356*, 358–361.

27 Abovich, N., Rosbash, M., Cross-intron bridging interactions in the yeast commitment complex are conserved in mammals. *Cell* **1997**, *89*, 403–412.

28 Legrain, P., Rosbash, M., Some cis- and trans-acting mutants for splicing target pre-mRNA to the cytoplasm. *Cell* **1989**, *57*, 573–583.

29 Fong, Y. W., Zhou, Q., Stimulatory effect of splicing factors on transcriptional elongation. *Nature* **2001**, *414*, 929–933.

30 Kay, B. K., Williamson, M. P., Sudol, M., The importance of being proline: the interaction of proline-rich motifs in signaling proteins with their cognate domains. *FASEB J.* **2000**, *14*, 231–241.

31 Chan, D. C., Bedford, M. T., Leder, P., Formin binding proteins bear WWP/WW domains that bind proline-rich peptides and functionally resemble SH3 domains. *EMBO J.* **1996**, *15*, 1045–1054.

32 Pufall, M. A., Graves, B. J., Autoinhibitory domains: modular effectors of cellular regulation. *Annu. Rev. Cell Dev. Biol.* **2002**, *18*, 421–462.

6
PTB Domains

Ben Margolis and Linton M. Traub

6.1
Introduction

The phosphotyrosine binding (PTB) domain also known as the phosphotyrosine interaction domain (PID) was the second domain (after the SH2 domain) found to bind phosphotyrosine-containing peptides. Over time, scientists have come to realize that this domain does more than mediate phosphotyrosine-related signaling, because its binding is not always dependent on the presence of phosphotyrosine. The domain was first identified in the amino terminus of the Shc protein [1–3], where it was found to bind to an Asp-Pro-*any amino acid*-pTyr (NPx*po*Y) motif found in many activated growth factor receptors. Simultaneously, a binding domain on insulin receptor substrate 1 (IRS1) and insulin receptor substrate 2 (IRS2) was identified that also bound the NPx*po*Y motif [2, 4]. Although the NPx*po*Y binding regions in the IRS and Shc proteins do not have significant primary sequence homology, they have similar binding specificity and 3D structure. PTB domains with sequence homology to those found in IRS proteins are referred to as PTBI for PTB-domain-IRS-like to differentiate them from PTB domains that have sequence homology with the Shc PTB domain (Simple Modular Architecture Research Tool; http://smart.embl-heidelberg.de). A relatively small number of proteins have been identified that contain PTBI domains, and these primarily function in tyrosine kinase signal transduction. Many more PTB domains with sequence similarity to the Shc PTB domain have been identified [5], and they have more diverse cellular functions. Evolutionarily, PTB domains are not found in yeast or plants but appear in *Drosophila* and *Caenorhabditis elegans* genomes. This chapter discusses the function of proteins containing PTB and PTBI domains and examines the structural basis of their interactions with peptide and phospholipid ligands.

Modular Protein Domains. Edited by G. Cesareni, M. Gimona, M. Sudol, M. Yaffe

ISBN: 3-527-30813-X

6.2
Function of PTB Domain Proteins

Proteins with PTB and PTBI domains are involved in multiple cellular functions. The following sections discuss their role in tyrosine kinase-dependent and -independent signaling, protein trafficking, and cell adhesion. A schematic representation of several PTB and PTBI domain proteins is displayed in Figure 6.1.

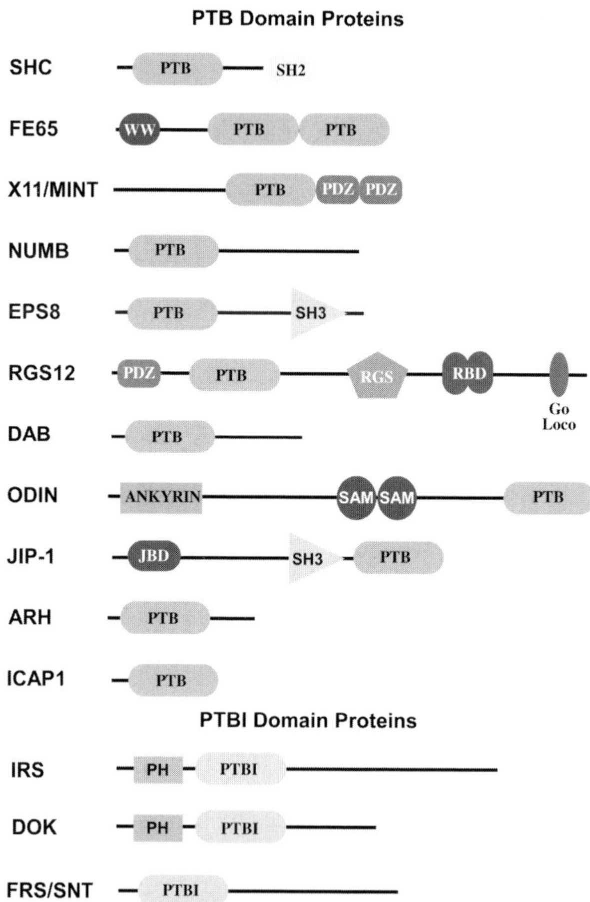

Figure 6.1 Domain architecture of several representative PTB domain-containing proteins and other modular domains within the PTB and PTBI protein families. Many of the other modular domains are discussed in other chapters in this volume. More details on those not covered in this book can be found at one of the numerous domain databases such as SMART (http://smart.embl-heidelberg.de). PTB = phosphotyrosine binding domain; PTBI = phosphotyrosine binding domain IRS-like; SH2 = Src homology 2 domain; WW = domain with two conserved tryptophans; PDZ = postsynaptic density 95/discs large/zona occludens-1 domain; SH3 = Src homology 3 domain; RGS = regulator of G protein signaling domain, RBD = Raf-like Ras-binding domain; Go Loco = $G\alpha_{I/O}$-Loco motif; SAM : sterile alpha motif; JBD = JNK binding domain; PH = pleckstrin homology domain.

6.2.1 Role of PTB Domain Proteins in Tyrosine Kinase Signaling

6.2.1.1 Shc

As described above, PTB domains are known to play an important role in tyrosine kinase signal transduction. This of course relates to the initial identification of the PTB domain in the tyrosine kinase substrate, ShcA, and has been reviewed previously [6]. The classical role of the Shc PTB domain is to allow Shc to bind to growth factor receptors, facilitating Shc phosphorylation [7, 8]. Once Shc is phosphorylated, it can bind to other downstream signal-transduction molecules. For example, the nerve growth factor receptor, TrkA, cannot bind efficiently to Grb2 and its binding partner, the Ras guanyl nucleotide exchange factor, Son of Sevenless (Sos). Grb2–Sos is the primary protein complex used by growth factor receptors to activate the Ras G protein. However, TrkA does have an NPxY motif that, once phosphorylated, can bind the PTB domain of Shc [8]. Once TrkA binds Shc, it mediates Shc tyrosine phosphorylation, allowing it to bind to the Grb2–Sos complex and activate Ras. Many growth factor receptors are known to utilize Shc and its paralogs as a crucial intermediary in signal transduction [9]. One of the better examples for the role of the PTB domain is found in *Drosophila* Shc [10]. In a genetic screen, Nüsslein-Volhard and coworkers identified Shc as a maternal gene important for embryonic development. They found that Shc signaled downstream of the *Drosophila* EGF and TOR tyrosine kinase receptors. Interestingly, one Shc mutant isolated had a mutation in the PTB domain and functioned similar to a null allele, confirming the crucial role of the PTB domain in Shc function.

A second aspect to the function of the Shc PTB domain and other PTB domains is the ability to bind to phospholipids [11]. The PTB domain is highly related in structure to the pleckstrin homology (PH) domain, a domain known to bind to phospholipids [12]; indeed, the protein binding capabilities of the PTB domain may have evolved from its initial ability to bind phospholipids as a PH domain [13]. The lipid binding region of the PTB domain is distinct from the peptide binding region, indicating that the PTB domain can bind both lipids and proteins simultaneously. This can assist the targeting of Shc to the membrane while binding to tyrosine-phosphorylated receptors. It has been shown that specific mutations within the PTB domain that disrupt phospholipid but not phosphotyrosine binding impair Shc tyrosine phosphorylation by growth factor receptors [11]. A unique aspect to Shc is that it also contains an SH2 domain in addition to a PTB domain (Figure 6.1). This organization is conserved throughout evolution, suggesting some important role for the SH2 domain. Yet there are very few examples of the SH2 domain rather than the PTB domain coupling Shc to tyrosine phosphorylated proteins. This suggests that the Shc SH2 domain may have an alternative role in Shc function, but the exact nature of this is unclear [9].

6.2.1.2 Proteins with PTBI Domains

Proteins with PTBI domains play a prominent role in signaling by receptor tyrosine kinases. As discussed previously, PTBI domains have a 3D structure similar to that of PTB domains, but there is limited primary sequence similarity between these domains. The classic family of proteins having this domain is the insulin receptor substrates (IRS). It had been known for several years that the insulin receptor had an NPxY motif centered on Tyr960 that was phosphorylated. It was also known that this motif was necessary for insulin-mediated signal transduction as well as phosphorylation of a 185-kDa substrate protein [14]. This substrate protein was later identified as IRS-1 [15] and shown to have a domain that can interact with the insulin receptor [16]. A similar domain was also identified in the related IRS-2 protein [4], and this domain is now referred to as the PTBI domain.

The IRS family of proteins functions as adapters similar to Shc. However, although both Shc PTB and the IRS PTBI domains bind NPx*po*Y motifs, there are differences in the sequences surrounding the NPx*po*Y motif that are required for high-affinity binding [17]. IRS proteins contain multiple tyrosine-phosphorylation sites and can mediate multiple downstream signaling pathways. A complete discussion of IRS signaling is beyond the scope of this review, but it has been described in other publications [18]. However, a few points specifically related to the PTBI domains of these proteins can be noted. One of the surprises is that the role of PTBI in signaling by IRS proteins has been difficult to demonstrate. The amino terminus of IRS proteins has one PTBI domain that sits just carboxy-terminal to the PH domain (Figure 6.1). As mentioned previously, a primary function of PH domains is to bind phospholipid headgroups. In IRS-1, the PH domain is necessary for IRS tyrosine phosphorylation after insulin stimulation [19]. The PTB domain seems to be essential for high-affinity interactions between IRS-1 and insulin receptor, but the actual necessity for this interaction in insulin signaling has been difficult to demonstrate [19]. The picture becomes more confusing when one considers that the PTB domain of IRS proteins can bind phospholipids [20] and that the PH domain of IRS-1 also has protein binding partners [21]. The most reasonable explanation for the function of these domains comes from structural studies, which suggest that cooperation between PTB and PH domains mediates a high-affinity interaction between insulin receptor and IRS-1 [22]. Many of the studies using mutagenesis of these domains relied on overexpression and thus may miss the cooperative nature of domain interactions when the proteins are expressed at physiologic levels.

Two other families of proteins with PTBI domains are also important in growth factor receptor signal transduction. One family is referred to as the fibroblast growth factor receptor substrate (FRS) family [23]. These proteins are also called Suc-associated neurotrophic factor-induced tyrosine-phosphorylated target (SNT). This protein family was first identified by its binding to p13suc1 agarose, an affinity reagent that binds to certain cyclin-dependent kinases [24]. This suggested that this tyrosine-phosphorylated substrate of fibroblast growth factor (FGF) and Trk receptors might directly control the cell cycle, but later purification indicated it was a large scaffold protein that bound to growth factor receptors at the cell surface [23]. Proteins in this family have a single PTBI domain at their amino terminus that mediates interactions with FGF and Trk receptors, as well as other growth factor receptors [23, 25, 26]. Like Shc and IRS proteins, the FRS proteins become tyrosine-phosphorylated after association with growth factor receptors and then recruit signaling molecules onto the tyrosine phosphorylation sites [27, 28]. The binding specificity of the FRS PTBI domain is unique. On TrkA it binds to phosphorylated Y490, a classic NPx*po*Y site, and can compete with Shc for binding to this site [29]. In contrast, the binding to FGF receptor appears to be constitutive and not dependent on receptor phosphorylation [25, 27]. The peptide from the juxtamembrane domain of the FGF receptor binds to the FRS PTBI domain in a fashion completely different from that seen with the NPx*po*Y motif from the TrkA receptor [30] and represents a novel peptide–PTB domain interaction. One trend seen in previously discussed PTB domain proteins is also seen with the FRS family of proteins: both Shc and IRS proteins have the ability to bind membranes and protein peptide motifs either with their PTB domain alone or with the PTB domain in combination with another domain. The FRS proteins address this problem, not with an additional domain, but by adding a myristate group at the amino terminus that links them to membranes [23].

The other family of PTBI domain proteins is the DOK (downstream of kinase) proteins (Figure 6.1). The first DOK family member was identified as a 62-kDa tyrosine-phosphorylated protein that binds to the p120 RasGAP protein and contains an amino-terminal PH and PTB domain [31, 32]. Since then, an additional four members of the DOK family have been identified [33–40]. The PTB domain of DOK proteins can bind to the canonical NPx*po*Y motifs and to highly related motifs present in EGF receptor [41] and other tyrosine-phosphorylated proteins [38, 42]. However, there also appears to be unique phosphotyrosine-containing motifs that can bind these PTB domains, such as those found in Tie1 receptor [43]. As in IRS-1, there appears to be a need for cooperation between the PH and PTB domains to mediate the growth factor-induced tyrosine phosphorylation of DOK proteins [41]. However, in contrast to IRS and FRS family members, the primary role of DOK proteins focuses on the inhibition of growth-promoting signals. This is particularly evident for DOK signaling in cells of the hematopoietic lineage, where DOK family members inhibit signaling from tyrosine kinases to the MAP kinase–Erk pathway [36, 44]. However, there have also been reports of positive roles for DOK proteins in signal transduction pathways [38] and cell migration [45].

6.2.1.3 Additional PTB Domain Proteins Involved in Tyrosine Kinase Signaling

Other proteins with PTB domains appear to play a role in tyrosine kinase signaling, although the role of the PTB domain in these processes is not as clear. The EPS8 protein was first identified as a protein that became tyrosine-phosphorylated after EGF stimulation [46]. EPS8 appears to bind to the EGF receptor, but surprisingly, the PTB domain does not appear to be involved in this process [47]. EPS8 protein also contains an SH3 domain (Figure 6.1). An important insight came when it was determined that EPS8 can localize to sites of actin polymerization [48]. When EPS8 is knocked out in mice, the mice appear to develop normally, but there are defects in actin polymerization in response to growth factor receptors in knockout fibroblasts [49]. Based on several studies, it was concluded that EPS8 is in a complex with several proteins, including Sos, and mediates Rac activation after growth factor receptor stimulation in a PI-3 kinase-dependent fashion [50, 51]. EPS8 may also play a role in the endocytosis of growth factor receptors [52].

Another PTB domain known to be phosphorylated after EGF stimulation is Odin [53]. Odin also has SAM domains as well as multiple ankyrin repeats (Figure 6.1). It appears in the protein database in multiple isoforms with and without ankyrin repeats. The function of the PTB domain in this protein is unclear, because its removal does not affect Odin tyrosine phosphorylation, and it is not certain that this protein even associates with growth factor receptors [53]. A highly related gene called EB-1 was identified as a gene that is induced in hematopoietic cells transformed by the E2A-PBX oncoprotein, but the function of EB-1, as well as of Odin, is unknown [54]. Regulator of G protein signaling 12 (RGS12) is another signaling molecule that contains a PTB domain and is involved in tyrosine kinase signaling (Figure 6.1). RGS proteins function as GTPase activating proteins for the α subunits of heterotrimeric G proteins [55]. In RGS12, it has been reported that the PTB domain of this protein can bind to tyrosine-phosphorylated calcium channels [56]. The binding of the PTB domain to the calcium channel may modulate calcium channel activity as well as regulate the heterotrimeric G proteins that control the channel.

6.2.2 PTB Domain Proteins That Function Independent of Phosphotyrosine

The above signaling pathways involve the binding of PTB and PTBI domains to tyrosine-phosphorylated proteins. Indeed, these were the early signaling pathways that defined the function of PTB and PTBI domains. However, soon after the identification of this domain, it was found that PTB domains can also bind NPxY motifs independent of tyrosine phosphorylation [57]. This led to an expanded appreciation of the role of PTB domains in signaling, cell adhesion, and protein processing and trafficking.

6.2.2.1 PTB Domain Proteins That Bind APP

The realization that PTB domains can bind nonphosphorylated NPxY motifs came with studies of amyloid precursor protein (APP). APP is of great interest because it is processed via proteases to amyloid beta, a protein deposited in the brains of Alzheimer disease patients. Evidence obtained over the past several years has strongly pointed to the deposition of amyloid beta as an important causative factor in Alzheimer's disease [58]. Despite the known role of APP processing in Alzheimer's disease, the physiological role of APP and highly related proteins is unclear. However, the binding of PTB domains to APP has provided unique insights into the normal function of this protein.

Initial studies pointed to the binding of two PTB domain proteins, FE65 and X11/Mint, to the NPxY motif of APP [57, 59, 60]. Later, several other PTB domain proteins, including Disabled (Dab), JIP-1, and Numb, were identified as proteins that can bind APP [61–66]. There are also reports of APP becoming tyrosine-phosphorylated and binding Shc via an NPx*po*Y motif, but the physiologic relevance of this interaction is uncertain [67, 68]. A large number of studies have examined the effect of PTB domain proteins on APP processing to amyloid beta [64, 69–72]. A weakness of all these studies is that they employ overexpression of the PTB domain proteins, and thus it is not clear that these proteins physiologically regulate APP processing normally in the brain. Nonetheless, a large amount of data has been obtained on the effects of X11/Mint proteins on the processing of APP. X11/Mint proteins have one PTB domain and two PDZ domains (Figure 6.1; see also Chapter 13). X11/Mint protein overexpression slows APP processing and reduces amyloid beta production, both in tissue culture and in vivo [69, 70, 73]. The mechanism of X11/Mint action in this regard is unclear; however, studies on the *C. elegans* homolog of X11/Mint proteins, known as Lin-10, are instructive [74, 75]. Lin-10 plays a role in proper trafficking of proteins in worm epithelia and brain, suggesting that X11/Mint might have a similar role in mammalian cells. The localization of X11/Mint in the Golgi fits with such a role for this protein [76, 77]. In addition, X11/Mint proteins also interact with Munc-18, a protein that modulates SNARE interactions and thus possibly vesicle fusion [78]. Munc-18 interaction also modulates the effects of X11/Mint on APP processing, adding to the hypothesis that altered trafficking of APP is at the core of X11 actions on APP processing [79, 80]. The processing of APP is known to occur in discrete intracellular compartments, and rerouting of APP in the cell by X11/Mint proteins could alter amyloid beta production [81].

A more exciting story has emerged on the interaction of the FE65 family of proteins with APP. FE65 family members have two PTB domains and an amino-terminal WW domain (Figure 6.1; see also Chapter 3). The effects of FE65 on APP processing have also been studied, and it appears that FE65 usually increases amyloid beta production, but the effects may be variable depending on the situation [71, 82]. FE65 also interacts with another transmembrane protein called lipoprotein receptor-related protein (LRP), a member of the lipoprotein receptor family [83]. The amino-terminal PTB domain of FE65 interacts with LRP while the carboxy-terminal PTB binds to APP, allowing FE65 to act as a bridge linking LRP to APP [83, 84]. This is of significance, because a large body of literature suggests that LRP can control APP processing, and FE65 is an important link between these two proteins [85, 86].

Further studies revealed that FE65 is a nuclear protein and that its localization to the nucleus is restricted by binding to APP [87]. A breakthrough study by Cao and Sudhof [88] then demonstrated that processing of APP leads to translocation of FE65 to the nucleus bound to the cleaved intracellular domain of APP. It was demonstrated that this complex of FE65 bound to the intracellular domain of APP can affect gene transcription by binding to the Tip60 protein, a histone acetyltransferase [88]. This pathway can be modified by the interaction of APP with X11/Mint [89] and by the interaction of FE65 with LRP [90]. The concept of FE65 and APP as a unit that can control transcriptional activity has been confirmed by several other groups [91, 92]. These findings suggest a signal-transduction pathway in which the processing of APP and translocation of its intracellular domain to the nucleus leads to control of gene expression. Further work will be necessary to define what mechanisms are utilized to control the processing of APP and the nature of the genes induced. Other roles have also been invoked for FE65, including the control of actin bundling and cell migration [93], but these findings still await confirmation.

APP has also been reported to interact with the PTB-domain proteins JNK interacting protein-1 (JIP-1) and JNK interacting protein-2 (JIP-2). JIPs were first identified as scaffolding proteins that can regulate the MAP kinase pathway that leads to JNK activation [94, 95]. There are several theories as to how JIP might control JNK activity, and it may have both inhibitory and activating effects on JNK activation [94–96]. JIP-1 and JIP-2 contain a PTB domain that can interact with APP [63, 64] and a carboxy terminus that can interact with the light chain of conventional kinesin, kinesin-1 [97]. In this fashion, APP, a transmembrane protein, can connect transport vesicles to a kinesin motor and assist vesicle movement along microtubules. Thus, another physiological role for APP, in addition to control of transcription, may involve vesicle trafficking along microtubules in neurons. Indeed, several reports have linked APP to kinesins, both directly and indirectly [98, 99]. The exact role of JIPs, and more specifically, the associated MAP pathway kinases in control of kinesin transport is an area of active investigation. It is interesting to note that another APP binding partner, X11/MINT, has also been implicated as binding to kinesins [100], indicating that APP may have a general role in coupling transport vesicles to kinesin motors. Thus, in conclusion, several PTB-containing proteins interact with the NPxY motifs of APP and give important clues to the physiological role of APP and its paralogs in transcription control and vesicle trafficking. The studies may also provide important additional clues to the pathogenesis of Alzheimer's disease.

6.2.2.2 **PTB Domain Proteins That Bind Integrins**

Another family of transmembrane proteins known to contain NPxY motifs are beta integrins, which serve to link the intracellular actin cytoskeleton with the extracellular matrix [101]. NPxY motifs are found in multiple copies in several of the beta integrins' cytoplasmic tails [102, 103]. Recent studies have shown that several PTB-domain proteins can interact with integrins [103]. However, the best data are found for the interaction of integrin NPxY motifs with two proteins, integrins' cytoplasmic domain associated protein-1 (ICAP1) [104, 105] and talin [106]. ICAP1 was first identified by yeast two-hybrid screening using the intracellular domain of beta integrins as bait [104, 105]. ICAP1α is a small phosphoprotein of 200 amino acids (Figure 6.1), with a C-terminal PTB domain and an N terminus that is the site of threonine phosphorylation [104, 107]. Several roles have been ascribed to ICAP1. One set of studies found that ICAP1 can function as a guanyl dissociation inhibitor, sequestering the small G proteins, Cdc42 and Rac, in the cytoplasm and leading to altered actin dynamics [108]. In this regard it has been demonstrated that modulation of small G protein function can regulate ICAP phosphorylation [104]. ICAP may also alter cell adhesion and cell spreading by modulating the interactions of integrins with other members of the actin cytoskeleton [109].

One of the members of the actin cytoskeleton whose binding to integrins may be affected by ICAP1 is talin. Talin is a large protein (> 2500 amino acids) that interacts with integrins via its FERM domain [110]. FERM domains were originally identified in members of the Band 4.1 family of actin-binding proteins [111]. The FERM domain of talin interacts with the NPxY motifs in beta integrins [106]. It is of great interest that a crystal structure of this interaction reveals that this FERM domain interaction is very similar to PTB domain interactions and that a region of the FERM domain has a structure very similar to that of PTB and PH domains [112, 113]. Thus, some FERM-domain proteins may have interactions that mimic PTB domains and may greatly broaden the proteins that can interact with NPxY motifs.

6.2.2.3 PTB Domain Proteins That Control Endocytosis

Nearly a decade before the discovery of the PTB domain, the Goldstein and Brown laboratory [114] showed that a single point mutation (the JD mutation) in the gene encoding the low density lipoprotein (LDL) receptor can prevent receptor internalization. The LDL receptor governs the internalization of plasma LDL particles, primarily in the liver, and targets internalized LDL for lysosomal degradation. The JD mutation substitutes a tyrosine for a cysteine residue within the cytosolic domain of the receptor [114], causing stagnation of the receptor on the cell surface and leading to hypercholesterolemia and coronary artery disease. More complete mapping revealed that an FxNPxY internalization sequence is required to drive the efficient endocytic uptake of the LDL receptor and that this sequence is both autonomous and transplantable [102]. Most other members of the LDL receptor superfamily contain at least one FxNPxY motif within the cytosolic portion but, because the PTB domain was originally presumed to be pTyr-specific, a link between PTB domain proteins and endocytosis from the cell surface was not immediately appreciated. It is now known that several PTB domain proteins, including Disabled-1 (Dab1), Disabled-2 (Dab2), FE65, and JIP-1/2, can bind to these nonphosphorylated FxNPxY sequences directly [61, 83, 115–117].

The PTB domain–endocytosis connection became apparent when Dab2 was found to colocalize with clathrin at internalization sites on the cell surface [117]. The C-terminal region of Dab2, following the PTB domain, is essentially unstructured and contains multiple interaction motifs that bind physically to clathrin, the clathrin-associated AP-2 adaptor complex, and several other endocytic components [117–119]. Clathrin and AP-2 are the two principal components of a sorting coat that assembles on a class of endocytic transport vesicles termed clathrin-coated vesicles [120]. Like the signaling PTB-domain scaffolding proteins, Dab2 appears to mesh LDL receptor cargo with assembling coated vesicles by simultaneously binding to an NPxY internalization signal, PtdIns(4,5)P_2, and the clathrin coat machinery. In so doing, Dab2 expands the cargo selective capacity of the major sorting adaptor AP-2. Indeed, overexpression of a tandem Dab2 PTB domain prevents uptake of LDL but not of transferrin, because the transferrin receptor uses a distinct internalization sequence that binds directly to the AP-2 adaptor [118]. In mice, targeted disruption of the Dab2 gene is lethal, but conditional disruption only in the embryo leads to viable animals with a proteinuria similar to, albeit milder than, that seen in megalin nullizygous mice [121, 122]. A member of the LDL receptor superfamily, megalin is a scavenger receptor that plays a vital role in the recovery of protein and vitamins from the renal glomerular filtrate [122].

Although there is evidence for Dab2 participating in signal transduction [121, 123–125], the trafficking activity appears phylogenetically conserved. Ce-Dab-1, the single Disabled ortholog in *C. elegans*, acts together with two LDL receptor-related proteins (Ce-LRP-1 and Ce-LRP-2) in the proper biosynthetic delivery of Egl17, a fibroblast growth factor [126].

The recent identification of a gene responsible for a rare form of hypercholesterolemia characterized by a recessive pattern of inheritance has fortified the connection between PTB domains and endocytosis. Autosomal recessive hypercholesterolemia patients have a clinical phenotype remarkably similar to that of familial hypercholesterolemia patients, but have no mutations in the LDL receptor. Instead, the defective gene encodes a 308-amino-acid adaptor protein, termed autosomal recessive hypercholesterolemia (ARH), with an N-terminal PTB domain related to the Dab1/2 and Numb PTB domains (Figure 6.1) [127, 128]. ARH patient lymphoblasts exhibit defective LDL uptake despite normal levels of LDL receptor, but retroviral-mediated expression of exogenous ARH rescues LDL receptor activity in these cells [128]. The hypercholesterolemic phenotype of $ARH^{-/-}$ mice fed a high-fat diet and the prominent accumulation of the LDL receptor on the sinusoidal surface of hepatocytes in these animals also support a role for ARH in regulating the endocytic activity of the LDL receptor [129]. The ARH PTB domain interacts with FxNPxY sequences while the C-terminal region physically binds to AP-2 and clathrin [130–132]. In *Xenopus* oocytes, production of mutant ARH that cannot engage the endocytic machinery severely blunts the uptake of vitellogenin [132], a lipoprotein endocytosed by the vitellogenin receptor, a member of the LDL receptor superfamily with an FxNPxY internalization signal. Because ARH patients do not exhibit defects in LDL uptake in all tissues [133, 134] and the $Dab2^{-/-}$ phenotype is rather mild [121], it is possible that Dab2 and ARH are functionally redundant to some extent.

The function of Numb is perhaps best understood in terms of the development of external sensory organs in *Drosophila*. Sensory bristles on the fly head are composed of four different cell types that all originate from a single sensory organ precursor (SOP) cell by asymmetric cell division. This results in the formation of two nonequivalent daughter cells with different cell fates. Numb is a modular PTB domain protein (Figure 6.1) that antagonizes signaling by the transmembrane receptor Notch [135]. The bristle phenotype of *notch* mutants is similar to that seen on Numb overexpression [136], and there is some evidence for Numb binding directly to Notch [137, 138]. During SOP cell mitosis, Numb partitions asymmetrically, repressing Notch activity in the cell preferentially enriched in Numb, which is instrumental in establishing the different identities of the daughter cells. The PTB domain of Numb is necessary and sufficient for membrane association and asymmetric localization [139]. As in Dab2, the C-terminal region of Numb is likely disordered, and there are binary associations between Numb and the AP-2 adaptor

[136, 140]. In fact, certain AP-2 (α-adaptin) mutant alleles phenocopy *numb* mutations, and Numb is necessary to drive asymmetric partitioning of AP-2 in the mitotic SOP cell [136]. Epistasis analysis reveals that AP-2 acts between Numb and Notch, leading to a model in which Numb-dependent endocytosis and degradation of Notch underlies the different levels of Notch signaling that cause distinct cell fates. The asymmetric distribution of Numb in mitotic cells may be controlled by NIP (Numb-interacting protein), a transmembrane protein that binds the PTB domain via two redundant NPxF-type sequences [141]. That the PTB domain is important in asymmetric cell division is also demonstrated by the gene dosage-dependent effect of overexpression of Numb-associated kinase (Nak), which engages the PTB domain via an FSNMSF sequence [142].

A significant complication of the Notch down-regulation model is that it has not been convincingly shown that the levels of Notch differ significantly between the two daughter cells. In a modified model, Numb may regulate Notch activity by binding to the multispanning transmembrane protein Sanpodo and promoting its internalization [143]. Endocytosis of Sanpodo prevents the protein from localizing to the plasma membrane where it can bind and assist Notch in signal transduction [143]. It is not yet known whether endocytic uptake of Sanpodo requires the Numb PTB domain, but the YTNPAF sequence at the N terminus of Sanpodo is highly suggestive. Irrespective, it seems clear that Numb, ARH, and Dab2 are globally similar adaptors that use the PTB domain to select designated cargo while synchronously interacting with the endocytic machinery.

6.3
PTB Domain Structure

The high-resolution atomic structures of the PTB domain from eight proteins (Shc [13], IRS-1 [144, 145], Numb [146], X11 [147], SNT-1 [30], Disabled-1 (Dab1) [148, 149], Disabled-2 (Dab2) [149], and Dok1 [150]) currently available reveal that, despite low primary sequence identity and conservation, all adopt a common basic fold. The canonical PTB domain core is composed of seven central β strands, folded into two antiparallel β sheets oriented nearly orthogonally to one another and capped by an α helix at one end of the β sandwich [151]. Typically, one sheet is composed of strands β1–β4 and the other of β5–β7, although β1 can be long and arched, allowing this strand to contribute to the β5–β7 sheet as well (Figures 6.2 and 6.3). As mentioned previously, the overall topological arrangement is homologous to that of a phosphoinositide interaction module, the PH domain [12], making the PTB domain a member of the PH superfold. Indeed, the core C_α atoms of the IRS-1 PTB domain and the dynamin, phospholipase C-δ1, and spectrin PH domains superimpose with an rms deviation of ~1.0 Å [144].

In the PTB domain, the common flanking amphipathic C-terminal helix, along with the abutting β5 strand, play a central role in binding the NPx(*po*)Y peptide ligand. These two structural elements, in part, generate an elongated groove in the PTB domain that accounts for several of the generic features of PTB–NPx(*po*)Y

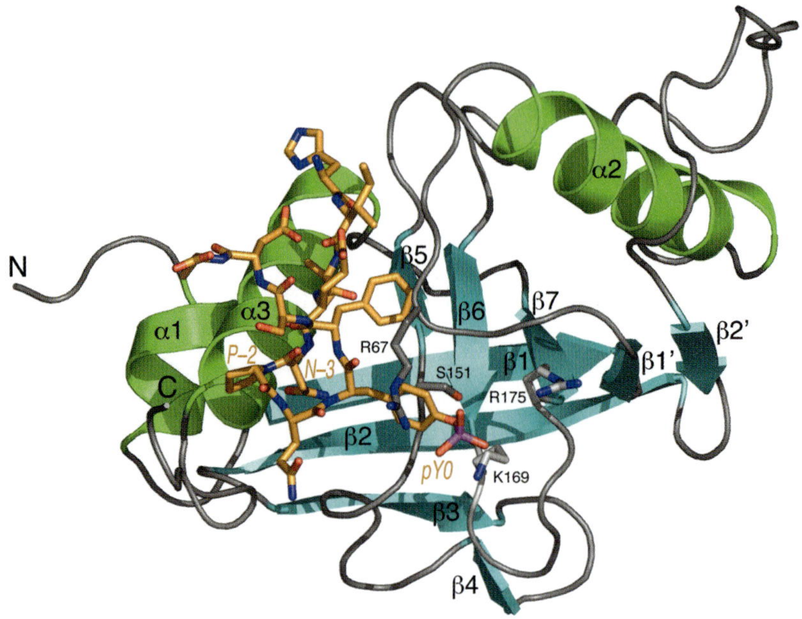

Figure 6.2 Schematic representation of the structure of the Shc PTB domain bound to a TrkA NPx*po*Y phosphopeptide. Ribbon diagram with α helices colored green, β strands cyan, and connecting loops gray. The TrkA phosphopeptide (HIIENPQ*po*YFSDA, gold) is shown in stick representation, with nitrogen atoms colored blue, oxygen red, and phosphorus magenta. The position of PTB domain sidechains involved in coordinating the pTyr 0 (*pY0*) residue of the peptide are also shown in stick representation and colored according to the secondary structural elements from which they emanate. The locations of the NPx*po*Y peptide residues N–3, P–2, and pY0 are indicated in gold italic type. Notation of phosphotyrosine at position ϕ, indicated by pY0, should not be confused with the Seefeld convention for phosphotyrosine, *po*Y, as described in Chapter 22.

engagement. In most cases, the ligand residues proximal to the Asn –3 residue are extended, participating in a β augmentation via backbone hydrogen bonds. This melds an extra antiparallel strand derived from the binding partner into one sheet of the β sandwich, between β5 and the C-terminal helix. The sidechain amide of the NPxY Asn –3 hydrogen bonds to the PTB domain, and the carbonyl oxygen is involved in the formation of the characteristic type I β turn. An intramolecular hydrogen bond to the mainchain amide of the –1 residue, and the propensity of Pro for tight turn formation, reorients the NPxY peptide backbone roughly 90°, bringing the Tyr into proximity with the loops connecting the β4–β5 and β6–β7 strands of the sandwich and stabilizing electrostatic interactions (Figures 6.2 and 6.3). Thus, the ligand is positioned in an L-shaped conformation and the generally conserved mode of NPxY sequence recognition and engagement is illustrated by the extremely similar backbone conformations of a nonphosphorylated NGYNPTY peptide, derived from APP, in the X11 PTB and the ApoER2-derived FNFDNPVY peptide in the Dab1 PTB domain [148].

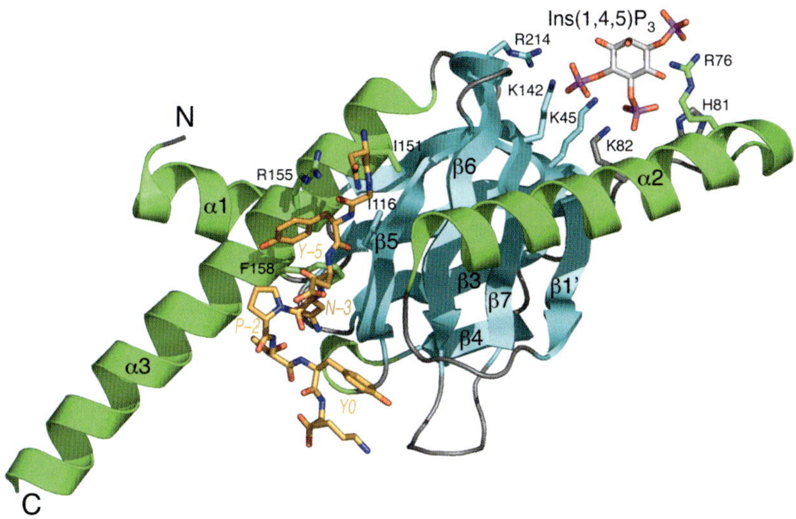

Figure 6.3 Ribbon diagram illustrating the 3D structure of a ternary complex of the Dab1 PTB domain, an APP-derived NPxY peptide, and inositol 1,4,5-trisphosphate (Ins(1,4,5)P_3). Helices are green, β strands cyan, and connecting loops gray. The positions of PTB domain sidechains involved in accommodating the Tyr–5 (*Y–5*) residue of the YxNPxY peptide and in coordination of the Ins(1,4,5)P_3 phosphates are shown in stick representation colored according to the secondary structural elements from which they emanate. The APP peptide (NGYENPTYK, gold) and Ins(1,4,5)P_3 are shown in stick form with nitrogen atoms blue, oxygen red, and phosphorus magenta. The locations of the NPxY peptide residues Y–5, N–3, P–2, and Y0 are indicated in gold italic type.

6.3.1 Broad Binding Specificity

As discussed above, PTB domains are capable of binding a variety of NPx(*po*)Y-type sequences, discriminating between tyrosine-phosphorylated and -nonphosphorylated versions of this motif, as well as engaging sequences unrelated to the NPxY consensus. Flexibility in recognition comes in part from specializations and variations in the basic PTB domain architecture in the form of additional strands and helices. A common addition, which differentiates the PTB from the PTBI domain, is the insertion of a β1′ strand and following α helix between β1 and β2 (found in Shc, Numb, X11, Dab1, and Dab2). The insertion extends considerably the loop between the β1 and β2 strands and typically results in a second helix positioned over the β5/6/7 sheet, although the precise orientation differs among the PTB domains (Figures 6.2 and 6.3). In X11, residues from the N-terminal segment of the α1 helix make hydrophobic and electrostatic contributions to interactions with the N-terminal –5 and –8 residues of the bound, nonphosphorylated APP NPxY peptide [147]. The X11–APP interaction is also unusual because the NPxY tight turn is followed immediately by a short 3_{10} helix that orients two aromatic sidechains (Phe +2 and Phe +3) for optimal placement in a hydrophobic pocket created by sidechains from the longer C-terminal region of the capping C-terminal α2 helix [147]. Both these interactions contribute to the binding energy, because removal of the –7 and –8 residues of the APP peptide or substitution of Phe +2 or +3 for Ala, substantially diminish the affinity of the peptide for the X11 PTB domain [147]. Thus, X11 uses extended contacts to bind a nonphosphorylated NPxY sequence with high affinity ($K_d \approx 300$ nM) [147]. A nonphosphorylatable Numb PTB domain ligand, GFS**NMSF**EDFP derived from the Numb binding partner Numb-associated kinase (Nak), uses a somewhat similar mechanism to complex with Numb [152].

6.3.2
Diverse Modes of Engagement

As alluded to above, the FRS2/SNT1 PTBI domain binds either NPx*po*Y in TrkA receptor or a completely unrelated sequence in FGF receptor. NMR analysis of FRS2 complexed with a 22-residue peptide (HSQMAVHKLAKSIPLRRQVTVS) corresponding to a segment of the human FGF receptor uncovers a novel extended mode of ligand engagement [30] for this canonical PTBI domain scaffold. Despite the absence of the NPxY motif, the C-terminal portion of the FGF receptor peptide (QVTVS) forms the usual additional antiparallel β strand, but between β5 and β8, an extra strand unique to FRS2/SNT1 [30] that folds back over the capping C-terminal helix and assumes some of the function of the lower flanking α1 helix. Turning the peptide backbone nearly 90°, the proximal part of the FGF receptor ligand peptide extends over the face of the β5/6/7 sheet and turns again, embedding the N-terminal MAVH sequence in a hydrophobic depression created by sidechains projecting off the three loops joining β1–β2, β3–β4, and β6–β7. Together, this wraps the FGF receptor peptide around the PTB domain, employing extensive hydrophobic contacts all along the interaction surface as well as electrostatic interactions mainly at the turns [30]. Interestingly, deletion of β8 abolishes FGF receptor binding without disrupting the binding of NPx*po*Y-bearing neurotrophin receptors such as TrkA. The thermodynamics governing FGF receptor and neurotrophin peptide engagement are different [151], yet FRS2/SNT-1 does use the same general surface between β5 and α1/β8 to bind both, because FGF receptor and TrkB peptides compete with each other for PTB binding [30]. The high selectivity of FRS2 for phosphotyrosine is most probably due to two arginine residues structurally equivalent to the two sidechains (Arg212 and Arg227) in IRS-1 that compensate for the negative charge of the phosphate [144, 145]. In FRS2, Arg63, located at the beginning of β5, and Arg78, projecting off the loop connecting β6 to β7, likely coordinate the phosphotyrosine, because mutation of FRS2 Arg63 or Arg78 disrupts TrkB binding without affecting FGF receptor interactions [30]. Similarly positioned basic sidechains for phosphotyrosine coordination are also found in Shc (Arg67, Lys169, Arg175; Figure 6.2), which likewise selects NPx*po*Y ligands over NPxY [13, 153, 154], and in the Dok1 PTB domain [150]. Altogether, these structures provide a molecular explanation for phosphotyrosine-specific binding as well as for NPxY-unrelated sequence engagement.

Another aspect of PTB domain versatility has been uncovered by NMR studies on the Shc PTB domain. Compared to the NPx*po*Y-liganded structure, the free Shc PTB domain is relatively unfolded: 25 N-terminal and 16 C-terminal residues of the domain are not ordered in the unliganded domain [155]. In addition, the vital β7 strand (equivalent to the canonical PTB domain β5 strand) is unstructured, and the α1 and α3 helices are partly unwound and separated, severely dismembering the NPx*po*Y engagement groove (Figure 6.2). Local disorder in the absence of bound ligand also splays the two β sheets apart, altering the location of important sidechains required for proper engagement of the phosphopeptide ligand [155]. This information suggests either an induced-fit type of ligand-dependent folding or equilibrium between the disordered and folded states, with ligand binding to and stabilizing only the folded conformation, or some degree of both [155]. It is not yet clear whether partial unfolding also occurs in other unliganded PTB domains.

Although the Shc, IRS-1, Dok1, and FRS2 PTB domains preferentially bind phosphorylated partners, the Dab1 and Dab2 PTB domains have 10–100-fold higher selectivity for a nonphosphorylated NPxY peptide [61, 117]. And although X11 is Tyr-neutral, in that both Ala and pTyr are accommodated at the Tyr (0) position, these modifications to an APP peptide ligand severely inhibit binding to the Dab1 PTB domain [61]. The crystal structures of the Dab1 and Dab2 PTB domains nicely explain how the NPxY motif is selected over NPx*po*Y sequences in a phylogenetically distinct class of PTB-domain proteins including Numb, ARH [127] and the apoptosis protein CED-6/GULP [156]. An NPxY peptide derived from either ApoER2 [148] or APP [149] adopts the typical antiparallel β strand/tight turn conformation on the Dab1 PTB domain, and the orientation of the APP peptide upon the Dab1 or Dab2 PTB domain is very similar [149]. Strikingly, there is a lack of surface-exposed arginine residues in the immediate vicinity of the Tyr (0) hydroxyl group and, instead, a hydrogen bond between the Tyr sidechain and the mainchain carbonyl oxygen of Gly131 in Dab1, or Gly139 in Dab2, occurs [148, 149]. The β6 strand in Dab1 and 2 is also longer than in X11 and terminates closer to Tyr (0) and, together with the short loop joining β4–β5 and the tight turn connecting β6–β7, confines the aromatic portion of Tyr (0) with nearby PTB sidechains making hydrophobic contacts [148, 149] (Figure 6.3). Consequently, steric clashes would prevent the larger pTyr from engaging appropriately, allowing these PTB domains to effectively discriminate between NPxY and NPx*po*Y. Loop length and/or mobility in this region might allow some PTB domains to accommodate either form of tyrosine. In the X11–APP structure, the β6–β7 connecting loop is mobile and lacks good density [147]. This flexibility would allow more space to house the bulkier pTyr.

Different PTB domains bind NPx(*po*)Y sequences with clear specificity; the Dok1 PTB domain, for example, does not bind NPx(*po*)Y peptides derived from IL-4 receptor or from TrkA, despite binding the related sequence WIENKL*po*YGM derived from RET [150]. Additional ligand specificity/selectivity comes from peripheral alterations to the surface topology of the major interaction site. The marked preference for bulky hydrophobic sidechains at the –5 (Shc, X11, Dab1/2) or –6 position (Dok1) of the ligand reflects hydrophobic pockets appropriately positioned to accommodate these residues. For example, in Dab1 (and Dab2) Phe

−5 packs into a hydrophobic site generated by Ile116, Ile151, Phe158, and the aliphatic portion of Arg155 [148, 149] (Figure 6.3). The surface of the Dok1 PTB domain is optimized for a sequence containing Leu at −1 [150]. All together, these features explain how different PTB domains preferentially recognize distinct peptide sequences despite an overall similar mode of engagement.

6.3.3
Phospholipid Binding

As previously discussed, the PTB domains of Shc [11], Dab1 [61], and Dab2 [118] also bind the acidic phospholipid phosphatidylinositol 4,5-bisphosphate (PtdIns(4,5)P_2) noncompetitively with the NPxY ligand. The phosphoinositide interaction surface in the Disabled proteins is located at the opposite end of the second β sheet, allowing the crystallization of ternary PTB–NPxY–PtdIns(4,5)P_2/Ins(1,4,5)P_3 complexes [148, 149]. The lipid-binding site is spatially related to the phosphoinositide-binding surface in the PH domain, where the negatively charged headgroup is coordinated by basic sidechains positioned to optimize hydrogen bonding to the appropriate phosphatidylinositol phosphate form [12]. Similarly, Lys, Arg, and His residues (Lys45, Arg76, His81, Lys82, Arg124, and Lys142; Figure 6.3) cluster to generate a basic surface patch on the Dab1 and 2 PTB domains involved in coordinating the phosphate groups [148, 149]. This coherent positively charged interaction surface is not found at the equivalent position in either X11 or Shc [148], but is likely to occur in Numb and ARH. Indeed, the PTB domains from both these proteins bind phosphoinositides in vitro [131, 157].

6.4
Conclusions

In summary, the PTB domain utilizes the basic PH-domain architectural scaffold to create a topologically conserved NPx(*po*)Y ligand binding cleft absent from the phosphoinositide binding PH domain. The peripheral mode of pTyr engagement, utilizing surface-exposed sidechains not absolutely conserved in primary sequence, is fundamentally different from that of the SH2 domain, which bears no structural relationship to the PTB domain. The mode of NPxY binding has allowed the evolution of different PTB domains specifically selecting either NPx*po*Y or NPxY, or displaying no bias for pTyr over Tyr, making the original PTB designation now a misnomer for this interaction module.

Due to this ability to bind to diverse ligands, PTB-domain proteins are involved in multiple cellular functions. As described in Section 6.2, major categories of activity include tyrosine kinase signaling, cell adhesion, and protein trafficking. However, there is still much to be learned about the functions of these proteins and their possible role in diseases. Already, the identification of the ARH protein has yielded important insights into cholesterol trafficking, and the multiple APP binding proteins are likely to lead to new discoveries in the pathogenesis of Alzheimer's disease. Additionally, there are many PTB domains that we know very little about, such as Odin. Future work will focus on delineating the function of these proteins, and no one will be surprised if unique types of interactions are described for their PTB domains. Overall, the PTB domain is a good example of how studies based on the simple analysis of protein binding can evolve so as to yield many important discoveries in unrelated fields of study.

References

1 Blaikie, P., Immanuel, D., Wu, J., Li, N., Yajnik, V., Margolis, B., A region in Shc distinct from the SH2 domain can bind tyrosine phosphorylated growth factor receptors. *J. Biol. Chem.* **1994**, *269*, 32031–32034.

2 Gustafson, T. A., He, W., Craparo, A., Schaub, C. D., O'Neill, T. J., Phosphotyrosine-dependent interaction of Shc and IRS-1 with the NPEY motif of the insulin receptor via a novel (non-SH2) domain. *Mol. Cell. Biol.* **1995**, *15*, 2500–2508.

3 Kavanaugh, W. M., Williams, L. T., An alternative to SH2 domains for binding tyrosine-phosphorylated proteins. *Science* **1994**, *266*, 1862–1865.

4 Sun, X. J., et al., Role of IRS-2 in insulin and cytokine signalling. *Nature* **1995**, *377*, 173–177.

5 Bork, P., Margolis, B., A phosphotyrosine interaction domain. *Cell* **1995**, *80*, 693–694.

6 Borg, J. P., Margolis, B., Function of PTB domains. *Curr. Top. Microbiol. Immunol.* **1998**, *228*, 23–38.

7 Batzer, A. G., Blaikie, P., Nelson, K., Schlessinger, J., Margolis, B., The phosphotyrosine interaction domain of Shc binds an LXNPXY motif on the epidermal growth factor receptor. *Mol. Cell. Biol.* **1995**, *15*, 4403–4409.

8 Dikic, I., Batzer, A. G., Blaikie, P., Obermeier, A., Ullrich, A., Schlessinger, J., Margolis, B., Shc binding to nerve growth factor receptor is mediated by the phosphotyrosine interaction domain. *J. Biol. Chem.* **1995**, *270*, 15125–15129.

9 Ravichandran, K. S., Signaling via Shc family adapter proteins. *Oncogene* **2001**, *20*, 6322–6330.

10 Luschnig, S., Krauss, J., Bohmann, K., Desjeux, I., Nusslein-Volhard, C., The *Drosophila* SHC adaptor protein is required for signaling by a subset of receptor tyrosine kinases. *Mol. Cell* **2000**, *5*, 231–241.

11 Ravichandran, K. S., Zhou, M. M., Pratt, J. C., Harlan, J. E., Walk, S. F., Fesik, S. W., Burakoff, S. J., Evidence for a requirement for both phospholipid and phosphotyrosine binding via the Shc phosphotyrosine-binding domain in vivo. *Mol. Cell. Biol.* **1997**, *17*, 5540–5549.

12 Lemmon, M. A., Phosphoinositide recognition domains. *Traffic* **2003**, *4*, 201–213.

13 Zhou, M. M., et al., Structure and ligand recognition of the phosphotyrosine binding domain of Shc. *Nature* **1995**, *378*, 584–592.

14 White, M. F., Livingston, J. N., Backer, J. M., Lauris, V., Dull, T. J., Ullrich, A., Kahn, C. R., Mutation of the insulin receptor at tyrosine 960 inhibits signal transmission but does not affect its tyrosine kinase activity. *Cell* **1988**, *54*, 641–649.

15 Sun, X. J., et al., Structure of the insulin receptor substrate IRS-1 defines a unique signal transduction protein. *Nature* **1991**, *352*, 73–77.

16 He, W., O'Neill, T. J., Gustafson, T. A., Distinct modes of interaction of SHC and insulin receptor substrate-1 with the insulin receptor NPEY region via non-SH2 domains. *J. Biol. Chem.* **1995**, *270*, 23258–23262.

17 Wolf, G., et al., PTB domains of IRS-1 and Shc have distinct but overlapping binding specificities. *J. Biol. Chem.* **1995**, *270*, 27407–27410.

18 White, M. F., IRS proteins and the common path to diabetes. *Am. J. Physiol. Endocrinol. Metab.* **2002**, *283*, E413–22.

19 Yenush, L., Makati, K. J., Smithhall, J., Ishibashi, O., Myers, M. G., White, M. F., The pleckstrin homology domain is the principle link between insulin receptor and IRS-1. *J. Biol. Chem.* **1996**, *271*, 24300–24306.

20 Takeuchi, H., et al., PTB domain of insulin receptor substrate-1 binds inositol compounds. *Biochem. J.* **1998**, *334*, 211–218.

21 Farhang-Fallah, J., Randhawa, V. K., Nimnual, A., Klip, A., Bar-Sagi, D., Rozakis-Adcock, M., The pleckstrin homology (PH) domain-interacting protein couples the insulin receptor substrate 1 PH domain to insulin signaling pathways leading to mitogenesis and GLUT4 translocation. *Mol. Cell. Biol.* **2002**, *22*, 7325–7336.

22 Dhe-Paganon, S., Ottinger, E. A., Nolte, R. T., Eck, M. J., Shoelson, S. E., Crystal structure of the pleckstrin homology-phosphotyrosine binding (PH-PTB) targeting region of insulin receptor substrate 1. *Proc. Natl. Acad. Sci. USA* **1999**, *96*, 8378–8383.

23 Kouhara, H., Hadari, Y. R., Spivak-Kroizman, T., Schilling, J., Bar-Sagi, D., Lax, I., Schlessinger, J., A lipid-anchored Grb2-binding protein that links FGF-receptor activation to the Ras/MAPK signaling pathway. *Cell* **1997**, *89*, 693–702.

24 Rabin, S. J., Cleghon, V., Kaplan, D. R., SNT, a differentiation-specific target of neurotrophic factor-induced tyrosine kinase activity in neurons and PC12 cells. *Mol. Cell. Biol.* **1993**, *13*, 2203–2213.

25 Xu, H., Lee, K. W., Goldfarb, M., Novel recognition motif on fibroblast growth factor receptor mediates direct association and activation of SNT adapter proteins. *J. Biol. Chem.* **1998**, *273*, 17987–17990.

26 Kurokawa, K., Iwashita, T., Murakami, H., Hayashi, H., Kawai, K., Takahashi, M., Identification of SNT/FRS2 docking site on RET receptor tyrosine kinase and its role for signal transduction. *Oncogene* **2001**, *20*, 1929–1938.

27 Ong, S. H., Guy, G. R., Hadari, Y. R., Laks, S., Gotoh, N., Schlessinger, J., Lax, I., FRS2 proteins recruit intracellular signaling pathways by binding to diverse targets on fibroblast growth factor and nerve growth factor receptors. *Mol. Cell. Biol.* **2000**, *20*, 979–989.

28 Hadari, Y. R., Gotoh, N., Kouhara, H., Lax, I., Schlessinger, J., Critical role for the docking-protein FRS2 alpha in FGF receptor-mediated signal transduction pathways. *Proc. Natl. Acad. Sci. USA* **2001**, *98*, 8578–8583.

29 Meakin, S. O., MacDonald, J. I., Gryz, E. A., Kubu, C. J., Verdi, J. M., The signaling adapter FRS-2 competes with Shc for binding to the nerve growth factor receptor TrkA. A model for discriminating proliferation and differentiation. *J. Biol. Chem.* **1999**, *274*, 9861–9870.

30 Dhalluin, C., et al., Structural basis of SNT PTB domain interactions with distinct neurotrophic receptors. *Mol. Cell* **2000**, *6*, 921–929.

31 Yamanashi, Y., Baltimore, D., Identification of the Abl- and rasGAP-associated 62-kDa protein as a docking protein, Dok. *Cell* **1997**, *88*, 205–211.

32 Carpino, N., Wisniewski, D., Strife, A., Marshak, D., Kobayashi, R., Stillman, B., Clarkson, B., p62(dok): a constitutively tyrosine-phosphorylated, GAP-associated protein in chronic myelogenous leukemia progenitor cells. *Cell* **1997**, *88*, 197–204.

33 Jones, N., Dumont, D. J., The Tek/Tie2 receptor signals through a novel Dok-related docking protein, Dok-R. *Oncogene* **1998**, *17*, 1097–1108.

34 Nelms, K., Snow, A. L., Hu-Li, J., Paul, W. E., FRIP, a hematopoietic cell-specific rasGAP-interacting protein phosphorylated in response to cytokine stimulation. *Immunity* **1998**, *9*, 13–24.

35 Di Cristofano, A., et al., Molecular cloning and characterization of p56dok-2 defines a new family of RasGAP-binding proteins. *J. Biol. Chem.* **1998**, *273*, 4827–4830.

36 Cong, F., Yuan, B., Goff, S. P., Characterization of a novel member of the DOK family that binds and modulates Abl signaling. *Mol. Cell. Biol.* **1999**, *19*, 8314–8325.

37 Lemay, S., Davidson, D., Latour, S., Veillette, A., Dok-3, a novel adapter molecule involved in the negative regulation of immunoreceptor signaling. *Mol. Cell. Biol.* **2000**, *20*, 2743–2754.

38 Grimm, J., Sachs, M., Britsch, S., Di Cesare, S., Schwarz-Romond, T., Alitalo, K., Birchmeier, W., Novel p62dok family members, dok-4 and dok-5, are substrates of the c-Ret receptor tyrosine kinase and mediate neuronal differentiation. *J. Cell Biol.* **2001**, *154*, 345–354.

39 Cai, D., Dhe-Paganon, S., Melendez, P. A., Lee, J., Shoelson, S. E., Two new substrates in insulin signaling, IRS5/DOK4 and IRS6/DOK5. *J. Biol. Chem.* **2003**, *278*, 25323–25330.

40 Favre, C., Gerard, A., Clauzier, E., Pontarotti, P., Olive, D., Nunes, J. A., DOK4 and DOK5: new Dok-related genes expressed in human T cells. *Genes Immunol.* **2003**, *4*, 40–45.

41 Jones, N., Dumont, D. J., Recruitment of Dok-R to the EGF receptor through its PTB domain is required for attenuation of Erk MAP kinase activation. *Curr. Biol.* **1999**, *9*, 1057–1060.

42 Songyang, Z., Yamanashi, Y., Liu, D., Baltimore, D., Domain-dependent function of the rasGAP-binding protein p62Dok in cell signaling. *J. Biol. Chem.* **2001**, *276*, 2459–2465.

43 Jones, N., Chen, S. H., Sturk, C., Master, Z., Tran, J., Kerbel, R. S., Dumont, D. J., A unique autophosphorylation site on Tie2/Tek mediates Dok-R phosphotyrosine binding domain binding and function. *Mol. Cell. Biol.* **2003**, *23*, 2658–2668.

44 Yamanashi, Y., Tamura, T., Kanamori, T., Yamane, H., Nariuchi, H., Yamamoto, T., Baltimore, D., Role of the rasGAP-associated docking protein p62(dok) in negative regulation of B cell receptor-mediated signaling. *Genes. Dev.* **2000**, *14*, 11–16.

45 Master, Z., Jones, N., Tran, J., Jones, J., Kerbel, R. S., Dumont, D. J., Dok-R plays a pivotal role in angiopoietin-1-dependent cell migration through recruitment and activation of Pak. *EMBO J.* **2001**, *20*, 5919–5928.

46 Fazioli, F., Minichiello, L., Matoska, V., Castagnino, P., Miki, T., Wong, W. T., Di Fiore, P. P., Eps8, a substrate for the epidermal growth factor receptor kinase, enhances EGF-dependent mitogenic signals. *EMBO J.* **1993**, *12*, 3799–3808.

47 Castagnino, P., Biesova, Z., Wong, W. T., Fazioli, F., Gill, G. N., Di Fiore, P. P., Direct binding of eps8 to the juxtamembrane domain of EGFR is phosphotyrosine- and SH2-independent. *Oncogene* **1995**, *10*, 723–729.

48 Provenzano, C., Gallo, R., Carbone, R., Di Fiore, P. P., Falcone, G., Castellani, L., Alema, S., Eps8, a tyrosine kinase substrate, is recruited to the cell cortex and dynamic F-actin upon cytoskeleton remodeling. *Exp. Cell Res.* **1998**, *242*, 186–200.

49 Scita, G., et al., EPS8 and E3B1 transduce signals from Ras to Rac. *Nature* **1999**, *401*, 290–293.

50 Scita, G., et al., An effector region in Eps8 is responsible for the activation of the Rac-specific GEF activity of Sos-1 and for the proper localization of the Rac-based actin-polymerizing machine. *J. Cell Biol.* **2001**, *154*, 1031–1044.

51 Innocenti, M., Frittoli, E., Ponzanelli, I., Falck, J. R., Brachmann, S. M., Di Fiore, P. P., Scita, G., Phosphoinositide 3-kinase activates Rac by entering in a complex with Eps8, Abi1, and Sos-1. *J. Cell Biol.* **2003**, *160*, 17–23.

52 Lanzetti, L., Rybin, V., Malabarba, M. G., Christoforidis, S., Scita, G., Zerial, M., Di Fiore, P. P., The Eps8 protein coordinates EGF receptor signalling through Rac and trafficking through Rab5. *Nature* **2000**, *408*, 374–377.

53 Pandey, A., et al., Cloning of a novel phosphotyrosine binding domain containing molecule, Odin, involved in signaling by receptor tyrosine kinases. *Oncogene* **2002**, *21*, 8029–8036.

54 Fu, X., McGrath, S., Pasillas, M., Nakazawa, S., Kamps, M. P., EB-1, a tyrosine kinase signal transduction gene, is transcriptionally activated in the t(1;19) subset of pre-B ALL, which express oncoprotein E2a-Pbx1. *Oncogene* **1999**, *18*, 4920–4929.

55 Neubig, R. R., Siderovski, D. P., Regulators of G-protein signalling as new central nervous system drug targets. *Nat. Rev. Drug Discov.* **2002**, *1*, 187–197.

56 Schiff, M. L., et al., Tyrosine-kinase-dependent recruitment of RGS12 to the N-type calcium channel. *Nature* **2000**, *408*, 723–727.

57 Borg, J. P., Ooi, J., Levy, E., Margolis, B., The phosphotyrosine interaction domains of X11 and FE65 bind to distinct sites on the YENPTY motif of amyloid precursor protein. *Mol. Cell. Biol.* **1996**, *16*, 6229–6241.

58 Hardy, J., Selkoe, D. J., The amyloid hypothesis of Alzheimer's disease: progress and problems on the road to therapeutics. *Science* **2002**, *297*, 353–356.

59 Fiore, F., Zambrano, N., Minopoli, G., Donini, V., Duilio, A., Russo, T., The regions of the Fe65 protein homologous to the phosphotyrosine interaction/phosphotyrosine binding domain of Shc bind the intracellular domain of the Alzheimer's amyloid precursor protein. *J. Biol. Chem.* **1995**, *270*, 30853–30856.

60 McLoughlin, D. M., Miller, C. C. J., The intracellular cytoplasmic domain of the Alzheimer's disease amyloid precursor protein interacts with phosphotyrosine-binding domain proteins in the yeast two-hybrid system. *FEBS Lett.* **1996**, *397*, 197–200.

61 Howell, B. W., Lanier, L. M., Frank, R., Gertler, F. B., Cooper, J. A., The disabled-1 phosphotyrosine-binding domain binds to the internalization signals of transmembrane glycoproteins and to phospholipids. *Mol. Cell. Biol.* **1999**, *19*, 5179–5188.

62 Homayouni, R., Rice, D. S., Sheldon, M., Curran, T., Disabled-1 binds to the cytoplasmic domain of amyloid precursor-like protein 1. *J. Neurosci.* **1999**, *19*, 7507–7515.

63 Matsuda, S., et al., c-Jun N-terminal kinase (JNK)-interacting protein-1b/islet-brain-1 scaffolds Alzheimer's amyloid precursor protein with JNK. *J. Neurosci.* **2001**, *21*, 6597–6607.

64 Taru, H., Kirino, Y., Suzuki, T., Differential roles of JIP scaffold proteins in the modulation of amyloid precursor protein metabolism. *J. Biol. Chem.* **2002**, *277*, 27567–27574.

65 Roncarati, R., et al., The gamma-secretase-generated intracellular domain of beta-amyloid precursor protein binds Numb and inhibits Notch signaling. *Proc. Natl. Acad. Sci. USA* **2002**, *99*, 7102–7107.

66 Scheinfeld, M. H., Roncarati, R., Vito, P., Lopez, P. A., Abdallah, M., D'Adamio, L., Jun NH2-terminal kinase (JNK) interacting protein 1 (JIP1) binds the cytoplasmic domain of the Alzheimer's beta-amyloid precursor protein (APP). *J. Biol. Chem.* **2002**, *277*, 3767–3775.

67 Russo, C., Dolcini, V., Salis, S., Venezia, V., Zambrano, N., Russo, T., Schettini, G., Signal transduction through tyrosine-phosphorylated C-terminal fragments of amyloid precursor protein via an enhanced interaction with Shc/Grb2 adaptor proteins in reactive astrocytes of Alzheimer's disease brain. *J. Biol. Chem.* **2002**, *277*, 35282–35288.

68 Tarr, P. E., Roncarati, R., Pelicci, G., Pelicci, P. G., D'Adamio, L., Tyrosine phosphorylation of the beta-amyloid precursor protein cytoplasmic tail promotes interaction with Shc. *J. Biol. Chem.* **2002**, *277*, 16798–16804.

69 Borg, J. P., Yang, Y., de Taddeo-Borg, M., Margolis, B., Turner, R. S., The X11alpha protein slows cellular amyloid precursor protein processing and reduces Abeta40 and Abeta42 secretion. *J. Biol. Chem.* **1998**, *273*, 14761–14766.

70 Sastre, M., Turner, R. S., Levy, E., X11 Interaction with beta-amyloid precursor protein modulates its cellular stabilization and reduces amyloid beta-protein secretion. *J. Biol. Chem.* **1998**, *273*, 22351–22357.

71 Sabo, S. L., Lanier, L. M., Ikin, A. F., Khorkova, O., Sahasrabudhe, S., Greengard, P., Buxbaum, J. D., Regulation of beta-amyloid secretion by FE65, an amyloid protein precursor-binding protein. *J. Biol. Chem.* **1999**, *274*, 7952–7957.

72 Chang, Y., et al., Generation of the beta-amyloid peptide and the amyloid precursor protein C-terminal fragment gamma are potentiated by FE65L1. *J. Biol. Chem.* **2003**, *278*, 51100–51107.

73 Lee, J. H., et al., The neuronal adaptor protein X11alpha reduces Abeta levels in the brains of Alzheimer's APPswe Tg2576 transgenic mice. *J. Biol. Chem.* **2003**, *278*, 47025–47029.

74 Kaech, S. M., Whitfield, C. W., Kim, S. K., The LIN-2/LIN-7/LIN-10 complex mediates basolateral membrane localization of the *C. elegans* EGF receptor LET-23

in vulval epithelial cells. *Cell* **1998**, *94*, 761–771.

75 Rongo, C., Whitfield, C. W., Rodal, A., Kim, S. K., Kaplan, J. M., LIN-10 is a shared component of the polarized protein localization pathways in neurons and epithelia. *Cell* **1998**, *94*, 751–759.

76 Borg, J. P., Lopez-Figueroa, M. O., de Taddeo-Borg, M., Kroon, D. E., Turner, R. S., Watson, S. J., Margolis, B., Molecular analysis of the X11-mLin-2/CASK complex in brain. *J. Neurosci.* **1999**, *19*, 1307–1316.

77 Whitfield, C. W., Benard, C., Barnes, T., Hekimi, S., Kim, S. K., Basolateral localization of the *Caenorhabditis elegans* epidermal growth factor receptor in epithelial cells by the PDZ protein LIN-10. *Mol. Biol. Cell* **1999**, *10*, 2087–2100.

78 Okamoto, M., Sudhof, T. C., Mints, Munc18-interacting proteins in synaptic vesicle exocytosis. *J. Biol. Chem.* **1997**, *272*, 31459–31464.

79 Ho, C. S., Marinescu, V., Steinhilb, M. L., Gaut, J. R., Turner, R. S., Stuenkel, E. L., Synergistic effects of Munc18a and X11 proteins on amyloid precursor protein metabolism. *J. Biol. Chem.* **2002**, *277*, 27021–27028.

80 Hill, K., et al., Munc18 interacting proteins: ADP-ribosylation factor-dependent coat proteins that regulate the traffic of beta-Alzheimer's precursor protein. *J. Biol. Chem.* **2003**, *278*, 36032–36040.

81 King, G. D., Perez, R. G., Steinhilb, M. L., Gaut, J. R., Turner, R. S., X11alpha modulates secretory and endocytic trafficking and metabolism of amyloid precursor protein: mutational analysis of the YENPTY sequence. *Neuroscience* **2003**, *120*, 143–154.

82 Ando, K., Iijima, K. I., Elliott, J. I., Kirino, Y., Suzuki, T., Phosphorylation-dependent regulation of the interaction of amyloid precursor protein with Fe65 affects the production of beta-amyloid. *J. Biol. Chem.* **2001**, *276*, 40353–40361.

83 Trommsdorff, M., Borg, J. P., Margolis, B., Herz, J., Interaction of cytosolic adaptor proteins with neuronal apolipoprotein E receptors and the amyloid precursor protein. *J. Biol. Chem.* **1998**, *273*, 33556–33560.

84 Kinoshita, A., Whelan, C. M., Smith, C. J., Berezovska, O., Hyman, B. T., Direct visualization of the gamma secretase-generated carboxyl-terminal domain of the amyloid precursor protein: association with Fe65 and translocation to the nucleus. *J. Neurochem.* **2002**, *82*, 839–847.

85 Hyman, B. T., Strickland, D., Rebeck, G. W., Role of the low-density lipoprotein receptor-related protein in beta-amyloid metabolism and Alzheimer disease. *Arch. Neurol.* **2000**, *57*, 646–650.

86 Pietrzik, C. U., Busse, T., Merriam, D. E., Weggen, S., Koo, E. H., The cytoplasmic domain of the LDL receptor-related protein regulates multiple steps in APP processing. *EMBO J.* **2002**, *21*, 5691–5700.

87 Minopoli, G., de Candia, P., Bonetti, A., Faraonio, R., Zambrano, N., Russo, T., The beta-amyloid precursor protein functions as a cytosolic anchoring site that prevents Fe65 nuclear translocation. *J. Biol. Chem.* **2001**, *276*, 6545–6550.

88 Cao, X., Sudhof, T. C., A transcriptionally [correction of transcriptively] active complex of APP with Fe65 and histone acetyltransferase Tip60. *Science* **2001**, *293*, 115–120.

89 Biederer, T., Cao, X., Sudhof, T. C., Liu, X., Regulation of APP-dependent transcription complexes by Mint/X11s: differential functions of Mint isoforms. *J. Neurosci.* **2002**, *22*, 7340–7351.

90 Kinoshita, A., Shah, T., Tangredi, M. M., Strickland, D. K., Hyman, B. T., The intracellular domain of the low density lipoprotein receptor-related protein modulates transactivation mediated by amyloid precursor protein and Fe65. *J. Biol. Chem.* **2003**, *278*, 41182–41188.

91 Baek, S. H., Ohgi, K. A., Rose, D. W., Koo, E. H., Glass, C. K., Rosenfeld, M. G., Exchange of N-CoR corepressor and Tip60 coactivator complexes links gene expression by NF-kappaB and beta-amyloid precursor protein. *Cell* **2002**, *110*, 55–67.

92 Zhao, Q., Lee, F. S., The transcriptional activity of the APP intracellular domain–Fe65 complex is inhibited by activation of the NF-kappaB pathway. *Biochemistry* **2003**, *42*, 3627–3634.

93 Sabo, S. L., Ikin, A. F., Buxbaum, J. D., Greengard, P., The Alzheimer amyloid precursor protein (APP) and FE65, an APP-binding protein, regulate cell movement. *J. Cell Biol.* **2001**, *153*, 1403–1414.

94 Whitmarsh, A. J., Cavanagh, J., Tournier, C., Yasuda, J., Davis, R. J., A mammalian scaffold complex that selectively mediates MAP kinase activation. *Science* **1998**, *281*, 1671–1674.

95 Dickens, M., et al., A cytoplasmic inhibitor of the JNK signal transduction pathway. *Science* **1997**, *277*, 693–696.

96 Nihalani, D., Meyer, D., Pajni, S., Holzman, L. B., Mixed lineage kinase-dependent JNK activation is governed by interactions of scaffold protein JIP with MAPK module components. *EMBO J.* **2001**, *20*, 3447–3458.

97 Verhey, K. J., Meyer, D., Deehan, R., Blenis, J., Schnapp, B. J., Rapoport, T. A., Margolis, B., Cargo of kinesin identified as JIP scaffolding proteins and associated signaling molecules. *J. Cell Biol.* **2001**, *152*, 959–970.

98 Inomata, H., Nakamura, Y., Hayakawa, A., Takata, H., Suzuki, T., Miyazawa, K., Kitamura, N., A scaffold protein JIP-1b enhances amyloid precursor protein phosphorylation by JNK and its association with kinesin light chain 1. *J. Biol. Chem.* **2003**, *278*, 22946–22955.

99 Kamal, A., Almenar-Queralt, A., LeBlanc, J. F., Roberts, E. A., Goldstein, L. S., Kinesin-mediated axonal transport of a membrane compartment containing beta-secretase and presenilin-1 requires APP. *Nature* **2001**, *414*, 643–648.

100 Setou, M., Nakagawa, T., Seog, D. H., Hirokawa, N., Kinesin superfamily motor protein KIF17 and mLin-10 in NMDA receptor-containing vesicle transport. *Science* **2000**, *288*, 1796–1802.

101 Hynes, R. O., Integrins: bidirectional, allosteric signaling machines. *Cell* **2002**, *110*, 673–687.

102 Chen, W. J., Goldstein, J. L., Brown, M. S., NPXY, a sequence often found in cytoplasmic tails, is required for coated pit-mediated internalization of the low density lipoprotein receptor. *J. Biol. Chem.* **1990**, *265*, 3116–3123.

103 Calderwood, D. A., et al., Integrin beta cytoplasmic domain interactions with phosphotyrosine-binding domains: a structural prototype for diversity in integrin signaling. *Proc. Natl. Acad. Sci. USA* **2003**, *100*, 2272–2277.

104 Chang, D. D., Wong, C., Smith, H., Liu, J., ICAP-1, a novel beta1 integrin cytoplasmic domain-associated protein, binds to a conserved and functionally important NPXY sequence motif of beta1 integrin. *J. Cell Biol.* **1997**, *138*, 1149–1157.

105 Zhang, X. A., Hemler, M. E., Interaction of the integrin beta1 cytoplasmic domain with ICAP-1 protein. *J. Biol. Chem.* **1999**, *274*, 11–19.

106 Calderwood, D. A., Zent, R., Grant, R., Rees, D. J., Hynes, R. O., Ginsberg, M. H., The talin head domain binds to integrin beta subunit cytoplasmic tails and regulates integrin activation. *J. Biol. Chem.* **1999**, *274*, 28071–28074.

107 Bouvard, D., Block, M. R., Calcium/calmodulin-dependent protein kinase II controls integrin alpha5beta1-mediated cell adhesion through the integrin cytoplasmic domain associated protein-1 alpha. *Biochem. Biophys. Res. Commun.* **1998**, *252*, 46–50.

108 Degani, S., et al., The integrin cytoplasmic domain-associated protein ICAP-1 binds and regulates Rho family GTPases during cell spreading. *J. Cell Biol.* **2002**, *156*, 377–387.

109 Bouvard, D., et al., Disruption of focal adhesions by integrin cytoplasmic domain-associated protein-1 alpha. *J. Biol. Chem.* **2003**, *278*, 6567–6574.

110 Horwitz, A., Duggan, K., Buck, C., Beckerle, M. C., Burridge, K., Interaction of plasma membrane fibronectin receptor with talin: a transmembrane linkage. *Nature* **1986**, *320*, 531–533.

111 Chishti, A. H., et al., The FERM domain: a unique module involved in the linkage of cytoplasmic proteins to the membrane. *Trends Biochem. Sci.* **1998**, *23*, 281–282.

112 Garcia-Alvarez, B., et al., Structural determinants of integrin recognition by talin. *Mol. Cell* **2003**, *11*, 49–58.

113 Pearson, M. A., Reczek, D., Bretscher, A., Karplus, P. A., Structure of the ERM protein moesin reveals the FERM domain fold masked by an extended actin binding tail domain. *Cell* **2000**, *101*, 259–270.

114 Davis, C. G., Lehrman, M. A., Russell, D. W., Anderson, R. G., Brown, M. S., Goldstein, J. L., The J. D. mutation in familial hypercholesterolemia: amino acid substitution in cytoplasmic domain impedes internalization of LDL receptors. *Cell* **1986**, *45*, 15–24.

115 Gotthardt, M., et al., Interactions of the low density lipoprotein receptor gene family with cytosolic adaptor and scaffold proteins suggest diverse biological functions in cellular communication and signal transduction. *J. Biol. Chem.* **2000**, *275*, 25616–25624.

116 Oleinikov, A. V., Zhao, J., Makker, S. P., Cytosolic adaptor protein Dab2 is an intracellular ligand of endocytic receptor gp600/megalin. *Biochem. J.* **2000**, *347 Pt 3*, 613–621.

117 Morris, S. M., Cooper, J. A., Disabled-2 colocalizes with the LDLR in clathrin-coated pits and interacts with AP-2. *Traffic* **2001**, *2*, 111–123.

118 Mishra, S. K., Keyel, P. A., Hawryluk, M. J., Agostinelli, N. R., Watkins, S. C., Traub, L. M., Disabled-2 exhibits the properties of a cargo-selective endocytic clathrin adaptor. *EMBO J.* **2002**, *21*, 4915–4926.

119 Morris, S. M., Arden, S. D., Roberts, R. C., Kendrick-Jones, J., Cooper, J. A., Luzio, J. P., Buss, F., Myosin VI binds to and localises with Dab2, potentially linking receptor-mediated endocytosis and the actin cytoskeleton. *Traffic* **2002**, *3*, 331–341.

120 Conner, S. D., Schmid, S. L., Regulated portals of entry into the cell. *Nature* **2003**, *422*, 37–44.

121 Morris, S. M., Tallquist, M. D., Rock, C. O., Cooper, J. A., Dual roles for the Dab2 adaptor protein in embryonic development and kidney transport. *EMBO J.* **2002**, *21*, 1555–1564.

122 Leheste, J. R., et al., Megalin knockout mice as an animal model of low molecular weight proteinuria. *Am. J. Pathol.* **1999**, *155*, 1361–1370.

123 Zhou, J., Hsieh, J. T., The inhibitory role of DOC-2/DAB2 in growth factor receptor-mediated signal cascade. DOC-2/DAB2-mediated inhibition of ERK phosphorylation via binding to Grb2. *J. Biol. Chem.* **2001**, *276*, 27793–27798.

124 Hocevar, B. A., Smine, A., Xu, X. X., Howe, P. H., The adaptor molecule Disabled-2 links the transforming growth factor beta receptors to the Smad pathway. *EMBO J.* **2001**, *20*, 2789–2801.

125 Hocevar, B. A., Mou, F., Rennolds, J. L., Morris, S. M., Cooper, J. A., Howe, P. H., Regulation of the Wnt signaling pathway by disabled-2 (Dab2). *EMBO J.* **2003**, *22*, 3084–3094.

126 Kamikura, D. M., Cooper, J. A., Lipoprotein receptors and a Disabled family cytoplasmic adaptor protein regulate EGL–17/FGF export in *C. elegans*. *Genes. Dev.* **2003**, *17*, 2798-2811.

127 Garcia, C. K., et al., Autosomal recessive hypercholesterolemia caused by mutations in a putative LDL receptor adaptor protein. *Science* **2001**, *292*, 1394–1398.

128 Eden, E. R., et al., Restoration of LDL-receptor function in cells from patients with autosomal recessive hypercholesterolemia by retroviral expression of *ARH1*. *J. Clin. Invest.* **2002**, *110*, 1695–1702.

129 Jones, C., Hammer, R. E., Li, W. P., Cohen, J. C., Hobbs, H. H., Herz, J., Normal sorting, but defective endocytosis of the LDL receptor in mice with autosomal recessive hypercholesterolemia. *J. Biol. Chem.* **2003**, *278*, 29024–29030.

130 He, G., Gupta, S., Michaely, P., Hobbs, H. H., Cohen, J. C., ARH is a modular adaptor protein that interacts with the LDL receptor, clathrin and AP-2. *J. Biol. Chem.* **2002**, *277*, 44044–44049.

131 Mishra, S. K., Watkins, S. C., Traub, L. M., The autosomal recessive hypercholesterolemia (ARH) protein interfaces directly with the clathrin-coat machinery. *Proc. Natl. Acad. Sci. USA* **2002**, *99*, 16099–16104.

132 Zhou, Y., Zhang, J., King, M. L., Xenopus ARH couples lipoprotein receptors with the AP-2 complex in oocytes and embryos and is required for vitellogenesis. *J. Biol. Chem.* **2003**, *278*, 44584–44592.

133 Cohen, J. C., Kimmel, M., Polanski, A., Hobbs, H. H., Molecular mechanisms of autosomal recessive hypercholesterolemia. *Curr. Opin. Lipidol.* **2003**, *14*, 121–127.

134 Soutar, A. K., Naoumova, R. P., Traub, L. M., Genetics, clinical phenotype, and molecular cell biology of autosomal recessive hypercholesterolemia. *Arterioscler. Thromb. Vasc. Biol.* **2003**, *23*, 1963–1970.

135 Jafar-Nejad, H., Norga, K., Bellen, H., Numb. "Adapting" notch for endocytosis. *Dev. Cell* **2002**, *3*, 155–156.

136 Berdnik, D., Torok, T., Gonzalez-Gaitan, M., Knoblich, J., The endocytic protein α-adaptin is required for Numb-mediated asymmetric cell division in Drosophila. *Dev. Cell* **2002**, *3*, 221–231.

137 Wakamatsu, Y., Maynard, T. M., Jones, S. U., Weston, J. A., Numb localizes in the basal cortex of mitotic avian neuroepithelial cells and modulates neuronal differentiation by binding to Notch-1. *Neuron* **1999**, *23*, 71–81.

138 Zhong, W., Feder, J. N., Jiang, M. M., Jan, L. Y., Jan, Y. N., Asymmetric localization of a mammalian numb homolog during mouse cortical neurogenesis. *Neuron* **1996**, *17*, 43–53.

139 Knoblich, J. A., Jan, L. Y., Jan, Y. N., The N terminus of the *Drosophila* Numb protein directs membrane association and actin-dependent asymmetric localization. *Proc. Natl. Acad. Sci. USA* **1997**, *94*, 13005–13010.

140 Santolini, E., Puri, C., Salcini, A. E., Gagliani, M. C., Pelicci, P. G., Tacchetti, C., Di Fiore, P. P., Numb is an endocytic protein. *J. Cell Biol.* **2000**, *151*, 1345–1352.

141 Qin, H., Percival-Smith, A., Li, C., Jia, C. Y., Gloor, G., Li, S. S., A novel transmembrane protein recruits numb to the plasmic membrane in asymmetric cell division. *J. Biol. Chem.* **2003**, in press.

142 Chien, C. T., Wang, S., Rothenberg, M., Jan, L. Y., Jan, Y. N., Numb-associated kinase interacts with the phosphotyrosine binding domain of Numb and antagonizes the function of Numb in vivo. *Mol. Cell. Biol.* **1998**, *18*, 598–607.

143 O'Connor-Giles, K. M., Skeath, J. B., Numb inhibits membrane localization of Sanpodo, a four-pass transmembrane protein, to promote asymmetric divisions in Drosophila. *Dev. Cell* **2003**, *5*, 231–243.

144 Eck, M. J., Dhe-Paganon, S., Trub, T., Nolte, R. T., Shoelson, S. E., Structure of the IRS-1 PTB domain bound to the juxtamembrane region of the insulin receptor. *Cell* **1996**, *85*, 695–705.

145 Zhou, M. M., et al., Structural basis for IL-4 receptor phosphopeptide recognition by the IRS-1 PTB domain. *Nat. Struct. Biol.* **1996**, *3*, 388–393.

146 Li, S. C., Zwahlen, C., Vincent, S. J., McGlade, C. J., Kay, L. E., Pawson, T., Forman-Kay, J. D., Structure of a Numb PTB domain–peptide complex suggests a basis for diverse binding specificity. *Nat. Struct. Biol.* **1998**, *5*, 1075–1083.

147 Zhang, Z., Lee, C. H., Mandiyan, V., Borg, J. P., Margolis, B., Schlessinger, J., Kuriyan, J., Sequence-specific recognition of the internalization motif of the Alzheimer's amyloid precursor protein by the X11 PTB domain. *EMBO J.* **1997**, *16*, 6141–6150.

148 Stolt, P. C., Jeon, H., Song, H. K., Herz, J., Eck, M. J., Blacklow, S. C., Origins of peptide selectivity and phosphoinositide binding revealed by structures of Disabled-1 PTB domain complexes. *Structure (Camb)* **2003**, *11*, 569–579.

149 Yun, M., et al., Crystal structures of the dab homology domains of mouse disabled 1 and 2. *J. Biol. Chem.* **2003**, *278*, 36572–36581.

150 Shi, N., et al., Structural basis for the specific recognition of RET by the Dok1 PTB domain. *J. Biol. Chem.* **2003**, in press.

151 Yan, K. S., Kuti, M., Yan, S., Mujtaba, S., Farooq, A., Goldfarb, M. P., Zhou, M. M., FRS2 PTB domain conformation regulates interactions with divergent neurotrophic receptors. *J. Biol. Chem.* **2002**, *277*, 17088–17094.

152 Zwahlen, C., Li, S. C., Kay, L. E., Pawson, T., Forman-Kay, J. D., Multiple modes of peptide recognition by the PTB domain of the cell fate determinant Numb. *EMBO J.* **2000**, *19*, 1505–1515.

153 Songyang, Z., Margolis, B., Chaudhuri, M., Shoelson, S. E., Cantley, L. C., The phosphotyrosine interaction domain of SHC recognizes tyrosine-phosphorylated NPXY motif. *J. Biol. Chem.* **1995**, *270*, 14863–14866.

154 Trub, T., Choi, W. E., Wolf, G., Ottinger, E., Chen, Y., Weiss, M., Shoelson, S. E., Specificity of the PTB domain of Shc for beta turn-forming pentapeptide motifs amino-terminal to phosphotyrosine. *J. Biol. Chem.* **1995**, *270*, 18205–18208.

155 Farooq, A., Zeng, L., Yan, K. S., Ravichandran, K. S., Zhou, M. M., Coupling of folding and binding in the PTB domain of the signaling protein Shc. *Structure (Camb)* **2003**, *11*, 905–913.

156 Su, H. P., Nakada-Tsukui, K., Tosello-Trampont, A. C., Li, Y., Bu, G., Henson, P. M., Ravichandran, K. S., Interaction of CED-6/GULP, an adapter protein involved in engulfment of apoptotic cells, with CED-1 and CD91/LRP. *J. Biol. Chem.* **2001**, *277*, 11772–11779.

157 Dho, S. E., French, M. B., Woods, S. A., McGlade, C. J., Characterization of four mammalian numb protein isoforms. Identification of cytoplasmic and membrane-associated variants of the phosphotyrosine binding domain. *J. Biol. Chem.* **1999**, *274*, 33097–33104.

7
The FHA Domain

Daniel Durocher

7.1
Introduction

One of the major and universal roles of reversible protein phosphorylation is the regulation of protein–protein interactions [1]. Protein phosphorylation can influence protein–protein interactions in multiple ways. For example, phosphorylation of a serine, threonine, or tyrosine residue can result in the masking of an epitope necessary for the interaction of a protein with another. Conversely, phosphorylation can induce a conformational change that allows two proteins to interact. Finally, the phosphorylated residue can be an essential and integral part of an epitope used in a protein–protein interaction. This latter mode of influencing protein function via protein phosphorylation is a recurring theme in many cellular processes and in almost all signal transduction pathways of eukaryotes [2, 3]. Phospho-dependent protein–protein interactions are so versatile and powerful that, during the course of evolution, protein domains specialized in mediating such interactions have arisen multiple times [2, 3]. The main advantage of using protein modules resides in their 'transferability', since they are independently folding units for which the coding DNA can be shuffled around the genome [4]. This shuffling results in the generation of multidomain proteins with a diverse array of functions. The class of phospho-recognition modules is composed of the SH2, PTB, BRCT, and WW domains, Polo-box, 14-3-3 proteins [3, 5–8], and the subject of this chapter, the forkhead-associated (FHA) domain (reviewed in [9–12]).

The FHA domain was discovered in 1995 by Hofmann and Bucher, who recognized a protein sequence motif in a subset of forkhead-type transcription factors [13]. A set of approximately 20 proteins containing this domain signature was identified by profile searches, and a first consensus alignment of the FHA domain was derived [13]. Today, using other methods of database searching such as hidden Markov models (HMMs) and iterative BLAST searches, the FHA domain can be found in more than 400 different proteins (see SMART and PFAM databases) from eukaryotic, eubacterial, and some archeal species. For humans, the ENSEMBL database lists 23 unique FHA-containing proteins encoded by the human genome.

Modular Protein Domains. Edited by G. Cesareni, M. Gimona, M. Sudol, M. Yaffe

ISBN: 3-527-30813-X

However, this number is likely to be higher, given that some additional FHA-containing proteins are clearly being missed by current HMM searches (Table 7.1). For example, the mammalian DNA repair enzyme polynucleotide kinase (PNK) was recently proposed by Caldecott and colleagues to have a divergent FHA domain that shares homology with the histidine triad-containing protein, aprataxin (APTX) [14]. The gene encoding APTX is mutated in the neurological disease ataxia with occular apraxia (AOA) [15, 16] and, given the high degree of similarity between APTX and PNK, is proposed to be involved in the repair of single-strand breaks as is PNK. Recent work from our laboratory suggests that the divergent FHA domain of PNK does function as a phosphopeptide-binding domain (A. Koch, R. Agyei, and D. Durocher, unpublished data). Interestingly, iterative PSI-BLAST searches readily identify a third member of this subgroup (MGC47799, gi:27734905) thus, in total, adding three new FHA-containing proteins to the 23 reported in the ENSEMBL database. Since the three noted FHA domains are not detected by many of the bioinformatics tools used at present, it is tempting to wonder how many additional FHA domains exist unnoticed in protein databases.

Table 7.1 Human FHA-containing proteins.

Name	*Accession number*	*Indigene cluster*	*Other domains*	*Role*	*Associated disease*
Chk2	GI:6005850	Hs.146329	Kinase	Cell cycle checkpoint	variant Li–Fraumeni syndrome; breast cancer
Nbs1/Nibrin	GI:4505339	Hs.25812	BRCT	DNA repair/ checkpoint	Nijmegen-breakage syndrome
MDC1	GI:7661966	Hs.433653	BRCT	Checkpoint	None reported
Ki-67	GI:4505189	Hs.80976	PEST	Mitosis	None reported
Chfr	GI:8922675	Hs.23794	Ring finger	Prophase checkpoint	None reported
KAB1/KIAA0470	GI:7662142	Hs.25132	Similar to KIAA0284	Interaction with KARP-1	None reported
KIAA0284	GI:37546019	Hs.182536	Similar to KAB1	?	None reported
NIPP1	GI:13699256	Hs.356590	ND	Protein phosphatase I regulatory subunit	None reported
AF-6/MLLT4	GI:5174575	Hs.100469	RalGDS, DIL, PDZ	Cellular adhesion/signal transduction	None reported

Table 7.1 (continued)

Name	*Accession number*	*Indigene cluster*	*Other domains*	*Role*	*Associated disease*
Foxk1/MNF/ILF	GI:3183529	Hs.439387	Forkhead	Transcription factor	None reported
TCF19	GI:6005892	Hs.512706	PHD	Chromatin (?)	None reported
MCRS1	GI:29893564	Hs.25313	ND	Nucleolar, interacts with Herpes simplex protein	None reported
KIF1A/ATSV	GI:2497523	Hs.389765	Kinesin, PH domain	Motor protein	None reported
KIF1B	GI:17377940	Hs.444757	Kinesin	Motor protein	Charcot–Marie–Tooth disease type 2A
KIF1C	GI:3913961	Hs.525871	Kinesin	Motor protein	None reported
KIF13B	GI:318739	Hs.15711	Kinesin	Motor protein	None reported
KIF14	GI:7661878	Hs.3104	Kinesin, ERM	Motor protein	None reported
Smad nuclear interacting protein	GI:20072537	Hs.99601	ND	Interacts with Smad proteins	None reported
FLJ10283/ HSU84971	GI:8922326	Hs.104530	RRM motifs	RNA metabolism	None reported
KIAA0638 protein; LL5alpha	GI:38424073	Hs.149653	Kinesin domain PH domain	Motor protein	None reported
TIF2A /T2BP	GI:33563370	Hs.310640	ND	Innate immunity (?)	None reported
HSU84971	GI:9558745	Hs.104530	G-patch	RNA binding (?)	None reported
RNF8	GI:34304336	Hs.24439	RING finger	Protein ubiquitination	None reported
PNK	GI:6005836	Hs.78016	DNA kinase and phos-phatases	DNA repair	None reported
APTX	GI:8923156	Hs.444529	Histidine triad	DNA repair (?)	Ataxia with ocular apraxia 1 (AOA1)
MGC47799	GI:27734905	Hs.258941	ND	?	None reported

Prior to the bioinformatics work that led to the recognition of the FHA domain signature, work from the Walker laboratory [17] gave the first clues as to what the FHA domain might do. In a search for proteins that could interact with the plant receptor-like protein kinase RLK5, they identified a PP2C-type phosphatase that they named KAPP, for kinase-associated protein phosphatase [17]. The region of KAPP required for the interaction was termed the kinase-interaction (KI) domain, and it was shown that only the phosphorylated form of the kinase can be bound by KAPP [17]. This observation suggested a mechanism analogous to that of other eukaryotic signaling systems, in which tyrosine-phosphorylated receptor tyrosine kinases are bound by effectors and regulators via phospho-dependent interactions.

Following the demonstration by Hoffmann and Bucher [13] that the minimal KI region of KAPP has homology to the FHA domain, Walker and colleagues [18] sought to investigate the relevance of the FHA homology for KI activity. They employed site-directed mutagenesis to disrupt amino acid residues conserved among FHA domains. Mutation of these conserved residues abolished KI activity, suggesting that the FHA domain of KAPP mediates phospho-dependent protein–protein interactions [18]. However, this result did not resolve the important issue of whether these interactions directly involve the recognition of a phosphorylated residue or whether they are initiated by a conformational change triggered by phosphorylation.

In a parallel set of experiments, studies on the budding yeast DNA damage checkpoint brought in the last elements to complete the picture of FHA domain function. In eukaryotes, a conserved protein-phosphorylation cascade is initiated in response to DNA damage detection [19]. This cascade is governed by ATM-like kinases, and the DNA damage signal is propagated downstream to members of the Chk1 and Chk2 kinase families (reviewed in [19]). Chk2 and its homologs are recognizable by the presence of one or two FHA domains flanking a kinase domain of the Ca^{2+}/calmodulin kinase family [20]. In budding yeast, two Rad53 paralogs are also observed: Dun1 and Mek1/Mre4 are both protein kinases that play roles in the mitotic and meiotic checkpoint responses, and for which the FHA domain plays an essential role [21, 22]. Early on, the presence of FHA domains on Rad53 suggested a role in protein–protein interactions and, consistent with this, Rad53 was found to interact with Rad9 following DNA damage [23–26]. The interaction between Rad9 and Rad53 is dependent on Rad9 phosphorylation and was shown to require the C-terminal FHA domain (FHA2) of Rad53, although the N-terminal FHA domain (FHA1) has now been shown to play an important role in binding to Rad9 ([27] and D. Durocher and F. Sweeney, unpublished data).

To determine whether the FHA-dependent interaction involved direct recognition of a phospho epitope on Rad9, Durocher et al. [25] tested a series of phosphopeptides for their ability to compete with the Rad9–Rad53^{FHA1} interaction. A phosphothreonine-containing peptide derived from the N terminus of p53, but not its unphosphorylated counterpart, was found to efficiently compete with the Rad9–Rad53^{FHA1} interaction [25]. In addition, this p53-derived phosphopeptide was shown to directly bind Rad53^{FHA1}, in experiments making use of a variety of assays such as pulldowns, surface plasmon resonance, and isothermal titration calorimetry.

Mutation of the conserved Arg70 and His88 residues, but not of the variable Asp117 residue, in Rad53^{FHA1} abolished its interaction with both synthetic phosphopeptides and phosphorylated Rad9 [25]. Furthermore, the Rad53^{FHA1} domain was shown to protect the phosphorylated residue from phosphatase treatment, suggesting that the phosphate moiety is in intimate contact with the FHA domain [25]. Intriguingly, substitution of the phosphothreonine with a phosphoserine residue abolished the interaction, suggesting that the FHA1 domain of Rad53 is specific for phosphothreonine-containing epitopes [25].

These latter results indicated that the Rad53^{FHA1} domain recognizes phosphothreonine-containing epitopes. In an effort to test whether most if not all FHA domains possess the same binding capacity, Durocher et al. [25] tested the binding of various FHA domains from various species (including plants and bacteria) to degenerate peptide libraries using surface plasmon resonance. This assay unequivocally indicated that all the FHA domains tested could recognize the phosphothreonine peptide library but not the unphosphorylated counterpart [25]. Collectively, these results strongly suggested that the FHA domain directly recognizes threonine-phosphorylated epitopes on proteins.

7.2 FHA Domain Structure

7.2.1 Topology

In their bioinformatics work describing the FHA domain, Hoffmann and Bucher [13] predicted that the FHA domain would consist mainly of β strands. Their prediction was confirmed with the structure determinations of the Rad53 FHA1 and FHA2 domains by X-ray crystallography and NMR, respectively [28–30]. These structures revealed that the FHA domain is composed of an 11-stranded β sandwich consisting of two twisted β sheets (Figure 7.1) [28–30]. To date, four other FHA domain structures have been reported (Chfr, Ki67, Chk2, KAPP), and the above topology is shared by all four of these FHA domains [31–34]. Small variations between structures include the length and the spatial arrangements of the loops connecting the β strands and the presence of small helical insertions. Furthermore, the Chfr FHA domain lacks the eleventh β strand, since the domain forms a segment-swapped dimer under crystallographic conditions (discussed in more detail below [33]).

The FHA domain, as a folded unit, is approximately 100–120 amino acids long, much larger than the FHA homology region used for profile or HMM searches (which is around 75 amino acids) [28, 29, 35]. The core homology motif encompasses six of the 11 β strands (β3 to β8) and loops β3/4 to β7/8. As explained below, the homology motif contains the three loops that directly contact the phosphopeptide, as well as the structural framework needed to organize these loops in a spatial arrangement conducive to phosphopeptide binding [28, 31].

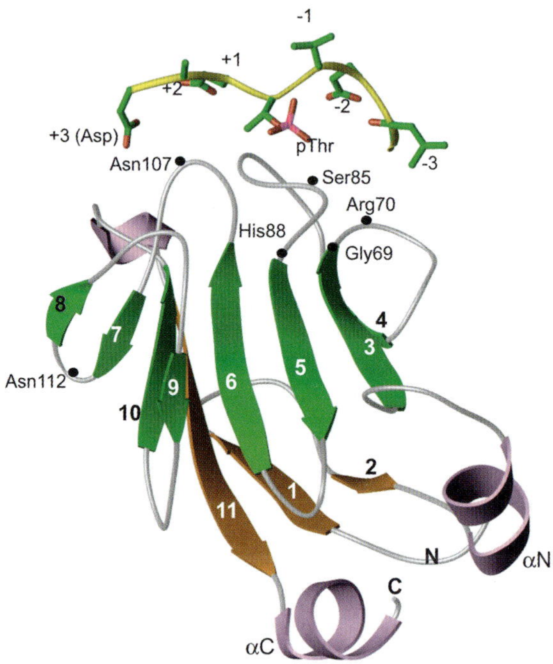

β1 β2 αN β3 β4
Rad53 (34) CRVIC TTGQIPI RDL SADIS QVLK EKRSIK KVWTF GRNPACD YH LGNI

β5 β6 β7 β8 β9
(82) SRLSNKH FQILL GEDGN LLLNDI STNG TWL NG QK EKNSNQLL SQG

β10 β11
(128) D EITV GVGVESDI LSLVIF IN

Figure 7.1 Upper panel: ribbon representation of the Rad53^{FHA1} domain in complex with a phosphothreonine-containing peptide (shown in ball-and-stick style). The core FHA homology (sheets β3–β10) is colored green. Conserved residues on the FHA domain discussed in the chapter are also indicated. Lower panel: the amino acid sequence of the Rad53FHA1 domain. The secondary structure elements are boxed following the same color scheme as in the ribbons diagram. Helix αC is omitted.

FHA domains are usually monomeric, but a structure determination of the Chfr FHA domain revealed that, under some conditions, the FHA domain can form a segment-swapped dimer in which the four C-terminal β strands and an α helix are exchanged between two FHA domains [33]. Although this dimer may form as a consequence of crystallization conditions, a number of reports have indicated that segment-swapping may be physiologically relevant (discussed in [33]). Further studies are required to determine whether this is true for Chfr and whether other FHA domain can undergo segment swapping under physiological conditions. If this type of dimer can exist under physiological conditions, it raises the possibility that segment-swapped FHA domain heterodimers may also exist.

7.2.2
FHA–Phosphopeptide Interaction

The FHA–phosphopeptide interaction buries between 600 Å^2 (Rad53^{FHA1}) and 1000 Å^2 (Chk2) of the FHA domain's solvent-accessible surface [28, 31]. The phosphopeptide, when bound to the FHA domain, adopts an extended conformation and interacts via the loops connecting the β strands (loops β3/4, β4/5 and β6/7; Figure 7.1 [28, 31]). Interestingly, this mode of binding is reminiscent of the binding of the complementary determining region (CDR) of antibodies (also β sandwiches) to antigen epitopes [36]. In the latter interaction, the antigen epitope is also bound in an extended conformation and interacts with the CDR via loops connecting the β strands [36]. As mentioned above, the determinants of phosphopeptide recognition by the FHA domain are all embedded in the FHA domain homology. Most of the conserved residues within the FHA domain are involved in binding the phosphopeptide. In the Rad53^{FHA1} domain, six amino acid residues of the phosphopeptide (–2, –1, phosphothreonine, +1, +2, +3) are contacted via either main-chain interactions (for residues –2, +1, and +3) or side-chain interactions (for the phosphothreonine and +3 residue). The conserved residues involved in this network of interaction includes, on Rad53^{FHA1}: Arg70, Ser82, and Asn107 located on loops β3/4, β4/5, and β6/7, respectively (Figure 7.1).

Unlike many phospho-dependent interactions, such as those mediated by SH2 domains or 14-3-3 domains, the phosphothreonine residue does not make direct salt interactions with arginine residues but is rather bound through an array of hydrogen bonds involving Arg70 and Ser85 in Rad53^{FHA1} [28]. Arg70 in Rad53^{FHA1} (Arg117 in Chk2) is the only conserved arginine residue involved in a direct contact with the phosphoamino acid residue, and it contacts it solely via a charged hydrogen bond with the γ oxygen of phosphothreonine. The second conserved interaction between the FHA domain and the phosphate moiety is the hydrogen bond between the hydroxyl group of Ser85 and the ε oxygen of the phosphate [28, 31].

The two other functionally important and strongly conserved residues in the FHA domain, Gly69 and His88 in Rad53^{FHA1}, do not play a role in directly contacting the phosphopeptide, despite being essential for phosphopeptide binding activity [18, 25]. The function of His88 appears to be to stabilize the architecture of the phosphopeptide binding site [28]. In particular, the imidazole ring of His88 tethers the β4/5 and β6/7 loops through hydrogen bonds with the main-chain atoms of Ser85 [28].

A unique feature of the FHA domain is its ability to discriminate phosphothreonine from phosphoserine residues [25]. Threonine differs from serine only by the presence of a γ methyl group on the threonine side-chain. The crystal structure of the Chk2 FHA and Rad53^{FHA1} domains indicate that phosphothreonine is selected as a consequence of the docking of the γ methyl group to a small pocket consisting of a conserved Asn residue (Asn166 in Chk2, Asn107 in Rad53) and surrounding amino acid residues, via van der Waals interactions [28, 31]. These interactions facilitate the orientation of the phosphothreonine residue in a position suitable for establishing a network of hydrogen bonds necessary to contact the FHA domain.

Several FHA domains seem to be selective for particular amino acid residues at the +3 position relative to the phosphothreonine. For example, Rad53^{FHA1} selects for an Asp residue at this position, whereas the Chk2 FHA domain selects for Ile/Leu, and the mycobacterial protein Rv1827 strongly selects for Tyr [28, 30, 37]. Some light on the structural basis of this +3 selectivity was shed by comparing the structures of the Rad53^{FHA1} and Chk2 FHA domains bound to high-affinity-binding phosphopeptides [28, 31]. Sequence selectivity in the Rad53^{FHA1} domain is imparted by the formation of a salt bridge between the guanidinium moiety of Arg83 (β4/5 loop) and the carbonyl group of the +3 Asp side-chain [28]. However, in Chk2, the hydrophobic selectivity at +3 can be explained by a complementary shallow hydrophobic pocket formed by the side-chain atoms of Thr138, equivalent to the position occupied by Arg83 in Rad53, along with the side-chain atoms of Asn166 (β6/7 loop) and Ser192/Leu193 (β10/11 loop) [31]. Since the β6/7–β10/11 loops are poorly conserved in sequence and size among various FHA domains, these findings suggest that each FHA domain may display its own specificity.

7.2.3
A Second Binding Interface?

Recent work suggests that the FHA domain may be able to bind proteins outside the phosphopeptide binding surface or that an ancillary binding surface cooperates with the phosphopeptide binding activity. A strong argument for an ancillary binding interface first emerged from studying the impact of the Ile157–Thr mutation of Chk2 found in families displaying a rare variant of the Li–Fraumeni syndrome [38], a highly penetrant familial cancer disease. Ile157 is located far from the phosphopeptide binding site, at the C-terminal end of the β5/6 loop, some 25 Å away from the phosphothreonine-binding area [31]. Accordingly, mutation of Ile157 to Thr does not affect binding to phosphopeptides in vitro nor does it affect protein stability in vitro and in vivo [31]. Surprisingly however, this mutation abrogates the binding of the Chk2 FHA domain to a number of proteins, including Brca1 and Cdc25C [31, 39]. The interaction with Brca1 also requires Arg117, indicating that the interaction is phospho-dependent [31]. Thus, the Arg117- and Ile157-dependent surfaces act cooperatively to bind certain proteins. Li et al. [34] proposed that Ile157 lies at the heart of a hydrophobic surface that stabilizes an interaction initiated by the phospho-dependent engagement of Arg117 [31]. Interestingly, evolutionary trace (ET) analysis performed on a large number of FHA domains suggests that the surface surrounding Ile157 in Chk2 may be functionally conserved in a large number of FHA domains [34].

The recent characterization Ki-67 and its interaction with hNIFK provides a second argument supporting the possibility that some FHA domains require additional binding surfaces for optimal binding [32]. The human NIFK protein was discovered as a binding partner of the mitotic/chromosome condensation regulator Ki-67 in a two-hybrid screen using the FHA domain of Ki-67 [40]. The interaction region of hNIFK was mapped to a peptide of 44 amino acid residues containing two threonine residues that are required for binding [32]. Tsai's group [32] detected high-affinity phospho-dependent binding between synthetic peptides corresponding to the minimal interaction region of hNIFK and the Ki-67 FHA domain, as expected [32]. Furthermore, they deduced that the phosphothreonine at a position equivalent to Thr234 in hNIFK was required for the interaction. However, two unexpected binding properties were noted: first, using HSQC NMR experiments, a weak interaction was detected between the unphosphorylated 44-mer and the Ki-67 FHA domain (with a K_D of approximately 100 μM), indicating that a binding determinant outside the canonical phosphothreonine-recognition determinants may exist [32]. Furthermore, the interaction requires a phosphopeptide longer than the 6 amino acid residue phosphopeptide normally sufficient for binding to the Rad53, Chk2, Rv1827, and other FHA domains [32]. This requirement for a longer phosphopeptide indicates either that Ki-67 binds structured epitopes or that a second set of binding determinants is present on the hNIFK-derived peptide. In support of the latter possibility, NMR analysis of chemical shift changes on the Ki-67 FHA domain indicated that residues outside the phosphopeptide-binding loops are involved in binding the hNIFK 44-mer [32]. This potential binding surface includes the β1/2 loop and residues located on β10, a strand neighboring β1 and β2. If this surface is experimentally confirmed to be required for binding, it would be a distinct surface from the ancillary binding surface of Chk2 which, as mentioned above, is located at the C terminus of the β5/6 loop.

7.2.4 The FHA Domain is Part of a Domain Superfamily

Structural homology searches place the FHA domain in a superfamily of domains with similar topology. The topology of the FHA domain is similar to that of the Smad MH2 domain and the recently described IRF-3 activation domain (IAD) [9, 10, 41–43]. All three share an 11-stranded β sandwich core of nearly identical topology. The IRF IAD and SMAD MH2 domains are more closely related to each other, having additional structural features that surround the β sandwich not present in the FHA domain [41].

Interestingly, all three domains can act as transcriptional activators (see below) and two of them (the MH2 and FHA domains) have the capacity to engage in phospho-dependent protein–protein interactions [28, 44]. IRF-3 IAD also engages in phospho-dependent dimerization, although it is yet unclear if this occurs via phospho-epitope recognition [41]. As noted previously, this structural and functional homology suggests that the FHA/MH2/IAD superfamily shares a common ancestor [9, 44], and it is tempting to speculate that this ancestor may have been involved in the regulation of transcription. As structural genomics programs progress, novel members of this superfamily may emerge. If so, it will be interesting to determine if they too are involved in phospho-dependent interactions and/or in the regulation of transcription.

7.3
Molecular and Signaling Function

To date, the only molecular function of the FHA domain is its phosphopeptide binding activity. However, this function is used in a number of different ways to regulate the activity of the proteins containing it. Below, I briefly review some of the ways FHA-mediated protein–protein interactions are used to accomplish this task.

7.3.1
FHA Domain Can Regulate Protein Localization

Protein interaction modules such as the PDZ and PH domains often regulate the activity of proteins by localizing them to distinct subcellular compartments [2]. The FHA domain is no exception and has also been shown to act in this manner. For example, the FHA-containing protein NIPP1 is a regulatory subunit of protein phosphatase 1. NIPP1 localizes to nuclear speckles, which are sites of pre-mRNA processing where it recruits PP1 [45, 46]. The Bollen group [45] has shown that the FHA domain NIPP1 is critical for targeting the NIPP1 to nuclear speckles and for nuclear retention. The spliceosomal protein that binds the FHA domain of NIPP1 remains unknown, but both CDC5L and SAP155 are excellent candidates, because both can interact with NIPP1 in an FHA- and phosphorylation-dependent manner [47, 48]. Interestingly, the interaction between NIPP1 and SAP155 is controlled by mitotic phosphorylation of SAP155 [48]. It will be interesting to determine whether the proline residue, characteristic of mitotic phosphorylation sites, is an integral part of the motif recognized by the FHA domain of NIPP1.

Another striking example of FHA-dependent protein localization is the localization of Dma1 to spindle pole bodies (SPBs). Dma1 is an FHA-containing ring finger protein from *Schizosaccharomyces pombe* analogous to human Chfr, a protein often mutated in tumors [49, 50]. Dma1 is a component of the septation-initiation network (S/N), a signaling pathway that links the end of mitosis with cytokinesis (called septation in fission yeast) [49, 51]. Elegant work by Guertin and colleagues [52] recently showed that Dma1 localizes to SPBs in an FHA-dependent manner via an interaction with the spindle pole component Sid4 [52]. The localization of Dma1 to SPBs is essential for its function in regulating SIN by modulating the activity of Plo1, the fission yeast polo kinase [52]. Interestingly, this function of Dma1 is also conserved in humans, since Chfr regulates the destruction and activity of human Polo kinase, Plk1 [53].

7.3.2
FHA Domain Binding to Enzyme Substrates

A number of FHA-containing proteins also display catalytic activity. These include kinases, phosphatases, ubiquitin ligases, protein motors, prolyl isomerases, and oxidases. It is therefore not surprising to find that the FHA domain can serve to recruit substrates. The first and most characterized example concerns the plant PP2C-type phosphatase KAPP [17]. As mentioned above, Walker and colleagues [17] identified KAPP in 1994 by virtue of its ability to interact with the phosphorylated RLK5 receptor, suggesting that KAPP may modulate its activity [17]. More recent genetic studies in *Arabidopsis* indicate that KAPP is a negative regulator of CLAVATA 1 (CLV1)-dependent signaling. As with the KAPP–RLK5 interaction, KAPP was found to directly bind to activated CLV1, a plant receptor-like kinase [54]. These genetic and physical interactions suggest that recruitment of KAPP by activated receptors during plant signal transduction acts in an analogous manner to the recruitment of SH2-containing tyrosine phosphatases to activated receptor tyrosine kinases, and that phosphorylation-dependent attenuation of receptor-mediated signaling is a common feature of both plant and animal cells [10].

The cis–trans prolyl isomerase PinA in the slime mould *Dictyostelium discoideum* may represent a particularly interesting example of FHA-dependent recruitment of substrates. PinA is the slime-mould homolog of Pin1, a cis–trans prolyl isomerase with specificity for proline residues preceded by phosphoserine or phosphothreonine. In most metazoans, the phospho dependence of Pin1 substrate recognition requires a specialized WW domain [55]. However, in PinA, a WW domain is not apparent, but the presence of an FHA domain in its place indicates that it may serve an equivalent function. The FHA domain of PinA could therefore help to recruit phosphorylated proteins for proline isomerization. If this is true, PinA may represent an instance of convergent evolution or domain shuffling.

7.3.3
FHA Domain Binding to Regulators

The function of FHA domain-containing proteins can also be modulated upon FHA-dependent phosphoprotein binding. In fact, it is likely that a large proportion of FHA-dependent interactions will be found to regulate the function of the FHA-containing protein. For example, the FHA-dependent Rad53–Rad9 interaction, which led to the discovery of the FHA domain as a phosphothreonine binding domain, is essential for the activation of Rad53 in response to DNA damage [56]. However, the exact mechanism by which Rad9 regulates Rad53 activity remains poorly understood and is still under investigation.

7.3.4
Reversible Protein–Protein Interactions

Intuitively, the presence of a phosphopeptide binding domain is suggestive of an ability to form protein–protein interactions in response to signals. However, in some instances, the interaction between the FHA domain and the phosphoprotein appears constitutive under normal conditions but is rapidly inhibited under perturbing conditions. In these examples, the phospho-dependent interaction may function as a switch to enforce rapid reversibility. This variation on the theme of dynamic interactions appears to underlie the interaction between Rad53 and Asf1

[27, 57]. In normally cycling budding yeast cells, Rad53 binds constitutively to Asf1 in an FHA-dependent manner [27]. However, after DNA damage, Rad53 is released from Asf1 by an as-yet-uncharacterized mechanism. Thus, Rad53 appears to be 'released' from Asf1 upon detection of DNA damage, allowing it to accomplish its functions at the DNA replication fork or to signal DNA damage to the transcription and cell cycle machineries. Hence, in this particular instance, the main advantage of an FHA-dependent interaction may reside in the ability to break the interaction in response to a stimulus.

7.3.5 FHA Domain as a Transcriptional Activator Domain

The forkhead-type transcription factor Fkh2 of budding yeast is a key regulator of the transcription program governing mitosis [58–60]. Fkh2 binds alongside the MADS-box Mcm1 protein to the promoter region of mitosis-specific genes. Fkh2 binds to regulatory sequences known as SFF elements, which act as positive elements during mitosis and repressor elements during interphase [58–60]. Fkh2 binds the SFF element throughout the cell cycle, indicating that an additional level of regulation must exist to allow SFF-dependent transcriptional activation in M phase and SFF-dependent repression of M-phase genes in interphase. This regulation is imparted by the binding of the coactivator Ndd1 during mitosis (Figure 7.2) [58, 61, 62]. The Ndd1–Fkh2 interaction is controlled by Cdk phosphorylation and was recently demonstrated to be mediated by a phospho-dependent interaction between the FHA domain of Fkh2 and phThr239 of Ndd1 [61, 62]. These results demonstrate that the FHA domain of Fkh2 can act as a transcriptional activator. Indeed, mutation of the critical Arg residue in the Fkh2 FHA domain mimics an *NDD1* deletion [61, 62]. In the absence of Ndd1 binding, the FHA domain of Fkh2 appears to act as repressor. In support of this idea, *NDD1* deletion is lethal, whereas the double mutation *ndd1Δ fkh2Δ* is viable [58]. This genetic interaction is consistent with the model whereby Thr239-phosphorylated Ndd1 acts to convert Fkh2 from a repressor to an activator of the G2/M transcription program [61, 62].

Fkh2 is part of a eukaryotic subfamily of forkhead transcription factors. In budding yeast, the Fkh2 paralogs Fkh1 and Fhl1 have a similar FHA domain–forkhead DNA-binding domain structure. Fkh1 plays a role in the regulation of G2/M transcription but is also critical for regulation of the recombination enhancer of the mating-type locus, whereas Fhl1 is a transcriptional regulator of ribosome biogenesis [63]. Interestingly, Fhl1 binds to a small regulator called Ifh1 and the *IFH1* gene is essential only when *FHL1* is present, suggesting a mode of transcriptional regulation essentially identical to the Fkh2–Ndd1 regulation of G2/M transcription (Figure 7.2) [64]. It will be interesting to determine if the interaction of Fhl1 to Ifh1 is phospho- and FHA-dependent and if it behaves similarly in other respects to the Fkh2–Ndd1 interaction.

The parallel between Fkh2–Ndd1 and Fhl1–IFh1 is striking (Figure 7.2). Given that at least one Fkh2/Fhl1 homologue is present in most eukaryotes sequenced to date, it will be interesting to determine whether this mode of phospho-dependent

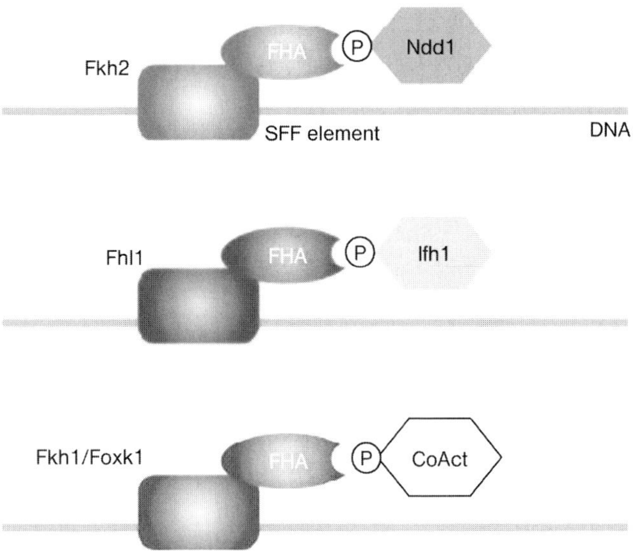

Figure 7.2 The FHA–forkhead paradigm. In budding yeast, FHA-containing forkhead-type transcription factors Fkh2 and Fhl1 (upper and middle panels) appear to act in a similar way. In conditions under which the FHA domain is not engaged in a phospho-dependent interaction, they act as transcriptional repressors. Upon FHA-dependent binding to their coactivator proteins, they are converted into transcriptional activators. See text for further details.

transcriptional activation is also conserved. In mammals, MNF/ILF/Foxk1 is the sole Fkh2-like protein present and was identified as a regulator of muscle gene expression and of the IL-2 promoter [65–67]. Interestingly, the N terminus of MNF/ILF/Foxk1 binds to the transcriptional corepressor mSIN3A [68]. The mSin3A-binding domain encompasses the FHA domain, perhaps indicating that phospho-dependent engagement of this forkhead transcription factor may convert it from a repressor to an activator. Further studies will be required to test this possibility. Nevertheless, these results indicate that Foxk1 may act in a manner similar to Fkh2/Fhl1-dependent regulation of transcription (Figure 7.2).

7.4 Emerging Research Direction

7.4.1 Bacterial FHA Domains

Prokaryotic FHA domains contain all the amino acid residues required for phosphothreonine binding and, indeed, two mycobacterial FHA domains (Rv1827 and Rv0020c) have been shown to bind to phosphothreonine libraries in vitro.

This result raised the exciting possibility that modular phosphopeptide recognition is widespread in bacteria [69]. Recent analysis of bacterial genomes has lent further support to this possibility. Pallen et al. [70] demonstrated a near-perfect correlation between the presence of genes encoding for FHA domain-containing proteins and the presence of genes coding for eukaryotic-type kinases and phosphatases [70], extending the earlier observations of Ponting et al [71]. This striking association suggests that bacterial FHA-containing proteins are engaging in processes in which phosphorylation-dependent protein–protein interactions play an important role. Such a possibility has very recently gained further support after the demonstration that EmbR, a putative transcriptional regulator, contacts the eukaryote-type protein kinase pknH in a phospho-dependent manner via its FHA domain [72]. Whether this interaction occurs in vivo remains to be demonstrated, but this recent observation supports the notion that 'eukaryotic-type' signaling systems operate in bacteria.

7.4.2
A Potential Role for FHA Domains During Innate Immunity?

The presence of bacterial FHA-containing proteins in type III secretion systems indicates that FHA-containing proteins may play a role during pathogenic bacterial infections [70]. The recent identifycation of T2BP/TIFA as a TRAF2/6-interacting FHA-containing protein suggests that at least one FHA-containing protein may play a role in innate immunity by influencing IL-1 and/or TNF signaling [73, 74]. T2BP/TIFA appears to play a positive role in NFκB and MAP kinase signaling, because its overexpression potentiates both signaling processes [73, 74]. The presence of FHA-containing proteins in these important signaling processes, as well as their presence in pathogenic bacteria, may suggest a role for bacterial FHA domains in pathogen–host interactions. Although this possibility remains speculative, some pathogenic bacteria may use their FHA domains to remodel TRAF2/6 complexes during bacterial infection. This hypothesis should be relatively straightforward to assess.

7.4.3
FHA Domain and Phosphothreonine-Proline Motifs

The characterization of the Rad53 and Chk2 FHA domains has led to a model whereby the main determinants for the FHA domain–phosphopeptide interaction are the phosphorylated residue along with the +3 residue [28]. Recent work on a number of FHA domains is changing this view of FHA-dependent phosphopeptide recognition. Indeed, the FHA domains of NIPP1, Ki-67, and possibly Fkh2/Fkh1 appear to specifically require a proline residue immediately C-terminal to the phosphothreonine residue [32, 48, 62]. As mentioned above, recent work from the Tsai laboratory indicates that, in the FHA domain of Ki67, the +1 proline residue may play an essential and direct role in phosphopeptide recognition. This striking characteristic places the FHA domain in a growing class of domains that recognize pSer/pThr-Pro motifs and which include the WD40 domain, the WW domain, and the polo box domain [8, 55, 75, 76]. It will be interesting to determine whether all the above FHA domains employ a common mechanism of proline recognition at the +1 position. If so, such determinants could prove valuable for predicting whether a given FHA-containing protein participates in processes regulated by proline-directed kinases.

7.4.4 FHA Domain Chimeras as Phosphorylation Biosensors

The ability of FHA domains to recognize different phosphothreonine-containing epitopes makes it an attractive tool for the development of genetically encoded phosphorylation reporters. In an elegant study, Newton's group [77] developed an FHA-based reporter that monitors the activation of protein kinase C (PKC) in vivo and in real time. The design of the reporter is illustrated in Figure 7.3 and is based on the alteration of fluorescence resonance energy transfer (FRET) between two fluorescent proteins, cyan fluorescent protein (CFP) and yellow fluorescent protein (YFP) [77]. Both fluorescent proteins flank the FHA2 domain of Rad53, an FHA domain with a preference for the bulky hydrophobic residues Ile and Leu at the +3 position C-terminal to the phosphothreonine [28]. The FHA domain is fused to both fluorescent proteins by flexible linkers, and a consensus phosphorylation site for PKC is introduced in the flexible linker between the FHA domain and the C-terminal YFP protein. Since the specificity of PKC phosphorylation is determined mainly by amino acid residues N-terminal to the phosphate acceptor threonine, it was possible to engineer a reporter protein that has optimal specificity for both PKC phosphorylation and FHA2-domain phosphothreonine recognition [77].

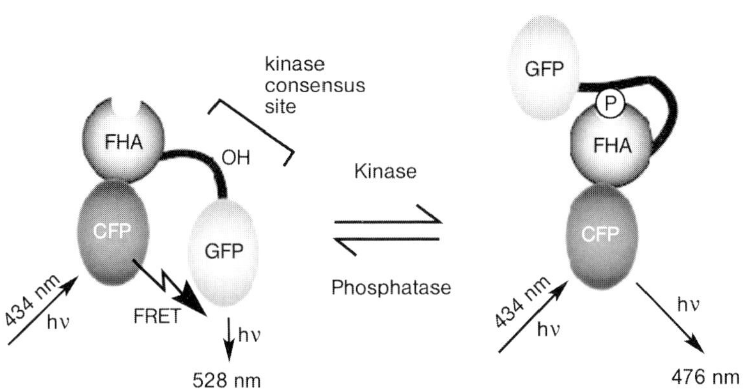

Figure 7.3 The FHA domain-based kinase activity reporter. The FHA domain is flanked by the CFP and YFP proteins and a consensus phosphorylation site for PKC (or any other kinase) is introduced in the flexible linker between the FHA domain and the C-terminal YFP protein. Upon kinase activation, the flexible linker is phosphorylated, creating an optimal FHA2 binding site. The FHA2 domain then engages the phosphorylated linker, resulting in a loss of FRET signal [77].

Upon PKC activation, the flexible linker is phosphorylated by PKC, creating an optimal FHA2 binding site [77]. The FHA2 domain then engages the phosphorylated linker in an intramolecular interaction, thereby moving one fluorescent protein away from the other, leading to a loss of FRET signal [77]. Interestingly, the relatively low affinity of the FHA2 domain for its optimal binding site appears to be a major factor in enabling the monitoring of dynamic phosphorylation events in real time. Indeed, the lower binding affinity enables the access of phosphatase to the phosphothreonine residue, with concomitant restoration of the FRET signal [77].

In addition to the PKC system, it is easy to imagine how this design could be applicable to the study of a wide variety of phosphorylation events. For example, FHA domains with proline-directed phospho recognition could be used to monitor Cdk and MAP kinase activation. Furthermore, since FHA2 domain phosphopeptide recognition is mainly dependent on the presence of an Ile residue three residues C-terminal to the phosphorylated residue, the reporter described by Newton et al. can be easily adapted to monitor the regulation of a number of kinases, including ATM-like kinases and basic-directed kinases.

7.5 Concluding Remarks

The discovery of the FHA domain as a phosphothreonine binding module was a significant step toward gaining a better understanding of Ser/Thr-based signaling systems in prokaryotes and eukaryotes. The understanding of FHA domain function is also helping us to decipher the molecular basis of some diseases, such as the variant Li–Fraumeni and Nijmegen-breakage syndromes. In the years to come, I expect that the study of FHA-containing proteins will continue to further of our understanding of various cellular processes and that the discovery of novel phospho-dependent protein interaction modules will have a similar impact as that of the FHA domain.

Acknowledgements

I thank the members of my laboratory for their help and support. I would also like to thank Frank Sicheri for his critical reading of this manuscript. I am a recipient of the Hitchings–Elion fellowship of the Burroughs–Wellcome Fund and I hold a Canada Research Chair (tier II) in Bioinformatics, Proteomics and Functional Genomics. Work in my laboratory is supported by grants from the Canadian Institutes for Health Research and the National Cancer Institute of Canada.

References

1 Pawson, T., Nash, P., Protein–protein interactions define specificity in signal transduction. *Genes. Dev.* **2000**, *14*, 1027–1047.
2 Pawson, T., Nash, P., Assembly of cell regulatory systems through protein interaction domains. *Science* **2003**, *300*, 445–452.
3 Yaffe, M. B., Elia, A. E., Phosphoserine/threonine-binding domains. *Curr. Opin. Cell Biol.* **2001**, *13*, 131–138.
4 Ponting, C. P., Russell, R. R., The natural history of protein domains. *Annu. Rev. Biophys. Biomol. Struct.* **2002**, *31*, 45–71.
5 Manke, I. A., Lowery, D. M., Nguyen, A., Yaffe, M. B., BRCT repeats as phosphopeptide-binding modules involved in protein targeting. *Science* **2003**, *302*, 636–639.
6 Rodriguez, M., Yu, X., Chen, J., Songyang, Z., Phosphopeptide binding specificities of BRCA1 COOH-terminal (BRCT) domains. *J. Biol. Chem.* **2003**, *278*, 52914–52918.
7 Yu, X., Chini, C. C., He, M., Mer, G., Chen, J., The BRCT domain is a phospho-protein binding domain. *Science* **2003**, *302*, 639–642.
8 Elia, A. E., Cantley, L. C., Yaffe, M. B., Proteomic screen finds pSer/pThr-binding domain localizing Plk1 to mitotic substrates. *Science* **2003**, *299*, 1228–1231.
9 Durocher, D., Jackson, S. P., The FHA domain. *FEBS Lett.* **2002**, *513*, 58–66.
10 Li, J., Lee, G. I., Van Doren, S. R., Walker, J. C., The FHA domain mediates phosphoprotein interactions. *J. Cell Sci.* **2000**, *113 Pt 23*, 4143–4149.
11 Tsai, M. D., FHA: a signal transduction domain with diverse specificity and function. *Structure (Camb)* **2002**, *10*, 887–888.
12 Hammet, A., Pike, B. L., McNees, C. J., Conlan, L. A., Tenis, N., Heierhorst, J., FHA domains as phospho-threonine binding modules in cell signaling. *IUBMB Life* **2003**, *55*, 23–27.
13 Hofmann, K., Bucher, P., The FHA domain: a putative nuclear signalling domain found in protein kinases and transcription factors. *Trends Biochem. Sci.* **1995**, *20*, 347–349.
14 Caldecott, K. W., DNA single-strand break repair and spinocerebellar ataxia. *Cell* **2003**, *112*, 7–10.
15 Date, H., et al., Early-onset ataxia with ocular motor apraxia and hypoalbuminemia is caused by mutations in a new HIT superfamily gene. *Nat. Genet.* **2001**, *29*, 184–188.
16 Moreira, M. C., et al., The gene mutated in ataxia-ocular apraxia 1 encodes the new HIT/Zn-finger protein aprataxin. *Nat. Genet.* **2001**, *29*, 189–193.
17 Stone, J. M., Collinge, M. A., Smith, R. D., Horn, M. A., Walker, J. C., Interaction of a protein phosphatase with an *Arabidopsis* serine-threonine receptor kinase. *Science* **1994**, *266*, 793–795.
18 Li, J., Smith, G. P., Walker, J. C., Kinase interaction domain of kinase-associated protein phosphatase, a phosphoprotein-binding domain. *Proc. Natl. Acad. Sci. USA* **1999**, *96*, 7821–7826.
19 Rouse, J., Jackson, S. P., Interfaces between the detection, signaling, and repair of DNA damage. *Science* **2002**, *297*, 547–551.
20 Bartek, J., Lukas, J., Chk1 and Chk2 kinases in checkpoint control and cancer. *Cancer Cell* **2003**, *3*, 421–429.
21 Bashkirov, V. I., Bashkirova, E. V., Haghnazari, E., Heyer, W. D., Direct kinase-to-kinase signaling mediated by the FHA phosphoprotein recognition domain of the Dun1 DNA damage checkpoint kinase. *Mol. Cell Biol.* **2003**, *23*, 1441–1452.
22 Wan, L., De Los Santos, T., Zhang, C., Shokat, K., Hollingsworth, N. M., Mek1 kinase activity functions downstream of RED1 in the regulation of meiotic double strand break repair in budding yeast. *Mol. Biol. Cell* **2004**, *15*, 11–23.
23 Sun, Z., Hsiao, J., Fay, D. S., Stern, D. F., Rad53 FHA domain associated with phosphorylated Rad9 in the DNA damage checkpoint. *Science* **1998**, *281*, 272–274.
24 Vialard, J. E., Gilbert, C. S., Green, C. M., Lowndes, N. F., The budding yeast Rad9 checkpoint protein is subjected to Mec1/Tel1-dependent hyperphosphory-

lation and interacts with Rad53 after DNA damage. *EMBO J.* **1998**, *17*, 5679–5688.

25 Durocher, D., Henckel, J., Fersht, A. R., Jackson, S. P., The FHA domain is a modular phosphopeptide recognition motif. *Mol. Cell* **1999**, *4*, 387–394.

26 Emili, A., MEC1-dependent phosphorylation of Rad9p in response to DNA damage. *Mol. Cell* **1998**, *2*, 183–189.

27 Schwartz, M. F., Lee, S. J., Duong, J. K., Eminaga, S., Stern, D. F., FHA domain-mediated DNA checkpoint regulation of Rad53. *Cell Cycle* **2003**, *2*, 384–396.

28 Durocher, D., Taylor, I. A., Sarbassova, D., Haire, L. F., Westcott, S. L., Jackson, S. P., Smerdon, S. J., Yaffe, M. B., The molecular basis of FHA domain: phosphopeptide binding specificity and implications for phospho-dependent signaling mechanisms. *Mol. Cell* **2000**, *6*, 1169–1182.

29 Liao, H., Byeon, I. J., Tsai, M. D., Structure and function of a new phosphopeptide-binding domain containing the FHA2 of Rad53. *J. Mol. Biol.* **1999**, *294*, 1041–1049.

30 Liao, H., et al., Structure of the FHA1 domain of yeast Rad53 and identification of binding sites for both FHA1 and its target protein Rad9. *J. Mol. Biol.* **2000**, *304*, 941–951.

31 Li, J., et al., Structural and functional versatility of the FHA domain in DNA-damage signaling by the tumor suppressor kinase Chk2. *Mol. Cell* **2002**, *9*, 1045–1054.

32 Li, H., Byeon, I. J., Ju, Y., Tsai, M. D., Structure of human Ki67 FHA domain and its binding to a phosphoprotein fragment from hNIFK reveal unique recognition sites and new views to the structural basis of FHA domain functions. *J. Mol. Biol.* **2004**, *335*, 371–381.

33 Stavridi, E. S., Huyen, Y., Loreto, I. R., Scolnick, D. M., Halazonetis, T. D., Pavletich, N. P., Jeffrey, P. D., Crystal structure of the FHA domain of the Chfr mitotic checkpoint protein and its complex with tungstate. *Structure (Camb)* **2002**, *10*, 891–899.

34 Lee, G. I., Ding, Z., Walker, J. C., Van Doren, S. R., NMR structure of the forkhead-associated domain from the *Arabidopsis* receptor kinase-associated protein phosphatase. *Proc. Natl. Acad. Sci. USA* **2003**, *100*, 11261–11266.

35 Hammet, A., Pike, B. L., Mitchelhill, K. I., Teh, T., Kobe, B., House, C. M., Kemp, B. E., Heierhorst, J., FHA domain boundaries of the Dun1p and Rad53p cell cycle checkpoint kinases. *FEBS Lett.* **2000**, *471*, 141–146.

36 Davies, D. R., Cohen, G. H., Interactions of protein antigens with antibodies. *Proc. Natl. Acad. Sci. USA* **1996**, *93*, 7–12.

37 Qin, D., Lee, H., Yuan, C., Ju, Y., Tsai, M. D., Identification of potential binding sites for the FHA domain of human Chk2 by in vitro binding studies. *Biochem. Biophys. Res. Commun.* **2003**, *311*, 803–808.

38 Bell, D. W., et al., Heterozygous germ line hCHK2 mutations in Li–Fraumeni syndrome. *Science* **1999**, *286*, 2528–2531.

39 Falck, J., Mailand, N., Syljuasen, R. G., Bartek, J., Lukas, J., The ATM–Chk2–Cdc25A checkpoint pathway guards against radioresistant DNA synthesis. *Nature* **2001**, *410*, 842–847.

40 Takagi, M., Sueishi, M., Saiwaki, T., Kametaka, A., Yoneda, Y., A novel nucleolar protein, NIFK, interacts with the forkhead associated domain of Ki-67 antigen in mitosis. *J. Biol. Chem.* **2001**, *276*, 25386–25391.

41 Moustakas, A., Heldin, C. H., The nuts and bolts of IRF structure. *Nat. Struct. Biol.* **2003**, *10*, 874–876.

42 Qin, B. Y., Liu, C., Lam, S. S., Srinath, H., Delston, R., Correia, J. J., Derynck, R., Lin, K., Crystal structure of IRF-3 reveals mechanism of autoinhibition and virus-induced phosphoactivation. *Nat. Struct. Biol.* **2003**, *10*, 913–921.

43 Takahasi, K., et al., X-ray crystal structure of IRF-3 and its functional implications. *Nat. Struct. Biol.* **2003**, *10*, 922–927.

44 Wu, J. W., et al., Crystal structure of a phosphorylated Smad2: recognition of phosphoserine by the MH2 domain and insights on Smad function in TGF-beta signaling. *Mol. Cell* **2001**, *8*, 1277–1289.

45 Jagiello, I., Van Eynde, A., Vulsteke, V., Beullens, M., Boudrez, A., Keppens, S., Stalmans, W., Bollen, M., Nuclear and

subnuclear targeting sequences of the protein phosphatase-1 regulator NIPP1. *J. Cell Sci.* **2000**, *113 Pt 21*, 3761–3768.

46 Trinkle-Mulcahy, L., Ajuh, P., Prescott, A., Claverie-Martin, F., Cohen, S., Lamond, A. I., Cohen, P., Nuclear organisation of NIPP1, a regulatory subunit of protein phosphatase 1 that associates with pre-mRNA splicing factors. *J. Cell Sci.* **1999**, *112*, 157–168.

47 Boudrez, A., et al., NIPP1-mediated interaction of protein phosphatase-1 with CDC5L, a regulator of pre-mRNA splicing and mitotic entry. *J. Biol. Chem.* **2000**, *275*, 25411–25417.

48 Boudrez, A., Beullens, M., Waelkens, E., Stalmans, W., Bollen, M., Phosphorylation-dependent interaction between the splicing factors SAP155 and NIPP1. *J. Biol. Chem.* **2002**, *277*, 31834–31841.

49 Murone, M., Simanis, V., The fission yeast *dma1* gene is a component of the spindle assembly checkpoint, required to prevent septum formation and premature exit from mitosis if spindle function is compromised. *EMBO J.* **1996**, *15*, 6605–6616.

50 Scolnick, D. M., Halazonetis, T. D., Chfr defines a mitotic stress checkpoint that delays entry into metaphase. *Nature* **2000**, *406*, 430–435.

51 Bardin, A. J., Amon, A., Men and sin: what's the difference? *Nat. Rev. Mol. Cell Biol.* **2001**, *2*, 815–826.

52 Guertin, D. A., Venkatram, S., Gould, K. L., McCollum, D., Dma1 prevents mitotic exit and cytokinesis by inhibiting the septation initiation network (SIN). *Dev Cell* **2002**, *3*, 779–790.

53 Kang, D., Chen, J., Wong, J., Fang, G., The checkpoint protein Chfr is a ligase that ubiquitinates Plk1 and inhibits Cdc2 at the G2 to M transition. *J. Cell Biol.* **2002**, *156*, 249–259.

54 Stone, J. M., Trotochaud, A. E., Walker, J. C., Clark, S. E., Control of meristem development by CLAVATA1 receptor kinase and kinase-associated protein phosphatase interactions. *Plant Physiol* **1998**, *117*, 1217–1225.

55 Lu, P. J., Zhou, X. Z., Shen, M., Lu, K. P., Function of WW domains as phosphoserine- or phosphothreonine-binding modules. *Science* **1999**, *283*, 1325–1328.

56 Durocher, D., Smerdon, S. J., Yaffe, M. B., Jackson, S. P., The FHA domain in DNA repair and checkpoint signaling. *Cold Spring Harbor Symposia on Quantitative Biology, Vol. LXV.* **2000**, 423–431.

57 Emili, A., Schieltz, D. M., Yates, J. R. 3rd, Hartwell, L. H., Dynamic interaction of DNA damage checkpoint protein Rad53 with chromatin assembly factor Asf1. *Mol. Cell* **2001**, *7*, 13–20.

58 Koranda, M., Schleiffer, A., Endler, L., Ammerer, G., Forkhead-like transcription factors recruit Ndd1 to the chromatin of G2/M-specific promoters. *Nature* **2000**, *406*, 94–98.

59 Kumar, R., Reynolds, D. M., Shevchenko, A., Goldstone, S. D., Dalton, S., Forkhead transcription factors, Fkh1p and Fkh2p, collaborate with Mcm1p to control transcription required for M-phase. *Curr. Biol.* **2000**, *10*, 896–906.

60 Pic, A., et al., The forkhead protein Fkh2 is a component of the yeast cell cycle transcription factor SFF. *EMBO J.* **2000**, *19*, 3750–3761.

61 Darieva, Z., et al., Cell cycle-regulated transcription through the FHA domain of Fkh2p and the coactivator Ndd1p. *Curr. Biol.* **2003**, *13*, 1740–1745.

62 Reynolds, D., Shi, B. J., McLean, C., Katsis, F., Kemp, B., Dalton, S., Recruitment of Thr 319-phosphorylated Ndd1p to the FHA domain of Fkh2p requires Clb kinase activity: a mechanism for CLB cluster gene activation. *Genes. Dev.* **2003**, *17*, 1789–1802.

63 Hermann-Le Denmat, S., Werner, M., Sentenac, A., Thuriaux, P., Suppression of yeast RNA polymerase III mutations by FHL1, a gene coding for a fork head protein involved in rRNA processing. *Mol. Cell Biol.* **1994**, *14*, 2905–2913.

64 Cherel, I., Thuriaux, P., The IFH1 gene product interacts with a fork head protein in *Saccharomyces cerevisiae*. *Yeast* **1995**, *11*, 261–270-

65 Nirula, A., Moore, D. J., Gaynor, R. B., Constitutive binding of the transcription factor interleukin-2 (IL-2) enhancer binding factor to the IL-2 promoter. *J. Biol. Chem.* **1997**, *272*, 7736–7745.

66 Li, C., Lai, C. F., Sigman, D. S., Gaynor, R. B., Cloning of a cellular factor, interleukin binding factor, that binds to NFAT-like motifs in the human immunodeficiency virus long terminal repeat. *Proc. Natl. Acad. Sci. USA* **1991**, *88*, 7739–7743.
67 Bassel-Duby, R., Hernandez, M. D., Yang, Q., Rochelle, J. M., Seldin, M. F., Williams, R. S., Myocyte nuclear factor, a novel winged-helix transcription factor under both developmental and neural regulation in striated myocytes. *Mol. Cell Biol.* **1994**, *14*, 4596–4605.
68 Yang, Q., Kong, Y., Rothermel, B., Garry, D. J., Bassel-Duby, R., Williams, R. S., The winged-helix/forkhead protein myocyte nuclear factor beta (MNF-beta) forms a co-repressor complex with mammalian sin3B. *Biochem. J.* **2000**, *345 Pt 2*, 335–343.
69 Durocher, D., Bacterial signal transduction: a FHAscinating glimpse at the origins of phospho-dependent signal transduction. *Trends Microbiol* **2003**, *11*, 67–68.
70 Pallen, M., Chaudhuri, R., Khan, A., Bacterial FHA domains: neglected players in the phospho-threonine signaling game? *Trends Microbiol* **2002**, *10*, 556–563.
71 Ponting, C. P., Aravind, L., Schultz, J., Bork, P., Koonin, E. V., Eukaryotic signalling domain homologues in archaea and bacteria: ancient ancestry and horizontal gene transfer. *J. Mol. Biol.* **1999**, *289*, 729–745.
72 Molle, V., Kremer, L., Girard-Blanc, C., Besra, G. S., Cozzone, A. J., Prost, J. F., An FHA phosphoprotein recognition domain mediates protein EmbR phosphorylation by PknH, a Ser/Thr protein kinase from *Mycobacterium tuberculosis*. *Biochemistry* **2003**, *42*, 15300–15309.
73 Takatsuna, H., et al., Identification of TIFA as an adapter protein that links tumor necrosis factor receptor-associated factor 6 (TRAF6) to interleukin-1 (IL-1) receptor-associated kinase-1 (IRAK-1) in IL-1 receptor signaling. *J. Biol. Chem.* **2003**, *278*, 12144–12150.
74 Kanamori, M., Suzuki, H., Saito, R., Muramatsu, M., Hayashizaki, Y., T2BP, a novel TRAF2 binding protein, can activate NF-kappaB and AP-1 without TNF stimulation. *Biochem. Biophys. Res. Commun.* **2002**, *290*, 1108–1113.
75 Nash, P., et al., Multisite phosphorylation of a CDK inhibitor sets a threshold for the onset of DNA replication. *Nature* **2001**, *414*, 514–521.
76 Elia, A. E., et al., The molecular basis for phosphodependent substrate targeting and regulation of Plks by the Polo-box domain. *Cell* **2003**, *115*, 83–95.
77 Violin, J. D., Zhang, J., Tsien, R. Y., Newton, A. C., A genetically encoded fluorescent reporter reveals oscillatory phosphorylation by protein kinase C. *J. Cell Biol.* **2003**, *161*, 899–909.

Websites related to FHA domains

SMART database: http://smart.embl-heidelberg.de/

Pfam database: http://www.sanger.ac.uk/Software/Pfam/

Ensembl genome browser: http://www.ensembl.org/

FHA domain entry, Pawson laboratory: http://www.mshri.on.ca/pawson/fha.html

8
Phosphoserine/Threonine Binding Domains

Andrew E. H. Elia and Michael B. Yaffe

8.1
Introduction

The orchestration of complex cellular events requires the timely assembly of multimolecular signaling complexes at specific locations within the cell. This temporal and spatial control is often achieved through phosphorylation-dependent binding and activation. Prior to the discovery of SH2 domains, protein phosphorylation was thought to regulate substrate binding and catalytic activity primarily by inducing allosteric changes in protein tertiary structure. A new view emerged in 1990, however, with the realization that binding of SH2 domains to tyrosine residues occurred only when they carried a phosphate moiety, introducing the idea that phosphorylation could function as a direct regulatory switch for protein–protein interactions [1–3]. This view was not immediately applied to serine/threonine kinase signaling, since SH2 and subsequently identified PTB domains were highly specific for phosphotyrosine [4]. Speculation ensued based on the idea that basic signaling mechanisms were different for Tyr and Ser/Thr phosphorylation events. The unanticipated finding in 1996 that 14-3-3 proteins recognize phosphorylated serine- and threonine-based motifs, however, rapidly dispelled this idea [5, 6].

Additional phosphoserine (pSer)/phosphothreonine (pThr)-binding modules have been discovered since 1996 and currently include five additional family members: WW domains, FHA domains, WD40 repeats, the Polo-box domain, and BRCT repeats [7, 8–10]. These domains comprise a diverse structural group, demonstrating that numerous divergent tertiary folds have been capable of acquiring a phospho-dependent binding function through evolution. Importantly, these domains all recognize phosphoserine or phosphothreonine within a unique consensus motif that directs the specificity of ligand binding. This chapter provides an overview of the identification, cellular function, and structural basis of binding for current members of the pSer/pThr-binding family. Their expanding repertoire is leading to a more general appreciation of the role of phosphopeptide recognition domains in regulating the reversible assembly of multiprotein complexes.

Modular Protein Domains. Edited by G. Cesareni, M. Gimona, M. Sudol, M. Yaffe

ISBN: 3-527-30813-X

8.2
The 14-3-3 Proteins

8.2.1
History and Functions

The first pSer/pThr-binding molecules to be identified were 14-3-3 proteins, a family of abundant polypeptides found in all eukaryotic cells. Initially discovered with no known function, the observation that ligand phosphorylation might be critical for 14-3-3 binding emerged from work on tryptophan hydroxylase [11], an enzyme involved in neurotransmitter biosynthesis. Widespread interest in 14-3-3 proteins subsequently grew when they were found to interact with Raf, the upstream activator of the classical MAP kinase pathway, and polyoma middle T antigen [12–14] and to play an essential role in the DNA damage checkpoint of fission yeast [15]. Investigation of the 14-3-3 binding sites on Raf [5] and in-vitro peptide-library screening [6] led to the identification of two optimal phosphoserine/threonine-containing motifs – RSx*po*S/TxP and Rxxx*po*S/TxP – that are recognized by all 14-3-3 isotypes [6].

Over 100 proteins interact with 14-3-3, including various protein kinases and phosphatases, apoptotic factors, transcription factors, cell surface receptors, ion channels, cytoskeletal proteins, and metabolic proteins [16–18]. 14-3-3 regulates the function of its bound partners through a number of general mechanisms, including increasing or decreasing the ligand's catalytic activity, facilitating or blocking molecular interactions between the ligand and other molecules, and regulating the subcellular localization of the bound protein.

One of the earliest elucidated functions of 14-3-3 was the regulation of tryptophan hydroxylase activity. 14-3-3 was found to bind tryptophan hydroxylase upon calmodulin kinase II phosphorylation and to thereby activate the enzyme [19, 20]. Shortly thereafter, 14-3-3 proteins were found to play inhibitory roles toward other enzymes, including PKC [21] and apoptosis signal-regulating kinase 1 (ASK1) [22]. In some instances, 14-3-3 plays a dynamic role with different functions during the course of catalytic activation. For example, 14-3-3 helps to maintain the mitogen-stimulated kinase Raf in an inactive but activatable conformation in resting cells. Upon cellular stimulation with growth factors, 14-3-3 is partly displaced from Raf to allow Raf activation, but 14-3-3 binding is not completely eliminated, because its continued presence is necessary for full catalytic activity [23–27].

Besides influencing enzymatic function, 14-3-3 proteins regulate some biological processes by modulating the interaction between two protein binding partners. A well studied example is the binding of 14-3-3 to the pro-apoptotic factor BAD [28, 29]. Exposure of cells to survival factors such as IL-3 results in the phosphorylation of BAD at three sites by the kinases Akt1, RSK1, and PKA [30–33]. These phosphorylations cooperatively mediate 14-3-3 binding, which ultimately interferes with the ability of BAD to bind and inhibit the anti-apoptotic factor Bcl-2. The net outcome is Bcl2 release and the prevention of apoptosis. Another example of molecular interference by 14-3-3 involves its binding to IRS-1, which functions to inhibit the interaction between IRS-1 and phosphatidylinositol-3-kinase (PI3-K), causing a

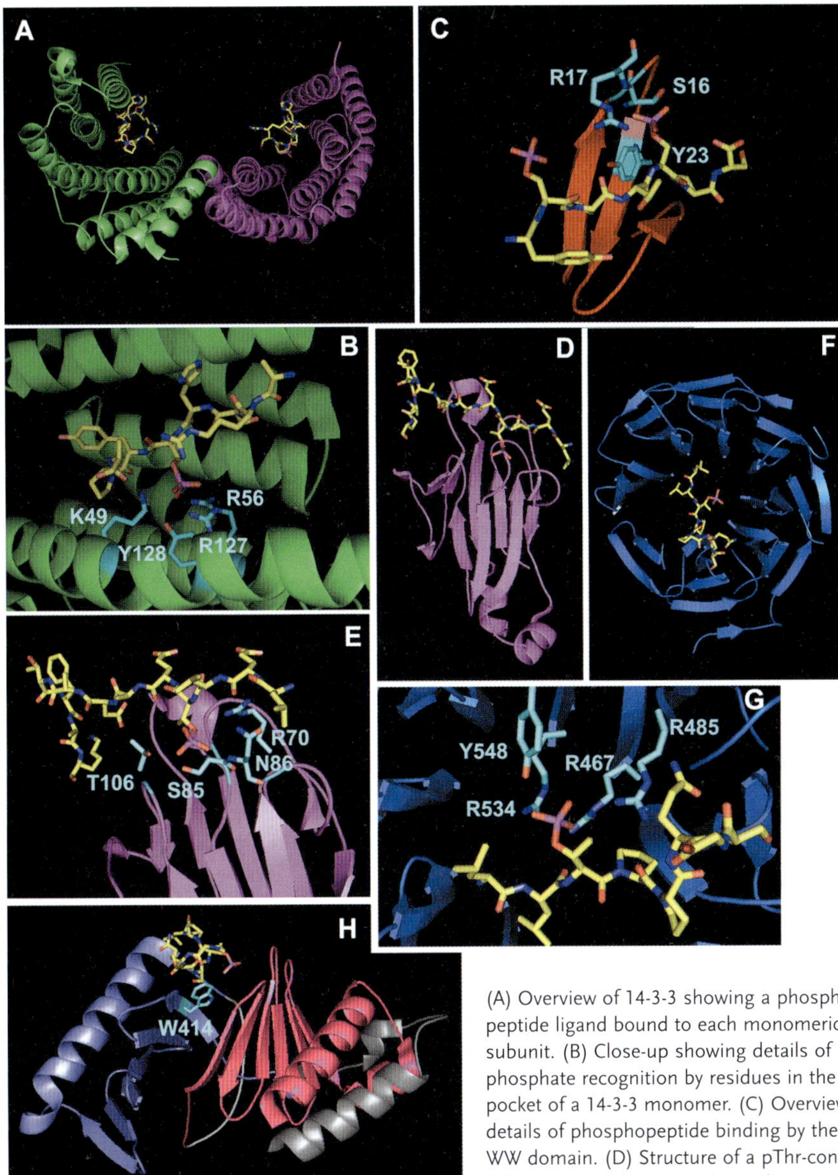

Figure 8.1 Structural diversity among phosphoserine/threonine binding domains. In each panel, the bound phosphopeptide ligand is shown in stick representation with carbons colored yellow, nitrogens blue, oxygens red, and phosphate purple. Selected sidechains from each of the structures that interact with the phosphopeptide ligand are also shown in ball-and-stick representation with their carbon atoms colored cyan and the oxygen and nitrogen atoms colored as above. (A) Overview of 14-3-3 showing a phosphopeptide ligand bound to each monomeric subunit. (B) Close-up showing details of phosphate recognition by residues in the basic pocket of a 14-3-3 monomer. (C) Overview and details of phosphopeptide binding by the Pin1 WW domain. (D) Structure of a pThr-containing peptide bound to the Rad53 N-terminal FHA domain. (E) Close-up showing the residues involved in the Rad53 N-terminal FHA–pThr phosphate interaction. (F) Overview of the Cdc4 WD40 repeats binding to a pThr-containing phosphopeptide.
(G) Close-up showing WD40 residue sidechains important for pThr binding.
(H) Structure of the Plk1 Polo-box domain bound to an optimal phosphopeptide. The critical Trp414 residue involved in phosphopeptide binding and Ser-1 selection is shown.

decrease in insulin-stimulated PI3-K activity [34]. Although 14-3-3 functions as a molecular 'blocker' in these instances, it promotes protein binding in others by exploiting the presence of two phosphopeptide binding sites within its dimeric structure. Thus, a single dimeric 14-3-3 molecule can function as an adaptor by simultaneously binding two distinct ligands to bridge them together. Reports suggest that such 14-3-3–mediated complexes exist between Bcr and Raf [35] and between the kinases PKCζ and Raf [36].

14-3-3 proteins regulate subcellular localization by inducing the cytoplasmic sequestration of some proteins and the nuclear translocation of others. A well characterized example of the former includes cytoplasmic retention of the mitotic phosphatase Cdc25C upon phosphorylation in response to DNA damage. Subsequent 14-3-3 binding and cytoplasmic tethering prevents access of Cdc25C to nuclear Cdc2-cyclin B and thereby inhibits mitotic entry [37–39]. Another example includes the 14-3-3–induced cytoplasmic sequestration of the pro-apoptotic transcription factor FKHRL-1, which occurs upon Akt phosphorylation in response to external cell survival stimuli [40, 41]. The general notion that 14-3-3 always promotes cytoplasmic localization, however, is contradicted by the observation that binding to the homeodomain transcription factor Tlx-2 induces its nuclear translocation [42]. Such diverse effects of 14-3-3 binding on subcellular localization may result from the fact that 14-3-3 proteins, by themselves, do not contain any functional nuclear export sequences (NES) or nuclear localization sequences (NLS). Instead, 14-3-3 molecules may regulate nucleo–cytoplasmic trafficking by exposing or masking such sequences within the bound target molecule, as has been observed for FKHRL-1 [41]. In this way, 14-3-3 proteins act as molecular 'chauffeurs', with the final subcellular destination of the 14-3-3-bound complex controlled by the bound protein passenger [41, 43].

8.2.2
Structure and Binding

The X-ray structures of 14-3-3τ and ζ in the absence of bound ligand revealed the molecule to be a cup-shaped dimer (Figure 8.1a) [44, 45], with each monomeric subunit consisting of nine α helices. The dimer interface is formed from hydrophobic and salt-bridge interactions that conceal a large surface area of over 2000 $Å^2$. In all ligand-bound 14-3-3 structures [6, 46, 47], the peptide occupies one of two amphipathic grooves that line the central channel formed by interaction of the monomers. In phosphopeptide structures, the phosphate oxygens form ionic and hydrogen bonds with three basic residues, Lys49, Arg56, and Arg127, which form a positively charged pocket, and with Tyr128 (Figure 8.1b). The entire phosphopeptide mainchain is held in an extended conformation up to two residues after the phosphoserine, at which point there is a sudden change in chain direction. This ligand geometry is required to exit the 14-3-3 binding cleft and rationalizes why optimal 14-3-3-binding sites contain a proline at this position. This general mode of binding has been validated for a physiologic protein substrate by the recent structure of 14-3-3ζ in complex with serotonin N-acetyl transferase [48].

14-3-3 proteins bind to their ligands with high affinity, having dissociation constants typically in the nanomolar range. A phosphopeptide binding site present within each monomer suggests that a dimeric 14-3-3 molecule might engage two distinct motifs within a single molecule to make use of an avidity effect. This type of bidentate interaction has been directly observed in 14-3-3 binding to serotonin N-acetyl transferase [48] and may occur in other 14-3-3 substrates that contain two or more phosphorylated 14-3-3 binding sites, such as Raf and BAD [5, 18, 26–28]. A synthetic phosphopeptide that contains tandem 14-3-3 consensus motifs binds to 14-3-3 over 30 times more tightly than the same peptide containing only one phospho-motif [6].

8.3 WW Domains

WW domains, named for the presence of two conserved tryptophan residues within their 40-residue sequence, can be grouped into six classes, whose members all recognize proline-rich sequences [49–51]. Only class IV WW domains exhibit a phospho-dependent binding function [52]. Three proteins with class IV domains have been described: the mitotic prolyl isomerase Pin1/Ess1, the splicing factor Prp40, and the HECT domain E3 ubiquitin ligase Nedd4/Rsp5. Phosphospecific binding to the WW domain of Pin1 has been most extensively studied. In addition to its WW domain, Pin1 contains a C-terminal prolyl isomerase domain, the X-ray crystal structure of which was solved prior to the recognition of its WW domain phospho-specificity. The structure of the isomerase domain in complex with an Ala-Pro dipeptide revealed a sulfate ion located 5 Å from the C_β of Ala, suggesting that the isomerase might prefer phosphorylated substrates [53]. Indeed, Pin1 was found to catalyze the specific cis–trans isomerization of pSer/Thr-Pro bonds [54]. Furthermore, Pin1 was shown to interact with numerous mitotic phosphoproteins, including the important mitotic regulators Cdc25, Myt1, Wee1, Plk1/Plx1, and Cdc27 [55]. This phospho-dependent recognition was subsequently found to occur, in part, through the WW domain, defining it as the first modular phosphoserine/threonine-binding domain [8]. All class IV WW domains show specific binding to phosphoserine-proline or phosphothreonine-proline motifs that are created upon substrate phosphorylation by proline-directed kinases such as cyclin-dependent kinases and MAPKs.

The effects of Pin1 on mitotic progression are complex, since it functions to delay mitotic entry but is also required for proper passage through mitosis [56]. The molecular mechanism underlying these functions is not completely understood but seems to involve regulation of the phosphatase Cdc25. Normal mitotic entry involves dephosphorylation and consequent activation of Cdc2 kinase, which is present within a complex containing cyclin B and a small regulatory subunit called Cks1, by a phosphorylated form of Cdc25. The activated Cdc2 complex, in turn, further phosphorylates Cdc25, increasing its activity and creating a positive feedback loop. Upon binding to phosphorylated Ser/Thr-Pro sites in Cdc25, Pin1 likely serves two functions to regulate this process. First, it catalytically induces a conformational change in Cdc25 that facilitates its dephosphorylation [57, 58], and second, it competes with Cks1 for binding to Cdc25, thereby limiting Cdc25 phosphorylation by Cdc2 [59, 60]. The cumulative effect is a decrease in Cdc25 phosphorylation and inhibition of Cdc25 activity during early mitotic entry.

Pin1 plays additional roles outside of mitosis, one of which involves the regulation of β-catenin turnover and nuclear translocation. The WW domain of Pin1 inhibits interaction of adenomatous polyposis coli protein (APC) with β-catenin by binding to a phosphorylated Ser-Pro motif near the APC-binding site of β-catenin [61]. Since APC is necessary for the cytoplasmic assembly of β-catenin into a multimolecular complex that triggers its degradation, Pin1 binding serves to increase β-catenin stability in the nucleus [61]. Pin1 is also involved in activating the tumor-suppressor protein p53 upon DNA damage. Multiple types of genotoxic stress induce binding of Pin1 to three phosphorylated Ser/Thr-Pro sites in p53 [62, 63]. Pin1 then appears to effect a conformational change in p53 that increases both its stability and its transactivation function. Pin1-deficient cells are defective in both p53 activation and stabilization and in checkpoint control in response to DNA damage. The Pin1-induced conformational changes in p53 may function to activate p53 by protecting it from interaction with the protein HDM2, which regulates p53 stability, and/or by directly influencing the ability of p53 to act as a transcription factor [62, 63].

WW domains fold into three antiparallel β strands, forming a single groove that recognizes proline-rich ligands in the context of a type II polyproline helix (Figure 8.1c) [64, 65]. Specificity for different proline-rich motifs is determined largely by residues within the loop regions that connect the β1/β2 and β2/β3 strands [65], somewhat akin to the mechanism of ligand binding utilized by FHA domains. The structure of the Pin1 WW domain bound to a Y*po*SPT*po*SPS peptide from the C-terminal region of RNA polymerase II shows that phospho binding occurs through four hydrogen bonds between the peptide's second phosphoserine and two residues in the β1/β2 loop (Arg17 and Ser16), along with one in the β2 strand (Tyr23) (Figure 8.1c) [66]. Because the majority of WW domains lack an Arg residue in the β1 loop, pSer/pThr binding by WW domains may be the exception rather than the rule. Pin1 WW domain selection for proline at the pSer/Thr +1 position is explained by the presence of a hydrophobic pocket formed by the aromatic amino acids Tyr23 and Trp34. The pSer-Pro backbone inserts into this pocket, where it is sterically clamped in a trans conformation.

8.4
FHA Domains

FHA (forkhead associated) domains were originally identified through sequence profiling as a region of homology in forkhead transcription factors [67]. They have since been found in a wide diversity of both prokaryotic and eukaryotic proteins. FHA domains are extensively reviewed by Durocher in Chapter 7 of this book, and therefore only salient points regarding FHA domain–phosphopeptide binding are briefly summarized below.

The necessity of ligand phosphorylation for FHA binding initially emerged from studies in *Arabidopsis* showing that a region encompassing the FHA domain of the protein KAPP (kinase-associated protein phosphatase) bound exclusively to the phosphorylated form of the receptor-like protein kinase RLK5 [68]. Interest in FHA domains increased in 1998 when Sun et al. [69] demonstrated that phospho-selective binding of the C-terminal FHA domain of Rad53 to the BRCT-containing protein Rad9 was necessary for DNA damage-induced G2/M arrest in *Saccharomyces cerevisiae* [69]. Firm evidence for FHA domains as phosphopeptide docking modules was secured when Durocher et al. [70] demonstrated that selective binding of the Rad53 FHA domain to phosphorylated Rad9 could be inhibited by exogenous peptides. These authors also showed that FHA domains from other proteins could also bind directly to isolated phosphopeptides. Oriented peptide library screening to discern consensus motifs for phosphothreonine peptide binding has allowed a tentative grouping of FHA domains into discrete classes based on the specificity at the pThr +3 position (three residues C-terminal from the phosphothreonine) [71].

FHA domain-containing proteins have been most extensively investigated in cell cycle control and in the cellular response to genotoxic damage. Three such proteins are the kinases Chk2, Rad53, and Dun1, all of which mediate cell cycle arrest at multiple checkpoints in response to DNA damage. The simple concept that these proteins employ their FHA domains for targeting substrates to their kinase domains, however, has not been supported by the available data. Instead, it appears that the FHA domains of these kinases are more important for targeting the kinases to scaffolds, where their own activating phosphorylation can occur. Mutations in the C-terminal FHA domain of Rad53 impair its ability to bind phosphorylated Rad9 after DNA damage and result in a loss of Rad53 activation [69]. Rad9 is thought to promote Rad53 activation by acting as an adaptor that either recruits Rad53 to a complex containing the activating kinase Mec1 [72] or brings separate Rad53 molecules close together to facilitate their trans-autophosphorylation [73]. In this regard, the FHA domain of Chk2 appears to regulate trans-autophosphorylation of its kinase domain by mediating direct homodimerization through interaction with ATM-phosphorylated T68 of Chk2 [74, 75]. Similarly, Dun1's FHA domain is critical for direct phosphorylation and activation of Dun1 by Rad53 [76].

Two non-kinase mediators of checkpoint signaling that contain FHA domains are Nbs1 and MDC1. Nbs1 is a component of the Mre11–Rad50–Nbs1 (MRN) complex, which localizes to sites of DNA damage in order to coordinate DNA repair and checkpoint signaling. It contains an FHA domain and an adjacent BRCT domain, both of which are involved in targeting this complex to nuclear DNA damage foci [77, 78]. Recombinant FHA/BRCT domain of Nbs1 binds directly in vitro to ATM-phosphorylated histone H2AX, which likely mediates MRN localization to DNA damage sites in vivo [77]. Another recently identified mediator of checkpoint signaling containing an FHA domain is the molecule MDC1, which functions in the intra-S and G2–M checkpoints. Within minutes after ionizing radiation, MDC1 associates with the MRN complex and with numerous other DNA damage-response proteins, including 53BP1 and BRCA1. Like Nbs1, it forms nuclear foci that colocalize with H2AX foci induced by DNA damage, and peptide binding data suggest that its FHA domain may also mediate direct binding to phosphorylated histone H2AX [79–81].

The X-ray crystal structure of the N-terminal Rad 53 FHA exhibits a core fold consisting of an 11-stranded β sandwich, with a strand topology essentially identical to that of the MH2 domain from SMAD signaling molecules (Figure 8.1d) [71]. Phosphopeptide binding occurs at one end of the domain, through interactions between selected residues in the phosphopeptide and loops that connect the β3/4, β4/5, and β6/7 strands. The phosphate moiety is held by five hydrogen bonds to Arg70, Ser85, Asn86, and Thr106 of Rad53 FHA1 (Figure 8.1e). Phospho-independent contacts between FHA domain sidechains and peptide backbone atoms maintain the peptide in an extended conformation. In the Rad53 FHA domain complex, the pThr +3 specificity is derived from salt bridging interactions of Arg83 with an aspartate at the pThr +3 residue of the bound phosphopeptide.

FHA domains differ from other phosphothreonine-docking modules in that replacement of pThr with pSer completely eliminates phosphopeptide binding [71]. Curiously, though, the C-terminal Rad53 FHA domain binds phosphotyrosine-containing peptides [82]. The biological significance of this finding is unclear, since yeast have limited tyrosine kinase signaling. However, it raises the interesting possibility that FHA domains in higher eukaryotes may function as dual-specificity phosphopeptide-binding modules. Interestingly, FHA domains appear to have an additional phospho-independent binding surface on the opposite side of the phosphopeptide-interacting groove. For the FHA domain of human Chk2, this surface is necessary, in conjunction with the phosphopeptide binding surface, for binding BRCA1 [83].

8.5
WD40 Repeats of F-box Proteins

Phospho-dependent substrate recognition has proven to play a key role in the regulation of protein ubiquitination. Skp1–Cullin–F-box (SCF) complexes are E3 ubiquitin ligases that facilitate the transfer of ubiquitin from a ubiquitin-conjugating enzyme (E2) to lysine residues on a substrate, ultimately forming polyubiquitylation chains that target the substrate for degradation by the proteasome. SCF complexes contain Skp1, Cul1/Cdc53, Roc1, and an F-box-containing protein [84–87]. In addition to an N-terminal F-box motif, most F-box proteins typically contain either WD40 domains or leucine-rich repeats (LRRs) that mediate phospho-specific substrate targeting. Whereas WD40 domains have been definitively shown to directly recognize phosphorylated motifs within substrates (reviewed in [7]), the exact role of LRRs in phospho-dependent substrate recognition is less clear, since they appear to additionally require the adaptor protein Csk1 [88]. The earliest studies implicating WD40 domains in phospho-binding came in 1997 from Ben-Neriah and coworkers, who discovered that phosphopeptides corresponding to sites within the NF-κB inhibitor, IκBα, specifically inhibit IκBα ubiquitination and degradation mediated by the WD40-containing F-box protein β-TrCP [89, 90]. Subsequent work showed that β-TrCP in cellular lysates interacts with immobilized peptides in a phosphospecific fashion [91]. However, definitive proof was not attained until Sicheri, Tyers, and colleagues [92] crystallized the WD40 repeat of the F-box protein Cdc4 directly bound to a phosphopeptide ligand and Pavletich and colleagues [93] crystallized β-TRCP bound to a phosphopeptide from β-catenin.

Phosphospecific binding by F-box proteins has been most extensively analyzed for Cdc4 and β-TrCP, two F-box proteins containing WD40 domains. In *S. cerevisiae*, studies on Cdc4 have focused on the substrate Sic1, a Cdk inhibitor that is specific for the S-phase kinase Clb5-Cdc28 kinase and whose degradation is necessary for S-phase entry. Phosphorylation of Sic1 by Cln1/2-Cdc28 kinases during G1 leads to its Cdc4-mediated ubiquitination and subsequent proteasomal degradation. Consequently, Clb5-Cdc28 becomes activated and drives S-phase entry [94]. The necessity of multisite phosphorylation for Sic1 recognition by Cdc4 has proven to play an important role in the kinetics of S-phase entry [95]. The optimal binding motif for Cdc4, which is LI-L/I-*po*T-P-$^\wedge(RK)_4$ (where $^\wedge(RK)_4$ denotes selection against basic residues in the next four positions), is not found in Sic1. Rather, nine suboptimal sites are present, six of which must be phosphorylated for Cdc4 binding. This multisite requirement makes the binding of Cdc4 to Sic1 an ultrasensitive process with a maximum theoretical Hill coefficient (n_H) of 6. When Cln-Cdc28 levels are subthreshold, phosphorylation is not sufficient to drive Sic1–Cdc4 complex formation, but when Cln-Cdc28 activity meets a critical threshold level, virtually all of the Sic1 is rapidly phosphorylated and degraded within a short time [95–97].

Another F-box protein with WD40 repeats, β-TrCP, plays roles in both developmental and inflammatory signaling pathways. It mediates the ubiquitination of phosphorylated β-catenin to down-regulate Wnt signaling [98] and of phosphorylated IκBα to induce nuclear translocation of the transcription factor NFκB for induction of immunological genes [99]. β-TrCP recognizes a common D*po*SGxx*po*S motif, which is generated in β-catenin by GSK-3 phosphorylation and in IκBα by IκK phosphorylation [99].

An X-ray crystal structure of a ternary complex consisting of Skp1, Cdc4, and a high-affinity phosphopeptide provides direct visualization of phosphoepitope binding to the WD40 domain of Cdc4 (Figure 8.1f) [92]. The WD40 repeats form an eight-bladed β propeller, a somewhat unusual finding since all previously solved WD40 structures possess only seven blades. The phosphopeptide binds in an extended conformation across one blade with the N terminus oriented toward the central pore of the WD40 domain and the C terminus oriented toward the outer rim. The phosphate moiety is bound by electrostatic interactions with the guanidinium groups of three arginines and with the hydroxyl group of a tyrosine residue (Figure 8.1g). The pThr +1 proline selection derives from insertion of the proline into a hydrophobic pocket whose Trp426 engages the proline pyrrolidine ring through a coplanar interaction in a manner strikingly similar to that which occurs in the Pin1 WW domain binding to a pThr +1 proline. The pThr –1 and –2 aliphatic selections are rationalized by hydrophobic binding, and the disfavor of basic residues in the C-terminal peptide stretch arises from repulsive forces delivered by a cluster of positively charged basic residues adjacent to the phosphate binding pocket [92].

The mechanism for cooperative binding between Cdc4 and phospho-Sic1 has yet to be determined. In principle, multiple phosphorylated sites could enhance binding through an avidity effect by engaging more than one pocket on Cdc4. Examining the Cdc4 surface, though, reveals no obvious additional sites for phosphate binding, because its most conserved region overlaps with the known basic phosphate binding pocket. The authors, therefore, prefer a model in which phospho-Sic1 engages a single phosphopeptide binding groove of Cdc4 [100, 101]. The presence of multiple Sic1 sites would strengthen binding in this model by increasing the local concentration of Sic1 once an initial contact is made. In this scenario, the rate of Sic1 diffusion away from Cdc4 upon dissociation of a single site would be overwhelmed by rebinding at a second site. Alternatively, cooperativity could be achieved by avidity binding of multiply phosphorylated Sic1 to multimerized Cdc4, for which some evidence exists [100].

8.6
Polo-box Domains

Polo-like kinases play important roles in cell cycle progression and in checkpoint pathways activated by DNA damage or mitotic spindle disruption. They are characterized by the presence of a conserved Ser/Thr kinase domain and a noncatalytic C-terminal region composed of two homologous ~70–80-residue segments called Polo-boxes [102, 103]. For nearly 10 years, this C-terminal region was recognized as essential for the in vivo function of Plks [104–108]. Evidence suggested that it targeted polo kinases to substrates at particular locations within the cell. However, the molecular mechanism through which such localization occurred remained mysterious until identification of the Plk1 C terminus in a proteomic screen for novel phosphopeptide binding domains [9]. In this screen, an immobilized library of partially degenerate phosphopeptides was biased toward the phosphorylation motif for cyclin-dependent kinases (Cdks) and used to isolate phospho-binding domains that bind to proteins phosphorylated by Cdks. This screen revealed that both Polo-boxes of Plk1 function together as a single phosphoserine/threonine-binding domain, which has been termed the Polo-box domain (PBD). Oriented peptide library screening with PBDs from all canonical human Plks (Plk1 1, Plk2, Plk3) and from *S. cerevisiae* and *Xenopus* has demonstrated that PBDs recognize the common consensus motif S-[*po*T/*po*S]-(P/x) [109]. Peptide array studies have shown that serine at the (pThr or pSer) –1 position is absolutely necessary and that proline at the (pThr or pSer) +1 position is preferred but not required [9].

In contrast to other, more ubiquitous, phosphopeptide binding domains, PBDs occur only in Polo-like kinases, where they perform two critical functions regulating the activity of their adjacent kinase domains. In the basal state, they inhibit the catalytic activity of the kinase through intramolecular binding, but upon activation, they target the kinase domain to previously phosphorylated (primed) substrates or docking proteins. Plk1, which has been the most extensively studied among members of the polo kinase family, has a distinct subcellular localization pattern during mitosis, originating at centrosomes and kinetochores in prophase and moving to the spindle midzone during late stages of mitosis (reviewed in [110–113]). The PBD is necessary and sufficient for this localization pattern. Mutation of the phosphopeptide binding pocket in the PBD [109] or injection of its optimal phosphopeptide ligand into permeabilized cells [9] disrupts centrosomal localization, demonstrating that this process relies on phospho binding by the PBD.

Evidence that phosphopeptide binding by the Plk1 PBD is necessary for mitotic progression emerged from the finding that mutation of its phospho-binding pocket prevents G2–M arrest when a dominant negative PBD construct is overexpressed in mammalian cells [109]. The mitotic phosphatase Cdc25 is one particular protein targeted by the PBD of Plk1 to regulate mitotic entry. During mitosis, Cdc25 is phosphorylated at five Ser/Thr-Pro sites in its N terminus [114, 115] by Cdks. The Plk1 PBD interacts selectively with the phosphorylated form of one of these sites, Thr130, which contains a conserved PBD consensus motif. Both Cdc2-cyclin B and Plk1 were previously shown to cooperate in phosphorylating and activating Cdc25 in mitotic entry [115–119]. It is attractive to envision a model for this cooperation by which low amounts of Cdc2/cyclinB activity during prophase are insufficient to fully activate Cdc25 but provide priming phosphorylation of Cdc25, creating a PBD docking site. Subsequent recruitment of Plk1 would further phosphorylate and activate Cdc25, which would dephosphorylate Cdc2/cyclin B to increase its activity, priming additional Cdc25 molecules for activation by Polo-like kinases, and resulting in a positive feedback loop [9].

Multiple lines of evidence suggest that a mutually inhibitory interaction exists between the PBD and the kinase domain in full-length Plk1. Deletion of the PBD increases the kinase activity of Plk1 about three fold [108, 120], and the isolated PBD interacts with and inhibits the isolated kinase domain in trans [107]. Furthermore, addition of the optimal PBD phosphopeptide to full-length Plk1 increases its kinase activity by a factor of about three, suggesting that binding of the PBD to primed phosphorylation sites not only serves to target the kinase domain to substrates but simultaneously activates the kinase domain by relieving an inhibitory intramolecular interaction [109]. Interaction of the PBD with the kinase domain does not appear to involve the phosphopeptide binding pocket, since its mutation does not affect the level of binding [109]. Furthermore, the kinase domain does not contain a PBD consensus motif, and mutation of Thr210 [121] within the kinase domain to Asp, as a mimic of phosphorylation, actually inhibits interaction of the PBD with the kinase domain [107]. Thus, disruption of the PBD–kinase interaction likely opens the Plk1 molecule up, to facilitate phosphorylation within its activation loop at Thr210 by upstream kinases [121–123]. The phosphorylated Plk1 would then become locked in the active state throughout the remainder of mitosis by preventing rebinding of the PBD to the kinase.

X-ray crystal structures of the human Plk1 PBD bound to its optimal phosphopeptide ligand show that the two Polo-boxes of the PBD each comprise $\beta_6\alpha$ structures (Figure 8.1h) [109, 124], resembling the single Polo-box found in Sak [125]. Together, both Plk1 Polo-boxes form a novel 12-stranded β-sandwich domain, to which the phosphopeptide binds within a conserved, positively charged cleft located at the edge of the Polo-box interface. Binding at this interface rationalizes the requirement of both Polo-boxes for efficient subcellular localization of Plk1 in vivo and also an observed 1:1 stoichiometry of PBD–ligand binding. Preceding the Polo-boxes is a 45-residue region, termed the Polo cap, which wraps around Polo-box 2 like a hook tethering it to Polo-box 1 [109]. The phosphate group participates in eight hydrogen-bonding interactions, explaining the critical dependence on peptide phosphorylation for binding. Only two residues (His538 and Lys540) directly contact the phosphate, accounting for three of the hydrogen bonds, with the remaining hydrogen bonds formed by an extensive lattice of water molecule bridges. The structural basis for the high serine selectivity at the pThr –1 position results from three hydrogen bonds to the serine hydroxyl group, two of which arise from interactions with Trp414 mainchain atoms. This critical role of Trp414 in ligand binding explains prior observations that a W414F mutation eliminates both the centrosomal localization of Plk1 and its ability to complement a temperature-sensitive mutation in the Plk1 ortholog, Cdc5, in *S. cerevisiae* [104].

8.7
Conclusions and Future Directions

A remarkable amount of functional and structural diversity is observed in these examples of phosphoserine/phosphothreonine-binding domains. Most of these domains play at least some role in regulating normal cell progression or in halting the cell cycle after DNA damage. These observations underscore the critical role of protein phosphorylation-dependent cell signaling in the regulation of many aspects of cell division. The recent identification of tandem BRCT repeats as the newest member of the phosphoserine/threonine binding domain superfamily [10, 126] and the demonstration that mutations in the BRCT domains that eliminate phosphopeptide binding are also associated with an increased risk of breast and ovarian cancer further illustrate this point. The observation that so many different tertiary protein folds can adopt a phosphoserine/threonine binding function suggests that this property has been repeatedly 'rediscovered' during evolution and strongly argues that many additional phosphoserine/threonine binding domains remain to be found.

Acknowledgements

This work benefited from helpful discussions with members of the Yaffe laboratory. We apologize to the many researchers whose work was not cited here due to space limitations. Financial assistance from the NIH (grant GM60594) (MBY), and a Career Development Award from the Burroughs–Wellcome Fund (MBY), and the NIH Medical Scientist Training Program (AE) is gratefully acknowledged.

References

1 Pawson, T., Gish, G. D., SH2 and SH3 domains: from structure to function. *Cell* **1992**, *71*, 359–362.

2 Mayer, B. J., Gupta, R., Functions of SH2 and SH3 domains. *Curr. Top. Microbiol. Immunol.* **1998**, *228*, 1–22.

3 Kuriyan, J., Cowburn, D., Modular peptide recognition domains in eukaryotic signaling. *Annu. Rev. Biophys. Biomol. Struct.* **1997**, *26*, 259–288.

4 Yaffe, M. B., Phosphotyrosine-binding domains in signal transduction. *Nat. Rev. Mol. Cell Biol.* **2002**, *3*, 177–186.

5 Muslin, A. J., et al., Interaction of 14-3-3 with signaling proteins is mediated by the recognition of phosphoserine. *Cell* **1996**, *84*, 889–897.

6 Yaffe, M. B., et al., The structural basis for 14-3-3:phosphopeptide binding specificity. *Cell* **1997**, *91*, 961–971.

7 Yaffe, M. B., Elia, A. E., Phosphoserine/threonine-binding domains. *Curr. Opin. Cell Biol.* **2001**, *13*, 131–138.

8 Lu, P. J., et al., Function of WW domains as phosphoserine- or phosphothreonine-binding modules. *Science* **1999**, *283*, 1325–1328.

9 Elia, A. E., Cantley, L. C., Yaffe, M. B., Proteomic screen finds pSer/pThr-binding domain localizing Plk1 to mitotic substrates. *Science* **2003**, *299*, 1228–1231.

10 Manke, I. A., et al., BRCT repeats as phosphopeptide-binding modules involved in protein targeting. *Science* **2003**, *302*, 636–639.

11 Furukawa, Y., et al., Demonstration of the phosphorylation-dependent interaction of tryptophan hydroxylase with the 14-3-3 protein. *Biochem. Biophys. Res. Commun.* **1993**, *194*, 144–149.

12 Fantl, W. J., et al., Activation of Raf-1 by 14-3-3 proteins. *Nature* **1994**, *371*, 612–614.

13 Freed, E., et al., Binding of 14-3-3 proteins to the protein kinase Raf and effects on its activation. *Science* **1994**, *265*, 1713–1716.

14 Pallas, D. C., et al., Association of polyomavirus middle tumor antigen with 14-3-3 proteins. *Science* **1994**, *265*, 535–537.

15 Ford, J. C., et al., 14-3-3 protein homologs required for the DNA damage checkpoint in fission yeast. *Science* **1994**, *265*, 533–535.

16 Fu, H., Subramanian, R. R., Masters, S. C., 14-3-3 proteins: structure, function, and regulation. *Ann. Rev. Pharmacol. Toxicol.* **2000**, *40*, 617–647.

17 Tzivion, G., Shen, Y. H., Zhu, J., 14-3-3 proteins; bringing new definitions to scaffolding. *Oncogene* **2001**, *20*, 6331–6338.

18 van Hemert, M. J., Steensma, H. Y., van Heusden, G. P., 14-3-3 proteins: key regulators of cell division, signalling and apoptosis. *BioEssays* **2001**, *23*, 936–946.

19 Ichimura, T., et al., Brain 14-3-3 protein is an activator protein that activates tryptophan 5-monooxygenase and tyrosine 3-monooxygenase in the presence of Ca^{2+},calmodulin-dependent protein kinase II. *FEBS Lett.* **1987**, *219*, 79–82.

20 Ichimura, T., et al., Molecular cloning of cDNA coding for brain-specific 14-3-3 protein, a protein kinase-dependent activator of tyrosine and tryptophan hydroxylases. *Proc. Natl. Acad. Sci. USA* **1988**, *85*, 7084–7088.

21 Toker, A., et al., Protein kinase C inhibitor proteins: purification from sheep brain and sequence similarity to lipocortins and 14-3-3 protein. *Eur. J. Biochem.* **1990**, *191*, 421–429.

22 Zhang, L., Chen, J., Fu, H., Suppression of apoptosis signal-regulating kinase 1-induced cell death by 14-3-3 proteins. *Proc. Natl. Acad. Sci. USA* **1999**, *96*, 8511–8515.

23 Roy, S., et al., 14-3-3 facilitates Ras-dependent Raf-1 activation in vitro and in vivo. *Mol. Cell Biol.* **1998**, *18*, 3947–3955.

24 Light, Y., Paterson, H., Marais, R., 14-3-3 antagonizes Ras-mediated Raf-1 recruitment to the plasma membrane to maintain signaling fidelity. *Mol. Cell Biol.* **2002**, *22*, 4984–4996.

25 Thorson, J. A., et al., 14-3-3 proteins are required for maintenance of Raf-1 phosphorylation and kinase activity. *Mol. Cell Biol.* **1998**, *18*, 5229–5238.

26 Tzivion, G., Luo, Z., Avruch, J., A dimeric 14-3-3 protein is an essential cofactor for Raf kinase activity. *Nature* **1998**, *394*, 88–92.

27 Ory, S., et al., Protein phosphatase 2A positively regulates Ras signaling by dephosphorylating KSR1 and Raf-1 on critical 14-3-3 binding sites. *Curr. Biol.* **2003**, *13*, 1–20.
28 Zha, J., et al., Serine phosphorylation of death agonist BAD in response to survival factor results in binding to 14-3-3 not BCL-X(L). *Cell* **1996**, *87*, 619–628.
29 Datta, S. R., et al., Akt phosphorylation of BAD couples survival signals to the cell-intrinsic death machinery. *Cell* **1997**, *91*, 231–241.
30 Lizcano, J. M., Morrice, N., Cohen, P., Regulation of BAD by cAMP-dependent protein kinase is mediated via phosphorylation of a novel site, Ser155. *Biochem. J.* **2000**, *349*, 547–557.
31 Tan, Y., et al., BAD Ser-155 phosphorylation regulates BAD/Bcl-XL interaction and cell survival. *J. Biol. Chem.* **2000**, *275*, 25865–25869.
32 Datta, S. R., et al., 14-3-3 proteins and survival kinases cooperate to inactivate BAD by BH3 domain phosphorylation. *Mol. Cell* **2000**, *6*, 41–51.
33 Zhou, X. M., et al., Growth factors inactivate the cell death promoter BAD by phosphorylation of its BH3 domain on Ser155. *J. Biol. Chem.* **2000**, *275*, 25046–25051.
34 Kosaki, A., et al., 14-3-3beta protein associates with insulin receptor substrate 1 and decreases insulin-stimulated phosphatidylinositol 3′-kinase activity in 3T3L1 adipocytes. *J. Biol. Chem.* **1998**, *273*, 940–944.
35 Braselmann, S., McCormick, F., Bcr and Raf form a complex in vivo via 14-3-3 proteins. *EMBO J.* **1995**, *14*, 4839–4848.
36 Van Der Hoeven, P. C., et al., 14-3-3 isotypes facilitate coupling of protein kinase C-zeta to Raf-1: negative regulation by 14-3-3 phosphorylation. *Biochem. J.* **2000**, *345*, 297–306.
37 Kumagai, A., Dunphy, W. G., Binding of 14-3-3 proteins and nuclear export control the intracellular localization of the mitotic inducer Cdc25. *Genes. Dev.* **1999**, *13*, 1067–1072.
38 Lopez-Girona, A., et al., Nuclear localization of Cdc25 is regulated by DNA damage and a 14-3-3 protein. *Nature* **1999**, *397*, 172–175.
39 Zeng, Y., Piwnica-Worms, H., DNA damage and replication checkpoints in fission yeast require nuclear exclusion of the Cdc25 phosphatase via 14-3-3 binding. *Mol. Cell Biol.* **1999**, *19*, 7410–7419.
40 Brunet, A., et al., Akt promotes cell survival by phosphorylating and inhibiting a forkhead transcription factor. *Cell* **1999**, *96*, 857–868.
41 Brunet, A., et al., 14-3-3 transits to the nucleus and participates in dynamic nucleocytoplasmic transport. *J. Cell Biol.* **2002**, *156*, 817–828.
42 Tang, S. J., et al., Association of the TLX-2 homeodomain and 14-3-3eta signaling proteins. *J. Biol. Chem.* **1998**, *273*, 25356–25363.
43 Muslin, A. J., Xing, H., 14-3-3 proteins: regulation of subcellular localization by molecular interference. *Cell Signal* **2000**, *12*, 703–709.
44 Liu, D., et al., Crystal structure of the zeta isoform of the 14-3-3 protein. *Nature* **1995**, *376*, 191–194.
45 Xiao, B., et al., Structure of a 14-3-3 protein and implications for coordination of multiple signalling pathways. *Nature* **1995**, *376*, 188–191.
46 Petosa, C., et al., 14-3-3zeta binds a phosphorylated Raf peptide and an unphosphorylated peptide via its conserved amphipathic groove. *J. Biol. Chem.* **1998**, *273*, 16305–16310.
47 Rittinger, K., et al., Structural analysis of 14-3-3 phosphopeptide complexes identifies a dual role for the nuclear export signal of 14-3-3 in ligand binding. *Mol. Cell* **1999**, *4*, 153–166.
48 Obsil, T., et al., Crystal structure of the 14-3-3zeta:serotonin N-acetyltransferase complex. a role for scaffolding in enzyme regulation. *Cell* **2001**, *105*, 257–267.
49 Sudol, M., et al., Characterization of a novel protein-binding module: the WW domain. *FEBS Lett.* **1995**, *369*, 67–71.
50 Sudol, M., Sliwa, K., Russo, T., Functions of WW domains in the nucleus. *FEBS Lett.* **2001**, *490*, 190–195.
51 Otte, L., et al., WW domain sequence activity relationships identified using ligand recognition propensities of 42 WW domains. *Protein Sci.* **2003**, *12*, 491–500.

52 Sudol, M., Hunter, T., NeW wrinkles for an old domain. *Cell* **2000**, *103*, 1001–1004.

53 Ranganathan, R., et al., Structural and functional analysis of the mitotic rotamase Pin1 suggests substrate recognition is phosphorylation dependent. *Cell* **1997**, *89*, 875–886.

54 Yaffe, M. B., et al., Sequence-specific and phosphorylation-dependent proline isomerization: a potential mitotic regulatory mechanism. *Science* **1997**, *278*, 1957–1960.

55 Shen, M., et al., The essential mitotic peptidyl-prolyl isomerase Pin1 binds and regulates mitosis-specific phosphoproteins. *Genes. Dev.* **1998**, *12*, 706–720.

56 Lu, K. P., Hanes, S. D., Hunter, T., A human peptidyl-prolyl isomerase essential for regulation of mitosis. *Nature* **1996**, *380*, 544–547.

57 Zhou, X. Z., et al., Pin1-dependent prolyl isomerization regulates dephosphorylation of Cdc25C and tau proteins. *Mol. Cell* **2000**, *6*, 873–883.

58 Stukenberg, P. T., Kirschner, M. W., Pin1 acts catalytically to promote a conformational change in Cdc25. *Mol. Cell* **2001**, *7*, 1071–1083.

59 Landrieu, I., et al., p13(SUC1) and the WW domain of PIN1 bind to the same phosphothreonine-proline epitope. *J. Biol. Chem.* **2001**, *276*, 1434–1438.

60 Patra, D., et al., The *Xenopus* Suc1/Cks protein promotes the phosphorylation of G(2)/M regulators. *J. Biol. Chem.* **1999**, *274*, 36839–36842.

61 Ryo, A., et al., Pin1 regulates turnover and subcellular localization of beta-catenin by inhibiting its interaction with APC. *Nat. Cell Biol.* **2001**, *3*, 793–801.

62 Zacchi, P., et al., The prolyl isomerase Pin1 reveals a mechanism to control p53 functions after genotoxic insults. *Nature* **2002**, *419*, 853–857.

63 Zheng, H., et al., The prolyl isomerase Pin1 is a regulator of p53 in genotoxic response. *Nature* **2002**, *419*, 849–853.

64 Macias, M. J., et al., Structure of the WW domain of a kinase-associated protein complexed with a proline-rich peptide. *Nature* **1996**, *382*, 646–649.

65 Zarrinpar, A., Lim, W. A., Converging on proline: the mechanism of WW domain peptide recognition. *Nat. Struct. Biol.* **2000**, *7*, 611–613.

66 Verdecia, M. A., et al., Structural basis for phosphoserine-proline recognition by group IV WW domains. *Nat. Struct. Biol.* **2000**, *7*, 639–643.

67 Hofmann, K., Bucher, P., The FHA domain: a putative nuclear signalling domain found in protein kinases and transcription factors. *Trends Biochem. Sci.* **1995**, *20*, 347–349.

68 Stone, J. M., et al., Interaction of a protein phosphatase with an *Arabidopsis* serine-threonine receptor kinase. *Science* **1994**, *266*, 793–795.

69 Sun, Z., et al., Rad53 FHA domain associated with phosphorylated Rad9 in the DNA damage checkpoint [*see comments*]. *Science* **1998**, *281*, 272–274.

70 Durocher, D., et al., The FHA domain is a modular phosphopeptide recognition motif. *Mol. Cell* **1999**, *4*, 387–394.

71 Durocher, D., et al., The molecular basis of FHA domain:phosphopeptide binding specificity and implications for phosphodependent signaling mechanisms. *Mol. Cell* **2000**, *##*, *##–##*.

72 Schwartz, M. F., et al., Rad9 phosphorylation sites couple Rad53 to the *Saccharomyces cerevisiae* DNA damage checkpoint. *Mol. Cell* **2002**, *9*, 1055–1065.

73 Gilbert, C. S., Green, C. M., Lowndes, N. F., Budding yeast Rad9 is an ATP-dependent Rad53 activating machine. *Mol. Cell* **2001**, *8*, 129–136.

74 Ahn, J. Y., et al., Phosphorylation of threonine 68 promotes oligomerization and autophosphorylation of the Chk2 protein kinase via the forkhead-associated domain. *J. Biol. Chem.* **2002**, *277*, 19389–19395.

75 Xu, X., Tsvetkov, L. M., Stern, D. F., Chk2 activation and phosphorylation-dependent oligomerization. *Mol. Cell Biol.* **2002**, *22*, 4419–4432.

76 Bashkirov, V. I., et al., Direct kinase-to-kinase signaling mediated by the FHA phosphoprotein recognition domain of the Dun1 DNA damage checkpoint kinase. *Mol. Cell Biol.* **2003**, *23*, 1441–1452.

77 Kobayashi, J., et al., NBS1 localizes to gamma-H2AX foci through interaction with the FHA/BRCT domain. *Curr. Biol.* **2002**, *12*, 1846–1851.

78 Tauchi, H., et al., The forkhead-associated domain of NBS1 is essential for nuclear foci formation after irradiation but not essential for hRAD50–hMRE11–NBS1 complex DNA repair activity. *J. Biol. Chem.* **2001**, *276*, 12–15.
79 Goldberg, M., et al., MDC1 is required for the intra-S-phase DNA damage checkpoint. *Nature* **2003**, *421*, 952–956.
80 Lou, Z., et al., MDC1 is coupled to activated CHK2 in mammalian DNA damage response pathways. *Nature* **2003**, *421*, 957–961.
81 Stewart, G. S., et al., MDC1 is a mediator of the mammalian DNA damage checkpoint. *Nature* **2003**, *421*, 961–966.
82 Wang, P., et al., II. Structure and specificity of the interaction between the FHA2 domain of rad53 and phosphotyrosyl peptidesdagger [In Process Citation]. *J. Mol. Biol.* **2000**, *302*, 927–940.
83 Li, J., et al., Structural and functional versatility of the FHA domain in DNA-damage signaling by the tumor suppressor kinase Chk2. *Mol. Cell* **2002**, *9*, 1045–1054.
84 Peters, J. M., SCF and APC: the yin and yang of cell cycle regulated proteolysis. *Curr. Opin. Cell Biol.* **1998**, *10*, 759–768.
85 Willems, A. R., et al., SCF ubiquitin protein ligases and phosphorylation-dependent proteolysis. *Philos. Trans R Soc. London B Biol. Sci.* **1999**, *354*, 1533–1550.
86 Winston, J. T., et al., A family of mammalian F-box proteins. *Curr. Biol.* **1999**, *9*, 1180–1182.
87 Deshaies, R. J., SCF and Cullin/Ring H2-based ubiquitin ligases. *Annu. Rev. Cell Dev. Biol.* **1999**, *15*, 435–467.
88 Harper, J. W., Protein destruction: adapting roles for Cks proteins. *Curr. Biol.* **2001**, *11*, R431–435.
89 Yaron, A., et al., Inhibition of NF-kappa-B cellular function via specific targeting of the I-kappa–B-ubiquitin ligase. *EMBO J.* **1997**, *16*, 6486–6494.
90 Yaron, A., et al., Identification of the receptor component of the IkappaBalpha–ubiquitin ligase. *Nature* **1998**, *396*, 590–594.
91 Winston, J. T., et al., The SCFbeta–TRCP–ubiquitin ligase complex associates specifically with phosphorylated destruction motifs in IkappaBalpha and beta-catenin and stimulates IkappaBalpha ubiquitination in vitro [erratum appears in *Genes. Dev.* **1999**, *13*, 1050]. *Genes. Dev.* **1999**, *13*, 270–283.
92 Orlicky, S., et al., Structural basis for phosphodependent substrate selection and orientation by the SCFCdc4 ubiquitin ligase. *Cell* **2003**, *112*, 243–256.
93 Wu, G., et al., Structure of a beta-TrCP1–Skp1–beta-catenin complex: destruction motif binding and lysine specificity of the SCF(beta-TrCP1) ubiquitin ligase. *Mol. Cell* **2003**, *11*, 1445–1456.
94 Tyers, M., Jorgensen, P., Proteolysis and the cell cycle: with this RING I do thee destroy. *Curr. Opin. Genet. Dev.* **2000**, *10*, 54–64.
95 Nash, P., et al., Multisite phosphorylation of a CDK inhibitor sets a threshold for the onset of DNA replication. *Nature* **2001**, *414*, 514–521.
96 Deshaies, R. J., Ferrell, J. E. Jr., Multisite phosphorylation and the countdown to S phase. *Cell* **2001**, *107*, 819–822.
97 Harper, J. W., A phosphorylation-driven ubiquitination switch for cell-cycle control. *Trends Cell Biol.* **2002**, *12*, 104–107.
98 Maniatis, T., A ubiquitin ligase complex essential for the NF-kappaB, Wnt/Wingless, and Hedgehog signaling pathways. *Genes. Dev.* **1999**, *13*, 505–510.
99 Karin, M., Ben-Neriah, Y., Phosphorylation meets ubiquitination: the control of NF-[kappa]B activity. *Annu. Rev. Immunol.* **2000**, *18*, 621–663.
100 Jackson, P. K., Ubiquitinating a phosphorylated Cdk inhibitor on the blades of the Cdc4 beta-propeller. *Cell* **2003**, *112*, 142–144.
101 Klein, P., Pawson, T., Tyers, M., Mathematical modeling suggests cooperative interactions between a disordered polyvalent ligand and a single receptor site. *Curr. Biol.* **2003**, *13*, 1669–1678.
102 Seong, Y. S., et al., A spindle checkpoint arrest and a cytokinesis failure by the dominant-negative polo-box domain of Plk1 in U-2 OS cells. *J. Biol. Chem.* **2002**, *277*, 32282–32293.

103 Sonnhammer, E. L., et al., Pfam: multiple sequence alignments and HMM-profiles of protein domains. *Nucleic Acids Res.* **1998**, *26*, 320–322.

104 Lee, K. S., et al., Mutation of the polo-box disrupts localization and mitotic functions of the mammalian polo kinase Plk. *Proc. Natl. Acad. Sci. USA* **1998**, *95*, 9301–9306.

105 Lee, K. S., Song, S., Erikson, R. L., The polo-box-dependent induction of ectopic septal structures by a mammalian polo kinase, plk, in *Saccharomyces cerevisiae*. *Proc. Natl. Acad. Sci. USA* **1999**, *96*, 14360–14365.

106 May, K. M., et al., Polo boxes and Cut23 (Apc8) mediate an interaction between polo kinase and the anaphase-promoting complex for fission yeast mitosis. *J. Cell Biol.* **2002**, *156*, 23–28.

107 Jang, Y. J., et al., Functional studies on the role of the C-terminal domain of mammalian polo-like kinase. *Proc. Natl. Acad. Sci. USA* **2002**, *99*, 1984–1989.

108 Mundt, K. E., et al., On the regulation and function of human polo-like kinase 1 (PLK1): effects of overexpression on cell cycle progression. *Biochem. Biophys. Res. Commun.* **1997**, *239*, 377–385.

109 Elia, E. A. H., et al., The molecular basis for phosphodependent substrate targeting and regulation of Plks by the Polo-box domain. *Cell* **2003**, *##*, *##–##*.

110 Nigg, E. A., Polo-like kinases: positive regulators of cell division from start to finish. *Curr. Opin. Cell Biol.* **1998**, *10*, 776–783.

111 Glover, D. M., Ohkura, H., Tavares, A., Polo kinase: the choreographer of the mitotic stage? *J. Cell Biol.* **1996**, *135*, 1681–1684.

112 Glover, D. M., Hagan, I. M., Tavares, A. A., Polo-like kinases: a team that plays throughout mitosis. *Genes. Dev.* **1998**, *12*, 3777–3787.

113 Donaldson, M. M., et al., The mitotic roles of Polo-like kinase. *J. Cell Sci.* **2001**, *114*, 2357–2358.

114 Kumagai, A., Dunphy, W. G., Regulation of the cdc25 protein during the cell cycle in *Xenopus* extracts. *Cell* **1992**, *70*, 139–151.

115 Izumi, T., Maller, J. L., Elimination of cdc2 phosphorylation sites in the cdc25 phosphatase blocks initiation of M-phase. *Mol. Biol. Cell* **1993**, *4*, 1337–1350.

116 Karaiskou, A., et al., MPF amplification in *Xenopus* oocyte extracts depends on a two-step activation of cdc25 phosphatase. *Exp. Cell Res.* **1998**, *244*, 491–500.

117 Karaiskou, A., et al., Phosphatase 2A and polo kinase, two antagonistic regulators of cdc25 activation and MPF auto-amplification. *J. Cell Sci.* **1999**, *112*, 3747–3756.

118 Qian, Y. W., et al., The polo-like kinase Plx1 is required for activation of the phosphatase Cdc25C and cyclin B-Cdc2 in *Xenopus* oocytes. *Mol. Biol. Cell* **2001**, *12*, 1791–1799.

119 Qian, Y. W., et al., Activated polo-like kinase Plx1 is required at multiple points during mitosis in *Xenopus laevis*. *Mol. Cell Biol.* **1998**, *18*, 4262–4271.

120 Lee, K. S., Erikson, R. L., Plk is a functional homolog of *Saccharomyces cerevisiae* Cdc5, and elevated Plk activity induces multiple septation structures. *Mol. Cell Biol.* **1997**, *17*, 3408–3417.

121 Jang, Y. J., et al., Phosphorylation of threonine 210 and the role of serine 137 in the regulation of mammalian polo-like kinase. *J. Biol. Chem.* **2002**, *277*, 44115–44120.

122 Kelm, O., et al., Cell cycle-regulated phosphorylation of the *Xenopus* polo-like kinase Plx1. *J. Biol. Chem.* **2002**, *277*, 25247–25256.

123 Ellinger-Ziegelbauer, H., et al., Ste20-like kinase (SLK), a regulatory kinase for polo-like kinase (Plk) during the G2/M transition in somatic cells. *Genes. Cells* **2000**, *5*, 491–498.

124 Cheng, K. Y., et al., The crystal structure of the human polo-like kinase-1 polo box domain and its phospho-peptide complex. *EMBO J.* **2003**, *22*, 5757–5768.

125 Leung, G. C., et al., The Sak polo-box comprises a structural domain sufficient for mitotic subcellular localization. *Nat. Struct. Biol.* **2002**, *9*, 719–724.

126 Yu, X., et al., The BRCT domain is a phospho-protein binding domain. *Science* **2003**, *302*, 639–642.

9
The Eukaryotic Protein Kinase Domain

Arvin C. Dar, Leanne E. Wybenga-Groot, and Frank Sicheri

9.1
Introduction

The eukaryotic protein kinases serve as molecular switches to control diverse biological processes such as metabolism, development, the response to DNA damage, and cell cycle progression. Over 500 members of this superfamily are encoded in the human genome [1], and most share a common ability to catalyze the transfer of the gamma-phosphate moiety of adenosine triphosphate (ATP) to hydroxyl groups in target proteins. The phosphorylation of target proteins can cause conformational changes that influence protein function or can serve as points of nucleation for the formation of macromolecular assemblies. The latter occurs through the action of specialized phospho-recognition modules, which are the focus of other chapters of this book. Although protein kinases catalyze the same basic chemical reaction, great diversity of function is imparted by differences in catalytic switching and substrate recognition mechanisms. Indeed, it is the great potential for functional diversification that accounts for the pervasiveness of protein kinases as regulatory switches within the cell.

In this chapter we review the general architecture of the protein kinase catalytic domain and then outline specific examples of protein kinase systems that highlight the diverse mechanisms by which catalytic switching and substrate recognition are controlled. In particular, we focus on representative systems for which mechanisms of regulation have been characterized by high-resolution X-ray crystallographic methods.

9.2
Architecture of the Kinase Domain

The eukaryotic protein kinases employ a highly conserved catalytic core, referred to here as the protein kinase domain, consisting of ~250–300 amino acid residues [2]. This domain is found in cytoplasmic proteins or within the cytoplasmic

Modular Protein Domains. Edited by G. Cesareni, M. Gimona, M. Sudol, M. Yaffe

ISBN: 3-527-30813-X

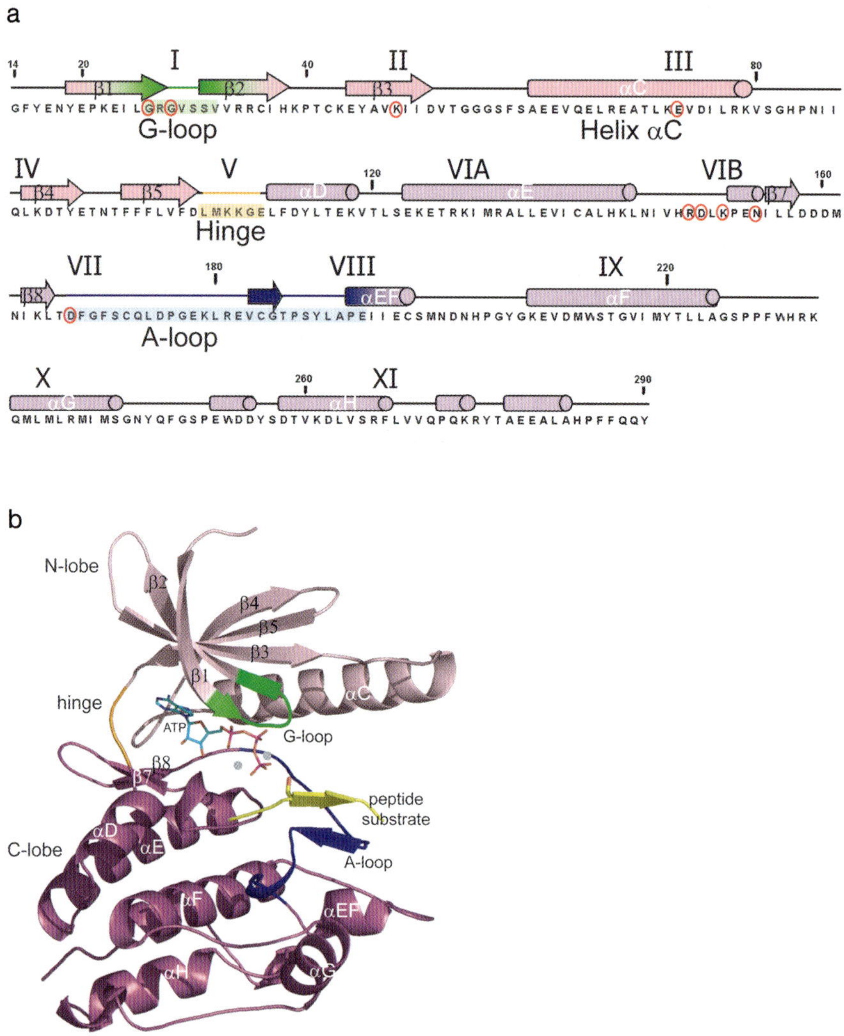

Figure 9.1 Primary and tertiary structure of a representative protein kinase domain. (a) The primary structure of phosphorylase kinase (PDB Id: 2PHK). Conserved secondary structural elements are highlighted, with β strands and α helices as arrows and cylinders, respectively. Protein kinase subdomains are denoted according to the nomenclature of Hanks and Hunter, 1995 [2]. N-lobe, C-lobe, hinge, G-loop, and A-loop elements are colored pink, purple, orange, green, and blue, respectively. Highly conserved protein kinase domain residues involved in catalysis and ATP binding are circled in red. (b) Ribbon representation of the protein kinase domain of phosphorylase kinase in complex with ATP, magnesium, and a canonical peptide substrate. The N lobe and C lobe are colored pink and purple, respectively. The hinge region, the G loop, and A loop are colored orange, green, and blue, respectively. The ATP molecule is shown in stick representation and colored according to chemical properties. Two magnesium ions are represented as silver dots. A canonical peptide substrate is colored yellow, with the phospho-acceptor residue represented in stick style.

component of transmembrane glycoproteins. Based on a comprehensive comparative analysis of human sequences in genomic, cDNA, and EST databases, the eukaryotic protein kinase superfamily can be divided into 10 groups (AGC, CAMK, CMGC, TK, TKL, STE, CKI, RGC, atypical, and others) and further subdivided into numerous families (134 total) and subfamilies (201 total) [1]. Twelve sequences of primary structure, termed subdomains, are highly conserved or invariant throughout the protein kinase superfamily (Figure 9.1a) [2]. Most play crucial roles in enzyme function, including ATP binding, peptide substrate binding, and catalysis [3]. Given the high degree of homology, all protein kinase domains are expected to adopt similar 3D structures (see Figure 9.1b for representative ribbon schematic of phosphorylase kinase) and to mediate phosphotransfer by a common mechanism [2]. This prediction is borne out by the large number of protein kinase structures (≥ 246 redundant structures at last count) deposited in the Protein Data Bank.

The first protein kinase visualized by X-ray crystallography was the cyclic AMP-dependent protein kinase (cAPK) or protein kinase A [4, 5]. Its structure revealed a bilobal architecture, with an ATP binding pocket and active site located within a deep cleft between its component lobes. The upper and smaller N-terminal lobe (or N lobe) is characterized by a β-sheet architecture, and the lower and larger C-terminal lobe (or C lobe) is predominantly α-helical. A single stretch of polypeptide, referred to as the hinge region, connects the two lobes. Minimally, the N lobe consists of a twisted 5-stranded antiparallel β sheet (denoted β1 to β5) and a single helix αC, inserted between β strands 3 and 4, flanking one side of the β sheet. The larger C lobe is composed minimally of two antiparallel β strands (β7 and β8) and six α helices (αD, αE, αEF, αF, αG, αH), with the β strands located on the top surface of the C lobe at the inter-lobe cleft [4]. Variability in the relative orientation of the two protein kinase lobes has been observed, representing both open and closed conformations. However, a closed conformation is generally accepted to be required for enzymatic activity, because this forms the ATP binding pocket and brings catalytic elements into proper orientation for catalysis [6]. Rotation of the N and C lobes as semi-rigid bodies is achieved by a pivot motion at a well-conserved glycine residue within the hinge region [7, 8].

Many protein kinase domains contain inserts of variable length, sequence, and position. For instance, a kinase insert is located between helices αD and αE in the platelet-derived growth factor receptors (PDGFRs), and the SR protein kinase family (which phosphorylate the SR family of serine/arginine-rich pre-mRNA splicing factors) contain a unique insert between strands β7 and β8 [2, 9, 10]. The lengths of kinase inserts vary greatly from 10 to 300 residues, depending on the individual protein kinase [2, 9, 11] and do not appear to be required for intrinsic catalytic function [9, 12].

9.2.1
ATP Binding Pocket

The majority of conserved subdomain elements localize to the cleft region of the protein kinase domain and serve in some capacity to bind and orient ATP. Four of the five conserved subdomains in the N lobe function in this regard. Subdomain I, termed the G loop, with the consensus sequence Gly-x-Gly-x-x-Gly-x-Val (where x is any amino acid), forms the C and N termini of strands β1 and β2, respectively, and an intervening flexible linker. This structural element serves to anchor and coordinate the ATP phosphate groups from a top position [4, 5, 13, 14]. The G loop also contributes in part to a hydrophobic pocket that contacts the adenine ring of ATP [13]. Two invariant N-lobe residues corresponding to a Lys residue (subdomain II) in strand β3 and a Glu (subdomain III) in helix αC, function to coordinate the nonhydrolyzable phosphates of ATP from a lateral position. The Lys sidechain, which is positioned by a salt interaction with the Glu sidechain, directly coordinates the α- and β-phosphoryl groups of ATP [5, 13, 14]. Mainchain atoms of the kinase hinge region (subdomain V) hydrogen bond with the adenine nitrogen atoms [13–15], while sidechains of the hinge region contribute to the hydrophobic pocket surrounding the adenine ring [14].

A large number of conserved C-lobe residues also serve in some capacity to bind ATP. Included are residues immediately C-terminal to the hinge region and residues adjacent to or contained within strand β7 (subdomain VIB). The latter residues hydrogen bond to the hydroxyl groups of the ribose moiety, either directly or via bound water molecules [13–15]. Furthermore, C-lobe residues coordinate two magnesium ions, which in turn coordinate the ATP phosphate groups from a bottom position. Of greatest importance in this regard is the aspartic acid sidechain in the highly conserved Asp-Phe-Gly triplet (subdomain VII), located C-terminal to strand β8 [4, 13, 14].

9.2.2
Peptide Binding and Catalytic Residues

In addition to composing half of the ATP binding pocket, the C lobe contributes key residues implicated in catalysis, the vast majority of the peptide substrate-binding infrastructure, and a key structural element termed the activation segment which is implicated in autoregulation. A conserved sequence element called the catalytic loop (subdomain VIB), located N-terminal to strand β7, plays a critical role in both the phosphotransfer mechanism and in orienting peptide substrates in the vicinity of the active site [4, 5]. Conserved residues in this loop hydrogen bond with the phospho-acceptor group of the peptide substrate, positioning it optimally for catalysis. In addition, an invariant Asp residue is thought to function as the catalytic base, acting to abstract a proton from the target hydroxyl group of the protein substrate. The resulting alcoholate or phenolate ion is then well positioned to attack the γ-phosphate of ATP [3, 4, 6, 13–16]. Other regions of the C lobe also participate in peptide substrate binding through a combination of electrostatic and hydrophobic interactions [5, 15]. The nature of these contacts, which typically do not extend beyond the three positions flanking each side of the phospho-acceptor site, dictate the preferred and/or tolerated peptide sequences recognized by the protein kinase [17]. A more detailed discussion of substrate recognition is provided in Section 9.4.

Also contained within the C-terminal lobe of the protein kinase domain is a lengthy polypeptide sequence called the activation segment, or A loop (Figure 9.1a and 9.2a). This segment, located between strand β8 and helix αEF (starting and finishing at the conserved triplets Asp-Phe-Gly of subdomain VII and Ala-Pro-Glu of subdomain VIII), varies considerably in sequence and length, from typically 20 to 25 amino acids to greater than 100 residues in some protein kinases [2, 6]. The A loop is a focal point of regulatory control in many protein kinases. For protein kinases in their active states, the A loop adopts an extended conformation that sprawls laterally away from the active site along the lower kinase lobe before redirecting back beyond the active site to the C-lobe helix αEF. Considerable variability exists in the sequence of the A loop, its precise conformation, and the interactions that maintain it in productive and nonproductive conformations. The productive conformation of the A loop plays an important role in orienting the peptide substrate for phosphotransfer, providing a platform for peptide substrate binding. Specifically, a β-sheet–like hydrogen-bonding pattern is commonly observed between residues at the C-terminal end of the A loop and substrate residues C-terminal to the phospho-acceptor position (Figure 9.1b). In addition, the A loop governs the distance between the mainchain position of the phosphate-acceptor residue and the catalytic base, thereby defining a preference for Ser/Thr- versus Tyr-containing targets [11, 13, 15].

Far greater variability is observed in the conformation of A loops in protein kinase structures representing down-regulated states. In such structures, the A loop is commonly found disordered [18–24] or adopts a conformation incompatible with catalysis due to a perturbation of the ATP or peptide binding pockets or the position of catalytic residues. Many protein kinases have one or more regulatory sites within their A loops that can be phosphorylated either by a distinct 'activating' kinase or by autophosphorylation. These events serve to restructure the activation segment into a catalytically productive conformation. The structural basis for phospho-regulation by the A loop is described in more detail in the following review of protein kinase catalytic switching mechanisms.

9.3 Catalytic Switching Mechanisms

9.3.1 Kinase Regulation by the A Loop

The ability to alternate between catalytically active and inactive states in response to specific stimuli represents a critical point of diversity and control for protein kinase function. The crystal structure of the insulin receptor kinase (IRK) in its down-regulated state provides a prime example of the regulation of protein kinase catalytic function by mechanisms involving A-loop interference with the active site (Figure 9.2b). In this structure, the activation segment adopts a conformation that

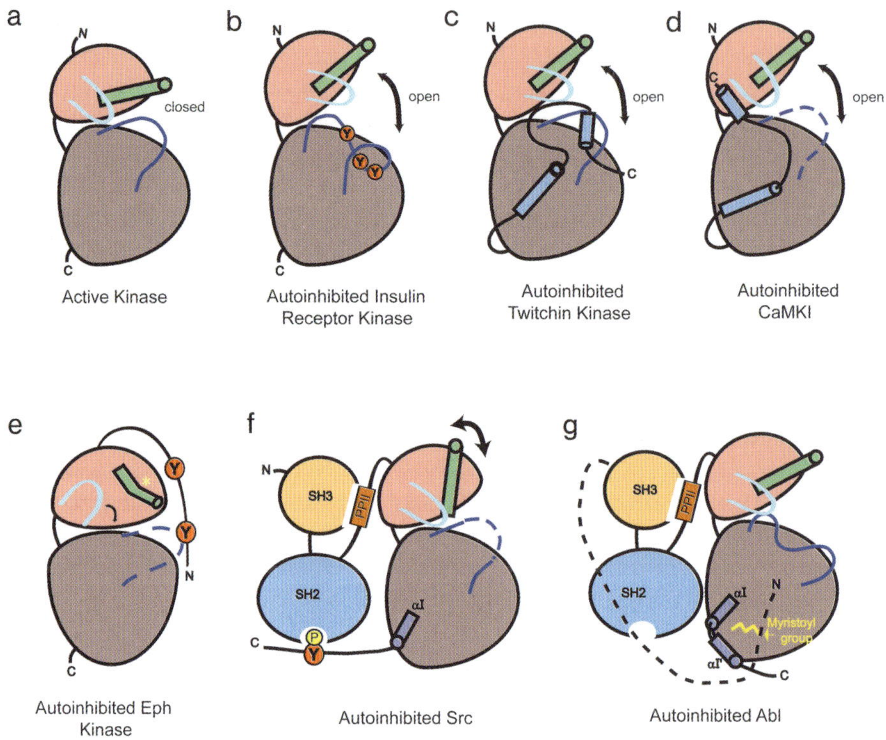

Figure 9.2 Protein kinase catalytic switching mechanisms. Cartoon representations of various protein kinase domains with N lobe, C lobe, G loop, A loop, and helix αC colored pink, purple, light blue, dark blue, and green, respectively. Additional domains and regulatory motifs are labeled. Phospho-regulatory tyrosine residues are highlighted as red circles and phosphates are highlighted as yellow dots. (a) Productive closed conformation. (b) Autoinhibited insulin receptor kinase domain. The A loop sterically blocks ATP and peptide substrate binding. In its phosphorylated state, the A loop transitions to a catalytically competent conformation. (c) Pseudosubstrate regulation of twitchin protein kinase is characterized by a C-terminal regulatory segment that extends through the active site of the kinase domain. This causes a relative opening between the N and C lobes and prevents ATP and peptide substrate binding. (d) Pseudosubstrate regulation of CaMKI. The C-terminal regulatory region of CaMKI associates with the N lobe to stabilize an open, noncatalytic conformation of the kinase domain. Also, the G-loop conformation is severely distorted. (e) Juxtamembrane region regulation of the Eph receptor tyrosine kinases. In its unphosphorylated state, the juxtamembrane region associates with the N lobe of the kinase domain, inducing a kink in helix αC (yellow asterisk). As a result, the G loop is shifted up and the A loop is disordered. (f) Autoinhibition of the Src family kinases. Intramolecular engagement of the SH2 domain in blue with the phosphorylated C-terminal tail and the SH3 domain in orange with a polyproline II (PPII; red box) helical motif stabilizes an autoinhibited kinase domain conformation involving the lateral displacement of helix αC. (g) Autoinhibitory model of Abl tyrosine kinase. Like the Src family kinases, the SH3 domain of Abl engages a polyproline type II helical motif in the SH2-kinase domain linker, and both SH3 and SH2 domains are similarly positioned on the back of the kinase domain. Shown in yellow is a myristoyl group bound to the lower kinase domain lobe.

prevents both ATP and substrate binding [11]. Specifically, one of three A-loop phospho-regulatory residues, Tyr1162 [25], engages the substrate acceptor site, thereby competitively inhibiting substrates from binding [11]. Simultaneously, the Phe-Gly dipeptide of subdomain VII within the N-terminal end of the A loop occupies the ATP binding site, so that cis autophosphorylation of the optimally placed Tyr1162 sidechain is precluded. Additionally, steric clashes between the Asp-Phe dipeptide of subdomain VII and residues within the G loop maintain the N and C lobes in a nonproductive open conformation.

The crystal structure of triply phosphorylated IRK reveals how phosphorylation of its A loop relieves structural distortions characteristic of its autoinhibited state. In response to autophosphorylation of three A-loop tyrosine residues (Tyr1158, Tyr1162, and Tyr1163), the A loop undergoes a dramatic rearrangement to a conformation stabilized by both phosphotyrosine (pTyr) and non-pTyr A-loop interactions [15]. This allows the N and C lobes to reorient to a closed conformation with unrestricted access to the ATP and peptide substrate binding sites [15] (Figure 9.2a). Structures of the related insulin-like growth factor 1 receptor tyrosine kinase demonstrate that the IRK A-loop regulatory mechanism is conserved across a larger group of receptor tyrosine kinases [16, 26].

A subtly different mechanism of A-loop regulation is employed by the fibroblast growth factor receptor 1 (FGFR1) kinase. As revealed by its autoinhibited structure, both the active site and ATP binding site are unobstructed by the A loop; instead, the C-terminal end of the A loop selectively interferes with peptide substrate binding [27]. Additionally, steric interactions between residues in helix αC and the Asp-Phe-Gly tripeptide motif of subdomain VII appear to impede lobe closure upon ATP binding. Presumably, autophosphorylation of two A-loop regulatory tyrosines (Tyr653 and Tyr654), which is required for the up-regulation of FGFR1 kinase activity [28], would result in rearrangement of the A loop, allowing productive substrate binding and interlobe closure.

The activation segment has been shown to regulate the activity of numerous other protein kinases. Examples for which structures are available include the muscle-specific receptor tyrosine kinase MuSK, vascular endothelial growth factor receptor 2 kinase, Tie2 receptor tyrosine kinase, aurora-2 kinase, Abelson leukemia virus tyrosine kinase Abl, cyclin-dependent kinase 2 CDK2, the mitogen-activated protein kinase ERK2, human lymphocyte kinase Lck, and the Src family kinases [19, 24, 29–36]. Many protein kinases do not require A-loop phosphorylation for full activation and, although some of these kinases are simply constitutively active, others employ alternative mechanisms to regulate kinase activity. A survey of alternative mechanisms of kinase regulation follows.

9.3.2
Regulation of Catalysis by Elements External to the Kinase Domain

The positioning of active-site elements within the catalytic domain serves as points of regulation for many protein kinases. In many instances, complex regulatory mechanisms are composed through the association of the kinase domain with structural elements external to the kinase domain. These elements can be part of the same protein, giving rise to intramolecular modes of regulation, or they can be components of other proteins, giving rise to intermolecular modes of regulation. Furthermore, external structural elements can be linear polypeptide sequences devoid of structure when not in complex with a protein kinase domain, or they can adopt autonomous folds with additional kinase-independent functions.

9.3.2.1 **Pseudosubstrate Regulation**

A number of protein kinases are negatively regulated by linear sequences C-terminal to the protein kinase domain that engage the kinase domain in an extended manner across the ATP and/or peptide substrate binding sites. This mode of regulation, termed pseudosubstrate inhibition, is exemplified by twitchin kinase, a large 753-kDa protein from *Caenorhabditis elegans* and a member of the CaMK protein kinase group and myosin lightchain kinase subfamily [2, 7] (Figure 9.2c). Its domain structure consists of multiple copies of N-terminal fibronectin type III domains and immunoglobulin-like domains followed by a Ser/Thr kinase domain and a C-terminal 60-residue autoregulatory segment [37–39]. In a crystal structure representing its autoinhibited state, the C-terminal regulatory segment of twitchin adopts a conformation consisting of an α helix, a large extended loop, a second α helix, and then a terminating short β strand [7]. The extended-loop portion of the regulatory segment extends through the kinase active site, wedging between the two kinase lobes. This association causes a 30° rotation of the N lobe relative to the C lobe, giving rise to an open conformation. The first α helix and extended loop also interact with determinants of peptide substrate binding. In addition, the loop region occupies the equivalent position of the ribose 5′-diphosphate of ATP and forms electrostatic interactions with the Asp and Lys sidechains of subdomains VIB and II, respectively. As a result, both ATP and peptide substrate binding sites are occluded. Structural and functional analyses of titan kinase, the human homolog of *C. elegans* twitchin kinase, revealed a nearly identical mechanism of pseudosubstrate regulation [40].

Interestingly, both ATP and peptide substrate binding are precluded in the autoregulatory mechanism of the Ca^{2+}/calmodulin-dependent protein kinase I (CaMKI) (Figure 9.2d). However, the C-terminal regulatory sequence of CaMKI, which contains both autoinhibitory and Ca^{2+}/calmodulin binding sites [41], adopts an α-helix, extended-loop, α-helix structure that differs in many respects from that of twitchin kinase [18]. Overall, the C-terminal regulatory segment of CaMKI associates with the catalytic domain to stabilize an open, nonproductive conformation. However, the molecular details of intramolecular association differ significantly. Specifically, while the first helix and extended loop of the regulatory segment form ATP and peptide substrate-like interactions with the protein kinase domain, the remaining structural elements do not enter the active site. Instead, they redirect away from the active site and bind to the outer surface of the kinase N lobe, wedging between the G loop and hinge region. As a consequence, the G-loop conformation is severely deformed [18].

The binding of a modulator to the C-terminal regulatory segment serves to activate both twitchin and CaMKI by displacing the intramolecular association of the regulatory segments with the protein kinase domains. The specific modulator of each kinase differs, however, as does the requirement for an additional phosphorylation event for full catalytic activation. In twitchin, the activator S100A1 binds to a portion of the C-terminal regulatory segment to fully activate the kinase [42–44]. In contrast, calmodulin binds to the C-terminal autoregulatory segment of CaMKI in a calcium-dependent manner. When coupled with phosphorylation of an A-loop Thr residue, full activation of CaMKI is achieved [18, 45, 46]. Interestingly, although the structural basis for autoinhibition of titan and twitchin kinases are strikingly similar, titan, like CaMKI, requires A-loop phosphorylation for maximum activation.

The crystal structure of unphosphorylated human MAPKAPK2 (mitogen-activated protein kinase-activated protein kinase 2) provides a third example of pseudosubstrate regulation by a C-terminal segment [23]. This example is notable in that its down-regulated state does not appear to preclude ATP binding. The domain structure of MAPKAPK2 consists, from its N terminus, of two proline-rich segments followed by a Ser/Thr kinase domain and a C-terminal regulatory segment containing both nuclear localization and export signal motifs [47, 48]. As observed for CaMK1 and twitchin, the C-terminal regulatory segment of MAPKAPK2 adopts a familiar α-helix, extended-loop, α-helix conformation [23]. Similarly, the first α helix and extended loop of the regulatory segment act like a pseudosubstrate by occupying the presumed peptide substrate binding pocket. However, in contrast to CaMKI and twitchin, the ATP binding pocket and interlobe orientation appear unaffected. Instead, the second α helix of the regulatory segment sterically impedes the A loop from adopting a productive conformation, which is disordered in the MAPKAPK2 structure. Activation of MAPKAPK2 occurs by phosphorylation of a Thr residue located between the kinase domain and the C-terminal regulatory segment by the p38 MAP kinase [49, 50]. This event is predicted to displace the C-terminal regulatory region from the kinase domain, thereby activating catalytic function and, in a coordinated manner, altering its subcellular localization.

A last variation on the theme of pseudosubstrate regulation, with additional levels of regulatory complexity, has been uncovered from the X-ray crystal structure of the p21-activated kinase PAK1 [51]. In the autoinhibited state, a kinase inhibitor segment of PAK1 binds within its active site in a manner similar to the pseudosubstrate interactions of the above-mentioned kinase systems. However, as opposed to the C-terminal regulatory sequences of those kinases, the kinase inhibitor segment of PAK1 is positioned N-terminal to the protein kinase domain adjacent to a p21 binding domain and a dimerization motif. Intriguingly, the autoinhibited state of PAK1 is dimeric, which facilitates a trans inhibitory mechanism [52]. Specifically, the kinase inhibitor segment of one PAK molecule inhibits the kinase domain of a second PAK molecule within a dimeric configuration. Upon binding to the GTPase Cdc42 or Rac, the autoinhibitory interactions between the N-terminal regulatory segment and the kinase domain of PAK1 disengage so that the A loop is released for phosphorylation and activation [51, 52].

9.3.2.2 Receptor Tyrosine Kinase Regulation by the Juxtamembrane Region

In general, the domain structure of receptor tyrosine kinases (RTKs) consists, from the N terminus, of an extracellular ligand binding region, a transmembrane spanning segment, and a cytoplasmic region containing a tyrosine kinase domain. RTKs are typically activated by interaction with their ligands, which induces oligomerization or reorientation of monomeric receptors and leads to the intermolecular autophosphorylation of cytoplasmic tyrosine residues [53]. Frequently, tyrosine residues located in the receptor's juxtamembrane region, the sequence between the transmembrane segment and the kinase domain, become phosphorylated in this manner. This event can serve to create passive recruitment sites for SH2 and PTB domain-containing proteins or can play a direct role in regulating receptor catalytic function. The latter mechanism of action is exemplified by the Eph receptor tyrosine kinases.

The Eph receptors are the largest family of RTKs in humans, with 14 members. Their domain structure is highly conserved from worms to humans, consisting, from their N terminus, of an extracellular ligand binding domain, a cysteine-rich region, two fibronectin type III repeats, a single membrane-spanning segment, a juxtamembrane segment, a tyrosine kinase domain, a sterile alpha motif (SAM) domain, and a C-terminal PDZ domain binding motif. Within the juxtamembrane region is a highly conserved motif ($\mathbf{Y}_{604}$IDPFT$\mathbf{Y}_{610}$EDP in EphB2) containing two key phospho-regulatory sites [54, 55]. In the absence of phosphorylation, the juxtamembrane region serves to repress the catalytic function of the adjacent protein kinase domain [55]. Once phosphorylated, this repressive function is relieved. The crystal structure of a cytoplasmic fragment of murine EphB2, consisting of the conserved juxtamembrane segment and the protein kinase domain, revealed the mechanism by which autoregulation is achieved [20] (Figure 9.2e). In its dephosphorylated state, the juxtamembrane region associates intimately with helix αC of the kinase N lobe, inducing a 14° kink. In consequence, the conformation of the G loop and the salt interaction between the Lys and Glu sidechains of subdomains II and III are distorted, thereby perturbing the coordination of the ATP ribose and phosphate groups. Although the integrity of the adenine ring binding pocket is maintained, the resultant mode of binding ATP is nonproductive. In addition to perturbing the conformation of N-lobe elements, Tyr604 of the juxtamembrane region also prevents the A loop from attaining a productive conformation. This is predicted to disrupt the binding of peptide substrates. Based on modeling and mutagenesis data, phosphorylation of the juxtamembrane Tyr residues was predicted to disengage the juxtamembrane region from the kinase domain through electrostatic and steric effects. This would allow for restoration of N- and C-lobe structural elements to productive conformations. This prediction has been confirmed by structural analyses of activated forms of EphA4 and EphB2 receptor kinases (L. E. W.-G. and F. S., unpublished data). Given the highly conserved nature of the juxtamembrane–kinase domain interface, the mechanism of autoregulation characterized for EphB2 is likely to apply to all Eph receptor family members [20].

The highly conserved juxtamembrane region of the platelet-derived growth factor receptor family of RTKs, which includes PDGFβ, Kit, and Flt3 receptors, plays an analogous role in receptor regulation to that observed for the Eph RTKs. However, the structural basis for juxtamembrane regulation by this RTK family has yet to be characterized. In support of a regulatory role for their juxtamembrane regions, autophosphorylation of juxtamembrane tyrosines 579 and 581 in the PDGFβ receptor is required for full kinase activation, and mutation to phenylalanine abolishes receptor activation [56]. In addition, juxtamembrane mutations N-terminal to the phospho-regulatory sites give rise to constitutive receptor activation [57]. Similarly, some oncogenic variants of the Kit receptor, identified in gastrointestinal stromal tumors or mast cell leukemia cell lines, contain mutations within the juxtamembrane region that result in ligand-independent activation [58–61]. Detailed biochemical and kinetic analyses of Kit have revealed an ability of the juxtamembrane region to inhibit the catalytic domain as separate polypeptides through a direct binding interaction. This activity of the juxtamembrane region is perturbed by oncogenic mutation [62]. A third PDGFβ receptor family member, Flt3, has been shown to be mutated in approximately 20% of acute myeloid leukemias, with most mutations mapping to conserved sites in the juxtamembrane region [63–65]. Although less well characterized, an autoregulatory role for the juxtamembrane region of the muscle-specific receptor tyrosine kinase (MuSK) has also been identified [30, 66, 67]. The lack of significant sequence similarity between the juxtamembrane regions of MuSK, the PDGFβ receptor family, and the Eph family of RTKs suggest that the underlying structural basis for juxtamembrane regulation differs considerably.

9.3.2.3 Intramolecular Regulation Involving Autonomously Folded Domains

Members of the Src family tyrosine kinases share a common domain architecture consisting of an N-terminal myristylation signal, an SH3 domain, an SH2 domain, a tyrosine kinase domain, and a C-terminal regulatory tail containing a key tyrosine phospho-regulatory site. Together, the SH3, SH2, catalytic domain, and C-terminal regulatory tail play interdependent roles in down-regulating protein kinase activity [68]. The structural basis for this behavior was uncovered by crystallographic analyses of the Src family kinase Hck (hematopoietic cell kinase) and Src itself [34, 35]. Down-regulated states for both family members are achieved by the intramolecular engagement of the SH2 domain with the phosphorylated C-terminal tail and the SH3 domain with a polyproline II helical motif located in the SH2 domain–protein kinase domain linker (Figure 9.2f). These intramolecular modes of engagement localize the SH2 and SH3 domains on the back of the kinase domain, away from the active site. Although remotely positioned on the kinase domain, the SH2 and SH3 domains still influence the active site by stabilizing the lateral displacement of the N-lobe helix αC. A displaced helix αC is also stabilized in part by a nonproductive conformation of the kinase-domain A loop in its dephosphorylated form [34]. As a consequence of helix-αC displacement, the invariant Glu residue of subdomain III is rotated outward into a nonproductive position, thereby perturbing catalytic function.

The lateral displacement of helix αC is a key component of the cyclin-dependent kinase regulatory mechanism. As shown by an X-ray crystallographic analysis of the down-regulated state of CDK2 [33], helix αC adopts a laterally displaced position in the absence of cyclin binding and A-loop phosphorylation. This feature appears to be stabilized in part by the conformation of the A loop, which forms a large helical structure in the cleft region of the kinase. Both CDKs and Src kinases employ phosphorylation of their A loops as part of their activation mechanism. However, in contrast to the Src family kinases, which require dissociation of SH2 and SH3 domains from the kinase domain (through competitive binding to high-affinity phosphopeptide and proline-rich ligands, respectively) for full activation, the positioning of helix αC in CDK2 is regulated by the intermolecular association of a cyclin regulatory subunit [69].

The Abl tyrosine kinase possesses a similar domain architecture to that of the Src family kinases in possessing an SH3 and SH2 domain preceding a tyrosine kinase catalytic domain [70–73]. Importantly, Abl lacks the characteristic phospho-regulatory tail characteristic of the Src family kinases [74] and instead possesses a unique N-terminal myristylated 'cap' of ~80 residues preceding the SH3 domain that serves an analogous functional role [75, 76]. Despite these differences, Abl adopts a strikingly similar autoinhibitory structure to that of the Src family kinases. In particular, the SH3 domain of Abl engages a polyproline type II helical motif in the SH2-kinase domain linker, and both SH3 and SH2 domains are similarly positioned on the back of the kinase domain, remote from the active site [77] (Figure 9.2g). Significant differences between the Src-family and Abl autoinhibited structures include the absence of an intramolecular phosphopeptide ligand for the SH2 domain of Abl and the mechanism by which the higher-order association of domains serves to influence kinase domain activity.

In the Abl structure, the lack of an SH2–phosphorylated tail interaction may be compensated for by a more robust direct interaction between the SH2 domain and the back of the kinase C lobe. Although the equivalent direct point of contact in the Src family kinase structures is almost exclusively electrostatic in nature and hence easily solvated, the interface on the Abl structure is composed of a mixture of electrostatic, hydrogen bonding, and hydrophobic interactions. Very close to this interface on the kinase C lobe of Abl resides a binding site for the N-terminal myristoylation cap sequence. Comparison of the isolated Abl kinase domain structure with the higher-order structure of autoinhibited Abl reveals that binding of a myristoyl group to the kinase C lobe induces a conformational change compatible with the intramolecular association of the SH2 domain. Specifically, the terminal helix αI of the isolated kinase domain undergoes rearrangement to two shorter α helices, αI and αI′, that compose part of the SH2 binding surface. Hence, the binding of a myristoyl group may serve as a modulator of Abl kinase activity. In contrast to the Src kinases, the autoinhibitory mechanism of Abl is not characterized by a lateral displacement of the kinase domain helix αC. Instead, the higher-order structure is hypothesized to act as a clamp that impedes N- and C-lobe motions critical for repositioning the kinase domain A loop to a productive conformation in response to A-loop phosphorylation. In the absence of phosphorylation, the N-terminal portion of the A loop folds into the active site to prevent peptide substrate binding in a manner similar to that observed for the insulin receptor tyrosine kinase [31, 77, 78].

9.4 Protein Kinase Substrate Recognition

The ability to recognize and discriminate substrates represents a second critical point for diversity and control of protein kinase functions. Insight into the mechanisms that underlie protein kinase substrate recognition has added to our understanding of protein kinase function under normal physiological conditions. Moreover, these insights have provided a framework for the development of specific therapeutics for protein kinases implicated in human diseases.

The study of substrate recognition by the catalytic domain of protein kinases has focused largely on their ability to discriminate target phosphorylation sites (Ser/Thr vs. Tyr) and the flanking amino acid sequences in short polypeptides. However, it has become increasingly evident that the determinants that underlie protein kinase substrate specificity within the cell are complex and extend beyond the immediate site of phosphorylation in the target substrate. In this section we examine the underlying principles and the diverse mechanisms for which substrate recognition by a protein kinase is defined, with a focus on systems for which structural insight is available.

9.4.1 Canonical Peptide Substrate Recognition

The most basic mechanism employed by all protein kinases to discriminate substrates centers on their ability to distinguish between Ser/Thr and Tyr acceptor sites and the flanking sequences in short polypeptide sequences. This level of substrate specificity, referred to here as canonical peptide substrate recognition or active-site-directed specificity, is defined in the immediate vicinity of the protein kinase active site (Figures 9.1b and 9.3a). Optimal polypeptide sequences, which are recognized by the active site of a kinase, can be determined empirically by using short synthetic peptide libraries [79], phage-display methods [80], or comparing characterized phosphorylation sites in known protein kinase substrates (Note: P0 denotes the acceptor site position, and P –x and P +x denote relative positions N- and C-terminal to the P0 position). The structural basis for canonical peptide substrate recognition has been studied by X-ray crystallographic analyses of protein kinase catalytic domains in complex with short peptide substrates [3, 15, 81, 82].

As first exemplified by the X-ray crystal structure of cAPK and its pseudosubstrate inhibitor PKI, the determinants that anchor a peptide substrate into the kinase active site predominantly reside in the C lobe of the kinase domain [3, 5, 13]. Binding pockets within the C lobe, defined in part by the A loop, accommodate residues adjacent to the P0 phospho-acceptor site and appear to dictate the tolerated and preferred amino acids for each position. For example, as revealed by an insulin receptor kinase (IRK) domain and peptide substrate cocrystal structure, two hydrophobic pockets within the C lobe of the catalytic domain accommodate two methionine residues in the P +1 and P +3 positions of a bound peptide substrate

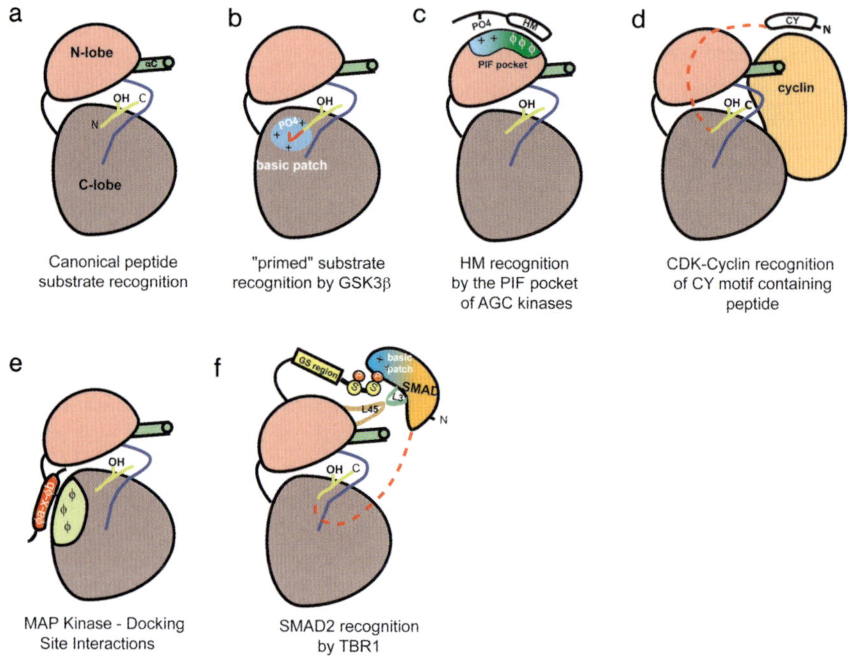

Figure 9.3 Cartoon representations of various protein kinase domains colored as in Figure 9.2. (a) A canonical peptide substrate shown in yellow engages a binding site on the lower C lobe of the kinase domain. (b) 'Primed' substrate recognition by GSK3β. A phosphate group (corresponding to the P +4 position) of a primed substrate binds to a basic patch, colored blue. (c) The hydrophobic motif of AGC kinases serves as a recruitment signal and/or activation motif. The hydrophobic motif (HM) engages AGC kinases through the PIF pocket (blue-green) in the N lobe. (d) Bipartite substrate recognition by the cyclin-dependent protein kinases. Regulatory cyclins (orange) compose a binding site for the CY substrate motif, which complements the canonical peptide substrate specificity of the kinase domain. (e) The docking peptides from MAPK substrates and activators interact with a binding site in the kinase domain C lobe (green). (f) Upon phosphorylation, the GS region (yellow, with phosphoserines as red and yellow circles) of TGFβ type 1 receptors, together with the L45 loop (orange), compose a Smad substrate binding site. A basic patch and the L3 loop compose the complementary binding site on Smad2.

[15]. Furthermore, a preferred acidic residue at the P –1 position forms a water-mediated hydrogen bond with a basic patch on the protein kinase domain. Similarly, in a crystallographic analysis of the phosphorylase Ser/Thr kinase, numerous ion pairs, hydrogen bonds, and hydrophobic contacts are made between residues in the peptide substrate and the C lobe of the kinase domain [81]. As in IRK, the nature of these contacts allows for a rationalization of the optimal canonical peptide substrate specificity of this kinase (Lys-Gln-Met-Ser-Phe-Arg-Leu [79, 81]).

In their bound state, canonical peptide substrates adopt an extended conformation and, to some extent, form an antiparallel alignment with respect to the C-terminal portion of the A loop of the protein kinase domain (Figures 9.1b and 9.3a).

The length and path of the A loop appear to distinguish the ability of a protein kinase to phosphorylate Ser or Thr vs. Tyr residues. Specifically, the A loops of Tyr protein kinases are spaced farther from the catalytic site relative to those of Ser/Thr kinases [15]. This is critical for accommodating the larger size of tyrosine relative to the serine and threonine sidechains [17]. The conformation of the activation segment can also influence specificity for residues surrounding the target site of phosphorylation. For example, an A-loop residue within the cyclin dependent kinase, CDK2, adopts an unusual conformation that is essential for accommodating a proline at the P +1 position of the substrate [82]. Indeed, this feature enforces an absolute requirement for proline at the P +1 position of the CDK substrate consensus sequence, Ser/Thr-Pro-x-Lys/Arg [83, 84]. Similar A-loop determinants impose a requirement for proline at the P +1 substrate position for the mitogen-activated protein kinase, ERK2 [85]. Interestingly, for both ERK2 and CDK2, A-loop phosphorylation facilitates formation of the peptide binding groove. As deduced from comparisons of phosphorylated and unphosphorylated forms of ERK2 and CDK2, phosphorylation within their A loops relieves steric blockage of the peptide binding groove by residues within the A loop itself [29, 33, 82, 85].

Based on Ser/Thr kinase domain structures in complex with short peptide substrates, together with empirically derived peptide substrate specificities for a large set of protein kinases, an algorithm has been devised for predicting canonical peptide substrate specificities [86]. Based solely on the input of the primary structure of a Ser/Thr protein kinase domain of interest, this algorithm has demonstrated significant utility in predicting both in vitro and in vivo protein kinase substrates. However, as with empirically determined canonical substrate specificities, the number of predicted substrates typically far exceeds the total number of real substrates. Hence, this predictive tool still requires experimental validation. The significant error rates that are associated with all predictive methods may be due to the fact that canonical substrate specificity is supplemented by other, more specialized, mechanisms of substrate selection. In the following sections we now survey several more specialized mechanisms that influence protein kinase substrate selection.

9.4.2
'Phospho-priming'–Dependent Substrate Recognition

A key element of substrate recognition by glycogen synthase kinase 3β (GSK3β) is the requirement for prior phosphorylation or 'priming' of its substrate. In response to insulin receptor tyrosine kinase signaling, phosphorylation of glycogen synthase at Ser656 by casein kinase II (CK2) serves as the priming event to allow GSK3β to sequentially phosphorylate GS at Ser652, Ser648, Ser644, and Ser640 [87, 88]. Similarly, Wnt signaling leads to the phosphorylation of β-catenin by casein kinase I (CKIα) at residue Ser45, which then serves as a priming event for GSK3β to sequentially phosphorylate β-catenin at residues Thr41, Ser37, Ser33, and Ser29 [89]. In a general sense, once a substrate is primed by phosphorylation by another kinase, GSK3β can recognize and progressively hyperphosphorylate its substrate, with each subsequent phosphorylation dependent on prior phosphorylation at the P +4 position [87].

X-ray crystal structures of GSK3β allow for a rationalization of the unique 'phospho-priming' requirement of this protein kinase [90, 91]. In these structures, three basic residues remotely positioned in the primary structure (Arg96, Arg180, and Lys205) project to form a basic patch on the front surface of the protein kinase, close to the predicted P +4 position of a bound canonical peptide substrate (Figure 9.3b). In the absence of a bound peptide substrate, the basic region of GSK3β is neutralized by a negatively charged phosphate [90] or sulfonate ion [91] supplied by the crystallization solution. Upon binding a primed peptide substrate, the basic patch is predicted to engage the 'priming' phosphate moiety and thereby position a Ser or Thr residue located four positions N-terminal for acceptance of a phosphate group in a catalytic cycle [90, 91]. The newly phosphorylated site could then serve as the priming site for a subsequent round of catalysis.

Interestingly, the basic patch on the catalytic domain of GSK3β is also predicted to play a role in stabilizing an autoinhibited state [91, 92]. The phosphorylation of GSK3β at Ser9 in its N-terminal tail by the Ser/Thr protein kinase B (PKB/Akt) causes inactivation of the enzyme [93]. Competition analysis with a primed phosphorylated peptide and mutagenesis data suggest that the N-terminal segment of GSK3β occupies the P +4 binding site when phosphorylated on Ser9. This would serve to block the binding of exogenous substrates [91, 92].

9.4.3
Regulation of AGC Kinases by the Hydrophobic Motif

Over 40 distinct human members comprise the AGC protein kinase group that includes cAPK, 3-phosphoinositide-dependent protein kinase-1 (PDK1), and PKB/Akt. In addition to sharing a high level of overall sequence similarity within their catalytic domains and a common phospho-regulatory site within their A loop, most AGC kinases contain a regulatory sequence C-terminal to the catalytic domain, referred to as the hydrophobic motif [94]. Often, this hydrophobic motif contains a key phospho-acceptor residue (either Ser or Thr), flanked by large aromatic residues such as Phe or Tyr (consensus sequence: Phe-x-x-Phe-Ser/Thr-Phe/Tyr) [94, 95]. In other members of the AGC group, an acidic residue such as aspartic or glutamic acid substitutes for the phospho-regulatory position of the hydrophobic motif [96]. PDK1, however, is unique among AGC kinases in possessing a truncated hydrophobic motif with no phospho-regulatory or phospho-substituting position. Although external to the kinase domain, the hydrophobic motif plays important regulatory roles in the AGC group of protein kinases. Specifically, association of the hydrophobic motif with the protein kinase domain of AGC group members can serve to stabilize an active conformation and recruit substrates. A common hydrophobic motif binding site in the kinase domain of the AGC group members serves both functional roles.

The protein kinase PDK1 recognizes the hydrophobic motif of other AGC protein kinases, making them substrates of PDK1. The ability of PDK1 to recruit p70 ribosomal S6 kinase, S6K [97, 98], p90 ribosomal S6 kinase, RSK [99], and the serum- and glucocorticoid-induced kinase, SGK [100, 101] is promoted by phosphorylation of these kinases within their hydrophobic motifs [102–104]. Upon recruitment, PDK1 regulates these kinases through phosphorylation of sites within their A loops. However, not all PDK1 substrates are recruited in this manner. For example, phosphorylation of PKB/Akt [105, 106] by PDK1 is stimulated by the production of phosphatidylinositol-3,4,5-triphosphate, which serves to colocalize the two protein kinases to the plasma membrane via interactions mediated by their respective pleckstrin homology (PH) domains [104, 107, 108].

A conserved pocket within the catalytic domain of PDK1, referred to as the PIF region, forms the complementary binding site for the hydrophobic motif in its substrates [103, 104]. As revealed by a crystallographic analysis of PDK1, the PIF region localizes to the N lobe of the kinase [109], forming a 5-Å pocket with residues from helices αB and αC and strand β5 [109] (Figure 9.3c). Part of the PIF region, composed of two conserved basic residues (Arg131 and Lys76), functions to coordinate the phosphoserine or threonine position of the hydrophobic motif [110]. Consistent with a substrate recruitment role for the PIF region of PDK1, mutations within this region greatly diminish the ability of PDK1 to recruit and phosphorylate its substrates [104, 109, 110].

In addition to serving as a recruitment signal for substrates, the hydrophobic motif may function to regulate the catalytic activity of certain AGC kinase family members [111, 112]. This role is illustrated by the behavior of PKB/Akt, which requires phosphorylation both within its A loop at Thr308 and within its hydrophobic motif at Ser474 for maximum catalytic activation [106, 113, 114]. In the absence of Ser474 phosphorylation but in the presence of A-loop phosphorylation at Thr308, the kinase domain adopts an inactive conformation, which is characterized by a large degree of structural disorder within the A loop and in helices αB and αC [112]. Substitution of Ser474 with Asp or Glu to mimic a phosphorylation event allows the hydrophobic motif to associate with the top surface of the kinase N lobe, causing an ordering of the A loop and helices αB and αC [111]. This should serve to activate PKB/Akt's catalytic function.

The hydrophobic motif binding site on PKB and the PIF recruitment region in PDK1 map to the same position on the kinase domain N lobe. In addition, a similar area in cAPK is bound by its own variant of the hydrophobic motif, which contains the Phe-x-x-Phe sequence but lacks the following phospho-regulatory or analogous acidic amino acid position [4]. Finally, the same region in the AGC group member, aurora-2, is occupied by a linear peptide sequence of its regulatory partner TPX-2 [115]. These findings suggest that peptide engagement at a common site in the N-terminal lobe is a general feature underlying the regulation of and substrate recognition by this group of protein kinases. As such, this region provides a potential target for the design of small molecules that could selectively activate or inhibit AGC kinase group members.

9.4.4 CDK–Cyclin Interactions with Substrates Mediated through the CY Motif

The CDKs, which act in concert with cyclin regulatory subunits, employ a bipartite mechanism for substrate selection. Each subunit of the CDK–cyclin pair participates in substrate selection through the binding of epitopes remotely positioned in the primary structure of the substrate [116–119]. The bipartite elements of CDK substrates are unstructured linear peptide sequences with variable intervening spacing. The peptide motif recognized by the CDK catalytic domain consists of the tetrameric sequence Ser/Thr-Pro-x-(Lys/Arg), where x is any amino acid and Ser/Thr is the phospho-acceptor site. The peptide motif recognized by the cyclin (referred to as the CY motif) consists minimally of the trimeric sequence Arg-X-Leu. The presence of both motifs is required for efficient substrate phosphorylation in vitro and in vivo. The CY motif is also present in CDK inhibitor proteins, including $p27^{Kip1}$, where it acts in concert with a linear inhibitory motif that substitutes for a phosphorylation-site consensus sequence. In the crystal structure of CDK2–cyclinA in complex with a fragment of the p107 substrate [82] and the structure of CDK2–cyclin A in complex with the $p27^{Kip1}$ inhibitor [120], the CY motif binds to the cyclin at coordinates far removed ($\cong$ 35 Å) from the canonical peptide substrate binding site in the catalytic domain (Figure 9.3d). In the CDK2–cyclin A–$p27^{Kip1}$ inhibitor complex, the inhibitor sequence binds in an extended manner across the cyclin and the N-terminal lobe of the kinase domain, inducing conformational changes that perturb catalytic function and ATP binding. Interestingly, the γ-herpes virus has exploited the substrate-targeting mechanism of the CDKs by encoding a viral cyclin with the ability to activate mammalian CDK2 and recognize native substrates to drive cell cycle progression [121]. Yet, as revealed by the structure of the γ-herpes virus cyclin bound to CDK2, this complex differs sufficiently to be resistant to $p27^{Kip}$ inhibition [121].

The ability to predict CDK substrates based on consensus sequence matches of bipartite elements is very useful for generating testable hypotheses of kinase–substrate relationships. However, the variable spacing of bipartite elements in CDK substrates and the short and degenerate nature of each consensus sequence make the accuracy of predictions less than ideal. These same characteristics, however, afford a large degree of flexibility as to how the motifs are incorporated into the primary structure of a substrate and where the phosphorylation event takes place.

9.4.5 MAPK Docking Site Interactions: Common Recognition Mechanisms for Substrates, Activators, and Scaffolds

The MAP kinases operate as modules composed of three protein kinases – a MAP kinase, a MAP kinase kinase (MKK), and a MAP kinase kinase kinase (MKKK) – that phosphorylate and activate each other sequentially [122]. At the end of the cascade, the MAP kinases are activated by phosphorylation at sites within their A loops by an upstream MKK and, in response, phosphorylate a variety of proteins, including transcription factors and cytosolic proteins to influence cellular behavior [123]. In most instances, the assembly of MAP kinase modules occurs through the organizing function of scaffold proteins [124].

MAP kinases recognize their substrates and interact with upstream MAP kinase activators and organizing scaffolds in part through an ability to recognize a simple linear docking sequence present in each of these molecules [125]. The docking sequence consists of a positively charged amino acid followed by a hydrophobic position and having the consensus sequence Arg/Lys-x_n-Φ_A-X-Φ_B, where $\Phi_{A/B}$ are hydrophobic residues and $x_n = 1 - 6$ amino acids [126, 127]. This recognition mechanism supplements the canonical phosphorylation site-specificity of the Erk1/2 MAP kinases, which, like the cyclin-dependent kinases, is proline-directed (consensus: Pro-Leu-Ser/Thr-Pro [79, 128, 129]).

Crystal structures of the MAP kinase, p38, in complex with two different docking site peptides, including one derived from a transcription factor substrate MEF2A and one from its activating MKK, MKK3b, has revealed a common binding site on the back of the C-terminal kinase lobe [130] (Figure 9.3e). In both complexes, the docking peptide binds in an extended manner with strong electron density observed for the Φ_A-X-Φ_B portion of the peptide. Electron density corresponding to the basic portion of the motif, however, was not apparent. This observation is consistent with mutational studies, which indicate that the hydrophobic residues are the major determinants of binding [131]. The composition of residues defining the docking site on the lower kinase lobe are relatively well conserved among MAPK subfamilies [130]. Small differences in composition across subfamilies are predicted to contribute to the selectivity of each MAPK subfamily for its respective docking partner [130, 132].

9.4.6
Substrate Recognition through a Phosphorylated Epitope in TGFβ

The transforming growth factor β (TGFβ) family of proteins regulate cellular responses such as growth, differentiation, and cell fate specification [133]. This family includes the prototype TGFβ proteins, the activin proteins, and the bone morphogenic proteins (BMPs), among others. Signaling by these proteins is mediated through cell surface receptor protein kinases (referred to collectively as the TGFβ receptors), their substrates the Smad family of proteins, and a variety of DNA-binding proteins that regulate transcription in the nucleus. The TGFβ receptor family consists of the related type 1 (TβR-1) and type 2 (TβR-2) classes, which share a common domain architecture consisting of an N-terminal ligand binding domain, a transmembrane region, and a C-terminal serine/threonine kinase domain. The TβR-1 class differs from the TβR-2 class in possessing a regulatory motif, referred to as the GS region, preceding the protein kinase domain [134]. In response to ligand binding, TβR-1 signaling is activated by TβR-2 through trans phosphorylation of the GS region [135]. Upon activation, TβR-1 phosphorylates a subset of Smad proteins (the receptor-regulated class), which then oligomerize with other Smad proteins (coregulator class) and translocate to the nucleus to modulate transcription [136, 137]. An intact TβR-1–Smad protein complex has yet to be visualized directly, but a wealth of biochemical and mutational data, together with structures of the individual TβR-1 and Smad2 proteins, suggest that the two proteins bind using extensive complementary surfaces [138–140] (Figure 9.3f).

Phosphorylation of the GS region of TβR-1 does not affect its intrinsic catalytic efficiency [139]. Rather, in its phosphorylated state, the GS region serves as a recruitment motif for Smad substrates [139]. In addition to the GS motif, a second region of TβR-1 participates in Smad2 binding, namely, the L45 loop within the N lobe of the kinase domain [141, 142]. Situated between strands β4 and β5, the L45 loop lies close to the GS region and, together, the two are postulated to form an extended binding surface for Smad substrates [138, 139]. The structure of Smad2 revealed a characteristic basic surface conserved within a subset of Smad proteins that selectively bind TβR-1 [140]. This basic patch, along with a proximal sequence called the L3 loop, are hypothesized to form the complementary binding surface for the phosphorylated GS region and L45 loop of TβR-1 [141, 143]. Interestingly, the TβR-1 binding surface on Smad proteins resides ~30 Å from the site of TβR-1 phosphorylation [140, 141, 143].

In the absence of phosphorylation, the GS region serves as the binding site for the TβR-1 inhibitor protein, FKBP12. As revealed by the crystal structure of TβR-1 in complex with FKBP12, the inhibitor binds to the GS region of TβR-1 so that phospho-regulatory sites within the GS region are inaccessible to TβR-2 [138]. This enforces an intricate coordination of TβR-1 catalytic activation and substrate recognition.

9.4.7
Substrate Recognition by the eIF2α Protein Kinases: Recognition of a Complex Epitope Presented by Globular Fold

The majority of substrate-recognition mechanisms detailed above relate to linear peptide motifs (e.g., phosphorylation acceptor sites, hydrophobic motifs, CY motifs, docking peptides) binding to specialized binding sites on a structured protein domain. In the context of their unbound states, these peptide epitopes are typically unstructured and map to flexible, highly dynamic regions of a protein. Substrate recognition by the eIF2α (eukaryotic protein translation initiation factor 2α) protein kinase family appears to involve a domain–domain interaction more akin to the TβR-1–Smad2 interaction.

The eIF2α protein kinases, which include the RNA-dependent protein kinase (PKR), GCN2, the heme-regulated eIF2α kinase (HRI), and the pancreatic eIF2α kinase (PEK), respond to distinct stress stimuli within the cell but share a common ability to phosphorylate eIF2α at an identical site, Ser51, to potently inhibit protein translation. The recognition of eIF2α by the eIF2α protein kinases is defined by two sets of determinants encoded within their catalytic domains. Canonical peptide substrate specificity consists of a preference for Ser/Thr residues flanked by basic residues, which closely matches the (ILLSELS$_{51}$RRIR) sequence flanking the Ser51 phosphorylation site in eIF2α [144].

Insight into the secondary determinant of substrate specificity has been provided by structures of eIF2α and its structural mimic K3L, a viral inhibitor protein that subverts the antiviral function of the PKR. The regions of highest sequence identity between K3L and the PKR's natural substrate eIF2α are dispersed throughout the primary structure of the proteins remote from the Ser51 phospho-regulatory site [145]. As revealed by the crystal structures of K3L [146] and eIF2α [147, 148], these conserved elements project to a single highly structured surface located 21.5 Å from Ser51 in eIF2α [146]. This surface, which has been shown by mutagenesis to be critical for recognition in vivo and in vitro, has been termed the PKR recognition motif [145, 146].

The complementary binding site for the PKR-recognition motif on the kinase domain of the eIF2α kinases has been elusive. This may be due to the recent discovery that dimerization of the PKR catalytic domain functions as a switch that allows for both catalytic activation and substrate recognition [146, 149]. An understanding of how the eIF2α/K3L binding site is composed or exposed in response to dimerization awaits the structures of inactive and active conformations of the PKR catalytic domain and its complexes with eIF2α or K3L.

9.5 Conclusions

The structural analysis of protein kinases in catalytically active and repressed states and in complex with their substrates has revealed highly specialized and complex mechanisms for catalytic switching and substrate recognition. These studies have contributed greatly to our understanding of the rules that govern protein kinase function within the cell. However, since the protein kinase systems studied in fine detail represent but a small fraction of the 518 human kinases encoded in the genome, there is still very much to be learned. Indeed, accumulating data suggest that large variations on the currently uncovered themes of regulation and altogether novel mechanisms of regulation have yet to be uncovered. The importance of kinases as regulatory switches within the cell and the involvement of kinase dysregulation in human disease warrant further study and characterization of protein kinase regulatory mechanisms.

References

1 Manning, G., Whyte, D. B., Martinez, R., Hunter, T., Sudarsanam, S., The protein kinase complement of the human genome. *Science* **2002**, *298*, 1912–1934.

2 Hanks, S. K., Hunter, T., Protein kinases 6. The eukaryotic protein kinase superfamily: kinase (catalytic) domain structure and classification. *FASEB J.* **1995**, *9*, 576–596.

3 Madhusudan, Trafny, E. A., Xuong, N. H., Adams, J. A., ten Eyck, L. F., Taylor, S. S., Sowadski, J. M., cAMP-dependent protein kinase: crystallographic insights into substrate recogni-

tion and phosphotransfer. *Protein Sci.* **1994**, *3*, 176–187.

4 Knighton, D. R., Zheng, J. H., ten Eyck, L. F., Ashford, V. A., Xuong, N. H., Taylor, S. S., Sowadski, J. M., Crystal structure of the catalytic subunit of cyclic adenosine monophosphate-dependent protein kinase. *Science* **1991**, *253*, 407–414.

5 Knighton, D. R., Zheng, J. H., ten Eyck, L. F., Xuong, N. H., Taylor, S. S., Sowadski, J. M., Structure of a peptide inhibitor bound to the catalytic subunit of cyclic adenosine monophosphate-dependent protein kinase. *Science* **1991**, *253*, 414–420.

6 Johnson, L. N., Noble, M. E., Owen, D. J., Active and inactive protein kinases: structural basis for regulation. *Cell* **1996**, *85*, 149–158.

7 Hu, S. H., Parker, M. W., Lei, J. Y., Wilce, M. C., Benian, G. M., Kemp, B. E., Insights into autoregulation from the crystal structure of twitchin kinase. *Nature* **1994**, *369*, 581–584.

8 Olah, G. A., Mitchell, R. D., Sosnick, T. R., Walsh, D. A., Trewhella, J., Solution structure of the cAMP-dependent protein kinase catalytic subunit and its contraction upon binding the protein kinase inhibitor peptide. *Biochemistry* **1993**, *32*, 3649–3657.

9 Nolen, B., Yun, C. Y., Wong, C. F., McCammon, J. A., Fu, X. D., Ghosh, G., The structure of Sky1p reveals a novel mechanism for constitutive activity. *Nat. Struct. Biol.* **2001**, *8*, 176–183.

10 Siebel, C. W., Feng, L., Guthrie, C., Fu, X. D., Conservation in budding yeast of a kinase specific for SR splicing factors. *Proc. Natl. Acad. Sci. USA* **1999**, *96*, 5440–5445.

11 Hubbard, S. R., Wei, L., Ellis, L., Hendrickson, W. A., Crystal structure of the tyrosine kinase domain of the human insulin receptor. *Nature* **1994**, *372*, 746–754.

12 Taylor, G. R., Reedijk, M., Rothwell, V., Rohrschneider, L., Pawson, T., The unique insert of cellular and viral Fms protein tyrosine kinase domains is dispensable for enzymatic and transforming activities. *EMBO J.* **1989**, *8*, 2029–2037.

13 Bossemeyer, D., Engh, R. A., Kinzel, V., Ponstingl, H., Huber, R., Phosphotransferase and substrate binding mechanism of the cAMP-dependent protein kinase catalytic subunit from porcine heart as deduced from the 2.0 Å structure of the complex with Mn^{2+} adenylyl imidodiphosphate and inhibitor peptide PKI(5-24). *EMBO J.* **1993**, *12*, 849–859.

14 Zheng, J., Knighton, D. R., ten Eyck, L. F., Karlsson, R., Xuong, N., Taylor, S. S., Sowadski, J. M., Crystal structure of the catalytic subunit of cAMP-dependent protein kinase complexed with MgATP and peptide inhibitor. *Biochemistry* **1993**, *32*, 2154–2161.

15 Hubbard, S. R., Crystal structure of the activated insulin receptor tyrosine kinase in complex with peptide substrate and ATP analog. *EMBO J.* **1997**, *16*, 5572–5581.

16 Favelyukis, S., Till, J. H., Hubbard, S. R., Miller, W. T., Structure and autoregulation of the insulin-like growth factor 1 receptor kinase. *Nat. Struct. Biol.* **2001**, *8*, 1058–1063.

17 Johnson, L. N., Lowe, E. D., Noble, M. E., Owen, D. J., The Eleventh Datta Lecture. The structural basis for substrate recognition and control by protein kinases. *FEBS Lett.* **1998**, *430*, 1–11.

18 Goldberg, J., Nairn, A. C., Kuriyan, J., Structural basis for the autoinhibition of calcium/calmodulin-dependent protein kinase I. *Cell* **1996**, *84*, 875–887.

19 McTigue, M. A., et al., Crystal structure of the kinase domain of human vascular endothelial growth factor receptor 2: a key enzyme in angiogenesis. *Structure Fold Des* **1999**, *7*, 319–330.

20 Wybenga-Groot, L. E., Baskin, B., Ong, S. H., Tong, J., Pawson, T., Sicheri, F., Structural basis for autoinhibition of the Ephb2 receptor tyrosine kinase by the unphosphorylated juxtamembrane region. *Cell* **2001**, *106*, 745–757.

21 Pautsch, A., Zoephel, A., Ahorn, H., Spevak, W., Hauptmann, R., Nar, H., Crystal structure of bisphosphorylated IGF-1 receptor kinase: insight into domain movements upon kinase activation. *Structure (Camb)* **2001**, *9*, 955–965.

22 Nowakowski, J., et al., Structures of the cancer-related aurora-A, FAK, EphA2 protein kinases from nanovolume crystallography. *Structure (Camb)* **2002**, *10*, 1659–1667.

23 Meng, W., et al., Structure of mitogen-activated protein kinase-activated protein (MAPKAP) kinase 2 suggests a bi-functional switch that couples kinase activation with nuclear export. *J. Biol. Chem.* **2002**, *277*, 37401–37405.

24 Cheetham, G. M., et al., Crystal structure of aurora-2, an oncogenic serine/threonine kinase. *J. Biol. Chem.* **2002**, *277*, 42419–42422.

25 White, M. F., Shoelson, S. E., Keutmann, H., Kahn, C. R., A cascade of tyrosine autophosphorylation in the beta-subunit activates the phosphotransferase of the insulin receptor. *J. Biol. Chem.* **1988**, *263*, 2969–2980.

26 Munshi, S., et al., Crystal structure of the Apo, unactivated insulin-like growth factor-1 receptor kinase: implication for inhibitor specificity. *J. Biol. Chem.* **2002**, *277*, 38797–38802.

27 Mohammadi, M., Schlessinger, J., Hubbard, S. R., Structure of the FGF receptor tyrosine kinase domain reveals a novel autoinhibitory mechanism. *Cell* **1996**, *86*, 577–587.

28 Mohammadi, M., Dikic, I., Sorokin, A., Burgess, W. H., Jaye, M., Schlessinger, J., Identification of six novel autophosphorylation sites on fibroblast growth factor receptor 1 and elucidation of their importance in receptor activation and signal transduction. *Mol. Cell Biol.* **1996**, *16*, 977–989.

29 Zhang, F., Strand, A., Robbins, D., Cobb, M. H., Goldsmith, E. J., Atomic structure of the MAP kinase ERK2 at 2.3 Å resolution. *Nature* **1994**, *367*, 704–711.

30 Till, J. H., et al., Crystal structure of the MuSK tyrosine kinase: insights into receptor autoregulation. *Structure (Camb)* **2002**, *10*, 1187–1196.

31 Schindler, T., Bornmann, W., Pellicena, P., Miller, W. T., Clarkson, B., Kuriyan, J., Structural mechanism for STI-571 inhibition of Abelson tyrosine kinase. *Science* **2000**, *289*, 1938–1942.

32 Shewchuk, L. M., et al., Structure of the Tie2 RTK domain: self-inhibition by the nucleotide binding loop, activation loop, C-terminal tail. *Structure Fold Des* **2000**, *8*, 1105–1113.

33 De Bondt, H. L., Rosenblatt, J., Jancarik, J., Jones, H. D., Morgan, D. O., Kim, S. H., Crystal structure of cyclin-dependent kinase 2. *Nature* **1993**, *363*, 595–602.

34 Sicheri, F., Moarefi, I., Kuriyan, J., Crystal structure of the Src family tyrosine kinase Hck. *Nature* **1997**, *385*, 602–609.

35 Xu, W., Harrison, S. C., Eck, M. J., Three-dimensional structure of the tyrosine kinase c-Src. *Nature* **1997**, *385*, 595–602.

36 Yamaguchi, H., Hendrickson, W. A., Structural basis for activation of human lymphocyte kinase Lck upon tyrosine phosphorylation. *Nature* **1996**, *384*, 484–489.

37 Kemp, B. E., Pearson, R. B., Intrasteric regulation of protein kinases and phosphatases. *Biochim Biophys Acta* **1991**, *1094*, 67–76.

38 Benian, G. M., Kiff, J. E., Neckelmann, N., Moerman, D. G., Waterston, R. H., Sequence of an unusually large protein implicated in regulation of myosin activity in *C. elegans*. *Nature* **1989**, *342*, 45–50.

39 Benian, G. M., L'Hernault, S. W., Morris, M. E., Additional sequence complexity in the muscle gene, *unc-22*, and its encoded protein, twitchin, of *Caenorhabditis elegans*. *Genetics* **1993**, *134*, 1097–1104.

40 Mayans, O., et al., Structural basis for activation of the titin kinase domain during myofibrillogenesis. *Nature* **1998**, *395*, 863–869.

41 Yokokura, H., Picciotto, M. R., Nairn, A. C., Hidaka, H., The regulatory region of calcium/calmodulin-dependent protein kinase I contains closely associated autoinhibitory and calmodulin-binding domains. *J. Biol. Chem.* **1995**, *270*, 23851–23859.

42 Heierhorst, J., Kobe, B., Feil, S. C., Parker, M. W., Benian, G. M., Weiss, K. R., Kemp, B. E., Ca^{2+}/S100 regulation of giant protein kinases. *Nature* **1996**, *380*, 636–639.

43 Kobe, B., Heierhorst, J., Feil, S. C., Parker, M. W., Benian, G. M., Weiss, K. R., Kemp, B. E., Giant protein kinases: domain interactions and structural basis of autoregulation. *EMBO J.* **1996**, *15*, 6810–6821.

44 Kobe, B., Kemp, B. E., Active site-directed protein regulation. *Nature* **1999**, *402*, 373–376.

45 Matsushita, M., Nairn, A. C., Characterization of the mechanism of regulation of Ca^{2+}/calmodulin-dependent protein kinase I by calmodulin and by Ca^{2+}/calmodulin-dependent protein kinase kinase. *J. Biol. Chem.* **1998**, *273*, 21473–21481.

46 Haribabu, B., et al., Human calcium-calmodulin dependent protein kinase I: cDNA cloning, domain structure and activation by phosphorylation at threonine-177 by calcium-calmodulin dependent protein kinase I kinase. *EMBO J.* **1995**, *14*, 3679–3686.

47 Stokoe, D., Caudwell, B., Cohen, P. T., Cohen, P., The substrate specificity and structure of mitogen-activated protein (MAP) kinase-activated protein kinase-2. *Biochem. J.* **1993**, *296*, 843–849.

48 Engel, K., Kotlyarov, A., Gaestel, M., Leptomycin B-sensitive nuclear export of MAPKAP kinase 2 is regulated by phosphorylation. *EMBO J.* **1998**, *17*, 3363–3371.

49 Zu, Y. L., Wu, F., Gilchrist, A., Ai, Y., Labadia, M. E., Huang, C. K., The primary structure of a human MAP kinase activated protein kinase 2. *Biochem. Biophys. Res. Commun.* **1994**, *200*, 1118–1124.

50 Ben-Levy, R., Leighton, I. A., Doza, Y. N., Attwood, P., Morrice, N., Marshall, C. J., Cohen, P., Identification of novel phosphorylation sites required for activation of MAPKAP kinase-2. *EMBO J.* **1995**, *14*, 5920–5930.

51 Lei, M., Lu, W., Meng, W., Parrini, M. C., Eck, M. J., Mayer, B. J., Harrison, S. C., Structure of PAK1 in an autoinhibited conformation reveals a multistage activation switch. *Cell* **2000**, *102*, 387–397.

52 Parrini, M. C., Lei, M., Harrison, S. C., Mayer, B. J., Pak1 kinase homodimers are autoinhibited in trans and dissociated upon activation by Cdc42 and Rac1. *Mol. Cell* **2002**, *9*, 73–83.

53 Schlessinger, J., Cell signaling by receptor tyrosine kinases. *Cell* **2000**, *103*, 211–225.

54 Dodelet, V. C., Pasquale, E. B., Eph receptors and ephrin ligands: embryogenesis to tumorigenesis. *Oncogene* **2000**, *19*, 5614–5619.

55 Binns, K. L., Taylor, P. P., Sicheri, F., Pawson, T., Holland, S. J., Phosphorylation of tyrosine residues in the kinase domain and juxtamembrane region regulates the biological and catalytic activities of Eph receptors. *Mol. Cell Biol.* **2000**, *20*, 4791–4805.

56 Baxter, R. M., Secrist, J. P., Vaillancourt, R. R., Kazlauskas, A., Full activation of the platelet-derived growth factor beta-receptor kinase involves multiple events. *J. Biol. Chem.* **1998**, *273*, 17050–17055.

57 Irusta, P. M., DiMaio, D., A single amino acid substitution in a WW-like domain of diverse members of the PDGF receptor subfamily of tyrosine kinases causes constitutive receptor activation. *EMBO J.* **1998**, *17*, 6912–6923.

58 Kitayama, H., et al., Constitutively activating mutations of c-kit receptor tyrosine kinase confer factor-independent growth and tumorigenicity of factor-dependent hematopoietic cell lines. *Blood* **1995**, *85*, 790–798.

59 Tsujimura, T., et al., Activating mutation in the catalytic domain of c-kit elicits hematopoietic transformation by receptor self-association not at the ligand-induced dimerization site. *Blood* **1999**, *93*, 1319–1329.

60 Nakahara, M., et al., A novel gain-of-function mutation of c-kit gene in gastrointestinal stromal tumors. *Gastroenterology* **1998**, *115*, 1090–1095.

61 Hirota, S., et al., Gain-of-function mutations of c-kit in human gastro-intestinal stromal tumors. *Science* **1998**, *279*, 577–580.

62 Chan, P. M., Ilangumaran, S., La Rose, J., Chakrabartty, A., Rottapel, R., Autoinhibition of the kit receptor tyrosine kinase by the cytosolic juxtamembrane region. *Mol. Cell Biol.* **2003**, *23*, 3067–3078.

63 Yokota, S., et al., Internal tandem duplication of the *FLT3* gene is preferentially seen in acute myeloid leukemia and myelodysplastic syndrome among various hematological malignancies: a study on a large series of patients and cell lines. *Leukemia* **1997**, *11*, 1605–1609.

64 Hayakawa, F., Towatari, M., Kiyoi, H., Tanimoto, M., Kitamura, T., Saito, H., Naoe, T., Tandem-duplicated Flt3 constitutively activates STAT5 and MAP kinase and introduces autonomous cell growth in IL-3–dependent cell lines. *Oncogene* **2000**, *19*, 624–631.

65 Nakao, M., et al., Internal tandem duplication of the *flt3* gene found in acute myeloid leukemia. *Leukemia* **1996**, *10*, 1911–1918.

66 Herbst, R., Burden, S. J., The juxtamembrane region of MuSK has a critical role in agrin-mediated signaling. *EMBO J.* **2000**, *19*, 67–77.

67 Watty, A., Neubauer, G., Dreger, M., Zimmer, M., Wilm, M., Burden, S. J., The in vitro and in vivo phosphotyrosine map of activated MuSK. *Proc. Natl. Acad. Sci. USA* **2000**, *97*, 4585–4590.

68 Brown, M. T., Cooper, J. A., Regulation, substrates and functions of src. *Biochim Biophys Acta* **1996**, *1287*, 121–149.

69 Jeffrey, P. D., Russo, A. A., Polyak, K., Gibbs, E., Hurwitz, J., Massague, J., Pavletich, N. P., Mechanism of CDK activation revealed by the structure of a cyclinA–CDK2 complex. *Nature* **1995**, *376*, 313–320.

70 Barila, D., Superti-Furga, G., An intramolecular SH3-domain interaction regulates c-Abl activity. *Nat. Genet.* **1998**, *18*, 280–282.

71 Franz, W. M., Berger, P., Wang, J. Y., Deletion of an N-terminal regulatory domain of the c-Abl tyrosine kinase activates its oncogenic potential. *EMBO J.* **1989**, *8*, 137–147.

72 Jackson, P., Baltimore, D., N-terminal mutations activate the leukemogenic potential of the myristoylated form of c-Abl. *EMBO J.* **1989**, *8*, 449–456.

73 Muller, A. J., Pendergast, A. M., Parmar, K., Havlik, M. H., Rosenberg, N., Witte, O. N., En bloc substitution of the Src homology region 2 domain activates the transforming potential of the c-Abl protein tyrosine kinase. *Proc. Natl. Acad. Sci. USA* **1993**, *90*, 3457–3461.

74 Superti-Furga, G., Courtneidge, S. A., Structure–function relationships in Src family and related protein tyrosine kinases. *BioEssays* **1995**, *17*, 321–330.

75 Pluk, H., Dorey, K., Superti-Furga, G., Autoinhibition of c-Abl. *Cell* **2002**, *108*, 247–259.

76 Hantschel, O., Nagar, B., Guettler, S., Kretzschmar, J., Dorey, K., Kuriyan, J., Superti-Furga, G., A myristoyl/phosphotyrosine switch regulates c-Abl. *Cell* **2003**, *112*, 845–857.

77 Nagar, B., et al., Structural basis for the autoinhibition of c-Abl tyrosine kinase. *Cell* **2003**, *112*, 859–871.

78 Nagar, B., et al., Crystal structures of the kinase domain of c-Abl in complex with the small molecule inhibitors PD173955 and imatinib (STI-571). *Cancer Res* **2002**, *62*, 4236–4243.

79 Songyang, Z., et al., A structural basis for substrate specificities of protein Ser/Thr kinases: primary sequence preference of casein kinases I and II, NIMA, phosphorylase kinase, calmodulin-dependent kinase II, CDK5, and Erk1. *Mol. Cell Biol.* **1996**, *16*, 6486–6493.

80 Schmitz, R., Baumann, G., Gram, H., Catalytic specificity of phosphotyrosine kinases Blk, Lyn, c-Src and Syk as assessed by phage display. *J. Mol. Biol.* **1996**, *260*, 664–677.

81 Lowe, E. D., Noble, M. E., Skamnaki, V. T., Oikonomakos, N. G., Owen, D. J., Johnson, L. N., The crystal structure of a phosphorylase kinase peptide substrate complex: kinase substrate recognition. *EMBO J.* **1997**, *16*, 6646–6658.

82 Brown, N. R., Noble, M. E., Endicott, J. A., Johnson, L. N., The structural basis for specificity of substrate and recruitment peptides for cyclin-dependent kinases. *Nat. Cell Biol.* **1999**, *1*, 438–443.

83 Higashi, H., Suzuki-Takahashi, I., Taya, Y., Segawa, K., Nishimura, S., Kitagawa, M., Differences in substrate specificity between Cdk2-cyclin A and Cdk2-cyclin E in vitro. *Biochem. Biophys. Res. Commun.* **1995**, *216*, 520–525.

84 Songyang, Z., Blechner, S., Hoagland, N., Hoekstra, M. F., Piwnica-Worms, H., Cantley, L. C., Use of an oriented

peptide library to determine the optimal substrates of protein kinases. *Curr. Biol.* **1994**, *4*, 973–982.

85 Canagarajah, B. J., Khokhlatchev, A., Cobb, M. H., Goldsmith, E. J., Activation mechanism of the MAP kinase ERK2 by dual phosphorylation. *Cell* **1997**, *90*, 859–869.

86 Brinkworth, R. I., Breinl, R. A., Kobe, B., Structural basis and prediction of substrate specificity in protein serine/threonine kinases. *Proc. Natl. Acad. Sci. USA* **2003**, *100*, 74–79.

87 Fiol, C. J., Wang, A., Roeske, R. W., Roach, P. J., Ordered multisite protein phosphorylation: analysis of glycogen synthase kinase 3 action using model peptide substrates. *J. Biol. Chem.* **1990**, *265*, 6061–6065.

88 Wang, Y., Roach, P. J., Inactivation of rabbit muscle glycogen synthase by glycogen synthase kinase-3: dominant role of the phosphorylation of Ser-640 (site-3a). *J. Biol. Chem.* **1993**, *268*, 23876–23880.

89 Liu, C., et al., Control of beta-catenin phosphorylation/degradation by a dual-kinase mechanism. *Cell* **2002**, *108*, 837–847.

90 ter Haar, E., Coll, J. T., Austen, D. A., Hsiao, H. M., Swenson, L., Jain, J., Structure of GSK3beta reveals a primed phosphorylation mechanism. *Nat. Struct. Biol.* **2001**, *8*, 593–596.

91 Dajani, R., Fraser, E., Roe, S. M., Young, N., Good, V., Dale, T. C., Pearl, L. H., Crystal structure of glycogen synthase kinase 3 beta: structural basis for phosphate-primed substrate specificity and autoinhibition. *Cell* **2001**, *105*, 721–732.

92 Frame, S., Cohen, P., Biondi, R. M., A common phosphate binding site explains the unique substrate specificity of GSK3 and its inactivation by phosphorylation. *Mol. Cell* **2001**, *7*, 1321–1327.

93 Cross, D. A., Alessi, D. R., Cohen, P., Andjelkovich, M., Hemmings, B. A., Inhibition of glycogen synthase kinase-3 by insulin mediated by protein kinase B. *Nature* **1995**, *378*, 785–789.

94 Pearson, R. B., Dennis, P. B., Han, J. W., Williamson, N. A., Kozma, S. C., Wettenhall, R. E., Thomas, G., The principal target of rapamycin-induced p70s6k inactivation is a novel phosphorylation site within a conserved hydrophobic domain. *EMBO J.* **1995**, *14*, 5279–5287.

95 Keranen, L. M., Dutil, E. M., Newton, A. C., Protein kinase C is regulated in vivo by three functionally distinct phosphorylations. *Curr. Biol.* **1995**, *5*, 1394–1403.

96 Parekh, D. B., Ziegler, W., Parker, P. J., Multiple pathways control protein kinase C phosphorylation. *EMBO J.* **2000**, *19*, 496–503.

97 Pullen, N., Dennis, P. B., Andjelkovic, M., Dufner, A., Kozma, S. C., Hemmings, B. A., Thomas, G., Phosphorylation and activation of p70s6k by PDK1. *Science* **1998**, *279*, 707–710.

98 Alessi, D. R., Kozlowski, M. T., Weng, Q. P., Morrice, N., Avruch, J., 3-Phosphoinositide-dependent protein kinase 1 (PDK1) phosphorylates and activates the p70 S6 kinase in vivo and in vitro. *Curr. Biol.* **1998**, *8*, 69–81.

99 Jensen, C. J., Buch, M. B., Krag, T. O., Hemmings, B. A., Gammeltoft, S., Frodin, M., 90-kDa Ribosomal S6 kinase is phosphorylated and activated by 3-phosphoinositide-dependent protein kinase-1. *J. Biol. Chem.* **1999**, *274*, 27168–27176.

100 Kobayashi, T., Cohen, P., Activation of serum- and glucocorticoid-regulated protein kinase by agonists that activate phosphatidylinositide 3-kinase is mediated by 3-phosphoinositide-dependent protein kinase-1 (PDK1) and PDK2. *Biochem. J.* **1999**, *339*, 319–328.

101 Park, J., Leong, M. L., Buse, P., Maiyar, A. C., Firestone, G. L., Hemmings, B. A., Serum and glucocorticoid-inducible kinase (SGK) is a target of the PI 3-kinase–stimulated signaling pathway. *EMBO J.* **1999**, *18*, 3024–3033.

102 Balendran, A., Biondi, R. M., Cheung, P. C., Casamayor, A., Deak, M., Alessi, D. R., A 3-phosphoinositide–dependent protein kinase-1 (PDK1) docking site is required for the phosphorylation of protein kinase Czeta (PKCzeta) and PKC-related kinase 2 by PDK1. *J. Biol. Chem.* **2000**, *275*, 20806–20813.

103 Biondi, R. M., Cheung, P. C., Casamayor, A., Deak, M., Currie, R. A.,

Alessi, D. R., Identification of a pocket in the PDK1 kinase domain that interacts with PIF and the C-terminal residues of PKA. *EMBO J.* **2000**, *19*, 979–988.

104 Biondi, R. M., Kieloch, A., Currie, R. A., Deak, M., Alessi, D. R., The PIF-binding pocket in PDK1 is essential for activation of S6K and SGK, but not PKB. *EMBO J.* **2001**, *20*, 4380–4390.

105 Alessi, D. R., James, S. R., Downes, C. P., Holmes, A. B., Gaffney, P. R., Reese, C. B., Cohen, P., Characterization of a 3-phosphoinositide–dependent protein kinase which phosphorylates and activates protein kinase Balpha. *Curr. Biol.* **1997**, *7*, 261–269.

106 Stokoe, D., et al., Dual role of phosphatidylinositol-3,4,5-trisphosphate in the activation of protein kinase B. *Science* **1997**, *277*, 567–570.

107 Scheid, M. P., Woodgett, J. R., PKB/AKT: functional insights from genetic models. *Nat. Rev. Mol. Cell Biol.* **2001**, *2*, 760–768.

108 Collins, B. J., Deak, M., Arthur, J. S., Armit, L. J., Alessi, D. R., In vivo role of the PIF-binding docking site of PDK1 defined by knock-in mutation. *EMBO J.* **2003**, *22*, 4202–4211.

109 Biondi, R. M., Komander, D., Thomas, C. C., Lizcano, J. M., Deak, M., Alessi, D. R., van Aalten, D. M., High resolution crystal structure of the human PDK1 catalytic domain defines the regulatory phosphopeptide docking site. *EMBO J.* **2002**, *21*, 4219–4228.

110 Frodin, M., Antal, T. L., Dummler, B. A., Jensen, C. J., Deak, M., Gammeltoft, S., Biondi, R. M., A phosphoserine/threonine-binding pocket in AGC kinases and PDK1 mediates activation by hydrophobic motif phosphorylation. *EMBO J.* **2002**, *21*, 5396–5407.

111 Yang, J., Cron, P., Good, V. M., Thompson, V., Hemmings, B. A., Barford, D., Crystal structure of an activated Akt/protein kinase B ternary complex with GSK3-peptide and AMP-PNP. *Nat. Struct. Biol.* **2002**, *9*, 940–944.

112 Yang, J., Cron, P., Thompson, V., Good, V. M., Hess, D., Hemmings, B. A., Barford, D., Molecular mechanism for the regulation of protein kinase B/Akt by hydrophobic motif phosphorylation. *Mol. Cell* **2002**, *9*, 1227–1240.

113 Stephens, L., et al., Protein kinase B kinases that mediate phosphatidylinositol 3,4,5-trisphosphate–dependent activation of protein kinase B. *Science* **1998**, *279*, 710–714.

114 Alessi, D. R., Andjelkovic, M., Caudwell, B., Cron, P., Morrice, N., Cohen, P., Hemmings, B. A., Mechanism of activation of protein kinase B by insulin and IGF-1. *EMBO J.* **1996**, *15*, 6541–6551.

115 Bayliss, R., Sardon, T., Vernos, I., Conti, E., Structural basis of aurora-A activation by TPX2 at the mitotic spindle. *Mol. Cell* **2003**, *12*, 851–862.

116 Adams, P. D., Sellers, W. R., Sharma, S. K., Wu, A. D., Nalin, C. M., Kaelin, W. G. Jr., Identification of a cyclin-cdk2 recognition motif present in substrates and p21-like cyclin-dependent kinase inhibitors. *Mol. Cell Biol.* **1996**, *16*, 6623–6633.

117 Chen, J., Saha, P., Kornbluth, S., Dynlacht, B. D., Dutta, A., Cyclin-binding motifs are essential for the function of p21CIP1. *Mol. Cell Biol.* **1996**, *16*, 4673–4682.

118 Dynlacht, B. D., Moberg, K., Lees, J. A., Harlow, E., Zhu, L., Specific regulation of E2F family members by cyclin-dependent kinases. *Mol. Cell Biol.* **1997**, *17*, 3867–3875.

119 Zhu, L., Harlow, E., Dynlacht, B. D., p107 uses a p21CIP1-related domain to bind cyclin/cdk2 and regulate interactions with E2F. *Genes. Dev.* **1995**, *9*, 1740–1752.

120 Russo, A. A., Jeffrey, P. D., Patten, A. K., Massague, J., Pavletich, N. P., Crystal structure of the p27Kip1 cyclin-dependent-kinase inhibitor bound to the cyclin A–Cdk2 complex. *Nature* **1996**, *382*, 325–331.

121 Card, G. L., Knowles, P., Laman, H., Jones, N., McDonald, N. Q., Crystal structure of a gamma-herpesvirus cyclin–cdk complex. *EMBO J.* **2000**, *19*, 2877–2888.

122 Kolch, W., Meaningful relationships: the regulation of the Ras/Raf/MEK/ERK pathway by protein interactions. *Biochem. J.* **2000**, *351 Pt 2*, 289–305.

123 Payne, D. M., et al., Identification of the regulatory phosphorylation sites in pp42/mitogen-activated protein kinase (MAP kinase). *EMBO J.* **1991**, *10*, 885–892.

124 Pawson, T., Scott, J. D., Signaling through scaffold, anchoring, and adaptor proteins. *Science* **1997**, *278*, 2075–2080.

125 Holland, P. M., Cooper, J. A., Protein modification: docking sites for kinases. *Curr. Biol.* **1999**, *9*, R329–331.

126 Sharrocks, A. D., Yang, S. H., Galanis, A., Docking domains and substrate-specificity determination for MAP kinases. *Trends Biochem. Sci.* **2000**, *25*, 448–453.

127 Bardwell, L., Thorner, J., A conserved motif at the amino termini of MEKs might mediate high-affinity interaction with the cognate MAPKs. *Trends Biochem. Sci.* **1996**, *21*, 373–374.

128 Gonzalez, F. A., Raden, D. L., Davis, R. J., Identification of substrate recognition determinants for human ERK1 and ERK2 protein kinases. *J. Biol. Chem.* **1991**, *266*, 22159–22163.

129 Alvarez, E., et al., Pro-Leu-Ser/Thr-Pro is a consensus primary sequence for substrate protein phosphorylation. Characterization of the phosphorylation of c-myc and c-jun proteins by an epidermal growth factor receptor threonine 669 protein kinase. *J. Biol. Chem.* **1991**, *266*, 15277–15285.

130 Chang, C. I., Xu, B. E., Akella, R., Cobb, M. H., Goldsmith, E. J., Crystal structures of MAP kinase p38 complexed to the docking sites on its nuclear substrate MEF2A and activator MKK3b. *Mol. Cell* **2002**, *9*, 1241–1249.

131 Barsyte-Lovejoy, D., Galanis, A., Sharrocks, A. D., Specificity determinants in MAPK signaling to transcription factors. *J. Biol. Chem.* **2002**, *277*, 9896–9903.

132 Tanoue, T., Maeda, R., Adachi, M., Nishida, E., Identification of a docking groove on ERK and p38 MAP kinases that regulates the specificity of docking interactions. *EMBO J.* **2001**, *20*, 466–479.

133 Roberts, A. B., Sporn, M. B., The transforming growth factors-βs. In *Peptide Growth Factors and Their Receptors* (Eds.: Sporn, M. B., Roberts, A. B.) **1990**, Heidelberg, Germany: Springer-Verlag.

134 Wieser, R., Wrana, J. L., Massague, J., GS domain mutations that constitutively activate T beta R-I, the downstream signaling component in the TGF-beta receptor complex. *EMBO J.* **1995**, *14*, 2199–2208.

135 Wrana, J. L., Attisano, L., Wieser, R., Ventura, F., Massague, J., Mechanism of activation of the TGF-beta receptor. *Nature* **1994**, *370*, 341–347.

136 Heldin, C. H., Miyazono, K., ten Dijke, P., TGF-beta signalling from cell membrane to nucleus through SMAD proteins. *Nature* **1997**, *390*, 465–471.

137 Massague, J., TGF-beta signal transduction. *Annu. Rev. Biochem.* **1998**, *67*, 753–791.

138 Huse, M., Chen, Y. G., Massague, J., Kuriyan, J., Crystal structure of the cytoplasmic domain of the type I TGF beta receptor in complex with FKBP12. *Cell* **1999**, *96*, 425–436.

139 Huse, M., Muir, T. W., Xu, L., Chen, Y. G., Kuriyan, J., Massague, J., The TGF beta receptor activation process: an inhibitor- to substrate-binding switch. *Mol. Cell* **2001**, *8*, 671–682.

140 Wu, G., et al., Structural basis of Smad2 recognition by the Smad anchor for receptor activation. *Science* **2000**, *287*, 92–97.

141 Chen, Y. G., Hata, A., Lo, R. S., Wotton, D., Shi, Y., Pavletich, N., Massague, J., Determinants of specificity in TGF-beta signal transduction. *Genes. Dev.* **1998**, *12*, 2144–2152.

142 Feng, X. H., Derynck, R., A kinase subdomain of transforming growth factor-beta (TGF-beta) type I receptor determines the TGF-beta intracellular signaling specificity. *EMBO J.* **1997**, *16*, 3912–3923.

143 Lo, R. S., Chen, Y. G., Shi, Y., Pavletich, N. P., Massague, J., The L3 loop: a structural motif determining specific interactions between SMAD proteins and TGF-beta receptors. *EMBO J.* **1998**, *17*, 996–1005.

144 Mellor, H., Proud, C. G., A synthetic peptide substrate for initiation factor-2 kinases. *Biochem. Biophys. Res. Commun.* **1991**, *178*, 430–437.

145 Kawagishi-Kobayashi, M., Silverman, J. B., Ung, T. L., Dever, T. E., Regulation

of the protein kinase PKR by the vaccinia virus pseudosubstrate inhibitor K3L is dependent on residues conserved between the K3L protein and the PKR substrate eIF2alpha. *Mol. Cell Biol.* **1997**, *17*, 4146–4158.

146 Dar, A. C., Sicheri, F., X-ray crystal structure and functional analysis of vaccinia virus K3L reveals molecular determinants for PKR subversion and substrate recognition. *Mol. Cell* **2002**, *10*, 295–305.

147 Dhaliwal, S., Hoffman, D. W., The crystal structure of the N-terminal region of the alpha subunit of translation initiation factor 2 (eIF2alpha) from *Saccharomyces cerevisiae* provides a view of the loop containing serine 51, the target of the eIF2alpha-specific kinases. *J. Mol. Biol.* **2003**, *334*, 187–195.

148 Nonato, M. C., Widom, J., Clardy, J., Crystal structure of the N-terminal segment of human eukaryotic translation initiation factor 2alpha. *J. Biol. Chem.* **2002**, *21*, 21.

149 Ung, T. L., Cao, C., Lu, J., Ozato, K., Dever, T. E., Heterologous dimerization domains functionally substitute for the double-stranded RNA binding domains of the kinase PKR. *EMBO J.* **2001**, *20*, 3728–3737.

10
Structure, Specificity, and Mechanism of Protein Lysine Methylation by SET Domain Enzymes

James H. Hurley and Raymond C. Trievel

10.1
Discovery and Biology of SET Domains

The SUV39/enhancer of zeste/trithorax (SET) domain is a ~110 amino acid motif found in more than 700 regulatory proteins distributed throughout all kingdoms of life [1]. The SET domain was first discovered as a region of homology in proteins involved in gene silencing and transcriptional regulation [2], but its biochemical function was unknown. Subsequent to the identification of the domain, more sensitive searches for sequence homology identified SET domains in several plant rubisco large-subunit methyltransferases (LSMTs) [3]. Rubisco LSMTs catalyze the *S*-adenosylmethionine (AdoMet)-dependent N-methylation of a single Lys residue in the flexible N-terminal tail of rubisco large subunits. The homology between plant rubisco LSMTs and regulatory proteins of gene silencing led to the inference that the latter might also be protein methyltransferases. This led in turn to the breakthrough discovery in 2000 that the one of the latter protein families, SUVAR3-9, was a site-specific histone methyltransferase (HMTs) [3]. Although it had long been known that histones were methylated in vivo, the importance of this modification was not understood until it was linked to the regulatory action of the SUV39 family of gene-silencing proteins. The discovery of many other HMTs quickly followed.

Over the past three years, lysine ε-N–methylation of histones joined phosphorylation, acetylation, and ubiquitination as a key post-translational modification of histone tails, which together comprise the so-called histone code [4–6]. In the N-terminal tail of histone H3, lysines 4, 9, 27, and 36 have been shown to be methylated in vivo, and Lys20 is methylated in histone H4 [4–6]. The effects of histone methylation in chromatin remodeling and transcriptional regulation vary depending on the lysine modified and the context in which the methylation occurs. Methylation of histone H3 Lys9 and Lys27 is strongly correlated with transcriptional silencing. Silencing requires the recruitment of chromodomain-bearing proteins that specifically recognize and bind the methylated forms of these two lysines. Methylation of H3 Lys9 occurs in both heterochromatin [7–10] and transcriptionally active euchromatin [11–14] through recruitment of heterochromatin protein 1 (HP1)

Modular Protein Domains. Edited by G. Cesareni, M. Gimona, M. Sudol, M. Yaffe

ISBN: 3-527-30813-X

via its methyl-Lys9–specific chromodomain [15, 16]. Histone H3 Lys27 methylation recruits Polycomb through its methyl-Lys27–specific chromodomain to repress transcription in euchromatic regions [17, 18]. Methylation of histone H3 Lys4 plays diverse roles in chromatin remodeling, including transcriptional activation at loci that are trimethylated at Lys4 [19, 20], transcriptional elongation [21, 22], and rDNA and telomere silencing in yeast [23–26]. Histone H3 Lys36 methylation is also associated with the elongation phase of transcription [27–31], although it was originally reported to function in repression [32]. Methylation of histone H4 at Lys20 is associated with cell cycle-dependent transcriptional repression, which peaks during mitosis [33–35]. Finally, the rubisco large subunit is methylated at Lys14 in its N-terminal tail in a variety of plant species [36–39], though the function of this modification has not been elucidated. Although other protein substrates of SET domain enzymes have been identified to date, the large number and widespread distribution of these methyltransferases suggests that their range will extend beyond just histones and the Rubisco large subunit.

10.2
Structure of the SET Domain

The recent intense interest in protein lysine methylation has spawned an equally intense effort to determine the structures of SET domain proteins. Structures of five different SET domains emerged from seven different laboratories between October 2002 and March 2003. Structures have been determined for the rubisco LSMT of the garden pea [40] and the HMTs DIM-5 of *Neurospora crassa* [41], human SET7/9 [42–44], *Schizosaccharomyces pombe* Clr4 [45], and the *Chlorella bursaria* virus 'vSET' enzyme [46] (Table 10.1).

Table 10.1 Reported SET domain structures. Bound substrate and cofactor molecules are shown if they are ordered and included in the deposited PDB coordinates.

Enzyme	*PDB entry*	*Ordered ligands*	*References*
DIM-5	1ml9		41
	1peg	SAH, H3 7-13	51
SET7/9	1h3i		42
	1muf	SAH	43
	1mt6	SAM	43
	1n6a	SAM	44
	1n6c	SAH, K4-Me H3 1-10	44
	1o9s		50
Rubisco LSMT	1mlv	SAH, HEPES	40
	1ozv	SAH, lysine	52
	1p0y	SAH, MeLys	52
Clr4	1mvh		45
	1mvx		45
vSET	1n3j		46

10.2.1
The SET Domain Fold

The SET domain fold consists of 12 β strands organized into five sheets. A single one-turn 3_{10} helix precedes the eighth β strand (Figure 10.1). There were minor disparities in the initial publications reporting the structures, owing to the small size of some of the secondary structure elements and ambiguity about the definition of the conserved structural core. Now that the dust has settled, it is clear that the 12 β strand fold occurs in all of the structurally characterized SET domains. The organization of the five β sheets represents a novel structural fold not previously seen in any other methyltransferase, nor indeed in any previously determined structure. The arrangement of the sheets has been compared to a box [41]. Significant stretches of nonconserved sequence are inserted between β5 and β6 and between β7 and β8. The most peculiar feature of the fold is that the C terminus of the SET domain is threaded through the loop connecting strands 8 and 9. The unusual topology engendered considerable discussion about whether it constitutes a knot [43, 47–49]. An apparent resolution to the issue has been reached by Taylor and colleagues [49], who classify the structure as a 'pseudoknot', borrowing terminology from the RNA field.

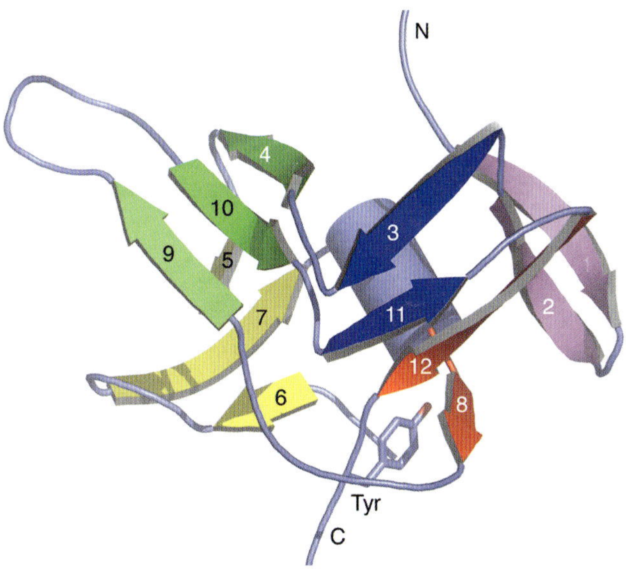

Figure 10.1 The SET domain fold. Conserved core elements common to all SET domains are shown, using the core of the Rubisco LSMT structure as an example. β sheets I–V are colored violet, blue, green, yellow, and red, respectively. The sidechain of the absolutely conserved Tyr at the C terminus is shown. The Tyr marks the location where the C-terminal loop threads under the β8–β9 junction, forming the pseudoknot.

10.2.2
The Active Site

The active site of the SET domain is formed at a small pore joining two clefts, one on each side of the domain. The AdoMet binding cleft is formed by the β1–β2 turn, β6, and β8. The reaction product *S*-adenosylhomocysteine (AdoHcy) is more stable than the methyl donor AdoMet, and all but one of the cofactor-bound complexes determined so far contain AdoHcy instead of AdoMet. The AdoHcy-bound complexes of SET7/9 [43] and Rubisco LSMT [40] and the AdoMet-bound structure of SET7/9 [44] showed that the cofactor interacts with all of these SET domains in the same way. One early report of a different binding mode [42] was later ascribed to poorly defined electron density, rather than a real biochemical difference [50]. A conserved Asn-His (NH) sequence motif on β8 is the pivotal recognition element for the cofactor. The mainchain amide and carbonyl of the His residue in the NH motif recognize the adenine base of the cofactor via hydrogen bonds with the adenine N6 and N7. The sidechain Oδ of the NH motif Asn accepts a hydrogen bond from the amino group of the amino acid moiety of the cofactor. In addition, the amine nitrogen and one oxygen of the carboxylate of the cofactor engage in hydrogen bonds with the backbone of residues from the conserved G-x-G motif in the β2 sheet. The other oxygen interacts with different nonconserved basic sidechains at different positions in the primary sequences of different SET domains (Figure 10.2).

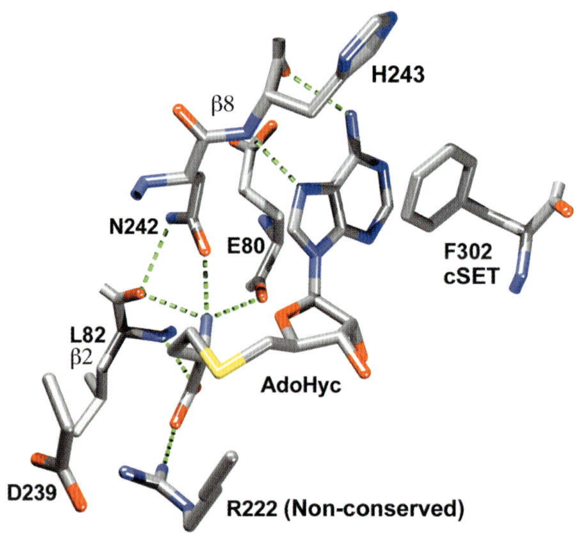

Figure 10.2 Cofactor binding site. The desmethyl form of the cofactor *S*-adenosylhomocysteine (AdoHcy) is shown bound to Rubisco LSMT. Glu80 and Leu82 are part of or adjoin the conserved G-x-G motif at the start of β2. Asn243 and His243 represent the conserved NH motif. The Arg222, and the Asp239 with which it interacts, are conserved in the LSMTs but not in other SET domains. The basic sidechain interaction with the AdoHcy carboxylate is replaced by equivalent interactions using different positions in other SET domains. Phe302 in the LSMT cSET region makes hydrophobic contacts that are replaced in HMTs by interaction with other nonconserved cSET or post-SET domains (modified from [40]).

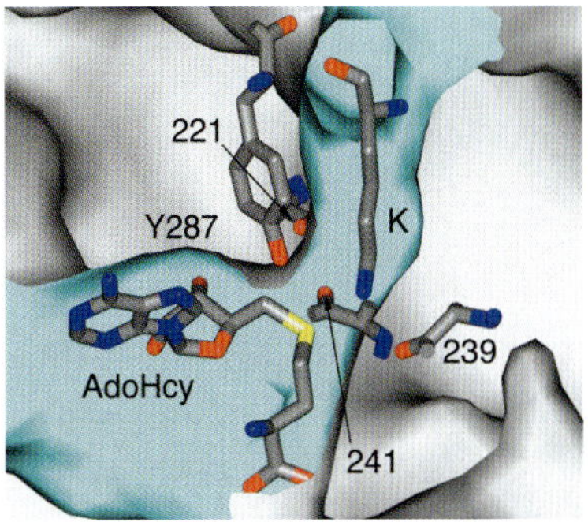

Figure 10.3 The active site pore. The role of the conserved Tyr and the carbonyl cage at the pore between the Lys and AdoMet binding sites is illustrated using the LSMT structure (modified from [40]).

The substrate binding site was first inferred from the binding of a HEPES buffer ion to the putative Lys-binding pocket of Rubisco LSMT [40] and from molecular modeling of DIM-5 [41]. These structures were followed by structures of SET7/9 and DIM-5 bound to methylated [50] and unmethylated Lys peptides [51], respectively, and by structures of Rubisco LSMT bound to Lys and MeLys [52]. The structures show a consistent picture of the substrate Lys residue bound in a cleft that is deeper and narrower than the AdoMet binding cleft. This cleft is formed by the β6 and β12 strands, as well as by nonconserved regions inserted between β5 and β6 and just C-terminal to the SET domain. The cleft is closed by nonconserved helical regions C-terminal to the SET domain in SET7/9 and LSMT and by the post-SET domain in DIM-5 and Clr4. The Lys-binding cleft meets the AdoMet-binding cleft at a connecting pore, and the ε-amino group of the Lys points into the pore. The sides of the upper part of the cleft are formed by hydrophobic residues. The pore at the bottom of the cleft is formed by an absolutely conserved Tyr residue close to the C terminus of the SET domain and by four mainchain carbonyl groups (Figure 10.3). Two of the carbonyl groups come from the section immediately preceding β6, and the other two project from the C-terminal end of the 3_{10} helix.

10.2.3
Interactions with Other Domains

SET domains generally occur in close juxtaposition with other domains. Some of these have been visualized along with the SET domain in recent structures. The first family of HMTs to be characterized contains Cys-rich pre-SET and post-SET domains. The structures of DIM-5 [41] and Clr4 [45] showed that the pre-SET consists of a 3-Zn^{2+}/9-Cys cluster distal to the active site (Figure 10.4). The post-SET domain is disordered in the absence of substrate. In the presence of bound substrate, the post-SET region folds into a 1-Zn^{2+}/4-Cys cluster, with three Cys contributed by the post-SET domain and the fourth from the active-site region [51].

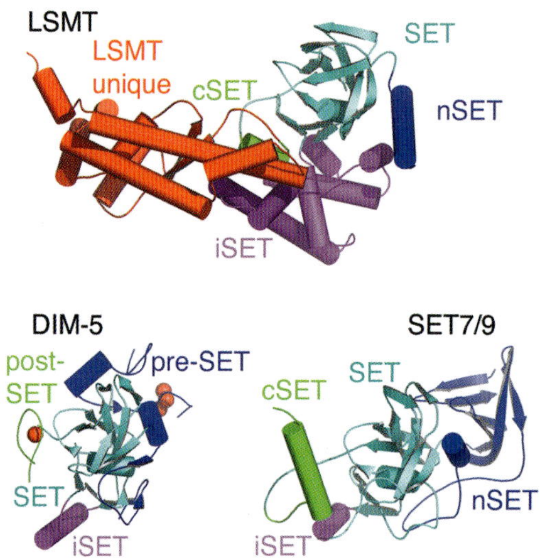

Figure 10.4 SET domains in context. The SET domain is colored cyan throughout. The nonconserved N-terminal regions (pre-SET in DIM-5, nSET in others) are blue. The nonconserved region inserted between β5 and β6 (iSET) is magenta. The immediate C-terminal extension of the SET domain (post-SET in DIM-5, cSET in others), which is closely associated with the active site, is green. The large C-terminal lobe unique to LSMT is red. The bound Zn^{2+} ions in DIM-5 are red spheres.

The folding of the post-SET domain is thus intimately coupled to the activity of the SET domain. Rubisco LSMTs contain a unique 110 amino acid helical 'iSET' domain inserted between β5 and β6 and a 290 amino acid helical C-terminal domain (Figure 10.4). Neither domain is similar to any other known structure. These domains are thought to be involved in stabilizing the SET domain and in interactions with Rubisco. Homologs of the C-terminal domain are present in *Neurospora, Drosophila*, mouse, and human ORFs, suggesting that these could be metazoan orthologs of the LSMTs. The identity of the substrates of these possible LSMT orthologs is an intriguing mystery, and the sequence homology shared by this SET domain subfamily suggests that the substrate(s) could be conserved among these various species. SMART and PSI–Blast searches show that the SET7/9 N-terminal domain consists primarily of MORN repeats. The MORN repeat is a little-characterized ~22 amino acid repeat also found in junctophilins, protein tyrosine kinases, Vps9-homology domain proteins, and plant phosphatidylinositol phosphate kinases. The structure of SET7/9 shows that each MORN repeat corresponds to one pair of antiparallel β strands (Figure 10.4). The MORN repeats form a domain that stabilizes the SET domain. It is not yet clear if this region has other roles, or what is the significance of the relationship to MORN repeats in other signaling proteins.

10.3
Substrate Specificity and Catalytic Mechanism

10.3.1
Substrate Specificity

SET domain enzymes, with a few exceptions, are remarkably specific for individual lysines in histones and other protein substrates (Table 10.2; see Sims et al., 2003 [53] for another recent review). Over the past several years, many groups have been involved in characterizing the various SET domain families that are responsible for methylation of these lysines. The founding members of the SET family, *Drosophila* SUVAR3-9 [3, 54], enhancer of zeste [55–57], and trithorax [55, 58, 59] represent three classes of SET methyltransferases with distinct substrate specificities for lysines 9, 9/27, and 4 in histone H3, respectively. Human SET7/9 [60, 61], G9A [62], and yeast SET2 [32] methylate histone H3 at lysines 4, 9/27, and 36, respectively. Human PRSET-7 [34] and its isoform SET8 [33] methylate histone H4 Lys20. Some SET methyltransferases appear to have expanded histone specificity, such as *Drosophila* ASH1, which methylates Lys4 and Lys9 in histone H3 and Lys20 in histone H4 [63].

Table 10.2 A list of currently identified SET methyltransferases and their protein substrates. Species listed include *At – Arabidopsis thaliana*; *Dm – Drosophila melanogaster*; *Hs – Homo sapiens*; *Mm – Mus musculus*; *Nc – Neurospora crassa*; *Nt – Nicotiana tabacum*; *Ps – Pisum sativum*; *Sc – Saccharomyces cerevisiae*; *So – Spinacia oleracea*; *Sp – Schizosaccharomyces pombe*.

Protein substrate	*SET methyltransferase*
Histone H3-K4	*Dm* trithorax [55] / *Hs* MLL/ALL-1 [58, 59], *Hs* ALR1/2 [67], *Hs* HALR [67] *Hs* SET1 [68] / *Sc* SET1 [23, 69, 70] *Hs* SET7/9 [60, 61] *Dm* ASH1 [63]
Histone H3-K9	*Dm* SU(VAR)3-9 [54] / *Hs, Mm* SUV39H1 [3] / SpCLR4 [3, 7] *Hs, Mm* G9A [62], *Hs* EuHMTase1 [71] *Dm* E(Z) [55, 57] *Hs* E(Z)/PRC2 [56] *Hs, Mm* ESET/SETDB1 [13, 72] *Nc* DIM-5 [73] *At* kryptonite [74] *Ds* ASH1 [63]
Histone H3-K27	*Hs, Mm* G9A [62] *Dm* E(Z) [55, 57] *Hs* E(Z)/PRC2 [56] vSET [46]
Histone H3-K36	*Sc* SET2 [32] *Hs* NSD1 [75]
Histone H4-K20	*Hs* PR-SET7 [34] *Dm, Hs* SET8 [33] *Dm* ASH1 [63] *Hs* NSD1 [75] (weak)
Rubisco LS-K14	*Ps, Nt, So* LSMT [37–39]

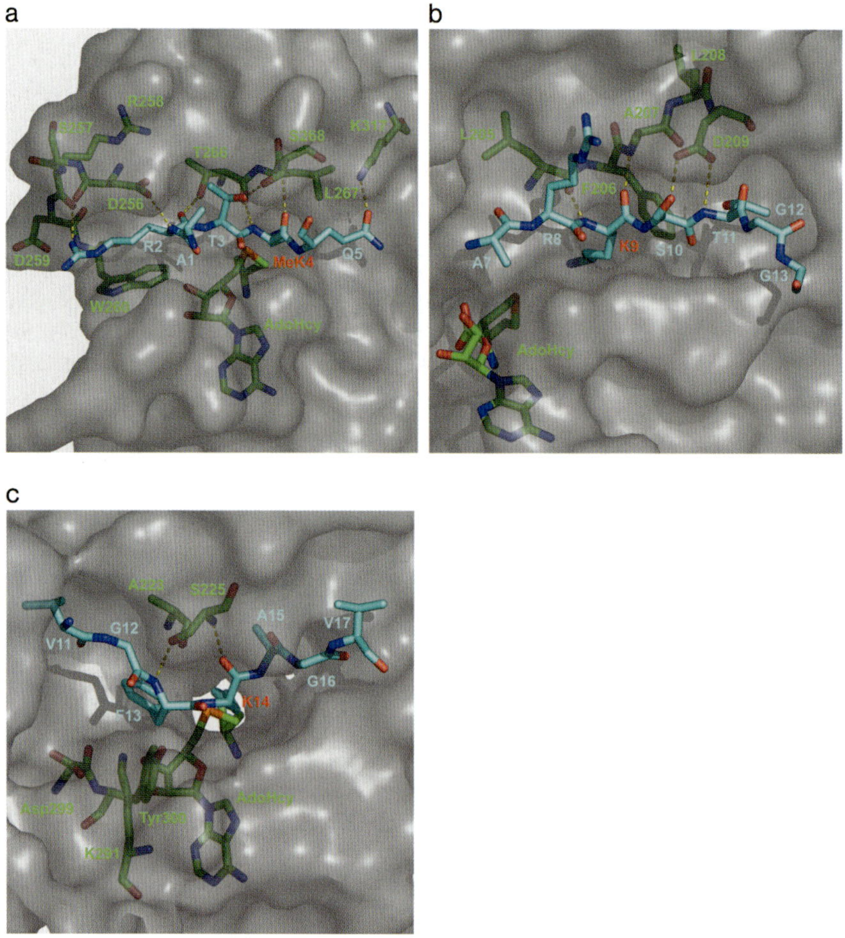

Figure 10.5 Substrate binding by SET domains.
(a) Solvent-accessible molecular surface of the substrate binding cleft of SET7/9 with a bound histone H3 peptide centered around Lys4. Residues in SET 7/9 that interact with histone H3 are highlighted, and the carbon atoms from SET7/9 and histone H3 are colored green and cyan, respectively, for clarity. Hydrogen bonds are denoted by yellow dashed lines.
(b) Molecular surface of the active site of DIM-5 bound to a histone H3 peptide centered around Lys9. The coloring is as in (a).
(c) Molecular surface of the substrate binding cleft of LSMT bound to a modeled Rubisco LS peptide centered around Lys14. The peptide was modeled on a consensus methylation sequence found in the N-terminal tails of the Rubisco LS in many plant species [52]. The sidechain of Arg 226 in the substrate binding cleft has been removed for clarity. The figure is colored as in (a).

Substrates bind to SET domains in a β conformation, forming an extensive parallel β-sheet interaction with β6 and a more limited interaction with β12 [50–52]. The bound substrate thus bridges two β sheets. It is interesting that the enzyme binds the substrate between the edges of two β sheets, because this could partly explain why such a complex β sheet topology evolved for the SET domain. By locking the substrate between two sheets, it can be held rigidly in place, perhaps enhancing sequence readout by these exquisitely specific enzymes. Only 5–7 peptide residues are ordered, hence sequence readout by SET domain active sites is local.

The HMTs that have been structurally characterized so far are uniquely specific for Lys4 or Lys9, but never both. The N-terminal tail of histone H3 starts with the sequence ARTKQTARKSTGGKA. Nonconserved iSET and cSET regions of SET7/9 form a binding pocket for Arg2 of histone H3, which is at the –2 position relative to the methylated Lys (Figure 10.5a). The Arg2 sidechain is sandwiched between two loops that project their carbonyl groups toward it. This Arg appears to be a particularly important determinant for Lys4 methylation, and other histone tail Lys residues do not have an Arg at the –2 position. Residues at the –1 and +1 positions probably contribute to specificity to a lesser degree. DIM-5 is a member of the class of HMTs that methylate Lys9 of histone H3. Methylation of this Lys by a HMTs is inhibited by phosphorylation of the adjacent Ser10 of histone H3 [3, 7, 13]. Ser10 of histone H3 forms a hydrogen bond with an Asp sidechain of DIM-5 (Figure 10.5b). Phospho-inhibition probably involves direct steric and electrostatic repulsion between the introduced phosphate group and the acidic Asp sidechain. Rubisco LSMT methylates Lys14 of the Rubisco large subunit. Lys14 is surrounded by the highly conserved sequence VGFKAGV. Molecular modeling based on the structure of free Lys complexed with Rubisco LSMT suggests that these are the residues that interact with the SET domain. A deep hydrophobic pocket is located adjacent to the Lys binding site and forms the putative binding site for Phe13 (Figure 10.5c). The surrounding groove is shallow and sterically restricted on either side of the Phe and Lys pockets, consistent with readout of a sequence consisting of small sidechains.

10.3.2
Catalytic Mechanism

The substrate Lys ε-amino group is poised for a direct in-line attack on the δ-methyl group of AdoMet, consistent with a S_N2 nucleophilic mechanism (Figure 10.6a). Such a mechanism requires that the Lys ε-amino group be deprotonated. SET domain enzymes have relatively low turnover numbers and high pH optima (> 9). The question arises of whether the Lys residue is already deprotonated when it binds or whether the enzyme provides a general base for this purpose. The high pH optimum might in principle be consistent with a Cys or Tyr residue acting as a general base. Indeed, the C-terminal conserved Tyr is the only conserved residue that presents its sidechain to the catalytic pore. The structures of substrate complexes bound to the active site show that the Tyr hydroxyl is near the substrate, but is not juxtaposed with it at an appropriate angle

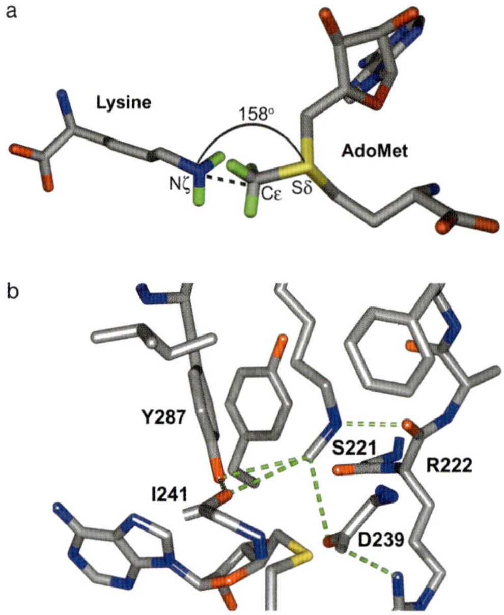

Figure 10.6 Catalysis by SET domain enzymes. (a) Putative reactive structure of the Lys–AdoMet ternary complex, showing geometry for nucleophilic attack. (b) Carbon–oxygen hydrogen bonds between the SET domain and MeLys in the LSMT active site (modified from [52]).

to abstract a proton. This seems to rule out a function for the Tyr in general base catalysis. It is more likely that the high pH optimum for the reaction reflects a requirement that the substrate be deprotonated in solution. Consistent with this, SET domain reaction rates are very low at neutral pH, where the great majority of Lys residues are protonated.

The AdoMet methyl group occupies the pore between the substrate and cofactor binding sites, sequestering it from solvent and poising it for transfer between the two reactive entities. The AdoMet methyl group lacks hydrophobic contacts, but interacts with surrounding carbonyl cage oxygens and the conserved Tyr hydroxyl group. The interatomic distances between the carbon and oxygen, 3.3–3.6 Å, are shorter than would be expected from ordinary van der Waals interactions. By the same token, the ζ-methyl group of the product MeLys makes similar polar interactions, but lacks hydrophobic interactions (Figure 10.6b). The AdoMet and MeLys methyl groups are unusually electropositive, due to electron withdrawal by the S or N atoms to which they are bonded. Methyl groups of this type can form exceptionally strong carbon–oxygen hydrogen bonds [64, 65]. The distances and angles between the methyl C and the carbonyl and Tyr oxygens are consistent with the formation of CH–O hydrogen bonds. This observation led us to suggest that one major role of the carbonyl cage and the conserved Tyr is to activate the AdoMet methyl for transfer by forming CH–O hydrogen bonds, promoting electron withdrawal and enhancing it as a target for nucleophilic attack.

10.3.3
Methylation Multiplicity

One reason Lys methylation is such a versatile regulatory mechanism is that one, two, or three methyl groups can be attached to the Lys ε-amino group. SET domains have the substrate and cofactor enter from opposite sides of the protein. This enables the cofactor to bind and dissociate without dissociation of the protein substrate. This led us to suggest that the substrate could remain bound during multiple rounds of cofactor association and dissociation, facilitating the transfer of up to three methyl groups prior to dissociation of the protein substrate. This theory was later supported by a thorough biochemical analysis of the methylation of Lys9 in histone H3 by DIM-5, which demonstrated that the enzyme catalyzes trimethylation of Lys9 without releasing the substrate during the consecutive methyl transfer reactions [51].

Some, but not all, SET domain enzymes trimethylate substrate Lys residues. Others, of which SET7/9 has become the archetype, are exclusively monomethyl transferases. Distinct patterns of mono and multiple methylation are observed in histones, suggesting there is probably some regulatory importance to differential methylation multiplicities [66]. Structural comparisons between the monomethyl transferase SET7/9 and the trimethyl transferases DIM-5 and LSMT reveal how methylation multiplicity is controlled. In SET7/9, two Tyr sidechains form hydrogen bonds with the Lys ε-amino group. One of these, Tyr245, is present in other HMTs, but the other, Tyr305, is absent in known trimethyl transferases. In the structure of LSMT, in which both Tyr are absent, MeLys binds with its ε-amino group about 1 Å closer to the active pore than in SET7/9. The additional hydrogen bonds between the SET7/9 Tyr residues and the ε-amino group appear to lock the MeLys into a nonproductive conformation in which further rounds of methyl transfer are blocked. The mutation Y305F converts SET7/9 into a multiple methyltransferase [50, 51]. A steric inhibition mechanism has also been suggested to explain how Tyr305 inhibits multiple methylation; however, the difference in size between the Tyr and Phe sidechains is small and seems unlikely to account for all of the effect. Despite observations suggesting that there may be a regulatory role for differential methylation multiplicity, tools have been lacking to precisely define its role. The discovery that a point mutation in SET domains can interconvert mono and multiple methyltransferases is an exciting advance, since it creates new tools for probing regulation in vivo.

10.4
Emerging Directions and Conclusions

SET domain and protein methylation research has witnessed explosive progress since the connection between the two was first made three years ago. The advances in regulation by protein methylation highlight how powerful the concept of modular domain families can be in biology. Prior to 2000, essentially nothing was known about SET domains or protein lysine-methylation–based gene regulatory mechanisms. In three years, numerous mechanisms have been identified and characterized. The structures and catalytic mechanisms of several SET domain enzymes have been elucidated.

The role of methylation multiplicity is likely to be a central area of inquiry in the near future. There is evidence to suggest that the number of methyl groups attached to a Lys residue has important regulatory consequences. Until recently, this was a difficult area to probe, but mutational tools for dissecting such mechanisms are now available, and rapid progress seems likely.

The known HMTs account for a minority of all SET domain proteins. Although the Rubisco large subunit is the only other established substrate for methylation by SET domain enzymes, it seems very unlikely that it is truly the only other substrate. The expansion of SET domain methylation to other areas of biology seems inevitable.

References

1 Schultz, J., Milpetz, F., Bork, P., Ponting, C. P., SMART, a simple modular architecture research tool: identification of signaling domains. *Proc. Natl. Acad. Sci. USA* **1998**, *95*, 5857–5864.

2 Jenuwein, T., Laible, G., Dorn, R., Reuter, G., SET domain proteins modulate chromatin domains in eu- and heterochromatin. *Cell Mol Life Sci* **1998**, *54*, 80–93.

3 Rea, S., et al., Regulation of chromatin structure by site-specific histone H3 methyltransferases. *Nature* **2000**, *406*, 593–599.

4 Rice, J. C., Allis, C. D., Gene regulation: code of silence. *Nature* **2001**, *414*, 258–261.

5 Felsenfeld, G., Groudine, M., Controlling the double helix. *Nature* **2003**, *421*, 448–453.

6 Fischle, W., Wang, Y. M., Allis, C. D., Binary switches and modification cassettes in histone biology and beyond. *Nature* **2003**, *425*, 475–479.

7 Nakayama, J., Rice, J. C., Strahl, B. D., Allis, C. D., Grewal, S. I., Role of histone H3 lysine 9 methylation in epigenetic control of heterochromatin assembly. *Science* **2001**, *292*, 110–113.

8 Peters, A., Mermoud, J. E., O'Carroll, D., Pagani, M., Schweizer, D., Brockdorff, N., Jenuwein, T., Histone H3 lysine 9 methylation is an epigenetic imprint of facultative heterochromatin. *Nat. Genet.* **2002**, *30*, 77–80.

9 Schotta, G., et al., Central role of *Drosophila* SU(VAR)3-9 in histone H3-K9 methylation and heterochromatic gene silencing. *EMBO J.* **2002**, *21*, 1121–1131.

10 Cheutin, T., McNairn, A. J., Jenuwein, T., Gilbert, D. M., Singh, P. B., Misteli, T., Maintenance of stable heterochromatin domains by dynamic HP1 binding. *Science* **2003**, *299*, 721–725.

11 Nielsen, A. L., Oulad-Abdelghani, M., Ortiz, J. A., Remboutsika, E., Chambon, P., Losson, R., Heterochromatin formation in mammalian cells: interaction between histones and HP1 proteins. *Mol. Cell* **2001**, *7*, 729–739.

12 Hwang, K. K., Eissenberg, J. C., Worman, H. J., Transcriptional repression of euchromatic genes by *Drosophila* heterochromatin protein 1 and histone modifiers. *Proc. Natl. Acad. Sci. USA* **2001**, *98*, 11423–11427.

13 Schultz, D. C., Ayyanathan, K., Negorev, D., Maul, G. G., Rauscher, F. J., SETDB1: a novel KAP-1–associated histone H3, lysine 9-specific methyltransferase that contributes to HP1-mediated silencing of euchromatic genes by KRAB zinc-finger proteins. *Genes. Dev.* **2002**, *16*, 919–932.

14 Ayyanathan, K., et al., Regulated recruitment of HP1 to a euchromatic gene induces mitotically heritable, epigenetic gene silencing: a mammalian cell culture model of gene variegation. *Genes. Dev.* **2003**, *17*, 1855–1869.

15 Bannister, A. J., Zegerman, P., Partridge, J. F., Miska, E. A., Thomas, J. O., Allshire, R. C., Kouzarides, T., Selective recognition of methylated lysine 9 on histone H3 by the HP1 chromo domain. *Nature* **2001**, *410*, 120–124.
16 Lachner, M., O'Carroll, N., Rea, S., Mechtler, K., Jenuwein, T., Methylation of histone H3 lysine 9 creates a binding site for HP1 proteins. *Nature* **2001**, *410*, 116–120.
17 Fischle, W., Wang, Y., Jacobs, S. A., Kim, Y., Allis, C. D., Khorasanizadeh, S., Molecular basis for the discrimination of repressive methyl-lysine marks in histone H3 by Polycomb and HP1 chromodomains. *Genes. Dev.* **2003**, *17*, 1870–1881.
18 Min, J., Zhang, Y., Xu, R. M., Structural basis for specific binding of Polycomb chromodomain to histone H3 methylated at Lys 27. *Genes. Dev.* **2003**, *17*, 1823–1828.
19 Bernstein, B. E., Humphrey, E. L., Erlich, R. L., Schneider, R., Bouman, P., Liu, J. S., Kouzarides, T., Schreiber, S. L., Methylation of histone H3 Lys 4 in coding regions of active genes. *Proc. Natl. Acad. Sci. USA* **2002**, *99*, 8695–8700.
20 Santos-Rosa, H., et al., Active genes are tri-methylated at K4 of histone H3. *Nature* **2002**, *419*, 407–411.
21 Ng, H. H., Robert, F., Young, R. A., Struhl, K., Targeted recruitment of set1 histone methylase by elongating pol II provides a localized mark and memory of recent transcriptional activity. *Mol. Cell* **2003**, *11*, 709–719.
22 Krogan, N. J., et al., The Paf1 complex is required for histone h3 methylation by COMPASS and Dot1p: linking transcriptional elongation to histone methylation. *Mol. Cell* **2003**, *11*, 721–729.
23 Briggs, S. D., Bryk, M., Strahl, B. D., Cheung, W. L., Davie, J. K., Dent, S. Y., Winston, F., Allis, C. D., Histone H3 lysine 4 methylation is mediated by Set1 and required for cell growth and rDNA silencing in *Saccharomyces cerevisiae*. *Genes. Dev.* **2001**, *15*, 3286–3295.
24 Bryk, M., Briggs, S. D., Strahl, B. D., Curcio, M. J., Allis, C. D., Winston, F., Evidence that SET1, a factor required for methylation of histone H3, regulates rDNA silencing in *S. cerevisiae* by a sir2-independent mechanism. *Curr. Biol.* **2002**, *12*, 165–170.
25 Krogan, N. J., Dover, J., Khorrami, S., Greenblatt, J. F., Schneider, J., Johnston, M., Shilatifard, A., COMPASS, a histone H3 (lysine 4) methyltransferase required for telomeric silencing of gene expression. *J. Biol. Chem.* **2002**, *277*, 10753–10755.
26 Kanoh, J., Francesconi, S., Collura, A., Schramke, V., Ishikawa, F., Baldacci, G., Geli, V., The fission yeast spSet1p is a histone H3-K4 methyltransferase that functions in telomere maintenance and DNA repair in an ATM kinase Rad3-dependent pathway. *J. Mol. Biol.* **2003**, *326*, 1081–1094.
27 Li, J., Moazed, D., Gygi, S. P., Association of the histone methyltransferase Set2 with RNA polymerase II plays a role in transcription elongation. *J. Biol. Chem.* **2002**, *277*, 49383–49388.
28 Li, B., Howe, L., Anderson, S., Yates, J. R. 3rd, Workman, J. L., The Set2 histone methyltransferase functions through the phosphorylated carboxyl-terminal domain of RNA polymerase II. *J. Biol. Chem.* **2003**, *278*, 8897–8903.
29 Xiao, T., Hall, H., Kizer, K. O., Shibata, Y., Hall, M. C., Borchers, C. H., Strahl, B. D., Phosphorylation of RNA polymerase II CTD regulates H3 methylation in yeast. *Genes. Dev.* **2003**, *17*, 654–663.
30 Schaft, D., Roguev, A., Kotovic, K. M., Shevchenko, A., Sarov, M., Neugebauer, K. M., Stewart, A. F., The histone 3 lysine 36 methyltransferase, SET2, is involved in transcriptional elongation. *Nucleic Acids Res.* **2003**, *31*, 2475–2482.
31 Krogan, N. J., et al., Methylation of histone H3 by Set2 in *Saccharomyces cerevisiae* is linked to transcriptional elongation by RNA polymerase II. *Mol. Cell Biol.* **2003**, *23*, 4207–4218.
32 Strahl, B. D., et al., Set2 is a nucleosomal histone H3-selective methyltransferase that mediates transcriptional repression. *Mol. Cell Biol.* **2002**, *22*, 1298–1306.
33 Fang, J., et al., Purification and functional characterization of SET8, a nucleosomal histone H4-lysine 20–specific

methyltransferase. *Curr. Biol.* **2002**, *12*, 1086–1099.

34 Nishioka, K., et al., PR-Set7 is a nucleosome-specific methyltransferase that modifies lysine 20 of histone H4 and is associated with silent chromatin. *Mol. Cell* **2002**, *9*, 1201–1213.

35 Rice, J. C., Nishioka, K., Sarma, K., Steward, R., Reinberg, D., Allis, C. D., Mitotic-specific methylation of histone H4 Lys 20 follows increased PR-Set7 expression and its localization to mitotic chromosomes. *Genes. Dev.* **2002**, *16*, 2225–2230.

36 Houtz, R. L., Stults, J. T., Mulligan, R. M., Tolbert, N. E., Post-translational modifications in the large subunit of ribulose bisphosphate carboxylase/oxygenase. *Proc. Natl. Acad. Sci. USA* **1989**, *86*, 1855–1859.

37 Klein, R. R., Houtz, R. L., Cloning and developmental expression of pea ribulose-1,5-bisphosphate carboxylase oxygenase large subunit N-methyltransferase. *Plant Mol Biol* **1995**, *27*, 249–261.

38 Ying, Z., Janney, N., Houtz, R. L., Organization and characterization of the ribulose-1,5-bisphosphate carboxylase/oxygenase large subunit epsilon N-methyltransferase gene in tobacco. *Plant Mol Biol* **1996**, *32*, 663–671.

39 Ying, Z., Mulligan, R. M., Janney, N., Houtz, R. L., Rubisco small and large subunit N-methyltransferases. Bi- and mono-functional methyltransferases that methylate the small and large subunits of rubisco. *J. Biol. Chem.* **1999**, *274*, 36750–36756.

40 Trievel, R. C., Beach, B. M., Dirk, L. M. A., Houtz, R. L., Hurley, J. H., Structure and catalytic mechanism of a SET domain protein methyltransferase. *Cell* **2002**, *111*, 91–103.

41 Zhang, X., Tamaru, H., Khan, S. I., Horton, J. R., Keefe, L. J., Selker, E. U., Cheng, X. D., Structure of the neurospora SET domain protein DIM-5, a histone H3 lysine methyltransferase. *Cell* **2002**, *111*, 117–127.

42 Wilson, J. R., Jing, C., Walker, P. A., Martin, S. R., Howell, S. A., Blackburn, G. M., Gamblin, S. J., Xiao, B., Crystal structure and functional analysis of the histone methyltransferase SET7/9. *Cell* **2002**, *111*, 105–115.

43 Jacobs, S. A., Harp, J. M., Devarakonda, S., Kim, Y., Rastinejad, F., Khorasanizadeh, S., The active site of the SET domain is constructed on a knot. *Nat. Struct. Biol.* **2002**, *9*, 833–838.

44 Kwon, T., Chang, J. H., Kwak, E., Lee, C. W., Joachimiak, A., Kim, Y. C., Lee, J. W., Cho, Y. J., Mechanism of histone lysine methyl transfer revealed by the structure of SET7/9-AdoMet. *EMBO J.* **2003**, *22*, 292–303.

45 Min, J. R., Zhang, X., Cheng, X. D., Grewal, S. I. S., Xu, R. M., Structure of the SET domain histone lysine methyltransferase Clr4. *Nat. Struct. Biol.* **2002**, *9*, 828–832.

46 Manzur, K. L., Farooq, A., Zeng, L., Plotnikova, O., Koch, A. W., Sachchidanand, Zhou, M. M., A dimeric viral SET domain methyltransferase specific to Lys27 of histone H3. *Nat. Struct. Biol.* **2003**, *10*, 187–196.

47 Yeates, T. O., Structures of SET domain proteins: protein lysine methyltransferases make their mark. *Cell* **2002**, *111*, 5–7.

48 Jacobs, S. A., Harp, J. M., Devarakonda, S., Kim, Y., Rastinejad, F., Khorasanizadeh, S., The active site of the SET domain is constructed on a knot. *Nat. Struct. Biol.* **2003**, *10*, 578.

49 Taylor, W. R., Xiao, B., Gamblin, S. J., Lin, K., A knot or not a knot? SETting the record 'straight' on proteins. *Comput. Biol. Chem.* **2003**, *27*, 11–15.

50 Xiao, B., et al., Structure and catalytic mechanism of the human histone methyltransferase SET7/9. *Nature* **2003**, *421*, 652–656.

51 Zhang, X., Yang, Z., Khan, S. I., Horton, J. R., Tamaru, H., Selker, E. U., Cheng, X. D., Structural basis for the product specificity of histone lysine methyltransferases. *Mol. Cell* **2003**, *12*, 177–185.

52 Trievel, R. C., Flynn, E. M., Houtz, R. L., Hurley, J. H., Mechanism of multiple lysine methylation by the SET domain enzyme rubisco LSMT. *Nat. Struct. Biol.* **2003**, *10*, 545–552.

53 Sims, R. J., Niskioka, K., Reinberg, D., Histone lysine methylation: a signature

for chromatin function. *Trends Genet.* **2003**, *19*, 629–639.

54 Czermin, B., Schotta, G., Hulsmann, B. B., Brehm, A., Becker, P. B., Reuter, G., Imhof, A., Physical and functional association of SU(VAR)3-9 and HDAC1 in *Drosophila*. *EMBO Rep* **2001**, *2*, 915–919.

55 Czermin, B., Melfi, R., McCabe, D., Seitz, V., Imhof, A., Pirrotta, V., *Drosophila* enhancer of Zeste/ESC complexes have a histone H3 methyltransferase activity that marks chromosomal Polycomb sites. *Cell* **2002**, *111*, 185–196.

56 Kuzmichev, A., Nishioka, K., Erdjument-Bromage, H., Tempst, P., Reinberg, D., Histone methyltransferase activity associated with a human multiprotein complex containing the Enhancer of Zeste protein. *Genes. Dev.* **2002**, *16*, 2893–2905.

57 Muller, J., et al., Histone methyltransferase activity of a *Drosophila* Polycomb group repressor complex. *Cell* **2002**, *111*, 197–208.

58 Milne, T. A., Briggs, S. D., Brock, H. W., Martin, M. E., Gibbs, D., Allis, C. D., Hess, J. L., MLL targets SET domain methyltransferase activity to Hox gene promoters. *Mol. Cell* **2002**, *10*, 1107–1117.

59 Nakamura, T., et al., ALL-1 is a histone methyltransferase that assembles a supercomplex of proteins involved in transcriptional regulation. *Mol. Cell* **2002**, *10*, 1119–1128.

60 Wang, H. B., Cao, R., Xia, L., Erdjument-Bromage, H., Borchers, C., Tempst, P., Zhang, Y., Purification and functional characterization of a histone H3-lysine 4-specific methyltransferase. *Mol. Cell* **2001**, *8*, 1207–1217.

61 Nishioka, K., Chuikov, S., Sarma, K., Erdjument-Bromage, H., Allis, C. D., Tempst, P., Reinberg, D., Set9, a novel histone H3 methyltransferase that facilitates transcription by precluding histone tail modifications required for heterochromatin formation. *Genes. Dev.* **2002**, *16*, 479–489.

62 Tachibana, M., Sugimoto, K., Fukushima, T., Shinkai, Y., SET domain-containing protein, G9a, is a novel lysine-preferring mammalian histone methyltransferase with hyperactivity and specific selectivity to lysines 9 and 27 of histone H3. *J. Biol. Chem.* **2001**, *276*, 25309–25317.

63 Beisel, C., Imhof, A., Greene, J., Kremmer, E., Sauer, F., Histone methylation by the *Drosophila* epigenetic transcriptional regulator Ash1. *Nature* **2002**, *419*, 857–862.

64 Derewenda, Z. S., Lee, L., Derewenda, U., The occurrence of C-H...O hydrogen bonds in proteins. *J. Mol. Biol.* **1995**, *252*, 248–262.

65 Scheiner, S., Kar, T., Gu, Y. L., Strength of the (CH)-H-alpha...O hydrogen bond of amino acid residues. *J. Biol. Chem.* **2001**, *276*, 9832–9837.

66 Santos-Rosa, H., et al., Active genes are tri-methylated at K4 of histone H3. *Nature* **2002**, *419*, 407–411.

67 Goo, Y. H., et al., Activating signal cointegrator 2 belongs to a novel steady-state complex that contains a subset of trithorax group proteins. *Mol. Cell Biol.* **2003**, *23*, 140–149.

68 Wysocka, J., Myers, M. P., Laherty, C. D., Eisenman, R. N., Herr, W., Human Sin3 deacetylase and trithorax-related Set1/Ash2 histone H3-K4 methyltransferase are tethered together selectively by the cell-proliferation factor HCF-1. *Genes. Dev.* **2003**, *17*, 896–911.

69 Roguev, A., Schaft, D., Shevchenko, A., Pijnappel, W., Wilm, M., Aasland, R., Stewart, A. F., The *Saccharomyces cerevisiae* Set1 complex includes an Ash2 homologue and methylates histone 3 lysine 4. *EMBO J.* **2001**, *20*, 7137–7148.

70 Nagy, P. L., Griesenbeck, J., Kornberg, R. D., Cleary, M. L., A trithorax-group complex purified from *Saccharomyces cerevisiae* is required for methylation of histone H3. *Proc. Natl. Acad. Sci. USA* **2002**, *99*, 90–94.

71 Ogawa, H., Ishiguro, K., Gaubatz, S., Livingston, D. M., Nakatani, Y., A complex with chromatin modifiers that occupies E2F- and Myc-responsive genes in G(0) cells. *Science* **2002**, *296*, 1132–1136.

72 Yang, L., Xia, L., Wu, D. Y., Wang, H., Chansky, H. A., Schubach, W. H., Hickstein, D. D., Zhang, Y., Molecular cloning of ESET, a novel histone H3-specific methyltransferase that interacts with ERG transcription factor. *Oncogene* **2002**, *21*, 148–152.

73 Tamaru, H., Selker, E. U., A histone H3 methyltransferase controls DNA methylation in *Neurospora crassa*. *Nature* **2001**, *414*, 277–283.

74 Jackson, J. P., Lindroth, A. M., Cao, X., Jacobsen, S. E., Control of CpNpG DNA methylation by the KRYPTONITE histone H3 methyltransferase. *Nature* **2002**, *416*, 556–560.

75 Rayasam, G. V., et al., NSD1 is essential for early post-implantation development and has a catalytically active SET domain. *EMBO J.* **2003**, *22*, 3153–3163.

11
The Structure and Function of the Bromodomain

Kelley S. Yan and Ming-Ming Zhou

11.1
Introduction

Covalent modifications of histones play a pivotal role in the control of chromatin structure, which in turn regulates a wide array of DNA-templated processes, including transcription, replication, recombination, and segregation [1–6]. Eukaryotic DNA is packaged in the form of chromatin, which is composed of a repeating nucleoprotein unit called the nucleosome. Within each nucleosome, chromosomal DNA of ~146 base pairs is wrapped around a histone octamer composed of two copies of each histone protein H2A, H2B, H3, and H4 [1]. Nucleosome cores are connected by short stretches of DNA bound to the linker histones H1 and H5 to form a nucleosomal filament, which is further folded into the higher-order chromatin fiber structure. Such dense packing and precise organization of DNA within chromatin structure is necessary for compaction of the genome into the nucleus of an eukaryotic cell. However, the question of how transcription or replication machineries gain access to the chromosomal DNA has been a subject of intense investigation. A rapidly growing body of knowledge has provided direct mechanistic links between the modification-induced dynamic modulation of chromatin architecture and the regulation of gene transcription [4, 6–8]. These modifications, including acetylation, methylation, phosphorylation, ubiquitination, ribosylation, and glycosylation, can occur on the conserved amino acid residues in the flexible N- and C-terminal sequences of histones and are directly linked to gene transcriptional activation and repression [8–10]. Such post-translational modifications, alone or in combination, of nucleosomal histones have been shown to be associated with a broad spectrum of distinct biological outcomes, a phenomenon that has been referred to as the 'histone code hypothesis' [3–6].

Although it was known that histones can be acetylated on specific lysine residues [11–14], the consequences of these modifications were not understood at a molecular level until more recently. Site-specific lysine acetylations of histones can serve as docking sites for the recruitment of chromatin remodeling complexes or simply alter electrostatic interactions between histones and DNA. The discovery by Dhalluin

Modular Protein Domains. Edited by G. Cesareni, M. Gimona, M. Sudol, M. Yaffe

ISBN: 3-527-30813-X

et al. [15] that the bromodomain may function as an acetyl-lysine binding domain by interacting with lysine-acetylated peptides derived from histones provided the first supporting evidence for the former possibility. The bromodomain is an evolutionarily conserved region of ~110 amino acids, which was first identified in the *Drosophila* protein brahma [16, 17] and named by analogy to the chromo domain, a chromatin-associated protein module [18]. The extensive bromodomain family contains members from over 500 eukaryotic chromatin-associated proteins and nuclear histone acetyltransferases (HATs), including ~128 human proteins [19, 20]. The suggested biological function of bromodomain binding to lysine-acetylated histones [15, 21] is analogous to that of SH2 (*Src* homology 2) [22] and PTB (phosphotyrosine binding) [23] domains of adaptor proteins binding to tyrosine-phosphorylated receptor tyrosine kinases in signal transduction [24]. Thus, histone lysine acetylation may serve as a regulatory modification to promote acetylation-dependent recruitment of proteins for chromatin remodeling or gene transcription. This mechanism agrees well with the histone code hypothesis, which postulates that distinct patterns of post-translational histone modifications function as a recognition code for the recruitment of different chromatin remodeling complexes [3–6].

The discovery of the bromodomain as an acetyl-lysine binding domain hinted at a mechanism for regulating protein–protein interactions via lysine acetylation. Such a mechanism has broad implications for the molecular events underlying a wide variety of cellular processes, including chromatin remodeling and transcriptional activation [3, 21, 25]. This mechanism also suggests that bromodomains may contribute to the observed hyperacetylated state of histones during transcription by tethering enzymatic activity of HATs to target chromosomal sites [12, 26, 27] so as to propagate the acetylation to neighboring nucleosomes. Moreover, the bromodomain may also assist in the directed assembly and activity of multiprotein chromatin remodeling complexes such as SAGA (Spt-Ada-Gcn5 acetyltransferase), RSC (remodeling the structure of chromatin), SWI/SNF, and NuA4 [28, 29]. Bromodomain disruption or deletion in various proteins across organisms results in pleiotrophic effects and may provide insights into the function of this module in vivo. For example, it has been shown that this module is indispensable for the function of GCN5p in the catalytic activity of the SAGA complex in *Saccharomyces cerevisiae* [30, 31]. Deletion of a bromodomain in HBRM, a protein component of the human SWI/SNF remodeling complex, causes both decreased stability and loss of nuclear localization [32, 33]. Bromodomains of Bdf1p, a *S. cerevisiae* protein, are required for sporulation and normal mitotic growth [34]. Finally, bromodomain deletion in Sth1, Rsc1, and Rsc2, three members of the nucleosome remodeling complex RSC, can cause a conditional lethal phenotype (in Sth1) [35] or a strong phenotypic inhibition of cell growth (in Rsc1 and Rsc2) [36]. Notably, the phenotypic effect observed in Rsc1 and Rsc2 results from deletion of only the second but not the first bromodomain, suggesting that these two bromodomains serve distinct functions through interactions with different biological ligands [36].

11.2
The Bromodomain Structure

The first 3D structure of the large bromodomain family was solved using the bromodomain from the transcriptional coactivator p300/CBP-associated factor (PCAF), as determined by nuclear magnetic resonance (NMR) spectroscopy [15]. The PCAF bromodomain adopts a left-handed four-helix bundle composed of amphipathic helices α_Z, α_A, α_B, and α_C (Figure 11.1a) [15]. At one end of the helical bundle, the N and C termini come together, emphasizing the modular architecture of this domain and underscoring the idea that the bromodomain acts as an independent functional unit for protein interactions [15, 18, 25]. At the opposite end of the bundle, a long intervening segment connecting helices α_Z and α_A (the 'ZA loop') packs against a relatively short segment connecting helices α_B and α_C (the 'BC loop') to form a surface-accessible hydrophobic pocket. Site-directed mutational analysis demonstrates that hydrophobic and aromatic tertiary contacts between the ZA and BC interhelical loops are important for stabilizing the 3D structure of this protein [15].

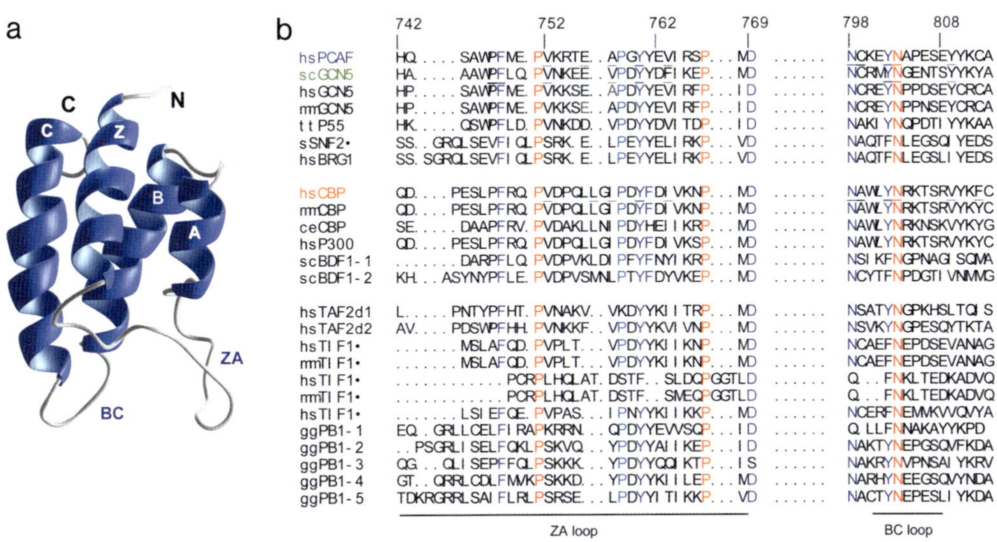

Figure 11.1 The bromodomain as an acetyl-lysine binding domain. (a) The 3D structure of the PCAF bromodomain in its free form, as determined by NMR spectroscopy [15]. (b) Sequence alignment of bromodomains highlighting amino acid variations in the ZA and BC loops. Bromodomains are grouped according to sequence similarities. Sequence numbers of PCAF are shown above the sequence. Absolutely and highly conserved residues in bromodomains are colored red and blue, respectively. Residues in the PCAF or CBP bromodomain that interact with p53 or HIV-1 Tat peptide, as shown by intermolecular NOEs, are underlined. Similarly, the residues of the yeast GCN5 bromodomain that directly contact the histone H4 peptide, as defined in the crystal structure, are also underlined.

This left-handed four-helix-bundle structural fold is highly conserved within the bromodomain family, as confirmed by more recently determined structures of bromodomains from human GCN5 [37], *S. cerevisiae* GCN5p [38], the double bromodomain module of human $TAF_{II}250$ [39], and the human transcriptional coactivator CBP (CREB-binding protein) [40]. Although the structural similarity shared by these bromodomains is very high for the four helices at the backbone level, structural differences do exist and are localized to the loop regions, particularly in the ZA and BC loops, which correspond to the segments of high amino acid sequence divergence [18] (Figure 11.1b). The conservation of 3D structure, as seen for the bromodomain fold, is another illustration of nature's general principle of exploiting a simple 3D scaffold to generate biological diversity. Although there may be only a limited number of evolutionarily conserved 3D structural folds, the functional use of these scaffolds can be amplified through amino acid sequence changes at their ligand binding sites to enable them to recognize a multitude of different biological targets.

11.3 The Bromodomain as an Acetyl-lysine Binding Domain

11.3.1 Acetyl-lysine Binding

The discovery of acetyl-lysine recognition by bromodomains is attributed to the unique ability of NMR spectroscopy to measure changes in local chemical environment and/or conformation of a protein induced upon binding to a ligand. Weak but highly specific interactions between a protein and a ligand (with a dissociation constant K_D in the micromolar-to-millimolar range) can be reliably detected with NMR, whereas most other techniques are limited to higher-affinity binding [41]. Furthermore, NMR spectroscopy provides insights into the location of the ligand-binding site within the protein through chemical shift mapping techniques, which were used to study the bromodomain from PCAF [15]. Ligand concentration-dependent NMR titrations of the PCAF bromodomain revealed that the protein can bind in a highly specific manner to lysine-acetylated peptides derived from major known acetylation sites on histones H3 or H4 [15]. The PCAF bromodomain failed to bind with either ligand in the absence of acetylation, demonstrating that the interaction is indeed dependent upon lysine acetylation. Chemical shift mapping using the titration data and NMR structural analysis of the PCAF bromodomain in complex with acetyl-histamine, a chemical analog of acetyl-lysine, showed that the acetyl-lysine binding site is localized to the hydrophobic cavity between the ZA and BC loops [15]. The methyl and methylene groups of acetyl-histamine make extensive contacts with the sidechains of V752, A757, Y760, Y802, N803, and Y809, which are highly conserved among the large bromodomain family [15, 18]. The observed acetyl-lysine dependence of the interactions and the location of the ligand-binding site were supported by another NMR study of the human GCN5 bromodomain binding to lysine-acetylated histone H4 peptides [37].

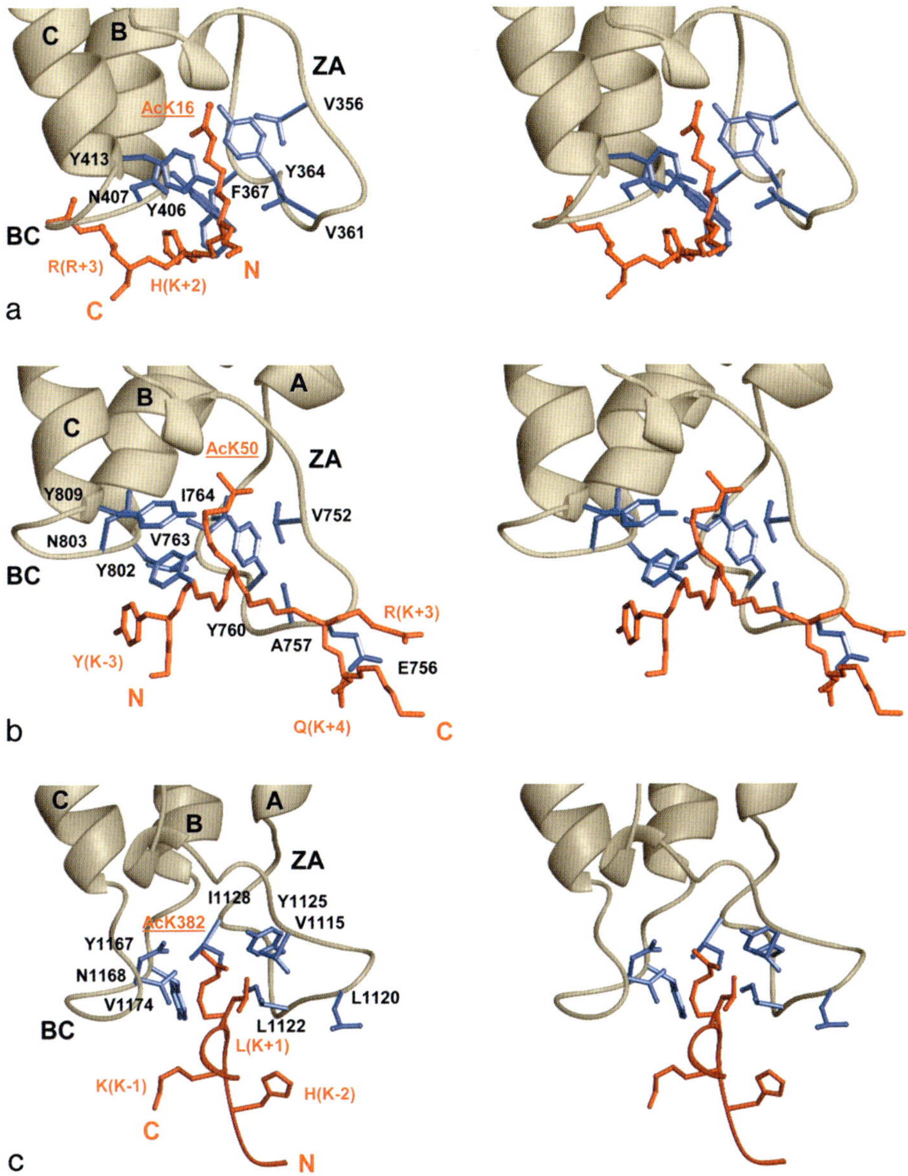

Figure 11.2 Differences in ligand selectivity of bromodomains. Stereoviews of the 3D structures of (a) scGCN5, (b) PCAF, and (c) CBP bromodomain in complex with the acetyl-lysine–containing peptide derived from histone H4 at K16 (A-*ac*K-RHRKILRNSIQGI), HIV-1 Tat at K50 (SYGR-*ac*K-KRRQR), or p53 at K382 (SHLKSKKGQSTSRHK-*ac*K-LMFK), respectively, showing interactions of the protein residues (blue) and the lysine-acetylated peptide residues (red) in the ligand binding sites. In all three bromodomain–ligand complex structures, the protein residues (blue) are numbered according to the sequences, and the peptide residues (red) are annotated according to their position with respect to the acetyl-lysine.

A crystal structure of *S. cerevisiae* GCN5p bromodomain solved in complex with an acetylated peptide derived from histone H4 at K16 (A-*ac*K-RHRKILRNSIQGI, where AcK represents N^{ε}-acetyl-lysine) reveals the molecular details of its acetyl-lysine recognition (Figure 11.2a) [38]. In addition to binding to the conserved hydrophobic and aromatic residues in the PCAF bromodomain, the acetyl-lysine forms a specific hydrogen bond between the oxygen of the acetyl carbonyl group and the sidechain amide nitrogen of an invariant asparagine residue in the bromodomain, N407 (corresponding to N803 in PCAF). A network of water-mediated hydrogen bonds with protein backbone carbonyl groups at the base of the cleft also contributes to acetyl-lysine binding. Site-directed mutagenesis confirmed the critical roles of these amino acid residues in binding to acetyl-lysine, suggesting that acetyl-lysine recognition is a general feature of bromodomains [15].

11.3.2
Molecular Determinants of Ligand Specificity

Although the major binding determinant in the GCN5p bromodomain–H4 peptide complex is the acetylated lysine itself, which sits in a deep hydrophobic pocket, the protein also has a limited number of contacts with residues C-terminal to the AcK at *ac*K +2 and *ac*K +3 in the H4 peptide that act as prongs plugged into two separate, shallower pockets (Figure 11.2a) [38]. Specifically, the aromatic ring of a histidine at *ac*K +2 interacts directly with aromatic sidechains of Y406 and F367, which are conserved in the bromodomain family. In addition to the GCN5p bromodomain–H4–AcK16 complex structure, the understanding of ligand specificity of bromodomains is further enhanced by the recent structural studies of two bromodomains in complex with biologically relevant nonhistone protein ligands. The first one is the highly selective association between the PCAF bromodomain and a lysine-acetylated trans activator protein Tat of human immunodeficiency virus type 1 (HIV-1) [42]. The second is the interaction between the bromodomain of the coactivator CBP (CREB binding protein) and a lysine-acetylated region in the C-terminal segment of the tumor suppressor protein p53 [40]. These structures also provide the first glimpses into structural features of bromodomain interactions with nonhistone proteins.

The viral Tat protein stimulates transcriptional activation of the integrated HIV-1 genome and promotes viral replication in infected host cells [43–47]. Tat trans-activation activity depends on acetylation at K50 by the HAT activity of p300/CBP and on its subsequent association with PCAF through a bromodomain-mediated interaction [48–50]. This bromodomain interaction results in the release of lysine-acetylated Tat from its association with the viral TAR RNA, leading to activation of HIV-1 transcription [51–53]. Deletion of the PCAF C-terminal region constituting the bromodomain potently abrogated Tat transactivation of integrated, but not of un-integrated, HIV-1 provirus [51].

The NMR structure of the PCAF bromodomain in complex with an acetylated Tat K50 peptide (SYGR-*ac*K-KRRQR) showed that, in addition to the acetyl-lysine, flanking residues both N- and C-terminal to the acetyl-lysine are important for this bromodomain interaction (Figure 11.2b). The Tat peptide adopts an extended conformation in the complex, in which its N-terminal Y(*ac*K −3) residue contacts Y802 and V763 and its C-terminal R(*ac*K +3) and Q(*ac*K +4) residues interact with E756. These specific interactions involving the acetyl-lysine moiety and its flanking residues, confirmed by site-directed mutagenesis, confer a highly selective association between the PCAF bromodomain and Tat [42]. The extensive number of contact points involved in ligand interactions may explain why the PCAF bromodomain binds to the Tat *ac*K50 peptide with a binding affinity (K_D ~10 μM) about 30 times higher that for a histone H4 *ac*K16 peptide (K_D ~300 μM) [15, 42].

The PCAF and GCN5p bromodomains share a high degree of sequence identity (~40%), yet the structures of these modules in their complexes with different ligands suggest that they possess different binding specificities. The differences in ligand selectivity are striking in both the location and orientation of the bound peptides in the PCAF and GCN5p bromodomains. The backbones of the Tat and H4 peptides both adopt an extended conformation, but are antiparallel in the two corresponding structures, with their N and C termini oriented nearly opposite to each other. Despite these differences, it is interesting to note that GCN5p binding of H4 H(*ac*K +2) residue is reminiscent of PCAF bromodomain recognition of Tat Y(*ac*K −3) residue via residues Y802 and V763, which are equivalent to residues Y406 and F367 in GCN5p. Because of this similar mode of molecular interaction, the two aromatic residues, which are located in very different positions in the Tat and H4 peptides with respect to the acetyl-lysine, are found, surprisingly, to be in a nearly identical position in the corresponding bromodomain complex structures. The high level of conservation of these ligand recognition residues in bromodomains suggests that selection of an aromatic or hydrophobic residue neighboring the acetyl-lysine is possibly a common mechanism used by this subgroup of the bromodomain family and that the ligand may be maneuvered into an orientation to accommodate this selection.

The human tumor suppressor p53 is another nonhistone protein that requires acetylation of its C-terminal lysine residues K320, K373, K382, and to a lesser extent K372 and K381, for its activity [54–57]. Recent in vivo studies show that acetylation-induced p53 activation as a transcription activator that controls expression of a large set of cellular genes in response to DNA damage does not result from an increase of its DNA binding activity, as hypothesized previously [58–60], but rather from its recruitment of coactivators and subsequent histone acetylation [54]. Despite the identification of these multiple acetylation sites, specific effects of single or combined acetylation of these lysine residues on p53 activity remain elusive. A recent study using structure-based functional analysis demonstrates that the bromodomain of CBP binds selectively to p53 at the acetylated K382 [40]. This molecular interaction is responsible for p53 acetylation-dependent coactivator recruitment after DNA damage, which is an essential step for p53-induced transcriptional activation of the cyclin-dependent kinase inhibitor p21 in G1 cell-cycle arrest [40].

The structure of the CBP bromodomain–p53 *ac*K382 peptide complex extends our knowledge of ligand selectivity of the bromodomain family (Figure 11.2c). Structural comparison of the CBP bromodomain–p53 *ac*K382 peptide

(SHLKSKKGQSTSRHK-*ac*K-LMFK), the PCAF bromodomain–Tat *ac*K50 peptide [42], and the GCN5p bromodomain–H4 AcK16 peptide [38] complexes further confirms that the mechanism of acetyl-lysine recognition is conserved. AcK recognition involves a nearly identical set of conserved residues in these different bromodomains, corresponding to V1115, Y1167, N1168, and V1174 in CBP. However, a different set of residues are used in the CBP bromodomain to recognize different amino acids flanking the AcK, including L(*ac*K +1), K(*ac*K −1), and H(*ac*K −2), to achieve its specificity. Notably, V763 in PCAF interacts with Y(*ac*K −3) in HIV-1 Tat [42], whereas mutation of the corresponding I1128 to alanine in CBP leads to only a partial reduction in p53 peptide binding. Moreover, E756 in PCAF, which is important for interactions with R(*ac*K +3) and Q(*ac*K +4) at the AcK50 in Tat, is changed to L1119 followed by the unique two-amino-acid insertion in CBP. The hydrophobic residues near this insertion are involved in CBP bromodomain binding to the L(*ac*K +1) and H(*ac*K −2) at the p53 AcK382 site. These distinct intermolecular interactions confer the binding preference of the CBP bromodomain for the *ac*K382 over *ac*K373 or *ac*K320 site in p53. Finally, the conformation of the bound peptides in CBP and PCAF bromodomains is also different, due to the differences in their modes of ligand interactions. The p53 peptide forms a β-turn–like conformation [40], whereas the HIV-1 Tat peptide adopts an extended conformation [42]. Taken together, these structural features of bromodomain–ligand complexes reinforce the notion that differences in ligand selectivity are attributed to just a few, although important, differences in bromodomain sequences, mostly in the ZA loop.

11.4 Emerging Developments

The structural and biochemical understanding of bromodomain–acetyl-lysine binding also facilitates the recent investigations of bromodomain functions in vivo. For instance, during transcription, p300 has been shown to bind directly to histones, preferentially to histone H3 [26, 61]. p300/CBP has been reported to interact through its bromodomain with lysine-acetylated myogenic factor MyoD [62]. Bromodomains of the catalytic subunits of SAGA and SWI/SNF may anchor these chromatin remodeling complexes to lysine-acetylated promoter nucleosomes [63]. Moreover, the fundamental importance of the bromodomain is further highlighted by the results of a systematic study that demonstrated its functional role in the recruitment of transcription complexes to lysine-acetylated histones [64].

The bromodomain modules from different proteins, or even from within the same proteins that contain two bromodomains, as exemplified by Bdf1p, are frequently found not to be biologically equivalent [65–67]. Using fluorescence resonance energy transfer, Kanno and coworkers recently demonstrated that bromodomain-containing proteins recognize different patterns of acetylated histones in the intact nuclei of living cells [68]. Specifically, they showed that the bromodomain protein Brd2 selectively interacts with acetylated K12 on histone

H4, whereas $TAF_{II}250$ and PCAF recognize H3 and other acetylated histones, indicating a high degree of specificity toward histone recognition exhibited by different bromodomains.

However, in other proteins, these double bromodomains found within the same protein may operate together to form a functional unit to act as a super-module. Brd4, one such protein containing tandem bromodomains, requires both bromodomains to interact with acetylated chromatin during mitosis and in transmitting transcriptional memory to daughter cells [69]. For the human $TAF_{II}250$, it has been proposed that the tandem bromodomains operate together in the cooperative binding of two neighboring acetyl-lysine sites in a histone protein that are separated by a distance of 25 Å, as suggested by the crystal structure of the bromodomains [39]. Additionally, some other bromodomains may function in combination with other modules to form heteromeric super-modules, such as the TIF1β bromodomain that is implicated in formation of a functional unit with its adjacent PHD finger [70], in transcriptional repression [71]. A number of other modular domains, including the BAH (bromo-adjacent homology) domain [72], are also frequently found to be adjacent to the bromodomain, and the juxtaposition of these two domains suggests that they could also operate as super-modules. Taken together, these findings suggest that, although bromodomains share a common basic biochemical function in acetyl-lysine binding that has been conserved throughout evolution, in vivo biological functions of individual bromodomains may be further modulated by the biological contexts in which they are found.

11.5 Concluding Remarks

Like the SH2-domain [22] and PTB-domain [73] recognition of tyrosine-phosphorylated proteins in signal transduction [24] (see Chapters 1 and 6), the bromodomain also binds with high selectivity to lysine-acetylated proteins through interactions with amino acid residues flanking the acetyl-lysine [25]. Since the residues important for acetyl-lysine recognition are largely conserved in bromodomains, binding of lysine-acetylated proteins is likely a general biochemical function of this family. The 3D left-handed four-helix bundle architecture provides a molecular framework for acetyl-lysine interaction with a hydrophobic cleft formed by the ZA and BC loops at one end of the bundle. Structural variations in the binding site are encoded by amino acid sequence variations in these loop regions, which allow for discrimination of different interaction targets. Differences in ligand selectivity may be attributed to a few important differences in bromodomain sequences. These include variations in the ZA and BC loops, which have relatively low sequence conservation and exhibit amino acid deletion or insertion in different bromodomains. These sequence variations enable individual bromodomains to use distinct sets of amino acids to interact with residues flanking the acetyl-lysine in a target protein. Due to the limited number of biologically relevant bromodomain–ligand complex structures currently available, a consensus understanding of ligand-binding specificity of different bromodomains is still lacking. Such new insights into ligand specificity will undoubtedly require additional structural analysis, which will help understand how functional diversity of this conserved structural fold is achieved through evolutionary modification of amino acid sequences that constitute the ligand-binding site. The emerging knowledge of the structure–function relationships of the bromodomain will enhance our mechanistic understanding of specific biological functions of bromodomain-containing proteins, which have been implicated in human diseases including Williams syndrome [74, 75], lymphoma, and leukemia [76], as well as in control of a large network of molecular interactions that regulate chromatin remodeling and gene transcription.

Acknowledgements

We thank all past and present members of the Zhou laboratory who have contributed to the studies discussed in this chapter. This work is supported by grants from the National Institutes of Health.

References

1 Wolffe, A. P., Hayes, J. J., Chromatin disruption and modification. *Nucleic Acids Res.* **1999**, *27*, 711–720.
2 John, S., Workman, J. L., Just the facts of chromatin transcription. *Science* **1998**, *282*, 1836–1837.
3 Strahl, B. D., Allis, C. D., The language of covalent histone modifications. *Nature* **2000**, *403*, 41–45.
4 Jenuwein, T., Allis, C. D., Translating the histone code. *Science* **2001**, *293*, 1074–1080.
5 Turner, B. M., Cellular memory and the histone code. *Cell* **2002**, *111*, 285–291.
6 Fischle, W., Wang, Y., Allis, C. D., Binary switches and modification cassettes in histone biology and beyond. *Nature* **2003**, *425*, 475–479.
7 Struhl, K., Histone acetylation and transcriptional regulatory mechanisms. *Genes. Dev.* **1998**, *12*, 599–606.
8 Mizzen, C., et al., Signaling to chromatin through histone modifications: how clear is the signal? Cold Spring Harbor Symposia on Quantitative Biology **1998**, *LXIII*, 469–481.
9 Luger, K., Mäder, A. W., Richmond, R. K., Sargent, D. F., Richmond, T. J., Crystal structure of the nucleosome core particle at 2.8 Å resolution. *Nature* **1997**, *389*, 251–260.
10 Grunstein, M., Histone acetylation in chromatin structure and transcription. *Nature* **1997**, *389*, 349–352.
11 Brownell, J. E., Zhou, J., Ranalli, T., Kobayashi, R., Edmondson, D. G., Roth, S. Y., Allis, C. D., *Tetrahymena* histone acetyltransferase A: a homolog to yeast Gcn5p linking histone acetylation to gene activation. *Cell* **1996**, *84*, 843–851.
12 Brownell, J. E., Allis, C. D., Special HATs for special occasions: linking histone acetylation to chromatin assembly and gene activation. *Curr. Opin. Genet. Dev.* **1996**, *6*, 176–184.
13 Filetici, P., Aranda, C., Gonzalez, A., Ballario, P., GCN5, a yeast transcriptional co-activator, induces chromatin reconfiguration of HIS3 promotor in vivo. *Biochem. Biophys. Res. Commun.* **1998**, *242*, 84–87.
14 Marcus, G. A., Silverman, N., Berger, S. L., Horiuchi, J., Guarente, L., Functional similarity and physical association between GCN5 and ADA2: putative transcriptional adaptors. *EMBO J.* **1994**, *13*, 4807–4815.

15 Dhalluin, C., Carlson, J. E., Zeng, L., He, C., Aggarwal, A. K., Zhou, M.-M., Structure and ligand of a histone acetyltransferase bromodomain. *Nature* **1999**, *399*, 491–496.

16 Tamkun, J. W., Deuring, R., Scott, M. P., Kissinger, M., Pattatucci, A. M., Kaufman, T. C., Kennison, J. A., brahma: a regulator of *Drosophila* homeotic genes structurally related to the yeast transcriptional activator SNF2/SWI2. *Cell* **1992**, *68*, 561–572.

17 Haynes, S. R., Dollard, C., Winston, F., Beck, S., Trowsdale, J., Dawid, I. B., The bromodomain: a conserved sequence found in human, *Drosophila* and yeast proteins. *Nucleic Acids Res.* **1992**, *20*, 2603–2603.

18 Jeanmougin, F., Wurtz, J. M., Le Douarin, B., Chambon, P., Losson, R., The bromodomain revisited. *Trends Biochem. Sci.* **1997**, *22*, 151–153.

19 Schultz, J., Milpetz, F., Bork, P., Ponting, C. P., SMART, a simple modular architecture research tool: identification of signaling domains. *Proc. Natl. Acad. Sci. USA* **1998**, *95*, 5857–5864.

20 Letunic, I., et al., Recent improvements to the SMART domain-based sequence annotation resource. *Nucleic Acids Res.* **2002**, *30*, 242–244.

21 Winston, F., Allis, C. D., The bromodomain: a chromatin-targeting module? *Nat. Struct. Biol.* **1999**, *6*, 601–604.

22 Schlessinger, J., Lemmon, M. A., SH2 and PTB domains in tyrosine kinase signaling. Sci. *STKE* **2003**, *2003*, RE12.

23 Yan, K. S., Kuti, M., Mujtaba, S., Farooq, A., Goldfarb, M. P., Zhou, M.-M., SNT PTB domain conformation regulates interactions with divergent neurotrophic receptors. *J. Biol. Chem.* **2002**, *277*, 17088–17094.

24 Pawson, T., Nash, P., Assembly of Cell regulatory systems through protein interaction domains. *Science* **2003**, *300*, 445–452.

25 Zeng, L., Zhou, M.-M., Bromodomain: an acetyl-lysine binding domain. *FEBS Lett.* **2001**, *513*, 124–128.

26 Manning, E. T., Ikehara, T., Ito, T., Kadonaga, J. T., Kraus, W. L., p300 forms a stable, template-committed complex with chromatin: role for the bromodomain. *Mol. Cell Biol.* **2001**, *21*, 3876–3887.

27 Travers, A., Chromatin modification: how to put a HAT on the histones. *Curr. Biol.* **1999**, *9*, 23–25.

28 Brown, C. E., Howe, L., Sousa, K., Alley, S. C., Carozza, M. J., Tan, S., Workman, J. L., Recruitment of HAT complexes by direct activator interactions with the ATM-related Tra1 subunit. *Science* **2001**, *292*, 2333–2337.

29 Sterner, D. E., et al., Functional organization of the yeast SAGA complex: distinct components involved in structural integrity, nucleosome acetylation, and TATA-binding protein interaction. *Mol. Cell Biol.* **1999**, *19*, 86–98.

30 Georgakopoulos, T., Gounalaki, N., Thireos, G., Genetic evidence for the interaction of the yeast transcriptional co-activator proteins GCN5 and ADA2. *Mol Gen Genet* **1995**, *246*, 723–728.

31 Syntichaki, P., Topalidou, I., Thireos, G., The Gcn5 bromodomain coordinates nucleosome remodelling. *Nature* **2000**, *404*, 414–417.

32 Muchardt, C., Bourachot, B., Reyes, J. C., Yaniv, M., ras transformation is associated with decreased expression of the brm/SNF2alpha ATPase from the mammalian SWI–SNF complex. *EMBO J.* **1998**, *17*, 223–231.

33 Muchardt, C., Yaniv, M., The mammalian SWI/SNF complex and the control of cell growth. *Semin Cell Dev Biol* **1999**, *10*, 189–195.

34 Chua, P., Roeder, G. S., Bdf1, a yeast chromosomal protein required for sporulation. *Mol. Cell Biol.* **1995**, *15*, 3685–3696.

35 Du, J., Nasir, I., Benton, B. K., Kladde, M. P., Laurent, B. C., Sth1p, a *Saccharomyces cerevisiae* Snf2p/Swi2p homolog, is an essential ATPase in RSC and differs from Snf/Swi in its interactions with histones and chromatin-associated proteins. *Genetics* **1998**, *150*, 987–1005.

36 Cairns, B. R., Schlichter, A., Erdjument-Bromage, H., Tempst, P., Kornberg, R. D., Winston, F., Two functionally distinct forms of the RSC nucleosome-remodeling complex, containing essential AT hook, BAH, and bromodomains. *Mol. Cell* **1999**, *4*, 715–723.

37 Hudson, B. P., Martinez-Yamout, M. A., Dyson, H. J., Wright, P. E., Solution structure and acetyl-lysine binding activity of the GCN5 bromodomain. *J. Mol. Biol.* **2000**, *304*, 355–370.

38 Owen, D. J., et al., The structural basis for the recognition of acetylated histone H4 by the bromodomain of histone acetyltransferase gcn5p. *EMBO J.* **2000**, *19*, 6141–6149.

39 Jacobson, R. H., Ladurner, A. G., King, D. S., Tjian, R., Structure and function of a human TAFII250 double bromodomain module. *Science* **2000**, *288*, 1422–1425.

40 Mujtaba, S., et al., Structural mechanism of the bromodomain of the coactivator CBP in p53 transcriptional activation. *Mol. Cell* **2004**, *13*, 251–263.

41 Hajduk, P. J., Measdows, R. P., Fesik, S. W., NMR-based screening in drug discovery. *Q Rev Biophys* **1999**, *32*, 211–240.

42 Mujtaba, S., He, Y., Zeng, L., Farooq, A., Carlson, J. E., Ott, M., Verdin, E., Zhou, M.-M., Structural basis of lysine-acetylated HIV-1 Tat recognition by P/CAF bromodomain. *Mol. Cell* **2002**, *9*, 575–586.

43 Cullen, B. R., HIV-1 auxiliary proteins: making connections in a dying cell. *Cell* **1998**, *93*, 685–692.

44 Jeang, K.-T., Xiao, H., Rich, E. A., Multifaceted activities of the HIV-1 transactivator of transcription, Tat. *J. Biol. Chem.* **1999**, *274*, 28837–28840.

45 Karn, J., Tackling Tat. *J. Mol. Biol.* **1999**, *293*, 235–254.

46 Adams, M., Sharmeen, L., Kimpton, J., Romeo, J. M., Garcia, J. V., Peterlin, B. M., Groudine, M., Emerman, M., Cellular latency in human immunodeficiency virus-infected individuals with high CD4 levels can be detected by the presence of promoter-proximal transcripts. *Proc. Natl. Acad. Sci. USA* **1994**, *91*, 3862–3866.

47 Garber, M. E., Jones, K. A., HIV-1 Tat: Coping with negative elongation factors. *Curr. Opin. Immunol.* **1999**, *11*, 460–465.

48 Kiernan, R. E., et al., HIV-1 Tat transcriptional activity is regulated by acetylation. *EMBO J.* **1999**, *18*, 6106–6118.

49 Ott, M., Schnolzer, M., Garnica, J., Fischle, W., Emiliani, S., Rackwitz, H.-R., Verdin, E., Acetylation of the HIV-1 Tat protein by p300 is important for its transcriptional activity. Curr. Biol. **1999**, *9*, 1489–1492.

50 Hottiger, M. O., Nabel, G. J., Interaction of human immunodeficiency virus type 1 Tat with the transcriptional coactivators p300 and CREB binding protein. *J. Virol.* **1998**, *72*, 8252–8256.

51 Benkirane, M., Chun, R. F., Xiao, H., Ogryzko, V. V., Howard, B. H., Nakatani, Y., Jeang, K.-T., Activation of integrated provirus requires histone acetyltransferase: p300 and P/CAF are co-activators for HIV-1 Tat. *J. Biol. Chem.* **1998**, *273*, 24898–24905.

52 Deng, L., et al., Acetylation of HIV-1 Tat by CBP/P300 increases transcription of integrated HIV-1 genome and enhances binding to core histones. *Virology* **2000**, *277*, 278–295.

53 Wei, P., Garber, M. E., Fang, S. M., Fischer, W. H., Jones, K. A., A novel CDK9-associated C-type cyclin interacts with HIV-1 Tat and mediates its high-affinity, loop-specific binding to TAR RNA. *Cell* **1998**, *92*, 451–462.

54 Barlev, N. A., Liu, L., Chehab, N. H., Mansfield, K., Harris, K. G., Halazonetis, T. D., Berger, S. L., Acetylation of p53 activates transcription through recruitment of coactivators/histone acetyltransferases. *Mol. Cell* **2001**, *8*, 1243–1254.

55 Ito, A., Lai, C. H., Zhao, X., Saito, S., Hamilton, M. H., Appella, E., Yao, T. P., p300/CBP-mediated p53 acetylation is commonly induced by p53-activating agents and inhibited by MDM2. *EMBO J.* **2001**, *20*, 1331–1340.

56 Li, M., Luo, J., Brooks, C. L., Gu, W., Acetylation of p53 inhibits its ubiquitination by Mdm2. *J. Biol. Chem.* **2002**, 50607–50611.

57 Ito, A., Kawaguchi, Y., Lai, C. H., Kovacs, J. J., Higashimoto, Y., Appella, E., Yao, T. P., MDM2-HDAC1–mediated deacetylation of p53 is required for its degradation. *EMBO J.* **2002**, *21*, 6236–6245.

58 Sakaguchi, K., Herrera, J. E., Saito, S., Miki, T., Bustin, M., Vassilev, A., Anderson, C. W., Appella, E., DNA damage activates p53 through

a phosphorylation–acetylation cascade. *Genes. Dev.* **1998**, *12*, 2831–2841.
59 Gu, W., Roeder, R. G., Activation of p53 sequence-specific DNA binding by acetylation of the p53 C-terminal domain. *Cell* **1997**, *90*, 595–606.
60 Liu, L., Scolnick, D. M., Trievel, R. C., Zhang, H. B., Marmorstein, R., Halazonetis, T. D., Berger, S. L., p53 sites acetylated in vitro by P/CAF and p300 are acetylated in vivo in response to DNA damage. *Mol. Cell Biol.* **1999**, *19*, 1202–1209.
61 An, W., Palhan, V. B., Karymov, M. A., Leuba, S. H., Roeder, R. G., Selective requirements for histone H3 and H4 N termini in p300-dependent transcriptional activation from chromatin. *Mol. Cell* **2002**, *9*, 811–821.
62 Polesskaya, A., Naguibneva, I., Duquet, A., Bengal, E., Robin, P., Harel-Bellan, A., Interaction between acetylated MyoD and the bromodomain of CBP and/or p300. *Mol. Cell Biol.* **2001**, *21*, 5312–5320.
63 Hassan, A. H., Prochasson, P., Neely, K. E., Galasinski, S. C., Chandy, M., Carrozza, M. J., Workman, J. L., Function and selectivity of bromodomains in anchoring chromatin-modifying complexes to promoter nucleosomes. *Cell* **2002**, *111*, 369–379.
64 Agalioti, T., Chen, G., Thanos, D., Deciphering the transcriptional histone acetylation code for a human gene. *Cell* **2002**, *111*, 381–392.
65 Matangkasombut, O., Buratowski, R. M., Swilling, N. W., Buratowski, S., Bromodomain factor 1 corresponds to a missing piece of yeast TFIID. *Genes. Dev.* **2000**, *14*, 951–962.
66 Matangkasombut, O., Buratowski, S., Different sensitivities of bromodomain factors 1 and 2 to histone H4 acetylation. *Mol. Cell* **2003**, *11*, 353–363.
67 Ladurner, A. G., Inouye, C., Jain, R., Tjian, R., Bromodomains mediate an acetyl-histone encoded antisilencing function at heterochromatin boundaries. *Mol. Cell* **2003**, *11*, 365–376.
68 Kanno, T., Kanno, Y., Siegel, R. M., Jang, M. K., Lenardo, M. J., Ozato, K., Selective recognition of acetylated histones by bromodomain proteins visualized in living cells. *Mol. Cell* **2004**, *13*, 33–43.
69 Dey, A., Chitsaz, F., Abbasi, A., Misteli, T., Ozato, K., The double bromodomain protein Brd4 binds to acetylated chromatin during interphase and mitosis. *Proc. Natl. Acad. Sci. USA* **2003**, *100*, 8758–8763.
70 Aasland, R., Gibson, T. J., Stewart, A. F., The PHD finger: implications for chromatin-mediated transcriptional regulation. *Trends Biochem. Sci.* **1995**, *20*, 56–59.
71 Schultz, D. C., Friedman, J. R., Rauscher, F. J. 3rd, Targeting histone deacetylase complexes via KRAB-zinc finger proteins: the PHD and bromodomains of KAP-1 form a cooperative unit that recruits a novel isoform of the Mi-2alpha subunit of NuRD. *Genes. Dev.* **2001**, *15*, 428–443.
72 Callebaut, I., Courvalin, J. C., Mornon, J. P., The BAH (bromo-adjacent homology) domain: a link between DNA methylation, replication and transcriptional regulation. *FEBS Lett.* **1999**, *446*, 189–193.
73 Yan, K. S., Kuti, M., Zhou, M.-M., PTB or not PTB – that is the question. *FEBS Lett.* **2002**, *513*, 67–70.
74 Lu, X., Meng, X., Morris, C. A., Keating, M. T., A novel human gene, WSTF, is deleted in Williams syndrome. *Genomics* **1998**, *54*, 241–249.
75 Bochar, D. A., Savard, J., Wang, W., Lafleur, D. W., Moore, P., Cote, J., Shiekhattar, R., A family of chromatin remodeling factors related to Williams syndrome transcription factor. *Proc. Natl. Acad. Sci. USA* **2000**, *97*, 1038–1043.
76 Greenwald, R. J., Tumang, J. R., Sinha, A., Currier, N., Cardiff, R. D., Rothstein, T. L., Faller, D. V., Denis, G. V., E mu-RD2 transgenic mice develop B-cell lymphoma and leukemia. *Blood* **2004**, *103*, 1475–1484.

12
Chromo and Chromo Shadow Domains

Joel C. Eissenberg and Sepideh Khorasanizadeh

12.1
Introduction and Brief History of the Module's Discovery

The chromo domain motif was first described and named when the *Polycomb* gene from *Drosophila* was cloned [1] and the gene product was found to have a 37 amino acid motif with 65% sequence identity to a sequence found in heterochromatin protein 1, HP1 [2, 3]. Both Polycomb and HP1 behave like transcriptional silencers in genetic assays, and both are chromosomal proteins, so it was inferred that their domain of homology conferred a homologous property underlying chromosome-based silencing. Thus, the name 'chromosome organization modifier' or chromo was coined to suggest a functional role for this motif. Subsequent molecular studies discussed here have demonstrated that chromo domains target modules that contribute to the organization of chromatin.

There are numerous examples of recognizable chromo domains in the eukaryotic kingdom [4, 5], nearly 500 proteins being identified in the nonredundant sequence database to date. Diverse proteins containing one or more chromo domains have been described in fungi, protozoa, nematodes, insects, plants, and animals. HP1, the prototypical chromo domain protein, was found to have a second C-terminal chromo domain motif, dubbed the chromo shadow domain. In all instances to date, chromo domain proteins are found to be involved in chromosome binding and chromatin metabolism. Chromo domain proteins have been divided into three classes [4]: (1) proteins having an N-terminal chromo domain followed by a chromo shadow domain; (2) proteins having a single chromo domain; and (3) proteins containing paired tandem chromo domains. In many instances, chromo domain proteins also encode catalytic activities, including DNA and protein methyltransferases (DNMT and SET domains), histone acetyltransferases (HAT domains), and ATPases [6].

In several instances, recognizable chromo domain homology is weak at the amino acid sequence level. An example is the HP1 chromo shadow domain, which shares less than 20% sequence identity with the chromo domain. However, 3D structural comparison confirms that the chromo and chromo shadow domains share a

Modular Protein Domains. Edited by G. Cesareni, M. Gimona, M. Sudol, M. Yaffe

ISBN: 3-527-30813-X

common fold [7–9]. For most chromo domains, however, the 3D structures have not been solved, and future studies will have to be carried out to establish their structures and any differences within the chromo domain fold of the HP1 family of proteins.

12.2
Structures of the Chromo and Chromo Shadow Domains

In this section we divide the chromo domain proteins into two main groups. Group 1 is the HP1 family, with one chromo domain followed by one chromo shadow domain (Figures 12.1 and 12.2), and group 2 includes nonHP1-family chromatin-associated proteins with one chromo domain or two chromo domains in tandem (Figure 12.3). Some members of the second group have a catalytic domain such as a SET, a HAT, or an ATPase domain C-terminal to the chromo domain. In the HP1 family of proteins, the chromo domain always occurs N-terminal to the chromo shadow domain, and the two domains are connected by a long (> 50 amino acids) nonconserved linker region. Limited proteolysis of the mammalian HP1β by subtilisin and trypsin has shown that these two conserved domains are the stable core regions of the HP1 proteins [7]. Gel filtration and analytical ultracentrifugation studies indicate that the intact HP1β and a truncated construct corresponding to only the chromo shadow domain are both dimeric, whereas the truncated construct corresponding to the chromo domain is monomeric [7, 10]. The high-resolution structures of the chromo domain [7] and the chromo shadow domain [8] of the mammalian HP1β have been determined by NMR spectroscopy. The high-resolution structure of the chromo shadow domain of the Swi6 protein, the HP1 homolog in *Schizosaccharomyces pombe*, has also been determined by X-ray crystallography [9]. In the following two paragraphs, we describe the 3D structures of each of these domains and discuss the unique attributes of the function of each domain.

The structure of the mammalian HP1β chromo domain is shown in Figure 12.2a. The chromo domain is a 50-amino acid module that consists of an N-terminal three-stranded antiparallel β sheet that packs against the C-terminal α helix. The sequence alignment of the chromo domains of the HP1 family (Figure 12.1a) indicates high homology across this family, suggesting that all members likely fold the same 3D structure. A set of conserved residues forms a groove on one surface of the β sheet (the same face shown in Figure 12.2a). The loops between the secondary structure units are long and consist of additional extended structure. Sequence alignment of chromo domains of nonHP1 proteins indicates that insertions and deletions in these connecting regions occur in different chromo domains that may not change the overall conserved fold (Figure 12.3). Ball et al. [7] observed that the chromo domain fold is homologous to two different classes of proteins. The first group includes the α/β chemokines, such as IL-8 and MCP1, which are involved in protein–protein interactions with receptors. The second group, which is highly related to the chromo domain fold, includes small DNA binding proteins, such as Sac7d and Sso7d. These proteins are found in the archaebacteria *Sulfolobus acidocaldarius* and *S. solfataricus*, respectively, and are also involved in the formation of chromatin structure. The HP1 chromo domain does not show appreciable nucleic acid binding activity, and, unlike Sac7d, it does not show a concentration of positively charged residues on its surface [7].

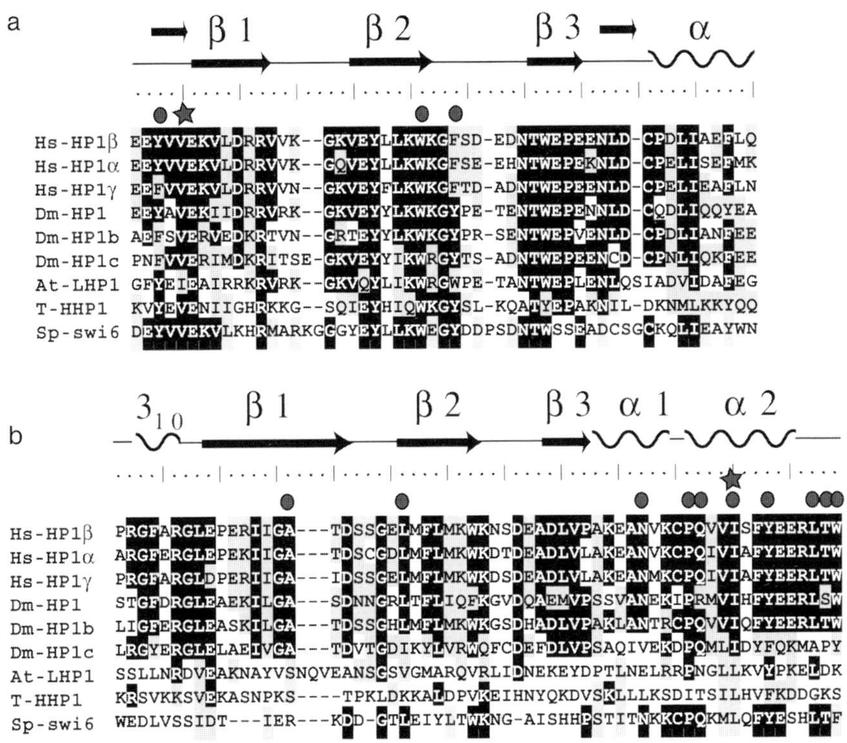

Figure 12.1 Sequences of the chromo domain (a) and the chromo shadow domain (b) of the HP1 proteins of *Homo sapiens* (Hs), *Drosophila melanogaster* (Dm), *Arabidopsis thaliana* (At), tetrahymena (T), and *S. pombe* (Sp). The accession codes in order of appearance are P23197, P45973, Q13185, P05205, Q9W396, Q9VCU6, AAL04059, O77159, and P40381. Conserved residues are highlighted. The secondary structure elements are indicated above the sequences based on the structures determined for each domain of Hs-HP1β (PDB accession codes 1AP0 and 1DZ1). Two additional β strands above the secondary structure in panel (a) indicate the locations of structures induced upon binding to histone H3 peptide, as shown in Figure 12.4a. The symbols above the sequence in panel (a) indicate the conserved residues of the aromatic cage (circles; see also Figure 12.4c) and the site of a valine-to-methionine missense mutation (star), which abolishes histone H3 interaction. The symbols above the sequence in panel (b) indicate the dimer interface (circles; see also Figure 12.2b) and the site of critical dimer interaction at which dimerization is interrupted by mutation to glutamate (star). Not shown are the sequences in the N- and C-terminal regions and the linker region in the HP1 family of proteins, which are not conserved and whose lengths vary significantly. To simplify the alignment in panel (b), a 22-residue insertion between α1 and α2 was removed from At-LHP1 sequence.

Moreover, it has been demonstrated that no combination of the naturally occurring amino acids gives rise to a motif that interacts with high affinity with the HP1 chromo domain [14]. As is discussed below, the function of some chromo domains, in particular HP1 and PC, is to bind histone H3 peptides containing a methyl-lysine, but in others, such as MOF and Mi-2, the chromo domain participates in nucleic acid binding.

The structure of the mammalian HP1β chromo shadow domain is shown in Figure 12.2b. Figure 12.1 shows that the composition of amino acid residues in the three β strands is different in the chromo domain and chromo shadow domain, suggesting unique functions for these domains. The chromo shadow domain is 70 amino acids long and forms a symmetrical dimer burying 687 $Å^2$ of surface area between its subunits [8, 9]. Interestingly, each monomer of the chromo shadow domain can be closely superimposed on the chromo domain, and the overall 3D fold of a chromo shadow domain monomer highly resembles that of the chromo domain. The residues that form the dimer interface are mostly hydrophobic and conserved only in the chromo shadow domain (Figure 12.1b). The dimerization constant for HP1β has been reported to be 2 μM [8]. One hydrophobic residue in α helix 2, an isoleucine, is at the heart of the dimer interface and plays a critical role in stabilization of the dimer (Figure 12.1b), and its mutation to a glutamate effectively disrupts the dimer [8].

The function of the chromo shadow domain is thought to be to establish numerous protein–protein interactions for the HP1 proteins. Random peptide phage-display libraries were probed with the chromo shadow domain of *Drosophila* HP1, and a consensus hydrophobic pentapeptide motif, [PL][WRY]V[MIV][MLV],

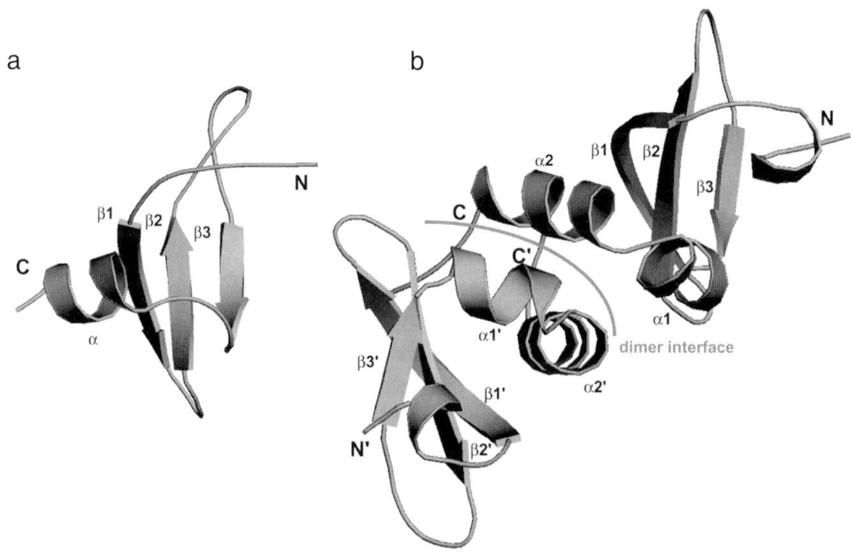

Figure 12.2 The 3D structure of HP1 chromo domain, PDB accession code 1AP0 (a) and of the HP1 chromo shadow domain dimer, PDB accession code 1DZ1 (b).

was identified for specific interaction [14]. Numerous nonhistone chromatin-associated proteins that interact with HP1 proteins contain this pentapeptide motif. Paradoxically, the chromo shadow domain also contains this motif in the region of α helix 2 and the dimer interface (Figure 12.1b) [14]. Figure 12.1 also shows that the chromo shadow domains in diverse HP1 proteins are not as conserved as the chromo domain, suggesting diversity in targeting by the chromo shadow domains of different HP1 proteins. For example, the last residue in the chromo shadow domain of some HP1 proteins is a tryptophan, which is not fully conserved. In mammalian HP1β, this residue was shown to be critical for interaction with peptides derived from CAFp150 or TIF (both have the consensus pentapeptide motif) [8]. It has also been shown that dimerization of the HP1β chromo shadow domain is necessary for interaction with CAFp150 or TIF peptides. The surface of the chromo shadow domain dimer that interacts with these peptides has been characterized by NMR spectroscopy and corresponds to the back of the dimer shown in Figure 12.2b toward which the C terminus points [8].

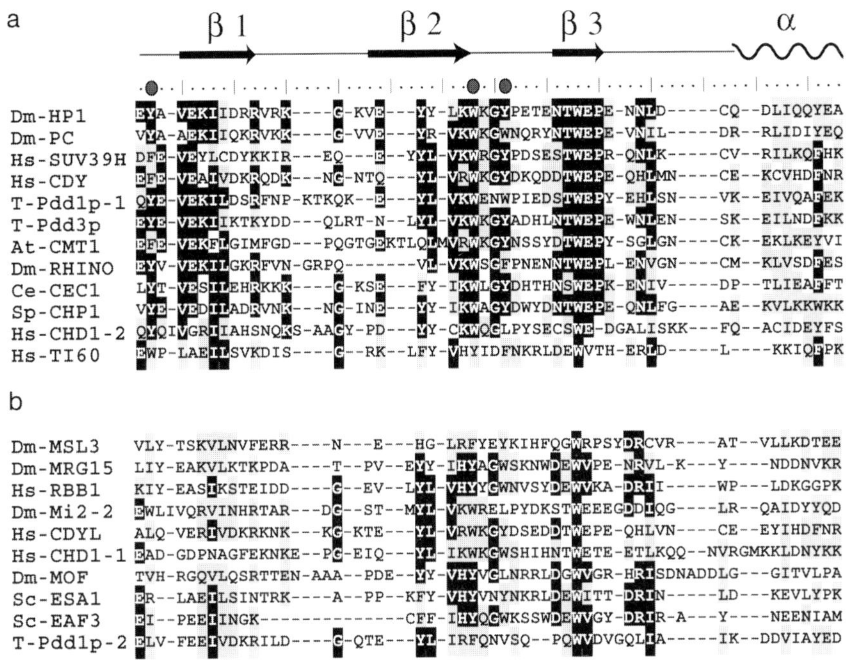

Figure 12.3 Structure-based sequence alignment and organization of diverse chromo domains. Abbreviations are Dm for *Drosophila melanogaster*, Hs for human, T for *Tetrahymena*, At for *Arabidopsis thaliana*, Ce for *Caenorhabditis elegans*, and Sp for *S. pombe*. The suffixes -1 and -2 indicate the N-terminal and C-terminal chromo domains, respectively, in tandem arrangements. Conserved residues are highlighted. The secondary structural elements are indicated above the sequence based on the structures determined for HP1. The sites of three aromatic cage residues that bind methyl-lysine in HP1 and polycomb are marked above the sequence with circles. All members of panel (b) lack one or more of the aromatic cage residues.

12.3
Function of the Chromo Domain

Although the structure of a chromo domain was first determined in 1997, the function of this domain remained elusive for a rather long period. In 2000, Suv39HI, which was known to interact with HP1, was shown to specifically methylate histone H3 at Lys9 via its SET domain [15]. Subsequently, several in vitro and in vivo studies showed that methylation at Lys9 in histone H3 produces a binding site for the chromo domain of HP1 protein [10,16–18]. The localization of HP1 to Lys9-methylated histone H3 has become a hallmark signal for epigenetic silencing in organisms as diverse as fission yeast and mammals. Characterization of the structure and energetic determinants of this complex are discussed below.

The high-resolution structures of mammalian HP1β and *Drosophila* HP1 were determined in complex with a histone H3 peptide containing methyl-lysine-9 [19, 20]. These structures show that a total of six residues of the peptide (Gln5–Ser10) directly bind to the chromo domain (Figure 12.4a). The structures involving dimethyl-lysine- and trimethyl-lysine–containing peptides are essentially superimposable. Moreover, the structures of the chromo domain in the free and complexed forms are very similar. The histone H3 peptide inserts as a β strand on the surface of the chromo domain at the conserved groove, which was predicted to be a protein binding site. Binding affinity is highly sensitive to mutations along the histone peptide binding groove. In particular, a missense mutation in *Drosophila*, the replacement of a valine with a methionine (see Figure 12.1a) [21], results in significant heterogeneity of structure in the peptide binding face of the HP1 chromo domain, and histone H3 interaction is completely abolished [10].

Whereas residues adjacent to residue 9 in histone peptide form important backbone hydrogen bonds and complementary surfaces with the sidechains of the chromo domain, the methyl-lysine recognition involves a special substructure referred to as an aromatic cage (Figure 12.4c) [19]. The aromatic residues of the chromo domain form a three-walled cage into which the methylammonium group inserts (Figure 12.4c); the fourth wall is solvent-exposed. The interaction between an amine (a cation) and a dense negatively charged aromatic ring (π electron system) is referred to as π–cation stabilization. Lysine methylation polarizes the Cε-Nζ bond in the lysine sidechain, lending a more cationic character to the methyl-lysine, resulting in stabilization of interactions with the aromatic cage. Experiments with isothermal titration calorimetry have shown that, upon going from dimethyl-lysine to trimethyl-lysine, binding improves about two-fold (K_D values of 7 μM and 2.5 μM, respectively) and there is favorable gain of enthalpy ($\Delta\Delta H = -0.4$ kcal mol^{-1}) and entropy [$\Delta(T\Delta S) = 0.2$ kcal mol^{-1}] associated with chromo domain binding. Therefore, binding improves polar interactions through π–cation interactions in the aromatic cage and also through van der Waals contacts. Interestingly, the aromatic cage of HP1 is highly superimposable with that found in the structure of the phosphatidylcholine transfer protein when bound to a phosphatidylcholine [22]. The quaternary amine of the phosphatidylcholine has a very similar π–cation interaction with three conserved aromatic residues of the protein.

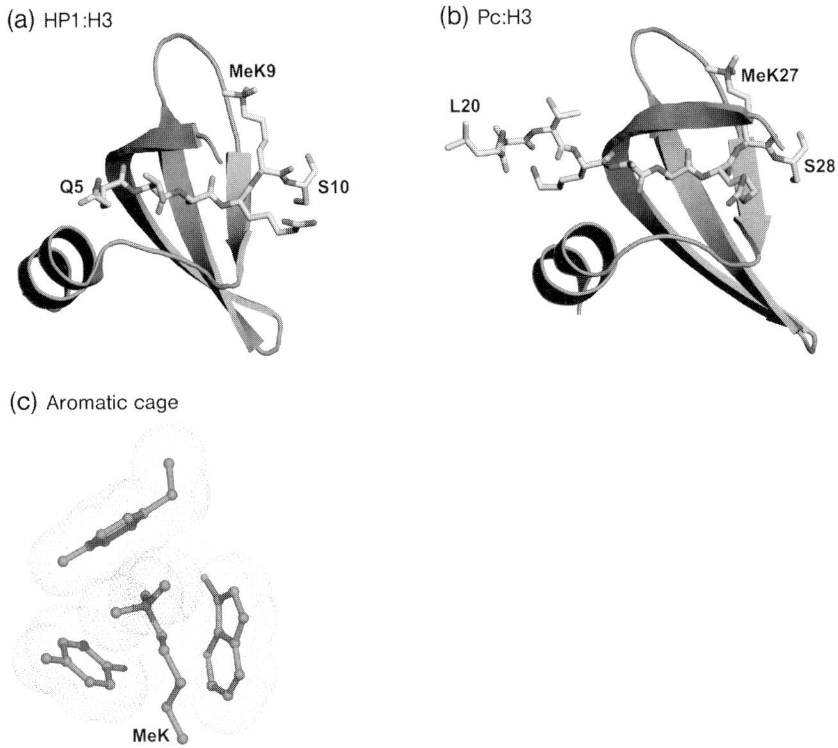

Figure 12.4 The 3D structure of HP1 chromo domain bound to histone H3 with methyl-lysine 9, PDB accession code 1KNE (a) and of polycomb chromo domain bound to histone H3 with methyl-lysine 27, PDB accession code 1PDQ (b). The chromo domains are shown in ribbon representation and their peptide partners are shown in stick style. Panel (c) zooms in on the aromatic cage of HP1 surrounding the methyl-lysine moiety. The structures in panels (a) and (b) are highly related and form similar aromatic cages around their methyl-lysine targets.

Very recently, the structure of the complex of the PC chromo domain with a histone H3 peptide containing methyl-lysine-27 revealed that highly related chromo domains can effectively discriminate highly related histone lysine-methylation sites to result in unique localizations in chromatin [23]. The chromo domains of *Drosophila* PC and HP1 proteins are 60% identical. The HP1 chromo domain binds to Lys9-methylated histone H3 16 times better than it binds to Lys27-methylated histone H3. Conversely, the PC chromo domain binds to Lys27-methylated histone H3 25 times better than it binds to Lys9-methylated histone H3. Although the HP1 chromo domain interacts with six residues of histone H3 ($QTARK^{9}S$) burying 1063 $Å^2$ of its surface area, the PC chromo domain interacts with nine residues of histone H3 ($LATKAARK^{27}S$) and buries 1482 $Å^2$ of its surface area [23, 24].

Besides the PC chromo domain, which binds to histone H3 having methyl-lysine-27, there are other nonHP1-class chromo domains that have an aromatic cage and are biochemically confirmed to bind histone H3 peptide having methyl-

Lys9. These include Suv39H1 [25], pdd1p-1, and pdd3p [26] chromo domains (Figure 12.3a). For other chromo domains having an aromatic cage, the targets appear to be other histone or nonhistone-derived peptides that have a methyl-lysine residue. Interestingly, a number of chromo domains have been implicated in nucleic acid binding. These include the MOF [27] and Mi2 [28] chromo domains, for which no high-resolution structures are available. Sequence analysis shows that the aromatic cage is not conserved in these two chromo domains (Figure 12.3). Moreover, there are numerous other chromo domains that have lost one or more of the residues of the aromatic cage. Additional biophysical studies are needed to establish the specific nature of interactions engaged by the chromo domains that lack an aromatic cage.

12.4
Genetic, Cytological, and Molecular Properties of the Chromo Domain

Mutations in the PC chromo domain [29] and the HP1 chromo domain [21] were identified by genetic screens for loss of silencing activity, specifically implicating these domains in the distinct silencing mechanisms of these proteins. Subsequent studies with the *S. pombe* HP1 family protein swi6 yielded similar results: both the chromo domain and the chromo shadow domain of swi6 are required for silencing at the mating type locus [30].

Although both PC and HP1 behave as transcriptional silencers in genetic assays, their genetic targets do not overlap. HP1 is primarily localized to pericentric heterochromatin and telomeres in higher eukaryotes. In *Drosophila*, HP1 is required for (1) silencing of euchromatic genes that come to lie next to heterochromatin by chromosome rearrangement or transposition [31], (2) activation of genes that lie in heterochromatin [32], (3) repression of certain euchromatic genes [33], and (4) telomere capping in mitotically active cells [34]. These domains of genetic activity are mirrored by the distribution of HP1, which is primarily concentrated in pericentric heterochromatin and telomeres, with limited euchromatic binding to the fourth chromosome and region 31 of the second chromosome [3,35]. In fission yeast, the HP1 family protein swi6 is required for silencing of the cryptic mating type cassettes [36] and for centromere function [37]. Chromatin immunoprecipitation data show that swi6 is enriched in paracentric chromatin and at the cryptic mating type [18, 38]. In *Drosophila*, PC is required for silencing of homeotic genes [39, 40]. The PC protein is found at > 100 sites in euchromatin [41] and colocalizes with regions containing Lys27-methylated histone H3 [23]. Double-label immunofluoresence staining reveals no detectable overlap with HP1 binding sites [21, 23]. Furthermore, mutation in HP1 results in none of the homeotic transformations seen with PC mutation, and mutation in PC has no effect on heterochromatic silencing.

Initial studies to determine chromo domain function employed fusion proteins to test chromosome binding activity. By this strategy, the PC chromo domain [29] and the HP1 chromo domain [21] were found to be sufficient to target specific chromosomal sites in vivo. Similar fusion protein assays with swi6 found that the chromo domain of this protein is required for chromosome targeting [30]. In contrast, only the chromo shadow domain of *Drosophila* HP1c is sufficient to target this HP1 family protein to euchromatin binding sites [42].

When the PC chromo domain is swapped for the HP1 protein chromo domain, the chimeric protein ($HP1^{PC-CD}$) targets both heterochromatin and the euchromatic sites of PC binding [21, 23]. Furthermore, the chimeric HP1/PC protein targets endogenous PC and other members of the PC complex to heterochromatin [23, 43], demonstrating that the PC chromo domain interacts via protein–protein contacts with one or more members of the PC complex assembly. Interestingly, the chimeric $HP1^{PC-CD}$ protein promotes silencing in transgenic flies [21].

In vitro, the chromo shadow domain of HP1 family of proteins mediates dimerization [8, 9, 14, 44]. Cytological studies revealed a heterochromatin targeting activity associated with the chromo shadow domains of *Drosophila* HP1 [45] and *S. pombe* swi6 [30]. This activity can be explained by heterodimerization with endogenous HP1 or swi6; in *S. pombe*, the chromosome binding activity of swi6 was lost for a chromo domain deletion expressed on a swi6 null background, but not on a swi6 wild-type background [30]. Similarly, expression of the $HP1^{PC-CD}$ chimeric protein results in the mislocalization of endogenous HP1 to euchromatic PC sites [21], a result best explained by heterodimerization between the endogenous and chimeric proteins through their chromo shadow domains.

In addition to self-association, chromo shadow domains have been reported to undergo a wide variety of heterologous interactions, including chromatin assembly factor 1 (CAF1; [46]), Ku70 [47], lamin B receptor [44], origin recognition complex protein 1 (ORC1; [48]), BRG1 [49], TIFα [50], KAP-1/TIF1β corepressor [51], Ki-67 antigen [52], and SP100 [53]. Whether and how many of these interactions are physiologically significant in vivo is unknown.

From the foregoing data, it seems very likely that H3K9 methylation confers specificity of binding to HP1 family proteins. It seems paradoxical then, that RNase A digestion of mammalian cell nuclei leads to reduced HP1α binding in the heterochromatin [54]. The RNA binding domain in HP1α lies at the C-terminal end of the linker domain connecting the chromo and chromo shadow domains [55]. One way to rationalize a controlling role for both RNA binding and methylated histone tail binding in HP1 protein targeting is that RNA binding may serve to link multiple HP1 molecules together, contributing cooperativity to HP1–nucleosome interaction. Since similar RNase A digestion in *Drosophila* cells has little or no effect on centromeric HP1 binding, this does not seem to be an evolutionarily conserved feature of HP1 targeting.

The HP1 chromo domain has also been implicated in RNA binding. Immunofluorescence localization of HP1 on *Drosophila* polytene chromosomes reveals significant staining over developmental puffs [3, 56], and upon heat shock, a significant amount of immunoreactive material accumulates over heat shock puffs [56]. In heat shock, it was shown that HP1 has dosage-dependent effects on Hsp70 steady-state mRNA levels. Polytene chromosome puffs are generally sites of intense transcription and massive RNA accumulation. RNase A digestion abolishes puff staining without affecting immunostaining in heterochromatic regions of the paracentric heterochromatin and telomeres. Similarly, puff staining is lost in a mutant HP1 containing a missense mutation in the chromo domain (valine to methionine, see Figure 12.1a). This same missense mutation was previously shown to interrupt binding to a methylated H3 tail peptide [10], suggesting that some aspects of chromo domain binding to histones and RNA may be shared. It also suggests that the structural basis for RNA binding by *Drosophila* HP1 and mammalian HP1α are different.

In *S. pombe*, genetic data strongly support a role for RNAi in the targeting of H3K9 methylation and swi6 binding to paracentric heterochromatin [57]. The mechanism of this targeting activity is unknown, but the unexpected observation that the chromo histone methylase clr4 is required for the RNAi mechanism as well as for histone methylation suggests the possibility of a chromo domain link [58].

The chromo domain of *Drosophila* MOF protein has been implicated in RNA binding. MOF, a chromo acetyltransferase [59], shows significant RNA binding activity in vitro, and point mutations in the chromo domain reduce or abolish detectable RNA binding [27]. A physiological significance of RNA binding by MOF is suggested by the observation that RNase A-digested nuclei lose both X-chromosome binding of MOF and histone H4 Lys16 acetylation, the product of MOF activity on chromatin [27].

In contrast to HP1 and PC, the chromo-helicase protein CHD1 is associated with transcriptionally active regions in *Drosophila*, as measured by immunolocalization of polytene chromosomes [60, 61], and with the body of transcriptionally active genes in yeast, as measured by chromatin immunoprecipitation [62]. Deletion of the chromo domains in the *Drosophila* CHD1 causes loss of chromosomal localization [63]. The chromo domains of yeast CHD1 have been shown to be required for functional complementation of a CHD1 deletion in the cold-sensitive phenotype of the *spt5-242* allele [62]. The biochemical function of the CHD1 chromo domains is unknown. However, it is interesting to note that transcriptionally active regions of chromatin are enriched in Lys4-methylated isoforms of histone H3 in yeast [18, 64, 65] and mammals [66]. The mechanistic role of this methyl marker in transcription is unknown, but it is tempting to speculate that a chromo domain protein associated with elongating RNA polymerase II on active chromatin, like CHD1, could target this methyl marker through its chromo domains.

12.5 Emerging Research Directions and Recent Developments

As discussed in detail above, the best-characterized chromo domain ligands are histone H3 tail regions containing a methyl-lysine, whereas the chromo shadow domain ligands include many proteins containing a consensus hydrophobic motif. For most chromo domains, the nature of ligand binding is unknown. In particular, for the chromo domains implicated in nucleic acid binding, the molecular basis for binding specificity is unknown. As the basis for binding specificity among the various chromo domain classes becomes understood, it should be possible to engineer chromo domains as targeting modules to deliver specific probes or enzyme activities to specific chromosomal regions.

Similarly, biochemical data suggest a large number of possible interactions between the HP1 chromo shadow domain and diverse protein complexes involved in transcriptional activation and repression, telomere stability and DNA repair, DNA replication, and nuclear organization. These in vitro interactions should be tested for physiological relevance. Authentic chromo shadow domain interactions should be characterized at the structural level, because this appears to be an important and versatile domain for organizing nuclear metabolism.

Although biochemical and biophysical data strongly support a model in which the chromo domains of HP1 family proteins and of PC specifically bind to methylated lysines in histone H3, these experiments have all been done using histone tail peptides and chromo domain peptides. To link these studies more directly with the biology of chromo domain proteins, it will be necessary to reconstitute chromo–histone interactions in the context of full-length chromo domain proteins binding to methylated histone assembled into nucleosome particles on DNA. Furthermore, self-association and heterologous associations will likely also contribute to binding affinity and targeting in vivo, and the contributions of these interactions remain unknown.

Acknowledgements

The work described here from the Khorasanizadeh lab has been supported by NIH grant GM 116635.

References

1 Paro, R., Hogness, D., The polycomb protein shares a homologous domain with a heterochromatin associated protein of *Drosophila*. *Proc. Natl. Acad. Sci. USA* **1991**, *88*, 263–267.

2 James, T. C., Elgin, S. C. R., Identification of a nonhistone chromosomal protein associated with heterochromatin in *Drosophila* and its gene. *Mol. Cell. Biol.* **1986**, *6*, 3862–3872.

3 James, T. C., Eissenberg, J. C., Craig, C., Dietrich, V., Hobson, A., Elgin, S. C. R., Distribution patterns of HP1, a heterochromatin-associated nonhistone chromosomal protein of *Drosophila*. *Eur. J. Cell Biol.* **1989**, *50*, 170–180.

4 Aasland, R., Stewart, A. F., The chromo shadow domain, a second chromo domain in heterochromatin-binding protein 1, HP1. *Nucleic Acids Res.* **1995**, *23*, 3163–3173.

5 Koonin, E. V., Zhou, S. B., Lucchesi, J. C., The chromo superfamily: new members, duplication of the chromo domain and possible role in delivering transcription regulators to chromatin. *Nucleic Acids Res.* **1995**, *3*, 4229–4233.

6 Eissenberg, J. C., Molecular biology of the chromo domain: an ancient chromatin module comes of age. *Gene* **2001**, *275*,19–29.

7 Ball, L. J., et al., Structure of the chromatin binding (chromo) domain from mouse modifier protein 1. *EMBO J.* **1997**,*16*, 2473–2481.

8 Brasher, S. V., et al., The structure of mouse HP1 suggests a unique mode of single peptide recognition by the shadow chromo domain dimer. *EMBO J.* **2000**, *19*, 1587–1597.

9 Cowieson, N. P., Partridge, J. F., Allshire, R. C., McLaughlin, P. J., Dimerisation of a chromo shadow domain and distinctions from the chromo domain as revealed by structural analysis. *Curr. Biol.* **2000**, *10*, 517–525.

10 Jacobs, S. A., Taverna, S. D., Zhang, Y., Briggs, S. D., Li, J., Eissenberg, J. C., Allis, C. D., Khorasanizadeh, S., Specificity of HP1 chromo domain for methylated N-terminus of histone H3. *EMBO J.* **2001**, *20*, 5232–5241.

11 Edmondson, S. P., Qiu, L., Shriver, J. W., Solution structure of the DNA-binding protein Sac7d from the hyperthermophile *Sulfolobus acidocaldarius*. *Biochemistry* **1995**, *34*, 13289–13304.

12 Baumann, H., Knapp, S., Lindbäck, T., Ladenstein, R., Härd, T., Solution structure and DNA-binding properties of a thermostable protein from the archaeon *Solfolobus solfataricus*. *Nature Struct. Biol.* **1994**, *1*, 808–819.

13 Gao, Y.-G., et al., The crystal structure of the hyperthermophile chromosomal protein Sso7d bound to DNA. *Nature Struct. Biol.* **1998**, *5*, 782–786.

14 Smothers, J. F., Henikoff, S., The HP1 chromo shadow domain binds a consensus peptide pentamer. *Curr. Biol.* **2000**, *10*, 27–30.

15 Rea, S., et al., Regulation of chromatin structure by site-specific histone H3 methyltransferases. *Nature* **2000**, *406*, 593–599.

16 Lachner, M., O'Carroll, D., Rea, S., Mechtler, K., Jenuwein, T., Methylation of histone H3 lysine 9 creates a binding site for HP1 proteins. *Nature* **2001**, *410*, 116–120.

17 Bannister, A. J., Zegerman, P., Partridge, J. F., Miska, E. A., Thomas, J. O., Allshire, R. C., Kouzarides, T., Selective recognition of methylated lysine 9 on histone 3 by the HP1 chromo domain. *Nature* **2001**, *410*, 120–124.

18 Noma, K., Allis, C. D., Grewal, S. I. S., Transitions in distinct histone H3 methylation patterns at the heterochromatin domain boundaries. *Science* **2001**, *293*, 1150–1155.

19 Jacobs, S. A., Khorasanizadeh, S., Structure of HP1 chromo domain bound to lysine 9-methylated histone H3 tail. *Science* **2002**, *295*, 2080–2083.

20 Nielsen, P. R., et al., Structure of the HP1 chromodomain bound to histone H3 methylated at lysine 9. *Nature* **2002**, *416*, 103–107.

21 Platero, J. S., Hartnett, T., Eissenberg, J. C., Functional analysis of the chromo domain of HP1. *EMBO J.* **1995**, *14*, 3977–3986.

22 Roderick, S. L., Chan, W. W., Agate, D. S., Olsen, L. R., Vetting, M. W., Rajashankar, K. R., Cohen, D. E., Structure of human phosphatidylcholine transfer protein in complex with its ligand. *Nature Struct. Biol.* **2002**, *9*, 507–511.

23 Fischle, W., Wang, Y., Jacobs, S. A., Kim, Y., Allis, C. D., Khorasanizadeh, S., Molecular basis for the discrimination of repressive methyl-lysine marks in histone H3 by Polycomb and HP1 chromo domains. *Genes. Dev.* **2003**, *17*, 1870–1881.

24 Min, J., Zhang, Y., Xu, R.-M., Structural basis for specific binding of Polycomb chromo domain to histone H3 methylated at Lys 27. *Genes. Dev.* **2003**, *17*, 1823–1828.

25 Jacobs, S. A., Fischle, W., Khorasanizadeh, S., Assays for the determination of structure and dynamics of the interaction of the chromodomain with histone peptides. *Methods Enzymol.* **2004**, *376*, 131–148.

26 Taverna, S. D., Coyne, R. S., Allis, C. D., Methylation of histone H3 at lysine 9 targets programmed DNA elimination in tetrahymena. *Cell* **2002**, *110*, 701–711.

27 Akhtar, A., Zink, D., Becker, P. B., Chromo domains are protein–RNA interaction modules. *Nature* **2000**, *407*, 405–408.

28 Bouazoune, K., Mitterweger, A., Längst, G., Imhof, A., Akhtar, A., Becker, P. B., Brehm, A., The dMi-2 chromo domains are DNA binding modules important for ATP-dependent nucleosome mobilization. *EMBO J.* **2002**, *21*, 2430–2440.

29 Messmer, S., Franke, A., Paro, R., Analysis of the functional role of the Polycomb chromo domain in *Drosophila melanogaster*. *Genes. Dev.* **1992**, *6*, 1241–1254.

30 Wang, G., Ma, A., Chow, C.-M., Horsley, D., Brown, N. R., Cowell, I. G., Singh, P. B., Conservation of heterochromatin protein 1 function. *Mol. Cell. Biol.* **2000**, *20*, 6970–6983.

31 Eissenberg, J. C., Morris, G. D., Reuter, G., Hartnett, T., The heterochromatin-associated protein HP-1 is an essential protein in *Drosophila* with dosage-dependent effects on position-effect variegation. *Genetics* **1992**, *131*, 345–352.

32 Lu, B. Y., Emtage, P. C. R., Duyf, B. J., Hilliker, A. J., Eissenberg, J. C., Heterochromatin protein 1 is required for the normal expression of two heterochromatin genes in *Drosophila*. *Genetics* **2000**, *155*, 699–708.

33 Hwang, K.-K., Eissenberg, J. C., Worman, H. J., Transcriptional repression of euchromatic genes by *Drosophila* heterochromatin protein 1 and histone modifiers. *Proc. Natl. Acad. Sci. USA* **2001**, *98*, 11423–11427.

34 Fanti, L., Giovinazzo, G., Berloco, M., Pimpinelli, S., The heterochromatin protein 1 prevents telomere fusions in *Drosophila*. *Mol. Cell* **1998**, *2*, 527–538.

35 Fanti, L., Berloco, M., Piacentini, L., Piminelli, S., Chromosomal distribution of heterochromatin protein 1 (HP1) in *Drosophila*: a cytological map of euchromatic HP1 binding sites. *Genetica* **2003**, *117*, 135–147.

36 Lorentz, A., Ostermann, K., Fleck, O., Schmidt, H., Switching gene *swi6*, involved in repression of silent mating-type loci in fission yeast, encodes a homologue of chromatin-associated proteins from *Drosophila* and mammals. *Gene* **1994**, *143*, 139–143.

37 Ekwall, K., Nimmo, E. R., Javerzat, J. P., Borgstrom, B., Egel, R., Cranston, G., Allshire, R., Mutations in the fission yeast silencing factors *clr4+* and *rik1+* disrupt localization of the chromo domain protein Swi6p and impair centromere function. *J. Cell Sci.* **1996**, *109*, 2637–2648.

38 Partridge, J. F., Borgstrom, B., Allshire, R. C., Distinct protein interaction domains and protein spreading in a complex centromere. *Genes. Dev.* **2000**, *14*, 783–791.

39 Duncan, I. M., Lewis, E. B., Genetic control of body segment differentiation in *Drosophila* In *Developmental Order: Its Origin and Regulation* (Eds.: Subtelny, S., Green, P. B.), pp. 533–554. Liss, New York **1982**.

40 Jürgens, G., A group of genes controlling the spatial expression of the bithorax complex in *Drosophila*. *Nature* **1985**, *316*, 153–155.

41 Zink, B., Paro, R., In vivo binding pattern of a trans-regulator of homoeotic genes in *Drosophila melanogaster*. *Nature* **1989**, *337*, 468–471.

42 Smothers, J. F., Henikoff, S., The hinge and chromo shadow domain impart distinct targeting of HP1-like proteins. *Mol. Cell. Biol.* **2001**, *21*, 2555–2569.

43 Platero, J. S., Sharp, E. J., Adler, P. N., Eissenberg, J. C., In vivo assay for protein–protein interactions using *Drosophila* chromosomes. *Chromosoma* **1996**, *140*, 393–404.

44 Ye, Q., Callebaut, I., Pezhman, A., Courvalin, J. C., Worman, H. J., Domain-specific interactions of human HP1-type chromo domain proteins and inner nuclear membrane protein LBR. *J. Biol. Chem.* **1997**, *272*, 14983–14989.

45 Powers, J., Eissenberg, J. C., Overlapping domains of the heterochromatin associated protein HP1 mediate nuclear localization and heterochromatin binding. *J. Cell Biol.* **1993**, *120*, 291–299.

46 Murzina, N., Verreault, A., Laue, E., Stillman, B., Heterochromatin dynamics in mouse cells: interaction between chromatin assembly factor 1 and HP1 proteins. *Mol. Cell* **1999**, *4*, 1–20.

47 Song, K., Jung, Y., Jung, D., Lee, I., Human Ku70 interacts with heterochromatin protein 1α. *J. Biol. Chem.* **2001**, *276*, 8321–8327.

48 Pak, D. T. S., Pflumm, M., I Chesnodov, Huang, D. W., Kellum, R., Marr, J., Romanowski, P., Botchan, M. R., Association of the origin recognition complex with heterochromatin and HP1

in higher eukaryotes. *Cell* **1997**, *91*, 311–323.

49 Nielsen, A. L., Sanchez, C., Ichinose, H., Cerviño, M., Lerouge, T., Chambon, P., Losson, R., Selective interaction between the chromatin-remodeling factor BRG1 and the heterochromatin-associated protein HP1α. *EMBO J.* **2002**, *21*, 5795–5806.

50 Le Douarin, B., Nielsen, A. L., Garnier, J.-M., Ichinose, H., Jeanmougin, F., Losson, R., Chambon, P., A possible involvement of TIF1α and TIF1β in the epigenetic control of transcription by nuclear receptors. *EMBO J.* **1996**, *15*, 6701–6715.

51 Lechner, M. S., Gegg, G. E., Speicher, D., Rauscher, F. J. III., Molecular determinants for targeting heterochromatin protein 1-mediated gene silencing: direct chromoshadow domain–KAP-1 corepressor interaction is essential. *Mol. Cell. Biol.* **2000**, *20*, 6449–6465.

52 Kametaka, A., Takagi, M., Hayakawa, T., Haraguchi, T., Hiraoka, Y., Yoneda, Y., Interaction of the chromatin compaction-inducing domain (LR domain) of Ki-67 antigen with HP1 proteins. *Genes. Cells* **2002**, *7*, 1231–1242.

53 Seeler, J.-S., Marchio, A., Sitterlin, D., Transy, C., Dejean, A., Interaction of SP100 with HP1 proteins: a link between the promyelocytic leukemia-associated nuclear bodies and the chromatin compartment. *Proc. Natl. Acad. Sci. USA* **1998**, *95*, 7316–7321.

54 Maison, C., et al., Higher-order structure in pericentric heterochromatin involves a distinct pattern of histone modification and an RNA component. *Nat. Genet.* **2002**, *30*, 329–334.

55 Muchardt, C., Guillemé, M., Seeler, J.-S., Trouche, D., Dejean, A., Yaniv, M., Coordinated methyl and RNA binding is required for chromatin localization of mammalian HP1α. *EMBO Rep.* **2002**, *3*, 975–981.

56 Piacentini, L., Fanti, L., Berloco, M., Perrini, B., Pimpinelli, S., Heterochromatin protein 1 (HP1) is associated with induced gene expression in *Drosophila* euchromatin. *J. Cell Biol.* **2003**, *161*, 707–714.

57 Volpe, T. A., Kidner, C., Hall, I. M., Teng, G., Grewal, S. I. S., Martienssen, R. A., Regulation of heterochromatic silencing and histone H3 lysine-9 methylation by RNAi. *Science* **2002**, *297*, 1833–1837.

58 Schramke, V., Allshire, R., Hairpin RNAs and retrotransposon LTRs effect RNAi and chromatin-based gene silencing. *Science* **2003**, *301*, 1069–1074.

59 Hilfiker, A., Hilfiker-Kleiner, D., Pannuti, A., Lucchesi, J. C., *mof*, a putative acetyl transferase gene related to the Tip60 and MOZ human gene and to the SAS genes of yeast, is required for dosage compensation in *Drosophila*. *EMBO J.* **1997**, *16*, 2054–2060.

60 Stokes, D. G., Perry, R. P., The DNA-binding and chromatin-localization properties of CHD1. *Mol. Cell. Biol.* **1995**, *15*, 2745–2753.

61 Stokes, D. G., Tartof, K. D., Perry, R. P., CHD1 is concentrated in interbands and puffed regions of *Drosophila* polytene chromosomes. *Proc. Natl. Acad. Sci. USA* **1996**, *93*, 7137–7142.

62 Simic, R., Lindstrom, D. L., Tran, H. G., Roinick, K. L., Costa, P. J., Johnson, A. D., Hartzog, G. A., Arndt, K. M., Chromatin remodeling protein Chd1 interacts with transcription elongation factors and localizes to transcribed genes. *EMBO J.* **2003**, *22*, 1846–1856.

63 Kelly, D. E., Stokes, D. G., Perry, R. P., CHD1 interacts with SSRP1 and depends on both its chromo domain and its ATPase/helicase-like domain for proper association with chromatin. *Chromosoma* **1999**, *108*, 10–25.

64 Bernstein, B. E., Humphrey, E. L., Erlich, R. L., Schneider, R., Bouman, P., Liu, J. S., Kouzarides, T., Schreiber, S. L., Methylation of histone H3 Lys 4 in coding regions of active genes. *Proc. Natl. Acad. Sci. USA* **2002**, *99*, 8695–8700.

65 Santos-Rosa, H., et al., Active genes are tri-methylated at K4 of histone H3. *Nature* **2002**, *419*, 407–411.

66 Litt, M. D., Simpson, M., Gaszner, M., Allis, C. D., Felsenfeld, G., Correlation between histone lysine methylation and developmental changes at the chicken β-globin locus. *Science* **2001**, *293*, 2453–2455.

Websites directly related to the subject of this chapter

http://www.uib.no/aasland/chromo.html

http://pfam.wustl.edu/cgi-bin/getdesc?name=chromo

http://smart.embl-heidelberg.de/smart/do_annotation.pl?BLAST=DUMMY&DOMAIN=CHROMO

http://bioinformatics.ccr.buffalo.edu/cgi-bin/pfam/getdesc?name=chromo

http://www.ii.uib.no/~inge/papers/mdl/chromo.top.html

http://us.expasy.org/cgi-bin/nicedoc.pl?PS00598

13
PDZ Domains: Intracellular Mediators of Carboxy-terminal Protein Recognition and Scaffolding

Laurence A. Lasky, Nicholas J. Skelton, and Sachdev S. Sidhu

13.1
Introduction

Protein–protein binding interfaces are involved with virtually every aspect of intra- and extracellular physiology. These interfaces range in size from the very large binding sites formed between antibody combining regions and their cognate antigens to the very small interfaces recognized by a variety of more compact motifs including, for example, the SH2 and SH3 domains (see Chapters 2 and 3). The vast majority of these protein modules bind to unmodified or post-translationally modified regions within proteins, as opposed to sites at the very amino- or carboxy-terminal ends of proteins. The potential benefits of recognition via the beginnings or ends of proteins are two-fold. First, recognition of these regions would allow for enhanced specificity by taking advantage of binding interactions involving the free ends of proteins. This might allow for a relatively small binding site, since specificity can be induced both by mainchain and sidechain interactions as well as by amino- or carboxy-type interactions. Second, recognition of terminal regions of proteins allows for the rest of the protein to be unencumbered by an internally bound protein-recognition motif. This would enable efficient scaffolding with other functionally related proteins together with simultaneous activity of the terminally-bound protein. It thus appeared likely to many investigators that scaffolding proteins that recognized the ends of other proteins were likely to exist. The discovery of a large family of proteins containing a compact domain, termed the PDZ domain, has fulfilled this expectation, at least with respect to C-terminal interactions.

Determination of the sequences of several large proteins in the early 1990s led to the discovery of a 90–100 amino acid long domain that was repeated within the proteins and was conserved in many other molecules derived from a variety of organisms and cell types. The initial discovery of this motif was in three polypeptides: the mammalian protein post-synaptic density-95 (PSD-95), the *Drosophila melanogaster* epithelial tumor suppressor protein, Discs Large (DLG), and the mammalian epithelial tight junction protein, zonula occludens-1 (ZO-1) [1–7]. Initial sequence comparisons revealed a reasonable degree of conservation between these diverse

Modular Protein Domains. Edited by G. Cesareni, M. Gimona, M. Sudol, M. Yaffe

ISBN: 3-527-30813-X

domains, including a highly conserved amino acid sequence, GLGF, and the motifs were initially referred to as GLGF domains. Subsequently, because this short sequence is not identical in all of these domains, as well as to honor the initial sites of discovery, the motifs were renamed PDZ domains (PSD-95, DLG, ZO-1).

Large-scale genomic analyses, together with in silico bioinformatics work, has allowed for the determination of the number of potential PDZ domains in the organisms for which we have reasonably complete genomic DNA sequence information. Interestingly, and in contrast to both total gene numbers and genome sizes, the numbers of PDZ domains in various organisms seem to increase dramatically with increasing complexity, although the number of genomes so far analyzed is far too small to be certain of the validity of this trend. However, there are ~90 domains in the nematode *Caenorhabditis elegans*, ~130 in the fly *D. melanogaster*, and over 400 in the human genome. Some of this increased complexity can be explained by increasingly larger gene families. For example, although the nematode appears to have only two members of the LAP (leucine-rich repeat and PDZ domain) family, higher organisms appear to have additional homologs encoded within their genomes. This appears to be so with many other PDZ domain-containing proteins, including DLG, ZO-type, the MAGI proteins, and LIN-7, among others. The potential reasons for this increased complexity, including tissue specificity, diversity of function, etc., remain to be fully elucidated.

PDZ domains are virtually always embedded in proteins that are assembled from multiple protein motifs (a complete list of PDZ-containing proteins together with other domains, functions, and annotations can be found at http://smart.embl-heidelberg.de/). These protein motifs can include other PDZ domains, and there are examples of proteins containing only PDZ domains, such as MUPP1, in which the protein consists of 13 PDZ motifs [8]. Other multi-PDZ domain-containing proteins include INAD, par-3, and NHERF. It seems likely that this type of multi-PDZ protein would be involved in the assembly of functionally related proteins via C-terminal recognition, and this has been proven both genetically and biochemically for the INAD protein of *Drosophila* (see below), as well as several others. A second large family of PDZ domain-containing proteins is the MAGUK (membrane-associated guanylate kinase) family. This large subgroup of PDZ-containing proteins contains the founding members of the PDZ family, and it is unified by the inclusion of a protein motif with distant homology to guanylate kinases (the GUK domain), although there is no evidence that this domain has enzymatic activity. Many of these proteins, including DLG, ZO-1, and the MAGI (membrane-associated guanylate kinases with inverted orientation) proteins, are associated with the tight junctions of epithelial cells and are presumably involved in assembly and maintenance of these important structures. This large family contains proteins with other potential protein interaction domains, including, for example, SH3, WW, (see chapters by Mayer and Saksela, and Sudol, respectively), calmodulin kinase (CAMK), and L27 motifs. Finally, a third large group of PDZ proteins contains a variety of other sequence motifs (but not a guanylate kinase-like domain) juxtaposed with one or more PDZ domains. Included in this family are proteins containing leucine-rich repeats (the LAP proteins), LIM, or crib motifs. A unifying feature of all of these PDZ-containing proteins is that the molecules contain a variety of protein interaction motifs with no clear evidence of enzymatic activity in any of the family members. These data, together with their localization to diverse intracellular sites, such as epithelial junctions and the neuronal post-synaptic density, suggest that the PDZ-containing proteins are predominately site-specific scaffolding proteins that assemble functional complexes, many of which appear to be important for the assembly, maintenance, and function of these subcellular anatomies.

13.2 Structural Analysis of PDZ Domains

Structural studies of many PDZ domains have identified a common 3D fold that undoubtedly provides the physical underpinning for the sequence motifs that can be used to identify the domains [2, 9]. Today, 34 independent structures of 20 different PDZ domains have been reported, and all consist of a six-stranded β barrel (strands βA through βF) capped by one short (α1) and one long (α2) helix (Figure 13.1) [10, 11]. Although the β-barrel core is common to all PDZ domain structures, there is sequence variation within the elements of regular secondary structure and also in the length and composition of the loops connecting them. In addition, a number of the domains also contain additional N- or C-terminal appendages that pack against the conserved core [9, 12, 13].

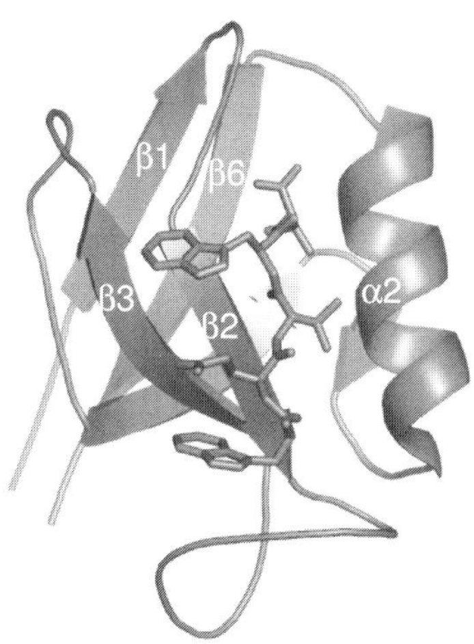

Figure 13.1 Schematic view of the PDZ domain fold. In this structure of the Erbin PDZ domain, β strands are shown as arrows and α helices as coils. The phage-derived peptide ligand (*ac*TGWETWV$_{COOH}$) is shown in stick form with the C-terminal carboxylate at the top of the view. Coordinates are taken from PDB accession number 1N7T.

Over half of the PDZ entries present in the structural database have a peptide ligand associated with the domain. In all instances, the GLGF motif of the domain wraps around the C-terminal carboxylate of the ligand [14]. The remainder of the ligand lies in the cleft between strand β2 and helix α2, with the peptide forming an antiparallel β-sheet interaction with β2. As a consequence of this fixed backbone hydrogen-bonding pattern, the ligand sidechains maintain a fixed register with respect to the PDZ domain and are able to contact only a limited subset of the domain sidechains (Figure 13.1). To standardize the categorization of all PDZ interactions, the ligand residues are numbered from 0 at the C terminus counting by increasingly negative integers toward the N terminus [15]. The importance of interactions at the 0 and –2 sites was appreciated from the earliest studies [16–19] and led to the classification of PDZ domains into two classes based on their ligand preference at the –2 site. Subsequent studies have shown that the –1 site can also make a crucial contribution to binding [20–23]. Moreover, PDZ domains having a preference for ligands with a particular –3 sidechain are common [16, 20–22, 24, 25], and preference for residues N-terminal to –3 have been observed, especially if the β2–β3 loop is long [13; 22; 26; 27]. Since the sidechains of ligand residues 0, –1, –2, and –3 all have relatively fixed interactions with their PDZ domain host, establishing the 'rules' for predicting ligand affinity and selectivity should be possible, given a suitable training set of PDZ domain–ligand interactions (e.g., see [28]).

In addition to the 'canonical' ligand binding mode, the co-complex between α-syntrophin and neuronal nitric oxide synthetase (nNOS) PDZ domains indicates that non–C-terminal interactions are also possible. In this particular example, an ancillary β hairpin at the N terminus of nNOS lies in the peptide binding groove of α-syntrophin with the reverse turn of the hairpin positioned approximately where a C-terminal carboxylate would ordinarily reside [29]. More recently, structural studies of the seventh PDZ domain from GRIP demonstrate that its peptide-binding groove is not well-formed [30]. NMR chemical shift mapping suggests that GRASP-1, the cognate ligand, binds to a hydrophobic surface formed by the strand β4–β5 hairpin and the face of helix α2 away from the peptide binding groove; precise details of the interaction were not reported. Finally, there have been reports that not all PDZ domains behave as independently folding units. Structures determined for tandem PDZ domains from PSD-95 and syntenin suggest that the domains can have a relatively fixed orientation with respect to each other, which may have implications for their function as scaffold proteins [31–33]. The quaternary structure observed for PDZ6 from GRIP-1 suggest alternative modes of such supramolecular assembly [34]. Finally, the PDZ4 and PDZ5 domains of GRIP-1 could not be expressed efficiently alone and had NMR spectra indicative of partial folding. However, the tandem domain expressed well and appeared highly folded by the same criteria [35]. Subsequent studies have shown that the two domains are intimately packed (with a different geometry than that observed in PSD-95 and syntenin tandem domains); the occluded peptide binding pocket of PDZ4 suggests that its primary function is not ligand binding, but rather, stabilization of the PDZ5 structure [36].

13.3 Analysis of PDZ Domain–Ligand Interactions with Mutagenesis and Synthetic Peptides

In vitro studies of PDZ domain–ligand interactions have proven invaluable for complementing and extending the knowledge acquired from structural analyses, and also for refining and even revising our views on physiologically relevant in vivo interactions. Studies with combinatorial peptide libraries, in particular, have been instrumental in defining the fine specificity of individual domains and revealing binding contributions from up to six C-terminal ligand residues. Early studies made use of synthetic peptide libraries [16] or a 'peptides-on-plasmids' library generated in *Escherichia coli* [24, 27–39], but more recently, C-terminal peptide libraries displayed on M13 [20] or lambda phage [40] have also been developed.

Peptides fused to the C terminus of the D-capsid protein of lambda phage resulted in high valency display that enabled the selection of even very low-affinity ligands for the seven PDZ domains of the human INAD-like (INADL) protein [40]. Different degrees of consensus were observed, with different PDZ domains selecting ligands with homology at two, three, or four C-terminal residues. Unfortunately, the resulting low homology data, which may have been hampered by the high display levels of the peptides, were not as enlightening as the genetic/biochemical data derived from the homologous *Drosophila* protein INAD (see below). In contrast, low valency display achieved through C-terminal fusion to the M13 major coat protein yielded high-affinity peptide ligands that bound to MAGI-3 PDZ2 with affinities in the submicromolar range and showed a clear consensus in all four C-terminal positions [20].

C-terminal M13 phage display was also used to study the binding specificity of the Erbin PDZ domain, which had originally been isolated as a putative binding partner for ErbB-2 in a yeast two-hybrid screen (see below) [41]. Interestingly, the Erbin PDZ-binding consensus defined by phage display ([D/E][T/S]WV$_{COOH}$) differed significantly from the C terminus of ErbB-2 (DVPV$_{COOH}$) and instead matched the conserved C termini of three p120-related catenins (DSWV$_{COOH}$). Subsequent in vitro affinity measurements with synthetic peptides showed that the Erbin PDZ domain binds to the catenin C-terminal sequence at least 100 times tighter than to the ErbB-2 C terminus. Although the yeast two-hybrid method is a powerful technology for discovering natural protein–protein interactions, experience with the Erbin PDZ domain shows that the method can be highly sensitive to even extremely low-affinity PDZ domain–ligand interactions and, thus, it is worthwhile corroborating yeast two-hybrid results with the results of other methods.

In a subsequent study, NMR spectroscopy, phage display, and in vitro affinity measurements were combined to provide a detailed analysis of the structure and function of the Erbin PDZ domain [22]. The NMR co-complex with a phage-optimized peptide revealed five distinct binding sites on the protein, which accommodated the five C-terminal ligand residues (Figure 13.2). Binding assays with a panel of synthetic peptides showed that each of the five ligand sidechains contributes to binding, with the last two sidechains and the C-terminal carboxylate providing the majority of the binding energy. A combinatorial mutagenesis strategy

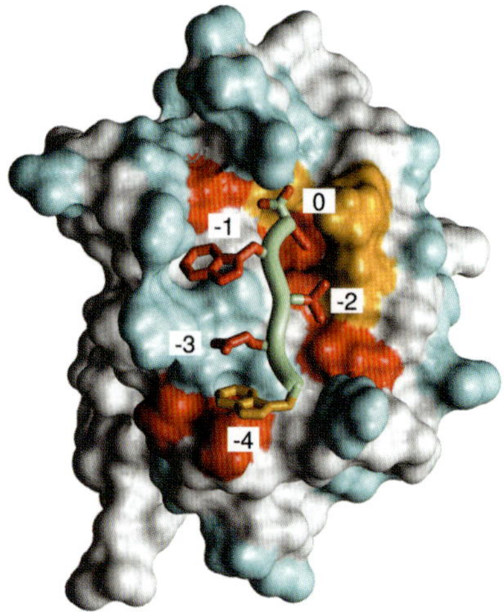

Figure 13.2 Summary of alanine-scanning mutagenesis data for Erbin PDZ domain binding to the phage-optimized peptide ligand. The solvent-accessible surface of Erbin residues that were mutated to alanine in a combinatorial phage library are colored according to their effect on ligand binding, with red, yellow, and blue indicating a large, moderate, or no effect on binding, respectively. Likewise, the ligand sidechains are colored according to the effect that peptide alanine substitutions had on binding to wild-type Erbin PDZ domain, with red and yellow indicating a greater than 100-fold or greater than 10-fold effect on IC_{50}, respectively.

was also used to assess the effects of alanine substitutions for 44 PDZ domain sidechains in and around the peptide-binding site. When the results of the peptide affinity and protein mutagenesis studies were mapped onto the NMR structure, they provided an extremely comprehensive view of the molecular elements involved in PDZ domain–ligand recognition (Figure 13.2).

More traditional point-mutation studies have been used to understand the binding of α-syntrophin to both C-terminal peptides and nNOS, and also to investigate the impact of charge substitutions on ligand binding [42, 43]. In an interesting extension of such an approach, Ranganathan and colleagues looked at sites of pairwise covariation among several hundred PDZ domains [44]. This analysis identified energetically coupled pathways emanating from the peptide-binding groove to sites on the opposite side of the domain. Experimental validation of these pathways was afforded by ligand-binding measurements in double-mutant cycles; their presence raises the possibility of functionally relevant allosteric processes associated with ligand binding [44]. Finally, progress has also been made on the development of computational algorithms for predicting and engineering PDZ domain specificities [28].

Given the importance of PDZ domains in mediating numerous protein assemblies, attempts have been made to antagonize their interactions so as to elicit a biological response. Optimal peptide ligands identified from phage display libraries have been used to such ends, to block the interaction between Erbin and δ-catenin [21]; such reagents were able to induce distinct phenotypes in neuron outgrowth assays (K. Kosik, personal communication). Small peptides have also been used to disrupt the in vivo interaction between an NMDA receptor and a PDZ domain of PSD-95 [45]. Recently, organic mimics of C-terminal peptide ligands have been proposed that can bind in a selective manner to MAGI-3 PDZ2 and inhibit formation of complexes with PTEN peptides [46]. In a further recent development, halothane anesthetics have been found to bind to PSD-95 and PSD-93 and to inhibit interaction with NMDA receptor or nNOS. The binding was localized to the peptide-binding groove, explaining the mechanism of antagonism of PDZ-mediated protein assembly, and possibly also explaining the mechanism by which these anesthetics exert their physiological response [47].

13.4 Molecular and Signaling Functions of PDZ Domains

Although an ever-increasing diversity of PDZ-mediated interactions has emerged in the literature (for example, see [7]), we focus here on a few examples of interactions that have been validated by biochemical as well as more physiological methods, particularly genetics. As outlined in the structure section above, many of the interactions determined by biochemical techniques, such as the yeast two-hybrid and coprecipitation methods, are encumbered with numerous potential artifacts, such as very low, probably nonphysiological, binding affinities. Thus, although it is clear that many of the interactions predicted by these approaches are undoubtedly correct, there are likely to be many incorrectly described interactions as well. Luckily, there are a variety of physiologically well characterized PDZ-mediated interactions that allow for a number of interesting conclusions regarding the diverse functions of these motifs.

13.4.1 INAD as a Molecular Scaffold

Anyone who has attempted to swat a fly cannot help but marvel at the insect's ability to escape certain death. This capability is in large part due to the rapid and efficient transduction of visual information. As in other organisms, vision in the fly *D. melanogaster* is initiated by the absorption of photons by the G protein-coupled receptor, rhodopsin. This results in a conformational change in the receptor, which induces the association with a Gq type of GTPase signaling protein. This interaction induces a modulation in downstream signaling molecules, including protein kinase C (PKC) and phospholipase C (PLC), which results in opening the transient receptor potential (TRP) and transient receptor potential-like (TRPL) calcium ion channels. The flow of ions through these channels is subsequently transmitted to the central nervous system, where the appropriate response to the initial visual input is engendered. Once the visual input signal is abolished, the system undergoes a negative feedback response to intracellular calcium concentrations by down-regulating the activities of rhodopsin and phospholipase C [48].

It is not surprising that the efficiency of visual signal transduction as regulated by this large number of signaling proteins would be enhanced by placing the proteins very close to one another. Genetic studies of strains of *Drosophila* carrying mutations at the *InaD* locus suggested that this gene was critical for proper function of the light-gathering ommatidial cells in the eye [49–52]. Morphological analysis of TRP channel localization demonstrated that this critical visual component appears to be mis-localized in the eyes of flies with various *InaD* mutations [53]. Isolation of the gene and analysis of the protein encoded by the *InaD* locus demonstrated that this gene encodes a molecule containing five PDZ domains [49–52]. This result immediately suggested the possibility that the INAD protein was involved in assembly of components of the visual signal transduction system into a macromolecular structure that enhanced signaling efficiency by closely associating these components. Biochemical analysis has demonstrated that at least seven proteins bind to the INAD protein, with many of the proteins binding via an interaction between their C termini and the INAD PDZ domains [38, 54–59]. These results suggest that INAD functions as a scaffold to assemble the components of the fly visual signal transduction system into a 'signalosome' (Figure 13.3). Functional analyses suggest, as expected, that the INAD signalosome functions to enhance signaling efficiency, since mutations in the scaffolding protein result in significantly increased latency between visual stimulus and cellular response. Furthermore, this macromolecular complex appears to be also involved in termination of the visual response [62, 63]. In summary, INAD meets the expectations that multi-PDZ proteins can function as scaffolds to assemble macromolecular complexes that enhance signaling.

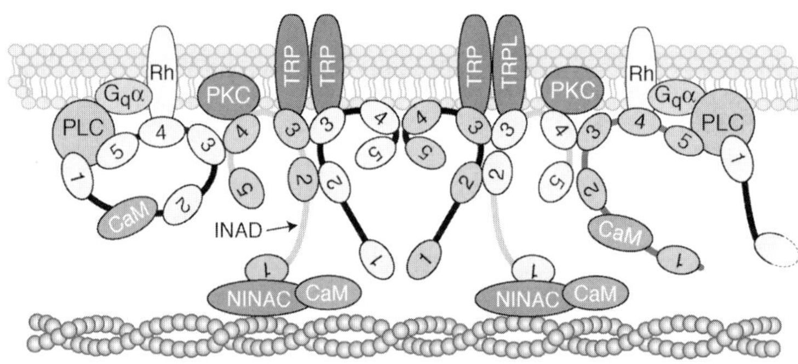

Figure 13.3 The *Drosophila* visual signal transduction complex is assembled by the multi-PDZ protein INAD (adapted from [48]). This figure illustrates that various cell-surface and intracellular proteins involved in signal transduction, including G proteins (Gqα), phospholipases (PLC), ion channels (TRP), are physically brought together by interactions between their C termini and the multi-PDZ protein INAD. The complex is maintained in an appropriate subcellular location by interactions between the NINAC protein and the cytoskeleton (F-actin). In addition, PDZ–PDZ interactions appear to hold separate complexes together into a supercomplex, although this type of interaction is not mediated by C-terminal binding.

13.4.2
LIN-7–Receptor Tyrosine Kinase Interactions and Subcellular Localization

Vulval development in *C. elegans* has provided an elegant system for analysis of the genetics and biochemistry of complex organ formation [62, 63]. Important early work in this system demonstrated that a homolog of the epidermal growth factor (EGF) receptor family, called LET-23, was critical for the formation of this intricate structure [64]. The activation of this receptor by LIN-3, an EGF-like molecule expressed by basolaterally localized gonadal anchor cells, suggested that the receptor was found on the basolateral surface of the vulval precursor cells, and in fact, this was found to be true [65]. The asymmetric basolateral orientation of this receptor was critical for completion of the vulval developmental program, since it allowed the receptor access to the adjacently produced activating factor. It appeared likely that the asymmetric localization of LET-23 was accomplished by a combination of both specific intracellular trafficking and retention mechanisms.

An elegant analysis by the Kim laboratory soon provided strong evidence that the basolateral localization of LET-23 was accomplished by an interaction with a PDZ domain-containing protein, LIN-7 [66, 67]. This protein was found to be associated in a complex with two other PDZ domain-containing proteins, LIN-2 and LIN-10 [66, 68]. The assembly of this complex did not involve PDZ-binding interactions, a finding that suggested that the PDZ domains within the complex were free to interact with the C termini of other proteins. Importantly, it was clearly demonstrated that the C terminus of LET-23 was a type 1 PDZ-binding motif (TCL_{COOH}), and it was established that this motif bound to the PDZ domain of LIN-7 [66–68]. Further genetic studies suggested that this interaction was critical for basolateral targeting of LET-23 and vulval development, providing the first molecular and genetic evidence for the role of PDZ-containing proteins in the subcellular targeting of signaling receptors in an epithelial cell [66–68] (Figure 13.4). Additional work suggested that the role of this interaction was to inhibit internalization of the bound receptor, consistent with the suggestion that the LIN-7–PDZ interaction was required for retention of the receptor at the basolateral surface of the polarized epithelial cell [66–68].

Because *C. elegans* LET-23 is a prototype of the EGF receptor family of higher organisms, it was important to establish the mechanisms by which these receptors, which include the oncogenically associated HER2-neu/ErbB-2 receptor, are asymmetrically maintained in polarized epithelial cells. An interesting initial study in this field was that of Borg and colleagues, who showed that Erbin, a member of the LAP family of PDZ-containing proteins, was involved in delivery and/or retention of the ErbB-2 receptor at the basolateral surface of epithelial cells [41]. However, as described above, further work demonstrated that the affinity of the ErbB-2–Erbin interaction was likely to be far too low to efficiently accomplish intracellular binding [21, 22]. In addition, several laboratories reported a high-affinity set of protein ligands for the Erbin PDZ domain, which included a variety of p120-type catenins [21, 69–71]. It thus appeared likely that Erbin was not involved in subcellular targeting of ErbB-2 [72], although initial data of Borg et al. did suggest that the ErbB-2

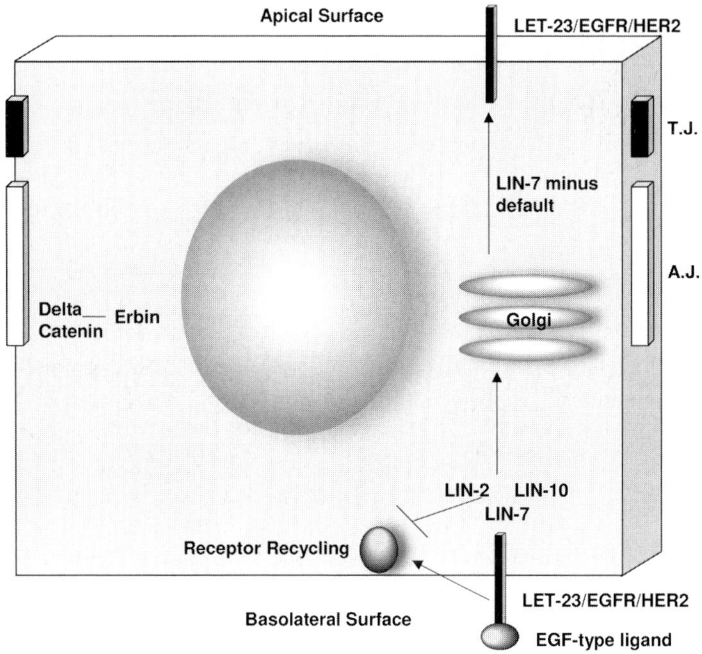

Figure 13.4 The EGF-like receptor LET-23 is localized in the basolateral region of the epithelial cell by the LIN-2, -7, -10 complex of PDZ proteins. This figure illustrates the role of the multi-PDZ protein complex containing LIN-2, -7, and -10 in mediating the subcellular localization of the *Caenorhabditis elegans* EGF-like receptor LET-23. LET-23 is kept in the basolateral part of the vulval epithelial progenitor cell, where it binds to an EGF-like ligand and induces differentiation. LIN-7 binds to both the C terminus of LET-23 and potentially, by analogy with the mammalian system, to a site in the kinase domain and determines trafficking by the golgi complex to the basolateral region. The PDZ complex is also involved in the inhibition of receptor–ligand endocytosis. In the absence of LIN-7, the default trafficking pathway for LET-23 is to the apical region of the cell. The figure also illustrates that Erbin, a PDZ protein previously thought to be involved in the subcellular localization of a LET-23–related receptor, HER2-neu (erb2), is now thought to interact with catenin-like proteins at the adherens junction (A. J.) of the epithelial cell (see Figure 13.5). Finally, the figure also suggests that a similar trafficking system is involved in the appropriate subcellular localization of other EGF-like receptors, including those in the EGFR family (i.e., erb2, EGFR, etc.).

C terminus was critical for basolateral localization [41]. This conundrum was recently resolved in a report that clearly demonstrated that the mammalian homolog of the LIN-7 protein was involved in the basolateral targeting and retention of the ErbB-2 protein [73]. Importantly, these investigators demonstrated that this targeting was accomplished by a bipartite signal in LIN-7, with an N-terminal domain involved in basolateral sorting and the PDZ domain involved in basolateral retention (Figure 13.4). Thus, the subcellular targeting of the EGF receptor tyrosine kinases, as well as, potentially, other receptor kinases, appears to involve PDZ domain interactions that are conserved from worms to humans [74]. Finally, it is likely that appropriate subcellular localization of other cell surface signaling molecules, such as for example, TGFα and the cystic fibrosis transmembrane regulator, is accomplished through interactions with PDZ-containing proteins such as GRASP and NHERF or CAP-70, respectively [75, 76].

13.4.3
PDZ Domain Proteins and Epithelial Polarity Induction and Maintenance

The contrast between the aesthetically beautiful structure of the polarized epithelium and the grossly malformed appearance of the oncogenically transformed tissue could not be greater. In general, oncogenic transformation of epithelial cells is accompanied by a loss of normal polarization accompanied by increased proliferation, loss of the normal flat, single-cell epithelial morphology, and invasion of adjacent normal tissues by tumor cells [77]. Epithelial cells separate their polarized apical and basolateral regions by using complex junctional structures composed of adherens junctions and tight (septate) junctions [78]. The molecular mechanisms involved in formation of the polarized epithelium are complex and involve a multitude of intracellular and extracellular proteins, including a variety of signaling, scaffolding, and adhesion molecules. Notably, PDZ domain-containing molecules play major roles in the appropriate assembly and maintenance of the epithelium of all metazoan organisms [79–82].

It is not surprising that the powerful genetics of lower metazoan organisms, such as the fly and the nematode, have played a critical part in elucidating the roles of various PDZ domain-containing molecules in the assembly and maintenance of the epithelium [78]. Interest in PDZ proteins was stimulated early on, when it was found that mutations in two different molecules, discs large (DLG), a MAGUK-type PDZ protein, and Scribble, a LAP-type PDZ protein, resulted in the loss of epithelial integrity and a tumorigenic phenotype in Drosophila [83–87]. Furthermore, of the ~50 known tumor suppressors in Drosophila, these two loci are the only places in which mutation leads to neoplasia in both imaginal discs and brain. These exciting results suggested that PDZ-containing molecules were likely to be involved in the assembly and/or maintenance of epithelial structures and that their loss correlated with a lack of morphological integrity as well as cellular hyperproliferation and invasion, all hallmarks of oncogenically transformed cells. Importantly, further analysis revealed that not every cell with a PDZ protein mutation gave rise to a tumor, suggesting the need for additional oncogenic insults.

Scribble and DLG both appear to be localized in the basolateral region of the epithelial cell. Another group of PDZ proteins, including the MAGUK protein Stardust (PALS-1), and the multi-PDZ protein Bazooka (Par-3) appear to be involved in formation of apical regions of the epithelial cell [88–90]. Elegant genetic and biochemical work from the Perrimon group has gone the farthest toward elucidating the roles of these various PDZ proteins in epithelial formation and maintenance [91]. These studies demonstrated that Bazooka is involved in the 'apicalization' of the cell, while the Scribble complex served to antagonize the Bazooka-induced formation of apical regions by repressing apicalization along the basolateral side of

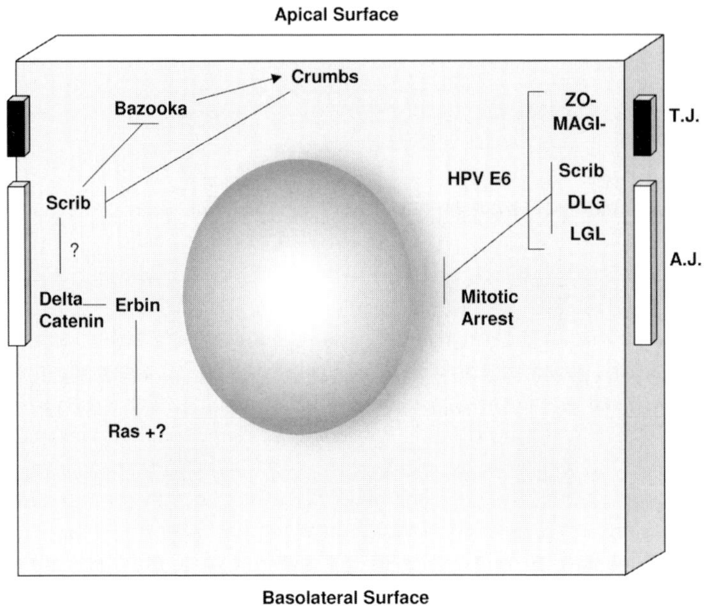

Figure 13.5 The apical–basolateral polarity of epithelial cells is in large part determined by a variety of PDZ proteins. Elegant genetic analyses in *Drosophila* and *Caenorhabditis* have demonstrated that a variety of PDZ-containing proteins, including Scribble (scrib), Discs Large (DLG), Lethal Giant Larvae (LGL), Erbin, and Bazooka are involved in the induction and maintenance of the apical–basal polarity of the epithelial cell. Scrib, LGL, and DLG are localized to the adherens junction, and their mutation leads to loss of apical–basal polarity and hyperproliferation that is tumor-like. Importantly, the oncogenic HPV viruses encode an E6 protein that is involved in mediating the degradation of these adherens junction components. Crumbs, an apical cell surface protein that appears to induce apicalization, interacts with Bazooka via a PDZ-mediated binding event. Scrib induces a basolateral polarity, and the balance between scrib and the Bazooka–Crumbs complex appears to control apical–basal polarization. Erbin interacts via a PDZ-mediated interaction with p120-like catenins (i.e., δ-catenin), and this interaction has been shown to regulate epithelial formation in *C. elegans*, possibly via an interaction with Ras-type GTPases. Finally, tight junction assembly and maintenance appears to be under the control of the ubiquitously expressed (in epithelia) ZO-1 and MAGI family of MAGUK-type PDZ proteins, which are also sensitive to E6-mediated degradation.

the cell. These data were consistent with the subcellular localization of each of these proteins. Earlier work had demonstrated that the overexpression of a cell-surface protein called Crumbs increased apical formation, and a PDZ-mediated complex between Crumbs and the Stardust MAGUK counteracted the basolateral induction by Scribble [92–95]. Together, these data suggest the importance of PDZ domain proteins in the determination of this critical subcellular morphology, although the molecular mechanisms by which this control is mediated are still poorly understood (Figure 13.5). Although some of the PDZ interactions involved in this important signaling pathway have been elucidated, it will be fascinating to use newer methods, such as C-terminal phage display [20], to further dissect the molecular interactions induced by these proteins. In addition, a variety of other proteins, including the ZO-type MAGUKs as well as the MAGI family of proteins (for example, see [96]), are localized to the tight junctions of virtually all epithelial cells and are therefore likely to play an important role in epithelialization. Finally, although genetic evidence for a role for these PDZ proteins in human tumor formation is still lacking, recent data suggest that oncogenic human papilloma viruses (HPV) target a variety of PDZ proteins, including Scribble, DLG, and the MAGI proteins, for ubiquitin-mediated proteasome degradation [97–101]. This targeting occurs via interactions between the C terminus of viral E6 protein, a ubiquitin ligase complex-associated protein, and various PDZ domains, suggesting that viruses can hijack cellular PDZ domains for their own nefarious purposes. A hallmark of HPV infection is loss of epithelial structure, hyperproliferation, and invasion again consistent with a role of these PDZ proteins in epithelial control.

Although Scribble, a LAP family member, has clearly been shown to be critical for epithelial formation, other members of this family, including Erbin, are also likely to be important for epithelialization. The strongest genetic data for the function of Erbin comes from C. elegans, where the LET-413 protein, a likely homolog of mammalian Erbin, was clearly shown to be involved in epithelial formation [102, 103]. As described above, the probable binding partners for the Erbin PDZ domain are a group of p120-like catenins, including p0071, δ-catenin, and ARVCF, all of which contain a conserved C terminus ($DSWV_{COOH}$), which was found to be a high-affinity Erbin PDZ-binding peptide [21, 22, 69–71]. Other important genetic data from C. elegans have recently been reported regarding JAC-1, the nematode homolog of the mammalian p120-related catenins. Mutation of this protein gives an epithelial phenotype that is reminiscent of the LET-413 mutations [104]. This result becomes even more interesting when viewed in the context of the work of Laura and colleagues, who predicted that the PDZ domain of LET-413 should bind with high affinity to the C terminus of JAC-1 ($DSWV_{COOH}$) [21]. Together, these data suggest that a complex between LET-413 and JAC-1 is likely to be involved in epithelial regulation. Although the mechanism of this regulation remains to be fully elucidated, recent work by Huang and colleagues has demonstrated that the leucine-rich repeats of Erbin appear to be involved in the regulation of Ras GTPase activity [105, 106]. Epithelial integrity and intracellular communication are, in part, mediated by the cadherin family of adhesion molecules. Because Erbin is linked to this adhesion system via its PDZ-mediated catenin binding, the data of Huang et al. suggest a mechanism whereby epithelial adhesion might regulate Ras subcellular localization and activity [107].

13.4.4 A Few Miscellaneous Examples: The Synapse, Disheveled, CARD MAGUKs, and Beta Adrenergic Receptors

In addition to the polarized epithelium, a second highly polarized cell type is the axon. In this cell type, the synapse corresponds to the highly polarized apical region of the epithelial cell, and it has been demonstrated that the synaptic and postsynaptic regions of neurons contain a variety of signaling molecules, including neural transmitters and the many channels that they bind to. Importantly, the subcellular localization, signaling efficiency, and signalosome assembly of the various synaptic and post-synaptic signaling proteins appear to be accomplished by a variety of PDZ proteins [11, 108, 109]. Many potential PDZ-mediated interactions in neurons have been identified, but we discuss only two examples for which there is some genetic analysis of their function. By biochemical studies, a variety of excitatory receptors, including NMDA and glutamate receptors, have been found to be associated with PSD-95, one of the prototypical MAGUK-type PDZ proteins, and it has been assumed that their clustering and/or localization to the post-synaptic density requires an interaction with this MAGUK. Murine gene knockout studies of PSD-95, however, revealed that these receptors were correctly localized, but their signaling pathways with respect to synaptic plasticity seemed affected, although subtly [110]. Other ion channels, such as the potassium channel, also appear to interact via a C-terminal–PDZ domain interaction [11, 108, 109]. This interaction appeared, in cell culture assays, to be critical for the clustering and activity of these channels. However, recent murine knockout experiments have shown that the localization of these channels to the juxtaparanodal regions of myelinated axons does not require PSD-95 [111]. Thus, although PSD-95 is colocalized to these sites with the potassium channels, its presence is clearly not required for the localization of these channels. Together, these two examples highlight the importance of in vivo analysis in assessing the true function of PDZ-containing proteins.

The signaling pathway regulated by the Wnt family of proteins has emerged as one of the most critical for numerous aspects of development, including the maintenance and expansion of stem cells. A critical component of this pathway is disheveled (Dsh/Dvl), a scaffolding protein containing a PDZ domain as well as a variety of other protein interaction motifs [112]. This protein is a critical component of the numerous pathways controlled by Wnt signaling, and mutations of the *Dsh/Dvl* gene result in various phenotypic changes in these pathways that mimic the loss of either Wnt protein or Wnt receptors [113–115]. Genetic analyses demonstrated that Dsh/Dvl was downstream from the Wnt/Wnt receptor but upstream from the β-catenin destruction complex, suggesting that Dsh/Dvl plays an important role in regulating this complex and its transcriptional activator, β-catenin. The role that the single Dsh/Dvl PDZ domain plays in regulating the Wnt pathway is controversial and appears to depend upon assay conditions, although much of this heterogeneity may be due to differences in the types of PDZ mutants that have been examined. Oddly, a bewildering array of proteins have been found to bind to the Dsh/Dvl PDZ domain [112]. Interestingly, these proteins appear to all have completely different C termini, suggesting that either this PDZ domain is unusually promiscuous or that most (or all) of these binding partners are artifactual. Recent data from our laboratory using the C-terminal phage-display method suggest that peptides that bind with high affinity to the Dsh/Dvl PDZ domain are quite different from the C-terminal sequences of these putative protein ligands (Zhang and Sidhu, unpublished observations). Thus, as with many other PDZ domain interactions reported in the literature, careful analysis of binding affinities is required to determine the validity of potential PDZ binding partners. Finally, a recently published study suggests that Dsh/Dvl is overexpressed in mesothelioma tumor cells, and this overexpression enhances β-catenin levels and cellular transformation [116].

As described above, the MAGUK-type PDZ domain-containing proteins are involved in the assembly of complexes that appear to be involved in polarization in both epithelial and neuronal cells. The CARD MAGUKs are a recently described family of proteins that contain MAGUK-type domains (including PDZ, SH3, and guanylate kinase domains) as well as a CARD (caspase recruitment domain) domain and a coiled-coil motif [117–121]. These proteins are encoded by multiple genes in mammals, and early data demonstrated that one CARMA-1 is expressed in cells of the immune system. Interesting early work demonstrated the function of CARMA-1 in the stabilization and activation of NFκB transcription factors, suggesting that this MAGUK may be involved in immune system function. This hypothesis was spectacularly supported by data from two groups who inactivated the CARMA-1 gene by using traditional knockout as well as a novel ENU-mediated point-mutation technique [122, 123]. Although some of the resultant phenotypes were different (likely due to the fact that one group introduced a point mutation in the coiled-coil domain and the other completely abolished expression), both types of mutant mice showed profound defects in immune system function, including loss of antigen receptor signaling in both B and T cells. Further analysis of the defects demonstrated that these mutant animals had defects in NFκB activation that appeared to be due to stabilization of the IκB inhibitor. Together, these data suggest that CARD MAGUKs are involved in antigen receptor signaling by controlling the stability of the inhibitor of NFκB. One possibility is that antigen receptor activation induces CARMA-1 to bring a protein degradation complex close to the IκB inhibitor. If the PDZ domain is critical to this event, inhibition of its binding activity by small-molecule antagonists may result in a decreased immune response, a potentially important result for diseases ranging from autoimmune syndromes to transplantation reactions.

Heart failure is a major unmet medical problem with little or no effective treatment. The three beta-adrenergic receptors are G-coupled proteins that are regulated by the sympathetic nervous system, and it is likely that their modulation plays a role in the pathogenesis of heart failure [124]. Both beta 1 and 2 adrenergic receptors appear to be localized to specific subcellular localizations and are associated with downstream intracellular signaling molecules including, for example, G proteins, beta arrestin, and various kinases. This is highly reminiscent of other proteins that utilize PDZ domain-containing molecules to accomplish subcellular localization and signalosome assembly, and it has been shown that the C-terminal PDZ binding motifs of these two G-coupled proteins associate with PSD-95/MAGI-2 (beta 1) and the sodium–hydrogen exchange regulatory factor (NHERF) (beta 2) [125, 126]. Two major functional aspects of these interactions appear to be the control of subcellular localization and endocytosis as well as coupling to various G proteins [127, 128]. Interestingly, mutation of the PDZ binding site in the beta 2 receptor or blocking the interaction with a C-terminally derived peptide both result in a change in intracellular coupling that is accompanied by increased contractility in response to ligand binding [129]. Increased contractility might be a useful response for heart failure patients, although other data suggest that modulation of intracellular coupling in this manner might also be detrimental to myocyte survival. In summary, these results provide for yet another interesting mechanism by which PDZ-containing scaffold or localization proteins play a critical role in cellular physiology and which might provide for a clinically useful application.

13.5 Concluding Remarks

PDZ domains and the proteins they are found embedded in have emerged as major components of mechanisms for subcellular localization, signaling, and polarity induction. Although this chapter has not exhaustively discussed the huge number of reported PDZ interactions, we have attempted to discuss representative examples that have strong biological and/or genetic support. In addition, a number of examples exist, with more being presented all the time, which suggest that PDZ domains may be involved in a variety of pathogenic situations. For example, disruption of the interaction between the C terminus of an NMDA receptor and a PDZ domain of PSD-95 by a small peptide results in an inhibition of glutamate-mediated damage in an ischemic stroke model [45], suggesting that the detrimental effects of excitotoxicity may depend on PDZ-mediated receptor localization and/or signaling. The structural data discussed here suggest that the interface between the PDZ domain and its ligand(s) is likely to be among the smallest of protein–protein binding sites. Thus, although the dream of disrupting protein–protein interactions appears quite daunting for most targets, the possibility of identifying and enhancing the binding activity of small-molecule antagonists of PDZ-mediated interactions seems quite likely. It will be interesting in the future to identify such small-molecule antagonists in both model and disease-related systems and to examine their effects on PDZ-mediated biology. If these studies are successful, these inhibitors might be among the first drugs to target pathogenic protein interfaces.

Acknowledgments

We thank David Wood for help with the figures.

References

1 Cho, K. O., Hunt, C. A., Kennedy, M. B., The rat brain postsynaptic density fraction contains a homolog of the *Drosophila* discs-large tumor suppressor protein. *Neuron* **1992**, *9*, 929–942.

2 Kennedy, M. B., Origin of PDZ (DHR, GLGF) domains. *Trends Biochem. Sci.* **1995**, *20*, 350.

3 Fanning, A. S., Anderson, J. M., Protein modules as organizers of membrane structure. *Curr. Opin. Cell Biol.* **1999**, *11*, 432–439.

4 Kornau, H. C., Seeburg, P. H., Kennedy, M. B., Interaction of ion channels and receptors with PDZ domain proteins. *Curr. Opin. Neurobiol.* **1997**, *7*, 368–373.

5 Woods, D. F., Bryant, P. J., Zo-1, DlgA and PSD95/SAP90: homologous proteins in tight, septate and synaptic cell junctions. *Mech. Dev.* **1993**, *44*, 889.

6 Kim, E., Niethammer, M., Rothschild, A., Jan, Y. N., Sheng, M., Clustering of Shaker-type K^+ channels by interaction with a family of membrane-associated guanylate kinases. *Nature* **1995**, *378*, 85–88.

7 Nourry, C., Grant, S. G. N., Borg, J. P., PDZ domain proteins: Plug and play! **2003**, *Sci. STKE* 2003, re7.

8 Ullmer, C., Schmuck, K., Figge, A., Lubbert, H., Cloning and characterization of MUPP1, a novel PDZ domain protein. *FEBS Lett.* **1998**, *424*, 63–68.

9 Morais Cabral, J. H., et al., Crystal structure of a PDZ domain. *Nature* **1996**, *382*, 649–652.

10 Harris, B. Z., Lim, W. A., Mechanism and role of PDZ domains in signaling complex assembly. *J. Cell Sci.* **2001**, *114*, 3219–3231.

11 Sheng, M., Sala, C., PDZ domains and the organization of supramolecular complexes. *Annu. Rev. Neurosci.* **2001**, *24*, 1–29.

12 Daniels, D. L., Cohen, A. R., Anderson, J. M., Brunger, A. T., Crystal structure of the hCASK PDZ domain reveals the structural basis of class II PDZ domain target recognition. *Nat. Struct. Biol.* **1998**, *5*, 317–325.

13 Tochio, H., Hung, F., Li, M., Bredt, D. S., Zhang, M., Solution structure and backbone dynamics of the second PDZ domain of postsynaptic density-95. *J. Mol. Biol.* **2000**, *295*, 225–237.

14 Doyle, D. A., Lee, A., Lewis, J., Kim, E., Sheng, M., MacKinnon, R., Crystal structures of a complexed and peptide-free membrane protein-binding domain: molecular basis of peptide recognition by PDZ. *Cell* **1996**, *85*, 1067–1076.

15 Aasland, R., et al., Normalization of nomenclature for peptide motifs as ligands of modular protein domains. *FEBS Lett.* **2002**, *513*, 141–144.

16 Songyang, Z., et al., Recognition of unique carboxyl-terminal motifs by distinct PDZ domains. *Science* **1997**, *275*, 73–77.

17 Niethammer, M., Kim, E., Sheng, M., Interaction between the C-terminus of NMDA receptor subunits and multiple members of the PSD-95 family of membrane-associated guanylate kinases. *J. Neurosci.* **1996**, *16*, 2157–2163.

18 Matsumine, A., et al., Binding of APC to the human homolog of the *Drosophila* discs large tumor suppressor protein. *Science* **1996**, *272*, 1020–1023.

19 Kornau, H. C., Schenker, L. T., Kennedy, M. B., Seeburg, P. H., Domain interaction between NMDA receptor subunits and the postsynaptic density protein PSD-95. *Science* **1995**, *269*, 1737–1740.

20 Fuh, G., Pisabarro, M. T., Li, Y., Quan, C., Lasky, L. A., Sidhu, S. S., Analysis of PDZ domain–ligand interactions using carboxyl-terminal phage display. *J. Biol. Chem.* **2000**, *275*, 21486–21491.

21 Laura, R. P., et al., The Erbin PDZ domain binds with high affinity and specificity to the carboxyl termini of delta-catenin and ARVCF. *J. Biol. Chem.* **2002**, *277*, 12906–12914.

22 Skelton, N. J., et al., Origins of PDZ domain specificity: structure determination and mutagenesis of the Erbin PDZ domain. *J. Biol. Chem.* **2003**, *278*, 7645–7654.

23 Karthikeyan, S., Leung, T., Ladias, J. A., Structural basis of the Na+/H+ exchanger regulatory factor PDZ1 interaction with the carboxyl-terminal region of the cystic fibrosis transmembrane conductance regulator. *J. Biol. Chem.* **2001**, *276*, 19683–19686.

24 Stricker, N. L., et al., PDZ domain of neuronal nitric oxide synthase recognizes novel C-terminal peptide sequences. *Nat. Biotech.* **1997**, *15*, 336–342.
25 Schultz, J., Hoffmuller, U., Krause, G., Ashurst, J., Macias, M. J., Schmieder, P., Schneider-Mergener, J., Oschkinat, H., Specific interactions between the syntrophin PDZ domain and voltage-gated sodium channels. *Nat. Struct. Biol.* **1998**, *5*, 19–24.
26 Kozlov, G., Banville, D., Gehring, K., Ekiel, I., Solution structure of the PDZ2 domain from cytosolic human phosphatase hPTP1E complexed with a peptide reveals contribution of the β2–β3 loop to PDZ domain–ligand interactions. *J. Mol. Biol.* **2002**, *320*, 813–820.
27 Birrane, G., Chung, J., Ladias, J. A., Novel mode of binding by the Erbin PDZ domain. *J. Biol. Chem.* **2003**, *278*, 1399–1402.
28 Reina, J., Lacroix, E., Hobson, S. D., Fernandez-Ballester, G., Rybin, V., Schwab, M. S., Serrano, L., Gonzalez, C., Computer-aided design of a PDZ domain to recognize new target sequences. *Nat. Struct. Biol.* **2002**, *9*, 621–627.
29 Hillier, B. J., Christopherson, K. S., Prehoda, K. E., Bredt, D. S., Lim, W. A., Unexpected modes of PDZ domain scaffolding revealed by structure of nNOS–syntrophin complex. *Science* **1999**, *284*, 812–815.
30 Feng, W., Fan, J.-S., Jiang, M., Shi, Y.-W., Zhang, M., PDZ7 of glutamate receptor interacting protein binds to its target via a novel hydrophobic surface area. *J. Biol. Chem.* **2002**, *277*, 41140–41146.
31 Kang, G., Cooper, D., Devedjiev, Y., Derewenda, U., Derewenda, Z., Molecular roots of degenerate specificity in syntenin's PDZ2 domain: reassessment of the PDZ recognition paradigm. *Structure* **2003**, *11*, 845–853.
32 Kang, G., Cooper, D., Jelen, F., Devedjiev, Y., Derewenda, U., Dauter, Z., Otlewski, J., Derewenda, Z. S., PDZ tandem of human syntenin: crystal structure and functional properties. *Structure* **2003**, *11*, 459–468.
33 Long, J.-F., Tochio, H., Wang, P., Fan, J.-S., Sala, C., Niethammer, M., Sheng, M., Zhang, M., Supramolecular structure and synergistic target binding of the N-terminal tandem PDZ domains of PSD-95. *J. Mol. Biol.* **2003**, *327*, 203–214.
34 Im, Y., Park, S., Rho, S.-H., Lee, J. H., Kang, G., Sheng, M., Kim, E., Eom, S., Crystal structure of GRIP1 PDZ6-peptide complex reveals the structural basis for class II PDZ target recognition and PDZ domain-mediated multimerization. *J. Biol. Chem.* **2003**, *278*, 8501–8507.
35 Zhang, Q., Fan, J.-S., Zhang, M., Interdomain chaperoning between PDZ domains of GRIP. *J. Biol. Chem.* **2001**, *276*, 43216–43220.
36 Feng, W., Shi, Y., Li, M., Zhang, M., Tandem PDZ repeats in glutamate receptor-interacting proteins have a novel mode of PDZ domain-mediated target binding. *Nat. Struct. Biol.* **2003**, *10*, 972–978.
37 Stricker, N. L., Schatz, P., Li, M., Using the Lac repressor system to identify interacting proteins. *Methods Enzymol.* **1999**, *303*, 451–468.
38 van Huizen, R., et al., Two distantly positioned PDZ domains mediate multivalent INAD–phospholipase C interactions essential for G protein-coupled signaling. *EMBO J.* **1998**, *17*, 2285–2297.
39 Wang, S., Raab, R. W., Schatz, P. J., Guggino, W. B., Li, M., Peptide binding consensus of the NHE-RF-PDZ1 domain matches the C-terminal sequence of cystic fibrosis transmembrane conductance regulator (CFTR). *FEBS Lett.* **1998**, *427*, 103–108.
40 Vaccaro, P., Brannetti, B., Montecchi-Palazzi, L., Philipp, S., Citterich, M. H., Cesareni, G., Dente, L., Distinct binding specificity of the multiple PDZ domains of INADL, a human protein with homology to INAD from *Drosophila melanogaster*. *J. Biol. Chem.* **2001**, *276*, 42122–42130.
41 Borg, J. P., et al., ERBIN: a basolateral PDZ protein that interacts with the mammalian ERBB2/HER2 receptor. *Nat. Cell Biol.* **2000**, *2*, 407–414.
42 Harris, B. Z., Hillier, B. J., Lim, W. A., Energetic determinants of internal motif recognition by PDZ domains. *Biochemistry* **2001**, *40*, 5921–5930.
43 Harris, B., Lau, F., Fujii, N., Guy, R., Lim, W. A., Role of electrostatic inter-

actions in PDZ domain ligand recognition. *Biochemistry* **2003**, *42*, 2797–2805.

44 Lockless, S. W., Ranganathan, R., Evolutionarily conserved pathways of energetic connectivity in protein families. *Science* **1999**, *286*, 295–299.

45 Aarts, M., et al., Treatment of ischemic brain damage by perturbing NMDA receptor–PSD-95 protein interactions. *Science* **2002**, *298*, 846–850.

46 Fujii, N., Haresco, J., Novak, K., Sokoe, D., Kuntz, I., Guy, R., A selective irreversible inhibitor targeting a PDZ protein interaction domain. *J. Am. Chem. Soc.* **2003**, *125*, 12074–12075.

47 Fang, M., et al., Synaptic PDZ domain-mediated protein interactions are disrupted by inhalation anesthetics. *J. Biol. Chem.* **2003**, *278*, 36669–36675.

48 Montell, C., Visual transduction in *Drosophila*. *Ann. Rev. Cell Dev. Biol.* **1999**, *15*, 231–268.

49 Huber, A., Sander, P., Gobert, A., Bahner, M., Hermann, R., Paulsen, R., The transient receptor potential protein (Trp), a putative store-operated Ca^{2+} channel essential for phosphoinositide-mediated photoreception, forms a signaling complex with NorpA, InaC and InaD. *EMBO J.* **1996**, *15*, 7036–7045.

50 Huber, A., Sander, P., Paulsen, R., Phosphorylation of the InaD gene product, a photoreceptor membrane protein required for recovery of visual excitation. *J. Biol. Chem.* **1996**, *271*, 11710–11717.

51 Saras, J., Heldin, C. H., PDZ domains bind carboxy-terminal sequences of target proteins. *Trends Biochem. Sci.* **1996**, *21*, 455–458.

52 Shieh, B. H., Niemeyer, B., A novel protein encoded by the InaD gene regulates recovery of visual transduction in *Drosophila*. *Neuron* **1995**, *14*, 201–210.

53 Chevesich, J., Kreuz, A. J., Montell, C., Requirement for the PDZ domain protein, INAD, for localization of the TRP store-operated channel to a signaling complex. *Neuron* **1997**, *18*, 95–105.

54 Huber, A., Sander, P., Bahner, M., Paulsen, R., The TRP Ca^{2+} channel assembled in a signaling complex by the PDZ domain protein INAD is phosphorylated through the interaction with protein kinase C (ePKC). *FEBS Lett.* **1998**, *425*, 317–322.

55 Scott, K., Zuker, C. S., Assembly of the *Drosophila* phototransduction cascade into a signalling complex shapes elementary responses. *Nature* **1998**, *395*, 805–808.

56 Shieh, B. H., Zhu, M. Y., Regulation of the TRP Ca^{2+} channel by INAD in *Drosophila* photoreceptors. *Neuron* **1996**, *16*, 991–998.

57 Shieh, B. H., Zhu, M. Y., Lee, J. K., Kelly, I. M., Bahiraei, F., Association of INAD with NORPA is essential for controlled activation and deactivation of *Drosophila* phototransduction in vivo. *Proc. Natl. Acad. Sci. USA* **1997**, *94*, 12682–12687.

58 Tsunoda, S., Sierralta, J., Sun, Y., Bodner, R., Suzuki, E., Becker, A., Socolich, M., Zuker, C. S., A multivalent PDZ-domain protein assembles signalling complexes in a G-protein–coupled cascade. *Nature* **1997**, *388*, 243–249.

59 Xu, X. Z., Choudhury, A., Li, X., Montell, C., Coordination of an array of signaling proteins through homo- and heteromeric interactions between PDZ domains and target proteins. *J. Cell Biol.* **1998**, *142*, 545–555.

60 Wes, P. D., Xu, X. Z., Li, H. S., Chien, F., Doberstein, S. K., Montell, C., Termination of phototransduction requires binding of the NINAC myosin III and the PDZ protein INAD. *Nature Neuroscience* **1999**, *2*, 447–453.

61 Li, H.-S., Porter, J. A., Montell, C., Requirement for the NINAC kinase/myosin for stable termination of the visual cascade. *J. Neurosci.* **1998**, *18*, 9601–9606.

62 Kornfeld, K., Vulval development in *Caenorhabditis elegans*. *Trends Genet.* **1997**, *13*, 55–61.

63 Eisenmann, D. M., Kim, S. K., Signal transduction and cell fate specification during *Caenorhabditis elegans* vulval development. *Curr. Opin. Genet. Dev.* **1994**, *4*, 508–516.

64 Aroian, R. V., Koga, M., Mendel, J. E., Ohshima, Y., Sternberg, P. W., The *let-23* gene necessary for *Caenorhabditis elegans* vulval induction encodes a

tyrosine kinase of the EGF receptor subfamily. *Nature* **1990**, *348*, 693–699.

65 Hill, R. J., Sternberg, P. W., The gene *lin-3* encodes an inductive signal for vulval development in *C. elegans*. *Nature* **1992**, *358*, 470–476.

66 Kaech, S. M., Whitfield, C. W., Kim, S. K., The LIN-2/LIN-7/LIN-10 complex mediates basolateral membrane localization of the *C. elegans* EGF receptor LET-23 in vulval epithelial cells. *Cell* **1998**, *94*, 761–771.

67 Simske, J. S., Kaech, S. M., Harp, S. A., Kim, S. K., LET-23 receptor localization by the cell junction protein LIN-7 during *C. elegans* vulval induction. *Cell* **1996**, *85*, 195–204.

68 Hoskins, R., Hajnal, A. F., Harp, S. A., Kim, S. K., The *C. elegans* vulval induction gene *lin-2* encodes a member of the MAGUK family of cell junction proteins. *Development* **1996**, *122*, 97–111.

69 Jaulin-Bastard, F., et al., Interaction between Erbin and a catenin-related protein in epithelial cells. *J. Biol. Chem.* **2002**, *277*, 2869–2875.

70 Izawa, I., Nishizawa, M., Tomono, Y., Ohtakara, K., Takahashi, T., Inagaki, M., ERBIN associates with p0071, an armadillo protein, at cell–cell junctions of epithelial cells. *Genes to Cells* **2002**, *7*, 475–485.

71 Ohno, H., Hirabayashi, S., Iizuka, T., Ohnishi, H., Fujita, T., Hata, Y., Localization of p0071-interacting proteins, plakophilin-related armadillo-repeat protein-interacting protein (PAPIN) and ERBIN, in epithelial cells. *Oncogene* **2002**, *21*, 7042–7049.

72 Dillon, C., Creer, A., Kerr, K., Kumin, A., Dickson, C., Basolateral targeting of ERBB2 is dependent on a novel bipartite juxtamembrane sorting signal but independent of the C-terminal ERBIN-binding domain. *Mol. Cell. Biol.* **2002**, *22*, 6553–6563.

73 Shelly, M., Mosesson, Y., Citri, A., Lavi, S., Zwang, Y., Melamed-Book, N., Aroeti, B., Yarden, Y., Polar expression of ErbB-2/HER2 in epithelia: bimodal regulation by Lin-7. *Developmental Cell* **2003**, *5*, 475–486.

74 Straight, S. W., Karnak, D., Borg, J. P., Kamberov, E., Dare, H., Margolis, B., Wade, J. B., mLin-7 is localized to the basolateral surface of renal epithelia via its NH_2 terminus. *Am. J. Physiol. Renal Physiol.* **2000**, *278*, F464–F475.

75 Kuo, A., Zhong, C., Lane, W. S., Derynck, R., Transmembrane transforming growth factor-alpha tethers to the PDZ domain-containing, golgi membrane-associated protein p59/GRASP55. *EMBO J.* **2000**, *19*, 6427–6439.

76 Bezprozvanny, I., Maximov, A., PDZ domains: more than just a glue. *Proc. Natl. Acad. Sci. USA* **2001**, *98*, 787–789.

77 Bissell, M. J., Radisky, D., Putting tumours in context. *Nat. Rev. Cancer* **2001**, *1*, 46–54.

78 Nagafuchi, A. Molecular architecture of adherens junctions. *Curr. Opin. Cell Biol.* **2001**, *13*, 600–603.

79 Bilder, D., PDZ proteins and polarity: functions from the fly. *Trends Genet.* **2001**, *17*, 511–519.

80 Humbert, P., Russell, S., Richardson, H., Dlg, Scribble and Lgl in cell polarity, cell proliferation and cancer. *BioEssays* **2003**, *25*, 542–553.

81 Wodarz, A., Tumor supressors: Linking cell polarity and growth control. *Curr. Biol.* **2000**, *10*, R624–R626.

82 Peifer, M., Tepass, U., Which way is up? *Nature* **2000**, *403*, 611–612.

83 Bilder, D., Perrimon, N., Localization of apical epithelial determinants by the basolateral PDZ protein Scribble. *Nature* **2000**, *403*, 676–680.

84 Bilder, D., Li, M., Perrimon, N., Cooperative regulation of cell polarity and growth by *Drosophila* tumor suppressors. *Science* **2000**, *289*, 113–116.

85 Woods, D. F., Hough, C., Peel, D., Callaini, G., Bryant, P. J., Dlg protein is required for junction structure, cell polarity, and proliferation control in *Drosophila* epithelia. *J. Cell Biol.* **1996**, *134*, 1469–1482.

86 Woods, D. F., Bryant, P. J., The discs-large tumor supressor gene of *Drosophila* encodes a guanylate kinase homolog localized at septate junctions. *Cell* **1991**, *66*, 451–464.

87 Jacob, L., Opper, M., Bernhard, M., Phannavong, B., Mechler, B. M., Structure of the *l(2)gl* gene of *Drosophila* and delimination of its tumor suppressor domain. *Cell* **1987**, *50*, 215–225.

88 Petronczki, M., Knoblich, J. A., DmPAR-6 directs epithelial polarity and asymmetric cell division of neuroblasts in *Drosophila*. *Nat. Cell Biol.* **2001**, *3*, 43–49.

89 Kuchinke, U., Grawe, F., Knust, E., Control of spindle orientation in *Drosophila* by the Par-3–related PDZ-domain protein Bazooka. *Curr. Biol.* **1998**, *8*, 1357–1365.

90 Muller, H. A., Wieschaus, E., Armadillo, bazooka, and stardust are critical for early stages in formation of the zonula adherens and maintenance of the polarized blastoderm epithelium in *Drosophila*. *J. Cell Biol.* **1996**, *134*, 149–163.

91 Bilder, D., Schober, M., Perrimon, N., Integrated activity of PDZ protein complexes regulates epithelial polarity. *Nat. Cell Biol.* **2003**, *5*, 53–58.

92 Hong, Y., Stronach, B., Perrimon, N., Jan, L. Y., Jan, Y. N., *Drosophila* Stardust interacts with Crumbs to control polarity of epithelia but not neuroblasts. *Nature* **2001**, *414*, 634–638.

93 Bachmann, A., Schneider, M., Theilenberg, E., Grawe, F., Knust, E., *Drosophila* Stardust is a partner of Crumbs in the control of epithelial cell polarity. *Nature* **2001**, *414*, 638–643.

94 Tepass, U., Theres, C., Knust, E., *Crumbs* encodes an EGF-like protein expressed on apical membranes of *Drosophila* epithelial cells and required for organization of epithelia. *Cell* **1990**, *61*, 787–799.

95 Wodarz, A., Hinz, U., Engelbert, M., Knust, E., Expression of Crumbs confers apical character on plasma membrane domains of ectodermal epithelia of *Drosophila*. *Cell* **1995**, *82*, 67–76.

96 Laura, R. P., Ross, S., Koeppen, H., Lasky, L. A., MAGI-1: a widely expressed, alternatively spliced tight junction protein. *Exp. Cell Res.* **2002**, *275*, 155–170.

97 Nakagawa, S., Huibregtse, J. M., Human scribble (Vartul) is targeted for ubiquitin-mediated degradation by the high-risk papillomavirus E6 proteins and the E6AP ubiquitin-protein ligase. *Mol. Cell. Biol.* **2000**, *20*, 8244–8253.

98 Kiyono, T., Hiraiwa, A., Fujita, M., Hayashi, Y., Akiyama, T., Ishibashi, M., Binding of high-risk human papillomavirus E6 oncoproteins to the human homologue of the Drosophila discs large tumor suppressor protein. *Proc. Natl. Acad. Sci. USA* **1997**, *94*, 11612–11616.

99 Gardiol, D., Kuhne, C., Glaunsinger, B., Lee, S. S., Javier, R., Banks, L., Oncogenic human papillomavirus E6 proteins target the discs large tumour suppressor for proteasome-mediated degradation. *Oncogene* **1999**, *18*, 5487–5496.

100 Pim, D., Thomas, M., Banks, L., Chimaeric HPV E6 proteins allow dissection of the proteolytic pathways regulating different E6 cellular target proteins. *Oncogene* **2002**, *21*, 8140–8148.

101 Mantovani, F., Banks, L., The human papillomavirus E6 protein and its contribution to malignant progression. *Oncogene* **2001**, *20*, 7874–7887.

102 Legouis, R., Gansmuller, A., Sookhareea, S., Bosher, J. M., Baillie, D. L., Labouesse, M., LET-413 is a basolateral protein required for the assembly of adherens junctions in *Caenorhabditis elegans*. *Nat. Cell Biol.* **2000**, *2*, 415–422.

103 McMahon, L., Legouis, R., Vonesch, J.-L., Labouesse, M., Assembly of *C. elegans* apical junctions involves positioning and compaction by LET-413 and protein aggregation by the MAGUK protein DLG-1. *J. Cell. Sci.* **2001**, *114*, 2265–2277.

104 Pettitt, J., Cox, E. A., Broadbent, I. D., Flett, A., Hardin, J., The *Caenorhabditis elegans* p120 catenin homologue, JAC-1, modulates cadherin-catenin function during epidermal morphogenesis. *J. Cell Biol.* **2003**, *162*, 15–22.

105 Huang, Y. Z., Zang, M., Xiong, W. C., Luo, Z., Mei, L., Erbin suppresses the MAP kinase pathway. *J. Biol. Chem.* **2003**, *278*, 1108–1114.

106 Li, W., Han, M., Guan, K.-L., The leucine-rich repeat protein SUR-8 enhances MAP kinase activation and forms a complex with Ras and Raf. *Genes. Dev.* **2000**, *14*, 895–900.

107 Kolch, W., Erbin: Sorting out ErbB2 receptors or giving Ras a break? *Sci. STKE* **2003**, *2003*, pe37.

108 Garner, C. C., Nash, J., Huganir, R. L., PDZ domains in synapse assembly and signalling. *Trends Cell Biol.* **2000**, *10*, 274–280.

109 Craven, S. E., Bredt, D. S., PDZ proteins organize synaptic signaling pathways. *Cell* **1998**, *93*, 495–498.

110 Migaud, M., et al., Enhanced long-term potentiation and impaired learning in mice with mutant postsynaptic density-95 protein. *Nature* **1998**, *396*, 433–439.

111 Rasband, M. N., Park, E. W., Zhen, D., Arbuckle, M. I., Poliak, S., Peles, E., Grant, S. G., Trimmer, J. S., Clustering of neuronal potassium channels is independent of their interaction with PSD-95. *J. Cell Biol.* **2002**, *159*, 663–672.

112 Wharton, K. A. Jr., Runnin' with the Dvl: proteins that associate with Dsh/Dvl and their significance to Wnt signal transduction. *Dev. Biol.* **2003**, *253*, 1–17.

113 Sokol, S. Y., Analysis of Dishevelled signalling pathways during *Xenopus* development. *Curr. Biol.* **1996**, *6*, 1456–1467.

114 Yanagawa, S., van Leeuwen, F., Wodarz, A., Klingensmith, J., Nusse, R., The dishevelled protein is modified by wingless signaling in *Drosophila*. *Genes. Dev.* **1995**, *9*, 1087–1097.

115 Krasnow, R. E., Wong, L. L., Adler, P. N., Dishevelled is a component of the frizzled signaling pathway in *Drosophila*. *Development* **1995**, *121*, 4095–4102.

116 Uematsu, K., Kanazawa, S., You, L., He, B., Xu, Z., Li, K., Peterlin, B. M., McCormick, F., Jablons, D. M., Wnt pathway activation in mesothelioma: evidence of Dishevelled overexpression and transcriptional activity of beta-catenin. *Cancer Res.* **2003**, *63*, 4547–4551.

117 Pomerantz, J. L., Denny, E. M., Baltimore, D., CARD11 mediates factor-specific activation of NF-kappaB by the T cell receptor complex. *EMBO J.* **2002**, *21*, 5184–5194.

118 Bertin, J., et al., CARD11 and CARD14 are novel caspase recruitment domain (CARD)/membrane-associated guanylate kinase (MAGUK) family members that interact with BCL10 and activate NF-kappa B. *J. Biol. Chem.* **2001**, *276*, 11877–11882.

119 McAllister-Lucas, L. M., et al., Bimp1, a MAGUK family member linking protein kinase C activation to Bcl10-mediated NF-κB induction. *J. Biol. Chem.* **2001**, *276*, 30589–30597.

120 Gaide, O., et al., CARMA1 is a critical lipid raft-associated regulator of TCR-induced NF-κB activation. *Nat. Immunol.* **2002**, *3*, 836–843.

121 Wang, D., et al., A requirement for CARMA1 in TCR-induced NF-κB activation. *Nat. Immunol.* **2002**, *3*, 830–835.

122 Jun, J. E., et al., Identifying the MAGUK protein Carma-1 as a central regulator of the humoral immune responses and atopy by genome-wide mouse mutagenesis. *Immunity* **2003**, *18*, 751–762.

123 Hara, H., et al., The MAGUK family protein CARD11 is essential for lymphocyte activation. *Immunity* **2003**, *18*, 763–775.

124 Lefkowitz, R. J., Rockman, H. A., Koch, W. J., Catecholamines, cardiac β-adrenergic receptors, and heart failure. *Circulation* **2000**, *101*, 1634–1637.

125 Hall, R. A., et al., The beta2-adrenergic receptor interacts with the Na+/H+-exchanger regulatory factor to control Na+/H+ exchange. *Nature* **1998**, *392*, 626–630.

126 Cao, T. T., Deacon, H. W., Reczek, D., Bretscher, A., von Zastrow, M., A kinase-regulated PDZ-domain interaction controls endocytic sorting of the beta2-adrenergic receptor. *Nature* **1999**, *401*, 286–290.

127 Xiang, Y., Devic, E., Kobilka, B., The PDZ binding motif of the beta 1 adrenergic receptor modulates receptor trafficking and signaling in cardiac myocytes. *J. Biol. Chem.* **2002**, *277*, 33783–33790.

128 Cong, M., Perry, S. J., Hu, L. A., Hanson, P. I., Claing, A., Lefkowitz, R. J., Binding of the β2 adrenergic receptor to *N*-ethylmaleimide–sensitive factor regulates receptor recycling. *J. Biol. Chem.* **2001**, *276*, 45145–45152.

129 Xiang, Y., Kobilka, B., The PDZ-binding motif of the β2-adrenorecptor is essential for physiologic signaling and trafficking in cardiac myocytes. *Proc. Natl. Acad. Sci. USA* **2003**, *100*, 10776–10781.

14
EH Domains and Their Ligands

Brian K. Kay, Michael D. Scholle, and Fred J. Stevens

14.1
Introduction

The Eps15 homology (EH) domain was first noted by DiFiore and colleagues during the analysis of a particular protein substrate of the epidermal growth factor receptor tyrosine kinase [1]. Within the N-terminal region of the epidermal growth factor receptor protein substrate 15 (Eps15) protein were three copies of a ~100 amino acid long region [2]. Computer searches with the repeated sequence revealed that it was present in a number of unrelated proteins, suggesting that it was an autonomous folding domain, much like the Src homology 2 (SH2) and 3 (SH3) domains. Since its discovery, a great deal of research has focused on the types of proteins that contain EH domains, the role of the domain in protein–protein interactions, the 3D structure of the domain, and its possible evolutionary origins. These topics are summarized below, and other recent reviews [3–5] contain additional information.

14.2
EH Domain-containing Proteins

Since the discovery of the EH domain in Eps15, it has been identified in numerous proteins. The SMART website [http://smart.embl-heidelberg.de], which annotates protein domain families utilizing sophisticated algorithms [6, 7] (see also Chapter 21 of this book), currently lists 243 EH domains in a total of 145 proteins. Although the domain has not been reported in bacteria or archae, the domain is present in many eukaryotes. In particular, the distributions (number of domains per number of proteins) in *Saccharomyces cerevisiae, Arabidopsis thaliana, Caenorhabditis elegans, Drosophila melanogaster, Mus musculus,* and *Homo sapiens* are 15/8, 6/4, 16/11, 25/13, 62/39, and 45/26, respectively. Comparison of the EH domain-containing proteins suggests that there are 35 architecturally distinct proteins in these six species [5]. As shown in Figure 14.1a, proteins can have one or more EH domains, in combination with many other protein interaction modules and ligand sequences.

Modular Protein Domains. Edited by G. Cesareni, M. Gimona, M. Sudol, M. Yaffe

ISBN: 3-527-30813-X

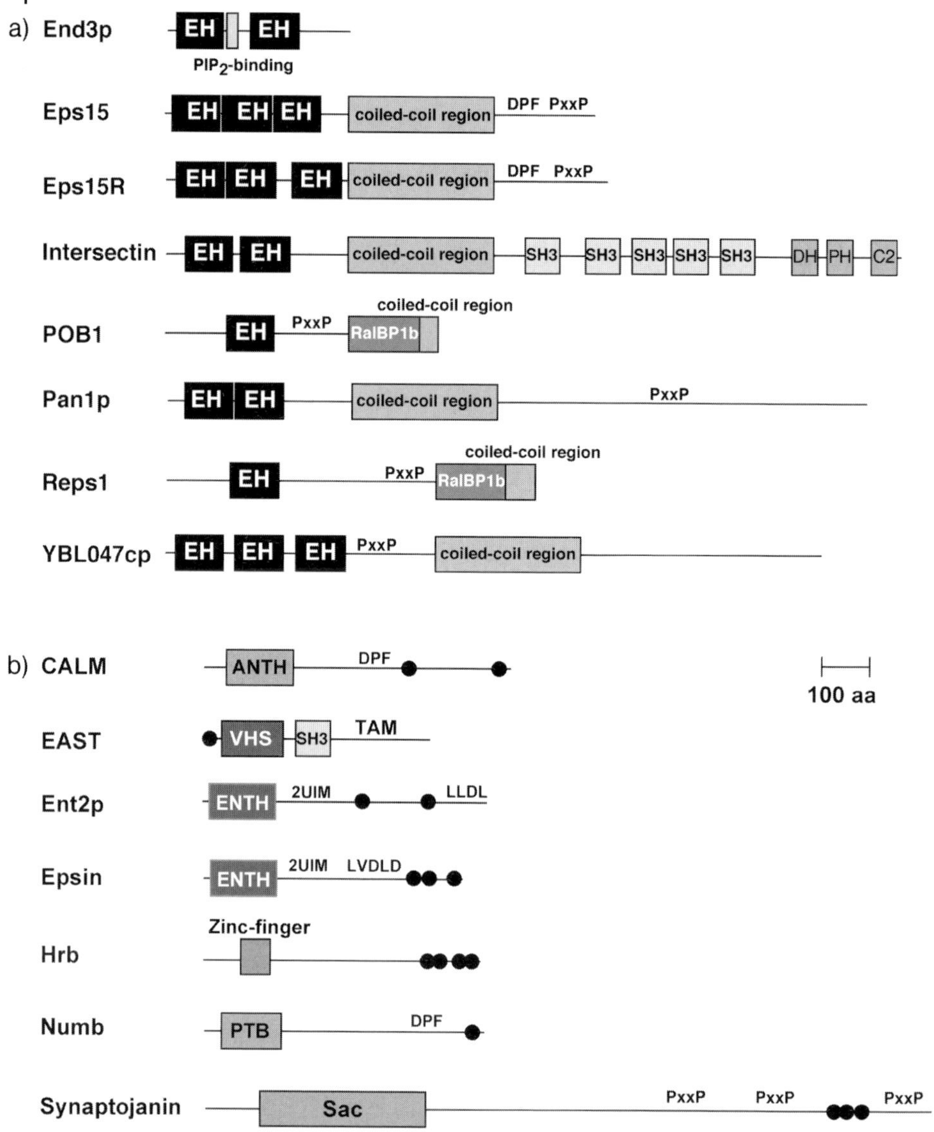

Figure 14.1 Proteins containing EH domains and their cellular ligands. (a) The architecture of eight proteins containing EH domains (black boxes) is shown. (b) The architecture of seven proteins demonstrated to be cellular ligands of EH domains is shown, with the ligand sites identified as black dots. All proteins are drawn to scale. Additional domain and motifs noted in these proteins include PIP_2 binding site, coiled-coil region, DPF binding motif for AP-2, PxxP binding motif for SH3 domains, SH3 domains, RalBP1b motif, ENTH domain, ANTH domain, VHS domain, ubiquitin-interacting domain (UIM), zinc finger, tyrosine activation motif (TAM), LLDL and LVDLD clathrin-binding motifs, phosphotyrosine binding domain (PTB), and the Sac domain. More information regarding these domains and ligand motifs can be found in this volume and elsewhere [3, 5, 8].

Analysis of the primary structures of the EH domains is shown in Figure 14.2a. Within the aligned sequences, the most conserved (i.e., > 95% identity) residues are Leu50 (L50) and Trp54 (W54). There are many other well conserved positions, with an average of 25% identity between any two EH domains. Interestingly, within one region (i.e., amino acids 58–69) there is a conserved cluster of acidic residues. As seen below, these residues serve to coordinate a calcium ion within some EH domains.

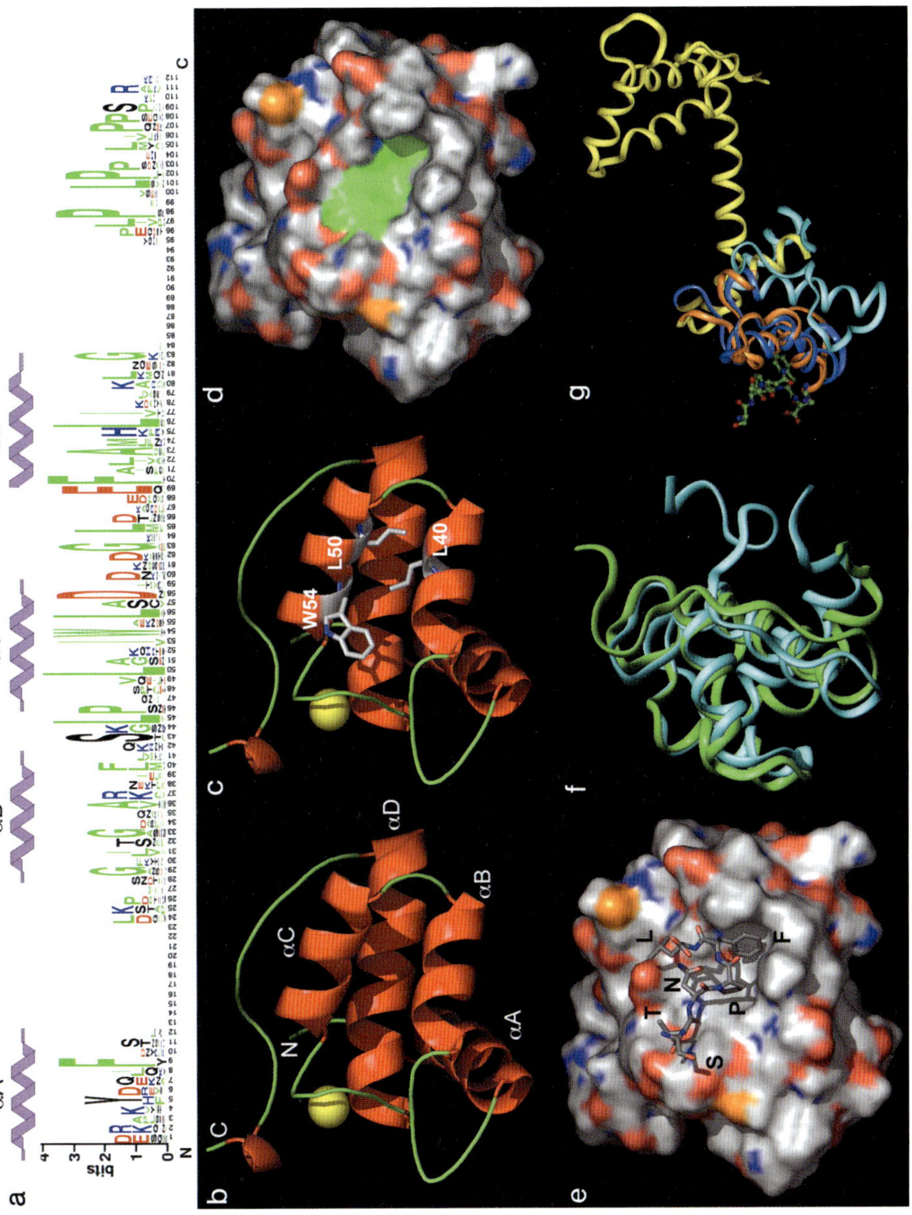

Figure 14.2 Primary, secondary, and tertiary structures of the EH domain (legend see also p. 282).

Figure 14.2 Primary, secondary, and tertiary structures of the EH domain.
(a) Alignment of the primary structures of EH domains. The sequences of 41 EH domains were aligned with the Logo program [12] on a website server [http://webLogo.berkeley.edu/]. To maximize the alignment among the primary structures, two gaps were introduced. Different amino acids are highlighted in four colors: blue (KRH), red (DE), green (GAVLIPWFM), and black (STCYNQ). The height (bits) of each letter corresponds to that amino acid's relative frequency at a given position. The numbering of the amino acid residues shown in this figure is used throughout this review. The four α-helical regions (i.e., αA, αB, αC, αD) are denoted above the aligned sequence.
(b) Ribbon diagram of the EH2 domain of Eps15. A diagram of the central EH domain of human Eps15 is shown with red helices (i.e., αA, αB, αC, and αD), green loops, and N and C termini labeled (i.e., N, C). The PyMOL Molecular Graphics System (DeLano Scientific, San Carlos, CA, USA) was used to generate this figure (and others) from the PDB coordinates for 1eh2 [13]. The calcium ion is shown as a yellow sphere bound in the loop between αC and αD.
(c) Location of three residues that form the hydrophobic pocket in EH2. The residues L40, L50, and W54 are drawn in stick style, with hydrogens not shown, in the ribbon diagram.
(d) Surface view of the EH2 domain and its hydrophobic pocket. The surface of the pocket formed by L40, L50, and W54 is shown in green. The elements C, N, O, and S are gray, blue, red, and orange, respectively.
(e) Structure of a peptide–EH domain complex. The second EH domain of human Eps15 is shown complexed with a peptide (STNPFL). In the stick representation of the peptide, C, N, and O are colored gray, blue, and red, respectively; hydrogens are omitted for clarity. The structure 1FF1 [23] was solved by NMR spectroscopy.
(f) Overlay of the 3D structures of two different EH domains. The carbon backbone of two EH domains, 1eh2 (blue) and 1fi6 (green), are superposed to demonstrate the conservation of the fold in two domains that share only 35% amino acid identity.
(g) Superposition of Eps15 homology domain (1f8h, blue) and calmodulin (4cln, yellow). Alpha carbons of residues 28–41 and 42–69 of EH2 are superimposed on alpha carbons 98–111 and 114–141 of calmodulin, respectively. The root mean square deviation (rmsd) of the superimposed alpha carbons is 4.58 Å. Superposition of all backbone atoms resulted in an rmsd of 4.36 Å, whereas superposition of the alpha carbons of the separate segments generated rms deviations of 3.30 and 3.88 Å, respectively, indicating a directional change between positions 111 and 114 in 4cln. For reference, the STNPFL peptide is shown docked in the EH2 pocket.

14.3 Peptide Ligands

When the EH domain was first discovered, it was unclear whether it had catalytic activity or participated in protein–protein interactions. From filter overlay and pulldown experiments, it appeared that the latter hypothesis was correct, since the EH domains of Eps15 were observed to interact selectively with a small set of proteins from mammalian cell lysates [2]. Like many other protein-interaction modules [8], the EH domain recognizes a short peptide motif within interacting proteins and is thus amenable to phage-display analysis. Screens of phage-displayed combinatorial peptide ligands [9–11] revealed that the optimal peptide ligand preference for most EH domains is Asn-Pro-Phe (NPF). Figure 14.3 shows a Logo plot [12] of the compiled peptide ligands for an assortment of EH domains. Interestingly, in some instances particular EH domains select for additional residues that flank the NPF motif, suggesting that the specificity of EH domains can differ subtly from each other, in agreement with cross-reactivity measurements [10, 11].

Through phage-display experiments, two other peptide motifs have been discovered for EH domains. This result has led to the classification of three different types of ligands: class I consists of NPF (as mentioned above); class II consists of aromatic and hydrophobic di- and tripeptide motifs, including the Phe-Trp (FW), Trp-Trp (WW), and Ser-Trp-Gly (SWG) motifs; and class III contains the His-(Thr/Ser)-Phe motif (HTF/HSF). Figure 14.3 displays examples of class II and III motifs, which appear to bind to their cognate EH domains through predominantly hydrophobic interactions, just like class I ligands (below). Interestingly, for some

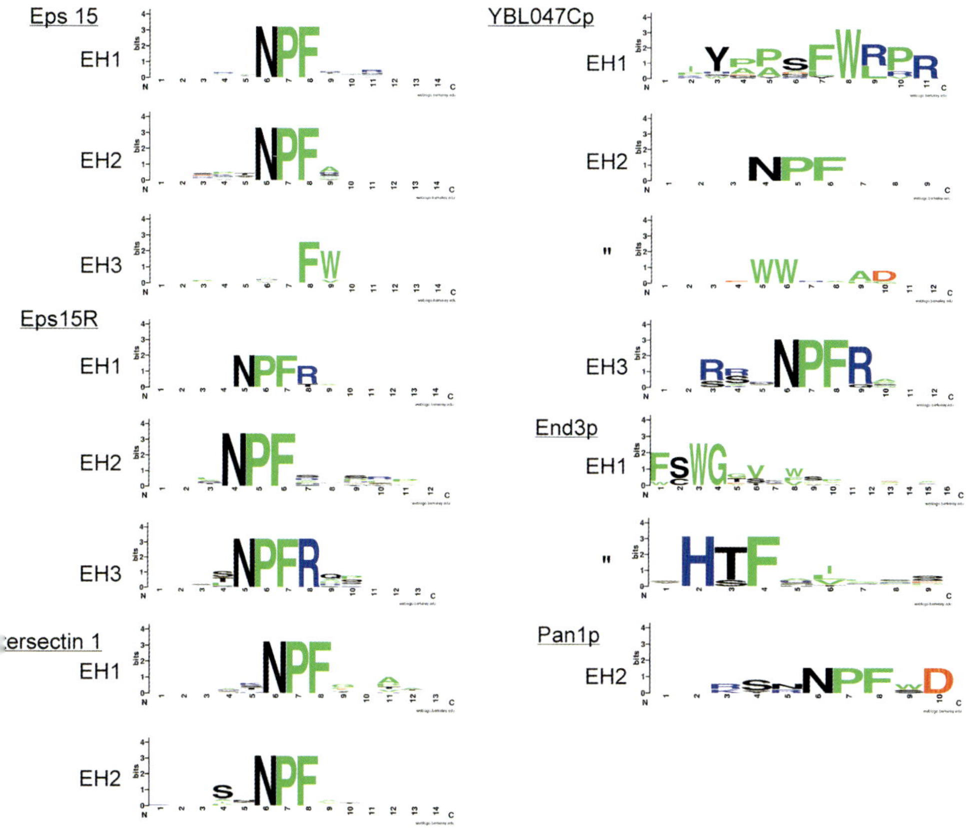

Figure 14.3 The optimal peptide ligands of various EH domains. Phage-displayed combinatorial peptide libraries were affinity selected with 14 individual EH domains [10, 11]. The peptide sequences of the recovered phage were aligned with the Logo program [12] on a website server [http://webLogo.berkeley.edu/]. Peptide ligands were selected for the three EH domains of Eps15 and Eps15R, the two EH domains of intersectin 1, the three EH domains of YBL047p, one of the two EH domains of End3p, and one of the EH domains of Pan1p. Several of the domains (denoted with commas) selected two different sets of peptide motifs. The height (bits) of the letter corresponds to that amino acid's relative frequency at a given position.

EH domains, there appears to be a dual specificity, but, as discussed below, it is likely that these motifs bind in the same hydrophobic pocket of the domain.

The affinity of EH domains for their peptide ligands is considered to be weak to modest. A dissociation constant of 560 μM was measured for the interaction of the EH2 domain of Eps15 with the PTGSSSTNPFL peptide by surface plasmon resonance [13]. The secondary structure of the peptide is likely to contribute to the binding strength, since NPF-containing peptides that are flanked by disulfide bonds bind up to 100 times stronger than their linear versions [11]. However, not all EH domains are the same in this respect. Careful analysis of the chemical shifts of the Reps EH domain by NMR spectroscopy, as increasing amounts of a linear or a cyclic peptide were added, indicated that this domain binds them with dissociation constants of 65 μM and 46 μM, respectively. Nevertheless, characterized EH domains bind their peptide ligands in the low- to mid-micromolar range, which is much weaker than SH2 [14] and SH3 domains [15], which typically bind peptide ligands with dissociation constants of 0.3–5 μM.

14.4 Cellular Ligands

The number of candidate proteins that interact with EH domains is steadily increasing. In the beginning, interacting proteins were identified by screening expression libraries with labeled EH domains [9, 11]. Such screens identified the human homolog of NUMB, the product of a developmentally regulated gene of *D. melanogaster*, Hrb, the HIV REV binding protein, and a number of novel proteins that were later named epsins [16]. Since each of these proteins contained one or more NPF motifs, this observation suggested a mechanism for protein–protein interaction that was later experimentally proven through mutagenesis of the NPF motif in Numb [9]. Now, with the optimal peptide ligand preferences of EH domains well documented through phage display, cellular ligands can be postulated based on the occurrence of any of the three classes of ligand motifs within a protein [17]. A total of 12 cellular ligands have been confirmed biochemically in eukaryotes, and several are diagrammed in Figure 14.1b. Genomic analysis of sequenced genomes has suggested that several hundred interacting proteins possibly exist [5].

Interestingly, many of the cellular ligands contain multiple (i.e., 2–7) copies of the NPF motif. One can speculate that the multiple occurrence of the motif within a protein is a mechanism by which it can simultaneously interact with multiple EH domain-containing proteins or that such a protein can interact very tightly with a single protein that carries multiple EH domains. Avidity is a very effective mechanism for converting weak-to-modest interactions into strong ones in the cell. One observation in support of this hypothesis is that the spacing of multiple EH domains, or their NPF peptide motifs, tends to be conserved between protein homologs from different species. We should note that the regulation of the EH domain–ligand interaction is yet to be elucidated, although phosphorylation of residues flanking the NPF motif has been demonstrated to inhibit the interaction [18].

14.5
Structures of the Domain and Its Ligands

To date, the structure of several EH domains has been solved by NMR spectroscopy. The three EH domains of murine or human Eps15 have been solved and their coordinates deposited in the Protein Database (PDB) as 1qjt [19], 1eh2 [13], and 1c07 [20], respectively. In addition, the structures of the EH domains of the Reps1 [21] and Pob1 [22] proteins have been solved and deposited in the PDB as 1iq3 and 1fi6, respectively. The EH domain consists of two helix-loop-helix motifs, also termed EF-hand motifs (Figure 14.2b). In three of the EH domains (Eps15 EH2, Pob1, Reps1), a calcium ion is bound in the second EF-hand (i.e., between the αC and αD helices); calcium is not bound in the first EF-hand, due to an absence of residues that coordinate calcium. In the other two EH domains, the calcium ion is coordinated by the first EF-hand of EH3 of Eps15, whereas EH1 of Eps15 does not bind calcium at all, due to substitutions. Thus, calcium ions are not required for the EH domain fold. In addition, there is no experimental evidence that calcium ions play any physiological role in regulating the ability of EH domains to bind other proteins [2].

The 3D structure of the EH domains is well conserved. Superimposition of the carbon backbone tracings of two EH domains (EH2 of Eps15, EH domain of Pob1) of modest amino acid identity (35%) demonstrates that the folds are highly conserved. Figure 14.2f compares the backbone folds of the structures 1eh2 and 1fi6 based on the superposition of the corresponding C-terminal segments (darker colors) of the two molecules, which yielded an α-carbon rmsd of 1.96 Å. Table 14.1 provides a comparison of all currently available EH domain structures.

Table 14.1 Dali comparisons of EH homology domains: Z score (rmsd).

gi	*Number of amino acids*	*PDB id*	*1f8h*	*1ff1*	*1c07*	*1fi6*	*1iq3*	*1qjt*	Z_{max}
5822057	106	1eh2	20.5 (1.0)	20.7 (0.7)	15.7 (1.5)	8.5 (2.4)	5.7 (2.3)	7.6 (2.5)	24.2
11514134	95	1f8h		21.2 (0.9)	15.6 (1.5)	8.6 (2.3)	5.6 (2.3)	7.2 (2.5)	24.2
11514159	95	1ff1			15.4 (1.5)	9.1 (2.3)	5.8 (2.2)	7.5 (2.6)	24.4
9955190	95	1c07				8.1 (2.6)	5.6 (2.9)	7.5 (2.5)	20.1
15826201	92	1fi6					8.3 (2.5)	6.3 (2.5)	21.5
14719573	110	1iq3						5.4 (3.0)	20.1
6980448	99	1qjt						–	

The structure of the second EH domain of Eps15 has been solved in complex with two different peptide ligands [23]. The peptides, PTGSSSTNPFR and PTGSSSTNPFL, were observed to bind in a hydrophobic pocket between two α helices (the αC and αD helices). The pocket is formed by two leucines (L155, L165) and a nearby tryptophan (W169) residue (Figure 14.2c). Mutational analysis of this domain demonstrated that the L165 and W169 residues were critical for binding [13]. As seen in Figure 14.2e, the STNPFL peptide segment adopts a type I Asn-Pro β turn within the hydrophobic pocket of the EH2 domain of Eps15 [23]. The binding pocket is located ~10 Å away from the calcium ion and, as in many other protein interaction modules [24], on a face opposite the N and C termini of the domain.

The details for NPF peptide ligand binding to the EH domain of Reps1 are very similar, except that the hydrophobic pocket is formed by F40, L50, and W54 [21]. By monitoring differential line broadening and chemical-shift changes in the NMR spectra of the Reps1 domain upon ligand binding, NPF peptide binding was found to affect residues F40, L50, Trp54, and S51, which are adjacent to the pocket [21]. In addition, two charged residues (K37, E55), which bracket the hydrophobic pocket, showed significant chemical-shift differences upon binding, suggesting that they act as a type of 'gate' through conformation changes.

NMR spectroscopic analysis of the third EH domain (EH3) of Eps15 interacting with NPF and FW peptide ligands has revealed that both motifs bind in the same hydrophobic pocket [20]. The chemical shifts in the resonances of the residues that line the pocket coincide when the EH3 domain interacts with 10-mer peptides containing either the NPF or FW motifs. Mutational analysis and surface plasmon resonance measurements have implicated the same residues in binding each peptide motif. However, none of these mutations completely abolishes binding to the FW motif, nor did introduction of such mutations into the EH2 domain induce recognition of the FW motif. Thus, the binding specificity of EH domains is still too complex to predict by examination of their primary structures.

14.6
Evolutionary Origins of the EH Domain

The calcium-binding motif known as the EF-hand is a common structural motif, with more than 70 EF-hand subfamilies known [25]. The question arises of whether the appearance of the EF-hand in EH domains reflects an evolutionary relationship that might contribute to insights on the functional origin of calcium binding or if it is a potential example of convergent evolution of function and structure. If the presence of an EF-hand in EH domains is the product of convergent evolution, then this would argue for a strong functional requirement for calcium binding, at least at some point during evolution. Since the regions of similarity extend beyond the calcium-binding loop, we speculate that the many representatives of the EF-hand family are the product of divergent evolution from an ancestral EH domain or EH domain-like structural motif.

To address this question, we performed computer searches with a representative EH domain (Eps15 EH2). (Qualitatively similar results are expected from sequences that have statistically significant similarity to each other.) Since the BLAST algorithm [26] is capable of identifying evolutionary relationships only between proteins whose divergence is relatively recent, so that the extent of primary structure identity retains statistical significance, we utilized the related algorithm, Psi-BLAST [27, 28]. Psi-BLAST significantly extends the range of evolutionary relationships that can be inferred, through the algorithm strategy itself and from the size of the NCBI nonredundant database, which is now approaching 2×10^6 sequence entries. The deposition of whole-genome data (i.e., unbiased data) is providing an increasingly diverse sequence repository from which to discover evolutionary relationships. Because Psi-BLAST results are dependent on the sequence profile that is developed, and the profile is intrinsically linked to the composition of the database, one can expect that the accumulating whole-genome data will enhance the ability of this algorithm to detect true homologs, despite extensive evolutionary distances.

When we used Psi-BLAST to find EH domain-related proteins, the first round of searching resulted in 200 matches, none of which were obvious EF-hand–containing proteins. In the second round of searching, more than 600 hits were obtained, including some with weak similarities to EF-hand proteins, such as troponin, calmodulin, fimbrin, calcineurin and others. More than 1000 potential matches were located in the third iteration, and > 3000 possible homologs were found in the fourth round. Included in the fourth round was calcyclin, an S100 family protein of known structure (1mho), which, in our opinion, is an EH domain-containing protein. As noted previously, the EH1 domain of Eps15 has itself been classified as a member of the S100 family [19].

The calcyclin structure supports the notion that an evolutionary relationship exists between EH domain proteins and the EF-hand proteins. Nevertheless, the region of alignment between the EH domain query sequence and the EF-hand representatives was consistently limited to the span of α helices B, C, and D. Additional support was provided by comparison of the structures. A Dali analysis [29, 30] of the EH2 domain of Eps15 (1f8h) and an arbitrarily selected calmodulin representative (4cln) resulted in a Z score of 3.7, consistent with structural similarity. Aligned segments of 1f8h and 4cln, as matched by Dali, are summarized in Table 14.2.

Table 14.2 Superposition of representative EH-domain and EF-hand proteins (Dali analysis – see text for explanation).

	1f8 h	*4cln*	*Number of amino acids*	*αC rmsd*	*Backbone rmsd*
a	9–21	81– 93	13	1.77	
b	23–26	94– 97	4	0.90	
c	28–41	98–111	14	3.30	
d	42–69	114–141	28	3.88	
e	93–96	144–147	4	1.29	
	c + d	c + d	42	4.58	4.36

Based on the rmsd of the alpha carbons of the superimposed segments, matches were found to be excellent (a, b, e) to good (c, d). The structural correspondence was particularly striking in the calcium-binding site; superposition of residues 58–60 in EH with 129–135 in the EF-hands of calmodulin yielded an rmsd of only 0.50 Å.

We propose that a distant evolutionary relationship links the EH domain proteins and the EF-hand proteins. The existence of a pair of EF-hands in the EH domain, one of which may or may not bind calcium, suggests that the progenitor protein of the EF-hand family may be related to a primitive EH domain, which, through repeated rounds of gene duplication, evolved into the large, functionally diverse EH domain and EF-hand families, which currently account for more than 6000 entries in the nonredundant database.

14.7 Functions of the EH Domain

The role of the EH domain is to mediate specific protein–protein interactions in the cytosol of eukaryotic cells. Many of the interactions function at various steps of receptor-mediated endocytosis [3] and synaptic vesicle recycling [31], and others participate in intracellular trafficking of endosomes to the golgi [32]. There is also evidence for the involvement of EH domain-containing proteins in other cellular processes, such as reorganization of the actin cytoskeleton [33, 34], transcriptional activation [35], mitogenic signaling [1, 36, 37], and nuclear shuttling [38, 39]. In the future, it will be interesting to determine the precise mechanism of how EH domain-containing proteins and their cellular ligands are involved in each of these processes and their interconnections. Biochemical and biophysical analyses will augment genome-wide analysis of EH domain–protein interactions and thereby increase our understanding of the role of the EH domain in cell function.

Acknowledgements

Our work at the Argonne National Laboratory was supported by the U. S. Department of Energy, Office of Science, Office of Basic Energy Sciences, under contract W-31-109-Eng-38.

References

1 Fazioli, F., Minichiello, L., Matoskova, B., Wong, W. T., Di Fiore, P. P., eps15, a novel tyrosine kinase substrate, exhibits transforming activity. *Mol. Cell. Biol.* **1993**, *13*, 5814–5828.

2 Wong, W. T., Schumacher, C., Salcini, A. E., Romano, A., Castagnino, P., Pelicci, P. G., Di Fiore, P., A protein-binding domain, EH, identified in the receptor tyrosine kinase substrate Eps15 and conserved in evolution. *Proc. Natl. Acad. Sci. USA* **1995**, *92*, 9530–9534.

3 Santolini, E., Salcini, A. E., Kay, B. K., Yamabhai, M., Di Fiore, P. P., The EH network. *Exp. Cell Res.* **1999**, *253*, 186–209.

4 Confalonieri, S., Di Fiore, P. P., The Eps15 homology (EH) domain. *FEBS Lett.* **2002**, *513*, 24–29.

5 Polo, S., Confalonieri, S., Salcini, A. E., Di Fiore, P. P., EH and UIM: endocytosis and more. *Sci STKE* **2003**, 2003, RE17.

6 Schultz, J., Milpetz, F., Bork, P., Ponting, C. P., SMART, a simple modular architecture research tool: identification of signaling domains. *Proc. Natl. Acad. Sci. USA* **1998**, *95*, 5857–5864.

7 Letunic, I., et al., Recent improvements to the SMART domain-based sequence annotation resource. *Nucleic Acids Res.* **2002**, *30*, 242–244.

8 Pawson, T., Nash, P., Assembly of cell regulatory systems through protein interaction domains. *Science* **2003**, *300*, 445–452.

9 Salcini, A. E., et al., Binding specificity and in vivo targets of the EH domain, a novel protein–protein interaction module. *Genes. Dev.* **1997**, *11*, 2239–2249.

10 Paoluzi, S., et al., Recognition specificity of individual EH domains of mammals and yeast. *EMBO J.* **1998**, *17*, 6541–6550.

11 Yamabhai, M., Hoffman, N. G., Hardison, N. L., McPherson, P. S., Castagnoli, L., Cesareni, G., Kay, B. K., Intersectin, a novel adaptor protein with two Eps15 homology and five Src homology 3 domains. *J. Biol. Chem.* **1998**, *273*, 31401–31407.

12 Schneider, T. D., Stephens, R. M., Sequence Logos: a new way to display consensus sequences. *Nucleic Acids Res.* **1990**, *18*, 6097–6100.

13 de Beer, T., Carter, R. E., Lobel-Rice, K. E., Sorkin, A., Overduin, M., Structure and Asn-Pro-Phe binding pocket of the Eps15 homology domain. *Science* **1998**, *281*, 1357–1360.

14 Ladbury, J. E., Lemmon, M. A., Zhou, M., Green, J., Botfield, M. C., Schlessinger, J., Measurement of the binding of tyrosyl phosphopeptides to SH2 domains: a reappraisal. *Proc. Natl. Acad. Sci. USA* **1995**, *92*, 3199–3203.

15 Chen, J. K., Lane, W. S., Brauer, A. W., Tanaka, A., Schreiber, S. L., Biased combinatorial libraries: novel ligands for the SH3 domain of phosphatidylinositol 3-kinase. *J. Am. Chem. Soc.* **1993**, *115*, 12591–12952.

16 Chen, H., Fre, S., Slepnev, V., Capua, M., Takei, K., Butler, M., Di Fiore, P., De Camilli, P., Epsin, an EH domain binding protein implicated in clathrin-mediated endocytosis. *Nature* **1998**, *394*, 793–798.

17 Kay, B. K., Kasanov, J., Knight, S., Kurakin, A., Convergent evolution with combinatorial peptides. *FEBS Lett.* **2000**, *480*, 55–62.

18 Kariya, K., Koyama, S., Nakashima, S., Oshiro, T., Morinaka, K., Kikuchi, A., Regulation of complex formation of POB1/epsin/adaptor protein complex 2 by mitotic phosphorylation. *J. Biol. Chem.* **2000**, *275*, 18399–18406.

19 Whitehead, B., Tessari, M., Carotenuto, A., van Bergen en Henegouwen, P. M., Vuister, G. W., The EH1 domain of Eps15 is structurally classified as a member of the S100 subclass of EF-hand–containing proteins. *Biochemistry* **1999**, *38*, 11271–11277.

20 Enmon, J. L., de Beer, T., Overduin, M., Solution structure of Eps15's third EH domain reveals coincident Phe-Trp and Asn-Pro-Phe binding sites. *Biochemistry* **2000**, *39*, 4309–4319.

21 Kim, S., Cullis, D. N., Feig, L. A., Baleja, J. D., Solution structure of the Reps1 EH domain and characterization of its binding to NPF target sequences. *Biochemistry* **2001**, *40*, 6776–6785.

22 Koshiba, S., Kigawa, T., Iwahara, J., Kikuchi, A., Yokoyama, S., Solution structure of the Eps15 homology domain of a human POB1 (partner of RalBP1). *FEBS Lett.* **1999**, *442*, 138–142.

23 de Beer, T., Hoofnagle, A. N., Enmon, J. L., Bowers, R. C., Yamabhai, M., Kay, B. K., Overduin, M., Molecular mechanism of NPF recognition by EH domains. *Nat. Struct. Biol.* **2000**, *7*, 1018–1022.

24 Pawson, T., Protein module and signalling networks. *Nature* **1995**, *373*, 573–560.

25 Kawasaki, H., Nakayama, S., Kretsinger, R. H., Classification and evolution of EF-hand proteins. *Biometals* **1998**, *11*, 277–295.

26 Altschul, S. F., Gish, W., Miller, W., Myers, E. W., Lipman, D. J., Basic local alignment search tool. *J. Mol. Biol.* **1990**, *215*, 403–410.

27 Altschul, S. F., Madden, T. L., Schaffer, A. A., Zhang, J., Zhang, Z., Miller, W., Lipman, D. J., Gapped BLAST and Psi-BLAST: a new generation of protein database search programs. *Nucleic Acids Res.* **1997**, *25*, 3389–3402.

28 Altschul, S. F., Koonin, E. V., Iterated profile searches with Psi-BLAST: a tool for discovery in protein databases. *Trends Biochem. Sci.* **1998**, *23*, 444–447.

29 Holm, L., Sander, C., Dali: a network tool for protein structure comparison. *Trends Biochem. Sci.* **1995**, *20*, 478–480.

30 Holm, L., Sander, C., Dali/FSSP classification of three-dimensional protein folds. *Nucleic Acids Res.* **1997**, *25*, 231–234.

31 Chen, H., Slepnev, V. I., Di Fiore, P. P., De Camilli, P., The interaction of epsin and Eps15 with the clathrin adaptor AP-2 is inhibited by mitotic phosphorylation and enhanced by stimulation-dependent dephosphorylation in nerve terminals. *J. Biol. Chem.* **1999**, *274*, 3257–3260.

32 Fernandez-Chacon, R., Achiriloaie, M., Janz, R., Albanesi, J. P., Sudhof, T. C., SCAMP1 function in endocytosis. *J. Biol. Chem.* **2000**, *275*, 12752–12756.

33 Duncan, M. C., Cope, M. J., Goode, B. L., Wendland, B., Drubin, D. G., Yeast Eps15-like endocytic protein, Pan1p, activates the Arp2/3 complex. *Nat. Cell Biol.* **2001**, *3*, 687–690.

34 Hussain, N. K., et al., Endocytic protein intersectin-l regulates actin assembly via Cdc42 and N-WASP. *Nat. Cell Biol.* **2001**, *3*, 927–932.

35 Mohney, R. P., Das, M., Bivona, T. G., Hanes, R., Adams, A. G., Philips, M. R., O'Bryan, J. P., Intersectin activates Ras but stimulates transcription through an independent pathway involving JNK. *J. Biol. Chem.* **2003**, *278*, 47038–47045.

36 Adams, A., Thorn, J. M., Yamabhai, M., Kay, B. K., O'Bryan, J. P., Intersectin, an adaptor protein involved in clathrin-mediated endocytosis, activates mitogenic signaling pathways. *J. Biol. Chem.* **2000**, *275*, 27414–27420.

37 Tong, X. K., et al., The endocytic protein intersectin is a major binding partner for the Ras exchange factor mSos1 in rat brain. *EMBO J.* **2000**, *19*, 1263–1271.

38 Doria, M., Salcini, A. E., Colombo, E., Parslow, T. G., Pelicci, P. G., Di Fiore, P. P., The eps15 homology (EH) domain-based interaction between eps15 and hrb connects the molecular machinery of endocytosis to that of nucleocytosolic transport. *J. Cell Biol.* **1999**, *147*, 1379–1384.

39 Hyman, J., Chen, H., Di Fiore, P. P., De Camilli, P., Brunger, A. T., Epsin 1 undergoes nucleocytosolic shuttling and its eps15 interactor NH(2)-terminal homology (ENTH) domain, structurally similar to armadillo and HEAT repeats, interacts with the transcription factor promyelocytic leukemia Zn(2)+ finger protein (PLZF). *J. Cell Biol.* **2000**, *149*, 537–546.

15
Ubiquitin Binding Modules: The Ubiquitin Network Beyond the Proteasome

Stefano Confalonieri and Pier Paolo Di Fiore

15.1
Introduction: The Ubiquitin System in Proteolysis and Beyond

Similar to other protein covalent modifications, such as phosphorylation, acetylation, or methylation, ubiquitination of substrates is used by the cell to regulate a variety of biochemical pathways and physiological functions. Ubiquitin (Ub) is a protein of 76 residues, extremely well conserved throughout evolution [1], that is appended to other proteins through an isopeptide bond between its terminal glycine residue (Gly76) and the ε-amino group of a lysine residue in the substrate, as the result of a complex chain of enzymatic reactions that culminates in the activation of distal effector enzymes, called Ub ligases or E3 enzymes (Figure 15.1) [2]. One peculiar characteristic of Ub is that it can itself serve as a substrate for further cycles of ubiquitination, leading to the formation of polyubiquitin chains. The Ub moiety has seven conserved lysine residue, which can form isopeptide bonds with other Ub molecules. Of these, four (Lys11, Lys29, Lys48, and Lys63) can be used to form polyubiquitin chains in vivo [3–6]. The most widely studied form of polyubiquitination is that in which chains branching from Lys48 are formed. This modification is known to target proteins for proteasomal degradation (reviewed in [2]). Indeed, it has been shown that multiubiquitin chains of at least four Ub molecules, linked through Lys48, are the optimal proteasome-targeting signal [7].

A recent body of evidence, however, has uncovered unexpected roles for ubiquitination beyond proteolysis, because Ub modification has been implicated in the regulation of processes as heterogeneous as cell-cycle progression, apoptosis, cellular differentiation, protein transport, endocytosis, and DNA repair (reviewed in [6]). It was discovered, for instance, that chains branching from Lys63 are not 'degradation' signals but seem to be important for the process of DNA repair and other functions (see [6, 8–10] for reviews). Proteins can also be monoubiquitinated, when a single Ub moiety is appended to a protein substrate. Monoubiquitination is a reversible nonproteolytic modification that regulates protein localization and function in endocytosis, DNA repair, histone activity, and virus budding (see [11] for a review).

Modular Protein Domains. Edited by G. Cesareni, M. Gimona, M. Sudol, M. Yaffe

ISBN: 3-527-30813-X

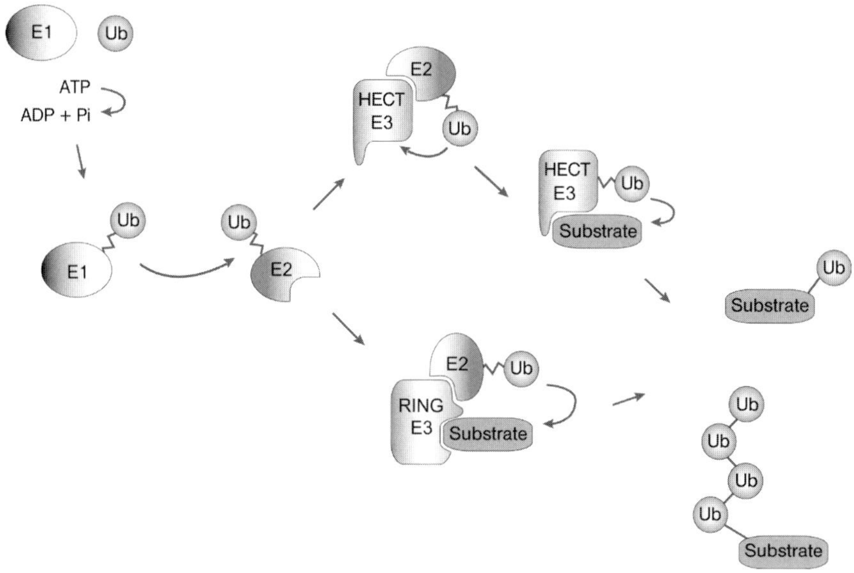

Figure 15.1 The ubiquitin conjugation pathway. Free ubiquitin (Ub) is activated in an ATP-dependent manner, with formation of a thiol–ester linkage between E1 and the carboxyl terminus of ubiquitin. Ubiquitin is then transferred to an E2 enzyme. E2 enzymes carrying a covalently bound Ub molecule can associate with E3 enzymes, which leads to ubiquitination of the substrate in two different ways. For HECT-domain–containing E3s, ubiquitin is first transferred to the active-site cysteine of the HECT domain and then to the substrate. RING-domain–containing E3s act instead as adaptors between the substrate and the E2 enzyme, which directly transfers the Ub molecule to the substrate. Ubiquitin can be added directly to the substrate once, producing monoubiquitinated substrates, or to a substrate-bound ubiquitin(s) to form Ub chains.

Substrate modifications or particular peptide motifs contained in proteins are recognized by specific protein modules or domains (for an exhaustive collection of reviews see the FEBS Letters Special Issue on Protein Domains, Vol. 513, Issue 1), thereby establishing networks of protein–protein interactions, which control many signaling pathways and regulatory systems. Polyubiquitin chains and monoubiquitin (mUb) moieties linked to substrate proteins are recognized by specific Ub binding domains contained in several protein families. The first indication of how Ub signals are recognized and interpreted came from the identification of a hydrophobic pocket on the Ub surface, defined mainly by the sidechains of Leu8, Ile44, and Val70, that is required for the binding to the S5a/Rpn10 subunit of the proteasome [12]. An extensive mutational analysis carried out in *Saccharomyces cerevisiae* has defined the structural features of Ub that are required for important phenotypes such as vegetative growth and endocytosis [13]. Only 16 of ubiquitin's 63 surface residues are indispensable for vegetative growth in yeast, and they cluster in three distinct regions. The first region maps near the hydrophobic pocket that participates in proteasome binding. Residues in this area, in particular Ile44, are essential for proteasome-mediated degradation and also critical for endocytosis [14]. The second region includes residues near Phe4, which is essential only for the endocytic process [14]. The third group resides in or near the tail region, which is crucial for ubiquitin conjugation and deubiquitination [15–18].

To date, at least six Ub binding domains have been found experimentally and/or by bioinformatics: CUE, UBA, and the UIM, which share some common structural features; UEV, which is a catalytic inactive version of the UBCc domain present in the E2 Ub-conjugating family of enzymes; and PAZ and NZF, which are zinc-finger domains. For five of them, the 3D structure has been solved, either alone or in complex with Ub. These studies have revealed that Ub-binding domains often interact with the hydrophobic pocket centered on Ile44 of Ub in a similar fashion.

In summary, novel modalities of regulation of protein function and intracellular signaling are emerging from studies of Ub, of protein domains or motifs that can bind to Ub, and of molecules that harbor such Ub-recognition devices (Ub receptors). In addition, many proteins are modified by Ub-like (UBL) modifiers [19–21], such as SUMO [22, 23] or Nedd8 [24], whose conjugation pathways closely resemble that of ubiquitin (Figure 15.1) but which do not seem to form multi-UBL chains [19, 21]. Many other proteins contain Ub-like domains (UDPs) embedded in their polypeptide chains, which do not form conjugates with cellular proteins [20] and may function as 'adaptor' molecules in the ubiquitin system, either by bridging the ubiquitination machinery to the proteasome or by regulating intracellular signaling by interacting with Ub-receptor proteins. Thus, Ub and Ub-like molecules constitute a vast functional 'domain' within the cell having projected vast impact on cellular homeostasis. In this review, we concentrate on the structural and functional features of Ub-binding devices and on some biological aspects of their networking with Ub.

15.2 CUE and UBA Domains

The CUE domain was identified by database searches for proteins of the endoplasmic-reticulum–associated degradation pathway, starting from the yeast Cue1p protein (CUE = coupling of ubiquitin conjugation to ER degradation), as a conserved module present in several eukaryotic proteins (Figure 15.2) [25]. CUEs have recently been shown to bind to mono- and polyubiquitin [26, 27], and the structures of the first CUE domain of Cue2p and of the CUE domain of Vps9p in complex with Ub have been solved [28, 29].

The UBA (ubiquitin associated) domain was identified experimentally as a Ub binding motif present in the C-terminal end of the p62 (SQSTM1) protein [30] (Figure 15.2) and was subsequently found in many other proteins of the ubiquitination pathway [31]. The UBA domain can bind both mono- and polyubiquitin [32, 33]. Recently, the 3D structures of the two UBA domains of RAD23A, the human homolog of yeast RAD23, and of the UBA domain of p62 (SQSTM1) have been solved [34–36].

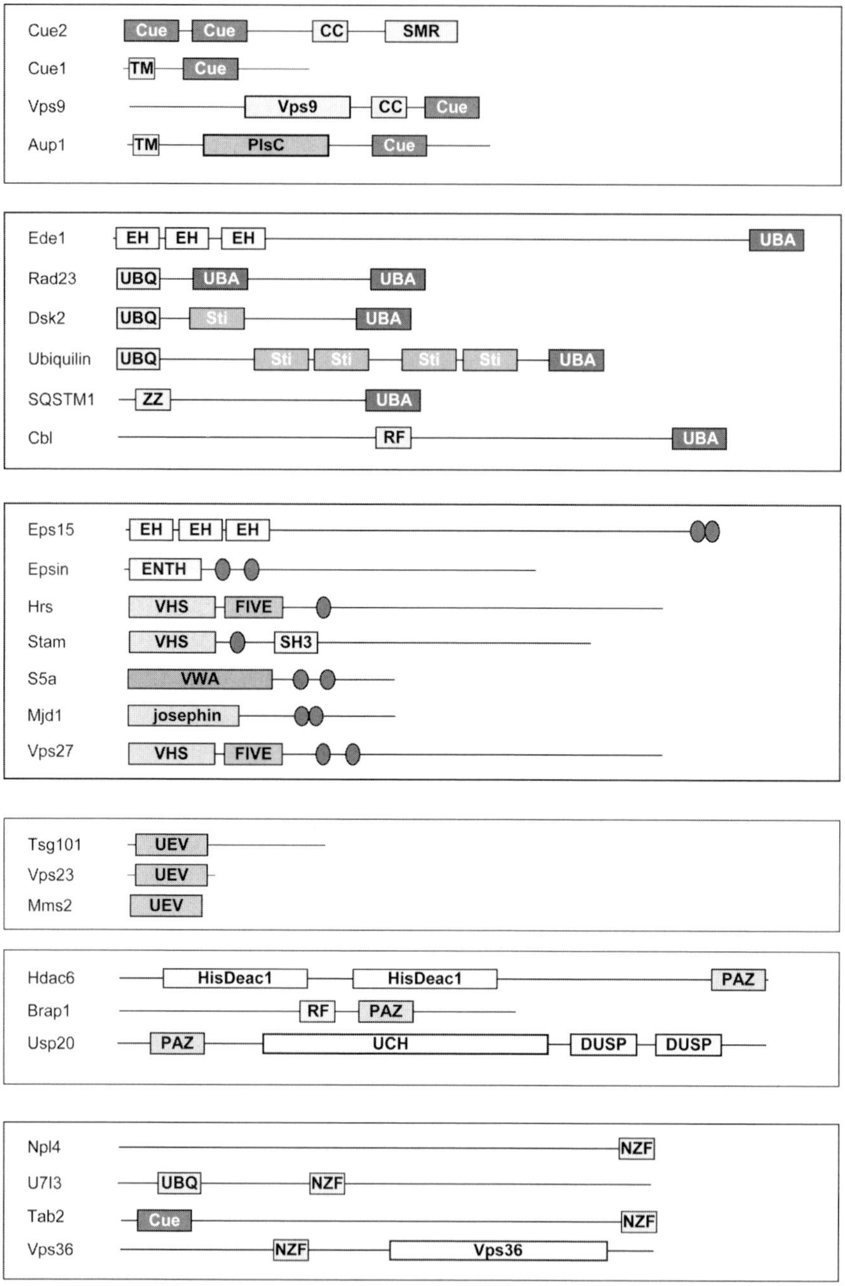

Figure 15.2 Proteins containing Ub-binding regions. The schematic diagram shows the domain architecture of selected proteins containing CUE, UBA, UIM, UEV, PAZ, and NZF. UIM is indicated by ovals. Other domains are drawn as rectangles and are labeled as follows: EH = Eps15-homology domain; ENTH = epsin N-terminal homology domain; VHS = Vps27/Hrs/Stam domain; FYVE = FYVE-finger domain; SH3 = Src-homology 3 domain; VWA = von Willebrand factor type A domain; Josephin = Josephin domain; HisDeac1 = histone deacetylase domain; UBQ = Ub-like domain; RF = RING finger domain; Sti = heat shock chaperonin-binding motif; UCH = Ub C-terminal hydrolase.

As shown in Figure 15.3, despite limited homology at the amino acid level, CUE and UBA domains share similar predicted secondary structures, a feature reflected by the overall similarity between their 3D structures (Figure 15.4a and b, respectively). Both domains consist of a compact three-helix bundle that is stabilized by a hydrophobic core and possesses an unusually large and conserved hydrophobic surface patch (Figure 15.4a–b).

The structure of the first CUE domain of Cue2p is available at high resolution also in complex with Ub (Figure 15.4c) [28] and reveals how the domain interacts with the hydrophobic core of Ub formed by the sidechains of Leu8, Ile44, and Val70. The CUE domain contains two highly conserved motifs, including the Met19-Phe20-Pro21 (MFP) motif, a dileucine motif Leu46-Leu47, and additional conserved hydrophobic residues including Leu39 and Ile43 (amino acid numbering corresponds to that of the Cue2p-1 CUE domain) (Figure 15.3a). Mutations at these positions in the CUE domains of Cue2p or Vps9p impair the interaction with Ub [26, 28]. These findings can be explained by analysis of the structure of the CUE-1 domain of Cue2p in complex with ubiquitin [28]. In the CUE domain the sidechain of Met19 fills the hydrophobic pocket of Ub, and this hydrophobic interaction is further stabilized by the interaction of Ile15, Leu39, and Ile43 with the rim of the pocket (Figure 15.4a and c). The complex is further stabilized by electrostatic interactions between the oppositely charged sidechains of Lys6 and Arg42 of Ub with Asp18 and Asp40 of the Cue2p CUE domain.

To date, there is no available high-resolution structure of a UBA–Ub complex. However, Kang and coworkers [28] modeled the UBA–Ub interaction, based on the structure of the CUE–Ub complex. They selected the first UBA domain of the human RAD23A, due to its high similarity to the first CUE domain of Cue2p. Although the three α-helices assemble in a slightly different manner in the two domains, the structural determinants that are essential for binding to Ub in the CUE domain are well conserved in the UBA domain (Figure 15.4a and b). The same hydrophobic and electrostatic interactions with Ub of the CUE domain are mimicked in the UBA domain by Met173, Tyr175, Val195, Leu199, and by Glu169 and Glu196, respectively. As can be seen from the alignments in Figure 15.3, the highly conserved MFP motif in CUE domains is replaced by the similar highly conserved MGF/Y (Met-Gly-Phe/Tyr) motif in UBA domains. At the sequence level, conservation between the two motifs is not readily noticeable, but it has important structural implications. In the model of the UBA–Ub complex [28], the glycine residue in the MGY motif of the first RAD23A UBA domain occupies a similar spatial position as the invariant proline in the MFP motif of the CUE domain of Cue2p. This glycine residue may be responsible for bringing the methionine and the phenylalanine/tyrosine residues of the MGF/MGY motifs close together, similar to what occurs in CUEs in which the methionine and phenylalanine residues of the MFP are adjacent. Thus, conserved functional and structural features allow the grouping of CUE and UBA domains into a superfamily of three-helical Ub-binding domains.

It is still not clear whether there is any specificity (or preference in binding) of CUE and UBA domains for mono- or polyubiquitin. The CUE domain of Vps9 binds efficiently to both mono- and polyubiquitin, but the CUE domain of Cue1p

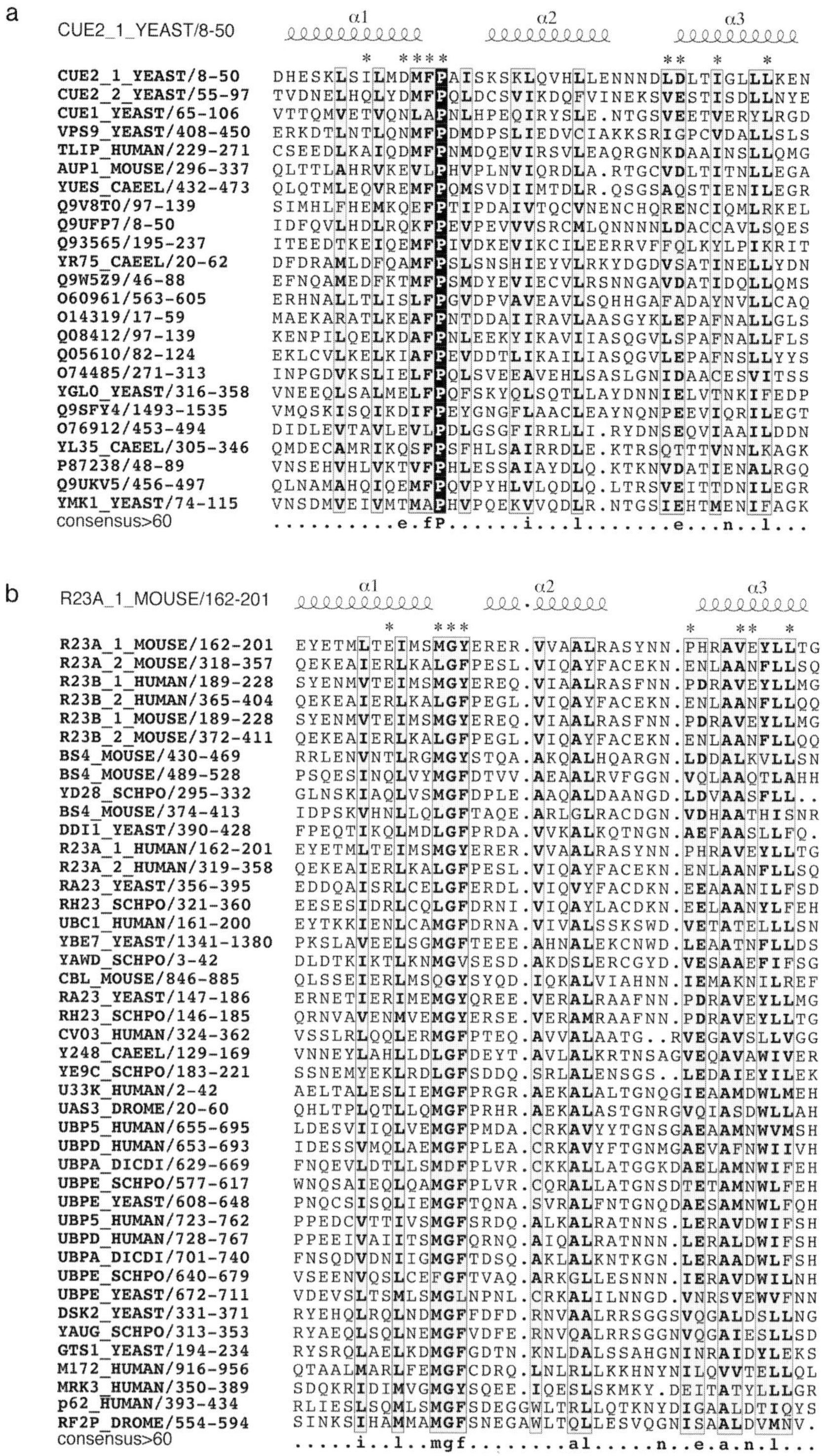

Figure 15.3 (legend see p. 297)

is apparently specific for monoubiquitin [26]. Thus, CUE domains might bind to both mono- and polyubiquitin, possibly with some domain-specific preference. This dual specificity seems confirmed by NMR studies that showed that, in a CUE–Ub complex, although Lys48 of Ub is quite protected, its ε-amino group is still accessible, and any covalent linkage at this position should not perturb the stability of the complex [28]. We should note, however, that available data indicate rather low affinity of CUEs for monoubiquitin. The CUE domain of Cue1p has a K_d of ~160 μM [26]. The Vps9p CUE shows a slightly better affinity (K_d ~20 μM), but this could be because it binds to Ubs as a dimer [29]. Indeed, the K_d of a monomeric Vps9p-CUE–Ub interaction was estimated to be around 170 μM [29]. It remains to be established, therefore, whether additional protein contacts between monoubiquitinated proteins and CUE-harboring proteins contribute to the interactions so as to achieve higher affinity. It should also be considered that relatively low affinities might have been selected for in those situations in which transient interactions and rapid dissociation are required, for instance, in the endocytic route (see below).

UBA domains, on the other hand, seem to preferentially interact with polyubiquitin. Despite reports that UBA domains of the *S. cerevisiae* Rad23p and Did1p proteins can bind to both mono- and polyubiquitin [37], in *Schizosaccharomyces pombe* the UBA domain-containing proteins Rhp23p and Mud1p (orthologs of Rad23p and Did1p, respectively) bind only to polyubiquitin [32]. In surface plasmon resonance experiments, Mud1p showed high affinity (K_d ~30 nM) for tetraubiquitin chains, but no binding to monoubiquitin was observed up to a monoubiquitin concentration of 1 mM [32]. Comparable results were obtained for other UBA-containing proteins [32], and in particular, the UBA domain of Dsk2p binds preferentially to K48-linked polyubiquitin chains [38]. Consistent with these observations, it was recently reported that the UBA domain of p62 (SQSTM1) can bind to polyubiquitin chains containing a minimum of two Ub molecules, but not to monoubiquitin [36], despite previous reports that showed binding to monomeric Ub–Sepharose beads [30].

Figure 15.3 Sequence comparison of selected CUE and UBA domains. Sequence alignment for the CUE and UBA domains was retrieved from the PFAM database (accession codes PF02845 and PF00627, respectively). The alignment was manually edited using the program Jalview (http://www.ebi.ac.uk/~michele/jalview/) and reproduced using the ESPript software [135]. Amino acids are grouped according to their physicochemical properties: H/K/R = polar positive; D/E = polar negative; S/T/N/Q = polar neutral; A/V/L/I/M = nonpolar aliphatic; F/Y/W = nonpolar aromatic; P/G = small residues; and C = cysteine (not groupable with other amino acids). Conserved residues are shaded gray when present at a plurality of > 60% and in black where conservation is 100%. Conserved equivalent residues are in bold. The consensus sequence was generated automatically by the ESPript software using criteria from MULTALIN [136]: uppercase indicates identity, and lowercase, consensus level > 60%. The meanings of the symbols are as follows: ! is any one of I/V, $ is any one of L/M, % is any one of F/Y, and # is any one of N/D/Q/E. Asterisks indicate all residues involved in Ub binding (for the CUE domain) or predicted to be involved (for the UBA domain). (a) Alignment of CUE domains with the secondary structure of the first CUE domain of yeast Cue2p indicated at the top of the alignment. (b) Alignment of UBA domains with the secondary structure of the first UBA domain of mouse RAD23A indicated at the top of the alignment.

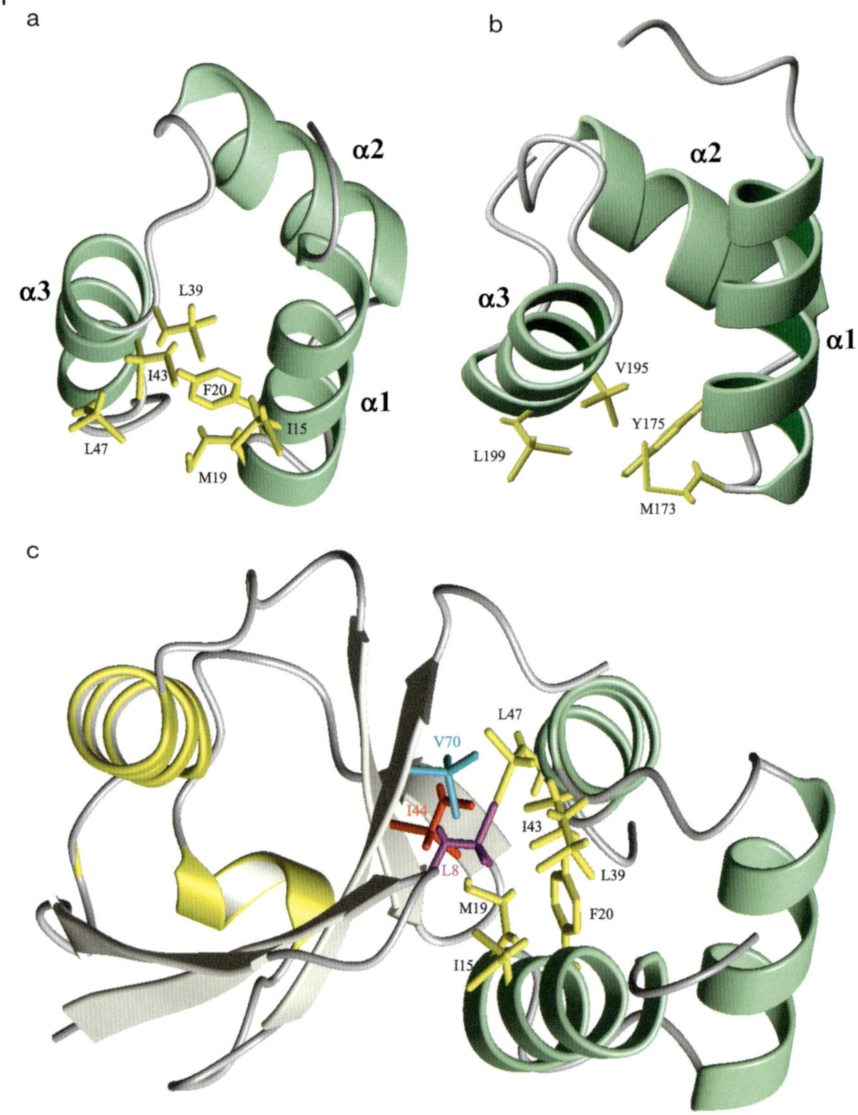

Figure 15.4 Structures of the first CUE domain of yeast Cue2p, the first UBA domain of human Rad23, and the CUE–Ub complex. (a) The first CUE domain of Cue2p, showing residues involved in Ub binding. (b) The first UBA domain of Rad23, showing key residues supposed to be involved in binding to Ub, as predicted by molecular modeling [28]. (c) The CUE–Ub complex. Both Ub (yellow) and CUE (green) are shown, together with key residues involved in the binding interface. PDB accession codes are 1IFY for the UBA domain and 1OTR for the CUE and CUE–Ub complex. Pictures of domain structures presented here and in subsequent figures were prepared with the program MOLMOL [137].

15.3 The Ubiquitin-interacting Motif

A short motif in the 26-S proteasome subunit 5a (S5a), composed of five alternating large and small hydrophobic residues flanked by negatively charged residues, was experimentally identified as the polyubiquitin binding site of this proteolytic organelle [39]. Based on its sequence, Hofman and Falquet identified and refined, through a bioinformatics approach, the ubiquitin interacting motif (UIM) [40]. Subsequently, several UIMs and UIM-containing proteins were experimentally validated as bona fide Ub binders [41–47]. UIM occurs in many proteins (Figure 15.2) and is frequently present in multiple copies within the same protein. It is a very short sequence motif, about 20 residues, and contains the conserved core x [–] [–] [–] [–] Ψ x x A x x x S x x [–], where Ψ represents an aliphatic residue and [–] an acidic residue (Figure 15.5a).

To date, several UIM structures are available, including the crystal structure of the second UIM of Vps27p [48] and the NMR spectra of the tandem UIMs of Vps27p. The first Vps27p-UIM has also been crystallized in complex with Ub [49], and the second UIM of S5a in complex with the Ub-like domain of RAD23A [50]. The UIM forms a short amphipathic α-helix in which all of the conserved residues are exposed on one face, creating a hydrophobic surface that binds to the hydrophobic pocket of Ub (Figure 15.5c). Leu262, Ile263, Ala266, and Ile267 (numbering according to the first UIM of Vps27p) directly contact the hydrophobic patch on Ub formed by Leu8, Ile44, and Val70. The sidechain of Ala266, a highly conserved residue among UIMs, is buried in a small pocket of Ub formed by Leu8, Ile44 His68, and Val70. Thus, structural studies confirm the predictions of functional studies, which have shown that mutagenesis of UIM-Ala266 (or corresponding residues in other UIMs) or Ub-Ile44 impair UIM–Ub interaction [39, 41, 42]. In addition, the conserved N-terminal acidic residues of UIM contact the positively charged patch on Ub formed by Arg42 and Arg72, and this interaction is thought to stabilize the complex, since mutation of these negatively charged residues strongly impairs the binding of the UIM of Hrs to Ub [48]. Interestingly, these hydrophobic and electrostatic interactions very closely resemble those observed in the CUE–Ub complex and predicted for the UBA–Ub complex. Finally, it is notable that the second UIM of Vps27p crystallized as an antiparallel four-helix bundle [48], in which most of the conserved residues involved in Ub recognition are buried into the middle of the bundle. Although this tetrameric assembly immediately suggests modes through which Ub binding might be regulated, it remains to be established whether it really occurs in vivo, and, if so, whether it is biologically relevant.

Structural data are also available for a complex involving a ubiquitin-like molecule, i.e., that between the second UIM of S5a/PSMD4 and the Ub-like domain of RAD23A [50]. As shown in the alignment in Figure 15.5b, there is limited homology between Ub and the Ub-like domain of Rad23 (22.3% and 33.7% identity in yeast and human, respectively); however, the two molecules show an overall high degree of structural similarity (Figure 15.5c and d). This is reflected in similar modalities of interaction of Ub or the Ub-like domain with UIMs. The core hydrophobic patch

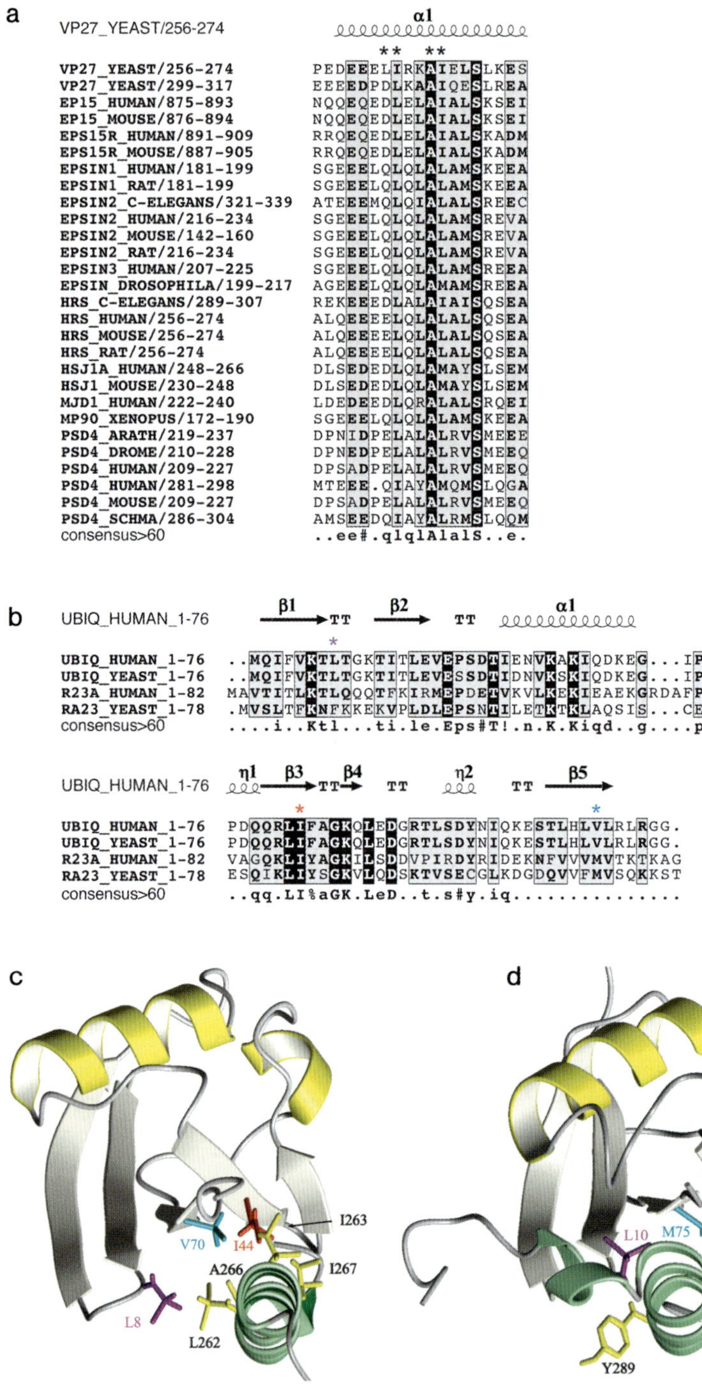

Figure 15.5 (legend see p. 301)

in Ub formed by Leu8, Ile44, and Val70 is formed by Leu10, Ile49, and Met75 in the Ub-like domain of RAD23A. As shown in Figure 15.5d, key interactions have been assigned to Tyr289, Ile287, Met291, and Leu295 of the UIM domain with Leu10, Met75, and Ile49 of the Ub-like domain of RAD23A and resemble those occurring between the UIM of Vps27p and Ub.

As in CUE domains, the interaction between UIMs and monoubiquitin appears to be low-affinity. As assessed by NMR [49, 51] and Biacore [45] studies, Ub binds to UIMs with K_d values in the high micromolar range (300 μM for the Hrs UIM, and ~277 μM for the first and ~177 μM for the second UIM of Vps27p, respectively). The affinity for the Ub-like domain of RAD23A of the second S5a UIM is ~30 fold higher (K_d of ~10 μM) [50], but still far from the affinity of a polyubiquitin chain (minimum of four Ub moieties) for the 26S proteasome (K_d < 200 nM) [52]. This can be explained by the fact that the dissociation constants measured in the former cases are representative of the interaction between isolated UIMs and one Ub molecule or Ub-like domain, but in the latter they represent the interaction of a polyubiquitin chain with the complete 26S proteasome particle. This may imply cooperative binding of the two UIMs of S5a to polyubiquitin or simply a lower affinity of the UIM for monoubiquitin relative to polyubiquitin [39, 53].

Despite the low affinity, almost all UIMs tested so far bind to monoubiquitin, in addition to polyubiquitin. This occurs for the UIMs of yeast Ent1p, Ent2p, and Vps27p and of mammalian Hrs, Epsin1, and Eps15 [41, 42]. We should note, however, that it has been reported that the two UIMs of ataxin-3 appear to bind exclusively to polyubiquitin chains, since no binding to free Ub, monoubiquitin fusion proteins, or ubiquitin chains containing fewer than four Ub moieties could be detected [47]. Here, the specificity for polyubiquitin might be directly linked to a putative function of ataxin-3 at the proteasomal level, since it has been proposed that ataxin-3 is a transiently associated multiubiquitin chain recognition subunit in the proteasome that receives ubiquitinated substrates through the concerted action of VCP/p97 and shuttle factors, such as Rad23 [54].

In summary, the issues of whether UIMs have specificity for mono- or polyubiquitin, of how meaningful stoichiometries of interaction can be obtained in vivo, given the apparent low affinity, and of whether tandem UIMs act in concert to bind to polyubiquitin efficiently are still a matter of debate and warrant further investigation.

Figure 15.5 Structures and modality of interaction between UIMs and Ub or Ub-like molecules. (a) Alignment of selected UIM domains from various proteins. Analysis and alignment are as in Figure 15.3. The secondary structure of the first UIM domain of yeast Vps27p is reported at the top of the alignment. Asterisks indicate residues involved in Ub binding. (b) Alignment between Ub and the Ub-like (UBQ) domain of Rad23 (both from yeast and human). The secondary structure of the human ubiquitin is reported at the top of the alignment. Asterisks indicate the key conserved residues that are involved in binding to UIMs and that form the hydrophobic pocket on the Ub surface. (c) Solution structure of the complex between the first UIM of Vps27p and ubiquitin (PDB accession code 1Q0W). Only key residues are shown. (d) Solution structure of the Ub-like domain of Rad23 in complex with the second UIM of S5a (PDB accession code 1P9D).

15.4 The UEV Domain

The UEV (ubiquitin E2 variant) domain, present in Tsg101 and Mms2 proteins (Figure 15.2), shares significant similarity with the catalytic domain of E2 enzymes, the UBCc (ubiquitin-conjugating catalytic) domain, at both the sequence and structural levels (Figure 15.6). At variance with UBCc domain of E2 enzymes, UEV domains are unable to catalyze Ub transfer because they lack the active-site cysteine that forms the transient thiol–ester bond with the C-terminal glycine (Gly76) of Ub [55, 56]. However, UEV domains retain the ability to bind to Ub, and they have also been shown, in Mms2, to act as cofactors in ubiquitination reactions [57–59]. The UEV domain fold resembles that of other E2 ligases and is composed of four helices packed against one side of a four-stranded antiparallel β-sheet. The UEV domains of both Mms2 and Tsg101 lack the two C-terminal helices found in all E2 proteins characterized so far (α4 and α5 in Figure 15.6d), a hallmark feature that differentiates UEVs from canonical UBCc domains [60]. The UEV of Tsg101 differs from that of Mms2 by the presence of an additional α-helix (α1 in Figure 15.6b), and also in that the first two β-strands, which project beyond the main body of the domain, form an extended β-hairpin tongue that contain residues essential for binding Ub [61]. These differences show how the Tsg101-UEV domain has diverged even further than Mms2 from the canonical E2/UBCc fold and reflect the different functional properties of the three domains, which are analyzed below.

Structural studies of yeast and human Mms2 have revealed how the Ub-binding abilities of their UEVs have been 'adapted' so as to serve as cofactors in a ubiquitination reaction leading to the formation of K63-linked polyubiquitin chains, instead of the canonical K48-linked chains. In both yeast and human, Mms2 forms a heterodimer with the E2 enzyme Ubc13 (Figure 15.7) [62, 63]. Both Mms2 and Ubc13 are essential in the formation of K63-linked polyubiquitin chains, and Ubc13 is the sole E2 that can bind to Mms2 and form an active complex [58, 62, 63]. The K63-chain assembly reaction performed by the Mms2/Ubc13 complex, as deduced from the molecular model proposed by VanDemark and colleagues [62], is achieved by the correct positioning of an Ub (acceptor) molecule, which is bound to Mms2, so as to present its Lys63 (and not Lys48) to the catalytic cysteine of the Ubc13 E2 enzyme carrying a 'donor' Ub molecule.

In the proposed model, Ubc13 carries a Gly76-linked donor Ub molecule bound to its catalytic cysteine and accommodates the C-terminal tail of ubiquitin in a hydrophobic channel near the active cysteine, similar to what has been shown for the canonical interaction between ubiquitin and the human Ubc2b E2 enzyme [64]. The correct orientation of donor Ub on the Ubc13 surface involves Ala110 of Ubc13, since mutation of this residue significantly inhibits chain assembly. This mutation does not, however, alter thiol–ester bond formation between Ub and the active Cys of Ubc13, suggesting that it is dispensable for the binding of the donor Ub to Ubc13 but required for correct positioning of the donor Ub [62]. The acceptor Ub lies in a channel formed by both the Mms2 and Ubc13 surfaces and contacts Mms2 (through Ub-Ile44) in a region centered around Ile57 of Mms2 that

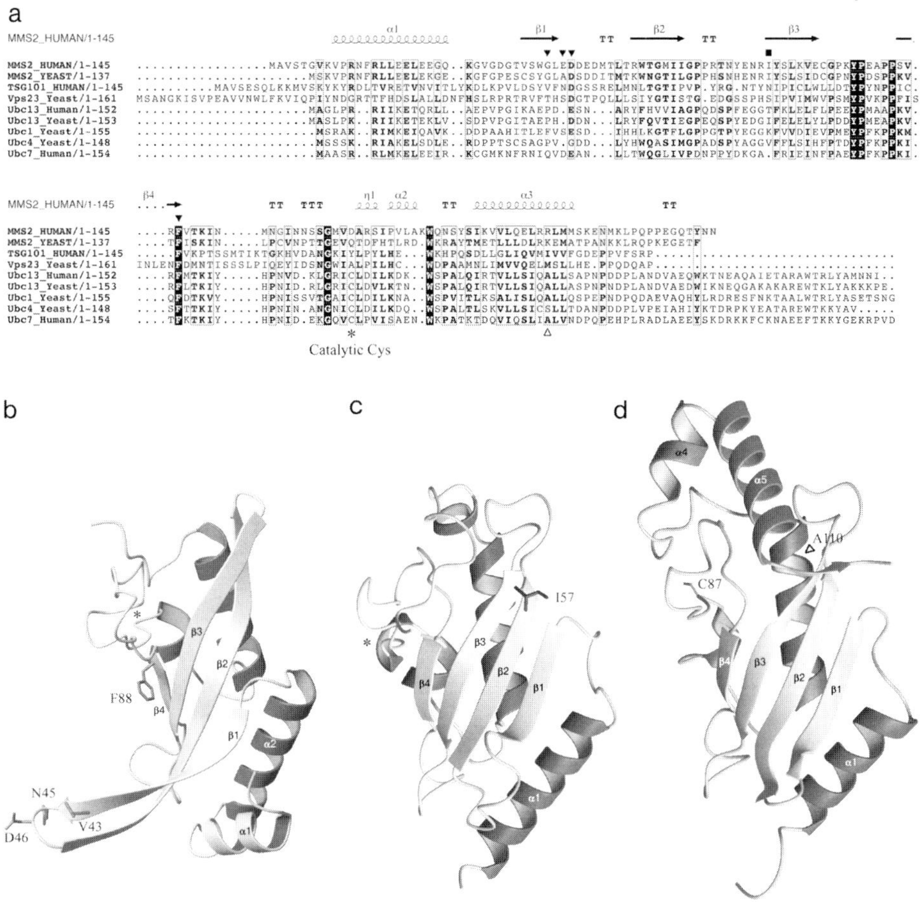

Figure 15.6 Sequences and structures of UBCc and UEV domains. (a) Alignment of selected UEV and UBCc domain-containing proteins from human and yeast. Analysis and alignment are as in Figure 15.3. The secondary structure of the UEV domain of human Mms2 is indicated at the top of the alignment. Black arrowheads indicate residues of Tsg101 involved in Ub binding. The black square indicates Ile57 of Mms2 which is supposed to interact directly with Ile44 of Ub. The open arrowhead indicates the position of Ala110 in Ubc13, which is not visible because it lies on the back of the structure shown in (d). The asterisk indicates the catalytic cysteine present in E2 enzymes but not conserved in UEV domains. (b) Structure of the Tsg101 UEV domain (PDB accession code 1KPQ). (c) Structure of the Mms2 UEV domain (PDB accession code 1J7D). (d) Structure of the UBCc domain of Ubc13 (PDB accession code 1J7D). For Ubc13, the active site cysteine is shown, and in Mms2 and Tsg101 UEV the asterisks indicate the positions of the inactive site. For all the structures residues involved in Ub binding are shown.

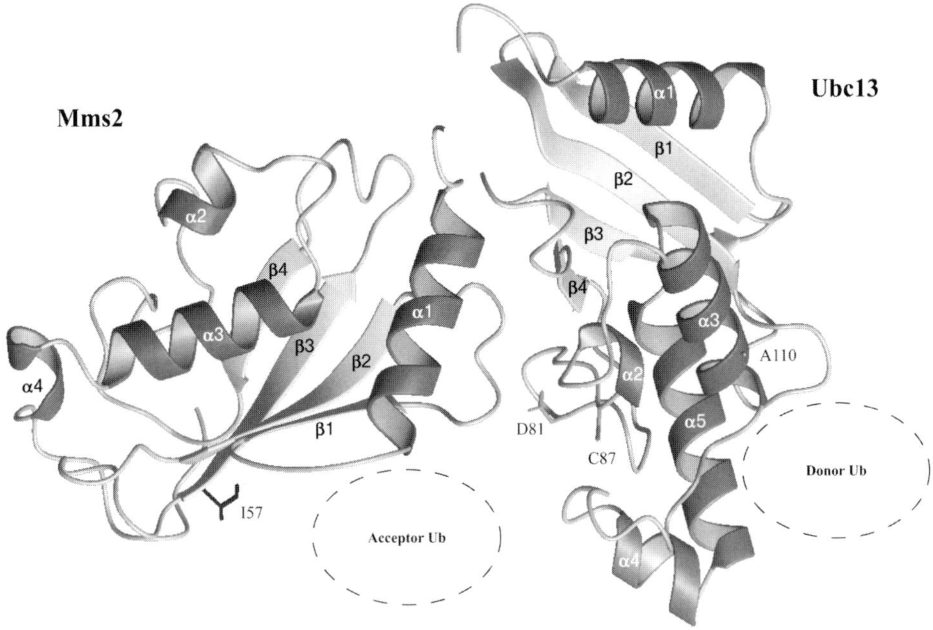

Figure 15.7 Crystal structure of the Mms2–Ubc13 complex. Mms2 is on the left and Ubc13 on the right; modeled ubiquitin molecules are not shown, but their positions are indicated. Acceptor ubiquitin lies near Ile57 of Mms2 in such a way as to present Lys63 close to the active Cys87 of Ubc13. The complex in stabilized by Asp81 of Ubc13. Donor ubiquitin binds to Ubc13 near α helix 3 and its C-terminal tail protrudes toward Cys87, to which it is covalently bound through Gly76. Ala110 of Ubc13 is involved in the correct orientation of donor ubiquitin relative to acceptor ubiquitin, which is not required for thiol–ester bond formation between Ub and the active Cys87 of Ubc13, but significantly inhibits chain assembly [62].

encompasses the C-terminal half of helix 1, the outer edge of β-strand 1, and the loop between β-strands 1 and 2 [62]. In this model, Ub is positioned so as to present the Lys63 sidechain to the active site Cys87 of Ubc13, and its position is stabilized by Ubc13 Asp81, whose mutation impairs diubiquitin synthesis [62] without affecting the catalytic activity of Ubc13 or its binding to Mms2. The specificity achieved is quite high, because neither manual nor computational approaches yielded the positioning of an acceptor Ub in a way suitable to exposing Lys48 to the active cysteine of Ubc13 in the Mms2–Ubc13 complex [62].

The UEV domain of Tsg101 behaves like a pure Ub-binding surface and is not involved in ubiquitination reactions [65]. As shown by chemical-shift mapping and mutagenesis experiments [61], it binds to Ub different than does Mms2–UEV and at low affinity (K_d ~500 μM) [57]. The binding site for Ub is formed by the lower half of the four-stranded sheet, along β-strands 1 and 2 and close to the vestigial active site, which flanks an extended surface hydrophobic patch formed by β-strands 3 and 4 (Figure 15.6b). Consistent with these observations, mutations of Tsg101-UEV Phe88, Val43, Asn45, and Asp46 (Figure 15.6a and b) severely reduce binding to Ub [61]. The region of Ub that interacts with Tsg101-UEV is mapped to β-strands 3 and 4 and the N-terminal end of β-strand 5 of Ub. This region of Ub includes Lys48 and contains several polar and charged residues, which may form electrostatic interactions with residues of Tsg101. This residue seems to be sequestered in the UEV–Ub interface, thus probably precluding the formation (and possibly the recognition) of K48-polyubiquitin chains. This hypothesis is in line with the demonstrated role of Tsg101 in the recognition of monoubiquitinated proteins in the endocytic pathway [59, 66] (see below).

Interestingly, the UEV domain of Tsg101 has been found to bind proline-rich sequences having the consensus tetrapeptide motif P(S/T)AP [57, 67–69] in a site distinct from that used to bind ubiquitin [61]. The binding site is shaped by the loop that connects strands 2 and 3, the C-terminal residues of the domain, and part of the vestigial active-site loop [61, 70]. The mechanism of proline recognition by the Tsg101 UEV domain is similar to that of SH3 and WW domains [70], making it part of the superfamily of proline-binding modules (see also Chapters 2–5 of this book). Through this kind of interaction Tsg101 can bind to Hrs, which contains a PSAP motif in its sequence [71], and together, they coordinate the endosomal trafficking of cargo receptors and the process of plasma membrane receptor down-regulation (see below).

In summary, the 'E2 fold' has achieved remarkable versatility in the cell and is used for a variety of functions, based on its ability to interact with Ub and, at least for the UEV domain of Tsg101, with proline-rich motifs. UBCc domains in canonical E2 enzymes are catalytically competent E2 folds that act as covalently attached Ub carrier modules along the ubiquitination pathway and participate to the transfer of Ub to the target substrate directly or in conjunction with an E3 protein ligase (reviewed in [72]) (see also Figure 15.1). The UEV domains are, in contrast, catalytically inactive E2 folds with a dual function: they can act as cofactors in the process of substrate ubiquitination, as in Mms2, or as 'pure' Ub receptors, as in the Tsg101 UEV domain.

15.5 The PAZ and NZF Domains

Recently, two additional Ub binding domains have been identified, NZF (novel zinc finger) and PAZ (polyubiquitin-associated zinc finger). As also indicated by their names, these modules are zinc finger domains endowed with Ub-binding properties.

The PAZ domain was initially identified as a Ub-binding module in mouse and human histone deacetylase 6 (HDAC6) [73, 74] and corresponds to the Znf-UBP and ZF-UBP domains from the SMART and PFAM databases, respectively. It is also present in several deubiquitinating enzymes and in the BRCA1-associated protein (BRAP) (Figure 15.2 and 8a). It remains to be established, however, whether these proteins can interact with Ub. Another domain named PAZ exists in the PFAM database (PF02170) but it is not related to the one described here.

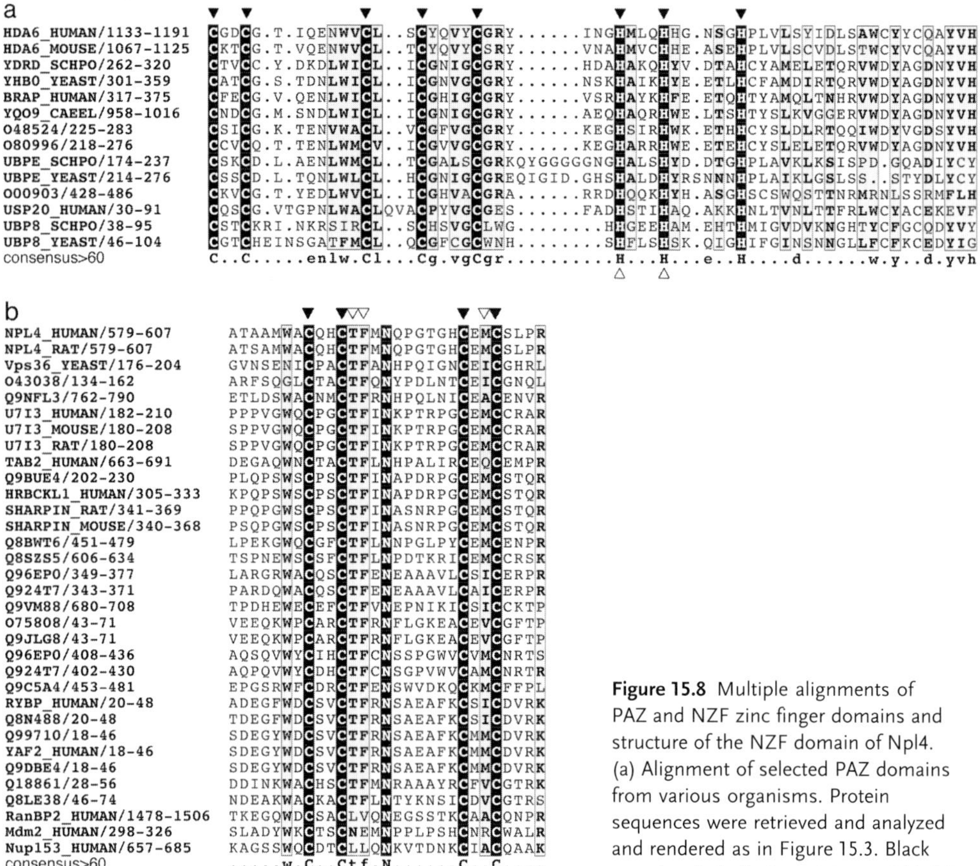

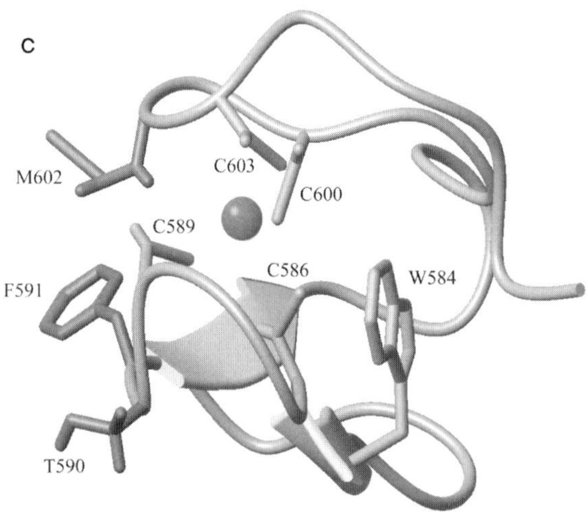

Figure 15.8 Multiple alignments of PAZ and NZF zinc finger domains and structure of the NZF domain of Npl4. (a) Alignment of selected PAZ domains from various organisms. Protein sequences were retrieved and analyzed and rendered as in Figure 15.3. Black arrowheads indicate cysteine and histidine residues conserved in the zinc finger domain, and open arrowheads indicate histidine residues whose mutation to alanine abolishes the binding of the PAZ domain of HDAC6 to Ub. (b) Alignment of selected NZF domains from various organisms. The zinc finger domains of Mdm2, RanBP2, and Nup153 are included for comparison. Black arrowheads indicate the four cysteine residues involved in zinc coordination, and open arrowheads point to residues involved in Ub binding. (c) Structure of the NZF domain of Npl4 (PDB accession code 1NJ3). Residues indicated in B are shown, together with Trp584; together they constitute the hydrophobic core of the domain. Zinc is indicated by a sphere.

There is no published structure of the PAZ domain so far. Seigneurin-Berny and coworkers have shown that mutations of two conserved histidines completely abolish the binding of HDAC6 to Ub [73]. We should point out, however, that these residues are supposed to be essential for formation of the zinc finger fold; thus, the engineered mutations might have disrupted the overall structure of the domain rather than just residues involved in Ub binding. The PAZ domain of HDAC6 can bind both monoubiquitin, in the form of GST-fusion protein, and polyubiquitin chains, although the reported results are not in perfect agreement [73, 74] and the isolated domain behaves differently than the full-length protein in the binding assays [73]. Thus, the issue of the binding specificity of the PAZ domain remains an open one.

The NZF domain (Figure 15.8b) was initially identified as a Ub-binding module in the human Npl4 and yeast Vps36p proteins [33, 75, 76]. Surprisingly, the human homolog of Vps36 and the yeast homolog of Npl4 do not have the domain. By database searches using consensus patterns, the NZF domain has been found in other proteins such as U7I3/RBCK1 (Ub-conjugating enzyme 7 binding protein 3), RYBP (YY1- and E4TF1-associated factor 1), Tab2, and Mdm2 [33] (Figure 15.8b).

NFZ domains are described in the Prosite database as part of a wider family called RanBP2 zinc finger, which includes zinc finger domains present in RanBP2 and Nup153, which are nuclear pore proteins that bind RanGDP via this motif. However, important features distinguish NZF domains from the zinc fingers of RanBP2 and Nup153 [77]. First, NZF domains from Npl4, Vps36, and U7I3/RBCK bind to Ub, but the zinc fingers of RanBP2 and Mdm2 do not [33]. Furthermore, the latter zinc fingers differ from 'authentic' NZF domains in the absence of the conserved TF dipeptide that follows the second cysteine residue in the domain (Figure 15.8b). We should also note that the NZF domain of Tab2 binds to Ub with lower efficiency than other NZFs. This is reflected at the sequence level by the fact that the conserved hydrophobic residue at position –1 relative to the fourth cysteine residue, present in all NZFs, is replaced by a polar residue in Tab2 (see below).

The structure of the NZF domain of Npl4 provides explanations for all these differences in binding to Ub. Mutagenesis analysis and chemical-shift mapping of the interaction surfaces of Npl4 NZF and Ub reveal that Thr590, Phe591, and Met602, which correspond to the conserved TF motif and to the hydrophobic residues at position –1 relative to the fourth cysteine in the Npl4 NZF domain (open arrowheads in Figure 15.8b), are probably involved in Ub binding (Figure 15.8c). On the Ub surface the most 'shifted' residues are Leu8, Ile44, Val70, Leu71, and Leu73. This interface between NZF and Ub resembles that formed by other Ub-binding domains, as discussed before [77]. The NZF domains of Npl4 and Vps36 can bind both mono- and polyubiquitin chains, as assessed by pulldown experiments [33]. The affinity of the binding to monoubiquitin, K_d ~122 μM for Npl4 and K_d ~199 μM for Vps36 NZF domains [77], is similar to those observed for other domains, as previously mentioned.

15.6 Ubiquitin-based Networks

The surprising abundance of Ub-binding modules predicts the establishment of extensive networks in the cell, involving Ub-modified proteins. The existence of ubiquitin-like domains biosynthetically present in several proteins [20] and of Ub-like modifiers such as SUMO [23] further projects a higher level of complexity on such networks. In this section we concentrate on some emerging functions of Ub-based networks.

Although the best-characterized function of Ub is to promote degradation of proteins, nonproteasomal functions of Ub are becoming increasingly apparent [2, 78]. A schematic representation of some of the connections involving Ub and Ub receptors in these pathways is depicted in Figure 15.9. In particular, monoubiquitination appears to serve as a signaling-inducible modification, involved in the regulation of many cellular processes. Monoubiquitination, for instance, is induced by extracellular signaling stimuli. Activated receptor tyrosine kinases (RTKs) induce their own monoubiquitination. This process requires the kinase activity of the receptor and is mediated by the E3 ubiquitin ligase Cbl [79]. It has been recently demonstrated that RTKs, which were long thought to be polyubiquitinated, are actually monoubiquitinated at multiple sites [80]. In addition, the monoubiquitination of RTKs has a signaling impact that is sufficient to promote receptor internalization and degradation [80]. A similar role for receptor monoubiquitination has been described in yeast [14]. Active RTKs, moreover, induce the monoubiquitination of several intracellular proteins, including Eps15, Eps15R, epsins, Hrs, and CIN85 [41, 81, 82]. In some instances, the combined analysis of the functions of these proteins in mammals and yeast indicates a role for their monoubiquitination in endocytosis and intracellular trafficking (extensively reviewed in [83–86]).

In the endosomal compartment the proteins Tsg101 and its yeast homolog Vps23p function as subunits of the ESCRT-1 protein complex and have a central role in late endosomal trafficking [66, 85]. Tsg101 is recruited to the endosomal membrane possibly by Hrs [71, 87] through the interaction of its UEV domain with the proline-rich motif PSAP present in Hrs. Hrs functions in the sorting of ubiquitinated proteins, coming from the endocytic route, to early endosomes. Tsg101 facilitates the further trafficking of ubiquitinated cargo-bearing endosomes to late endosomes and promotes their budding into the multivesicular bodies (MVBs) [66, 88], from where they are subsequently delivered to the lumen of the lysosome/vacuole.

There is a surprising relation between the cellular MVB machinery and the budding of viral particles form cells infected with the human immunodeficiency virus-1 (HIV-1). Human Tsg101 has been identified as the critical protein required for budding of HIV-1 and Ebola viruses from the cell surface [57, 67, 68]. These viruses do not encode their own machinery for accomplishing this function but recruit and reprogram specific cellular proteins. The HIV Gag protein, which directs virus assembly and budding (for review see [89–91]), provides a docking site for Tsg101 at the plasma membrane level, through a conserved PTAP motif present in its p6 'late domain' region. Pornillos and coworkers [87] have proposed a model in

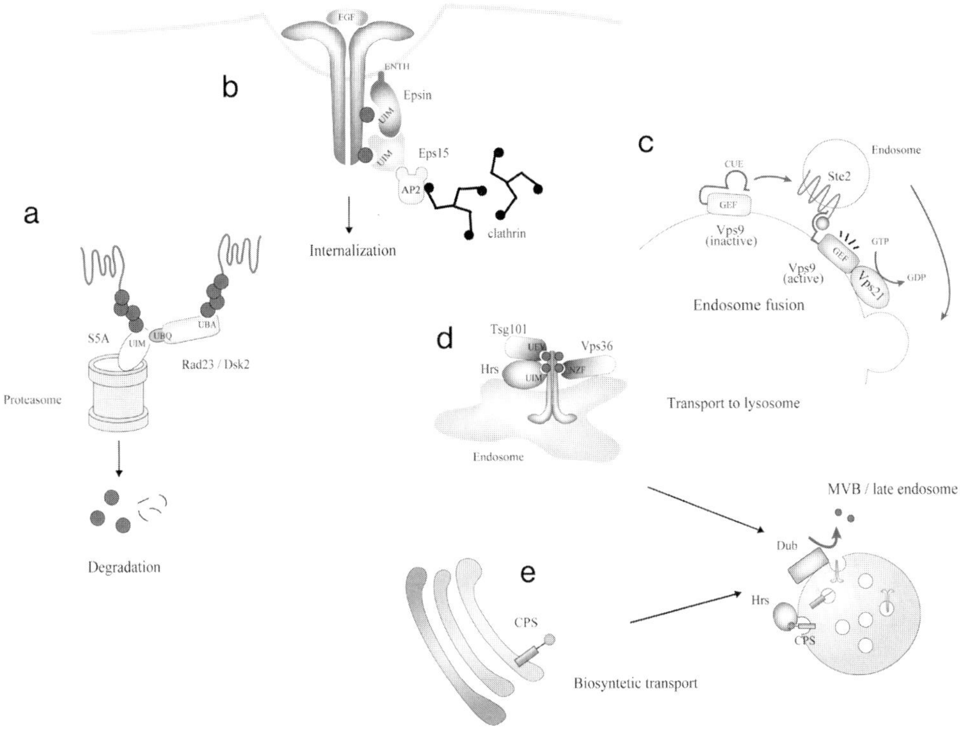

Figure 15.9 Ub–Ub-receptor interactions in proteasomal function, endocytosis, and vesicular trafficking. A schematic representation of the most relevant functions in the proteasome degradative pathway, in endocytosis, and in vesicular trafficking, mediated by Ub–Ub-receptors, is shown. The picture represents a conceptual integration of data obtained in mammals and in yeast. Monoubiquitin is represented by black circles. (a) Polyubiquitinated substrates are recognized by the S5a subunit of the proteasome via its UIM domains and degraded; free Ub is recycled. Rad23 and Dsk2 bind polyubiquitinated proteins via their UBA domains and, by associating with the proteasome via their UBQ domains, might help deliver substrates to the proteasome [38]. (b) Eps15 and epsin are adaptor proteins in EGFR internalization. They are most likely recruited to ubiquitinated plasma membrane receptors through their UIMs (reviewed in [86]). (c) The interaction between monoubiquitinated internalized receptors and a Ub receptor (Vps9p, displaying a CUE domain) is critical for endosome fusion in yeast. Vps9p (the yeast homolog of mammalian Rabex-5) is a GEF that regulates the activity of Vps21 (the yeast homolog of mammalian Rab5), which in turn promotes membrane fusion at the endosomal level. Biochemical and genetic data support the possibility that the CUE domain of Vps9p inhibits its GEF activity in cis. Upon interaction of Vps9p with ubiquitinated cargo receptors (in the picture, Ste2p), this inhibition is relieved, allowing activation of Vps21p and endosome fusion. The high degree of conservation of the entire system allows postulating a similar mechanism of regulation in mammals [26, 27]. (d) Hrs functions at the endosomal level, where internalized receptors are sorted to different destinations. Hrs recruits ubiquitinated receptors through the UIM and sorts them to the multivesicular body (MVB) and hence to the lysosome for degradation. The process also involves the ESCRT complex (not shown) and the Ub C-terminal hydrolase DoA4 (Dub) (reviewed in [85]). Tsg101 and Vps36 act in the same compartment, by binding to ubiquitinated cargo and promoting their sorting into the late endosome compartment [65] and their budding into the MVB. (e) Ubiquitinated biosynthetic cargoes, such as carboxypeptidase S (CPS), are also sorted to the MVB through the UIM of Hrs (reviewed in [85]).

which the HIV Gag protein mimics the Hrs recruiting of Tsg101 to the endosomal compartment by redirecting it to the plasma membrane. In support of this model, the affinity of the Tsg101 UEV domain for the PTAP element of HIV Gag is higher than its affinity for the PSAP sequence present in Hrs [87]. Moreover, overexpression of specific peptide sequences from the Gag protein relocalizes Tgs101 from the endosomes to the plasma membrane [67]. At the level of the plasma membrane, Tsg101 is thought to promote the budding of viral particles by recruiting all the components of the machinery involved in the budding process, which are normally exerted at the late endosomal compartment level (for recent reviews see [85, 91]). Accordingly, depletion of Tsg101 protein, mutant virus lacking the PTAP motif, or mutation in the UEV domain of Tsg101 inhibit HIV budding [61, 87]. We should note that the HIV Gag protein is monoubiquitinated [92]. This modification stabilizes and improves the interaction of Gag with Tsg101 [61] and correlates with the efficient budding of virus particles [93–96].

Another important function of monoubiquitination is regulation of histone function. Histone ubiquitination, which is conserved from yeast to mammals, has been known for a long time, but its mechanistic impact has been appreciated only recently, as a result of yeast studies that have elucidated an important function of monoubiquitinated histone H2B. The E2–E3 complex Rad6p/Bre1p ubiquitinates Lys123 in H2B. This event causes, in turn, methylation of Lys4 and Lys79 in histone H3, a modification directly involved in gene silencing [97–100]. How the ubiquitination of histone H2B directs the activity of H3 site-specific histone methylases is not known. Of course, in principle, not all Ub modifications must necessarily lead to Ub–Ub-receptor interactions. Indeed, monoubiquitination is a bulky modification that, especially in the context of chromatin, can lead to global changes in folding. However, the emergence of mUb as a protein networking device puts forward the intriguing possibility that the interaction of monoubiquitinated histones with yet-unidentified Ub receptors plays a role in transcriptional regulation.

Signaling stimuli of intracellular origin, such as DNA damage, also induce monoubiquitination. Molecular genetics studies of Fanconi anemia have revealed how, after DNA damage a macromolecular complex (the FA complex), containing five proteins associated with Ub ligase activity, induces monoubiquitination of the FANCD2 protein. This event leads to recruitment of FANCD2 into BRCA1-containing nuclear foci, which are in turn connected with DNA repair and checkpoint function [101, 102]. Recently, a new component of a Fanconi anemia protein complex, PHF9 (renamed FANCL), has been identified. It possesses E3 Ub ligase activity in vitro and is essential for FANCD2 monoubiquitination in vivo [103]. Interestingly, the FA-dependent monoubiquitination of FANCD2 is also required during the S phase of normal cell cycle progression [104].

Other pathways, such as those connected with regulation of transcription by NF-kB through IkB kinase β [8] or with regulation of DNA repair through PCNA (proliferating cell nuclear antigen) [105], appear to be regulated by monoubiquitination events as well. In reality, the real impact of monoubiquitination on protein function is only starting to be appreciated. The number of characterized monoubiquitinated proteins is quite low (Tables 15.1 and 15.2), due to lack of

Table 15.1 Monoubiquitinated proteins in mammals and their functions.

Protein name	*Function*	*Role of monoubiquitination*	*Reference*
p53	Cell cycle master regulator	Regulation/degradation?	109
Histone H2A	Structural nucleosomal component	Chromatin remodeling (associated to active transcription, repair)	110
Histone H2B	Structural nucleosomal component	Chromatin remodeling (associated to active transcription, repair)	111
Histone H3	Structural nucleosomal component	Chromatin remodeling (associated to active transcription, repair)	112
Histone H1	Structural nucleosomal component	Chromatin remodeling (associated to active transcription, repair)	113
PCNA	Adaptor in DNA repair and synthesis	Damage-induced PCNA ubiquitination is elementary for DNA repair	105
FANCD2	Involved in DNA repair And synthesis	Targeting to subnuclear Loci where repair and synthesis occur	102
EGFR	Receptor tyrosine kinase	Internalization and sorting signal	80
PDGFR	Receptor tyrosine kinase	Internalization and sorting signal	80
IL2R beta	Receptor tyrosine kinase	Sorting towards degradation	114
TCR	Receptor tyrosine kinase		115
CIN85	Endocytic adaptor		81
eps15	Endocytic adaptor		82
epsin 1	Endocytic adaptor		41
Hrs1	Endocytic adaptor		41
Jak2	Signaling molecule		116
Tyrosine hydroxylase	Rate-limiting enzyme in the biosynthesis of catecholamines		117
Phosphoglycerate mutase-B	Intramolecular transferase		118
Calmodulin	Calcium regulator	Decreases the biological activity of calmodulin towards phosphorylase kinase	119
Actin	Structural protein	Provide increased stability of the microfilaments, effector of plant response against microbes	120
(HIV, RSV, MuLV) GAG	Viral structural protein	Involved in the budding of the virus	121
Ro52/Trim21/SSA	Sjogren syndrome antigen A1		122

Table 15.2 Monoubiquitinated proteins in yeast and their functions.

Protein name	*Function*	*Role of monoubiquitination*	*Reference*
Ste2p	Pheromone α receptor	Internalization and sorting to MVB	123
Ste3p	Pheromone a receptor	Internalization and sorting to MVB	124
Ste6p	ABC protein	Internalization and sorting to MVB	125
Pdr5p	ABC protein	Internalization and sorting to MVB	126
Fur4p	Uracil permease	Internalization and sorting to MVB	127
Gap1p	Amino acid permease	Internalization and sorting to MVB	128
Tat2p	Tryptophan permease	Internalization and sorting to MVB	129
Gal2p	Galactose permease	Internalization and sorting to MVB	130
Zrt1p	Zinc transporter	Internalization and sorting to MVB	131
12-TMS	Maltose transporter	Internalization and sorting to MVB	132
CPS	Carboxypeptidase S	Sorting from golgi to MVB	59
Histone H2B	Structural nucleosomal component	Important in mitotic cell growth and meiosis	133
Vps9p	GEF for Rab		28
Calmodulin	Intracellular calcium receptor		134

systematic efforts so far. However, even this initial list reveals how many other biological functions might be regulated through Ub-mediated interactions. There is, therefore, great need for systematic approaches aimed at elucidating the monoubiquitin proteome. In principle, the approach is feasible, as demonstrated by the recent initial characterization of the yeast Ub proteome [106], and should enormously advance our knowledge of the impact of the nonproteolytic functions of ubiquitination on cellular homeostasis.

15.7
Conclusions

Ubiquitin modification constitutes a multifaceted mechanism to regulate many important physiological processes. Our understanding of the system is still quite rudimentary, and many outstanding questions await answers. First, how does ubiquitination, and in particular monoubiquitination, regulate protein function? The evidence reviewed here indicates that monoubiquitination should often be regarded as a signaling post-translational modification that is 'read' by Ub receptors intracellularly and that enables the formation of dynamic networks of protein–protein interactions. Ubiquitination is, however, a bulky modification, and thus it is projected to also have effects in cis on protein function, for example, by regulating the activity of ubiquitinated enzymes.

A second, and very important, question relates to the specificity of recognition between monoubiquitinated proteins and Ub receptors. It is unlikely that all Ub receptors recognize all monoubiquitinated proteins. But then, what does determine specificity? Although structural knowledge is being acquired on the modality of interaction between Ub and ubiquitin-binding modules, the issue of specificity remains unsolved. Another facet of the problem relates to how ubiquitin receptors discriminate between polyubiquitinated proteins, which are projected to be the vast majority of ubiquitin-containing proteins in the cell, and monoubiquitinated proteins. This question awaits answers as well.

Finally, since many E3 ligases are capable of both poly- and monoubiquitination, the issue of how the enzyme 'decides' its catalytic mode (whether to mono- or polyubiqutinate) is still obscure. Subversion of the ubiquitin pathway plays a role in many diseases [107, 108]. Our ability to harness and modulate the system for therapeutic purposes will largely depend on providing answers to the above questions.

Acknowledgments

Our work is supported by grants from AIRC (Italian Association for Cancer Research), Human Science Frontier Program, IARC (International Association for Cancer Research), the European Community (VI Framework), the Telethon Foundation, the Monzino Foundation, and the Italian Ministry of Health.

References

1 Wilkinson, K. D., Roles of ubiquitinylation in proteolysis and cellular regulation. *Annu. Rev. Nutr.* **1995**, *15*, 161–189.

2 Hershko, A., Ciechanover, A., The ubiquitin system. *Annu. Rev. Biochem.* **1998**, *67*, 425–479.

3 Spence, J., Sadis, S., Haas, A. L., Finley, D., A ubiquitin mutant with specific defects in DNA repair and multiubiquitination. *Mol. Cell Biol.* **1995**, *15*, 1265–1273.

4 Baboshina, O. V., Haas, A. L., Novel multiubiquitin chain linkages catalyzed by the conjugating enzymes E2EPF and RAD6 are recognized by 26 S proteasome subunit 5. *J. Biol. Chem.* **1996**, *271*, 2823–2831.

5 Chau, V., Tobias, J. W., Bachmair, A., Marriott, D., Ecker, D. J., Gonda, D. K., Varshavsky, A., A multiubiquitin chain is confined to specific lysine in a targeted short-lived protein. *Science* **1989**, *243*, 1576–1583.

6 Weissman, A. M., Themes and variations on ubiquitylation. *Nat. Rev. Mol. Cell Biol.* **2001**, *2*, 169–178.

7 Pickart, C. M., Targeting of substrates to the 26S proteasome. *FASEB J.* **1997**, *11*, 1055–1066.

8 Deng, L., et al., Activation of the IkappaB kinase complex by TRAF6 requires a dimeric ubiquitin-conjugating enzyme complex and a unique polyubiquitin chain. *Cell* **2000**, *103*, 351–361.

9 Spence, J., Gali, R. R., Dittmar, G., Sherman, F., Karin, M., Finley, D., Cell cycle-regulated modification of the

ribosome by a variant multiubiquitin chain. *Cell* **2000**, *102*, 67–76.

10 Schnell, J. D., Hicke, L., Non-traditional functions of ubiquitin and ubiquitin-binding proteins. *J. Biol. Chem.* **2003**, *278*, 35857–35860.

11 Hicke, L., Protein regulation by monoubiquitin. *Nat. Rev. Mol. Cell Biol.* **2001**, *2*, 195–201.

12 Beal, R., Deveraux, Q., Xia, G., Rechsteiner, M., Pickart, C., Surface hydrophobic residues of multiubiquitin chains essential for proteolytic targeting. *Proc. Natl. Acad. Sci. USA* **1996**, *93*, 861–866.

13 Sloper-Mould, K. E., Jemc, J. C., Pickart, C. M., Hicke, L., Distinct functional surface regions on ubiquitin. *J. Biol. Chem.* **2001**, *276*, 30483–30489.

14 Shih, S. C., Sloper-Mould, K. E., Hicke, L., Monoubiquitin carries a novel internalization signal that is appended to activated receptors. *EMBO J.* **2000**, *19*, 187–198.

15 Wilkinson, K. D., Tashayev, V. L., O'Connor, L. B., Larsen, C. N., Kasperek, E., Pickart, C. M., Metabolism of the polyubiquitin degradation signal: structure, mechanism, and role of isopeptidase T. *Biochemistry* **1995**, *34*, 14535–14546.

16 Hodgins, R. R., Ellison, K. S., Ellison, M. J., Expression of a ubiquitin derivative that conjugates to protein irreversibly produces phenotypes consistent with a ubiquitin deficiency. *J. Biol. Chem.* **1992**, *267*, 8807–8812.

17 Pickart, C. M., Kasperek, E. M., Beal, R., Kim, A., Substrate properties of site-specific mutant ubiquitin protein (G76A) reveal unexpected mechanistic features of ubiquitin-activating enzyme (E1). *J. Biol. Chem.* **1994**, *269*, 7115–7123.

18 Johnston, S. C., Riddle, S. M., Cohen, R. E., Hill, C. P., Structural basis for the specificity of ubiquitin C-terminal hydrolases. *EMBO J.* **1999**, *18*, 3877–3887.

19 Yeh, E. T., Gong, L., Kamitani, T., Ubiquitin-like proteins: new wines in new bottles. *Gene* **2000**, *248*, 1–14.

20 Jentsch, S., Pyrowolakis, G., Ubiquitin and its kin: how close are the family ties? *Trends Cell Biol.* **2000**, *10*, 335–342.

21 Hochstrasser, M., Evolution and function of ubiquitin-like protein-conjugation systems. *Nat. Cell Biol.* **2000**, *2*, E153–157.

22 Matunis, M. J., Coutavas, E., Blobel, G., A novel ubiquitin-like modification modulates the partitioning of the Ran-GTPase-activating protein RanGAP1 between the cytosol and the nuclear pore complex. *J. Cell Biol.* **1996**, *135*, 1457–1470.

23 Muller, S., Hoege, C., Pyrowolakis, G., Jentsch, S., SUMO, ubiquitin's mysterious cousin. *Nat. Rev. Mol. Cell Biol.* **2001**, *2*, 202–210.

24 Kamitani, T., Kito, K., Nguyen, H. P., Yeh, E. T., Characterization of NEDD8, a developmentally down-regulated ubiquitin-like protein. *J. Biol. Chem.* **1997**, *272*, 28557–28562.

25 Ponting, C. P., Proteins of the endoplasmic-reticulum–associated degradation pathway: domain detection and function prediction. *Biochem. J.* **2000**, *351 Pt 2*, 527–535.

26 Shih, S. C., Prag, G., Francis, S. A., Sutanto, M. A., Hurley, J. H., Hicke, L., A ubiquitin-binding motif required for intramolecular monoubiquitylation, the CUE domain. *EMBO J.* **2003**, *22*, 1273–1281.

27 Donaldson, K. M., Yin, H., Gekakis, N., Supek, F., Joazeiro, C. A., Ubiquitin signals protein trafficking via interaction with a novel ubiquitin binding domain in the membrane fusion regulator, Vps9p. *Curr. Biol.* **2003**, *13*, 258–262.

28 Kang, R. S., Daniels, C. M., Francis, S. A., Shih, S. C., Salerno, W. J., Hicke, L., Radhakrishnan, I., Solution structure of a CUE–ubiquitin complex reveals a conserved mode of ubiquitin binding. *Cell* **2003**, *113*, 621–630.

29 Prag, G., Misra, S., Jones, E. A., Ghirlando, R., Davies, B. A., Horazdovsky, B. F., Hurley, J. H., Mechanism of ubiquitin recognition by the CUE domain of Vps9p. *Cell* **2003**, *113*, 609–620.

30 Vadlamudi, R. K., Joung, I., Strominger, J. L., Shin, J., p62, a phosphotyrosine-independent ligand of the SH2 domain of p56lck, belongs to a new class of ubiquitin-binding proteins. *J. Biol. Chem.* **1996**, *271*, 20235–20237.

31 Hofmann, K., Bucher, P., The UBA domain: a sequence motif present in multiple enzyme classes of the ubiquitination pathway. *Trends Biochem. Sci.* **1996**, *21*, 172–173.

32 Wilkinson, C. R., Seeger, M., Hartmann-Petersen, R., Stone, M., Wallace, M., Semple, C., Gordon, C., Proteins containing the UBA domain are able to bind to multi-ubiquitin chains. *Nat. Cell Biol.* **2001**, *3*, 939–943.

33 Meyer, H. H., Wang, Y., Warren, G., Direct binding of ubiquitin conjugates by the mammalian p97 adaptor complexes, p47 and Ufd1–Npl4. *EMBO J.* **2002**, *21*, 5645–5652.

34 Dieckmann, T., Withers-Ward, E. S., Jarosinski, M. A., Liu, C. F., Chen, I. S., Feigon, J., Structure of a human DNA repair protein UBA domain that interacts with HIV-1 Vpr. *Nat. Struct. Biol.* **1998**, *5*, 1042–1047.

35 Mueller, T. D., Feigon, J., Solution structures of UBA domains reveal a conserved hydrophobic surface for protein–protein interactions. *J. Mol. Biol.* **2002**, *319*, 1243–1255.

36 Ciani, B., Layfield, R., Cavey, J. R., Sheppard, P. W., Searle, M. S., Structure of the ubiquitin-associated domain of p62 (SQSTM1) and implications for mutations that cause Paget's disease of bone. *J. Biol. Chem.* **2003**, *278*, 37409–37412.

37 Bertolaet, B. L., Clarke, D. J., Wolff, M., Watson, M. H., Henze, M., Divita, G., Reed, S. I., UBA domains of DNA damage-inducible proteins interact with ubiquitin. *Nat. Struct. Biol.* **2001**, *8*, 417–422.

38 Funakoshi, M., Sasaki, T., Nishimoto, T., Kobayashi, H., Budding yeast Dsk2p is a polyubiquitin-binding protein that can interact with the proteasome. *Proc. Natl. Acad. Sci. USA* **2002**, *99*, 745–750.

39 Young, P., Deveraux, Q., Beal, R. E., Pickart, C. M., Rechsteiner, M., Characterization of two polyubiquitin binding sites in the 26 S protease subunit 5a. *J. Biol. Chem.* **1998**, *273*, 5461–5467.

40 Hofmann, K., Falquet, L., A ubiquitin-interacting motif conserved in components of the proteasomal and lysosomal protein degradation systems. *Trends Biochem. Sci.* **2001**, *26*, 347–350.

41 Polo, S., et al., A single motif responsible for ubiquitin recognition and monoubiquitination in endocytic proteins. *Nature* **2002**, *416*, 451–455.

42 Shih, S. C., Katzmann, D. J., Schnell, J. D., Sutanto, M., Emr, S. D., Hicke, L., Epsins and Vps27p/Hrs contain ubiquitin-binding domains that function in receptor endocytosis. *Nat. Cell Biol.* **2002**, *4*, 389–393.

43 Katz, M., et al., Ligand-independent degradation of epidermal growth factor receptor involves receptor ubiquitylation and Hgs, an adaptor whose ubiquitin-interacting motif targets ubiquitylation by Nedd4. *Traffic* **2002**, *3*, 740–751.

44 Oldham, C. E., Mohney, R. P., Miller, S. L., Hanes, R. N., O'Bryan, J. P., The ubiquitin-interacting motifs target the endocytic adaptor protein epsin for ubiquitination. *Curr. Biol.* **2002**, *12*, 1112–1116.

45 Raiborg, C., Bache, K. G., Gillooly, D. J., Madshus, I. H., Stang, E., Stenmark, H., Hrs sorts ubiquitinated proteins into clathrin-coated microdomains of early endosomes. *Nat. Cell Biol.* **2002**, *4*, 394–398.

46 Bilodeau, P. S., Urbanowski, J. L., Winistorfer, S. C., Piper, R. C., The Vps27p Hse1p complex binds ubiquitin and mediates endosomal protein sorting. *Nat. Cell Biol.* **2002**, *4*, 534–539.

47 Burnett, B., Li, F., Pittman, R. N., The polyglutamine neurodegenerative protein ataxin-3 binds polyubiquitylated proteins and has ubiquitin protease activity. *Hum. Mol. Genet.* **2003**.

48 Fisher, R. D., Wang, B., Alam, S. L., Higginson, D. S., Robinson, H., Sundquist, W. I., Hill, C. P., Structure and ubiquitin binding of the ubiquitin-interacting motif. *J. Biol. Chem.* **2003**, *278*, 28976–28984.

49 Swanson, K. A., Kang, R. S., Stamenova, S. D., Hicke, L., Radhakrishnan, I., Solution structure of Vps27 UIM–ubiquitin complex important for endosomal sorting and receptor downregulation. *EMBO J.* **2003**, *22*, 4597–4606.

50 Mueller, T. D., Feigon, J., Structural determinants for the binding of ubiquitin-like domains to the

proteasome. *EMBO J.* **2003**, *22*, 4634–4645.

51 Shekhtman, A., Cowburn, D., A ubiquitin-interacting motif from Hrs binds to and occludes the ubiquitin surface necessary for polyubiquitination in monoubiquitinated proteins. *Biochem. Biophys. Res. Commun.* **2002**, *296*, 1222–1227.

52 Thrower, J. S., Hoffman, L., Rechsteiner, M., Pickart, C. M., Recognition of the polyubiquitin proteolytic signal. *EMBO J.* **2000**, *19*, 94–102.

53 Deveraux, Q., Ustrell, V., Pickart, C., Rechsteiner, M., A 26 S protease subunit that binds ubiquitin conjugates. *J. Biol. Chem.* **1994**, *269*, 7059–7061.

54 Doss-Pepe, E. W., Stenroos, E. S., Johnson, W. G., Madura, K., Ataxin-3 interactions with rad23 and valosin-containing protein and its associations with ubiquitin chains and the proteasome are consistent with a role in ubiquitin-mediated proteolysis. *Mol. Cell Biol.* **2003**, *23*, 6469–6483.

55 Sancho, E., et al., Role of UEV-1, an inactive variant of the E2 ubiquitin-conjugating enzymes, in in vitro differentiation and cell cycle behavior of HT-29-M6 intestinal mucosecretory cells. *Mol. Cell Biol.* **1998**, *18*, 576–589.

56 Ponting, C. P., Cai, Y. D., Bork, P., The breast cancer gene product TSG101: a regulator of ubiquitination? *J. Mol. Med.* **1997**, *75*, 467–469.

57 Garrus, J. E., et al., Tsg101 and the vacuolar protein sorting pathway are essential for HIV-1 budding. *Cell* **2001**, *107*, 55–65.

58 Hofmann, R. M., Pickart, C. M., Noncanonical MMS2-encoded ubiquitin-conjugating enzyme functions in assembly of novel polyubiquitin chains for DNA repair. *Cell* **1999**, *96*, 645–653.

59 Katzmann, D. J., Babst, M., Emr, S. D., Ubiquitin-dependent sorting into the multivesicular body pathway requires the function of a conserved endosomal protein sorting complex, ESCRT-I. *Cell* **2001**, *106*, 145–155.

60 Koonin, E. V., Abagyan, R. A., TSG101 may be the prototype of a class of dominant negative ubiquitin regulators. *Nat. Genet.* **1997**, *16*, 330–331.

61 Pornillos, O., Alam, S. L., Rich, R. L., Myszka, D. G., Davis, D. R., Sundquist, W. I., Structure and functional interactions of the Tsg101 UEV domain. *EMBO J.* **2002**, *21*, 2397–2406.

62 VanDemark, A. P., Hofmann, R. M., Tsui, C., Pickart, C. M., Wolberger, C., Molecular insights into polyubiquitin chain assembly: crystal structure of the Mms2/Ubc13 heterodimer. *Cell* **2001**, *105*, 711–720.

63 Moraes, T. F., Edwards, R. A., McKenna, S., Pastushok, L., Xiao, W., Glover, J. N., Ellison, M. J., Crystal structure of the human ubiquitin conjugating enzyme complex, hMms2–hUbc13. *Nat. Struct. Biol.* **2001**, *8*, 669–673.

64 Miura, T., Klaus, W., Gsell, B., Miyamoto, C., Senn, H., Characterization of the binding interface between ubiquitin and class I human ubiquitin-conjugating enzyme 2b by multidimensional heteronuclear NMR spectroscopy in solution. *J. Mol. Biol.* **1999**, *290*, 213–228.

65 Bishop, N., Horman, A., Woodman, P., Mammalian class E vps proteins recognize ubiquitin and act in the removal of endosomal protein–ubiquitin conjugates. *J. Cell Biol.* **2002**, *157*, 91–101.

66 Babst, M., Odorizzi, G., Estepa, E. J., Emr, S. D., Mammalian tumor susceptibility gene 101 (TSG101) and the yeast homologue, Vps23p, both function in late endosomal trafficking. *Traffic* **2000**, *1*, 248–258.

67 Martin-Serrano, J., Zang, T., Bieniasz, P. D., HIV-1 and Ebola virus encode small peptide motifs that recruit Tsg101 to sites of particle assembly to facilitate egress. *Nat. Med.* **2001**, *7*, 1313–1319.

68 Demirov, D. G., Ono, A., Orenstein, J. M., Freed, E. O., Overexpression of the N-terminal domain of TSG101 inhibits HIV-1 budding by blocking late domain function. *Proc. Natl. Acad. Sci. USA* **2002**, *99*, 955–960.

69 VerPlank, L., Bouamr, F., LaGrassa, T. J., Agresta, B., Kikonyogo, A., Leis, J., Carter, C. A., Tsg101, a homologue of ubiquitin-conjugating (E2) enzymes, binds the L domain in HIV type 1 Pr55(Gag). *Proc. Natl. Acad. Sci. USA* **2001**, *98*, 7724–7729.

70 Pornillos, O., Alam, S. L., Davis, D. R., Sundquist, W. I., Structure of the Tsg101 UEV domain in complex with the PTAP motif of the HIV-1 p6 protein. *Nat. Struct. Biol.* **2002**, *9*, 812–817.

71 Lu, Q., Hope, L. W., Brasch, M., Reinhard, C., Cohen, S. N., TSG101 interaction with HRS mediates endosomal trafficking and receptor downregulation. *Proc. Natl. Acad. Sci. USA* **2003**, *100*, 7626–7631.

72 Pickart, C. M., Mechanisms underlying ubiquitination. *Annu. Rev. Biochem.* **2001**, *70*, 503–533.

73 Seigneurin-Berny, D., Verdel, A., Curtet, S., Lemercier, C., Garin, J., Rousseaux, S., Khochbin, S., Identification of components of the murine histone deacetylase 6 complex: link between acetylation and ubiquitination signaling pathways. *Mol. Cell Biol.* **2001**, *21*, 8035–8044.

74 Hook, S. S., Orian, A., Cowley, S. M., Eisenman, R. N., Histone deacetylase 6 binds polyubiquitin through its zinc finger (PAZ domain) and copurifies with deubiquitinating enzymes. *Proc. Natl. Acad. Sci. USA* **2002**, *99*, 13425–13430.

75 Botta, A., Tandoi, C., Fini, G., Calabrese, G., Dallapiccola, B., Novelli, G., Cloning and characterization of the gene encoding human NPL4, a protein interacting with the ubiquitin fusion-degradation protein (UFD1L). *Gene* **2001**, *275*, 39–46.

76 Meyer, H. H., Shorter, J. G., Seemann, J., Pappin, D., Warren, G., A complex of mammalian ufd1 and npl4 links the AAA-ATPase, p97, to ubiquitin and nuclear transport pathways. *EMBO J.* **2000**, *19*, 2181–2192.

77 Wang, B., Alam, S. L., Meyer, H. H., Payne, M., Stemmler, T. L., Davis, D. R., Sundquist, W. I., Structure and ubiquitin interactions of the conserved zinc finger domain of Npl4. *J. Biol. Chem.* **2003**, *278*, 20225–20234.

78 Bonifacino, J. S., Weissman, A. M., Ubiquitin and the control of protein fate in the secretory and endocytic pathways. *Annu. Rev. Cell Dev. Biol.* **1998**, *14*, 19–57.

79 Levkowitz, G., et al., Ubiquitin ligase activity and tyrosine phosphorylation underlie suppression of growth factor signaling by c-Cbl/Sli-1. *Mol. Cell* **1999**, *4*, 1029–1040.

80 Haglund, K., Sigismund, S., Polo, S., Szymkiewicz, I., Di Fiore, P. P., Dikic, I., Multiple monoubiquitination of RTKs is sufficient for their endocytosis and degradation. *Nat. Cell Biol.* **2003**, *5*, 461–466.

81 Haglund, K., Shimokawa, N., Szymkiewicz, I., Dikic, I., Cbl-directed monoubiquitination of CIN85 is involved in regulation of ligand-induced degradation of EGF receptors. *Proc. Natl. Acad. Sci. USA* **2002**, *99*, 12191–12196.

82 van Delft, S., Govers, R., Strous, G. J., Verkleij, A. J., van Bergen en Henegouwen, P. M., Epidermal growth factor induces ubiquitination of Eps15. *J. Biol. Chem.* **1997**, *272*, 14013–14016.

83 Hicke, L., A new ticket for entry into budding vesicles – ubiquitin. *Cell* **2001**, *106*, 527–530.

84 Di Fiore, P. P., Polo, S., Hofmann, K., When ubiquitin meets ubiquitin receptors: a signalling connection. *Nat. Rev. Mol. Cell Biol.* **2003**, *4*, 491–497.

85 Katzmann, D. J., Odorizzi, G., Emr, S. D., Receptor downregulation and multivesicular-body sorting. *Nat. Rev. Mol. Cell Biol.* **2002**, *3*, 893–905.

86 Wendland, B., Epsins: adaptors in endocytosis? *Nat. Rev. Mol. Cell Biol.* **2002**, *3*, 971–977.

87 Pornillos, O., et al., HIV Gag mimics the Tsg101-recruiting activity of the human Hrs protein. *J. Cell Biol.* **2003**, *162*, 425–434.

88 Bache, K. G., Brech, A., Mehlum, A., Stenmark, H., Hrs regulates multivesicular body formation via ESCRT recruitment to endosomes. *J. Cell Biol.* **2003**, *162*, 435–442.

89 Freed, E. O., HIV-1 gag proteins: diverse functions in the virus life cycle. *Virology* **1998**, *251*, 1–15.

90 Gottlinger, H. G., The HIV-1 assembly machine. *Aids* **2001**, *15 Suppl 5*, S13–20.

91 Pornillos, O., Garrus, J. E., Sundquist, W. I., Mechanisms of enveloped RNA virus budding. *Trends Cell Biol.* **2002**, *12*, 569–579.

92 Ott, D. E., Coren, L. V., Chertova, E. N., Gagliardi, T. D., Schubert, U., Ubiquitination of HIV-1 and MuLV Gag. *Virology* **2000**, *278*, 111–121.

93 Harty, R. N., Brown, M. E., Wang, G., Huibregtse, J., Hayes, F. P., A PPxY motif within the VP40 protein of Ebola virus interacts physically and functionally with a ubiquitin ligase: implications for filovirus budding. *Proc. Natl. Acad. Sci. USA* **2000**, *97*, 13871–13876.

94 Harty, R. N., Brown, M. E., McGettigan, J. P., Wang, G., Jayakar, H. R., Huibregtse, J. M., Whitt, M. A., Schnell, M. J., Rhabdoviruses and the cellular ubiquitin–proteasome system: a budding interaction. *J. Virol.* **2001**, *75*, 10623–10629.

95 Schubert, U., et al., Proteasome inhibition interferes with gag polyprotein processing, release, and maturation of HIV-1 and HIV-2. *Proc. Natl. Acad. Sci. USA* **2000**, *97*, 13057–13062.

96 Vogt, V. M., Ubiquitin in retrovirus assembly: actor or bystander? *Proc. Natl. Acad. Sci. USA* **2000**, *97*, 12945–12947.

97 Briggs, S. D., Xiao, T., Sun, Z. W., Caldwell, J. A., Shabanowitz, J., Hunt, D. F., Allis, C. D., Strahl, B. D., Gene silencing: trans-histone regulatory pathway in chromatin. *Nature* **2002**, *418*, 498.

98 Wood, A., et al., Bre1, an E3 ubiquitin ligase required for recruitment and substrate selection of Rad6 at a promoter. *Mol. Cell* **2003**, *11*, 267–274.

99 Hwang, W. W., Venkatasubrahmanyam, S., Ianculescu, A. G., Tong, A., Boone, C., Madhani, H. D., A conserved RING finger protein required for histone H2B monoubiquitination and cell size control. *Mol. Cell* **2003**, *11*, 261–266.

100 Sun, Z. W., Allis, C. D., Ubiquitination of histone H2B regulates H3 methylation and gene silencing in yeast. *Nature* **2002**, *418*, 104–108.

101 Wang, Y., Cortez, D., Yazdi, P., Neff, N., Elledge, S. J., Qin, J., BASC, a super complex of BRCA1-associated proteins involved in the recognition and repair of aberrant DNA structures. *Genes. Dev.* **2000**, *14*, 927–939.

102 Gregory, R. C., Taniguchi, T., D'Andrea, A. D., Regulation of the Fanconi anemia pathway by monoubiquitination. *Semin. Cancer Biol.* **2003**, *13*, 77–82.

103 Meetei, A. R., et al., A novel ubiquitin ligase is deficient in Fanconi anemia. *Nat. Genet.* **2003**, *35*, 165–170.

104 Taniguchi, T., Garcia-Higuera, I., Andreassen, P. R., Gregory, R. C., Grompe, M., D'Andrea, A. D., S-phase–specific interaction of the Fanconi anemia protein, FANCD2, with BRCA1 and RAD51. *Blood* **2002**, *100*, 2414–2420.

105 Hoege, C., Pfander, B., Moldovan, G. L., Pyrowolakis, G., Jentsch, S., RAD6-dependent DNA repair is linked to modification of PCNA by ubiquitin and SUMO. *Nature* **2002**, *419*, 135–141.

106 Peng, J., et al., A proteomics approach to understanding protein ubiquitination. *Nat. Biotechnol.* **2003**, *21*, 921–926.

107 Vu, P. K., Sakamoto, K. M., Ubiquitin-mediated proteolysis and human disease. *Mol. Genet. Metab.* **2000**, *71*, 261–266.

108 Ciechanover, A., Schwartz, A. L., Ubiquitin-mediated degradation of cellular proteins in health and disease. *Hepatology* **2002**, *35*, 3–6.

109 Lai, Z., et al., Human mdm2 mediates multiple mono-ubiquitination of p53 by a mechanism requiring enzyme isomerization. *J. Biol. Chem.* **2001**, *276*, 31357–31367.

110 Goldknopf, I. L., French, M. F., Musso, R., Busch, H., Presence of protein A24 in rat liver nucleosomes. *Proc. Natl. Acad. Sci. USA* **1977**, *74*, 5492–5495.

111 Nickel, B. E., Allis, C. D., Davie, J. R., Ubiquitinated histone H2B is preferentially located in transcriptionally active chromatin. *Biochemistry* **1989**, *28*, 958–963.

112 Chen, H., Fre, S., Slepnev, V. I., Capua, M. R., Takei, K., Butler, M. H., Di Fiore, P. P., De Camilli, P., Epsin is an EH-domain–binding protein implicated in clathrin-mediated endocytosis. *Nature* **1998**, *394*, 793–797.

113 Pham, A. D., Sauer, F., Ubiquitin-activating/conjugating activity of TAFII250, a mediator of activation of gene expression in *Drosophila*. *Science* **2000**, *289*, 2357–2360.

114 Rocca, A., Lamaze, C., Subtil, A., Dautry-Varsat, A., Involvement of the ubiquitin/proteasome system in sorting of the interleukin 2 receptor beta chain to late endocytic compartments. *Mol. Biol. Cell* **2001**, *12*, 1293–1301.

115 Hou, D., Cenciarelli, C., Jensen, J. P., Nguygen, H. B., Weissman, A. M.,

Activation-dependent ubiquitination of a T cell antigen receptor subunit on multiple intracellular lysines. *J. Biol. Chem.* **1994**, *269*, 14244–14247.

116 Ungureanu, D., Saharinen, P., Junttila, I., Hilton, D. J., Silvennoinen, O., Regulation of Jak2 through the ubiquitin–proteasome pathway involves phosphorylation of Jak2 on Y1007 and interaction with SOCS-1. *Mol. Cell Biol.* **2002**, *22*, 3316–3326.

117 Doskeland, A. P., Flatmark, T., Ubiquitination of soluble and membrane-bound tyrosine hydroxylase and degradation of the soluble form. *Eur. J. Biochem.* **2002**, *269*, 1561–1569.

118 Usuba, T., Ishibashi, Y., Okawa, Y., Hirakawa, T., Takada, K., Ohkawa, K., Purification and identification of mono-ubiquitin–phosphoglycerate mutase B complex from human colorectal cancer tissues. *Int. J. Cancer* **2001**, *94*, 662–668.

119 Laub, M., Steppuhn, J. A., Bluggel, M., Immler, D., Meyer, H. E., Jennissen, H. P., Modulation of calmodulin function by ubiquitin–calmodulin ligase and identification of the responsible ubiquitylation site in vertebrate calmodulin. *Eur. J. Biochem.* **1998**, *255*, 422–431.

120 Dantan-Gonzalez, E., Rosenstein, Y., Quinto, C., Sanchez, F., Actin monoubiquitylation is induced in plants in response to pathogens and symbionts. *Mol. Plant Microbe Interact.* **2001**, *14*, 1267–1273.

121 Ott, D. E., Coren, L. V., Sowder, R. C., 2nd, Adams, J., Schubert, U., Retroviruses have differing requirements for proteasome function in the budding process. *J. Virol.* **2003**, *77*, 3384–3393.

122 Fukuda-Kamitani, T., Kamitani, T., Ubiquitination of Ro52 autoantigen. *Biochem. Biophys. Res. Commun.* **2002**, *295*, 774–778.

123 Hicke, L., Riezman, H., Ubiquitination of a yeast plasma membrane receptor signals its ligand-stimulated endocytosis. *Cell* **1996**, *84*, 277–287.

124 Roth, A. F., Davis, N. G., Ubiquitination of the yeast a-factor receptor. *J. Cell Biol.* **1996**, *134*, 661–674.

125 Kolling, R., Hollenberg, C. P., The ABC-transporter Ste6 accumulates in the plasma membrane in a ubiquitinated form in endocytosis mutants. *EMBO J.* **1994**, *13*, 3261–3271.

126 Egner, R., Kuchler, K., The yeast multidrug transporter Pdr5 of the plasma membrane is ubiquitinated prior to endocytosis and degradation in the vacuole. *FEBS Lett.* **1996**, *378*, 177–181.

127 Galan, J. M., Haguenauer-Tsapis, R., Ubiquitin lys63 is involved in ubiquitination of a yeast plasma membrane protein. *EMBO J.* **1997**, *16*, 5847–5854.

128 Springael, J. Y., Andre, B., Nitrogen-regulated ubiquitination of the Gap1 permease of *Saccharomyces cerevisiae*. *Mol. Biol. Cell* **1998**, *9*, 1253–1263.

129 Beck, T., Schmidt, A., Hall, M. N., Starvation induces vacuolar targeting and degradation of the tryptophan permease in yeast. *J. Cell Biol.* **1999**, *146*, 1227–1238.

130 Horak, J., Wolf, D. H., Catabolite inactivation of the galactose transporter in the yeast *Saccharomyces cerevisiae*: ubiquitination, endocytosis, and degradation in the vacuole. *J. Bacteriol.* **1997**, *179*, 1541–1549.

131 Gitan, R. S., Eide, D. J., Zinc-regulated ubiquitin conjugation signals endocytosis of the yeast ZRT1 zinc transporter. *Biochem. J.* **2000**, *346 Pt 2*, 329–336.

132 Lucero, P., Penalver, E., Vela, L., Lagunas, R., Monoubiquitination is sufficient to signal internalization of the maltose transporter in *Saccharomyces cerevisiae*. *J. Bacteriol.* **2000**, *182*, 241–243.

133 Robzyk, K., Recht, J., Osley, M. A., Rad6-dependent ubiquitination of histone H2B in yeast. *Science* **2000**, *287*, 501–504.

134 Jennissen, H. P., Botzet, G., Majetschak, M., Laub, M., Ziegenhagen, R., Demiroglou, A., Ca(2+)-dependent ubiquitination of calmodulin in yeast. *FEBS Lett.* **1992**, *296*, 51–56.

135 Gouet, P., Courcelle, E., Stuart, D. I., Metoz, F., ESPript: analysis of multiple sequence alignments in PostScript. *Bioinformatics* **1999**, *15*, 305–308.

136 Corpet, F., Multiple sequence alignment with hierarchical clustering. *Nucleic Acids Res.* **1988**, *16*, 10881–10890.

137 Koradi, R., Billeter, M., Wuthrich, K., MOLMOL: a program for display and analysis of macromolecular structures. *J. Mol. Graph.* **1996**, *14*, 51–55, 29–32.

16
The Calponin Homology (CH) Domain

Mario Gimona and Steven J. Winder

16.1
Introduction and Brief History

"Does Vav bind to F-actin through a CH domain?" This title of a 1995 paper by Castresana and Saraste [1] marks the birth of the calponin homology (CH) domain. With the help of structure-based sequence alignments the authors described a 100 residue long protein module that they found in a number of signaling and cytoskeletal proteins. The title of the manuscript prejudiced in a peculiar way the developments in the years to come. Not only did very few people know that the guanine nucleotide exchange factor Vav existed or even interacted with the actin cytoskeleton, but also the calponin community was surprised to find that the functionally almost dispensable N-terminal region of the calponin molecule would serve as the 'mother of actin-binding modules'. At any rate, once the CH domain was born it stirred up the fields of cytoskeleton research and signaling and made a significant contribution towards a greater mutual understanding of the importance of upstream and downstream targets, respectively.

Prior to this historical landmark, partial sequence similarities between the actin-binding domains of classical actin cross-linkers like α-actinin, filamin, spectrin, or fimbrin were noted [2–4] and the name-giving protein, calponin, likewise showed sequence similarities in parts of its N-terminal domain (see Figure 16.1 for sequence alignments and type classification). With the initial establishment of the CH domain, researchers began to look at actin-binding sites from a new perspective, and soon novel CH-domain family members were identified.

However, the early years of the CH domain were troublesome. A persistent misapprehension of CH domain function, based on an oversimplified interpretation of actin-binding data, delayed a more intellectual discourse with this fascinating protein module [5]. The CH domain was stuck in mediating actin association for all kinds of proteins, irrespective of their subcellular localization or molecular context. More recent and more careful annotations of additional functional sites present on all types of CH domains not only sharpened researchers' minds but also led to the revival of long-forgotten questions with respect to the regulation of

Modular Protein Domains. Edited by G. Cesareni, M. Gimona, M. Sudol, M. Yaffe

ISBN: 3-527-30813-X

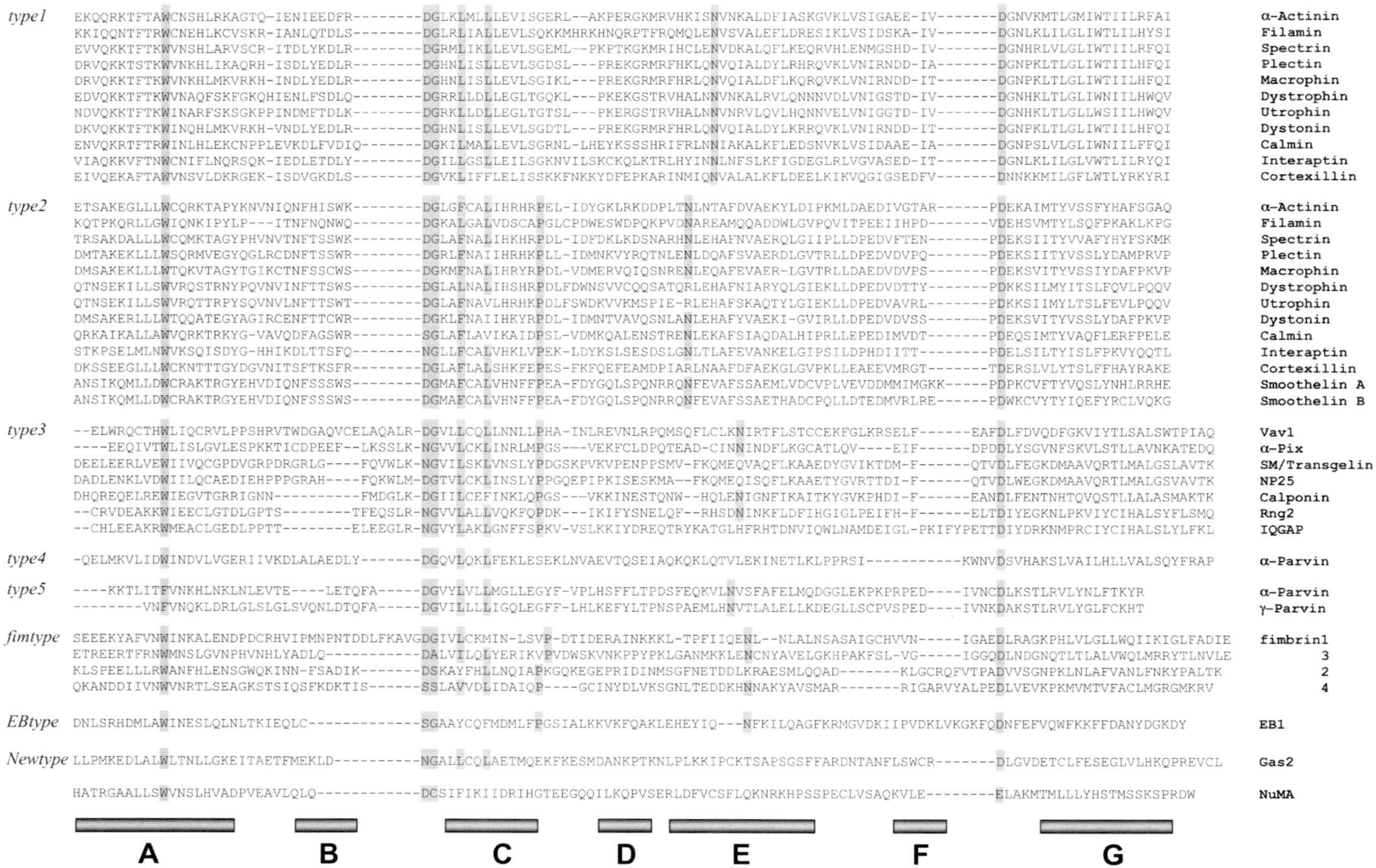

Figure 16.1 (legend see p. 323)

actin binding. Recent years have seen a greater appreciation for and a more liberal view of the functional polymorphism displayed by CH-domain–containing molecules. The realization that actin-binding CH domains can bind the filaments in multiple ways [6–9] considerably shaped the current view on actin-binding modules of the CH-domain family (reviewed in [10]). The final hits aiding in establishing the CH domain as a multifaceted tool were added by the unequivocal determination of the microtubule-anchoring function of the EB1 CH domain [11] and the identification of CH domains in proteins of the nuclear mitotic apparatus (NuMA) [12].

16.2 Structure of the Domain – The CH Domain Fold

The CH domain is a compact globular fold comprising four major (A, C, E, G) and two or three minor helices (B, D, F) interconnected by loops of variable lengths (Figure 16.2a). The four major helices of between 10 and 20 residues form the core of the domain, with the smaller helices making minor structural contributions. Helices C, E, and G run roughly parallel to each other, with the N-terminal helix A being roughly perpendicular to them. The A and G helices are the 'bread and butter' sandwiching the 'jam' represented by the hydrophobic helices C and E (Figure 16.2a). Among the 15 different CH domain structures currently known (Table 16.1), which range from those present in a single copy to those of the double tandem domain in fimbrin and represent actin-binding, signaling, and microtubule binding proteins, the structural unit is highly conserved with an rmsd of ~1 Å between Cα atoms of the four helices in all CH domains [13]. Despite strong structural conservation, CH domains can nonetheless be divided into distinct families, based on structural [7, 14] or sequence alignments [14]. In either situation, families arise due to differences in the lengths and positions of the minor helices and sequence variations in interconnecting loops. In all instances the integrity of the domain is maintained by hydrophobic interactions between the core helices, with only a tryptophan residue in helix A being absolutely conserved in all CH domains. This Trp residue generally forms nonpolar interactions with other aromatic or aliphatic residues in helices E and G, thus stabilizing helix A with respect to the triple-helical bundle formed by helices C, E, and G. The four or five residues that are conserved in character in all CH domains are mostly involved in helical packing of the core structure.

Figure 16.1 Sequence alignment of CH domains representing the individual types. Invariant key residues are highlighted, and the grey bars at the bottom indicate the positions of helices. The key residues tryptophan (W) at position 11 and the aspartate at the beginning of helix G are the most-conserved residues. Helix C usually follows the consensus DGXXLXXL. The proline (P) residue terminating helix C is invariant in type2 and type3 CH domains and also in the fimtype and EBtype CH domains, but is missing in type1, type4, and type5 CH domains. The asparagine (N) at or within helix E is conserved, but the position varies significantly among the different CH domain types.

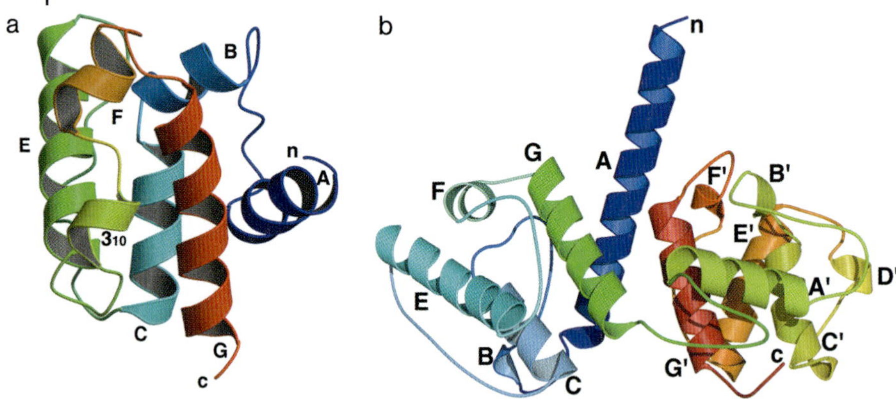

Figure 16.2 Representative structures of single and tandem CH domains. (a) The single CH domain of calponin [13] with major helices colored individually, from blue at the N terminus (n) to red at the C terminus (c), and labeled from A to G. A short 3_{10} helix unique to calponin is found between helices E and F, and there is no helix D. (b) The compact tandem CH-domain pair of plectin [20], colored and labeled as above. CH1 is on the left and CH2 on the right; as in calponin there is no helix D in CH1. In comparison to other tandem CH-domain structures, helix A in CH1 is very long. This is probably not a feature unique to plectin, since other studies have suggested a long N-terminal helix for utrophin [19], but the extended amino terminal part of the helix is not part of the true CH domain fold.

Table 16.1 CH domain structures.

Protein source	*Type of analysis*	*CH domain subtype*		*Reference*
Calponin	NMR	type3	1H67	13
EB1	crystal	EBtype	1PA7, 1UEG	17
Rng2	crystal	type3	1P2X, IP5S	18, 76
β-Spectrin	crystal	type2	1AA2, 1BKR	14, 16
Utrophin	crystal	type2	1BHD	7
α-Actinin	crystal	type1/type2 tandem	unpublished	22
Plectin	crystal	type1/type2 tandem	1MB8	20
Dystrophin	crystal	type1/type2 tandem	1DXX	23
Utrophin	crystal	type1/type2 tandem	1QAG	6
Fimbrin	crystal	type1/type2 tandem	1AOA	21

16.2.1
Structures of Single CH Domains

The first structures of single CH domains to be solved were of CH2 in spectrin and utrophin [7, 16]; however, these were single CH domains belonging to actin-binding domains containing tandem pairs of CH domains and are discussed in more detail below. The first structure of a true single CH domain, and to date the only solution structure of a CH domain, was that of the archetypal CH domain from chicken gizzard calponin [13] (Figure 16.2a); more recently this was followed by the crystal structures of EB1, a microtubule binding protein [17], and Rng2, an IQGAP protein from yeast [18]. In all instances the CH domains of these proteins are not thought to be necessary for actin binding, despite the association of Rng2 and calponin with actin or actin-containing structures. Calponin does not require its CH domain for actin binding and, as discussed below, the single CH domain is not an actin-binding domain per se. In EB1 however, this protein binds to microtubules and not to F-actin, but it is postulated that this interaction is mediated by hydrophobic residues in a similar manner to CH-domain interactions with F-actin [17]. It has been suggested that the interhelical loops, which in structural and sequence terms are the elements that show the most variation among CH domains, might confer the different properties on CH domains [19] and that flexibility in these regions might also confer unique properties, particularly in fimbrin. The solution structure of the calponin CH domain displayed very little if any flexibility in interhelical loops [13], and it is now accepted that it is probably conserved residues in the core helices that allow single CH domains to 'locate' on an actin filament.

16.2.2
Structures of Tandem CH Domains

The basic structure of individual CH domains within the tandem CH domain-containing proteins recapitulates the sequence-derived phylogeny, in that the CH1 domains are structurally more similar to CH1 domains in other proteins than to the CH2 domain in the same protein. The main differences stem from the relative lengths of the core helices and the number and position of the secondary helices, including short 3_{10} helices flanking helix C in CH2 domains and the presence of an additional helix D in CH2. These differences notwithstanding, the CH domains are remarkably similar in core structure [13]. The most striking and perhaps controversial feature of the tandem CH-domain structures so far elucidated is the positions of the first and second CH domains relative to each other. Fimbrin, plectin, and α-actinin all crystallized as compact monomers with a single molecule in the asymmetric unit [20–22], whereas dystrophin and utrophin crystallized in a more extended conformation and as antiparallel dimers [6, 23]. Biochemically, all the isolated actin-binding domains of these proteins are monomeric; furthermore, the interface between the antiparallel utrophin and dystrophin molecules was such that CH1 of one chain was juxtaposed with CH2 of the other chain in an orientation identical to the orientation of CH domains in fimbrin, α-actinin, and plectin. This sort of conservation of interactions that exists between domains in proteins that adopt two different states, here, crystallographic monomer versus crystallographic dimer, is known as 3D domain swapping [24]. As such, this is not unusual – the controversy arises, however, as to whether these tandem CH-containing actin-binding domains can adopt different conformations in solution and whether there is significant reorganization of the CH domains upon binding to actin [25]. To add fuel to the controversy, several cryoelectron microscopy reconstructions of tandem CH domains with F-actin have yielded different configurations for the CH-domain orientation on actin. Initially this difference was thought to be a consequence of the length of the inter-CH domain linker [6], because in the first two tandem CH-domain structures to be published, those of fimbrin and utrophin [6, 21], the long linker of fimbrin was believed to allow the two CH domains to fold back on each other and the shorter linker in utrophin resulted in a more open structure. α-Actinin however, which has one of the shortest interdomain linkers of all, crystallized as a compact monomer [22], effectively ruling out the linker hypothesis. An alternative view might be that the conformation of the actin-binding domain reflects the function of the whole protein, i.e., actin-bundling proteins like fimbrin and α-actinin require a compact formation because they bind at right angles to the actin filament [26, 27], and dystrophin and utrophin, which are more akin to side-binding proteins [28, 29], require an open conformation. This argument would seem to support available cryoelectron microscopic data (see [10]); however, the recent crystal structure of plectin [20], which is likely to also be an F-actin side-binding protein to some extent, was found to be a compact monomer (Figure 16.2b). Although Pereda and colleagues demonstrated rather elegantly [20], as had been suggested previously for utrophin [8, 30], that the two CH domains can undergo movement and can rearrange upon binding to actin [31], further direct experimentation is required to resolve this fascinating problem.

16.3
Molecular and Signaling Function

CH domains are found in a wide range of molecules, but the unifying theme among them is their involvement in cytoskeletal structure, dynamics, and signaling. The enormous functional plasticity displayed by the otherwise structurally highly conserved domain argues that the CH domain represents a platform for a plethora of functional sites.

16.3.1
Actin-binding Domains

Sequence- and structure-based profiling has led to the established classification of CH domains into at least five distinct subfamilies [15, 32]. By far the largest number of CH domains belongs to the type1 and type2 classes. With a single exception (smoothelin), these CH domains occur in tandem, and this dual module forms a high-affinity actin-binding domain (ABD) in a large variety of actin-binding and cross-linking proteins. The amino acid sequences of fimbrin-type CH domains are sufficiently different to place them in a separate subfamily, yet they follow the general consensus and arrangement of ABDs formed by the type1/type2 CH domains. Type1 and type2 CH domains differ not only in sequence, but also in their affinities for actin, as has been shown for the actin cross-linking protein α-actinin. Notably, the 'type1 fim' CH domains of fimbrin are able to associate additionally with the intermediate filament component vimentin [33] at a site encompassing residues 143–188 (corresponding to the type1 CH domain in the first ABD).

ABDs bind one actin monomer in the filament, with affinities typically in the low micromolar range. When analyzed in isolation, CH domains from ABDs have divergent actin-binding characteristics. All type1 CH domains bind actin with significant affinity, but for type2 CH domains actin binding is almost undetectable. Nevertheless, both N- and C-terminal CH domains are required for the generation of a fully functional ABD, in which the type2 CH domain appears to contribute to the overall stability of the module [34]. Similarly, the ABD in plectin requires only the CH1 for actin binding and dimerization [35]. Here, the crucial importance of the ABS2 residing in the most C-terminal helix of the CH1 was shown for the first time. Interestingly, the CH2 in plectin appears to have a negative influence on actin binding of the ABD, because deletion of the second half increases actin binding of the plectin ABD; however, this may be a reflection of alternatively spliced exons in the first CH domain because, depending on the spliced exon, the affinity for actin can vary considerably [36].

Structural studies of CH domain proteins and their interactions with actin filaments have suggested that a conserved hydrophobic surface is implicated in binding to actin filaments. It is therefore believed that the general mode of binding and the molecular interface employed for contacting the actin filament is conserved among actin-binding domains formed by a CH domain tandem. When the isolated ABD from the actin cross-linking protein α-actinin was used as a molecular targeting vehicle, a variety of otherwise cytoplasmic components (e.g., GFP, Vav) could be targeted to the thin filaments of transfected fibroblasts. However, the similarly arranged ABD from the actin-network–stabilizing molecule filamin failed to serve as a strong targeting vehicle, although the domain is undoubtedly required for filamin binding to actin [37]. This example illustrates that there are functional differences among ABDs from different subfamilies of CH-domain proteins, likely reflecting the differences displayed in the amino acid sequence of this module [38]. One may thus hypothesize that functional diversity occurs among type1/type2 CH domain ABDs and that these differences may account for the subtle differences in actin affinity, mode of cross-linking, and site of attachment along the actin filament.

16.3.2
Single EB-type CH Domains Function as Microtubule Anchors

In total contrast to the CaP-family and Vav-family CH domains, the CH domains of EB-family proteins (namely EB1, EB2, EB3, RB1, and the yeast EB1 homolog

Bim1) have been shown to contact growing microtubule ends. End-binding (EB) proteins are evolutionarily conserved proteins that modulate microtubule dynamics by regulating dynamic instability. In particular, EB1 targets growing microtubule ends, where it is favorably positioned to regulate microtubule polymerization [39] and to confer molecular recognition of the microtubule end [40, 41]. Immunodepletion of EB1 from *Xenopus* egg extracts has been shown to reduce microtubule length, and this effect was reversed by readdition of recombinant EB1 [42]. EB1 also decreased microtubule catastrophe, resulting in increased polymerization and stable microtubules in interphase cells. The effect of EB1 on microtubule dynamics is highly conserved, suggesting that this protein family belongs to a core set of regulatory factors conserved in higher organisms. In *Drosophila*, EB1 has been shown to play a crucial role in mitosis by its ability to promote the growth and interactions of microtubules within the central spindle and at the cell cortex [43]. Finally, in *Dictyostelium*, the largest known EB1 homolog (57 kDa), termed DdEB1, also localizes along microtubules and at microtubule tips, centrosomes, and protruding pseudopods and was found at the spindle, spindle poles, and kinetochores during mitosis. In addition, EB1 is involved in regulating the interaction of the tumor suppressor adenomatous polyposis coli (APC) with the microtubule apparatus [44, 45]. EB1 binds to the C terminus of the APC protein and may regulate accumulation in cortical clusters of extending membranes, but endogenous EB1 does not accumulate in the APC clusters [46].

16.3.3
Kinases, Phospholipids, and Other Cytoskeletal Components

Tandem CH domains are also present in the actopaxin/parvin family [47, 48], but here, both CH domains branch into separate subfamilies (type4, type5) and expose (overlapping) binding sites for F-actin and the focal adhesion proteins paxillin and integrin-linked kinase (ILK) [49, 50]. Thus, the CH domains in this family display functions different from those of type1, type2, or type3 CH domains. Unfortunately, more detailed molecular information about the residues involved in these interactions is lacking.

CH domains display different functions when present in single as compared to tandem motifs, and different proteins contain CH domains of different 'types'. All ABDs strictly follow the consensus type1/type2 arrangement. However, the relative contribution of each CH domain to actin binding is not known. Detailed correlative sequence analysis has revealed that CH domains in ABDs can harbor additional conserved binding motifs for phosphatidylinositol (type2 CH domains; see below) and also novel autonomous actin-binding sequences, like the recently discovered DFRxxL motif from myosin light chain kinase ([51], reviewed in [15]). These findings strongly suggest that even CH domains in ABDs may expose additional, as yet unidentified, features that contribute to the selectivity and specificity of binding to actin and possibly to other thin-filament–associated components.

A perfect example of this are the actin-binding domains from plectin and dystonin, which bind β4 integrin [52]. Mutations in the integrin binding pocket, formed by a sequence stretch unique to plectin and dystonin and corresponding to the region preceding and partially overlapping the ABS2 (residing in the C-terminal helix in the CH1 of plectin), significantly decrease β4 integrin binding but do not influence the F-actin binding ability of plectin's ABD. Interestingly, actin binding and integrin β4 association are mutually exclusive in plectin, suggesting that CH domain function can be switched 'on site' – the mechanism, however, remains obscure.

In contrast to the situation in type1 and type2 CH domains, the actin binding affinities of type3 CH domains are in the millimolar range, and researchers thus earlier questioned whether their physiological function is indeed related to actin binding. Although the contribution of the CH domain in calponin-family proteins is still an open question, we have seen an accumulation of data in recent years that support the original skepticism. It is clear now that the CH domain is neither sufficient nor necessary for actin binding activity in this protein family [53–56]. This development has helped to make manifest the view that type3 CH domains may serve a primarily regulatory function, and indeed several laboratories have identified a plethora of binding partners for type3 CH domains. In the name-giving protein calponin (CaP), for example, the type3 CH domain interacts with extracellular regulated kinases ERK1 and ERK2 in vitro [57] and is hypothesized to be modulated by an association with LIM kinase in vivo [58]. Work from the Gusev laboratory revealed that Hsp90 binds directly to the CH domain of smooth muscle *h*1CaP and affects actin binding. The authors hypothesized that, in the presence of Hsp90, CaP is trapped in a complex, which makes the molecule unavailable for interaction with G-actin. In this way, Hsp90 could decrease the CaP-induced polymerization of actin [59, 60]. Calponin interacts in vitro with the dimeric S100-family members calcyclin (S100A6), which exposes two functional Ca^{2+}-binding EF-hands per monomer, and S100A2. EF-hands have a similar amino acid arrangement as zinc fingers and require the presence of conserved Cys or His residues in a particular spatial arrangement. EF-hands of S100A2 can also bind Zn^{2+}, suggesting that Zn^{2+} binding is involved in the regulation of CaP via S100 molecules.

In agreement with this scenario, the type3 CH domain in the Rho family nucleotide exchange factor Vav binds intramolecularly to a zinc-finger–like domain [61] and has been shown to interact with the guanine nucleotide dissociation inhibitor (RhoGDI) in a two-hybrid interaction screen [62]. Vav proteins are activated by an N-terminal deletion that removes all (in Vav-2) or part (in Vav-1) of the CH domain, and this deletion occurs naturally in the *onco-vav* gene [63]. The CH domain is required for the modulation or down-regulation of Vav's GEF activity [64, 65] by mediating a conformational switch in the molecule, which involves interaction with the C-terminal zinc-finger domain and results in a steric block of the catalytic DH-PH domains [66]. Notably, the short loop connecting helices A and B in the Vav CH domain is responsible for this Vav-specific function, and replacement of the loop with the homologous region from CaP makes the molecule constitutively active [67].

Binding of the phosphoinositide PtdIns(4,5)-P_2 negatively regulates actin binding and bundling activity of the antiparallel actin filament cross linker α-actinin, likely by mediating changes in the molecular structure [68]. The conserved PiP_2 binding site resides in the CH2 of α-actinin, in good agreement with the supporting function of CH2 domains in ABDs. Site-directed mutagenesis in this region has revealed three critical basic residues. Mutant proteins carrying sequence alterations in these amino acids, leading to defective PiP_2 binding, display increased actin binding and bundling activity in vitro.

16.3.4 CH Domain-containing Proteins and Human Diseases

16.3.4.1 The Dystrophin ABD and Muscular Dystrophy

Dystrophin is a large cytoskeletal linker protein that connects the subsarcolemmal actin cytoskeleton of skeletal muscle to the transmembrane adhesion receptor dystroglycan [69]. Mutations in dystrophin give rise to the crippling and fatal X-linked disease Duchenne muscular dystrophy (DMD). The majority of mutations in the DMD gene give rise to premature stop codons, resulting in transcript instability and complete loss of protein. The milder allelic form of DMD, Becker muscular dystrophy, which is caused by in-frame deletion or missense point mutations, does allow the synthesis of mutated protein. Most point mutations in the CH domain-containing actin-binding region of dystrophin lead to a relatively severe phenotype [70–72], emphasizing the importance of the actin-binding domain for dystrophin function. However, it is doubtful that in the majority of cases the actin-binding domain is functional at all, because the four missense mutations; L54R, A171P, A168D, and Y321N and in-frame deletions of exon 3 (residues 32–62) and exon 5 (residues 89–119) are all expected to disrupt either the hydrophobic core of the protein or its overall structure [23]. Consequently, these mutations have not been particularly useful in determining function.

16.3.4.2 The Filamin ABD and Otopalatodigital Syndromes

Otopalatodigital (OPD) syndromes and related disorders are a diverse group of X-linked diseases affecting craniofacial, skeletal, brain, visceral, and urogenital structures. The affected gene encodes the cytoskeletal protein filamin A, with approximately half of the 17 mutations so far described being missense mutations residing in the CH2 domain of the filamin ABD [73]. No structure of a filamin CH domain is yet available, but mapping the mutations to a model of filamin based on the structures of spectrin or dystrophin CH2 suggested that some mutations were likely to not simply result in loss of actin-binding function [73], which points to additional and as-yet-unidentified roles for CH domains.

16.3.4.3 The α-Actinin ABD and Glomerulosclerosis

Mutations in α-actinin 4 have been described in familial focal segmental glomerulosclerosis (FSGS). FSGS is a common nonspecific renal lesion characterized by decreasing kidney function and often leading to end-stage renal failure. α-Actinin 4 has been implicated in some cases of autosomal dominant FSGS, with point mutations occurring in helix G of CH2 [74]. All three of the mutations characterized – K228E, T232I, and S235P – are on the solvent-accessible surface of helix G and are not expected to affect core structure, but are also not in a region implicated in direct interactions with actin, and so presumably are involved in some other as-yet-unidentified role of α-actinin in the kidney.

16.3.4.4 The β-Spectrin ABD and Spherocytosis

Hereditary spherocytosis (HS) includes a group of heterogeneous hemolytic anemias ranging in severity from asymptomatic to severe. In all cases the red blood cell has a distinct morphology, with varying degrees of surface area reduction leading to a spherocytic phenotype and osmotic fragility. Of the four characterized subsets of HS patients, two are characterized by a deficiency in β-spectrin. Several mutations in spectrin have been described and shown to be the molecular defect in HS with spectrin deficiency [75]. Of these mutations, two were found in the second CH domain, W182G and I220V, with both residues being important for maintaining the hydrophobic core of the CH domain. Changing Trp182 to Gly in the short helix B in particular would be expected to have a severe effect on the stability of the CH domain, with consequent effects on the function of the whole ABD.

Many other proteins that contain CH domains are implicated in diseases, for example EB1 is a tumor-suppressor protein, and plectin is involved in epidermolysis bullosa with muscular dystrophy, but to date, disease-causing mutations in these proteins have not been identified within the CH domains.

16.4 Emerging Research Directions and Recent Developments

The presence of a CH domain in any given protein is still taken as a strong indication that the molecule associates with the actin cytoskeleton, despite controversial interpretations of binding data and subcellular localization studies which call this simplified view into question. It is more than evident, from the diverse list of binding partners that have been identified for the various CH domains, that the module is a platform for a number of interaction sites with cytoskeleton and signaling components. The most interesting study of the past decade may be the identification of the EB1 CH domain. EB CH domains strictly follow the consensus of conserved residues and, even though the protein has not been ascribed any direct association with the actin cytoskeleton, are placed in the CH-domain family tree, representing a separate branch. The work of Hayashi and Ikura has demonstrated that the module folds almost identically to the CH domains described thus far for spectrin, fimbrin, and calponin – but the CH domain in EB1 is a microtubule anchor instead. It is difficult to envisage any similarity in the surface profiles between an actin filament and a microtubule. Hence, morphofunctional plasticity, established during a coevolutionary process of the diverse eukaryotic cytoskeleton filament systems, and the factors regulating their assembly and dynamics, may be the key to understanding this apparent paradox. Other important work, such as with the plectin ABD [52], should serve as future guidelines on how meaningful studies can be tailored to increase our knowledge of CH domain function and interactions. Future research activities might put greater emphasis on identifying the in vivo functions of the diverse CH domains and aim at determining in more detail the (sequence) parameters that drive functional plasticity in this family.

16.5 Concluding Remarks

Not even ten years have gone by since the CH domain was delineated in detail. Like all other domains and modules analyzed in this book, the CH domain obeys the rule of protein module definition. It is a stable fold, and the function can be transported to other molecules by simple fusion of the coding sequence. We must, however, be aware of the fact that protein linguistics and positional semantics may lead us into novel territory in which the strict functional definitions may not hold. There is rapidly increasing evidence for the existence of a novel class of modules that fulfill the criteria of autonomous function but not of folding (see also Chapter 21). These intrinsically unfolded protein (IUP) modules appear to fold exclusively at their ligands – and a good portion of these associate with the cytoskeleton! One should therefore keep in mind that the definition of domain borders is perhaps the most relevant parameter for assessing the proper in vivo function and that structurally dispensable flanking regions may contribute significantly to the fine-tuning of domain function.

Acknowledgements

Work cited in this chapter was funded by the BBSRC, MRC, and Wellcome Trust (SJW). We are grateful to Mike Broderick for critical reading of the manuscript and to Jose Pereda for providing Figure 16.1b. MG is recipient of the Marie Curie Excellence Grant MEXT-CT-2003-002573 of the European Union.

References

1 Castresana, J., Saraste, M., Does Vav bind to F-actin through a CH domain? *FEBS Lett.* **1995**, *374*, 149–151.

2 de Arruda, M. V., Watson, S., Lin, C. S., Leavitt, J., Matsudaira, P., Fimbrin is a homologue of the cytoplasmic phosphoprotein plastin and has domains homologous with calmodulin and actin gelation proteins. *J. Cell Biol.* **1990**, *111*, 1069–1079.

3 Matsudaira, P., Modular organization of actin crosslinking proteins. *Trends Biochem. Sci.* **1991**, *16*, 87–92.

4 Way, M., Pope, B., Weeds, A. G., Evidence for functional homology in the F-actin binding domains of gelsolin and alpha-actinin: implications for the requirements of severing and capping. *J. Cell Biol.* **1992**, *119*, 835–842.

5 Gimona, M., Winder, S. J., Single CH domains are not actin binding domains. *Current Biol.* **1998**, *8*, R674–675.

6 Keep, N. H., Winder, S. J., Moores, C. A., Walke, S., Norwood, F. L. M., Kendrick-Jones, J., Crystal structure of the actin-binding region of utrophin

reveals a head-to-tail dimer. *Struct. Fold. Design* **1999**, *7*, 1539–1546.

7 Keep, N. H., Norwood, F. L., Moores, C. A., Winder, S. J., Kendrick-Jones, J., The 2.0 Å structure of the second calponin homology domain from the actin-binding region of the dystrophin homologue utrophin. *J. Mol. Biol.* **1999**, *285*, 1257–1264.

8 Galkin, V. E., Orlova, A., Van Loock, M. S., Rybakova, I. N., Ervasti, J. M., Egelman, E. H., The utrophin actin-binding domain binds F-actin in two different modes: implications for the spectrin superfamily of proteins. *J. Cell Biol.* **2002**, *157*, 243–251.

9 Galkin, V. E., Orlova, A., Van Loock, M. S., Egelman, E. H., Do the utrophin tandem calponin homology domains bind F-actin in a compact or extended conformation? *J. Mol. Biol.* **2003**, *331*, 967–972.

10 Winder, S. J., Structural insights into actin-binding, branching and bundling proteins. *Curr. Opin. Cell Biol.* **2003**, *15*, 14–22.

11 Bu, W., Su, L.-K., Characterization of functional domains of human EB1 family proteins. *J. Biol. Chem.* **2003**, *278*, 49721–49731.

12 Novatchkova, M., Eisenhaber, F., A CH domain-containing N-terminus in NuMA? *Protein Sci.* **2002**, *10*, 2281–2284.

13 Bramham, J., Hodgkinson, J., Smith, B. O., Uhrín, D., Barlow, P. N., Winder, S. J., Solution structure of the calponin CH domain and fitting to the 3D helical reconstruction of F-actin: calponin. *Struct. Fold. Design.* **2002**, *10*, 249–258.

14 Banuelos, S., Saraste, M., Carugo, K. D., Structural comparisons of calponin homology domains: implications for actin binding. *Structure* **1998**, *6*, 1419–1431.

15 Gimona, M., Djinovic-Carugo, K., Kranewitter, W. J., Winder, S. J., Functional plasticity of CH domains. *FEBS Lett.* **2002**, *513*, 98–106.

16 Djinovic-Carugo, K. D., Bañuelos, S., Saraste, M., Crystal structure of a calponin homology domain. *Nat. Struct. Biol.* **1997**, *4*, 175–179.

17 Hayashi, I., Ikura, M., Crystal structure of the amino-terminal microtubule-binding domain of end-binding protein 1 (EB1). *J. Biol. Chem.* **2003**, *278*, 36430–36434.

18 Wang, C.-H., Walsh, M., Balasubramanian, M. K., Dokland, T., Expression, purification, crystallization and preliminary crystallographic analysis of the calponin-homology domain of Rng2. *Acta Crystallogr. D.* **2003**, *59*, 1809–1812.

19 Morris, G. E., Nguyen, T. M., Nguyen, T. N., Pereboev, A., Kendrick-Jones, J., Winder, S. J., Disruption of the utrophin–actin interface by monoclonal antibodies and prediction of an actin-binding surface of utrophin. *Biochem. J.* **1999**, *337*, 119–123.

20 Garcia-Alvarez, B., Bobkov, A., Sonnenberg, A., de Pereda, J. M., Structural and functional analysis of the actin binding domain of plectin suggests alternative mechanisms for binding to F-actin and integrin β4. *Structure* **2003**, *11*, 615–625.

21 Goldsmith, S. C., Pokala, N., Shen, W., Fedorov, A. A., Matsudaira, P., Almo, S. C., The structure of an actin-cross-linking domain from human fimbrin. *Nat. Struct. Biol.* **1997**, *4*, 708–712.

22 Djinovic-Carugo, K., personal communication.

23 Norwood, F. L. M., Sutherland-Smith, A. J., Keep, N. H., Kendrick-Jones, J., The structure of the N-terminal actin-binding domain of human dystrophin and how mutations in this domain may cause Duchenne or Becker muscular dystrophy. *Struct. Fold. Design* **2000**, *8*, 481–491.

24 Schulunegger, M., Bennet, M., Eisenberg, D., Oligomer formation by 3D domain swapping: a model for protein assembly and disassembly. *Adv. Protein Chem.* **1997**, *50*, 61–122.

25 Sutherland-Smith, A. J., Moores, C. A., F. L., Norwood, Hatch, V., Craig, R., Kendrick-Jones, J., Lehman, W., An atomic model for actin binding by the CH domains and spectrin repeat modules of utrophin and dystrophin. *J. Mol. Biol.* **2003**, *329*, 15–33.

26 Hanein, D., et al., An atomic model of fimbrin binding to F-actin and its implications for filament crosslinking

and regulation. *Nat. Struct. Biol.* **1998**, *5*, 787–792.

27 Ylänne, J., Scheffzek, K., Young, P., Saraste, M., Crystal Structure of the α-actinin rod reveals an extensive torsional twist. *Struct. Fold. Design* **2001**, *9*, 597–604.

28 Rybakova, I. N., Amman, K. J., Ervasti, J. M., A new model for the interaction of dystrophin with F-actin. *J. Cell Biol.* **1996**, *135*, 661–672.

29 Rybakova, I. N., Patel, J. R., Davies, K. E., Yurchenko, P. D., Ervasti, J. M., Utrophin binds laterally along actin filaments and can couple costameric actin with sarcolemma when over-expressed in dystrophin-deficient mice. *Mol. Biol. Cell* **2002**, *13*, 1512–1521.

30 Moores, C. A., Keep, N. H., Kendrick-Jones, J., Structure of the utrophin actin-binding domain bound to F-actin reveals binding by an induced fit mechanism. *J. Mol. Biol.* **2000**, *297*, 465–480.

31 Orlova, A., Rybakova, I. N., Prochniewicz, E., Thomas, D. D., Ervasti, J. M., Egelman, E. H., Binding of dystrophin's tandem calponin homology domain to F-actin is modulated by actin's structure. *Biophys. J.* **2001**, *80*, 1926–1931.

32 Korenbaum, E, Rivero, F., Calponin homology domains at a glance. *J. Cell Sci.* **2002** *115*, 3543–3545.

33 Correia, I., Chu, D., Chou, Y. H., Goldman, R. D., Matsudaira, P., Integrating the actin and vimentin cytoskeletons: adhesion-dependent formation of fimbrin–vimentin complexes in macrophages. *J. Cell Biol.* **1999**, *146*, 831–482.

34 McGough, A., Way, M., DeRosier, D., Determination of the alpha-actinin–binding site on actin filaments by cryoelectron microscopy and image analysis. *J. Cell Biol.* **1994**, *126*, 433–443.

35 Fontao, L., Geerts, D., Kuikman, I., Koster, J., Kramer, D., Sonnenberg, A., The interaction of plectin with actin: evidence for cross-linking of actin filaments by dimerization of the actin-binding domain of plectin. *J. Cell Sci.* **2001**, *114*, 2065–2076.

36 Fuchs, P., Zorer, M., Rezniczek, G. A., Spazierer, D., Oehler, S., Castanon, M. J., Hauptmann, R., Wiche, G., Unusual 5′ transcript complexity of plectin isoforms: novel tissue-specific exons modulate actin binding activity. *Hum. Mol. Genet.* **1999**, *8*, 2461–2472.

37 Gimona, M., unpublished.

38 Stradal, T., Kranewitter, W., Winder, S. J., Gimona, M., CH domains revisited. *FEBS Lett.* **1998**, *432*, 134–137.

39 Nakamura, M., Zhou, X. Z., Lu, K. P., Critical role for the EB1 and APC interaction in the regulation of microtubule polymerization. *Curr. Biol.* **2001**, *11*, 1062–1067.

40 Morrison, E. E., Moncur, P. M., Askham, J. M., EB1 identifies sites of microtubule polymerisation during neurite development. *Brain Res. Mol. Brain Res.* **2002**, *98*, 145–152.

41 Bienz, M., The subcellular destinations of APC proteins. *Nat. Rev. Mol. Cell Biol.* **2002** *3*, 328–338.

42 Tirnauer, J. S., Grego, S., Salmon, E. D., Mitchison, T. J., EB1–microtubule interactions in *Xenopus* egg extracts: role of EB1 in microtubule stabilization and mechanisms of targeting to microtubules. *Mol. Biol. Cell* **2002**, *13*, 3614–3626.

43 Rogers, S. L., Rogers, G. C., Sharp, D. J., Vale, R. D., *Drosophila* EB1 is important for proper assembly, dynamics, and positioning of the mitotic spindle. *J. Cell Biol.* **2002**, *158*, 873–884.

44 Mimori-Kiyosue, Y., Shiina, N., Tsukita, S., The dynamic behavior of the APC-binding protein EB1 on the distal ends of microtubules. *Curr. Biol.* **2000**, *10*, 865–868.

45 Askham, J. M., Moncur, P., Markham, A. F., Morrison, E. E., Regulation and function of the interaction between the APC tumour suppressor protein and EB1. *Oncogene* **2000**, *19*, 1950–1958.

46 Bienz, M., Spindle cotton on to junctions, APC and EB1. *Nat. Cell Biol.* **2001**, *3*, E67–E69.

47 Nikolopoulos, S. N., Turner, C. E., Actopaxin, a new focal adhesion protein that binds paxillin LD motifs and actin and regulates cell adhesion. *J. Cell Biol.* **2000**, *151*, 1435–1448.

48 Olski, T. M., Noegel, A. A., Korenbaum, E., Parvin, a 42 kDa focal adhesion protein, related to the alpha-actinin superfamily. *J. Cell Sci.* **2001**, *114*, 525–538.

49 Nikolopoulos, S. N., Turner, C. E., Integrin-linked kinase (ILK) binding to paxillin LD1 motif regulates ILK localization to focal adhesions. *J. Biol. Chem.* **2001**, *276*, 23499–23505.

50 Nikolopoulos, S. N., Turner, C. E., Molecular dissection of actopaxin-integrin–linked kinase–paxillin interactions and their role in subcellular localization. *J. Biol. Chem.* **2002**, *277*, 1568–1575.

51 Hatch, V., Zhi, G., Smith, L., Stull, J. T., Craig, R., Lehman, W., Myosin light chain kinase binding to a unique site on F-actin revealed by three-dimensional image reconstruction. *J. Cell Biol.* **2001**, *154*, 611–617.

52 Litjens, S. H. M., Koster, J., Kuikman, I., van Wilpe, S., de Pereda, J. M., Sonnenberg, A., Specific binding of the plectin actin-binding domain to β4 integrin. *Mol. Biol. Cell* **2003**, *14*, 4039–4050.

53 Gimona, M., Mital, R., The single CH domain of calponin is neither sufficient nor necessary for F-actin binding. *J. Cell Sci.* **1998** *111*, 1813–1821.

54 Fu, Y., Liu, H. W., Forsythe, S. M., Kogut, P., McConville, J. F., Halayko, A. J., Camoretti-Mercado, B., Solway, J., Mutagenesis analysis of human SM22: characterization of actin binding. *J. Appl. Physiol.* **2000**, *89*, 1985–1990.

55 Goodman, A., Goode, B. L., Matsudaira, P., Fink, G. R., The *Saccharomyces cerevisiae* calponin/transgelin homolog Scp1 functions with fimbrin to regulate stability and organization of the actin cytoskeleton. *Mol. Biol. Cell* **2003** *14*, 2617–2629.

56 Winder, S. J., Jess, T., Ayscough, K. R., SCP1 encodes an actin bundling protein in yeast. *Biochem. J.* **2003**, *375*, 287–295.

57 Leinweber, B. D., Leavis, P. C., Grabarek, Z., Wang, C. L., Morgan, K. G., Extracellular regulated kinase (ERK) interaction with actin and the calponin homology (CH) domain of actin-binding proteins. *Biochem. J.* **1999**, *344*, 117–123.

58 Grubinger, M., Gimona, M., unpublished.

59 Ma, Y., Bogatcheva, N. V., Gusev, N. B., Heat shock protein (hsp90) interacts with smooth muscle calponin and affects calponin binding to actin. *Biochim. Biophys. Acta* **2000**, *1476*, 300–310.

60 Bogatcheva, N. V., Ma, Y., Urosev, D., Gusev, N. B., Localization of calponin binding sites in the structure of 90 kDa heat shock protein (Hsp90). *FEBS Lett.* **1999**, *457*, 369–374.

61 Zugaza, J. L., Lopez-Lago, M. A., Caloca, M. J., Dosil, M., Movilla, N., Bustelo, X. R., Structural determinants for the biological activity of vav proteins. *J. Biol. Chem.* **2002**, *277*, 45377–45453.

62 Groysman, M., Shifrin, C., Russek, N., Katzav, S., Vav, a GDP/GTP nucleotide exchange factor interacts with GDIs, proteins that inhibit GDP/GTP dissociation. *FEBS Lett.* **2000**, *467*, 75–80.

63 Katzav, S., Cleveland, J. L., Heslop, H. E., Pulido, D., Loss of the amino-terminal helix–loop–helix domain of the *vav* proto-oncogene activates its transforming potential. *Mol. Cell. Biol.* **1991**, *11*, 1912–1920.

64 Abe, K., Whitehead, I. P., O'Bryan, J. P., Der, C. J., Involvement of NH_2-terminal sequences in the negative regulation of Vav signalling and transforming activity. *J. Biol. Chem.* **1999**, *274*, 30410–30418.

65 Yabana, N., Shibuya, M., Adaptor protein APS binds the NH_2-terminal autoinhibitory domain of guanine nucleotide exchange factor Vav3 and augments its activity. *Oncogene* **2002**, *21*, 7720–7729.

66 Aghazadeh, B., Lowry, W. E., Huang, X. Y., Rosen, M. K., Structural basis for relief of autoinhibition of the Dbl homology domain of proto-oncogene *Vav* by tyrosine phosphorylation. *Cell* **2000**, *102*, 625–633.

67 Kranewitter, W. J., Grubinger, M., Gimona, M., unpublished.

68 Fraley, T. S., Tran, T. C., Corgan, A. M., Nash, C. A., Hao, J., Critchley, D. R., Greenwood, J. A., Phosphoinositide binding inhibits α-actinin bundling activity. *J. Biol. Chem.* **2003**, *278*, 24039–24045.

69 Winder, S. J., The membrane–cytoskeleton interface: the role of dystrophin and utrophin. *J. Muscle Res. Cell Motil.* **1997**, *18*, 617–629.

74 Kaplan, J. M., et al., Mutations in *ACTN4*, encoding α-actinin-4, cause familial focal segmental glomerulosclerosis. *Nat. Genet.* **2000**, *24*, 251–256.

70 Beggs, A. H., et al., Exploring the molecular basis for variability among patients with Becker muscular dystrophy: dystrophin gene and protein studies. *Am. J. Hum. Genet.* **1991**, *49*, 54–67.

71 Prior, T. W., Bartolo, C., Pearl, D. K., Papp, A. C., Snyder, P. J., Sedra, M. S., Burghes, A. H. M., Mendell, J. R., Spectrum of small mutations in the dystrophin coding region. *Am. J. Hum. Genet.* **1995**, *57*, 22–33.

72 Roberts, R. G., Gardner, R. J., Bobrow, M., Searching for the 1 in 2,400,000: a review of dystrophin gene point mutations. *Hum. Mutat.* **1994**, *4*, 1–11.

73 Robertson, S. P., et al., Localized mutations in the gene encoding the cytoskeletal protein filamin A cause diverse malformations in humans. *Nat. Genet.* **2003**, *33*, 487–491.

75 Hassoun, H., et al., Characterization of the underlying molecular defect in hereditary spherocytosis associated with spectrin deficiency. *Blood* **1997**, *90*, 398–406.

76 Dokland, T., personal communication.

Websites Directly Related to the Domain

http://www.proteinmodules.org

General platform site for protein domains and modules and official web site of the Protein Modules Consortium.

17
PH Domains

Mark A. Lemmon and David Keleti

17.1
Introduction

The pleckstrin homology (PH) domain was first defined in two notes that were published in *Cell* and *Nature* in 1993 [1, 2], and a 1993 article in *Trends in Biochemical Sciences* [3] suggested that this domain is "a common piece in the structural patchwork of signaling proteins". PH domains were identified as stretches of 100–120 amino acids found in many (then) recently identified signaling molecules, occurring twice in the platelet protein pleckstrin. It would certainly have surprised the authors of these publications to learn – as was possible after the sequencing of the human genome nine years later [4] – that the PH domain they defined is the eleventh-most-common domain type in the human proteome, with some 252 examples. As we discuss in this chapter, the abundance of PH domains probably reflects the fact that their defining sequence characteristics are associated with a particularly stable protein fold that can have several different functions. In other words, there are large numbers of domains that are structurally related to PH domains (which is what the sequence analysis actually defines), and we are now appreciating that the functions of PH domains thus defined may be quite diverse. Indeed, unlike in the SH2 and SH3 domains, for example, there are no identifiable sequence motifs or even completely conserved residues in PH domains.

The identification of PH domains followed hard on the heels of the discovery and analysis of protein target recognition by Src homology domains 2 and 3 (SH2 and SH3) and the resulting conceptual leap in our understanding of intracellular signal transduction [5, 6]. Naturally, it was suggested that the PH domain might represent another small protein module that drives specific protein–protein interactions in cellular signaling, and many laboratories embarked upon searches to identify PH domain targets, using approaches that had borne fruit with SH2 and SH3 domains. Within a year, papers were published indicating that the βγ subunits of heterotrimeric G proteins [7, 8] and protein kinase C isoforms [9] are among the binding partners for PH domains. Although these interactions may

Modular Protein Domains. Edited by G. Cesareni, M. Gimona, M. Sudol, M. Yaffe

ISBN: 3-527-30813-X

well be relevant in some instances – certainly $G_{\beta\gamma}$ interaction is important for the PH domain of the β-adrenergic receptor kinase (βARK) [10, 11] – a large amount of effort failed to identify a common protein or peptide target of the PH domain. It thus became apparent that the PH domain is likely to differ significantly from SH2, SH3, or many subsequently identified domains in its function in intermolecular interactions.

A major advance in understanding at least some PH domains came in 1994, when Fesik's laboratory showed that the most N-terminal of the two PH domains from pleckstrin itself can bind to the lipid phosphatidylinositol-4,5-bisphosphate (PtdIns(4,5)P_2) [12]. This finding set the stage for a large number of subsequent studies that have established a role for PH domains in phosphoinositide-dependent recruitment of proteins to cellular membranes. There are well studied PH domains that specifically recognize PtdIns(4,5)P_2 and others that specifically recognize phosphoinositides that are phosphorylated at their 3-position and are therefore only recruited to membranes after signal-dependent activation of phosphatidylinositol 3-kinases [13]. In these instances, the biology and its structural basis are now quite well understood, and some PH domains are known to be membrane targeted in a signal-regulated manner. However, more recent studies indicate that relatively few of the 252 PH domains recognize phosphoinositides in this way. Most PH domains may combine nonspecific phosphoinositide binding and protein recognition to drive membrane targeting (or other events), and we still have a great deal to learn about how this is achieved (and what are the targets).

17.2 PH Domain Structure and Phosphoinositide Binding

Crystallographic or NMR structures have now been described for 16 different PH domains. For six of these, the crystal structure was determined in complex with a phosphoinositide headgroup [14–19]. Just one example of a PH domain–protein interaction – the PH domain-mediated interaction of βARK with $G_{\beta\gamma}$ – has been visualized crystallographically [11]. In this section we focus on phosphoinositide binding by PH domains and return to PH domain–protein interactions in Section 17.4.

17.2.1 Overall Structure – The PH Domain Fold

In all instances, the core structure of the PH domain is the same. It can be described as a β sandwich or a partly open β barrel having seven strands (Figure 17.1). Strands β1 through β4 form a β sheet that is almost orthogonal to a second sheet (containing strands β5 through β7). Both sheets have the topology of a β meander, and the contribution of strand β1 to both sheets gives the structure its opened-barrel appearance, as can be seen most clearly in Figure 17.1b. Because of their right-handed twist, the two β sheets in the sandwich contact one another closely at two (close) corners (left and right in Figure 17.1a), but are splayed apart at the other two (splayed) corners [20] (top and bottom in Figures 17.1a and b). One splayed corner is capped by a C-terminal α helix (α1) found in all PH domains. The other is covered by the β1/β2, β3/β4, and β6/β7 loops of the PH domain, which are the most variable in length and sequence among different PH domains and have been termed variable loops 1 through 3 [21]. These features of the PH domain fold have also been observed in five other protein domain families that were not identified by sequence analysis [22]. Each of these domains is involved in directing protein–protein interactions. They are the phosphotyrosine binding (PTB) domain [23], a Ran-binding domain [24], the enabled/VASP homology 1 (EVH1) domain [25], the third subdomain of the FERM (band four-point-one, ezrin, radixin, moesin) domain [26, 27], and a domain from neurobeachin [28].

A particular characteristic of PH domains that was noted early on [21, 29] is that they are often electrostatically polarized. With one exception (the PH domain from *Caenorhabditis elegans* UNC-89 [30]), all PH domains with known structure have a large area of positive electrostatic potential that surrounds variable loops 1 through 3 (marked in Figure 17.1) and regions of negative potential on other parts of their surface. The presence of the three variable loops in a region of positive potential is consistent with their constituting the binding site for negatively charged ligands such as phosphoinositides.

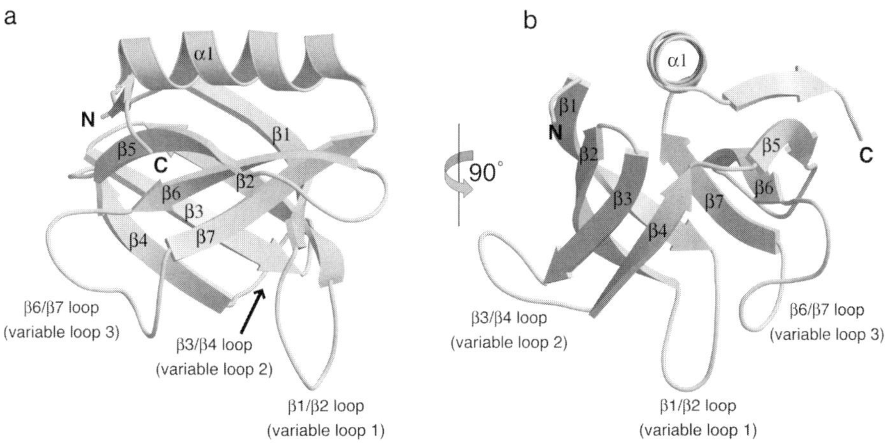

Figure 17.1 The PH domain fold. A ribbon representation of the PH domain from human dynamin-1 [21] is presented to illustrate the key points of the PH domain fold. The domain is shown from two orthogonal aspects, as indicated. Strands β1 through β7 are labeled, and the amphipathic C-terminal α helix that caps one splayed corner of the β sandwich is labeled α1. The N and C termini are marked, as are variable loops 1 to 3 that cap the splayed corner opposite α1. The variable loops correspond to the β1/β2, β3/β4, and β6/β7 loops, which are the most variable in length and sequence among PH domain sequences [21].

17.2.2 Structural Basis for Phosphoinositide Binding

Once phosphoinositides were identified as potential PH domain ligands [12], it became clear that certain PH domains specifically recognize a particular phosphoinositide (or subset of phosphoinositides) and bind with high affinity, but others bind more promiscuously and with much lower affinity. The PH domain from the N terminus of phospholipase C-δ_1 (PLC-δ_1) was the first shown to be capable of strong and specific phosphoinositide binding [31, 32]. The PLC-δ_1 PH domain binds to PtdIns(4,5)P_2 with a K_D in the 1–2 μM range, but at least 15-fold more weakly to any other phosphoinositide. By contrast, the N-terminal PH domain from pleckstrin and the β-spectrin PH domain are quite promiscuous in their phosphoinositide interactions and bind with K_D values in the 40 μM range [14, 33]. Structures of the PLC-δ_1 [15] and β-spectrin [14] PH domains with bound inositol-1,4,5-trisphosphate (Ins(1,4,5)P_3), the PtdIns(4,5)P_2 headgroup, provide insight into this specificity (or lack thereof).

17.2.2.1 High-affinity PtdIns(4,5)P_2 Binding

When bound to the PLC-δ_1 PH domain [15], Ins(1,4,5)P_3 makes direct contact with the β1/β2 and β3/β4 loops (variable loops 1 and 2), as well as a water-mediated hydrogen bond to the β6/β7 loop (variable loop 3). The bound Ins(1,4,5)P_3 molecule is located in the center of the positively charged surface of the PH domain, suggesting a mode of association with phosphoinositide-containing membranes that is illustrated in Figure 17.2. Phosphoinositide binding is driven largely by interactions between phosphate groups of the Ins(1,4,5)P_3 headgroup and (mostly) basic sidechains in the β1/β2 loop region. As discussed in more detail below, it was subsequently found that all PH domains that recognize phosphoinositides with high affinity and specificity share a sequence motif (with variations for different phosphoinositides) in the β1/β2 loop region [34–36]. The PLC-δ_1 PH domain recognizes the spatial array of phosphate groups in Ins(1,4,5)P_3 through a stereochemical cooperativity of interactions between primarily basic sidechains and the phosphates. Inspection of the structure [15] makes it clear how this cooperativity would be disrupted with inositol phosphate isomers other than Ins(1,4,5)P_3, thus significantly reducing the affinity of binding.

17.2.2.2 Low-affinity PtdIns(4,5)P_2 Binding

The crystal structure of the complex between the β-spectrin PH domain and Ins(1,4,5)P_3 [14] paints a different picture with regard to specificity, although the bound headgroup is again found in the center of the positively charged face of the domain. Whereas Ins(1,4,5)P_3 projects into a clear binding pocket when it is bound to the PLC-δ_1 PH domain, it appears to lie on the surface of the PH domain in β-spectrin, with a few hydrogen bonds (7, compared with 12 in PLC-δ_1) between its phosphate groups and primarily surface-located β-spectrin sidechains. The lack of stereospecificity in inositol phosphate (and phosphoinositide) binding by the β-spectrin PH domain and others suggests that binding in these examples is driven

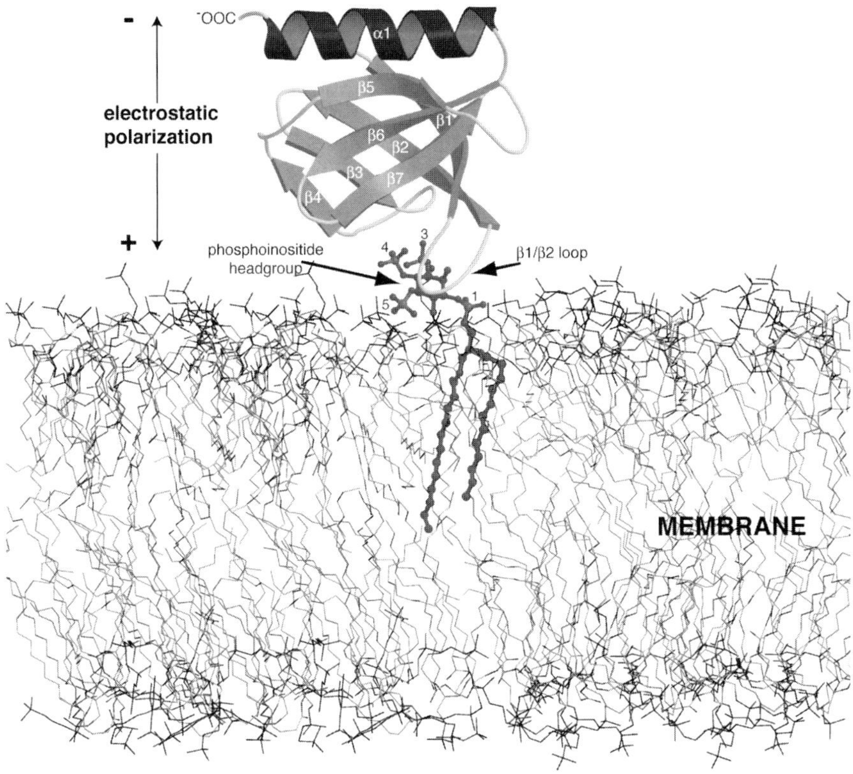

Figure 17.2 Hypothetical diagram of phosphoinositide-mediated membrane binding by a PH domain. The PH domain from DAPP1 [17] is used for illustration. This structure was determined in complex with Ins(1,3,4,5)P_4. Diacylglycerol has been added to the Ins(1,3,4,5)P_4 molecule to generate PtdIns(3,4,5)P_3, which has been placed in the context of a stick model of a phosphatidylcholine bilayer. The interaction and orientation of PtdIns(3,4,5)P_3 in the membrane is not intended to be accurate. This representation gives an impression of how the PH domain can interact with the phosphoinositide headgroup to drive its membrane association. The characteristic electrostatic polarization of PH domains, schematized to the left of DAPP1-PH, may also contribute to membrane association. The positively charged face (which includes the phosphoinositide-binding site) abuts the negatively charged membrane surface.

by delocalized electrostatic attraction between the positively charged surface of the PH domain and the highly negatively charged ligand. NMR studies of several PH domains in this class, including those from dynamin [37, 38], βARK [39], and pleckstrin-N [12], for example, also support this suggestion.

17.2.2.3 Specific Recognition of Phosphoinositide 3-Kinase Products

One of the best-studied PH domain functions is specific recognition of phosphoinositides having a phosphate group at their 3 position. Almost all cell-surface agonists activate one or other isoform of phosphoinositide 3-kinase (PI 3-kinase), leading to the phosphorylation of PtdIns(4,5)P_2 to yield PtdIns(3,4,5)P_3 [40].

$PtdIns(4,5)P_2$ is present constitutively in the plasma membrane of cells, and one estimate of its effective local concentration is approximately 5 mM [41]. By contrast, $PtdIns(3,4,5)P_3$ concentrations are estimated to be 1000 times lower (approximately 5 μM) prior to stimulation, rising to a maximum of around 200 μM following activation of PI 3-kinase by a cell-surface agonist [41]. A group of PH domains, including those from protein kinase B (PKB), Bruton's tyrosine kinase (Btk), and the general receptor for phosphoinositides-1 (Grp1), specifically recognize $PtdIns(3,4,5)P_3$ (or its immediate 5-dephosphorylation product $PtdIns(3,4)P_2$) and are directly recruited to the membrane as a result [13, 37, 42–45]. For these PH domains to be recruited to the plasma membrane only when $PtdIns(3,4,5)P_3$ is produced, their localization must be altered substantially by a local $PtdIns(3,4,5)P_3$ concentration of 200 μM, but not at all by a 25-fold higher local $PtdIns(4,5)P_2$ concentration of 5 mM. This feat is achieved with a selectivity for binding the $PtdIns(3,4,5)P_3$ headgroup ($Ins(1,3,4,5)P_4$) over the $PtdIns(4,5)P_2$ headgroup ($Ins(1,4,5)P_3$) of several hundred fold [13, 46], resulting from the addition of a single phosphate group.

Isakoff et al. [34] devised a convenient method for determining whether PH domains are capable of $PtdIns(3,4,5)P_3$-mediated (but not $PtdIns(4,5)P_2$-driven) membrane recruitment in yeast cells expressing a mammalian PI 3-kinase. Using

Figure 17.3 Sequence motif for high-affinity phosphoinositide binding to PH domains. (a) Sequences are shown for the β1 to β3 region of four PH domains specific for PI 3-kinase products and one (from PLC-δ_1) that is specific for $PtdIns(4,5)P_2$ as indicated. The positions of the lysine in strand β1, followed by the basic-X-basic pattern in strand β2 (see text), found in all PH domains that bind phosphoinositides with reasonably high affinity, are marked at the top. The motif used by Isakoff et al. [34] to predict which PH domains bind to PI 3-kinase products is shown below the sequences, and the residues found in Grp1-PH at each interacting position in the motif are marked. (b) Close-up of $Ins(1,3,4,5)P_4$ bound to the Grp1 PH domain [17]. Labels for residues in the sequence motif shown in panel A are boxed. The β1/β2 loop of Grp1-PH (variable loop 1) cradles the $Ins(1,3,4,5)P_4$ molecule, and motif residues fix the $Ins(1,3,4,5)P_4$ in position. Moving through the motif, the β1 leucine at the beginning of the motif projects into the hydrophobic core of the domain, helping to anchor the β1/β2 loop. The lysine in the middle of β1 (K273) forms hydrogen bonds with the both the 3- and 4-phosphate groups of the bound $Ins(1,3,4,5)P_4$. The amino acid immediately following strand β1 (G275) must have a small (or absent) sidechain to allow the orientation of the inositol ring shown here. R277 in the center of the β1/β2 loop closely approaches the 5-phosphate of the bound $Ins(1,3,4,5)P_4$ and is thought to contribute to $PtdIns(3,4,5)P_3$ specificity (this is absent from the PKB and DAPP1 PH domains). In strand β2, the first basic residue of the motif (K282 in Grp1-PH) makes a hydrogen bond with the 1-phosphate of $Ins(1,3,4,5)P_4$. The second (R284) hydrogen bonds extensively with the 3-phosphate. This conserved arginine corresponds to R40 in the PLC-δ_1 PH domain, R25 in PKB-PH, and R28 in Btk-PH (the site of XLA mutations). This interaction is critical for phosphoinositide binding. The aromatic sidechain at the end of β2 (F286 in Grp1-PH) anchors the β2 end of the β1/β2 loop into the hydrophobic core. Finally, the β3 tyrosine (Y295 in Grp1-PH) makes a hydrogen bond with the 4-phosphate of bound $Ins(1,3,4,5)P_4$. Specific to Grp1-PH, additional sidechains from a lysine, histidine, and asparagines (from the β6/β7 loop) interact with the 5-phosphate group to increase $PtdIns(3,4,5)P_3$ specificity.

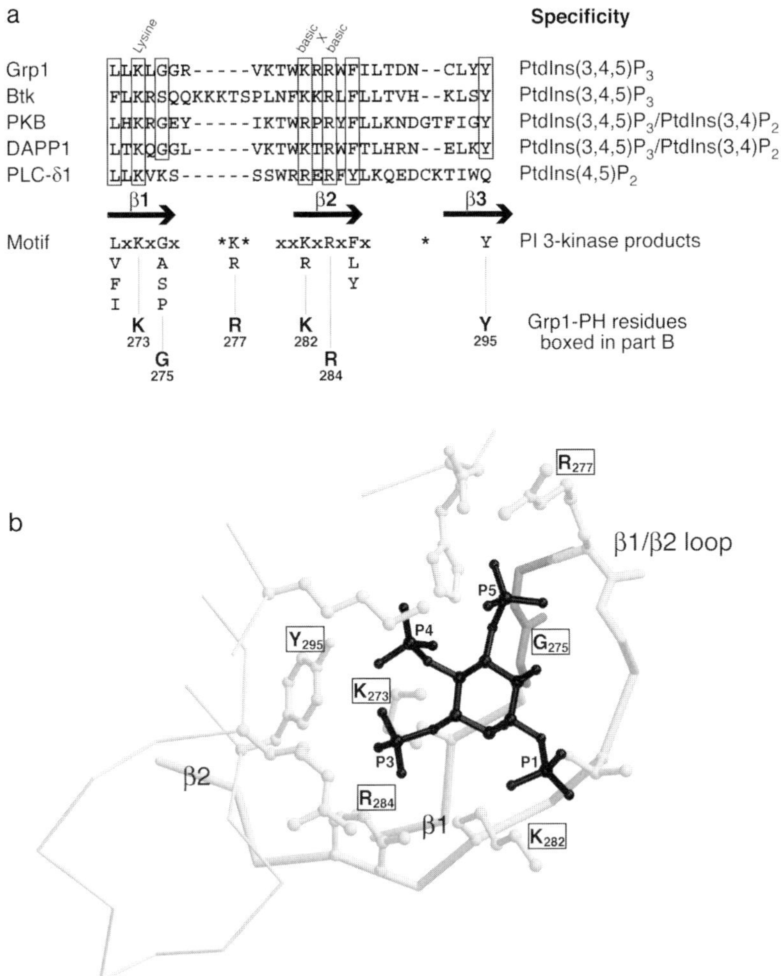

Figure 17.3 (legend see p. 342)

this approach they identified nearly all PH domains now known to have this property (and determined that many do not). Most interestingly, all PH domains that specifically recognize agonist-regulated PI 3-kinase products share a sequence motif in their β1/β2 loop region that is reminiscent of the corresponding region in the PLC-δ_1 PH domain. This motif is summarized in Figure 17.3a. Several subsequent X-ray crystal structures of PH domains bound to the headgroup of PtdIns(3,4,5)P_3 have provided a satisfying explanation of how this motif defines PH domain binding specificity [16–19]. The structural role of each conserved feature in the motif is illustrated in Figure 17.3b (and described in the figure legend).

The most critical elements of this motif are a lysine sidechain close to the end of strand β1 and two basic residues in the pattern 'basic-x-basic' at the beginning of strand β2. The last of these basic residues corresponds to R28 in Btk, which is mutated in X-linked agammaglobulinemia (XLA) [47, 48], and is the 'standard' residue that is mutated to impair phosphoinositide binding by PH domains experimentally (R25 in the PKB PH domain, R284 in the Grp1 PH domain). As seen in Figure 17.3b, when Ins(1,3,4,5)P_4 is bound to the Grp1, Btk, DAPP1, or PKB PH domains, the β1 lysine sidechain (K273 in Grp1-PH) makes hydrogen bonds with both the 3- and 4-phosphates of Ins(1,3,4,5)P_4, and the basic residues in β2 (K282 and R284 in Grp1-PH) hydrogen bond with the 1- and 3-phosphates, respectively. These hydrogen bonds form the 'core' set of interactions and are supplemented by additional contacts mediated by the loop region and/or other parts of the PH domain that define the precise binding specificity. In Grp1-PH, for example, selectivity for PtdIns(3,4,5)P_3 over PtdIns(3,4)P_2 appears to be determined in part by R277 in the β1/β2 loop, which is close to the 5-phosphate group, plus sidechains of a unique β6/β7 insert that also hydrogen bond with this phosphate group [17]. The absence of corresponding basic residues in the PKB and DAPP1 PH domains explains their ability to bind PtdIns(3,4)P_2 and PtdIns(3,4,5)P_3 with very similar affinities.

Interestingly, the headgroups of PtdIns(3,4,5)P_3 or PtdIns(3,4)P_2 can only be accommodated (with high affinity) in the configuration shown in Figure 17.3b if the last residue in strand β1 has a small (or absent) sidechain (G275 in Grp1-PH). Accordingly, this position is occupied by glycine, serine, alanine, or proline in all PI 3-kinase product-specific PH domains. A larger sidechain at this position would clash with the inositol ring of the bound Ins(1,3,4,5)P_4 or Ins(1,3,4)P_3, requiring a reorientation that would disrupt the ability of the PH domain to maximize cooperativity between interactions involving the 1-, 3-, and 4-phosphates. Alteration of the backbone configuration in this region of the Grp1 PH domain by insertion of a glycine reduces its PtdIns(3,4,5)P_3-binding affinity and specificity [49] – actually enhancing its PtdIns(4,5)P_2-binding affinity.

The PLC-δ_1 PH domain maintains most features of the motif shown in Figure 17.3 but does not have a residue with a small sidechain at the end of strand β1. Possibly as a result of this difference, Ins(1,4,5)P_3 binds to the PLC-δ_1 PH domain in an orientation that is rotated by 180° (about an axis between the 1- and 4-phosphates) compared with that seen for Ins(1,3,4,5)P_4 in the Grp1-PH binding site in Figure 17.3b. In the complex between the PLC-δ_1 PH domain and Ins(1,4,5)P_3, the 1- and 4-phosphates of Ins(1,4,5)P_3 are similar in position to those of Ins(1,3,4,5)P_4 in Figure 17.3b. However, because of the ~180° flip, the position occupied by the 3-phosphate in Figure 17.3b is instead occupied by the 5-phosphate of Ins(1,4,5)P_3 in the PLC-δ_1 complex. Thus, the β1 lysine of the PLC-δ_1 PH domain interacts with the 4- and 5-phosphates of Ins(1,4,5)P_3 (rather than the 3- and 4-phosphates of Ins(1,3,4,5)P_4), and the two basic sidechains in strand β2 interact with the 1- and 5-phosphates instead of the 1- and 3-phosphates. Thus, the principles that guide specific recognition of PI 3-kinase products by the PKB, Grp1, Btk, and other PH domains are similar to those that direct specific PtdIns(4,5)P_2 recognition by the PLC-δ_1 PH domain. The distinct specificities arise from small differences in stereochemistry.

17.2.2.4 PH Domains with Other Phosphoinositide-binding Specificities

No PH domains have been identified that bind with high affinity and specificity to any phosphoinositide other than PtdIns(4,5)P_2, PtdIns(3,4,5)P_3, or PtdIns(3,4,5)P_3/PtdIns(3,4)P_2. Dowler et al. [35] searched EST databases for uncharacterized PH domains with β1/β2 loop sequences similar to the motif described above and analyzed their phosphoinositide-binding properties. One, which they named TAPP1 (for tandem PH domain-containing protein-1), has two PH domains, the most C-terminal of which contains the motif described above and appeared to be PtdIns(3,4)P_2-specific in their studies. Subsequent crystallographic and mutational analyses [50] indicated that this PH domain has an alanine (rather than glycine) in its β1/β2 loop, which disfavors accommodation of the 5-phosphate of PtdIns(3,4,5)P_3. However, although this PH domain does bind more strongly to PtdIns(3,4)P_2 than to PtdIns(3,4,5)P_3, its PtdIns(3,4,5)P_3 binding affinity is significant [17, 50]. In headgroup competition studies, TAPP1 C-PH shows only a ~five-fold preference for the PtdIns(3,4)P_2 headgroup over the PtdIns(3,4,5)P_3 headgroup (Sankaran, V. G. and M. A. L., unpublished data). It is therefore not clear to what extent TAPP1 is truly PtdIns(3,4)P_2-specific. The evidence that the C-terminal TAPP1 PH domain prefers PtdIns(3,4)P_2 is strong, and it has been demonstrated that TAPP1 does relocate to the plasma membrane of cells in response to agonists that promote PtdIns(3,4)P_2 production [51]. However, the in vitro binding properties of the C-terminal TAPP 1 PH domain argue that TAPP1 is also likely to be regulated by PtdIns(3,4,5)P_3 under most conditions.

Dowler et al. [35] also identified PH domains with motifs related to those described in Figure 17.3 that appeared to be specific for PtdIns(3)P (proteins named PEPP1 and AtPH1), PtdIns(3,5)P_2 (centaurin-β2), or PtdIns(4)P. The interactions observed in this study (using a lipid-overlay approach) with PtdIns(3)P and PtdIns(3,5)P_2 appear to have rather low affinity, and their physiological significance is not yet clear. In the FAPP1 PH domain, although the lipid-overlay studies of Dowler et al. [35] indicate PtdIns(4)P specificity, our own analysis with multiple approaches (D. K. and M. A. L., unpublished data), as well as published studies by Levine and Munro [52], suggest it is in fact quite promiscuous in its phosphoinositide binding. FAPP1-PH appears to bind with a K_D of around 10 μM to all phosphoinositides tested.

17.2.2.5 Sequence Predictors of Phosphoinositide Binding

In a recent genome-wide study of *Saccharomyces cerevisiae* PH domains [36], only seven (of 33 in the genome) were found to bind phosphoinositides sufficiently strongly for the interaction to be measured by standard techniques (i.e., K_D values less than approximately 50 μM). All of these PH domains have patterns of basic residues in their β1/β2 loop regions that resemble (or match) those seen in the PLC-δ_1 PH domain and the PH domains that specifically recognize PI 3-kinase products. It is therefore likely that the mode of phosphoinositide headgroup binding shown in Figure 17.3 represents a model that is relevant to all other PH domains that bind these ligands with significant affinity. At least in the *S. cerevisiae* genome there do not appear to be alternative modes of phosphoinositide recognition by PH domains. Moreover, with one exception in the yeast genome, the presence of a lysine at the penultimate position in strand β1 plus the basic-X-basic pattern beginning at the second predicted residue of strand β2 are excellent predictors of significant phosphoinositide binding. Only one yeast PH domain having these features failed to bind phosphoinositides strongly in the studies of Yu et al [36]. Beyond these features, the characteristics described above and in the legend to Figure 17.3 can be used to predict whether the PH domain is likely to recognize PI 3-kinase products, and if so, whether it binds PtdIns(3,4,5)P_3 only or both PtdIns(3,4)P_2 and PtdIns(3,4,5)P_3.

It is important to appreciate that the majority of PH domains do not contain sequence patterns of this type. In yeast, 25 of the 33 identifiable PH domains (~75%) cannot be predicted to bind strongly to phosphoinositides according to these criteria, and experimental studies show this to be true [36]. Assuming a similar distribution in humans, there are probably approximately 190 human PH domains that bind very weakly (if at all) to phosphoinositides. Developing an understanding of the function of these PH domains, and its structural basis, is an important challenge for the future.

17.3 Molecular and Signaling Function of PH Domains

As should be clear from Section 17.2, only a small minority of PH domains (perhaps 20%) bind strongly to phosphoinositides, and these are the most well understood structurally and functionally. Indeed, in the past few years PH domains have become best known as modules that specifically recognize phosphoinositides and thus drive membrane targeting of their host proteins. This membrane targeting can be recapitulated in vivo using the isolated PH domain and has been well studied. We discuss this group of PH domains in Section 17.3.1, separating them into three groups according to their phosphoinositide binding specificity. However, it is very important to realize that this is a property or function of only a minority of PH domains. To put this into one perspective, there is only a single PH domain capable of specific high-affinity phosphoinositide binding among all 33 PH domains found in *S. cerevisiae*. This is a PtdIns(4,5)P_2-specific PH domain from Num1p, a protein involved in nuclear migration [53]. Most PH domains (75% or more) do not bind strongly to phosphoinositides and, moreover, do not show any tendency to be targeted to cellular membranes when studied in isolation. The function of these PH domains is much more enigmatic and is considered in Sections 17.3.2 and 17.3.4.

17.3.1 PH Domains as Phosphoinositide-dependent Membrane-targeting Domains

The description of the structural properties of phosphoinositide-specific PH domains provides a good conceptual introduction to their function. In essence, they target the protein that contains them to membranes containing the phosphoinositide(s) to which they bind with high affinity. Whether their membrane targeting is constitutive or signal-regulated depends on the phosphoinositide(s) that they recognize. Thus, PtdIns(4,5)P_2-specific PH domains are constitutively targeted to the plasma membrane and (to some extent) to other membranes, but PH domains that recognize only PI 3-kinase products remain cytoplasmic in unstimulated cells but are rapidly recruited to the plasma membrane upon activation of cell surface receptors. The clarity of our current view of phosphoinositide recognition by this class of PH domains has led to their use as in vivo probes of phosphoinositide production and location [54, 55]. The PLC-δ_1 PH domain fused to green fluorescent protein (GFP) has become an almost standard tool for observing the subcellular localization of PtdIns(4,5)P_2, and many studies have employed GFP fusions of PtdIns(3,4,5)P_3-specific or PtdIns(3,4,5)P_3/PtdIns(3,4)P_2-specific PH domains to monitor the accumulation and localization of PI 3-kinase products upon cell stimulation. It is generally assumed that phosphoinositides alone define the location of the GFP/PH domain fusions used in these studies, although other influences have not been excluded and there are several reasons to suspect that other (poorly defined) binding targets may also play a role [54].

17.3.1.1 PtdIns(4,5)P_2-specific PH Domains

PH domains from phospholipase C-δ isoforms (and related proteins) [31, 32, 56, 57] are the only known mammalian examples that are specific for PtdIns(4,5)P_2. In *S. cerevisiae*, as mentioned above, only the Num1p PH domain is PtdIns(4,5)P_2-specific [36]. The PH domain of PLC-δ_1 is located at its amino terminus and can bind both the substrate (PtdIns(4,5)P_2) and the product (Ins(1,4,5)P_3) of this enzyme. It is thought that binding of the PH domain to PtdIns(4,5)P_2 serves to anchor the whole enzyme to PtdIns(4,5)P_2-rich membranes, thus allowing processive or scooting-mode hydrolysis of substrate by PLC-δ_1 in these membranes [58, 59]. Binding of the soluble product (Ins(1,4,5)P_3) to the PH domain competes with PtdIns(4,5)P_2 binding [32], thus dissociating PLC-δ_1 from the membrane and inhibiting its activity [60, 61]. Through this mechanism, the PH domain influences PLC-δ_1 activity in a manner that depends on the ratio of PtdIns(4,5)P_2 and Ins(1,4,5)P_3 concentrations that it experiences.

The PtdIns(4,5)P_2-specific *S. cerevisiae* Num1p PH domain is required to anchor this protein to the mother cell cortex, so that it can serve as a cortical anchor for dynein as it drives nuclear migration through the bud neck during mitosis [62]. There is no reason to expect that Ins(1,4,5)P_3 binding is important for the Num1p PH domain.

Another very intriguing role for a PtdIns(4,5)P_2-binding PH domain was recently described for the Unc104 kinesin motor [63]. Unc104 has a PH domain, although one for which phosphoinositide binding specificity has yet to be determined. Klopfenstein et al. [63] found that Unc104 binds PtdIns(4,5)P_2-containing vesicles through this PH domain and uses its motor domain to transport these vesicles along microtubules. This finding illustrates the fact that we should bear in mind that PH domains can dock membranes (or lipid molecules) to proteins just as well as phosphoinositide-containing membranes can dock or recruit PH domain-containing proteins.

17.3.1.2 PI 3-kinase Product-binding PH Domains

Whereas membrane targeting by the very few PtdIns(4,5)P_2-specific PH domains is constitutive, it is directly signal-dependent for PH domains that recognize PI 3-kinase products and participate in PI 3-kinase signaling [40, 64, 65]. Conceptually, the role played by PH domains in responding to PI 3-kinase activation is very simple and is illustrated in Figure 17.4 for signaling via protein kinase B (PKB) and phosphoinositide-dependent kinase-1 (PDK1) [66]. After PI 3-kinase activation, PKB is rapidly recruited to the plasma membrane [67, 68] by virtue of the specific interaction of its N-terminal PH domain and the PtdIns(3,4,5)P_3 (and/or PtdIns(3,4)P_2) that transiently accumulates. This signal-dependent membrane recruitment can be reconstituted with the isolated PH domain and can be visualized directly (and dramatically) when the PH domain is fused to GFP [69–71]. Membrane recruitment of PKB brings it close to PDK1, which itself has a PI 3-kinase product-specific PH domain at its C terminus [72, 73]. There is disagreement as to whether PDK1 is recruited to the plasma membrane by PI 3-kinase products [74, 75]. However, binding of both PKB and PDK1 to the same phosphoinositides would certainly enhance their colocalization, and under these conditions PDK1-mediated phosphorylation of PKB, which is required for PKB activation, would be promoted. PKB is phosphorylated in its activation loop (at T308) by PDK1 and becomes fully activated when it is also phosphorylated at S473 by another uncharacterized kinase [66] (or possibly by autophosphorylation [76]). The role of the PH domains in this pathway is to colocalize two proteins transiently, and in a regulated manner, so that one can phosphorylate the other and cause its signal-dependent activation.

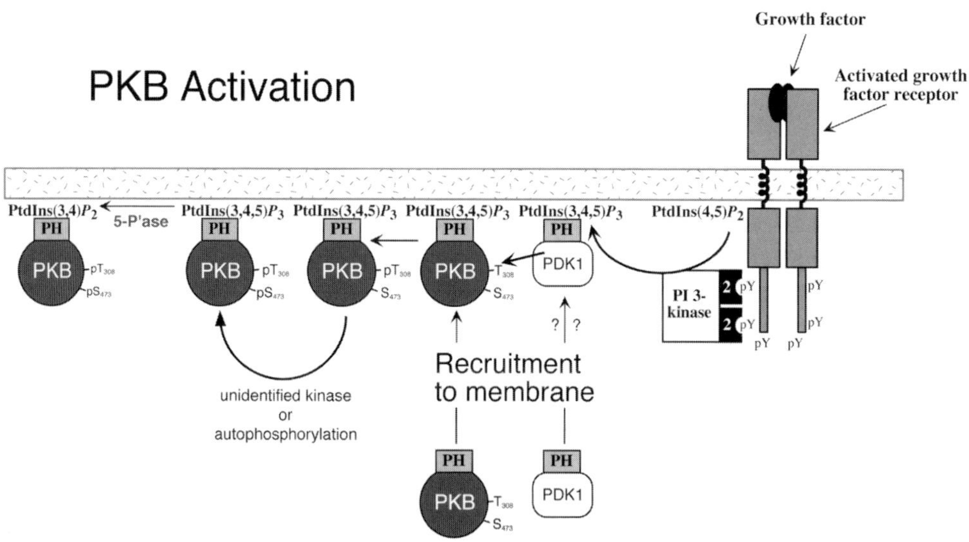

Figure 17.4 Role played by PH domains of PKB and PDK1 in PI 3-kinase signaling. A scheme for activation of PKB by growth factor-induced PI 3-kinase activation is shown. See text for details.

There are significant parallels between the roles that PH domains play in this pathway and those played by SH2 and PTB domains in other related pathways, as discussed in other chapters. Other PH domains that bind strongly and specifically to PI 3-kinase products function similarly. For example, by recognizing PtdIns(3,4,5)P_3, the PH domain of Btk brings this tyrosine kinase close to the membrane-associated Lyn tyrosine kinase that initiates Btk activation [77]. Grp1 is a guanine nucleotide exchanger for ADP ribosylation factor-6 (ARF6) at the plasma membrane. PtdIns(3,4,5)P_3-dependent recruitment of Grp1 to ARF6 at the plasma membrane activates this small G-protein, thus promoting cytoskeletal and membrane changes [77]. Other examples are detailed in the reviews of PI 3-kinase signaling cited above.

Although simple PI 3-kinase-dependent membrane recruitment seems sufficient to explain these signaling events, it has been speculated that an important conformational change in PKB is induced upon ligand binding to the PH domain. PKB mutants lacking the PH domain are reportedly more readily phosphorylated by PDK1 than is the intact protein [78], leading to the suggestion that the unliganded PH domain may shield T308 from PDK1 phosphorylation. A comparison of the crystal structures of unliganded and Ins(1,3,4,5)P_4-bound PKBα PH domain has suggested that this PH domain is unique in undergoing significant ligand-dependent conformational changes in the β1/β2, β3/β4 and β6/β7 loops [79]. The authors of this study suggested that these changes could disrupt intramolecular interactions that inhibit T308 phosphorylation and that this could be an important element of PI 3-kinase regulation of PKB. While this certainly is possible, we should note that the crystallization conditions used for the 'unliganded' and 'liganded' structure determinations began at pH values of 8.5 and 4.6, respectively, which could explain some of the observed structural differences. Moreover, a recent NMR study of the PKBβ PH domain (at pH 7.4) showed that the structures of all three of these loops (especially the β3/β4 and β6/β7 loops) are very ill defined – as seen for other PH domains – and no strong alteration in the dynamic behavior of these loops upon Ins(1,3,4,5)P_4 binding was detected by NMR [80]. It therefore remains unclear whether binding of PI 3-kinase products to the PKB PH domain promotes allosteric changes of the sort suggested by Milburn et al. [79] or whether these changes simply reflect different static structures trapped in different crystals that were grown at different pH values. However, even without significant, well defined ligand-induced changes in the PH domain itself, it does not seem unreasonable to suggest that reorientation of the PH and kinase domains at the membrane surface could promote the ability of PDK1 to phosphorylate PKB. Structural studies of the entire PKB molecule are required to address this question satisfactorily.

17.3.1.3 Membrane Targeting by PH Domains with Little Phosphoinositide-binding Specificity

A third group of PH domains that are clearly membrane targeted when analyzed in isolation (i.e., when expressed in cells as GFP-fusion proteins) has recently been identified. These bind phosphoinositides more than 10 times more weakly than the PLC-δ_1 PH domain or the PI 3-kinase product-specific PH domains. In addition, they show little or no stereospecificity in their phosphoinositide binding, interacting with all known phosphoinositides with approximately the same affinity. Levine and Munro were the first to identify a PH domain having these characteristics [52, 81] – that from the oxysterol binding protein (OSBP). The OSBP PH domain is specifically targeted to the golgi complex in both yeast and mammalian cells when analyzed as a GFP-fusion protein [81]. Several PH domains with closely related sequences, including those from FAPP1 [81, 82], and the *S. cerevisiae* OSBP homologs Osh1p [36, 83] and Osh2p [36] are all similarly localized to the golgi. Most interestingly, the specific localization of all of these PH domains is largely abolished if golgi production of PtdIns(4)P is impaired, in a *pik1*^{ts} yeast strain [36, 52, 81, 82]. By contrast, decreased concentrations of PtdIns(4,5)P_2 or other phosphoinositides, or (most intriguingly) even decreased PtdIns(4)P production at the plasma membrane (in an *stt4*^{ts} yeast strain [84]) have no influence on the golgi localization of these PH domains. All of these golgi-located PH domains bind equally well to all phosphoinositides in vitro [36, 52], yet in vivo they bind to PtdIns(4)P only at the golgi. The most reasonable explanation for these observations is that golgi localization requires two targets: PtdIns(4)P and a second unidentified binding partner. The second binding partner is likely to be present only in the golgi, thus defining the specific localization of these PH domains, and is thought to cooperate with golgi phosphoinositides (presumably mostly PtdIns(4)P) to recruit the PH domains from OSBP, FAPP1, and Osh proteins. Levine and Munro identified yeast Arf1p as one candidate for this second component that may cooperate with PtdIns(4)P in recruiting the OSBP PH domain to the golgi [52].

A recent genome-wide analysis of *S. cerevisiae* PH domains [36] identified at least four plasma membrane-targeted PH domains that may have similar characteristics. These PH domains (from Cla4p, Skm1p, Yil105cp, and Ynl047cp) are targeted to the plasma membrane in a manner that depends on PtdIns(4,5)P_2 production. However, the in vitro PtdIns(4,5)P_2 binding affinities of these PH domains are too weak for this to be solely responsible for the observed in vivo membrane targeting. We suggest that these PH domains are targeted to the plasma membrane by simultaneously binding PtdIns(4,5)P_2 and another, as yet unidentified, target. The probable existence of other such non-phosphoinositide targets is further suggested by one yeast PH domain identified in this study [36] (from Opy1p) that binds phosphoinositides very weakly in vitro, yet is efficiently targeted to the plasma membrane in a manner that does not depend on in vivo phosphoinositide production.

17.3.2
Function of Low-affinity PH Domains That Are Not Independently Membrane Targeted

As mentioned in several places in this chapter, the majority of PH domains neither bind tightly (or specifically) to phosphoinositides nor interact strongly with cellular membranes. Understanding the functions of these PH domains is currently the greatest challenge. Many have been reported to bind phosphoinositides with very low affinities (and usually no specificity). Indeed, in *S. cerevisiae*, all that is known about 17 of the 33 PH domains is that they bind phosphoinositides in lipid-overlay experiments (but not strongly enough to detect by other methods) and that they are cytoplasmic when expressed in yeast or mammalian cells as GFP fusions [36]. A similar conclusion has also been reached for the majority of mammalian PH domains that have been studied [46, 85], although several have also been reported to have protein targets [13].

The first question to address in light of these findings is whether or not the low-affinity phosphoinositide binding that is seen in vitro for nearly every PH domain has in vivo relevance. Indeed, it is troubling to realize that most PH domains (and many other proteins) can be purified very effectively by using cation exchange resins and that the binding matrices used to analyze phosphoinositide binding bear significant resemblance to these negatively charged resins. In other words, it actually seems quite likely that the low affinity and nonspecific phosphoinositide binding reported for many PH domains is actually an in vitro artifact. Contrary to this rather pessimistic view, it does appear that low-affinity phosphoinositide binding is important in at least some of these proteins – in particular for the PH domain from dynamin, a large GTPase involved in receptor-mediated endocytosis [86–88], and for the large group of PH domains that immediately follow Dbl homology (DH) domains in guanine nucleotide exchange factors (GEFs) for Rho GTPases.

17.3.2.1 The Dynamin PH Domain

Dynamin is a large GTPase (with one PH domain) that assembles at the necks of invaginated coated pits during receptor-mediated endocytosis and is intimately involved in the scission of these necks to produce endocytic vesicles [89]. PtdIns(4,5)P_2 and other phosphoinositides enhance the in vitro GTPase activity of dynamin [37, 90] in a way that appears to require its oligomerization into well defined assemblies [91, 92]. Moreover, although the monomeric dynamin PH domain has a very low affinity for phosphoinositides, it binds quite strongly to PtdIns(4,5)P_2-containing membranes when oligomerized (as in self-assembled dynamin) [93]. Thus, dynamin self-assembly and PH domain-mediated membrane binding appear to be thermodynamically coupled.

Expression of full-length dynamin with a phosphoinositide-binding–defective PH domain has a dominant negative effect on the function of endogenous dynamin in HeLa and other cells [86–88]. Dynamin assemblies containing some molecules with defective PH domains are therefore thought to be incapable of interacting sufficiently strongly with PtdIns(4,5)P_2-containing membranes for dynamin's function to be executed. In other words, overexpressed PH domain-mutated dynamin appears to 'poison' the avidity advantage afforded to the PH domain by self-assembly. Details of the step (or steps) in which PtdIns(4,5)P_2 binding to the dynamin PH domain is functionally important for its role in the scission of endocytic vesicles remain very poorly defined. However, the evidence for a requirement that multiple PH domains within a dynamin oligomer cooperate with one another to fulfill this role appears to be quite strong.

17.3.2.2 PH Domains of Dbl-family Proteins

The PH domains that follow Rho GEF/Dbl homology domains (referred to here as DH domains) in Dbl family proteins [94] are another group for which low-affinity, often promiscuous, phosphoinositide binding appears to be functionally important. The Dbl family proteins are guanine nucleotide exchange factors (GEFs) for Rho-family small GTPases, and the 200 amino-acid DH domain is necessary and sufficient for their GEF activity [94, 95]. In mammalian Dbl family proteins the DH domain is almost invariably followed by a PH domain, and the 46 or more examples for which this is true account for some 18% of all human PH domains. In all instances studied, DH-associated PH domains bind phosphoinositides with low affinity (K_D values > 10 μM), and in most examples, binding appears to be promiscuous [13, 96]. A series of recent publications have shown that mutations that impair the low-affinity in vitro phosphoinositide-binding interactions of these PH domains also impair the in vivo GEF activity of the full-length proteins [97–104]. Consistent with their low-affinity binding to phosphoinositides, neither isolated Dbl family PH domains nor DH/PH fragments of these proteins appear to be independently targeted to cellular membranes. By contrast, full-length Dbl family proteins are often found to be membrane-associated, most likely as a consequence of interactions mediated by the multiple other domains that they contain. It seems likely that phosphoinositide binding by the PH domain cooperates with these other domains in altering the extent or specific location of membrane targeting. PH domain mutations that impair the function of intact Dbl family proteins have been reported to alter subcellular localization in some mutants [101, 103] but not significantly in others [97, 98, 100].

PH domains that follow DH domains may also play a role that does not involve driving membrane targeting, and this may be their primary role. In the context of a membrane-associated multidomain protein, the PH domain could direct localization to membrane regions that are rich in phosphoinositides, which could be important for colocalizing the Dbl family proteins with its small GTPase targets. The specificity of the C-terminal Tiam-1 PH domain for PtdIns(3)P [96, 98] suggests such a 'lateral targeting' possibility, although this remains speculative. Structural studies of DH/PH fragments with and without bound Rho-family GTPases suggest an additional possibility [105–108]. In the structure of the Sos DH/PH fragment, the PH domain is positioned so that it prevents access of the small GTPase to its binding site on the DH domain [105]. In the structures of DH/PH fragments bound to small GTPases [106–108], the PH domain is oriented quite differently, so that the small GTPase has access to the DH domain. It has been proposed that phosphoinositide binding to the PH domain (at a membrane surface) promotes reorientation of the two domains so that the PH domain no longer interferes with access to the DH domain binding site, thus promoting exchange activity of the Dbl family protein [97, 105, 107]. The structure of the Dbs DH/PH fragment bound to cognate GTPases revealed additional unexpected direct contacts between the PH domain and the bound Cdc42 [106] or RhoA [108]. It is possible that phosphoinositide binding to the PH domain promotes an orientation of the DH and PH domains at the membrane surface that is ideal for these critical interactions. Thus, in the Dbl family proteins, PH domains may play roles in both membrane recruitment (either bulk localization or lateral targeting to specific regions of the membrane) and allosteric regulation by altering the relative orientations of adjacent domains.

17.3.3 Protein Targets of PH Domains

It is difficult to write about protein targets of PH domains – of which many have been suggested – without the risk of simply producing a list. However, two themes appear to be emerging. One was touched upon in the Introduction and involves PH domain interaction with the βγ subunits of heterotrimeric G proteins. The other was introduced in the previous section by the fact that the PH domain of Dbs directly contacts Cdc42 or RhoA bound to the DH/PH fragment of this Rho family GEF.

17.3.3.1 Small GTPases as PH Domain Targets

There have now been several reports of PH-domain interactions with small GTPases, and the surprising finding that the first Ran-binding domain (RanBD1) from Ran-binding protein-2 (RanBP2) closely resembles a PH domain [24] provides a conceptual framework for this. RanBP2 binds specifically to the GTP-bound form of the small GTPase Ran (involved in regulation of nucleocytoplasmic transport), using a binding site that contains its three PH-domain variable loops [24].

Three other reports provide compelling evidence for GTP-regulated interaction between PH domains and small GTPase targets. Jaffe et al. [109] recently showed that the PH domain of hCNK, the human homolog of connector enhancer of ksr (kinase suppressor of ras), interacts with Rho. Although the interaction appears to be relatively weak (and may require cooperation with other domains in CNK), it appears to be selective for the GTP-bound form of Rho (as in RanBP2–Ran interactions). Similarly, Kim et al. [110] presented evidence to suggest that Etk, a tyrosine kinase from the Btk family, interacts specifically with GTP-bound RhoA through the N-terminal Etk PH domain. A third example of this, suggested from genetic studies, was outlined above. In analyzing golgi localization of the OSBP PH domain, Levine and Munro [52] provided evidence that the small GTPase Arf1p may cooperate with phosphoinositides in recruiting PH domains to this organelle.

A more thorough and quantitative analysis has been applied to the interaction between the N-terminal PH domain of PLC-β_2 and Rac GTPases [111]. Sondek and colleagues analyzed PLC-β_2 binding by 17 different GTP-bound members of the Rho family and found that only the GTP-bound form of the three Rac GTPases bound significantly to (and activated) PLC-β_2. Binding of GTP-bound Rac GTPases to the isolated PLC-β_2 PH domain had essentially the same affinity (within a factor of two) as their binding to full-length PLC-β_2 in surface plasmon resonance studies, with K_D values in the 5–10 μM range. These more detailed studies add substantial credibility to the increasing number of other reports that certain PH domains bind to small GTPases and, by binding only to their GTP-bound state, may serve as Rac, Rho, or Arf1p effectors. This class of protein interaction is the closest that PH domains have come to having a common protein target.

17.3.3.2 Other Protein Targets of PH Domain Targets

We previously reviewed many of the proposed protein targets for PH domains that are based on one or two reports [13] and for which the physiological relevance is not completely clear. Rather than repeat this, here we summarize suggested targets by group and provide one or two examples. One of the difficulties in considering protein targets of PH domains is that the studies proposing them often rely solely on qualitative coprecipitation experiments and often use deletion mapping to implicate the PH domain – without accounting for the possible effects of deletions on overall protein conformation and stability.

The first class of proposed PH domain ligands are $G_{\beta\gamma}$ subunits of heterotrimeric G proteins [7], which initially attracted attention when it was appreciated that the site on βARK that interacts with $G_{\beta\gamma}$ is part of the βARK PH domain [112, 113]. The interaction of PH-domain–containing G protein-coupled receptor kinases (GRKs) (including βARK) with $G_{\beta\gamma}$ subunits has a well studied role in recruiting the kinase to the cell membrane after GPCR activation, so that it may phosphorylate and thus initiate down-regulation of the receptor [114]. Furthermore, a recent crystal structure has shown precisely how the GRK2/βARK PH domain associates with the $G_{\beta\gamma}$ subunit (Figure 17.5b), as discussed in Section 17.4. From this structure [11], it is clear that direct interaction of the PH domain with the membrane-associated $G_{\beta\gamma}$ subunit can cooperate with simultaneous PtdIns(4,5)P_2 binding by this PH domain [10] to drive its membrane targeting, possible also allosterically activating the GRK2 kinase domain. Apart from this clear and well-characterized example, however, no convincing physiological role for PH domain–$G_{\beta\gamma}$ interactions has been established, and no others have been characterized quantitatively.

The second class of potential PH domain protein targets that were suggested early on are protein kinase C (PKC) isoforms – appropriate, since pleckstrin is the major PKC substrate in platelets [115]. These examples are listed elsewhere [13], and their physiological roles remain to be established. In addition, a G_{α} subunit [116], filamentous actin [117], myosin II [118], the TCL1 oncogene product [119], and a transcription factor TFII-I [120] have all been reported to bind to one or more PH domains. However, no common theme emerges with regard to the structural basis or functional importance of these interactions, and much more investigation is needed. Finally, an intriguing set of results was reported from yeast two-hybrid studies of the IRS-1 PH domain, which demonstrated that this domain recognizes regions of protein sequence rich in acidic residues [121] (nucleolin being one example). In our own early studies of the Ras-GAP PH domain, a very similar sequence (in Nopp140) was identified as a preferred protein ligand in screens of bacterial expression libraries (M. A. L. and J. Schlessinger, unpublished). Although the relevance of these interactions is not clear, these studies do raise the possibility that there may be particular protein sequence motifs recognized by subsets of PH domains, which warrants further investigation.

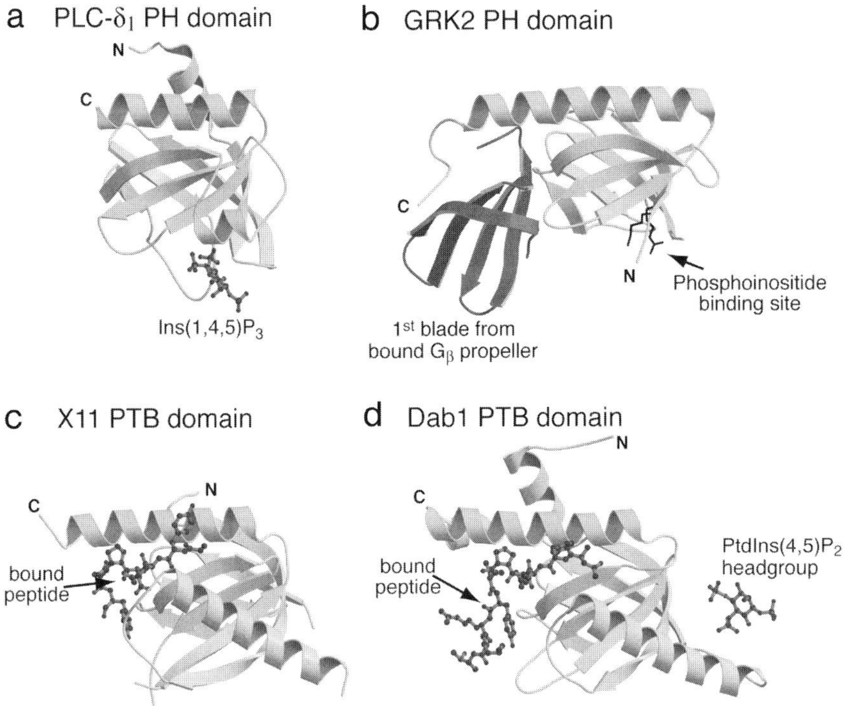

Figure 17.5 Examples of multiple ligands binding simultaneously to a single PH-related domain. (a) The PH domain from PLC-δ_1 [15] is shown bound to Ins(1,4,5)P_3 as an example of a singly liganded PH domain. (b) The PH domain from GRK2 is shown, with the first blade from the G_β propeller that interacts with the C-terminal part of the PH domain helix and strands β2 through β4 [11]. The phosphoinositide binding site on the GRK2 PH domain, identified by NMR [39], is also noted. This PH domain represents an example in which protein and phosphoinositide ligands can interact with the same PH domain simultaneously to drive membrane recruitment (see text). (c) The X11 PTB domain is shown [125], bound to a peptide modeled on a region of the amyloid precursor protein. The peptide lies in a groove between strand β5 and the C-terminal α helix, as seen for all PTB domain–peptide interactions [122]. This represents a 'classical' PTB domain interaction. (d) The structure of the disabled-1 PTB domain is shown [123], with a bound NPXY-containing peptide that interacts in a manner that is very similar to peptide binding to X11-PTB. In addition, the Dab1 PTB domain has a bound PtdIns(4,5)P_2 headgroup molecule, located relatively close to the phosphoinositide-binding site of a 'canonical' PH domain. This complex represents an example of how proteins and phosphoinositides can cooperate in driving membrane association of a PTB domain, which combines PTB-like and PH-like characteristics (see text).

17.4 Emerging Research Directions and Recent Developments

In our efforts to understand the function of PH domains, it is possible that things have been made more difficult by the (perhaps somewhat arbitrary) distinction between different subgroups of domains that contain the PH domain fold. It is true that sequence comparisons do not readily identify a PTB domain, for example, with PH domain sequence profiles. However, it is also true that there can be more sequence identity between a PTB domain and a PH domain than between two different PH domains. It is therefore worthwhile to consider domains with the PH domain fold as a whole when considering possible functions. Among the domains with the PH domain fold, there are subgroups that bind proline-rich sequences (EVH1 domains [25]), phosphorylated or β-turn peptides (PTB domains [122]), small GTPases (RanBD1 [24]), $G_{\beta\gamma}$ subunits (GRK2 [11]), and specific phosphoinositides (PH domains with the β1/β2 loop motif shown in Figure 17.3). There may well be other categories that have yet to be identified. However, there may be many examples that combine features from more than one of these categories. Indeed, several PH domains were discussed above, some identified in yeast, that appear to interact weakly with phosphoinositides and weakly with a second target – cooperation between the two targets being required for efficient membrane recruitment. In these instances it may be difficult to convincingly identify either target on its own because of low binding affinities, and new approaches for context-dependent screens may be needed. Figure 17.5 gives a structural impression – based on examples that are already well understood – of how such multi-ligand PH domains may function. On the left side is shown the PH domain of PLC-δ_1, with Ins(1,4,5)P_3 in its binding site (Figure 17.5a), and the X11 PTB domain with an NPXY peptide in its binding site that lies between strand β5 and the C-terminal α helix (Figure 17.5c). Recent studies of the PTB domain from disabled-1 (Dab1) and disabled-2 (Dab2) have shown how this domain with a PH fold can simultaneously exhibit features of a PTB domain and a PH domain [123, 124]. The Dab1 PTB domain binds its cognate peptide just as other PTB domains are seen to do (Figure 17.5d). In addition, it binds the headgroup of PtdIns(4,5)P_2 in a positively charged region that is similar in nature to the PtdIns(4,5)P_2 binding site of the PLC-δ_1 PH domain and is located broadly on the same face. The Dab PTB domains are thought to bind simultaneously to intracellular protein sequences in LDL receptor family members and to phosphoinositides to recruit them to the plasma membrane [123]. A second, more PH-like, solution to this problem is seen with the GRK2 PH domain that was discussed in Section 17.3.3.2. The structure of this PH domain, taken from the GRK2–$G_{\beta\gamma}$ complex structure [11] is shown in Figure 17.5b. The phosphoinositide-binding site, identified in NMR studies by Cowburn and colleagues [39], is marked. This corresponds well with the Ins(1,4,5)P_3 binding site in the PLC-δ_1 PH domain. Figure 17.5b also shows the first blade of the G_β propeller, which contains all the sites that contact the GRK2 PH domain. The binding site for G_β involves the C-terminal part of the PH domain helix (extended in GRK2-PH) and lies on the opposite side of this helix from the peptide binding site seen in PTB domains.

Whereas the peptide bound to PTB domains lies along the length of strand β5, G_β contacts the beginning of strand β1 and the C termini of strands β2 and β4. The locations of the binding sites for G_β and phosphoinositides on the GRK2 PH domain explain how these two ligands can cooperate to drive membrane targeting of this protein [10]. These examples represent one instance in which it could be argued that a phosphoinositide-binding PH domain has acquired the capacity to interact with additional protein targets (GRK2-PH), and another in which a PH-related protein binding domain has acquired phosphoinositide binding capacity (Dab1-PTB). We suggest that many other examples of this sort exist, in which phosphoinositide and protein (or two different proteins) are recognized simultaneously by the same domain. Identifying these domains and their binding targets will provide the next chapter in our quest to understand PH domain function.

References

1 Mayer, B. J., Ren, R., Clark, K. L., Baltimore, D., A putative modular domain present in diverse signaling molecules. *Cell* **1993**, *73*, 629–630.

2 Haslam, R. J., Koide, H. B., Hemmings, B. A., Pleckstrin domain homology. *Nature* **1993**, *363*, 309–310.

3 Musacchio, A., Gibson, T., Rice, P., Thompson, J., Saraste, M., The PH domain: a common piece in a patchwork of signalling proteins. *Trends Biochem. Sci.* **1993**, *18*, 343–348.

4 International Human Genome Sequencing Consortium. Initial sequencing and analysis of the human genome. *Nature* **2001**, *409*, 860–921.

5 Pawson, T., Gish, G. D., SH2 and SH3 domains: from structure to function. *Cell* **1992**, *71*, 359–362.

6 Schlessinger, J., SH2/SH3 signaling proteins. *Curr. Opin. Genet. Dev.* **1994**, *4*, 25–30.

7 Touhara, K., Inglese, J., Pitcher, J. A., Shaw, G., Lefkowitz, R. J., Binding of G protein beta gamma subunits to pleckstrin homology domains. *J. Biol. Chem.* **1994**, *269*, 10217–10220.

8 Tsukuda, S., Simon, M. I., Witte, O. N., Katz, A., Binding of beta gamma subunits of heterotrimeric G proteins to the PH domain of Bruton tyrosine kinase. *Proc. Natl. Acad. Sci. USA* **1994**, *91*, 11256–11260.

9 Yao, L., Kawakami, Y., Kawakami, T., The pleckstrin homology domain of Bruton tyrosine kinase interacts with protein kinase *Proc,* C., *Natl. Acad. Sci. USA* **1994**, *91*, 9175–9179.

10 Pitcher, J. A., Touhara, K., Payne, E. S., Lefkowitz, R. J., Pleckstrin homology domain-mediated membrane association and activation of the β-adrenergic receptor kinase requires coordinate interaction with $G_{\beta\gamma}$ subunits and lipid. *J. Biol. Chem.* **1995**, *270*, 11707–11710.

11 Lodowski, D. T., Pitcher, J. A., Capel, W. D., Lefkowitz, R. J., Tesmer, J. J., Keeping G proteins at bay: a complex between G protein-coupled receptor kinase 2 and Gbetagamma. *Science* **2003**, *300*, 1256–1262.

12 Harlan, J. E., Hajduk, P. J., Yoon, H. S., Fesik, S. W., Pleckstrin homology domains bind to phosphatidylinositol 4,5-bisphosphate. *Nature* **1994**, *371*, 168–170.

13 Lemmon, M. A., Ferguson, K. M., Signal-dependent membrane targeting by pleckstrin homology (PH) domains. *Biochem. J.* **2000**, *350*, 1–18.

14 Hyvönen, M., Macias, M. J., Nilges, M., Oschkinat, H., Saraste, M., Wilmanns, M., Structure of the binding site for inositol phosphates in a PH domain. *EMBO J.* **1995**, *14*, 4676–4685.

15 Ferguson, K. M., Lemmon, M. A., Schlessinger, J., Sigler, P. B., Structure of a high affinity complex between inositol-1,4,5-trisphosphate and a phospholipase C pleckstrin homology domain. *Cell* **1995**, *83*, 1037–1046.

16 Baraldi, E., et al., Structure of the PH domain from Bruton's tyrosine kinase in complex with inositol 1,3,4,5-tetrakisphosphate. *Structure* **1999**, *7*, 449–460.

17 Ferguson, K. M., Kavran, J. M., Sankaran, V. G., Fournier, E., Isakoff, S. J., Skolnik, E. Y., Lemmon, M. A., Structural basis for discrimination of 3-phosphoinositides by pleckstrin homology domains. *Mol. Cell* **2000**, *6*, 373–384.

18 Lietzke, S. E., Bose, S., Cronin, T., Klarlund, J., Chawla, A., Czech, M. P., Lambright, D. G., Structural basis of 3-phosphoinositide recognition by pleckstrin homology domains. *Mol. Cell* **2000**, *6*, 385–394.

19 Thomas, C., Deak, M., Alessi, D., van Aalten, D., High-resolution structure of the pleckstrin homology domain of protein kinase b/akt bound to phosphatidylinositol (3,4,5)-trisphosphate. *Curr. Biol.* **2002**, *12*, 1256–1262.

20 Chothia, C., Principles that determine the structure of proteins. *Annu. Rev. Biochem.* **1984**, *53*, 537–572.

21 Ferguson, K. M., Lemmon, M. A., Schlessinger, J., Sigler, P. B., Crystal structure at 2.2 Å resolution of the pleckstrin homology domain from human dynamin. *Cell* **1994**, *79*, 199–209.

22 Blomberg, N., Baraldi, E., Nilges, M., Saraste, M., The PH superfold: a structural scaffold for multiple functions. *Trends Biochem. Sci.* **1999**, *24*, 441–445.

23 Zhou, M.-M., et al., Structure and ligand recognition of the phosphotyrosine binding domain of Shc. *Nature* **1995**, *378*, 584–592.

24 Vetter, I. R., Nowak, C., Nishimotot, T., Kuhlmann, J., Wittinghofer, A., Structure of a Ran-binding domain complexed with Ran bound to a GTP analogue: implications for nuclear transport. *Nature* **1999**, *398*, 39–46.

25 Prehoda, K. E., Lee, D. J., Lim, W. A., Structure of the enabled/VASP homology 1 domain–peptide complex: a key component in the spatial control of actin assembly. *Cell* **1999**, *97*, 471–480.

26 Pearson, M. A., Reczek, D., Bretscher, A., Karplus, P. A., Structure of the ERM protein moesin reveals the FERM domain fold masked by an extended actin binding tail domain. *Cell* **2000**, *101*, 259–270.

27 Hamada, K., Shimizu, T., Matsui, T., Tsukita, S., Hakoshima, T., Structural basis of the membrane-targeting and unmasking mechanisms of the radixin FERM domain. *EMBO J.* **2000**, *19*, 4449–4462.

28 Jogl, G., Shen, Y., Gebauer, D., Li, J., Wiegmann, K., Kashkar, H., Kronke, M., Tong, L., Crystal structure of the BEACH domain reveals an unusual fold and extensive association with a novel PH domain. *EMBO J.* **2002**, *21*, 4785–4795.

29 Macias, M. J., Musacchio, A., Ponstingl, H., Nilges, M., Saraste, M., Oschkinat, H., Structure of the pleckstrin homology domain from β-spectrin. *Nature* **1994**, *369*, 675–677.

30 Blomberg, N., Baraldi, E., Sattler, M., Saraste, M., Nilges, M., Structure of a PH domain from the *C. elegans* muscle protein UNC-89 suggests a novel function. *Struct. Fold Des.* **2000**, *8*, 1079–1087.

31 Garcia, P., et al., The pleckstrin homology domain of phospholipase C-delta 1 binds with high affinity to phosphatidylinositol 4,5-bisphosphate in bilayer membranes. *Biochemistry* **1995**, *34*, 16228–16234.

32 Lemmon, M. A., Ferguson, K. M., O'Brien, R., Sigler, P. B., Schlessinger, J., Specific and high-affinity binding of inositol phosphates to an isolated pleckstrin homology domain. *Proc. Natl. Acad. Sci. USA* **1995**, *92*, 10472–10476.

33 Harlan, J. E., Yoon, H. S., Hajduk, P. J., Fesik, S. W., Structural characterization of the interaction between a pleckstrin homology domain and phosphatidylinositol 4,5-bisphosphate. *Biochemistry* **1995**, *34*, 9859–9864.

34 Isakoff, S. J., Cardozo, T., Andreev, J., Li, Z., Ferguson, K. M., Abagyan, R., Lemmon, M. A., Aronheim, A., Skolnik, E. Y., Identification and analysis of PH domain-containing targets of phosphatidylinositol 3-kinase using a novel in vivo assay in yeast. *EMBO J.* **1998**, *17*, 5374–5387.

35 Dowler, S., Currie, R. A., Campbell, D. G., Deak, M., Kular, G., Downes,

C. P., ALESSI, D. R., Identification of pleckstrin-homology-domain–containing proteins with novel phosphoinositide-binding specificities. *Biochem. J.* **2000**, *351*, 19–31.

36 YU, J. W., et al., Genome-wide analysis of membrane targeting by *S. cerevisiae* pleckstrin homology domains. *Mol. Cell* **2004**, *13*, 677–688.

37 SALIM, K., et al., Distinct specificity in the recognition of phosphoinositides by the pleckstrin homology domains of dynamin and Bruton's tyrosine kinase. *EMBO J.* **1996**, *15*, 6241–6250.

38 ZHENG, J., CAHILL, S. M., LEMMON, M. A., FUSHMAN, D., SCHLESSINGER, J., COWBURN, D., Identification of the binding site for acidic phospholipids on the PH domain of dynamin: implications for stimulation of GTPase activity. *J. Mol. Biol.* **1996**, *255*, 14–21.

39 FUSHMAN, D., NAJMABADI-KASKE, T., CAHILL, S., ZHENG, J., LEVINE, H., COWBURN, D., The solution structure and dynamics of the pleckstrin homology domain of G protein-coupled receptor kinase 2 (β-adrenergic receptor kinase 1): a binding partner of $G_{\beta\gamma}$ subunits. *J. Biol. Chem.* **1998**, *273*, 2835–2843.

40 TOKER, A., CANTLEY, L. C., Signaling through the lipid products of phosphoinositide 3-kinase. *Nature* **1997**, *387*, 673–676.

41 STEPHENS, L. R., JACKSON, T. R., HAWKINS, P. T., Agonist-stimulated synthesis of phosphatidylinositol 3,4,5-trisphosphate: a new intracellular signaling system? *Biochim. Biophys. Acta.* **1993**, *1179*, 27–75.

42 FRANKE, T. F., KAPLAN, D. R., CANTLEY, L. C., TOKER, A., Direct regulation of the Akt proto-oncogene product by phosphatidylinositol-3,4-bisphosphate. *Science* **1997**, *275*, 665–668.

43 RAMEH, L. E., et al., A comparative analysis of the phosphoinositide binding specificity of pleckstrin homology domains. *J. Biol. Chem.* **1997**, *272*, 22059–22066.

44 FUKUDA, M., KOJIMA, T., KABAYAMA, H., MIKOSHIBA, K., Mutation of the pleckstrin homology domain of Bruton's tyrosine kinase in immunodeficiency impaired inositol 1,3,4,5-tetrakisphosphate binding capacity. *J. Biol. Chem.* **1996**, *271*, 30303–30306.

45 KLARLUND, J. K., GUILHERME, A., HOLIK, J. J., VIRBASIUS, A., CZECH, M. P., Signaling by 3,4,5-phosphoinositide through proteins containing pleckstrin and Sec7 homology domains. *Science* **1997**, *275*, 1927–1930.

46 KAVRAN, J. M., KLEIN, D. E., LEE, A., FALASCA, M., ISAKOFF, S. J., SKOLNIK, E. Y., LEMMON, M. A., Specificity and promiscuity in phosphoinositide binding by pleckstrin homology domains. *J. Biol. Chem.* **1998**, *273*, 30497–30508.

47 RAWLINGS, D. J., et al., Mutation of unique region of Bruton's tyrosine kinase in immunodeficient XID mice. *Science* **1993**, *261*, 358–361.

48 THOMAS, J. D., SIDERAS, P., SMITH, C. I., VORECHOVSKY, I., CHAPMAN, V., PAUL, W. E., Colocalization of X-linked agammaglobulinemia and X-linked immunodeficiency genes. *Science* **1993**, *261*, 355–358.

49 KLARLUND, J. K., TSIARAS, W., HOLIK, J. J., CHAWLA, A., CZECH, M. P., Distinct polyphosphoinositide binding selectivities for pleckstrin homology domains of GRP1-like proteins based on diglycine versus triglycine motifs. *J. Biol. Chem.* **2000**, *275*, 32816–32821.

50 THOMAS, C. C., DOWLER, S., DEAK, M., ALESSI, D. R., VAN AALTEN, D. M., Crystal structure of the phosphatidylinositol 3,4-bisphosphate-binding pleckstrin homology (PH) domain of tandem PH-domain–containing protein 1 (TAPP1): molecular basis of lipid specificity. *Biochem. J.* **2001**, *358*, 287–294.

51 KIMBER, W. A., et al., Evidence that the tandem-pleckstrin-homology-domain–containing protein TAPP1 interacts with Ptd(3,4)P_2 and the multi-PDZ-domain–containing protein MUPP1 in vivo. *Biochem. J.* **2002**, *361*, 525–536.

52 LEVINE, T. P., MUNRO, S., Targeting of golgi-specific pleckstrin homology domains involves both PtdIns 4-kinase-dependent, and independent, components. *Curr. Biol.* **2002**, *12*, 695–704.

53 BLOOM, K., Nuclear migration: cortical anchors for cytoplasmic dynein. *Curr. Biol.* **2001**, *11*, R326–R329.

54 Balla, T., Bondeva, T., How accurately can we image inositol lipids in living cells? *Trends Pharmacol. Sci.* **2000**, *21*, 238–241.

55 Balla, T., Varnai, P., Visualizing cellular phosphoinositide pools with GFP-fused protein-modules. *Science STKE* **2002**.

56 Yagisawa, H., et al., Expression and characterization of an inositol 1,4,5-trisphosphate binding domain of phosphatidylinositol-specific phospholipase C-delta 1. *J. Biol. Chem.* **1994**, *269*, 20179–20188.

57 Takeuchi, H., et al., Localization of a high-affinity inositol 1,4,5-trisphosphate/inositol 1,4,5,6-tetrakisphosphate binding domain to the pleckstrin homology module of a new 130 kDa protein: characterization of the determinants of structural specificity. *Biochem. J.* **1996**, *318*, 561–568.

58 Cifuentes, M. E., Honkanen, L., Rebecchi, M. J., Proteolytic fragments of phosphoinositide-specific phospholipase C-delta 1: catalytic and membrane binding properties. *J. Biol. Chem.* **1993**, *268*, 11586–11593.

59 Rebecchi, M., Peterson, A., McLaughlin, S., Phosphoinositide-specific phospholipase C-δ1 binds with high affinity to phospholipid vesicles containing phosphatidylinositol 4,5-bisphosphate. *Biochemistry* **1992**, *31*, 12742–12747.

60 Kanematsu, T., Takeya, H., Watanabe, Y., Ozaki, S., Yoshida, M., Koga, T., Iwanaga, S., Hirata, M., Putative inositol 1,4,5-trisphosphate binding proteins in rat brain cytosol. *J. Biol. Chem.* **1992**, *267*, 6518–6525.

61 Cifuentes, M. E., Delaney, T., Rebecchi, M. J., D-myo-inositol 1,4,5-trisphosphate inhibits binding of phospholipase C-δ1 to bilayer membranes. *J. Biol. Chem.* **1994**, *269*, 1945–1994.

62 Farkasovsky, M., Kuntzel, H., Yeast Num1p associates with the mother cell cortex during S/G2 phase and affects microtubular functions. *J. Cell Biol.* **1995**, *131*, 1003–1014.

63 Klopfenstein, D. R., Tomishige, M., Stuurman, N., Vale, R. D., Role of phosphatidylinositol(4,5)bisphosphate organization in membrane transport by the Unc104 kinesin motor. *Cell* **2002**, *109*, 347–358.

64 Cantley, L. C., The phosphoinositide 3-kinase pathway. *Science* **2002**, *296*, 1655–1657.

65 Vanhaesebroeck, B., et al., Synthesis and function of 3-phosphorylated inositol lipids. *Annu. Rev. Biochem.* **2001**, *70*, 535–602.

66 Vanhaesebroeck, B., Alessi, D. R., The PI3K-PDK1 connection: more than just a road to PK*Biochem*, B., *J.* **2000**, *346*, 561–576.

67 Andjelkovic, M., et al., Role of translocation in the activation and function of protein kinase B. *J. Biol. Chem.* **1997**, *272*, 31515–31524.

68 Watton, S. J., Downward, J., Akt/PKB localisation and 3′ phosphoinositide generation at sites of epithelial cell–matrix and cell–cell interaction. *Curr. Biol.* **1999**, *9*, 433–436.

69 Gray, A., Van der Kaay, J., Downes, C. P., The pleckstrin homology domains of protein kinase B and GRP1 (general receptor for phosphoinositides-1) are sensitive and selective probes for the cellular detection of phosphatidylinositol 3,4-bisphosphate and/or phosphatidylinositol 3,4,5-trisphosphate in vivo. *Biochem. J.* **1999**, *344*, 929–936.

70 Servant, G., Weiner, O. D., Herzmark, P., Balla, T., Sedat, J. W., Bourne, H. R., Polarization of chemoattractant receptor signaling during neutrophil chemotaxis. *Science* **2000**, *287*, 1037–1040.

71 Jin, T., Zhang, N., Long, Y., Parent, C. A., Devreotes, P. N., Localization of the G protein beta gamma complex in living cells during chemotaxis. *Science* **2000**, *287*, 1034–1036.

72 Stephens, L., et al., Protein kinase B kinases that mediate phosphatidylinositol 3,4,5-trisphosphate–dependent activation of protein kinase B. *Science* **1998**, *279*, 710–714.

73 Alessi, D. R., James, S. R., Downes, C. P., Holmes, A. B., Gaffney, P. R., Reese, C. B., Cohen, P., Characterization of a 3-phosphoinositide–dependent protein kinase which phosphorylates and activates protein kinase B alpha. *Curr. Biol.* **1997**, *7*, 261–269.

74 Anderson, K. E., Coadwell, J., Stephens, L. R., Hawkins, P. T., Translocation of PDK-1 to the plasma membrane is important in allowing PDK-1 to activate protein kinase *Curr,* B., *Biol.* **1998**, *8*, 684–691.

75 Currie, R. A., et al., Role of phosphatidylinositol 3,4,5-trisphosphate in regulating the activity and localization of 3-phosphoinositide–dependent protein kinase-1. *Biochem. J.* **1999**, *337*, 575–583.

76 Toker, A., Newton, A. C., Akt/protein kinase B is regulated by autophosphorylation at the hypothetical PDK-2 site. *J. Biol. Chem.* **2000**, *275*, 8271–8274.

77 Li, Z., Wahl, M. I., Eguinoa, A., Stephens, L. R., Hawkins, P. T., Witte, O. N., Phosphatidylinositol 3-kinase-gamma activates Bruton's tyrosine kinase in concert with Src family kinases. *Proc. Natl. Acad. Sci. USA* **1997**, *94*, 13820–13825.

78 Stokoe, D., et al., Dual role of phosphatidylinositol-3,4,5-trisphosphate in the activation of protein kinase B. *Science* **1997**, *277*, 567–570.

79 Milburn, C. C., Deak, M., Kelly, S. M., Price, N. C., Alessi, D. R., van Aalten, D. M. F., Binding of phosphatidylinositol 3,4,5-trisphosphate to the pleckstrin homology domain of protein kinase B induces a conformational change. *Biochem. J.* **2003**, *375*, 531–538.

80 Auguin, D., Barthe, P., Auge-Senegas, M. T., Stern, M. H., Noguchi, M., Roumestand, C., Solution structure and backbone dynamics of the pleckstrin homology domain of the human protein kinase B (PKB/Akt): interaction with inositol phosphates. *J. Biomol. NMR* **2004**, *28*, 137–155.

81 Levine, T. P., Munro, S., The pleckstrin homology domain of oxysterol-binding protein recognizes a determinant specific to golgi membranes. *Curr. Biol.* **1998**, *8*, 729–739.

82 Stefan, C. J., Audhya, A., Emr, S. D., The yeast synaptojanin-like proteins control the cellular distribution of phosphatidylinositol (4,5)-bisphosphate. *Mol. Biol. Cell.* **2002**, *13*, 542–557.

83 Levine, T. P., Munro, S., Dual targeting of Osh1p, a yeast homologue of oxysterol-binding protein, to both the golgi and the nucleus–vacuole junction. *Mol. Biol. Cell.* **2001**, *12*, 1633–1644.

84 Audhya, A., Foti, M., Emr, S. D., Distinct roles for the yeast phosphatidylinositol 4-kinases, Stt4p and Pik1p, insecretion, cell growth, and organelle membrane dynamics. *Mol. Biol. Cell* **2000**, *11*, 2673–2689.

85 Takeuchi, H., et al., Distinct specificity in the binding of inositol phosphates by pleckstrin homology domains of pleckstrin, RAC-protein kinase, diacylglycerol kinase and a new 130 kDa protein. *Biochim. Biophys. Acta* **1997**, *1359*, 275 –285.

86 Achiriloaie, M., Barylko, B., Albanesi, J. P., Essential role of the dynamin pleckstrin homology domain in receptor-mediated endocytosis. *Mol. Cell. Biol.* **1999**, *19*, 1410–1415.

87 Lee, A., Frank, D. W., Marks, M. S., Lemmon, M. A., Dominant-negative inhibition of receptor-mediated endocytosis by a dynamin-1 mutant with a defective pleckstrin homology domain. *Curr. Biol.* **1999**, *9*, 261–264.

88 Vallis, Y., Wigge, P., Marks, B., Evans, P. R., McMahon, H. T., Importance of the pleckstrin homology domain of dynamin in clathrin-mediated endocytosis. *Curr. Biol.* **1999**, *9*, 257–260.

89 Warnock, D. E., Schmid, S. L., Dynamin GTPase, a force-generating molecular switch. *BioEssays* **1996**, *18*, 885–893.

90 Barylko, B., Binns, D., Lin, K. M., Atkinson, M. A., Jameson, D. M., Yin, H. L., Albanesi, J. P., Synergistic activation of dynamin GTPase by Grb2 and phosphoinositides. *J. Biol. Chem.* **1998**, *273*, 3791–3797.

91 Schmid, S. L., McNiven, M. A., De Camilli, P., Dynamin and its partners: a progress report. *Curr. Opin. Cell Biol.* **1998**, *10*, 504–512.

92 Warnock, D. E., Hinshaw, J. E., Schmid, S. L., Dynamin self-assembly stimulates its GTPase activity. *J. Biol. Chem.* **1996**, *271*, 22310–22314.

93 Klein, D. E., Lee, A., Frank, D. W., Marks, M. S., Lemmon, M. A., The pleckstrin homology domains of dynamin isoforms require oligomerization for high affinity phosphoinositide

binding. *J. Biol. Chem.* **1998**, *273*, 27725–27733.

94 Zheng, Y., Dbl family guanine nucleotide exchange factors. *Trends Biochem. Sci.* **2001**, *26*, 724–732.

95 Whitehead, I. P., Campbell, S., Rossman, K. L., Der, C. J., Dbl family proteins. *Biochim. Biophys. Acta* **1997**, *1332*, 1–23.

96 Snyder, J. T., Rossman, K. L., Baumeister, M. A., Pruitt, W. M., Siderovski, D. P., Der, C. J., Lemmon, M. A., Sondek, J., Quantitative analysis of the effect of phosphoinositide interactions on the function of Dbl family proteins. *J. Biol. Chem.* **2001**, *276*, 45868–45875.

97 Rossman, K. L., Cheng, L., Mahon, G. M., Rojas, R. J., Snyder, J. T., Whitehead, I. P., Sondek, J., Multifunctional roles for the PH domain of Dbs in regulating Rho GTPase activation. *J. Biol. Chem.* **2003**, *278*, 18393–18400.

98 Baumeister, M. A., Martinu, L., Rossman, K. L., Sondek, J., Lemmon, M. A., Chou, M. M., Loss of PtdIns-3-P binding by the C-terminal Tiam-1 pleckstrin homology (PH) domain prevents in vivo Rac1 activation without affecting membrane targeting. *J. Biol. Chem.* **2003**, *278*, 11457–11464.

99 Pruitt, W. M., et al., Role of the pleckstrin homology domain in intersectin-L Dbl homology domain activation of Cdc42 and signaling. *Biochim. Biophys. Acta* **2003**, *1640*, 61–68.

100 Palmby, T. R., Abe, K., Der, C. J., Critical role of the pleckstrin homology and cysteine-rich domains in Vav signaling and transforming activity. *J. Biol. Chem.* **2002**, *277*, 39350–39359.

101 Booden, M. A., Campbell, S. L., Der, C. J., Critical but distinct roles for the pleckstrin homology and cysteine-rich domains as positive modulators of Vav2 signaling and transformation. *Mol. Cell. Biol.* **2002**, *22*, 2487–2497.

102 Kubiseski, T. J., Culotti, J., Pawson, T., Functional analysis of the *Caenorhabditis elegans* UNC-73B PH domain demonstrates a role in activation of the Rac GTPase in vitro and axon guidance in vivo. *Mol. Cell. Biol.* **2003**, *23*, 6823–6835.

103 Fuentes, E. J., Karnoub, A. E., Booden, M. A., Der, C. J., Campbell, S. L., Critical role of the pleckstrin homology domain in Dbs signaling and growth regulation. *J. Biol. Chem.* **2003**, *278*, 21188–21196.

104 Russo, C., et al., Modulation of oncogenic DBL activity by phosphoinositol binding to pleckstrin homology domains. *J. Biol. Chem.* **2001**, *276*, 19524–19531.

105 Soisson, S. M., Nimnual, A. S., Uy, M., Bar-Sagi, D., Kuriyan, J., Crystal structure of the Dbl and pleckstrin homology domains from the human Son of Sevenless protein. *Cell* **1998**, *95*, 259–268.

106 Rossman, K. L., Worthylake, D. K., Snyder, J. T., Siderovski, D. P., Campbell, S. L., Sondek, J., A crystallographic view of interactions between Dbs and Cdc42: PH domain-assisted guanine nucleotide exchange. *EMBO J.* **2002**, *21*, 1315–1326.

107 Worthylake, D. K., Rossman, K. L., Sondek, J., Crystal structure of Rac1 in complex with the guanine nucleotide exchange region of Tiam1. *Nature* **2000**, *408*, 682–688.

108 Snyder, J. T., Worthylake, D. K., Rossman, K. L., Betts, L., Pruitt, W. M., Siderovski, D. P., Der, C. J., Sondek, J., Structural basis for the selective activation of Rho GTPases by Dbl exchange factors. *Nat. Struct. Biol.* **2002**, *9*, 468–475.

109 Jaffe, A. B., Aspenstrom, P., Hall, A., Human CNK1 acts as a scaffold protein, linking Rho and Ras signal transduction pathways. *Mol. Cell. Biol.* **2004**, *24*, 1736–1746.

110 Kim, O., Yang, J., Qiu, Y., Selective activation of small GTPase RhoA by tyrosine kinase Etk through its pleckstrin homology domain. *J. Biol. Chem.* **2002**, *277*, 30066–30071.

111 Snyder, J. T., Singer, A. U., Wing, M. R., Harden, T. K., Sondek, J., The pleckstrin homology domain of phospholipase C-beta2 as an effector site for Rac. *J. Biol. Chem.* **2003**, *278*, 21099–21104.

112 Pitcher, J. A., et al., Role of βγ subunits of heterotrimeric G proteins in targeting the β-adrenergic receptor kinase to membrane-bound receptors. *Science* **1992**, *257*, 1264–1267.

113 Koch, W. J., Inglese, J., Stone, W. C., Lefkowitz, R. J., The binding site for the βγ subunits of heterotrimeric G proteins on the β-adrenergic receptor kinase. *J. Biol. Chem.* **1993**, *268*, 8256–8260.

114 Pitcher, J. A., Freedman, N. J., Lefkowitz, R. J., G protein-coupled receptor kinases. *Annu. Rev. Biochem.* **1998**, *67*, 653–692.

115 Tyers, M., Rachubinski, R. A., Stewart, M. I., Varrichio, A. M., Shorr, R. G. L., Haslam, R. J., Harley, C. B., Molecular cloning and expression of the major protein kinase C substrate of platelets. *Nature* **1988**, *333*, 470–473.

116 Jiang, Y., Ma, W., Wan, Y., Kozasa, T., Hattori, S., Huang, X. Y., The G protein $G_{\alpha 12}$ stimulates Bruton's tyrosine kinase and a rasGAP through a conserved PH/BM domain. *Nature* **1998**, *395*, 808–813.

117 Yao, L., Janmey, P., Frigeri, L. G., Han, W., Fujita, J., Kawakami, Y., Apgar, J. R., Kawakami, T., Pleckstrin homology domains interact with filamentous actin. *J. Biol. Chem.* **1999**, *274*, 19752–19761.

118 Tanaka, M., Konishi, H., Touhara, K., Sakane, F., Hirata, M., Ono, Y., Kikkawa, U., Identification of myosin II as a binding protein to the PH domain of protein kinase *Biochem*, B., *Biophys. Res. Commun.* **1999**, *255*, 169–174.

119 Laine, J., Kunstle, G., Obata, T., Sha, M., Noguchi, M., The protooncogene TCL1 is an Akt kinase coactivator. *Mol. Cell* **2000**, *6*, 395–407.

120 Yang, W., Desiderio, S., BAP-135, a target for Bruton's tyrosine kinase in response to B cell receptor engagement. *Proc. Natl. Acad. Sci. USA* **1997**, *94*, 604–609.

121 Burks, D. J., Wang, J., Towery, H., Ishibashi, O., Lowe, D., Riedel, H., White, M. F., IRS pleckstrin homology domains bind to acidic motifs in proteins. *J. Biol. Chem.* **1998**, *273*, 31061–31067.

122 Schlessinger, J., Lemmon, M. A., SH2 and PTB domains in tyrosine kinase signaling. *Science STKE* **2003**, *Jul 15; 2003(191)*, RE12.

123 Stolt, P. C., Jeon, H., Song, H. K., Herz, J., Eck, M. J., Blacklow, S. C., Origins of peptide selectivity and phosphoinositide binding revealed by structures of disabled-1 PTB domain complexes. *Structure* **2003**, *11*, 569–579.

124 Yun, M., et al., Crystal structures of the Dab homology domains of mouse disabled 1 and 2. *J. Biol. Chem.* **2003**, *278*, 36572–36581.

125 Zhang, Z., Lee, C. H., Mandiyan, V., Borg, J. P., Margolis, B., Schlessinger, J., Kuriyan, J., Sequence-specific recognition of the internalization motif of the Alzheimer's amyloid precursor protein by the X11 PTB domain. *EMBO J.* **1997**, *16*, 6141–6150.

18
ENTH and VHS Domains

Vimal Parkash, Olli Lohi, Ismo Virtanen, and Veli-Pekka Lehto

18.1
Introduction

Endocytosis is a vital cellular process for nutrient uptake and down-regulation of the signaling activities of ligand–receptor complexes. It also provides a portal for virus entry and underlies the synaptic vesicle recycling in neurons. The best characterized of the multiple endocytic pathways is the clathrin-dependent, receptor-mediated endocytosis [1]. It involves recognition of the cytoplasmic tails of transmembrane receptors by an adaptor complex. Anchored adaptors, in turn, recruit clathrin and promote its polymerization into polyhedral lattices, which together form a coat on the surface of the nascent vesicle. The morphological correlates of these events are inward budding of the membrane with the formation, first, of clathrin-coated pits (CCPs) and then of clathrin coated vesicles (CCVs). CCVs are pinched off the membrane and released into the cytosol [2–4].

The principal adaptor complex in clathrin-dependent endocytosis is AP-2. It is a heterotetrameric molecule composed of four subunits, with the $\mu 2$ subunit containing the binding site for the internalization signal of the receptor (cargo) and the $\beta 2$ subunit binding directly to clathrin. Thus, AP-2 forms a bridge between clathrin and the receptors as they cluster at the budding site. It also recruits a class of proteins collectively called accessory proteins [5]. They form a heterogeneous group of mostly modular proteins, many of which bind to the 'ear' domain of the α subunit (α-adaptin) of AP-2 [6, 7]. Since they are required for endocytosis but are not enriched in purified CCVs, they are suggested to play a regulatory role in the process [5].

Accessory proteins constitute a variable group of mostly modular proteins. As a general strategy, they possess several protein–protein and/or protein–lipid interaction sites that enable formation of complex protein assemblies with the core endocytic components attached to the membrane at the site of the nascent bud formation [2, 8–10]. Among them are, for instance, both previously known and more novel ENTH-domain–containing proteins which, via the ENTH domain, seem to play decisive roles in some of the early events of endocytosis.

Modular Protein Domains. Edited by G. Cesareni, M. Gimona, M. Sudol, M. Yaffe

ISBN: 3-527-30813-X

CCVs also operate in vesicle transport between the trans-golgi network (TGN) and endosomes, especially in the selective ferrying of lysosomal hydrolases from TGN to endosomes and lysosomes [11]. The hydrolases carry a mannose-6-phosphate tag, which is recognized by the mannose-6-phosphate receptor (M6PR), a transmembrane receptor. The cytosolic tail of the receptor serves as a 'cargo' signal. Vesicle formation is initiated by an ADP-ribosylation factor (Arf). The cargo is then recruited by the adaptor complex AP-1, which is homologous and structurally similar to AP-2. One of the longstanding mysteries of this transport route has been the basis for its high selectivity, i.e., how the cargo that is destined to lysosomes is selected and transported separately from the other vesicle traffic from the TGN. This was largely solved by the discovery of the golgi-localized, γ-ear–containing, Arf-binding (GGA) proteins. They are a novel class of monomeric adaptor proteins, which link the cargo to the transport and clathrin-dependent vesiculation machinery [12, 13]. Some of the vital functions of these proteins are carried out by their VHS domains, a domain category having thus far poorly known functions, for instance, in some well studied endocytosis-associated proteins that belong to the STAM/EAST/Hbp family of proteins [14].

ENTH and VHS domains have distinct primary structures and functions. However, they share a similar fold and mostly occur in proteins involved in vesicular trafficking. Therefore, they are discussed together here.

18.2 History of ENTH

The first ENTH (epsin N-terminal homology)-containing protein in animal cells, epsin 1 (Eps15 interacting protein), was discovered and characterized on the basis of its interaction with Eps15, a main substrate for the tyrosine kinase activity of the EGF receptor [15]. An early link to the endocytic machinery was made when its identity with an until-then-unidentified 90-kDa band binding to the α-adaptin subunit of AP-2 was recognized [15, 16].

The interaction of epsin with Eps15 is mediated by the NPF motifs of epsin and by a short N-terminal module of Eps15 known as the EH (Eps15 homology) domain [17]. Other proteins similar or identical to epsin were also found on the basis of their interaction with EH domains of other EH-containing proteins or by direct cloning. These include, for instance, intersectin-binding protein 1 and 2 (Ibp1, Ibp2), epsin2, and the *Saccharomyces cerevisiae* homologs Ent1p and Ent2p [18–22].

The ENTH domain, identified and designated on the basis of sequence comparisons [15, 20, 23], typically occupies the very N-terminal locale. Epsins and their related proteins also contain other motifs, usually linear sequences that mediate interactions with the components of the endocytic machinery. They include DPF/W motifs for AP-2 interaction, clathrin boxes for clathrin binding, and UIM (ubiquitin interaction motif) motifs, which serve as binding sites for ubiquitin and ubiquitinated proteins. Epsin and its family members bind clathrin and the α subunit of AP-2 and stimulate clathrin polymerization [16, 20, 24–26]. Blocking its function inhibits receptor-mediated endocytosis of growth factor receptors [27]. This finding, together with mutational analysis and interaction studies with the *Drosophila* epsin homolog 'liquid facets' [28], conclusively defines epsin(s) as part of the endocytic machinery.

AP180 and CALM are homologous to nerve terminal-specific and ubiquitous clathrin-binding adaptors, respectively [29, 30]. Their N termini have only a low degree of similarity to ENTH. They were, nevertheless, considered to contain bona fide ENTH domains. Recent structural studies have, however, shown differences between the ENTH domains of epsin and AP180/CALM, and, consequently, the latter are often referred to as ANTH domains to emphasize their distinct nature and also their separate entourage of adjoining domains [31]. The same applies to

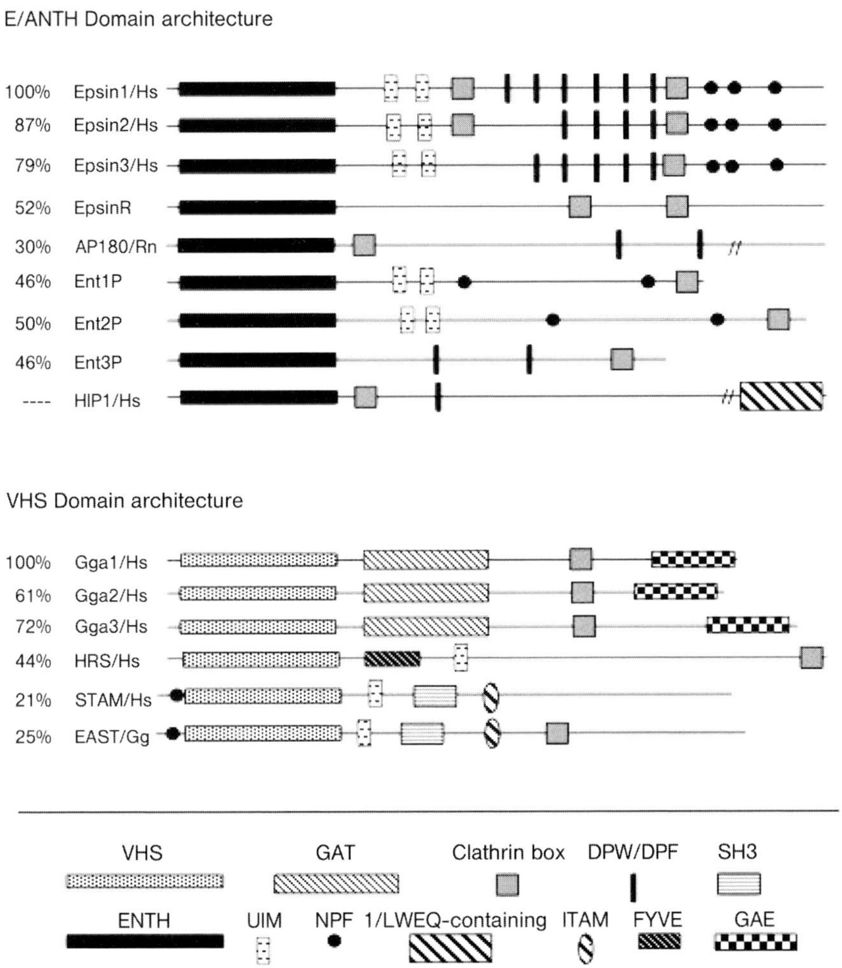

Figure 18.1 ENTH/ANTH- and VHS-containing proteins and their domain structures. Representatives of various subgroups from human, rat, chicken, and yeast are depicted.

HIP1 and HIP1r, related proteins that bind clathrin and, intriguingly, actin via a talin-homology region in the C terminus. They also carry N-terminal domains that are more similar to AP180/CALM ANTH than to ENTH domains [32, 33].

Figure 18.1 (upper panel) shows schematically the alignment and modular composition of some representatives of the various classes of ENTH proteins. Proteins containing a related VHS domain are shown in the lower panel.

Recently, the territory of ENTH-containing proteins has been expanded from the realm of endocytic machinery to the golgi apparatus by the work of four groups independently showing the presence, in TGN, of an epsin-related protein variedly named epsinR [34, 35], enthoprotin [36], and Clint [37]. Similarly, in yeast, epsin-related proteins Ent3p and Ent5 associate with the golgi [38].

A deeper insight into the workings of ENTH domains and, by extension, of ENTH-containing proteins in general, came with the observation that ENTH/ANTH domains of epsin 1 and API80/CALM bind to phosphoinositides, with PtdIns(4,5)P_2 as the strongly favored ligand [39, 40]. Phosphoinositides are phosphorylated (3, 4, or 5 positions) forms of phosphatidylinositol that occur in cells in various permutations and can be modified rapidly by headgroup phosphorylation and dephosphorylation. Moreover, certain types of phosphoinositides are restricted to certain membranes or organelles. Thus, they are well suited to forming transient membrane-targeting areas for proteins that have a capacity to bind them [41–43]. Table 18.1 lists the lipid ligands of ENTH and ANTH, along with their putative functions. For InsPtd(4,5)P_2 ligands, the structure of the lipid–domain complex has been solved [39, 44].

Table 18.1 ENTH/ANT proteins and their lipid ligands.

ENTH/ANTH	*Lipid*	*Suggested function*	*Structure of the complex (ref.)*	*References*
Epsin 1	**PI(4,5)P_2** PI(3,4,5)P_3	Anchorage to PM	39	40
AP180/CALM	PI(4,5)P_2 PI(3,4,5)P_3	Anchorage to PM	39, 44	39, 40
	PI(4,5)P_2 PI(3,4)P_2 PI(3,5)P_2	Anchorage to PM and endosomes		32 Hya et al.
Epsin R	PI(4)P	Anchorage to TGN		34, 35
Ent3p	PI(3,5)P_2	Anchorage to endosomes/MVB		58

Table 18.2 Structural data on ENTH/ANTH and their complexes with lipid ligands.

Protein	*Species*	*Pdb/ref.*	*Residues*	*Ligand*	*Technique*	*Resolution, Å*
Epsin 1	Rat	1EYH	15–158		X-ray	1.56
		1H0A/31	1–158	IP3	X-ray	1.70
		1EDU/49	2–150		X-ray	1.80
	Human	1INZ/47	1–148		NMR	
CALM	Rat	1HF8/39	19–281		X-ray	2
		1HG5/39	19–281	IP6	X-ray	2
		1HG2/39	19–281	IP2	X-ray	2
		1HFA/39	19–281	IP2	X-ray	2
AP180 (Lap)	*Drosophila melanogaster*	1HX8/44	22–299		X-ray	2.20

18.3 Structure of ENTH Domains

ENTH domains encompass 140–150 amino acid residues and invariably reside in the N terminus of their cognate proteins. The domain is highly conserved from plants, yeast, and nematodes to frogs and humans [20, 23, 45]. The structure of ENTH has been obtained for epsin 1, [40, 46] CALM [39], and *Drosophila* Lap (homolog of AP180) [44] (Table 18.2).

In the absence of a ligand, ENTH forms a superhelix of seven helices that are arranged into three helical hairpin repeats (Figure 18.2). They constitute a characteristic superhelical fold in which three helical hairpins are stacked with a right-handed twist. This arrangement corresponds to the originally defined domain boundary. In the analysis of extended domains encompassing also additional, less well conserved stretches, an eighth α helix was revealed, which is tethered on the side of the superhelical fold through only a few polar interactions and is not involved in the hydrophobic core of the domain. The most highly conserved amino acids are the internal residues involved in packing and the solvent-accessible residues. All in all, and also supported by recent NMR data on human epsin ENTH [47], the unliganded ENTH domain is defined by a seven-helix superhelical fold with a more variable C-terminal flanking α helix. The overall structural features of the epsin ENTH, on one hand, and of CALM ANTH [48] on the other, are closely similar despite their low similarity in primary structure. Outside the ENTH/ANTH family, the structure of epsin ENTH is most similar to the armadillo repeat segment of β-catenin and the HEAT repeat of karyopherin-β/importin-β [49].

The structure of ENTH/ANTH bound to the target inositol polyphosphates has been revealed for epsin 1 [50] and CALM [39]. Interestingly, comparison of the structures of the epsin ENTH in the unliganded and liganded state revealed that a new helix (α0; residues 3–15) becomes ordered at the N terminus upon PtdIns(4,5)P2 recognition and binding (Figure 18.2a, d). This induced helix becomes

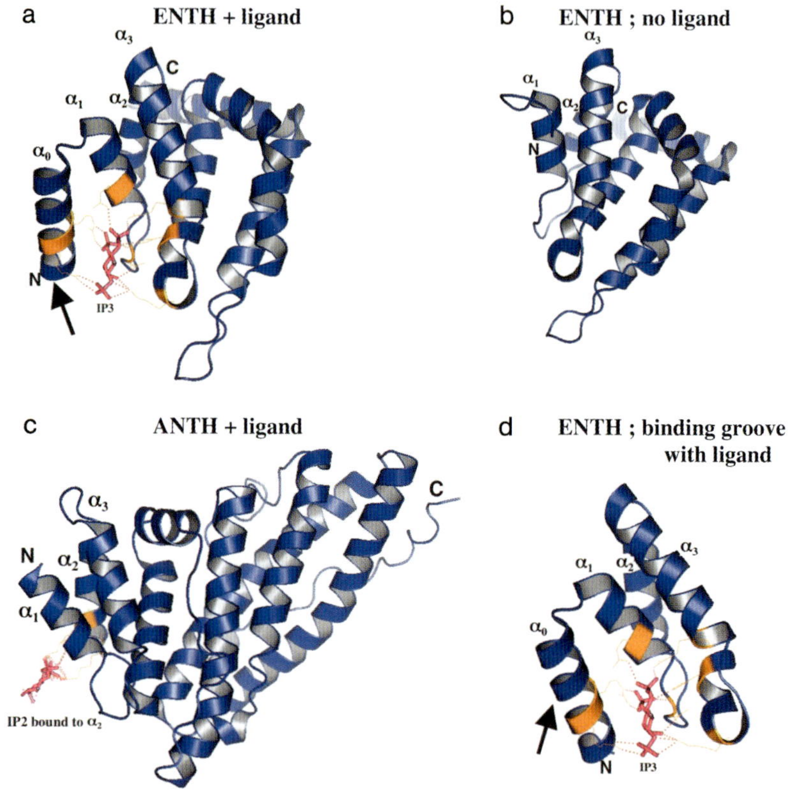

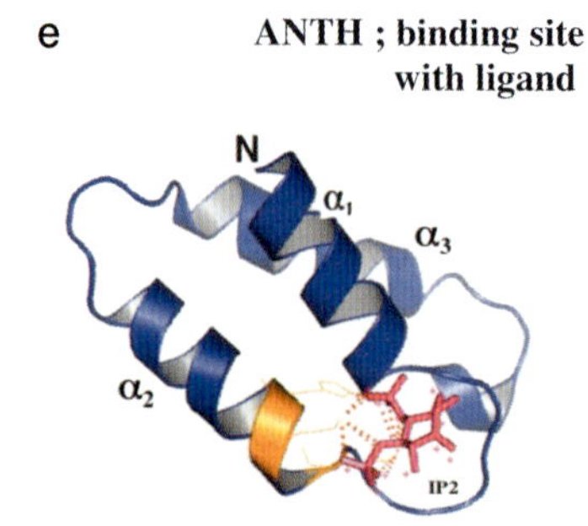

Figure 18.2 Structures of ENTH and ANTH with and without phosphoinositide ligand. (a) ENTH (N-terminal residues 1–158) of epsin with bound IP3 (PDB code 1HOA). (b) Unliganded ENTH of epsin (PDB code 1EDU). (c) ANTH (N-terminal residues 1–289) of CALM bound to IP2 (PDB code 1HG2). (d) and (e) show higher magnifications of the lipid-binding groove and surface of ENTH and ANTH, respectively. The arrow indicates the additional induced helix of ligand-bound ENTH.

a part of a ligand-binding groove in which, unusually for a soluble protein, the hydrophobic residues are on the outer surface and the polar residues form part of the phosphoinositide-binding groove (Figure 18.3). A similar amphipathic helix is predicted to be formed also in epsinR, based on modeling with epsin ENTH as a template [34, 37]. Intriguingly, such ordering of an additional amphipathic α helix is well suited for membrane penetration. Although unexpected, this mode of targeting is similar to what is seen in, e.g., the C1 domain of protein kinase C and

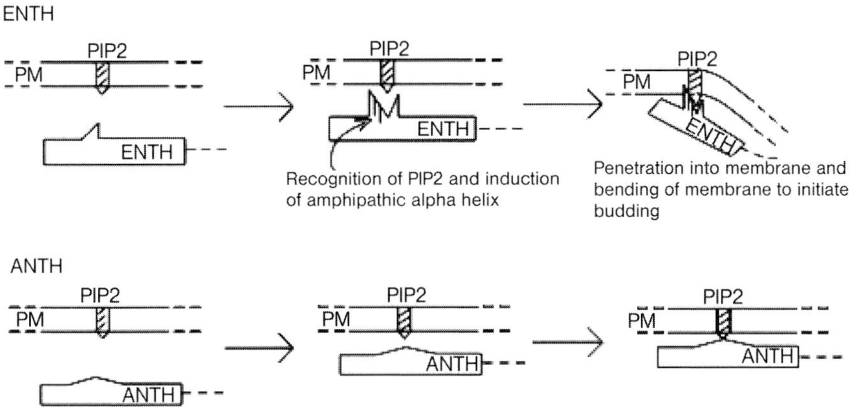

Figure 18.3 Schematic view of the different membrane binding strategies and effects of ENTH and ANTH.

in the FYVE domain; they also expose critical hydrophobic patches for membrane insertion upon ligand binding [41, 51, 52]. ANTH of AP180/CALM have the same lipid binding specificity but they drastically differ in their mode of interaction, which is based on a patch of lysines and histidines, with the phosphates perched on them like a ball balanced on the fingertips [39]. The corresponding site is not conserved in ENTH, and no hydrophobic surface like that of $\alpha0$ is generated in ANTH (Figure 18.2c, e, and Figure 18.3).

The rat, human, and *Drosophila* epsins share perfectly the critical eight sequences needed for lipid-binding in the induced groove [31]. In epsinR, on the other hand, despite the overall similar fold, the lipid-binding residues are poorly conserved. In fact, it has a different lipid specificity, preferentially binding PtdIns(4)P [34, 35]. The binding is rather weak but could play a role in targeting epsinR to TGN membranes, a location different from that of PtdIns(4,5)P_2-binding epsin.

The exact localization of the lipid binding sites also reveals the consensus sequence for the lipid binding in ANTH of AP180/CALM and their related proteins: (K/G)A(T/I)x 6(P/L/V)KxK(H/Y). In the corresponding region of ENTH there is a consensus sequence of (D/E)ATx2(D/E)PWGP, justifying their classification as related but distinct domains [31, 53].

18.4
Signaling and Molecular Functions of ENTH

The original discovery of ENTH domains in proteins, either as part of a coated pit or involved in endosome formation, strongly argued that ENTH played a role in clathrin-dependent events. This was later borne out by experiment, most importantly, by showing that ENTH proteins (1) harbor, in various combinations, motifs that serve as binding sites for clathrin (DLL motifs and clathrin boxes), for AP2

(DPF/W motifs), for EH domain (NPF), and/or for ubiquitinated proteins (UIM motifs) [53, 54] and (2) induce assembly of clathrin triskelia into clathrin cages [24, 29, 55]. Moreover, in immunofluorescence and electron microscopy studies, many ENTH-containing proteins have been shown to associate closely with CPs and CCVs.

Many of the ENTH-bearing proteins are predominantly cytosolic. Thus, their binding to location-specific phospholipids probably serves to anchor them to their target membranes [56, 57]. Once anchored by their N-terminal ENTH, their extended and modular C termini would be available for recruitment of other components needed for coat assembly and clathrin cage formation. Modulation and turnover of the lipid ligands, on the other hand, could lead to regulated release of the vesicle from its anchorage [43]. Such a scheme is substantiated by showing that clathrin polymerization on Ins(4,5)P_2 monolayers is driven by epsin 1 and AP180 [50]. On the other hand, mutations that block Ins(4,5)P_2 binding to ENTH also inhibit clathrin-mediated endocytosis [39, 40].

The conjecture of specific ENTH–lipid interactions underlying the specific targeting of the proteins is further strengthened by the observation that the yeast epsin-like protein Ent3p binds to PtdIns(3,5)P_2, a phosphopeptide concentrated in endosomes and lysosomes [58, 59] (Table 18.1). Interestingly, via this binding, Ent3 – and by extension its yet-to-be-identified mammalian homolog – could play a role in sorting proteins through multivesicular endosomes or bodies (MVBs) in their transit to lysosomes for degradation. The different lipid specificity of Ent3p as compared to epsins is in agreement with the poor conservation of the lipid-binding residues in its ENTH domain. Quite recently, it was shown that the ENTHs of HIP1 and HIPI preferentially bind PtdIns(3,4)P_2 and PtdIns(3, 5)P_2, which are distinctly localized to vesicles in the perinuclear sorting area [41, 60]. The accumulated structural data thus nicely validate the previously measured large differences in the binding affinities and specificities of different ENTHs toward different phosphoinositides. They also support the hypothesis that locally produced specific lipid species may serve as platforms on which ENTH/ANTH-containing proteins are recruited.

A new thematic development concerning the function of ENTH and epsins emerged when Ford et al. [50] found that the ENTH domain of epsin alone is able to cause tubulation of liposomes. This has led to questions regarding the validity of the classical view that it is clathrin that is responsible for the bending of the membrane at the site of the coated pit and of vesicle formation and that interactions and cooperation of multiple proteins are required to carry out such an energetically demanding task. An elegant alternative mechanism is now suggested, based on the ability of the induced amphipathic $\alpha 0$ helix of ENTH to be buried deep in the inner leaflet of the membrane. This, in turn, could induce a positive curvature, leading to bending of the membrane [50]. In fact, estimation of the energetic costs makes such a mechanism based on the perturbation of the lipid bilayer structure, presumably by membrane expansion, quite plausible [51, 61, 62].

ENTH/ANTH domains are best studied for their lipid interactions. Recently, however, interesting protein interactions have been described (Table 18.3). They include interaction with transcription factor PLZF (epsin 1 [49]), tubulin (epsin,

Table 18.3 ENTH/ANT proteins and their protein ligands.

Protein/Modif.	*E/ANTH*	*Binding site*	*Suggested function*	*References*
Tubulin	Epsin 1, 2	α1, α2, α7	Connects endocytic machinery to cytoskeleton	63
	AP180	α7		
	enthoprotin, HIP1, HIPR	N. D.		
Promyelocytic leukemia Zn^{+2} finger protein (PLZF)	Epsin 1	N. D.	Connects endocytic machinery to nuclear function	49
SNARE vti1b (mammalian)	Epsin R	N. D.	Connects vesicle formation to vesicle fusion and sorting	64
SNARE vti1p (yeast)	Ent 3			
Ubiquitin	Epsin 1	N. D.	Regulates PIP2 and/or PLZF binding	71

AP180, HIP1, HIP1r, enthoprotin [63]), and vtib, a SNARE (enthoprotin/clint/epsinR and Ent3 [64]). Binding to tubulin seems to be cooperatively mediated by several helical regions that are different from the lipid binding sites. For other proteins, the exact binding sites have not been determined.

The functional significance of the protein–protein interactions of ENTHs is still poorly known. Binding of PLZF suggests a nuclear function for epsin, especially since epsin was also shown to shuttle between the cytosol and the nucleus [49]. This at first glance was a surprising observation and suggestion, but it is now strongly supported by the dual cytoplasmic and nuclear functions of β-catenin, which also binds the transcription factor Kaiso [65] and which shares the armadillo repeat with epsin.

Binding of tubulin possibly points to a link between the endocytic system and the cytoskeletal machinery. Tubulin may, for instance, help recruit ENTH/ANTH-containing proteins to the vesicles. This is well in line with observations that microtubules can capture and actively transport endocytic vesicles [66, 67]. HIP1, on the other hand, may, by binding tubulin (by its ANTH domain) and actin (by its tail), link the microtubular system to actin microfilaments, which are important in restricting coated pit mobility [68].

Binding the SNARE protein vti1b is important because it raises the possibility that ENTH domains also operate in the selection of cargo, although so far, they have been associated only with initiation of the formation of transport vesicles. SNAREs have important roles in vesicle targeting and fusion, and therefore their locations and activities are carefully controlled. Binding of vti1b by epsinR/Ent3p is analogous to binding of some SNAREs to their locales by components of the budding machinery and adaptor complexes [69] and suggests that ENTH domains, via a SNARE partner, could have a role in cargo sorting in TGN.

The presence of a ubiquitin-interacting motif (UIM) means that epsin also belongs to the category of ubiquitin (Ub) receptors, which can bind ubiquitinated proteins or become monoubiquitinated [70]. Unexpectedly, epsin was found to be ubiquitinated in the ENTH domain, outside the UIM domain [71]. The physiological significance and meaning of this modification remains to be defined. The site of modification makes it is possible that it may alter the binding of the ENTH domain to PtdIns(4,5)P_2 and/or PLZF to ENTH [70].

18.5
History of VHS

The VHS domain was identified in a database screen, based on the multiple occurrence of stretches of sequences in signal transduction proteins [72]. It is ~140 residues long and is found in at least 130 proteins. In all of them, VHS domains occupy the N-terminal end of the polypeptide, suggesting that such a topology is important for its function [73].

The denotation VHS derives from its occurrence in Vps-27, Hrs, and STAM. VHS-containing proteins can be divided into four groups based on their modular composition. First, proteins of the STAM/EAST/Hbp family share the domain composition VHS-SH3-ITAM. The second group includes proteins with a FYVE domain C-terminal to VHS. The third group is GGA (golgi-localizing, γ-adaptin ear homology, Arf-binding) proteins with the domain composition VHS-GAT-GAE (GAT = GGA and TOM; GAE = γ-adaptin ear homology). The fourth group is proteins with a VHS domain alone or with domains other than those mentioned above [74]. Figure 18.1 shows the alignment and modular compositions of some of the representatives of the various classes of VHS proteins.

VHS domains were originally identified in proteins closely associated with receptor endocytosis (STAM/EAST [75–80]) and endosomal trafficking (Vps27, Hrs [81–84]). Later, GGA proteins (GGA1–3) were described as TGN components that bind to ADP-ribosylation factor and facilitate traffic between TGN and vacuole (in yeast) or lysosomes [85–92]. In the mouse, there are two GGA genes that are required for the correct sorting of hydrolases to a vacuolar organelle that is equivalent of the lysosome in animal cells [88, 93].

Despite the many interactions that these and many other VHS domain-containing proteins have, the binding partners and functions of the domain have remained unknown until quite recently. The breakthrough came when four laboratories independently showed that the VHS domains of mammalian GGAs recognize and bind transmembrane cargo proteins carrying a sorting signal of the acidic cluster dileucine (ACLL) family in their cytoplasmic domains [94–97]. This is specific, however, to GGAs, and the ligands of most other VHS domains still elude us.

18.6 Structure of VHS Domains

Crystal structures of VHS domains from *Drosophila* Hrs [48], human Tom1 [98], GGA1 [99, 100], GGA3 [101], and GGA2 [102] have been determined, with those of GGA1 and GGA3 in complex with a peptide corresponding to the ligand (Table 18.4).

The overall structure is a right-handed superhelix with eight α helices. The superhelix is double-layered, with the concave face containing helices α2, α4, and α7 (less well conserved), and the convex layer containing helices α1, α3, α6, and α8 (well conserved), with helix α5 connecting the two faces. A well conserved binding groove is formed by helices α6 and α8. The domain seems to be a chimera of two other motifs so that the first two helix hairpins (α1–α2 and α3–α4) closely resemble the HEAT motif, a repetitive sequence present, for instance, in huntingtin protein and importin β [103]. The third repeat consisting of helices α5–7 is reminiscent of the three-helix armadillo repeat, a motif seen, for instance, in β-catenin in which it serves as a binding site for cadherins, Tcf-family transcription factors, and the EGF-receptor tyrosine kinase domain [104, 105]. These repeats also form concave and convex primary faces, with the binding sites for interacting proteins residing on the concave face. The corresponding face of VHS domains contains large patches of conserved surface residues, suggesting that they also harbor protein binding sites.

The sequence similarity between the VHS and ENTH domains is low and cannot be detected by sequence analysis programs. Nevertheless, VHS and ENTH domains display a highly similar fold when superimposed, with a mean square deviation of only approximately 1.8 Å over the first seven helices (Figure 18.4). Some conserved, mostly hydrophobic residues are mapped to the channel created by the double helix layers. These make up the core of the domain and their distribution is the major determinant of the overall superhelical structure. The major difference is in the orientation of the C-terminal α8, which in VHS packs parallel to the α6 helix in the groove between α6 and α7, but in the ENTH packs perpendicularly across α4 and α2 [48, 74, 98]. The variable surface residues, on the other hand, probably account for the variable functions of these domains.

Table 18.4 Structural data on VHSs and VHS–peptide complexes.

Protein	*Species*	*Pdb/Ref.*	*Residues*	*Technique*	*Resolution, Å*
HRS	*Drosophila melanogaster*	1DVP/48	1–220	X-ray	2
TOM1	Human	1ELK/98	1–153	X-ray	1.50
GGA1	Human	1JWG/99	7–146	X-ray	2
		1JWF/99	7–145	X-ray	2.10
		1PY1/100	8–149	X-ray	2.60
GGA2	Human	1MHQ/102	25–167	X-ray	2.20
GGA3	Human	1JUQ/101	8–157	X-ray	2.20
		1LF8/107	1–166	X-ray	2.30
		1JPL/101	1–157	X-ray	2.40

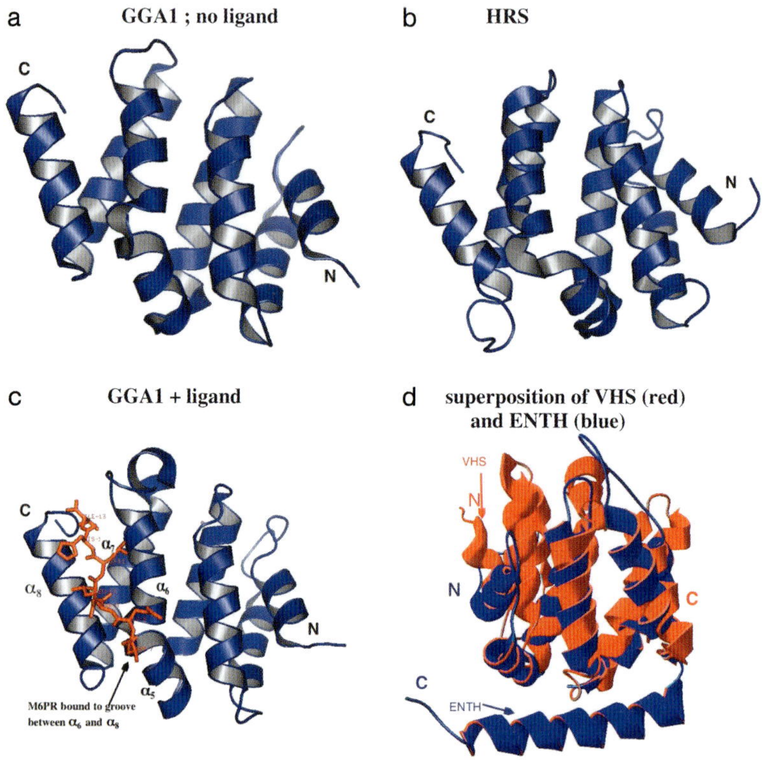

Figure 18.4 Structures of VHSs of GGA1 and Hrs with and without a ligand and comparison of the structures of the VHS of Hrs and ENTH of epsin. (a) VHS of GGA1 (PDB code 1JWF). (b) VHS of Hrs (PDB code 1DVP). (c) VHS of GGA1 with bound 7-amino-acid peptide corresponding to the DxxLL peptide of CI-M6PR (PDB code 1JWG). (d) Comparison of the VHS of Hrs and ENTH of epsin. Superposition: VHS red, ENTH blue.

The VHS domains of the GGAs are very similar in structure, reflecting their high degree of sequence similarity. The mean square of superimposed structures is in the range 0.8–1.1 Å. GGA2 VHS differs from the rest in the loop between helices 6 and 7 and in the fact that the end of helix 6 has an unwound rather than a helical conformation [102]. These differences probably underlie some of the distinct binding properties GGA2 as compared with those of GG1 and GGA3 [95]. A short isoform of GGA3 also exists, which lacks a region in the ligand-binding site in helix 6 and is unlikely to bind cargo [106].

As discussed above, the GGAs bind, via their VHS domains, the C-terminal ACLL sequences of some TGN-associated sorting receptors. The structural basis of this ligand binding has been resolved by studying crystals of complexes between GGA1 and GGA3 VHS domains and ligand peptides [99–101]. The peptide binds, in extended conformation, to a cleft between the α6 and α8 helices, in which a set of electrostatic and hydrophobic interactions guarantees the correct recognition (Table 18.5). The residues responsible for binding are conserved completely in

Table 18.5 VHSs and their protein ligands.

Protein	*VHS/GGA*	*Binding site*	*Function*	*References*
CD-M6PR	GGA1,GGA2,GGA3	cleft between α6 and α8 helices	Sorting of proteins at the TGN	95–97, 99–101, 140
CI-M6PR	GGA1,GGA2,GGA3			
SorLa	GGA1,GGA2			110
Sortilin	GGA1,GGA2			94, 97
LRP	GGA1			97
Memapsin 2 (BACE)	GGA1,GGA2,GGA3			97, 100, 109
Autoinhibitory site	GGA1, GGA3		Prevents cargo recognition dimerization?	112, 113
Ubiquitinated proteins	GGA3	VHS-GAT (cooperative)	Sorting from endosomes to MVB	114
TSG101	GGA3			114
Protein	***VHS/non-GGA***	***Binding site***	***Function***	***References***
Ubiquitinated protein	VHS+UIM	α2 and α4 helices	Sorting from endocytosis to MVB	119
	VHS+GAT			120
FYVE domain	HRS	α2 and α4 helices	Intramolecular interaction	98

human GGAs, to a high degree in their homologs in *Drosophila melanogaster* and *Caenorhabditis elegans*, only weakly in the yeast homologs, and not at all in non-GGA VSHs (TOM1, TOM1-like, Hrs and STAM). Phosphorylation of a serine residue upstream from an ACLL sequence further increases the affinity of the binding due to a slight movement of the sidechains in the cleft [107]. In response to the binding, extensive changes occur in the electrostatic potential and shape of the surface of the complex, which are probably important for downstream signaling.

18.7
Function of GGA-VHS Domains

GGAs were originally discovered as a result of a database search for novel proteins with specific domains known to be important in vesicular traffic [87, 88, 97, 108] or for effectors of Arfs [86]. Effectively, they are monomeric adaptor proteins, which are recruited to membranes through interactions with Arfs, interact with clathrin through specific sequences, bind to AP-1, and recognize peptide signals of their transmembrane cargo in TGN [12]. The cargo, the sorting receptors that traffic between TGN and the endosomal compartment, includes sortilin, a sorting receptor for cargo such as neurotensin [94, 97], the cation-dependent and -independent mannose-6-phosphate receptors (M6PR [95–97]), which act as carriers of the mannose-6-phosphate tagged hydrolases to be ferried to lysosomes, and the low-density lipoprotein receptor-related protein type 3 (LRP3 [97]). Binding of memapsin-2 (β-secretase BACE), a protein responsible for the β-site cleavage of the amyloid precursor protein and of SorLa, a homolog of sortilin, by GGA-VHS, was also demonstrated [97, 100, 109, 110].

The ACLL (also known as DxxLL) sequence is present near the C termini of the M6PR tails. They are instrumental in recognition of the receptors, to initiate a cycle in which the receptors carry lysosomal hydrolase precursors to lysosomes [111]. Attesting to the specificity of the interaction, many other transmembrane receptors that lack such complete recognition motifs fail to bind GGAs. SorLa, however, does not contain either an acidic cluster or dileucine in its recognition sequence [110]. Importantly, none of the known non-GGA VHS domains (TOM1, TOM1L1, Hrs, STAM) bind the ACLL sequences, an observation well in line with the known sequence and structural properties of the domains [101, 102].

GGA1 and GGA3 have an internal ACLL motif in their hinge regions. It binds in cis to the VHS domains of these polypeptides and thereby precludes ligand binding, thus acting in an autoinhibitory mode. Furthermore, the binding depends on phosphorylation of a serine residue just proximal to the sorting signal [112]. Intriguingly, GGA2 lacks this autoinhibitory sequence, suggesting that it has a different function in TGN-associated sorting and vesicle formation. Recent results show that the internal ACLL motif of GGA3 may also bind to the VHS domain of GGA2, thus contributing to the formation of complexes of GGAs on TGN [113].

All the cell biological and biochemical data presented above feature GGAs as adaptors important in the biosynthetic–secretory pathway from TGN to endosomes and lysosomes. Therefore, it came as a surprise that GGAs, and especially GGA3, also seem to have a role in endosomal trafficking. This relates to a study of Puertollano et al. [114], in which specific depletion of GGA3 was shown to inhibit the efficient transport of endocytosed and ubiquitinated growth factor receptors to multivesicular bodies (MVBs) and to lysosomes. Evidence was presented that GGAs are important in the sorting of ubiquitinated proteins at the early endosome stage. This intriguing sorting function was shown to be due to binding of both ubiquitin and TSG101, a component of the ubiquitin-dependent sorting machinery, by the VHS and GAT domains of GGA3. The exact binding site in VHS-GAT is still unresolved, but it lies outside the known ACLL interaction site. As a further twist to the evolving GGA saga, Scott et al. [115] showed that, first, ubiquitin-based sorting is not limited to endosomes but also occurs in the TGN, and, second, that it occurs in a GGA-dependent manner. Thus, new data are emerging that extend the realm of GGA functions beyond the already well known sorting events to less well defined pathways based on ubiquitin tags and their operational territory from the TGN to the endosomes [116].

18.8
Function of Non-GGA VHS Domains

The best-studied non-GGA VHS-containing proteins in mammals are Hrs, hepatocyte growth factor-regulated substrate, which is implicated in vesicular trafficking via the early endosomes [82–84]; STAM/EAST/Hbp proteins, which are associated with growth factor receptor endocytosis and cytokine-mediated intracellular signal transduction [76–78]; and Srcasm, a src-activating signaling molecule [117]. The presence of UIM in both STAM/EAST/Hbp and Hrs is a further indication that these proteins are important for endocytosis and in sorting for degradation at least some plasma membrane receptors via MVBs [118].

Table 18.5 lists the proteins known to bind VHS of GGA as well as non-GGA proteins. Known modifications of VHSs are indicated.

Unlike GGA-VHS, the VHS domains of the above non-GGA proteins do not have a variety of distinct binding partners. For instance, they are incapable of binding the ACLL signal, the favored ligand of GGAs, due to a lack of the critical amino acid residues in the binding pocket formed by helices α6 and α8. Neither do they bind to lipids, in contrast to the structurally related ENTH domains. In the absence of known ligands, the efforts to try to understand the functions of non-GGAs have been based mostly on studies in which isolated VHS domains have been overexpressed in cultured cells. Largely, their exact functions have remained a mystery.

The binding partners of non-GGA VHSs have remained elusive. This situation may be rapidly changing, however, by the discovery of Mizuno et al. [119] that STAM proteins bind to ubiquitinated proteins and that this interaction is mediated by the VHS domain and UIM motif. The binding is most probably cooperative and is suggested to be important in sorting cargo proteins, via the MVB pathway, to lysosomes. Similarly, Yamakami et al. [120] reported that TOM1, via its GAT domain and the C-terminal part of the VHS domain, binds to ubiquitinated proteins, with GAT probably playing a more important role than VHS. This, considering the importance of ubiquitin tagging in endocytosis, suggests, for the first time, a role for TOM1 in endosomal protein sorting. This is a novel function of non-GGA VHS domains and is only the second class of a family of proteins binding to VHS domains in general.

We can expect that the number of proteins binding to VHSs of non-GGAs is greater than currently known. This assessment is based, for instance, on a comparison between the VHSs of TOM1 (similar to the VHS of GGA) and Hrs. The VHS of Hrs interacts with the neighboring FYVE domain of the same polypeptide at a precisely delineated surface involving helices 2 and 4 of the VHS domain [98]. The key features of this binding surface are conserved in the structure of the VHS of TOM1, even though it has no FYVE domain. Thus, they could, in principle, constitute potential protein–protein interfaces that may be used for interaction with domains other than FYVE. This argument can be extended further to ENTH domains, in which the equivalent residues are also conserved.

18.9
Involvement of ENTH and VHS Domains in Human Disease

The vesicular transport mechanisms that the ENTH- and VHS-containing proteins are involved in serve many vital functions. Therefore, it is quite possible that closer scrutiny of the pathways could reveal pathogenetically important defects. Especially interesting in this respect are, for instance, the VHS-dependent endocytic events that serve to down-regulate active ligand–growth factor receptor complexes. Failure to do so could potentially lead, for instance, to sustained cell proliferation with subsequent developmental or growth derangements such as cancer [121]. Since STAM/EAST/Hbp are key regulators of endocytosis, it is quite feasible that they could be targets of cancer-promoting mutations. That this pathway is vulnerable to carcinogenic insults is evidenced by the presence of Eps15 as a fusion partner with HRX/ALL1 in some acute leukemias having translocations in chromosome 11q23 [122].

Another interesting target of pathogenetic studies is M6PR, the GGA-associated cargo receptor. It is a multifunctional receptor, which, besides delivering hydrolases from the TGN, also binds IGF-II on the cell surface. Hence, the alternative designation insulin-like growth factor II receptor (IGFII-R). Binding and down-regulating endocytosis of IGF-II prevents its potentially harmful effects, especially during fetal development. M6PR/IGFII-R also activates TGFβ-1 and prevents cathepsin secretion. All these capacities feature M6PR/IGFII-R as a presumptive tumor suppressor [111]. In fact, frequent loss of heterozygosity at M6PR/IGFII-R locus has been found in various types of cancers, especially in hepatomas. In regard to the critical role of GGAs in the trafficking of M6PR/IGFII-R, it will be important to look systematically at their potential role in these malignancies.

CALM, as an ANTH-containing protein, is an interesting example of how this class of proteins may even have a decisive role in carcinogenesis. It can be a target of the t(10;11)(p13;q14) translocation, which leads to almost full-length CALM fused to part of the AF-10 protein [30]. Recently, such a translocation has been shown to be more common than previously thought and also of prognostic significance in T-cell acute lymphocytic leukemia [123].

18.10
Emerging Research Directions

Thematically, there are several interesting developments in which further clarification of the roles of VHS and ENTH proteins are much needed. These include their involvement in cargo selection in endocytosis, in the bending of the membrane, and in fusion events in the endosomal system. Finally, at a higher level of hierarchy, these events need to be more rigorously related to the workings of the cytoskeletal system.

In a conventional view, ENTH-containing proteins are classified as accessory proteins, which merely assist in the mostly AP-2–driven selection of the cargo and orchestration of the events at the clathrin bud site. Recently, however, blocking studies using RNAi technology [124] and disruption of AP-2 function [125] have convincingly shown that AP-2 is not obligatory for all clathrin-mediated endocytosis, a corroboration of earlier observations in yeast [126, 127]. This leaves a void that seems to be conveniently filled by a group of proteins collectively called monomeric clathrin adaptors [12] or 'almost' adaptors [128] and which include, for instance, epsin, HIP1, Dab2, AP180/CALM, arrestin, and ARH [129]. By virtue of their lipid and clathrin binding properties, they could serve as alternative receptors for the AP-2–independent cargo [130]. This implies a new role for these proteins, in that they are supposed to participate in cargo recruitment in a discriminatory manner. An unresolved question is, however, how this selectivity is imparted and to what extent it is due to recognition of ubiquitin or other signals [130].

Bending of the membrane has emerged as a novel and surprising effect of ENTH domain binding to its target lipid. Since then, the BAR domain has been described as a new membrane binding and bending domain [131]. Interestingly, it is also found primarily in proteins associated with vesicular transport and endocytosis, for instance, amphiphysin, endophilins, and centaurins [132]. A curious feature, and one that sets it apart functionally from ENTH domains, is that BAR binding is sensitive to the curvature of the membrane, although ENTH binds irrespective of it. Amphiphysin and endophilin share AP-2 and clathrin binding properties with epsin but operate at a late stage of vesicle formation in preparation for scission of the vesicle from the membrane by dynamin [133]. This leads to a hypothetical (and verifiable) scenario in which epsin, amphiphysin, and endophilins, and possibly other membrane binding and bending adaptors, work in succession at sites dictated by the membrane curvature. Spatially restricted anchorage would also segregate the other regulatory elements that bind to the adaptors.

Sorting and fusion events at the level of endosomes, as a new field of action for VSH and ENTH proteins, has its origin in the discoveries that, first, ENTH proteins are present in TGN (epsinR) and, second, that cargo-recognizing VHS-containing proteins are found in endosomes (GGAs), both beyond their conventional subcellular territories. This brings up the question of their cooperative function in endosomal sorting, most probably at the level of MVBs ([111]. Closely involved with this issue is the question of how and to what extent do ubiquitination signals and the vesicle-associated SNARE contribute to this process.

A functional link between endocytosis and the cytoskeleton has been well established in numerous studies [134, 135]. It is, however, only through refined biochemical and cell biological studies based on domain–domain interactions and making use of novel imaging techniques that a sufficiently detailed understanding of these interactions can be gained [136, 137]. A third factor to be considered in this interplay are the membrane phosphoinositides that also regulate the activities and localizations of the cytoskeletal elements [138, 139].

18.11 Concluding Remarks

We have come a long way from characterizing VHS and ENTH domains originally on the basis of their multiple occurrences in similarly disposed proteins, to recognize them as the principal building blocks of the complex but harmoniously acting and reacting vesicular transport machinery involving membranes, lipids, proteins, and cytoskeleton. This deeper insight is, however, quite unevenly distributed, in that the GGA-associated events are known in great detail both structurally and biochemically, whereas even some fundamental aspects of the sorting machinery at the plasma membrane are still unresolved. On the other hand, rapid progress is in sight, enabled by new and powerful techniques for screening for interactions and spatial distributions and for imaging the targets in ever-greater detail in living cells.

Acknowledgements

The skillful secretarial assistance of Ms Arja Jumpponen is kindly acknowledged.

References

1 Schmid, S. L., Clathrin-coated vesicle formation and protein sorting: an integrated process. *Annu. Rev. Biochem.* **1997**, *66*, 511–548.

2 Brodsky, F. M., et al., Biological basket weaving: formation and function of clathrin-coated vesicles. *Annu. Rev. Cell Dev. Biol.* **2001**, *17*, 517–568.

3 Bonifacino, J. S., Lippincott-Schwartz, J., Coat proteins: shaping membrane transport. *Nat. Rev. Mol. Cell Biol.* **2003**, *4*, 409–414.

4 Scales, S. J., et al., Coat proteins regulating membrane traffic. *Int. Rev. Cytol.* **2000**, *195*, 67–144.

5 Slepnev, V. I., De Camilli, P., Accessory factors in clathrin-dependent synaptic vesicle endocytosis. *Nat. Rev. Neurosci.* **2000**, *1*, 161–172.

6 Traub, L. M., et al., Crystal structure of the alpha appendage of AP-2 reveals a recruitment platform for clathrin-coat assembly. *Proc. Natl. Acad. Sci. USA* **1999**, *96*, 8907–8912.

7 Owen, D. J., et al., A structural explanation for the binding of multiple ligands by the alpha-adaptin appendage domain. *Cell* **1999**, *97*, 805–815.

8 Kirchhausen, T., Clathrin. *Annu. Rev. Biochem.* **2000**, *69*, 699–727.

9 Robinson, M. S., Bonifacino, J. S., Adaptor-related proteins. *Curr. Opin. Cell Biol.* **2001**, *13*, 444–453.

10 Brett, T. J., et al., Accessory protein recruitment motifs in clathrin-mediated endocytosis. *Structure* **2002**, *(Camb) 10*, 797–809.

11 Traub, L. M., Kornfeld, S., The trans-golgi network: a late secretory sorting station. *Curr. Opin. Cell Biol.* **1997**, *9*, 527–533.

12 Bonifacino, J. S., The GGA proteins: adaptors on the move. *Nat. Rev. Mol. Cell Biol.* **2004**, *5*, 23–32.

13 Nakayama, K., Wakatsuki, S., The structure and function of GGAs, the traffic controllers at the TGN sorting crossroads. *Cell Struct. Funct.* **2003**, *28*, 431–442.

14 Lohi, O., Lehto, V. P., STAM/EAST/Hbp adapter proteins: integrators of signaling pathways. *FEBS Lett.* **2001**, *508*, 287–290.

15 Chen, H., et al., Epsin is an EH-domain–binding protein implicated in clathrin-mediated endocytosis. *Nature* **1998**, *394*, 793–797.

16 Wang, L. H., et al., The appendage domain of alpha-adaptin is a high affinity binding site for dynamin. *J. Biol. Chem.* **1995**, *270*, 10079–10083.

17 Confalonieri, S., Di Fiore, P. P., The Eps15 homology, (EH) domain. *FEBS Lett.* **2002**, *513*, 24–29.

18 Wendland, B., et al., Yeast epsins contain an essential N-terminal ENTH domain, bind clathrin and are required for endocytosis. *EMBO J.* **1999**, *18*, 4383–4393.

19 Wendland, B., Emr, S. D., Pan1p, yeast eps15, functions as a multivalent adaptor that coordinates protein–protein interactions essential for endocytosis. *J. Cell Biol.* **1998**, *141*, 71–84.

20 Rosenthal, J. A., et al., The epsins define a family of proteins that interact with components of the clathrin coat and contain a new protein module. *J. Biol. Chem.* **1999**, *274*, 33959–33965.

21 Yamabhai, M., et al., Intersectin, a novel adaptor protein with two Eps15 homology and five Src homology 3 domains. *J. Biol. Chem.* **1998**, *273*, 31401–31407.

22 Sengar, A. S., et al., The EH and SH3 domain Ese proteins regulate endocytosis by linking to dynamin and Eps15. *EMBO J.* **1999**, *18*, 1159–1171.

23 Kay, B. K., et al., Identification of a novel domain shared by putative components of the endocytic and cytoskeletal machinery. *Protein Sci.* **1999**, *8*, 435–438.

24 Kalthoff, C., et al., Unusual structural organization of the endocytic proteins AP180 and epsin 1. *J. Biol. Chem.* **2002**, *277*, 8209–8216.

25 Drake, M. T., et al., Epsin binds to clathrin by associating directly with the clathrin-terminal domain. Evidence for cooperative binding through two discrete sites. *J. Biol. Chem.* **2000**, *275*, 6479–6489.

26 Hussain, N. K., et al., Splice variants of intersectin are components of the endocytic machinery in neurons and nonneuronal cells. *J. Biol. Chem.* **1999**, *274*, 15671–15677.

27 Chen, H., et al., Epsin is an EH-domain–binding protein implicated in clathrin-mediated endocytosis. *Nature* **1998**, *394*, 793–797.

28 Cadavid, A. L., et al., The function of the *Drosophila* fat facets deubiquitinating enzyme in limiting photoreceptor cell number is intimately associated with endocytosis. *Development* **2000**, *127*, 1727–1736.

29 Ahle, S., Ungewickell, E., Purification and properties of a new clathrin assembly protein. *EMBO J.* **1986**, *5*, 3143–3149.

30 Dreyling, M. H., et al., The t(10;11)(p13;q14) in the U937 cell line results in the fusion of the *AF10* gene and *CALM*, encoding a new member of the AP-3 clathrin assembly protein family. *Proc. Natl. Acad. Sci. USA* **1996**, *93*, 4804–4809.

31 Ford, M. G., et al., Curvature of clathrin-coated pits driven by epsin. *Nature* **2002**, *419*, 361–366.

32 Mishra, S. K., et al., Clathrin- and AP-2–binding sites in HIP1 uncover a general assembly role for endocytic accessory proteins. *J. Biol. Chem.* **2001**, *276*, 46230–46236.

33 Metzler, M., et al., HIP1 functions in clathrin-mediated endocytosis through binding to clathrin and adaptor protein 2. *J. Biol. Chem.* **2001**, *276*, 39271–39276.

34 Hirst, J., et al., EpsinR: an ENTH domain-containing protein that interacts with AP-1. *Mol. Biol. Cell* **2003**, *14*, 625–641.

35 Mills, I. G., et al., EpsinR: an AP1/clathrin interacting protein involved in vesicle trafficking. *J. Cell Biol.* **2003**, *160*, 213–222.

36 Wasiak, S., et al., Enthoprotin: a novel clathrin-associated protein identified through subcellular proteomics. *J. Cell Biol.* **2002**, *158*, 855–862.

37 Kalthoff, C., et al., Clint: a novel clathrin-binding ENTH-domain protein at the golgi. *Mol. Biol. Cell* **2002**, *13*, 4060–4073.

38 Duncan, M. C., et al., Yeast epsin-related proteins required for golgi–endosome traffic define a gamma-adaptin ear-binding motif. *Nat. Cell Biol.* **2003**, *5*, 77–81.

39 Ford, M. G., et al., Simultaneous binding of PtdIns(4,5)P2 and clathrin by AP180 in the nucleation of clathrin lattices

on membranes. *Science* **2001**, *291*, 1051–1055.

40 Itoh, T., et al., Role of the ENTH domain in phosphatidylinositol-4,5-bisphosphate binding and endocytosis. *Science* **2001**, *291*, 1047–1051.

41 Lemmon, M. A., Phosphoinositide recognition domains. *Traffic* **2003**, *4*, 201–213.

42 DiNitto, J. P., et al., Membrane recognition and targeting by lipid-binding domains. *Sci. STKE* **2003**, *2003*, re16.

43 Overduin, M., et al., Signaling with phosphoinositides: better than binary. *Mol. Intervent.* **2001**, *1*, 150–159.

44 Mao, Y., et al., A novel all helix fold of the AP180 A-terminal domain for phosphoinositide binding and clathrin assembly in synaptic vesicle endocytosis. *Cell* **2001**, *104*, 433–440.

45 Lopez-Mendez, B., et al., NMR assignment of the hypothetical ENTH-VHS domain At3g16270 from *Arabidopsis thaliana*. *J. Biomol. NMR* **2004**, *29*, 205–206.

46 Hyman, J., et al., Epsin 1 undergoes nucleocytosolic shuttling and its eps15 interactor NH(2)-terminal homology, (ENTH) domain, structurally similar to armadillo and HEAT repeats, interacts with the transcription factor promyelocytic leukemia Zn(2)+ finger protein, (PLZF). *J. Cell Biol.* **2000**, *149*, 537–546.

47 Koshiba, S., et al., Solution structure of the epsin N-terminal homology, (ENTH) domain of human epsin. *J. Struct. Funct. Genomics* **2002**, *2*, 1–8.

48 Mao, Y., et al., Crystal structure of the VHS and FYVE tandem domains of Hrs, a protein involved in membrane trafficking and signal transduction. *Cell* **2000**, *100*, 447–456.

49 Hyman, J., et al., Epsin 1 undergoes nucleocytosolic shuttling, and its eps15 interactor NH(2)-terminal homology, (ENTH) domain, structurally similar to armadillo and HEAT repeats, interacts with the transcription factor promyelocytic leukemia Zn(2)+ finger protein, (PLZF). *J. Cell Biol.* **2000**, *149*, 537–546.

50 Ford, M. G., et al., Curvature of clathrin-coated pits driven by epsin. *Nature* **2002**, *419*, 361–366.

51 Hurley, J. H., Wendland, B., Endocytosis: driving membranes around the bend. *Cell* **2002**, *111*, 143–146.

52 Hurley, J. H., Misra, S., Signaling and subcellular targeting by membrane-binding domains. *Annu. Rev. Biophys. Biomol. Struct.* **2000**, *29*, 49–79.

53 Legendre-Guillemin, V., et al., ENTH/ANTH proteins and clathrin-mediated membrane budding. *J. Cell. Sci.* **2004**, *117*, 9–18.

54 Wendland, B., Epsins: adaptors in endocytosis? *Nat. Rev. Mol. Cell Biol.* **2002**, *3*, 971–977.

55 Morgan, J. R., et al., A conserved clathrin assembly motif essential for synaptic vesicle endocytosis. *J. Neurosci.* **2001**, *20*, 8667–8676.

56 Vanhaesebroeck, B., et al., Synthesis and function of 3-phosphorylated inositol lipids. *Annu. Rev. Biochem.* **2001**, *70*, 535–602.

57 Itoh, T., Takenawa, T., Regulation of endocytosis by phosphatidylinositol 4,5-bisphosphate and ENTH proteins. *Curr. Top. Microbiol. Immunol.* **2004**, *282*, 31–47.

58 Friant, S., et al., Ent3p Is a PtdIns(3,5)P2 effector required for protein sorting to the multivesicular body. *Dev. Cell* **2003**, *5*, 499–511.

59 Hicke, L., PtdIns(3,5)P2 finds a partner. *Dev. Cell.* **2003**, *5*, 363–364.

60 Hyun, T. S., et al., HIP1 and HIP1r stabilize receptor tyrosine kinases and bind 3-phosphoinositides via ENTH domains. *J. Biol. Chem.* **2004** (in press).

61 Nossal, R., Zimmerberg, J., Endocytosis: curvature to the ENTH degree. *Curr. Biol.* **2004**, *12*, R770–2.

62 Stahelin, R. V., et al., Contrasting membrane interaction mechanisms of AP180 N-terminal homology, (ANTH) and epsin N-terminal homology, (ENTH) domains. *J. Biol. Chem.* **2003**, *278*, 28993–28999.

63 Hussain, N. K., et al., A role for epsin N-terminal homology/AP180 N-terminal homology, (ENTH/ANTH) domains in tubulin binding. *J. Biol. Chem.* **2003**, *278*, 28823–28830.

64 Chidambaram, S., et al., Specific interaction between SNAREs and epsin N-terminal homology, (ENTH) domains of epsin-related proteins in trans-golgi network to endosome transport. *J. Biol. Chem.* **2004**, *279*, 4175–4179.

65 Daniel, J. M., Reynolds, A. B., The catenin p120(ctn) interacts with Kaiso, a novel BTB/POZ domain zinc finger transcription factor. *Mol. Cell. Biol.* **1999**, *19*, 3614–3623.

66 Goltz, J. S., et al., A role for microtubules in sorting endocytic vesicles in rat hepatocytes. *Proc. Natl. Acad. Sci. USA* **1992**, *89*, 7026–7030.

67 Oda, H., et al., Interaction of the microtubule cytoskeleton with endocytic vesicles and cytoplasmic dynein in cultured rat hepatocytes. *J. Biol. Chem.* **1995**, *270*, 15242–15249.

68 Gaidarov, I., et al., Spatial control of coated-pit dynamics in living cells. *Nat. Cell Biol.* **1999**, *1*, 1–7.

69 Ungar, D., Hughson, F. M., SNARE protein structure and function. *Annu. Rev. Cell Dev. Biol.* **2003**, *19*, 493–517.

70 Haglund, K., et al., Distinct monoubiquitin signals in receptor endocytosis. *Trends Biochem. Sci.* **2003**, *28*, 598–603.

71 Oldham, C. E., et al., The ubiquitin-interacting motifs target the endocytic adaptor protein epsin for ubiquitination. *Curr. Biol.* **2002**, *12*, 1112–1116.

72 Schultz, J., et al., SMART, a simple modular architecture research tool: identification of signaling domains. *Proc. Natl. Acad. Sci. USA* **1998**, *95*, 5857–5864.

73 Lohi, O., Lehto, V. P., VHS domain marks a group of proteins involved in endocytosis and vesicular trafficking. *FEBS Lett.* **1998**, *440*, 255–257.

74 Lohi, O., et al., VHS domain: a longshoreman of vesicle lines. *FEBS Lett.* **2002**, *513*, 19–23.

75 Schultz, J., et al., SMART: a web-based tool for the study of genetically mobile domains. *Nucleic Acids Res.* **2000**, *28*, 231–234.

76 Lohi, O., et al., EAST, an epidermal growth factor receptor- and Eps15-associated protein with Src homology 3 and tyrosine-based activation motif domains. *J. Biol. Chem.* **1998**, *273*, 21408–21415.

77 Takeshita, T., et al., Cloning of a novel signal-transducing adaptor molecule containing an SH3 domain and ITAM. *Biochem. Biophys. Res. Commun.* **1996**, *225*, 1035–1039.

78 Takata, H., et al., A Hrs binding protein having a Src homology 3 domain is involved in intracellular degradation of growth factors and their receptors. *Genes. Cells* **2000**, *5*, 57–69.

79 Takeshita, T., et al., STAM, signal transducing adaptor molecule, is associated with Janus kinases and involved in signaling for cell growth and c-myc induction. *Immunity* **1997**, *6*, 449–457.

80 Endo, K., et al., STAM2, a new member of the STAM family, binding to the Janus kinases. *FEBS Lett.* **2000**, *477*, 55–61.

81 Piper, R. C., et al., VPS27 controls vacuolar and endocytic traffic through a prevacuolar compartment in *Saccharomyces cerevisiae*. *J. Cell Biol.* **1995**, *131*, 603–617.

82 Komada, M., et al., Hrs, a tyrosine kinase substrate with a conserved double zinc finger domain, is localized to the cytoplasmic surface of early endosomes. *J. Biol. Chem.* **1997**, *272*, 20538–20544.

83 Komada, M., Kitamura, N., Growth factor-induced tyrosine phosphorylation of Hrs, a novel 115-kilodalton protein with a structurally conserved putative zinc finger domain. *Mol. Cell. Biol.* **1995**, *15*, 6213–6221.

84 Chin, L. S., et al., Hrs interacts with sorting nexin 1 and regulates degradation of epidermal growth factor receptor. *J. Biol. Chem.* **2001**, *276*, 7069–7078.

85 Poussu, A. M., et al., Vear, a novel golgi-associated protein, is preferentially expressed in type I cells in skeletal muscle. *Muscle Nerve* **2001**, *24*, 127–129.

86 Boman, A. L., et al., A family of ADP-ribosylation factor effectors that can alter membrane transport through the trans-Golgi. *Mol. Biol. Cell* **2000**, *11*, 1241–1255.

87 Dell'Angelica, E. C., et al., GGAs: a family of ADP ribosylation factor-binding proteins related to adaptors and associated with the Golgi complex. *J. Cell Biol.* **2000**, *149*, 81–94.

88 Hirst, J., et al., A family of proteins with gamma-adaptin and VHS domains that facilitate trafficking between the trans-golgi network and the vacuole/lysosome. *J. Cell Biol.* **2000**, *149*, 67–80.

89 Black, M. W., Pelham, H. R., A selective transport route from golgi to late endosomes that requires the yeast GGA proteins. *J. Cell Biol.* **2000**, *151*, 587–600.

90 Takatsu, H., et al., Adaptor gamma ear homology domain conserved in gamma-

adaptin and GGA proteins that interact with gamma-synergin. *Biochem. Biophys. Res. Commun.* **2000**, *271*, 719–725.

91 Costaguta, G., et al., Yeast Gga coat proteins function with clathrin in golgi to endosome transport. *Mol. Biol. Cell* **2001**, *12*, 1885–1896.

92 Zhdankina, O., et al., Yeast GGA proteins interact with GTP-bound Arf and facilitate transport through the golgi. *Yeast* **2001**, *18*, 1–18.

93 Mullins, C., Bonifacino, J. S., Structural requirements for function of yeast GGAs in vacuolar protein sorting, alpha-factor maturation, and interactions with clathrin. *Mol. Cell. Biol.* **2001**, *21*, 7981–7994.

94 Nielsen, M. S., et al., The sortilin cytoplasmic tail conveys golgi–endosome transport and binds the VHS domain of the GGA2 sorting protein. *EMBO J.* **2001**, *20*, 2180–2190.

95 Puertollano, R., et al., Sorting of mannose 6-phosphate receptors mediated by the GGAs. *Science* **2001**, *292*, 1712–1716.

96 Zhu, Y., et al., Binding of GGA2 to the lysosomal enzyme sorting motif of the mannose 6-phosphate receptor. *Science* **2001**, *292*, 1716–1718.

97 Takatsu, H., et al., Golgi-localizing, gamma-adaptin ear homology domain, ADP-ribosylation factor-binding, (GGA) proteins interact with acidic dileucine sequences within the cytoplasmic domains of sorting receptors through their Vps27p/Hrs/STAM, (VHS) domains. *J. Biol. Chem.* **2001**, *276*, 28541–28545.

98 Misra, S., et al., Structure of the VHS domain of human, Tom1 (target of myb 1): insights into interactions with proteins and membranes. *Biochemistry* **2000**, *39*, 11282–11290.

99 Shiba, T., et al., Structural basis for recognition of acidic-cluster dileucine sequence by GGA1. *Nature* **2002**, *415*, 937–941.

100 He, X., et al., Biochemical and structural characterization of the interaction of memapsin **2003**, 2, (beta-secretase) cytosolic domain with the VHS domain of GGA proteins. *Biochemistry 42*, 12174–12180.

101 Misra, S., et al., Structural basis for acidic-cluster-dileucine sorting-signal recognition by VHS domains. *Nature* **2002**, *415*, 933–937.

102 Zhu, G., et al., Crystal structure of GGA2 VHS domain and its implication in plasticity in the ligand binding pocket. *FEBS Lett.* **2003**, *537*, 171–176.

103 Groves, M. R., Barford, D., Topological characteristics of helical repeat proteins. *Curr. Opin. Struct. Biol.* **1999**, *9*, 383–389.

104 Huber, A. H., et al., Three-dimensional structure of the armadillo repeat region of beta-catenin. *Cell* **1997**, *90*, 871–882.

105 Peifer, M., et al., A repeating amino acid motif shared by proteins with diverse cellular roles. *Cell* **1994**, *76*, 789–791.

106 Wakasugi, M., et al., Predominant expression of the short form of GGA3 in human cell lines and tissues. *Biochem. Biophys. Res. Commun.* **2003**, *306*, 687–692.

107 Kato, Y., et al., Phosphoregulation of sorting signal–VHS domain interactions by a direct electrostatic mechanism. *Nat. Struct. Biol.* **2002**, *9*, 532–536.

108 Poussu, A., et al., Vear, a novel Golgi-associated protein with VHS and gamma-adaptin 'ear' domains. *J. Biol. Chem.* **2000**, *275*, 7176–7183.

109 He, X., et al., Memapsin 2 (beta-secretase) cytosolic domain binds to the VHS domains of GGA1 and GGA2: implications on the endocytosis mechanism of memapsin 2. *FEBS Lett.* **2002**, *524*, 183–187.

110 Jacobsen, L., et al., The sorLA cytoplasmic domain interacts with GGA1 and -2 and defines minimum requirements for GGA binding. *FEBS Lett.* **2002**, 511, 155–158.

111 Ghosh, P., et al., Mannose 6-phosphate receptors: new twists in the tale. *Nat. Rev. Mol. Cell Biol.* **2003**, *4*, 202–212.

112 Doray, B., et al., Autoinhibition of the ligand-binding site of GGA1/3 VHS domains by an internal acidic cluster-dileucine motif. *Proc. Natl. Acad. Sci. USA* **2002**, *99*, 8072–8077.

113 Ghosh, P., et al., Mammalian GGAs act together to sort mannose 6-phosphate receptors. *J. Cell Biol.* **2003**, 163, 755–766.

114 Puertollano, R., Bonifacino, J. S., Interactions of GGA3 with the ubiquitin sorting machinery. *Nat. Cell Biol.* **2004**, *6*, 244–251.

115 Scott, P. M., et al., GGA proteins bind ubiquitin to facilitate sorting at the trans-golgi network. *Nat. Cell Biol.* **2004**, *6*, 252–259.
116 Babst, M., GGAing ubiquitin to the endosome. *Nat. Cell Biol.* **2004**, *6*, 175–177.
117 Seykora, J. T., et al., Srcasm: a novel Src activating and signaling molecule. *J. Biol. Chem.* **2002**, 277, 2812–2822.
118 Hicke, L., A new ticket for entry into budding vesicles – ubiquitin. *Cell* **2001**, 106, 527–530.
119 Mizuno, E., et al., STAM proteins bind ubiquitinated proteins on the early endosome via the VHS domain and ubiquitin-interacting motif. *Mol. Biol. Cell* **2003**, *14*, 3675–3689.
120 Yamakami, M., et al., Tom1, a VHS domain-containing protein, interacts with tollip, ubiquitin, and clathrin. *J. Biol. Chem.* **2003**, *278*, 52865–52872.
121 Floyd, S., De Camilli, P., Endocytosis proteins and cancer: a potential link? *Trends Cell Biol.* **1998**, *8*, 299–301.
122 Rogaia, D., et al., The localization of the HRX/ALL1 protein to specific nuclear subdomains is altered by fusion with its eps15 translocation partner. *Cancer Res.* **1997**, *57*, 799–802.
123 Asnafi, V., et al., CALM-AF10 is a common fusion transcript in T-ALL and is specific to the TCRgammadelta lineage. *Blood* **2003**, 102, 1000–1006.
124 Motley, A., et al., Clathrin-mediated endocytosis in AP-2–depleted cells. *J. Cell Biol.* **2003**, 162, 909–918.
125 Conner, S. D., Schmid, S. L., Differential requirements for AP-2 in clathrin-mediated endocytosis. *J. Cell Biol.* **2003**, 162, 773–779.
126 Yeung, B. G., et al., Adaptor complex-independent clathrin function in yeast. *Mol. Biol. Cell* **1999**, *10*, 3643–3659.
127 Huang, K. M., et al., Clathrin functions in the absence of heterotetrameric adaptors and AP180-related proteins in yeast. *EMBO J.* **1999**, *18*, 3897–3908.
128 Watson, H. A., et al., In vivo role for actin-regulating kinases in endocytosis and yeast epsin phosphorylation. *Mol. Biol. Cell* **2001**, *12*, 3668–3679.
129 Gonzalez-Gaitan, M., Stenmark, H., Endocytosis and signaling: a relationship under development. *Cell* **2003**, 115, 513–521.
130 Traub, L. M., Sorting it out: AP-2 and alternate clathrin adaptors in endocytic cargo selection. *J. Cell Biol.* **2003**, 163, 203–208.
131 Peter, B. J., et al., BAR domains as sensors of membrane curvature: the amphiphysin BAR structure. *Science* **2004**, 303, 495–499.
132 Habermann, B., The BAR-domain family of proteins: a case of bending and binding? *EMBO Rep.* **2004**, *5*, 250–255.
133 Takei, K., et al., Interactions of dynamin and amphiphysin with liposomes. *Methods Enzymol.* **2001**, 329, 478–486.
134 Qualmann, B., Kessels, M. M., Endocytosis and the cytoskeleton. *Int. Rev. Cytol.* **2002**, 220, 93–144.
135 McPherson, P. S., The endocytic machinery at an interface with the actin cytoskeleton: a dynamic, hip intersection. *Trends Cell Biol.* **2002**, *12*, 312–315.
136 Merrifield, C. J., et al., Imaging actin and dynamin recruitment during invagination of single clathrin-coated pits. *Nat. Cell Biol.* **2002**, *4*, 691–698.
137 Rappoport, J. Z., et al., Movement of plasma-membrane–associated clathrin spots along the microtubule cytoskeleton. *Traffic* **2003**, *4*, 460–467.
138 Niggli, V., Structural properties of lipid-binding sites in cytoskeletal proteins. *Trends Biochem. Sci.* **2001**, *26*, 604–611.
139 Raucher, D., et al., Phosphatidylinositol 4,5-bisphosphate functions as a second messenger that regulates cytoskeleton–plasma membrane adhesion. *Cell* **2000**, 100, 221–228.
140 Doray, B., et al., Interaction of the cation-dependent mannose 6-phosphate receptor with GGA proteins. *J. Biol. Chem.* **2002**, 277, 18477–18482.

19
PX Domains

Matthew L. Cheever and Michael Overduin

19.1
Introduction and History of the PX Domain Discovery

It has become clear that the cell's membranes do much more than "function to organize biological processes by compartmentalizing them" [1]. Besides acting to separate one region of the cellular milieu from another, these complex biological systems are constantly undergoing changes in their compositions and locations. Membrane phospholipids are critical for cellular signaling and receptor trafficking and act as specific platforms for assembly of large complexes that mediate endocytosis and signal transduction. Although many proteins that provide membrane functionality are integrated into the lipid bilayer, a host of proteins have been discovered that function at the membrane–cytoplasm interface. These proteins can localize at the membrane in a constitutive fashion or can be brought to the membrane only when signaled to do so. Some of these proteins are directed to the membrane through protein–protein interactions with integral membrane proteins. More recently, a number of protein domains have been discovered that interact directly and often specifically with phospholipid headgroups, serving to target host proteins to subcellular membrane compartments. This group of proteins includes the phox homology (PX) and FYVE domains, whose primary specificity is for phosphatidylinositol 3-phosphate [PtdIns(3)P], as well as the ENTH and PH domains that are discussed in other chapters. The PX domain exemplifies the complexity of these modules, with a range of phospholipid specificities and modulatory protein interactions operating in a diverse array of intracellular pathways including cellular signaling, membrane fusion, microbial host defense, phospholipid metabolism, and receptor sorting.

The conserved sequence of the PX domain was identified within a number of eukaryotic proteins in 1996 [2]. These include $p47^{phox}$ and $p40^{phox}$, two subunits of its namesake, the superoxide producing phagocyte NADPH oxidase (phox) complex. In resting cells, these subunits and $p67^{phox}$ form a cytosolic complex which, when stimulated, assembles with the membrane-bound cytochrome b_{558}, a heterodimeric complex consisting of $p22^{phox}$ and $gp91^{phox}$, to produce superoxide. Phox complex

Modular Protein Domains. Edited by G. Cesareni, M. Gimona, M. Sudol, M. Yaffe

ISBN: 3-527-30813-X

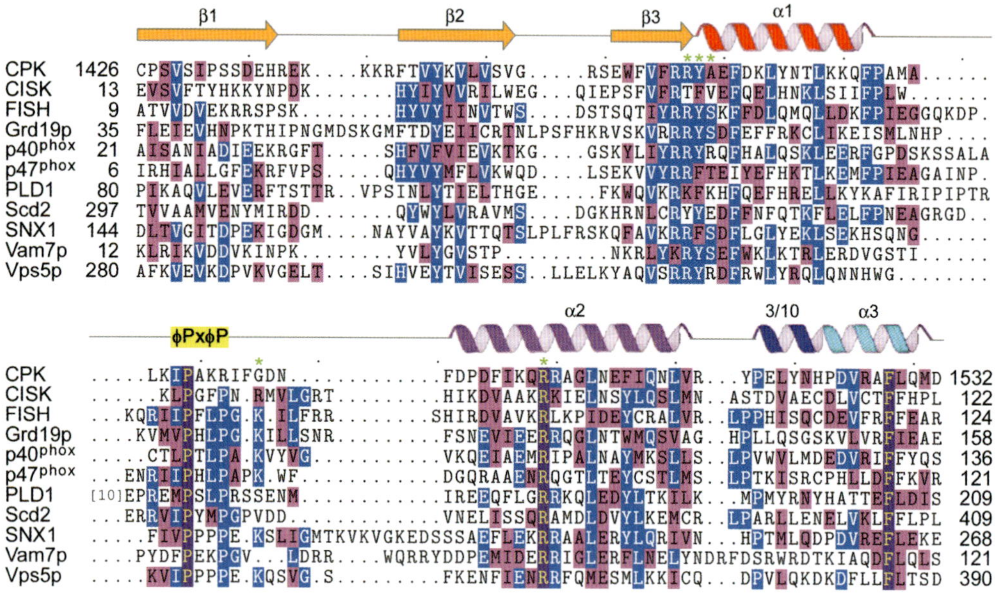

Figure 19.1 Amino acid sequence alignment of PX domains from several proteins. The positions of secondary structural elements and the SH3 binding motif (ϕPXϕP) of the p40^{phox} PX domain are indicated. Phospholipid binding residues of both Grd19p and p40^{phox} are marked with green asterisks. Absolutely conserved residues are shown with a purple background, partially conserved residues are shown in blue, and similar residues are shown in magenta. Each PX domain sequence is numbered according to the protein from which it is taken. The alignment was generated using the CLUSTAL W [75] program, and the coloring was done using the TEXshade [76] program from the Biology Workbench [77].

function is vital for phagocyte-mediated destruction of invading microbes, and mutations in phox components, including one found in the p47^{phox} PX domain, can lead to inherited chronic granulomatous disease (CGD). Patients with CGD show an increased susceptibility to bacterial and fungal infection due to lack of superoxide production [3, 4].

The PX domain, which is sometimes described as the PB2 (phox and Bem1p 2) domain, has subsequently been found in 48 human proteins, including protein kinases, PtdIns-kinases, and phospholipases (Figure 19.1). The proteins that contain PX domains can be classified into three types based on their modular architectures. Seven of the human sorting nexins (SNXs) consist of very little other than a PX domain. Thirteen other human SNXs also contain C-terminal coiled-coil regions that may mediate protein homo- and hetero-oligomerization. The third class contains a diverse array of protein and membrane-recognition modules in addition to their PX domains. These include domains that bind phospholipids (B41/ERM, C2, and PH domains), proline-rich motifs (SH3 domains), PC peptide motifs (PB1 domain), microtubules (MIT domain), and Gα subunits (RGS domain) [5] (Figure 19.2). The neighboring domains may have profound effects on PX domain function. For example, coiled-coil regions can increase membrane avidity by

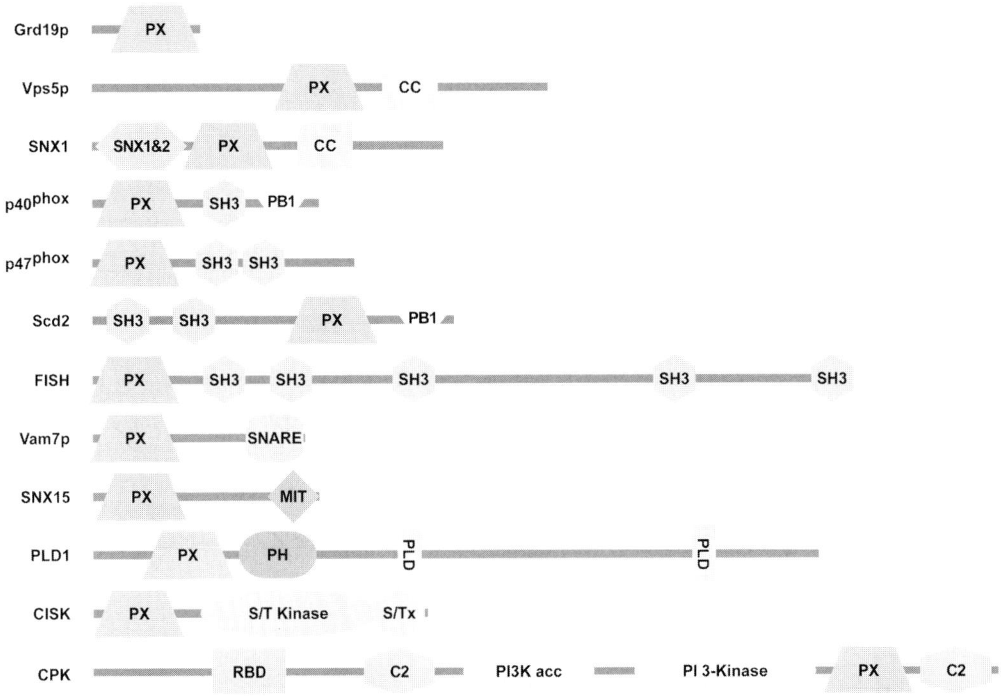

Figure 19.2 Schematic representation of the domain structure of several PX domain-containing proteins. Abbreviations: PX (phox homology), CC (coiled-coil), SNX1&2 (SNX1 and SNX2 N-terminal homology), SH3 (Src homology 3), PB1 (phox and Bem1p 1), SNARE (SNARE α helical), MIT (microtubule-interacting and trafficking), PH (pleckstrin homology), PLD (phospholipase D catalytic), S/T kinase (serine-threonine protein kinase catalytic), S/Tx (serine-threonine kinase extension), RBD (Ras binding), C2 (PKC conserved region 2), PI3K acc (PtdIns 3-kinase accessory), PI 3-kinase (PtdIns 3-kinase catalytic).

juxtaposing several PX domains within oligomeric protein complexes, and SH3 domains can provide regulatory interactions by interacting with proline-rich motifs contained within some PX domains [6–11].

The function of the PX domain remained elusive until 2001, when several groups reported its specific interaction with phosphoinositides (PIs) [12–16]. These phospholipids can be found in any of eight different forms, each distinguished by the phosphorylation state of the inositol ring. Positions 3, 4, and 5 on the inositol ring are phosphorylated or dephosphorylated by PI-specific kinases and phosphatases, thus providing a dynamic control over PI levels in various cellular membranes [17]. Several yeast PX domains bind specifically and with high affinity to PtdIns(3)P, which is concentrated in the yeast vacuole, as well as endocytic intermediate vesicles [18, 19]. Another set of PX domains bind specifically but weakly to PtdIns(3)P and appear to rely on other domains or multimeric complexes to associate stably with a particular membrane. Other PX domains bind preferentially to multiply phosphorylated PIs such as PtdIns(3,4)P_2, PtdIns(4,5)P_2, and PtdIns(3,4,5)P_3, as for $p47^{phox}$

[14, 20, 21], some PI 3-kinases [16], and SNX1 [7], respectively. The destinations of these proteins are diverse, with p47^{phox} targeting the plasma membrane, where levels of these multiply phosphorylated PIs increase in response to receptor activation, while SNX1 is found in the cytosol and endosomes [22], and the class II PI 3-kinases (CPK) distribute to the nucleus, cytosol, trans-golgi network, and clathrin-coated vesicles [23–25]. The range of PI affinities and specificities exhibited by PX domains suggests a plastic recognition mechanism that has been adapted to provide targeting to a range of cellular membranes in higher eukaryotes.

Concurrent with discovery of the function of the PX domain, Hiroaki and colleagues published the first PX domain structure, along with a demonstration that PX domains can interact with SH3 domains, albeit weakly [10]. Structural mechanisms have now been proposed to explain the abilities of PX domains to bind specifically to PtdIns(3)P and PtdIns(3,4)P_2, as well as the synergistic interaction with phosphatidic acid (PtdOH) and insertion into the membrane [26–28]. These structural analyses of PX domains by NMR and X-ray crystallography have been complemented by binding studies using analytical ultracentrifugation, blot overlays, fluorescence spectroscopy, isothermal titration calorimetry, and surface plasmon resonance, as well as extensive mutagenesis. Together, these studies have yielded a detailed understanding of PX domain function that illuminates its physiological roles.

19.2
Structure of the PX Domain

3D structures have been solved of the PX domains of the Grd19p sorting nexin homolog, the p40^{phox} and p47^{phox} subunits, and the Vam7p t-SNARE [10, 26–29]. The studies show that, despite sharing only eight strictly conserved residues, these domains possess a conserved core structure consisting of an N-terminal antiparallel β sheet draped across a C-terminal α helical subdomain (Figures 19.1 and 19.3). The twisted β sheet has three strands (β1, β2, β3), with a β bulge in β1 that helps to orient the N-terminal end of strand β2 to form one side of the PI binding pocket. The α-helical subdomain contains two long and structurally conserved helices (α1, α2), which are roughly parallel and are connected to each other by a long loop of varying sequence, length, and structure. In p40^{phox} and p47^{phox} this variable loop contains a type II polyproline helix (PP_{II}), and adjacent to this in p47^{phox} is a short 3_{10} helix [27, 28]. The portion of the PX domain C-terminal to α2 is also helical, with one helix in the NMR structure of p47^{phox} [10] and two helices in the X-ray structures of Grd19p [26], p40^{phox} [27], p47^{phox} [28], and the NMR structure of Vam7p [29] (Figures 19.2 and 19.3). Additional helices have been identified at the N and C termini of the p40^{phox} PX domain, and, although they are not conserved elements of the fold, they are structurally important appendages that may contribute to its functional specialization [27]. Grd19p consists of little more than a PX domain, having only 30 extra residues at its N-terminus, which appear to be unstructured, since they are vulnerable to proteolysis and are not apparent in the electron density.

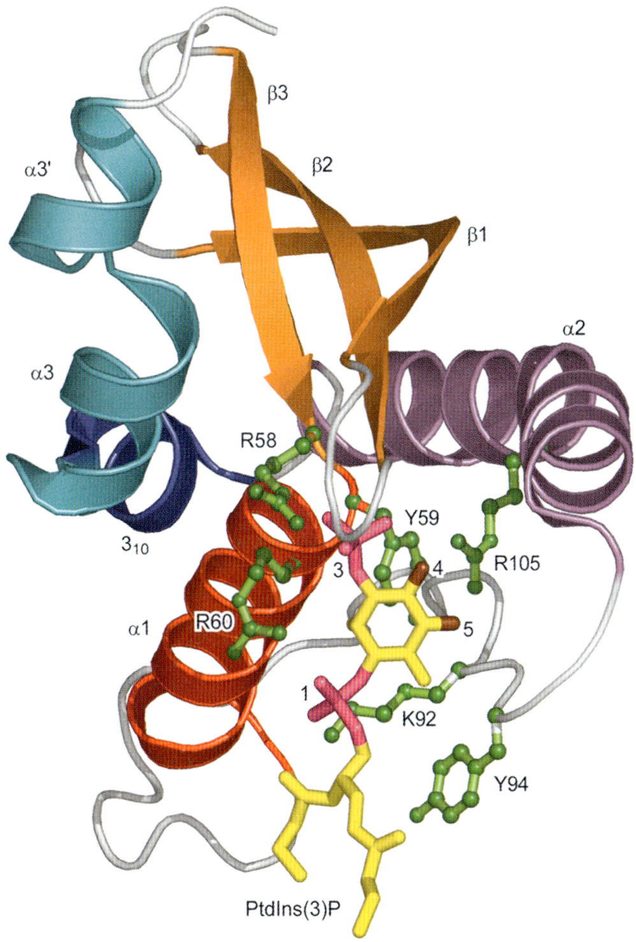

Figure 19.3 Structure of the PX domain of $p40^{phox}$ bound to dibutanoyl PtdIns(3)P. The β sheet is orange; helices α1, α2, and α3 are red, purple, and cyan; and the 3_{10} helix is blue. The residues involved in binding PtdIns(3)P are green, and dibutanoyl PtdIns(3)P is yellow. The 1- and 3-phosphates are magenta, and the 4- and 5-OH groups are brown. The various secondary structural elements and PtdIns(3)P-binding residues of the protein, as well as the PtdIns(3)P phosphates and OH groups are labeled. The figure was prepared using PyMOL [78]. PDB code 1H6H.

The similarity of the structures is indicated by the superposition of the Grd19p PX domain Cα atoms on those of the $p47^{phox}$, $p40^{phox}$, and Vam7p PX structures with root mean squared deviations (rmsd values) of 1.5, 2.2, and 2.1 Å, respectively. In contrast to the conserved core, the loops near the PI binding site adopt a wide variety of conformations, reflecting their contributions to specialized binding and regulatory functions.

19.2.1 Mechanism of PtdIns(3)P Coordination

PX domains bind a PI molecule in a positively charged pocket that is formed by five elements: the C-terminal end of strand β3, the N-terminal ends of strand β2 and helices α1 and α2, and residues in the variable loop (Figure 19.4a). The 3-phosphate of the inositol headgroup is hydrogen bonded by conserved basic and polar residues, which in Grd19p correspond to R81 and S83 [26]. In addition, the poorly conserved R118 forms a salt bridge with the 3-phosphate in one crystal form. The 1-phosphate is bound by K112 and E86 in the variable loop and α1, respectively. The latter residue also forms a hydrogen bond with the 2-OH group. The 4- and 5-OH groups are hydrogen bonded by R127, which is absolutely conserved in α2. The inositol ring is stabilized by stacking interactions with a highly conserved aromatic residue, Y82. These contacts allow the Grd19p protein to bind tightly to PtdIns(3)P, with no significant binding to the other PIs [18]. This specificity is explained by the steric clashes that would occur between a 5-phosphate and residues K112, I113, and R127 in the PtdIns(3)P binding pocket. It is less clear how PIs phosphorylated at the 4 position would be excluded from binding. Reorientation of R118, as seen in one of the two molecules in the asymmetric unit, would allow room for the 4-phosphate, and R127 could also provide favorable

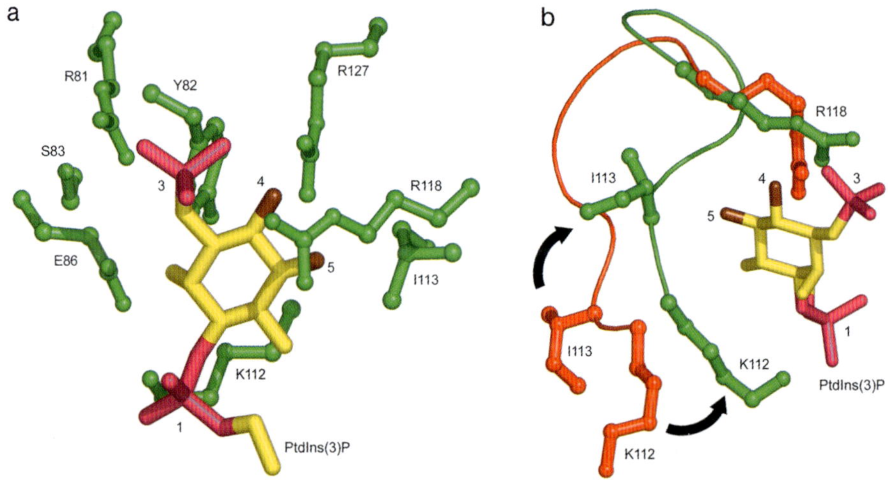

Figure 19.4 The PtdIns(3)P binding mechanism of the Grd19p PX domain. (a) View of the Grd19p residues that coordinate PtdIns(3)P. The coloring, labeling, and preparation are as described for Figure 19.3. Coordinates were provided by H. van Tilbeurgh. (b) Depiction of conformational changes in the variable-loop residues upon binding to dibutanoyl PtdIns(3)P. The residues and loop of the apo Grd19p PX domain structure are red, and those from chain B of the dibutanoyl PtdIns(3)P bound structure are green. Otherwise, colors, labels, and preparation are as described for Figure 19.3. Arrows indicate large sidechain movements in going from the unbound structure to the bound structure. The protein structures were superimposed by aligning the Cα atoms using SPOCK [79].

interactions. Thus, a predictive basis for specificity is complicated by induced fit and a preponderance of charge in the binding site and requires solving further structures of complexes for resolution.

The conformational changes associated with ligand binding are revealed by comparison of the structures of the Grd19p PX domain [26]. The apo and dibutanoyl PtdIns(3)P-bound forms have been refined at 2.0 and 2.3 Å resolution and superimpose with a backbone rmsd of 0.58 Å. The loop between $\alpha 1$ and $\alpha 2$ differs significantly between the bound and free states of Grd19p and forms a rim of the ligand binding pocket which moves towards the PtdIns(3)P in the bound structure (Figure 19.4b). The largest conformational change in the binding site involves K112, which moves 3.5 Å to accommodate the 1-phosphate, indicating the mobility of this variable loop (Figure 19.4b). Residue I113 also moves significantly, and like L114 and L115, is positioned to interact with the dibutanoyl tails of the dibutanoyl PtdIns(3)P co-crystal. However, the lack of solid electron density for L114, L115, and the PtdIns(3)P acyl chains prevents more precise positioning, although it is consistent with the mobility of these elements. The loop between strands $\beta 1$ and $\beta 2$ also differs in conformation between the two states, although these changes may be influenced by a crystal-induced dimer interaction seen only in the bound form [26].

A similar binding mechanism was inferred from the 1.7-Å crystal structure of the p40^{phox} domain bound to dibutanoyl PtdIns(3)P (Figure 19.3). In this complex, the 3-phosphate is coordinated by hydrogen bonds with R58, Y59, and R60 and by a salt link with R105. These residues superimpose closely with the corresponding Grd19p residues R81, Y82, S83, and R127, respectively, as does the PtdIns(3)P. The $\alpha 2$ helix dipole also provides a favorable interaction with the 3-phosphate. The 1-phosphate forms hydrogen bonds with R60 and K92 (which corresponds to Grd19p K112), while R105 (corresponding to Grd19p R127) also hydrogen bonds with the 4- and 5-hydroxyl groups of PtdIns(3)P. Also similar to Grd19p, the binding pocket is filled not only by the PtdIns(3)P headgroup but also by several ordered water molecules, which form a hydrogen-bond network between hydroxyls on the exposed face of the inositol ring and polar groups on the protein. The dibutanoyl tails of the bound PtdIns(3)P molecule can be seen in the p40^{phox} electron density and form hydrophobic interactions with Y94, a variable residue that precedes $\alpha 2$ [27]. The congruence of the Grd19p and p40^{phox} binding interactions suggests that all PtdIns(3)P-specific PX domains utilize a conserved binding mode.

NMR studies of the Vam7p PX domain reveal a similar PtdIns(3)P binding mechanism. Key ligand-coordinating residues from Grd19p and p40^{phox} are conserved as residues R41, Y42, S43, and R88 of the Vam7p PX domain. These residues exhibit significant chemical shift perturbations upon ligation of PtdIns(3)P, and their orientations resemble those of the bound conformations in the Grd19p and p40^{phox} complexes [12, 29]. The disorder of the binding-site residues and variable loop in the free state are consistent with the mobility inferred here in the Grd19p structures [29]. However, more extensive changes in conformation are possible, particularly in Vam7p's longer variable loop, given the extensive changes in the chemical shifts [12]. One PtdIns(3)P binding residue that is not conserved in

Vam7p's variable loop is Grd19p K112 (or p40^{phox} K92), which binds the 1-phosphate [26, 27]. However, the Vam7p residue K67 is located nearby and may play an analogous binding role, based on the local PtdIns(3)P-induced chemical shift changes [12, 29]. Another Vam7p residue that could substitute is the poorly conserved R77, which is also oriented toward the ligand-binding pocket. It has also been proposed that R77 may be involved in low-affinity interactions of Vam7p with PIs containing a 5-phosphate [29]. Within Vam7p's variable loop there are two hydrophobic residues, V70 and L71, which exhibit large dibutanoyl PtdIns(3)P-induced chemical shift changes and are located near the tails of PtdIns(3)P when overlaid with the p40^{phox} structure, suggesting potential hydrophobic interactions. The NMR studies show that the portion of the variable loop between V70 and Y79 interacts with membrane-mimicking dodecylphosphocoline micelles in the presence of dibutanoyl PtdIns(3)P, suggesting interactions between this element and the bilayer [12]. The heterogeneity in the variable loop is likely to be reflected by differences in the binding properties, such as in membrane affinity and insertion angle.

In contrast to the canonical PtdIns(3)P-specific PX domains of Grd19p, p40^{phox}, and Vam7p, the p47^{phox} PX domain is more promiscuous and prefers multiply phosphorylated PIs. Several reports suggest different PI specificities for this protein; however, PtdIns 3,4-bisphosphate [PtdIns(3,4)P_2] appears to be the consensus ligand [14, 20, 21, 28]. The structure of the p47^{phox} PX domain has been solved by NMR and X-ray crystallography and reveals that the variable loop contains a short helix following α1 and a distorted left-handed polyproline type II helix, with minor differences that can be attributed to crystal contacts [10, 28]. The p47^{phox} crystal structure includes two bound sulfates, one of which is located where the 3-phosphate of a PI would be expected to bind. This sulfate hydrogen bonds with R43 and is near R90, indicating a canonical 3-phosphate coordination. However, steric clashes may prevent the rest of the PI from binding in the same orientation as found in, for example, p40^{phox} [28]. The R90 sidechain blocks access by a 4-phosphate, and P78 occludes most of F44, the highly conserved aromatic residue that stacks with the inositol ring. Thus, it was proposed that the p47^{phox} PX domain binds a PI in a novel orientation in which the inositol ring is further rotated into a crevice between the C terminus of strand β2 and the N terminus of helix α2 [28]. Thus, although the PIs all appear to bind to the same pocket, the orientations of those other than PtdIns(3)P may differ significantly when bound to PX domains with divergent specificities. As with Vam7p, when the addition of PI to the p47^{phox} PX domain is monitored by NMR, chemical-shift changes are evident in residues distant from the binding pocket, providing support for a ligand-induced conformational change in the PX domain [8].

An accessory site that binds a second phospholipid molecule was identified in the p47^{phox} PX domain. This part of the PI binding pocket is occupied by a sulfate ion in this crystal structure, with another sulfate being located in the nearby 3-phosphate binding pocket [28]. In a physiological setting, this accessory pocket is thought to bind PtdOH or PtdSer, which can act synergistically with PtdIns(3,4)P_2 to provide plasma membrane targeting. However, this accessory site is not found in the p40^{phox}, Grd19p, or Vam7p PX domain structures, and the critical residues (R70, K55, and H51) are not conserved. Only the PX domains of the phospholipase D enzymes appear to share this accessory pocket, based on their conservation of these key basic residues.

19.3
Biological Function of the PX Domain

The primary role for PX domains is to localize proteins to cellular membranes by direct and reversible interactions with PIs such as PtdIns(3)P. These interactions occur with varying degrees of specificity and affinity and can be stabilized by the synergistic binding to other acidic phospholipids, as well as by hydrophobic insertion into the bilayer. PI binding can be negatively and positively modulated by SH3 domain interactions and protein oligomerization. These interaction networks are complex and not strictly conserved, thus contributing to the specialized functions of individual PX domains within the various proteins and pathways where they are found.

19.3.1
PI Binding Specificity

PX domains exhibit a range of PI specificities that reflect their subcellular localizations. The *Saccharomyces cerevisiae* genome codes for 15 proteins with PX domains, all but one of which show a preference for binding PtdIns(3)P over the other PIs. These yeast proteins can be divided into two groups based on PtdIns(3)P affinity, one in which four PX domains bind tightly with affinities around 2 or 3 μM, and the other in which the domains bind weakly [18]. The high-affinity group includes Grd19p and Vam7p, which localize to PtdIns(3)P-enriched endosomal and vacuole membranes, respectively [12, 30]. Membrane targeting by the low-affinity group may be enhanced by the formation of protein complexes and dimers, which would contain multiple PI binding sites with additive or synergistic binding activities.

The PI binding specificities of PX domains from higher eukaryotes are more divergent than those of yeast. For instance, the PX domains of CISK, CPK, p40^{phox}, p47^{phox}, PLD1, and SNX1 prefer PtdIns(3,4)P_2 [31], PtdIns(4,5)P_2 [16], PtdIns(3)P [14, 15, 20, 21], PtdIns(3,4)P_2) [14, 21], PtdIns(5)P [32], and PtdIns(3,4,5)P_3 [7], respectively. Some PX domain proteins are promiscuous in their PI binding, showing similar affinities for multiple PIs. Examples of this include the PX domains from SNX9, which binds PtdIns(3)P, PtdIns(3,4)P_2, PtdIns(4,5)P_2, and PtdIns(3,4,5)P_3 [33], and from RGS-PX1, which binds PtdIns(5)P and PtdIns(3)P [34]. This could reflect a need for localization at multiple membranes or may indicate a lower need for stringent localization. In some cases, the binding activity can be influenced by the method used. For example, two groups used a lipid blot binding assay to demonstrate that the SNX1 PX domain binds preferentially to

$PtdIns(3,4,5)P_3$, but liposome binding studies showed that this domain binds only PtdIns(3)P and $PtdIns(3,5)P_2$ [7, 35]. Further in vivo studies indicated that SNX1 localizes to membranes having high PtdIns(3)P concentrations but not high $PtdIns(3,4,5)P_3$ concentrations [35]. These results suggest that PX domains may exhibit different apparent PI specificities depending upon how a PI is presented, for example, as a soluble form or embedded in surfaces of various lipid compositions.

19.3.2
Synergistic Phospholipid Interactions

Although membrane targeting by PX domains is principally mediated by PIs, other acidic phospholipids also contribute. For example, the affinity of the PX domain of $p47^{phox}$ for lipid vesicles containing $PtdIns(3,4)P_2$ is 25 times stronger when PtdSer is supplemented [9]. This enhancement is due to occupancy of an accessory basic pocket by PtdSer or a similar molecule such as PtdOH [28]. The PtdSer interaction synergizes with the nearby $PtdIns(3,4)P_2$ binding by further reducing the domain's electrostatic potential and promoting partitioning of hydrophobic residues in the variable loop into the bilayer. Co-ligation of $PtdIns(3,4)P_2$ and another acidic phospholipid appears to be required for plasma membrane recruitment of $p47^{phox}$ PX constructs, which are normally cytosolic [9, 28]. Consistent with this idea, spiking membranes with negatively charged phospholipids, such as PtdSer, PtdOH, and phosphatidylglycerol, markedly increases NADPH oxidase activity [80], perhaps through interactions with the $p47^{phox}$ PX domain [36]. Furthermore, the in vitro stimulation of oxidase activity by the negatively charged amphiphile arachidonic acid may actually result from binding to the $p47^{phox}$ accessory site [36]. The conservation of this accessory site appears to be limited to two phospholipases, PLD1 and PLD2, and the membrane interaction of other PX domains, such as that of $p40^{phox}$, are not enhanced by PtdSer [9, 28]. Surprisingly, the PLD1 PX domain appears to bind PtdIns(5)P but not PtdOH or PtdSer [32]. It will be interesting to see whether other PX domains do indeed possess specificity for multiple phospholipids or whether $p47^{phox}$ is unique in this regard.

19.3.3
Membrane Insertion

Insertion of a PX domain into the hydrophobic portion of a bilayer is largely mediated by the variable loop that links helices $\alpha1$ and $\alpha2$. The extremities of this elongated loop contribute to the PI pocket and to the acidic phospholipid binding site found in $p47^{phox}$, respectively, and residues along its length are available for membrane interactions. Several hydrophobic residues in this loop, including I65 and W80 of $p47^{phox}$ and Y94 and V95 (as well as F35 between $\beta1$ and $\beta2$) of $p40^{phox}$, penetrate into the membrane [9]. Resonances of a corresponding region of the Vam7p PX domain spanning residues V70 to Y79 are specifically perturbed after association with lipid micelles, suggesting a similar insertion [12]. When the structure of a

PI-bound PX domain is modeled with a bilayer, this loop can be positioned near the membrane surface where it interacts with other lipid headgroups. This orientation not only directs the PtdIns(3)P acyl chains into the membrane, but it positions the PI and PtdSer/PtdOH binding sites of the p47^{phox} PX domain inside the membrane interface, thus explaining the ability of these lipids to independently induce membrane penetration [9]. Theoretical calculations indicate that the region of the PX domain predicted to insert into the membrane is surrounded by a strong positive electrostatic potential, which is greatly reduced when a PI is bound. A model was proposed whereby PI binding acts as an 'electrostatic switch' to decrease this electrostatic potential, thus allowing the hydrophobic residues to penetrate more readily into the interior of the bilayer [9].

19.3.4
Regulatory Protein Interactions

Like many other signaling modules, some PX domains are multifunctional, interacting with proteins such as SH3 domains, as originally hypothesized by Ponting [2]. Many, but not all, PX domains contain a class I or II SH3 binding motif in the variable loop, consisting of (R/K)xϕPXϕP or ϕPXϕPx(R/K) sequences, respectively, where x and ϕ refer to any residue and any hydrophobic residue, respectively [37]. In at least two proteins, intramolecular interactions between SH3 and PX domains provide regulation. For example, the ϕPXϕP motif of the p47^{phox} PX domain interacts with its C-terminal SH3 domain [10]. This is one of two intramolecular interactions that inhibit activation of the NADPH complex by preventing p47^{phox} from binding to PIs and the integral membrane subunit p22^{phox} [8, 28, 38, 81]. Phosphorylation of several of its C-terminal serine residues allows p47^{phox} to bind both PIs and p22^{phox} [8, 28, 39, 40, 81]. As well, when the SH3 domain is mutated to prevent its autoinhibition of the PX domain, the protein binds better to PtdIns(3,4)P_2-containing liposomes than does wild-type p47^{phox} [28]. These data are consistent with a model in which the closed conformation of p47^{phox} prevents superoxide production by blocking both membrane targeting and SH3-mediated interactions with membrane components of the oxidase. Phosphorylation then releases the intramolecular interactions, allowing the PX domain to interact with membrane lipids, including PIs and other acidic lipids, and bringing the SH3 domains into proximity with cytochrome b_{558}, thus activating the NADPH oxidase. In support of this, p47^{phox} shows cytoplasmic localization in resting cells but translocates to the membrane upon stimulation by the protein kinase C-activating compound PMA. Truncation of the PX domain or an R90K mutation that blocks PI binding impairs translocation and concomitant superoxide production [8]. With both its intramolecular SH3 domain autoinhibition and its phosphorylation-dependent membrane binding function, the p47^{phox} PX domain provides two levels at which the potentially cytotoxic function of the NADPH oxidase complex can be tightly regulated.

The Scd2 protein from *Schizosaccharomyces pombe* provides another example of a regulatory intramolecular interaction between SH3 and PX domains. This protein is a homolog of the *S. cerevisiae* protein Bem1p, which binds PtdIns(3)P and PtdIns(4)P through its PX domain [18, 21]. Scd2 is held in a closed conformation by an interaction between its PX and SH3 domains. The open conformation of Scd2 can be stabilized through binding Cdc42 and Scd1, where Cdc42 is a GTP-hydrolyzing switch protein that participates in controlling polarized cell growth and Scd1 is the nucleotide exchange factor that activates Cdc42. The Scd2 SH3 domain can then bind the protein kinase Shk1, allowing its activation by Cdc42 [11]. Beyond the examples of Scd2 and $p47^{phox}$, it is unclear how many other PX domains are involved in interactions with SH3 domains. However, the large number of PX domains containing canonical SH3 binding motifs and their partially exposed positions and potential mobility suggest that other PX-containing proteins will be found to use similar mechanisms of regulation.

19.3.5
Signaling Pathways of the PX Proteins

Within the family of PX proteins there exists a large subfamily of sorting nexins, so-named because of their roles in sorting receptors through membrane compartments such as endosomes and lysosomes (reviewed in [41]). Initially, this subfamily was more loosely defined, with the presence of a PX domain being the main prerequisite for inclusion as a sorting nexin. However, phylogenetic analyses suggest that the PX domains of SNX proteins are evolutionarily distinct and can be further classified into approximately six subclasses [41–43]. Although some sorting nexins, such as Grd19p and SNX3, consist of little more than a PX domain, the majority of these proteins contain a variety of other protein–protein interaction domains (Figure 19.2). Some of these additional protein domains interact with the receptors being sorted, and in sorting nexins 1, 2, 4, 6, and 15, coiled-coil regions mediate homo- or hetero-oligomeric interactions with other sorting nexins [7, 43, 44], It has been reported that the PX domains of sorting nexins 1, 2, 6, and 15 also play a role in oligomerization, and for SNX1, mutations that prevent vesicle localization and PI binding do not affect these interactions [7, 43, 44]. This indicates that PX domain-mediated oligomerization and PI binding rely on different parts of the protein and can function independently, although both are required for proper localization in vivo [7].

The first sorting nexin to be discovered was SNX1, and it remains the best-characterized member. It was first isolated as a binding partner for the epidermal growth factor receptor (EGFR), and overexpression of SNX1 was shown to decrease the amount of activated EGFR at the plasma membrane [45]. Subsequently, a number of other SNXs, including SNX2, SNX3, SNX6, SNX9, and SNX13, have been shown to interact with the EGFR or alter its endocytosis, transport, and lysosomal degradation [7, 13, 34, 44, 46, 47]. These sorting nexins may also help sort and regulate the cell surface receptors for cytokines, insulin, lepton, low-density lipoprotein, platelet-derived growth factor, thrombin, transferrin, and transforming growth factor receptor-β [43, 44, 46, 48–50], suggesting a broad importance in growth regulation.

Sorting nexins are also known to interact with a variety of other biologically important proteins. For instance, SNX5 binds to the Fanconi anemia complementation group A protein (FANCA), and its overexpression leads to increased cellular levels of FANCA [51]. The FANCA gene is mutated in the first of eight complementation groups of the autosomal-recessive syndrome Fanconi anemia, which is characterized by progressive bone-marrow failure, developmental abnormalities, and a predisposition to malignancy [52, 53]. The SNX6 protein interacts with the Pim1 proto-oncogene, and this interaction results in translocation of SNX6 to the nucleus [54]. An endocytic role for SNX9 is suggested by its interaction with the adaptor complex AP-2 [33]. The SNX13 protein, which is also known as RGS-PX1, is a regulator of G protein signaling that acts as a GTPase-activating protein for the $G\alpha_s$ class of heterotrimeric GTP-binding proteins [34]. Downstream responses regulated by these sorting nexins likely include transcription, endocytosis, cell growth, differentiation and proliferation, membrane trafficking, cardiac contraction and relaxation, hormone secretion, and learning and memory [33, 34, 54].

Another PX protein involved in growth regulation is PLD, an enzyme that hydrolyzes phosphatidylcholine to produce choline and phosphatidic acid, a lipid second messenger. Two mammalian forms, PLD1 and PLD2, are known to exist, with several splice variants. In addition to their PX domains, both PLDs contain pleckstrin homology (PH) domains, and PLD1 also contains a basic element, all of which may interact with PIs. Besides being a possible PI binding domain, the PH domain in PLD1 is palmitoylated, which would also likely promote insertion into the membrane. The exact interplay between each of these potential membrane-binding mechanisms is still unclear, although much progress has now been made [32, 55–57]. The ϕPXϕP motif in the PLD2 PX domain can be bound by the SH3 domain of phospholipase C-γ1 (PLC-γ1), allowing the potentiation of inositol 1,4,5-trisphosphate production by PLC-γ1 and Ca^{2+} release [58]. Depending on the cell type studied, PLD1 can cycle between perinuclear vesicular localization in resting cells and plasma membrane localization in activated cells. The PX domain appears to be required for perinuclear localization, but this does not depend on its PI binding activity and may involve PX–protein interactions. Although the PX domain does not seem to play a role in translocation to the plasma membrane in stimulated cells, it is required for proper internalization and endosomal relocalization. This effect depends on the PI binding activity of the PX domain, because mutation of residues likely involved in PI binding, such as R118 or F120/R179, have effects similar to deletion of the entire PX domain [32]. The perinuclear localization of PLD1 may be mediated through PX domain binding PtdIns(5)P on endocytic vesicles [32]. Although localized at the plasma membrane, PLD1 can be found in caveolin-enriched microdomains, where it is phosphorylated by protein kinase C on three residues, one of which, T147, is in the variable loop of the PX domain. Mutation of this residue reduced the activation of transiently expressed PLD1 after cellular stimulation [59, 60]. Thus, the function of the PLD1 PX domain is particularly complex, being influenced by an unusual PI specificity, a proposed accessory phospholipid site, protein interactions, phosphorylation, and the membrane interactions of neighboring domains.

A role for the PX domain in regulating cell survival is seen in the human protein serum- and glucocorticoid-inducible kinase 3 (SGK3) and its mouse homolog cytokine-independent survival kinase (CISK). As members of the Akt family of serine-threonine kinases, these proteins function by phosphorylating and inactivating pro-apoptotic proteins such as FKHRL1, a member of the forkhead family of transcription factors, which prevents cell cycle arrest and apoptosis [61]. Akt is activated through being recruited by its PH domain to the plasma membrane, where it is phosphorylated on its activation loop by another PI binding kinase, PDK1 [62, 63]. The binding of PIs by Akt not only brings it into proximity with PDK1, but also is thought to induce a conformational change that allows for its phosphorylation [64]. Although SGK3 and CISK do not have PH domains, it is thought that the PX domain performs a similar role in these kinases. The CISK PX domain has been shown to bind PIs by one group who used lipid blot and liposome binding studies to show a preference for PtdIns(3)P [65]; however, another study using only lipid blot methods demonstrated that PtdIns(3,5)P_2 and PtdIns(3,4,5)P_3 are the preferred ligands [31]. Regardless of PI specificity, the PI interactions of the PX domain are required for the proper targeting of CISK, since mutations that inhibit PI binding also prevent endosomal localization [31, 65]. Interestingly, mutations in the PX domain that prevent PI binding also prevent CISK activation, but deletion of the PX domain increases activation [31]. This suggests that, similar to the PH domain of Akt, the PX domain of CISK may have a role in negative regulation that can be relieved upon binding to PIs. CISK can interact with and be phosphorylated by PDK, further underscoring the similarities between these PX-domain–containing kinases and the more canonical member of the family, Akt [66]. Future studies may reveal why one member of this family evolved to contain a PH domain, while another selected the PX domain.

The Vam7p protein is unique among the yeast PX domain proteins, most of which can be classified as sorting nexins [67]. Although Vam7p also plays a role in protein sorting and membrane trafficking, it does not function as a sorting nexin, but instead helps mediate fusion of membrane vesicles with the vacuole [68]. In fact, Vam7p and two possible homologs in *S. pombe* and *Candida albicans* are currently the only known SNARE (soluble *N*-ethylmaleimide–sensitive fusion protein attachment protein receptor) proteins to contain PX domains [69]. SNAREs play a vital role in most types of intracellular membrane fusion, where they participate in docking two membranes together prior to the actual fusion event. This docking occurs through the interaction of α-helical regions from SNARE proteins on opposing membranes, thus forming a trans-SNARE complex. There is evidence to suggest that SNARE proteins can induce membrane fusion by themselves in vitro, but the actual in vivo process appears to be more complicated and requires a number of regulatory proteins [70]. Most SNAREs associate with membranes through transmembrane domains or through the posttranslational addition of lipid anchors, but for Vam7p, membrane localization is accomplished through PtdIns(3)P binding by its PX domain [12]. Deletion of the PX domain or a mutation that blocks PI binding prevents vacuolar localization of Vam7p, and loss of Vam7p vacuolar localization is also seen if PtdIns(3)P production is eliminated by inhibiting the function of the yeast PI 3-kinase, Vps34p. Vacuole fusion can be repressed by blocking Vam7p membrane anchoring sites with an exogenous PtdIns(3)P binding domain such as the PX domain or the FYVE domain. This repression also blocks enrichment of a number of other proteins required for downstream fusion events, such as the SNARE Vam3p, at the fusion site [71, 72]. It is clear that vacuole membrane fusion involving Vam7p requires a functional PX domain as a membrane anchor, but it is not yet clear why the PX domain has been selected for this purpose in *S. cerevisiae*. It has been suggested that PI binding is actually required for all membrane fusion events and that the proteins and domains involved in doing this vary between different sets of membrane fusion machineries and between different organisms [69].

19.4
Emerging Research Directions and Recent Developments

The mechanisms whereby PX domain function is regulated are certainly still in the process of emerging. At least one type of regulation comes through protein–protein interactions, such as seen in $p47^{phox}$. Other PX domains such as those from Scd2, SNX1, and SNX2 are also known to mediate protein–protein interactions, but the SNX6 interaction with Pim1 may be through its PX domain [54]. It is unclear what effect these interactions have on PI binding. As further characterization of PX domains from other proteins continues, we may see additional protein–protein interactions, some of which may have roles in regulating PI binding.

Phosphorylation of distant regulatory elements can control PI binding by a PX domain, as demonstrated for $p47^{phox}$ [8, 28]. For other proteins, direct phosphorylation of the PX domain may provide another mechanism of regulation. For example, PLD1 is phosphorylated by protein kinase C on T147 in the PX domain variable loop. It is unclear what effect, if any, this has on PX domain function, but mutation of this residue alters PLD1 activation [59, 60]. In the mouse protein FISH, unpublished data suggest an interaction between the PX domain and the third of five SH3 domains. This interaction is disrupted by Src phosphorylation; however, it is unknown where this phosphorylation occurs in vivo [73]. It is tempting to speculate that phosphorylation may provide a direct level of regulation of the function of these or other PX domains.

Finally, understanding the interplay between multiple membrane localization domains on one protein remains a challenge. For example, the PLD proteins employ several membrane interaction mechanisms. Other PX proteins, such as CPK (Figure 19.2) and zinc finger kinase (ZFK), contain additional putative membrane binding domains [2, 74]. Further study is needed to define the order and cooperativity of the respective binding events, as well as the implications of multivalent phospholipid binding for catalytic activity and protein positioning and dynamics on the membrane.

References

1 Voet, D., Voet, J. G., *Biochemistry.* Wiley, New York **1990**.

2 Ponting, C. P., Novel domains in NADPH oxidase subunits, sorting nexins, and PtdIns 3-kinases: binding partners of SH3 domains? *Protein Sci.* **1996**, *5*, 2353–2357.

3 Babior, B. M., NADPH oxidase: an update. *Blood* **1999**, *93*, 1464–1476.

4 Noack, D., Rae, J., Cross, A. R., Ellis, B. A., Newburger, P. E., Curnutte, J. T., Heyworth, P. G., Autosomal recessive chronic granulomatous disease caused by defects in NCF-1, the gene encoding the phagocyte p47-phox: mutations not arising in the NCF-1 pseudogenes. *Blood* **2001**, *97*, 305–311.

5 Schultz, J., Copley, R. R., Doerks, T., Ponting, C. P., Bork, P., SMART: a web-based tool for the study of genetically mobile domains. *Nucleic Acids Res.* **2000**, *28*, 231–234.

6 Hanson, B. J., Hong, W., Evidence for a role of SNX16 in regulating traffic between the early and later endosomal compartments. *J. Biol. Chem.* **2003**, *278*, 34617–34630.

7 Zhong, Q., et al., Endosomal localization and function of sorting nexin 1. *Proc. Natl. Acad. Sci. USA* **2002**, *99*, 6767–6772.

8 Ago, T., Kuribayashi, F., Hiroaki, H., Takeya, R., Ito, T., Kohda, D., Sumimoto, H., Phosphorylation of p47phox directs phox homology domain from SH3 domain toward phosphoinositides, leading to phagocyte NADPH oxidase activation. *Proc. Natl. Acad. Sci. USA* **2003**, *100*, 4474–4479.

9 Stahelin, R. V., Burian, A., Bruzik, K. S., Murray, D., Cho, W., Membrane binding mechanisms of the PX domains of NADPH oxidase p40phox and p47phox. *J. Biol. Chem.* **2003**, *278*, 14469–14479.

10 Hiroaki, H., Ago, T., Ito, T., Sumimoto, H., Kohda, D., Solution structure of the PX domain, a target of the SH3 domain. *Nat. Struct. Biol.* **2001**, *8*, 526–530.

11 Endo, M., Shirouzu, M., Yokoyama, S., The Cdc42 binding and scaffolding activities of the fission yeast adaptor protein Scd2. *J. Biol. Chem.* **2003**, *278*, 843–852.

12 Cheever, M. L., Sato, T. K., de Beer, T., Kutateladze, T. G., Emr, S. D., Overduin, M., Phox domain interaction with PtdIns(3)P targets the Vam7 t-SNARE to vacuole membranes. *Nat. Cell Biol.* **2001**, *3*, 613–618.

13 Xu, Y., Hortsman, H., Seet, L., Wong, S. H., Hong, W., Sorting nexin 3 (SNX3) regulates endosomal function via its PX domain-mediated interaction with phosphatidylinositol 3-phosphate. *Nat. Cell Biol.* **2001**, *3*, 658–666.

14 Kanai, F., Liu, H., Field, S. J., Akbary, H., Matsuo, T., Brown, G. E., Cantley, L. C., Yaffe, M. B., The PX domains of p47phox and p40phox bind to lipid products of phosphoinositide 3-kinase. *Nat. Cell Biol.* **2001**, *3*, 675–678.

15 Ellson, C. D., et al., PtdIns(3)P regulates the neutrophil oxidase complex by binding to the PX domain of p40(phox). *Nat. Cell Biol.* **2001**, *3*, 679–682.

16 Song, X., Xu, W., Zhang, A., Huang, G., Liang, X., Virbasius, J. V., Czech, M. P., Zhou, G. W., Phox homology domains specifically bind phosphatidylinositol phosphates. *Biochemistry.* **2001**, *40*, 8940–8944.

17 Vanhaesebroeck, B., et al., Synthesis and function of 3-phosphorylated inositol lipids. *Annu. Rev. Biochem.* **2001**, *70*, 535–602.

18 Yu, J. W., Lemmon, M. A., All phox homology (PX) domains from *Saccharomyces cerevisiae* specifically recognize phosphatidylinositol-3-phosphate. *J. Biol. Chem.* **2001**, *276*, 44179–44184.

19 Gillooly, D. J., Morrow, I. C., Lindsay, M., Gould, R., Bryant, N. J., Gaullier, J. M., Parton, R. G., Stenmark, H., Localization of phosphatidylinositol 3-phosphate in yeast and mammalian cells. *EMBO J.* **2000**, *19*, 4577–4588.

20 Zhan, Y., Virbasius, J. V., Song, X., Pomerleau, D. P., Zhou, G. W., The p40phox and p47phox PX domains of NADPH oxidase target cell membranes via direct and indirect recruitment by phosphoinositides. *J. Biol. Chem.* **2002**, *277*, 4512–4518.

21 Ago, T., Takeya, R., Hiroaki, H., Kuribayashi, F., Ito, T., Kohda, D., Sumimoto, H., The px domain as a novel phosphoinositide-binding module. *Biochem. Biophys. Res. Commun.* **2001**, *287*, 733–738.

22 Kurten, R. C., Eddington, A. D., Chowdhury, P., Smith, R. D., Davidson, A. D., Shank, B. B., Self-assembly and binding of a sorting nexin to sorting endosomes. *J. Cell Sci.* **2001**, *114*, 1743–1756.

23 Sindic, A., Aleksandrova, A., Fields, A. P., Volinia, S., Banfic, H., Presence and activation of nuclear phosphoinositide 3-kinase C2beta during compensatory liver growth. *J. Biol. Chem.* **2001**, *276*, 17754–17761.

24 Didichenko, S. A., Thelen, M., Phosphatidylinositol 3-kinase c2alpha contains a nuclear localization sequence and associates with nuclear speckles. *J. Biol. Chem.* **2001**, *276*, 48135–48142.

25 Domin, J., Gaidarov, I., Smith, M. E., Keen, J. H., Waterfield, M. D., The class II phosphoinositide 3-kinase PI3K-C2alpha is concentrated in the trans-golgi network and present in clathrin-coated vesicles. *J. Biol. Chem.* **2000**, *275*, 11943–11950.

26 Zhou, C. Z., et al., Crystal structure of the yeast PX-domain protein Grd19p complexed to phosphatidylinositol-3-phosphate. *J. Biol. Chem.* **2003**, in press.

27 Bravo, J., et al., The crystal structure of the PX domain from p40(phox) bound to phosphatidylinositol 3-phosphate. *Mol. Cell* **2001**, *8*, 829–839.

28 Karathanassis, D., Stahelin, R. V., Bravo, J., Perisic, O., Pacold, C. M., Cho, W., Williams, R. L., Binding of the PX domain of p47(phox) to phosphatidylinositol 3,4-bisphosphate and phosphatidic acid is masked by an intramolecular interaction. *EMBO J.* **2002**, *21*, 5057–5068.

29 Lu, J., Garcia, J., Dulubova, I., Sudhof, T. C., Rizo, J., Solution structure of the Vam7p PX domain. *Biochemistry* **2002**, *41*, 5956–5962.

30 Voos, W., Stevens, T. H., Retrieval of resident late-golgi membrane proteins from the prevacuolar compartment of *Saccharomyces cerevisiae* is dependent on the function of Grd19p. *J. Cell Biol.* **1998**, *140*, 577–590.

31 Xu, J., Liu, D., Gill, G., Songyang, Z., Regulation of cytokine-independent survival kinase (CISK) by the Phox homology domain and phosphoinositides. *J. Cell Biol.* **2001**, *154*, 699–705.

32 Du, G., Altshuller, Y. M., Vitale, N., Huang, P., Chasserot-Golaz, S., Morris, A. J., Bader, M. F., Frohman, M. A., Regulation of phospholipase D1 subcellular cycling through coordination of multiple membrane association motifs. *J. Cell Biol.* **2003**, *162*, 305–315.

33 Lundmark, R., Carlsson, S. R., Sorting nexin 9 participates in clathrin-mediated endocytosis through interactions with the core components. *J. Biol. Chem.* **2003**, *278*, 46772–46781.

34 Zheng, B., Ma, Y. C., Ostrom, R. S., Lavoie, C., Gill, G. N., Insel, P. A., Huang, X. Y., Farquhar, M. G., RGS-PX1, a GAP for GalphaS and sorting nexin in vesicular trafficking. *Science* **2001**, *294*, 1939–1942.

35 Cozier, G. E., Carlton, J., McGregor, A. H., Gleeson, P. A., Teasdale, R. D., Mellor, H., Cullen, P. J., The phox homology (PX) domain-dependent, 3-phosphoinositide–mediated association of sorting nexin-1 with an early sorting endosomal compartment is required for its ability to regulate epidermal growth factor receptor degradation. *J. Biol. Chem.* **2002**, *277*, 48730–48736.

36 Peng, G., Huang, J., Boyd, M., Kleinberg, M. E., Properties of phagocyte NADPH oxidase p47-phox mutants with unmasked SH3 (Src homology 3) domains: full reconstitution of oxidase activity in a semi-recombinant cell-free system lacking arachidonic acid. *Biochem. J.* **2003**, *373*, 221–229.

37 Mayer, B. J., SH3 domains: complexity in moderation. *J. Cell Sci.* **2001**, *114*, 1253–1263.

38 Sumimoto, H., et al., Role of Src homology 3 domains in assembly and activation of the phagocyte NADPH oxidase. *Proc. Natl. Acad. Sci. USA* **1994**, *91*, 5345–5349.

39 Ago, T., Nunoi, H., Ito, T., Sumimoto, H., Mechanism for phosphorylation-induced activation of the phagocyte

NADPH oxidase protein p47(phox): triple replacement of serines 303, 304, and 328 with aspartates disrupts the SH3 domain-mediated intramolecular interaction in p47(phox), thereby activating the oxidase. *J. Biol. Chem.* **1999**, *274*, 33644–33653.

40 Huang, J., Kleinberg, M. E., Activation of the phagocyte NADPH oxidase protein p47(phox): phosphorylation controls SH3 domain-dependent binding to p22(phox). *J. Biol. Chem.* **1999**, *274*, 19731–19737.

41 Worby, C. A., Dixon, J. E., Sorting out the cellular functions of sorting nexins. *Nat. Rev. Mol. Cell Biol.* **2002**, *3*, 919–931.

42 Teasdale, R. D., Loci, D., Houghton, F., Karlsson, L., Gleeson, P. A., A large family of endosome-localized proteins related to sorting nexin 1. *Biochem. J.* **2001**, *358*, 7–16.

43 Phillips, S. A., Barr, V. A., Haft, D. H., Taylor, S. I., Haft, C. R., Identification and characterization of SNX15, a novel sorting nexin involved in protein trafficking. *J. Biol. Chem.* **2001**, *276*, 5074–5084.

44 Parks, W. T., et al., Sorting nexin 6, a novel SNX, interacts with the transforming growth factor-beta family of receptor serine-threonine kinases. *J. Biol. Chem.* **2001**, *276*, 19332–19339.

45 Kurten, R. C., Cadena, D. L., Gill, G. N., Enhanced degradation of EGF receptors by a sorting nexin, SNX1. *Science* **1996**, *272*, 1008–1010.

46 Haft, C. R., de la Luz Sierra, M., Barr, V. A., Haft, D. H., Taylor, S. I., Identification of a family of sorting nexin molecules and characterization of their association with receptors. *Mol. Cell Biol.* **1998**, *18*, 7278–7287.

47 Lin, Q., Lo, C. G., Cerione, R. A., Yang, W., The Cdc42 target ACK2 interacts with sorting nexin 9 (SH3PX1) to regulate epidermal growth factor receptor degradation. *J. Biol. Chem.* **2002**, *277*, 10134–10138.

48 Wang, Y., Zhou, Y., Szabo, K., Haft, C. R., Trejo, J., Down-regulation of protease-activated receptor-1 is regulated by sorting nexin 1. *Mol. Biol. Cell* **2002**, *13*, 1965–1976.

49 Barr, V. A., Phillips, S. A., Taylor, S. I., Haft, C. R., Overexpression of a novel sorting nexin, SNX15, affects endosome morphology and protein trafficking. *Traffic* **2000**, *1*, 904–916.

50 Stockinger, W., Sailler, B., Strasser, V., Recheis, B., Fasching, D., Kahr, L., Schneider, W. J., Nimpf, J., The PX-domain protein SNX17 interacts with members of the LDL receptor family and modulates endocytosis of the LDL receptor. *EMBO J.* **2002**, *21*, 4259–4267.

51 Otsuki, T., Kajigaya, S., Ozawa, K., Liu, J. M., SNX5, a new member of the sorting nexin family, binds to the Fanconi anemia complementation group A protein. *Biochem. Biophys. Res. Commun.* **1999**, *265*, 630–635.

52 Liu, J. M., Buchwald, M., Walsh, C. E., Young, N. S., Fanconi anemia and novel strategies for therapy. *Blood* **1994**, *84*, 3995–4007.

53 Joenje, H., et al., Evidence for at least eight Fanconi anemia genes. *Am. J. Hum. Genet.* **1997**, *61*, 940–944.

54 Ishibashi, Y., Maita, H., Yano, M., Koike, N., Tamai, K., Ariga, H., Iguchi-Ariga, S. M., Pim-1 translocates sorting nexin 6/TRAF4-associated factor 2 from cytoplasm to nucleus. *FEBS Lett.* **2001**, *506*, 33–38.

55 Sugars, J. M., Cellek, S., Manifava, M., Coadwell, J., Ktistakis, N. T., Hierarchy of membrane-targeting signals of phospholipase D1 involving lipid modification of a pleckstrin homology domain. *J. Biol. Chem.* **2002**, *277*, 29152–29161.

56 Sciorra, V. A., Rudge, S. A., Wang, J., McLaughlin, S., Engebrecht, J., Morris, A. J., Dual role for phosphoinositides in regulation of yeast and mammalian phospholipase D enzymes. *J. Cell Biol.* **2002**, *159*, 1039–1049.

57 Powner, D. J., Wakelam, M. J., The regulation of phospholipase D by inositol phospholipids and small GTPases. *FEBS Lett.* **2002**, *531*, 62–64.

58 Jang, I. H., et al., The direct interaction of phospholipase C-gamma 1 with phospholipase D2 is important for epidermal growth factor signaling. *J. Biol. Chem.* **2003**, *278*, 18184–18190.

59 Kim, Y., et al., Phosphorylation and activation of phospholipase D1 by protein kinase C in vivo: determination of multiple phosphorylation sites. *Biochemistry* **1999**, *38*, 10344–10351.

60 Kim, Y., et al., Phospholipase D1 is phosphorylated and activated by protein kinase C in caveolin-enriched microdomains within the plasma membrane. *J. Biol. Chem.* **2000**, *275*, 13621–13627.

61 Liu, D., Yang, X., Songyang, Z., Identification of CISK, a new member of the SGK kinase family that promotes IL-3–dependent survival. *Curr. Biol.* **2000**, *10*, 1233–1236.

62 Stephens, L., et al., Protein kinase B kinases that mediate phosphatidylinositol 3,4,5-trisphosphate-dependent activation of protein kinase B. *Science* **1998**, *279*, 710–714.

63 Alessi, D. R., James, S. R., Downes, C. P., Holmes, A. B., Gaffney, P. R., Reese, C. B., Cohen, P., Characterization of a 3-phosphoinositide-dependent protein kinase which phosphorylates and activates protein kinase Balpha. *Curr. Biol.* **1997**, *7*, 261–269.

64 Stokoe, D., et al., Dual role of phosphatidylinositol-3,4,5-trisphosphate in the activation of protein kinase B. *Science* **1997**, *277*, 567–570.

65 Virbasius, J. V., Song, X., Pomerleau, D. P., Zhan, Y., Zhou, G. W., Czech, M. P., Activation of the Akt-related cytokine-independent survival kinase requires interaction of its phox domain with endosomal phosphatidylinositol 3-phosphate. *Proc. Natl. Acad. Sci. USA* **2001**, *98*, 12908–12913.

66 Nilsen, T., Slagsvold, T., Skjerpen, C. S., Brech, A., Stenmark, H., Olsnes, S., Peroxisomal targeting as a tool for assaying protein–protein interactions in the living cell: CISK binds PDK-1 in vivo in a phosphorylation-dependent manner. *J. Biol. Chem.* **2003**, in press.

67 Pelham, H. R., Insights from yeast endosomes. *Curr. Opin. Cell Biol.* **2002**, *14*, 454–462.

68 Sato, T. K., Darsow, T., Emr, S. D., Vam7p, a SNAP-25-like molecule, and Vam3p, a syntaxin homolog, function together in yeast vacuolar protein trafficking. *Mol. Cell Biol.* **1998**, *18*, 5308–5319.

69 Dietrich, L. E., Boeddinghaus, C., LaGrassa, T. J., Ungermann, C., Control of eukaryotic membrane fusion by N-terminal domains of SNARE proteins. *Biochim. Biophys. Acta* **2003**, *1641*, 111–119.

70 Jahn, R., Lang, T., Sudhof, T. C., Membrane fusion. *Cell* **2003**, *112*, 519–533.

71 Wang, L., Merz, A. J., Collins, K. M., Wickner, W., Hierarchy of protein assembly at the vertex ring domain for yeast vacuole docking and fusion. *J. Cell Biol.* **2003**, *160*, 365–374.

72 Boeddinghaus, C., Merz, A. J., Laage, R., Ungermann, C., A cycle of Vam7p release from and PtdIns 3-P–dependent rebinding to the yeast vacuole is required for homotypic vacuole fusion. *J. Cell Biol.* **2002**, *157*, 79–89.

73 Abram, C. L., Seals, D. F., Pass, I., Salinsky, D., Maurer, L., Roth, T. M., Courtneidge, S. A., The adaptor protein fish associates with members of the ADAMs family and localizes to podosomes of Src-transformed cells. *J. Biol. Chem.* **2003**, *278*, 16844–16851.

74 Vassella, E., Kramer, R., Turner, C. M., Wankell, M., Modes, C., van den Bogaard, M., Boshart, M., Deletion of a novel protein kinase with PX and FYVE-related domains increases the rate of differentiation of *Trypanosoma brucei*. *Mol. Microbiol.* **2001**, *41*, 33–46.

75 Thompson, J. D., Higgins, D. G., Gibson, T. J., CLUSTAL W: improving the sensitivity of progressive multiple sequence alignment through sequence weighting, position-specific gap penalties and weight matrix choice. *Nucleic Acids Res.* **1994**, *22*, 4673–4680.

76 Beitz, E., TEXshade: shading and labeling of multiple sequence alignments using LaTeX2 epsilon. *Bioinformatics* **2000**, *16*, 135–139.

77 Subramaniam, S., The Biology Workbench: a seamless database and analysis environment for the biologist. *Proteins* **1998**, *32*, 1–2.

78 DeLano, W. L., The PyMOL Molecular Graphics System. DeLano Scientific LLC. version 0.90. **2003**, http://www.pymol.org/

79 Christopher, J. A., SPOCK: The Structural Properties Observation and Calculation Kit. version 1.0b170. **1998**, http://mackerel.tamu.edu/spock/

80 Palicz, A., Foubert, T. R., Jesaitis, A. J., Marodi, L., McPhail, L. C., Phosphatidic acid and diacylglycerol directly activate NADPH oxidase by interacting with enzyme components. *J. Biol. Chem.* **2001**, *276*, 3090–3097.

81 Groemping, Y., Lapouge, K., Smerdon, S. J., Rittinger, K., Molecular basis of phosphorylation-induced activation of the NADPH oxidase. *Cell* **2003**, *113*, 343–355.

20
Peptide and Protein Repertoires for Global Analysis of Modules

Krzysztof Bialek, Andrzej Swistowski, and Ronald Frank

20.1
Introduction

In this chapter we give an overview on the experimental approaches that have been conceived and are being used to search genomes and proteomes for functional protein domains and their ligands as well as to assess their recognition specificities and functions on a molecular basis. With respect to the empirical nature of search approaches, most of these are very much concerned with the intelligent design and sophisticated preparation of competent domain and ligand repertoires that are suited to represent natural global pools of potential structures and which allow for efficient experimental exploration.

We focus particularly on the experimental design and concepts of the reports considered for this chapter and refer to the literature for more information on the types of domains and ligands that have been discovered this way, because these have been discussed in the preceding chapters of this book. We have organized our chapter according to the origin of the domain/ligand repertoires reported and the principles applied to prepare and screen them. This involves cell biology, biochemistry, molecular biology, and chemistry, which illustrates the interdisciplinary nature of the work in this field of research. Since the complete genome sequences of humans and many model organisms have become available, sequence homology search has become a potent tool for the recognition of domain structures and ligand families and has been used to design dedicated repertoires for further experimental analysis. The important contributions of the bioinformatics approach are well covered by Chapter 21.

This more technical overview is written to help the readers choose an appropriate approach and find useful material for their own research. We do not cover the recent efforts in developing nonnatural and small molecule drug-like ligands for use in therapeutic interference with protein domain function.

Modular Protein Domains. Edited by G. Cesareni, M. Gimona, M. Sudol, M. Yaffe

ISBN: 3-527-30813-X

20.2 Repertoires from Cell Extracts

A cell extract is the most natural source for a repertoire of the proteins to be investigated for function as modules or ligands. Cell extracts are relatively easy to prepare and they have the advantage that all proteins are correctly folded and post-translationally modified. Moreover, proteins in cell extracts may be easily metabolically labeled, for example, with ^{35}S or ^{32}P. Usually, protein extracts are probed with individual module or ligand molecules, and binding partners are isolated by affinity chromatography or pulldown of complexes via a tag at the probe molecule [1–7] or by immunoprecipitation [8]. The separated proteins are fractionated on polyacrylamide gels (one-or two-dimensional electrophoresis) and identified by mass spectrometric peptide fingerprinting, or they may be blotted and inspected with, e.g., antibodies. Such approaches were recently applied in a high-throughput systematic way to analyze the entire yeast proteome and yielded impressively comprehensive protein interaction networks [9, 10]. These networks obviously include many protein module interactions and may reveal novel module/ligand families.

20.3 Repertoires of Proteins Based on Expression Cloning of DNA Libraries

The major source for global domain and ligand repertoires is heterologous expression of coding DNA libraries. A common theme of approaches based on such sources is to construct a physical link between the individual member of the protein repertoire and its coding DNA fragment, as well as a suitable selection process for the module–ligand complexes. Usually, the complex partners are called bait and prey; either one or both may be represented by a global library. A 'DNA tag' always allows for efficient, easy amplification of even extremely minute amounts of module/ligand molecules. Identification is achieved through DNA sequence analysis (DNA sequencing or hybridization with respective complementary DNA probes). Examples are the yeast two-hybrid system [11] and the split-ubiquitin analog [12], phage expression library screens [13], the various molecular display systems [11, 14] used in iterative rounds of affinity enrichment (also called biopanning), and more recently, large-scale expressed protein collections in microtiter plate or microarray formats [15]. cDNA libraries are the most suitable source of global protein repertoires, rarely available in full length, but complementing the full proteome in the form of shorter redundant libraries of fragments. cDNA libraries can be chosen to represent specific cellular states, such as fetal or adult, brain or liver, healthy or oncogenic/pathogenic, etc. Genomic DNA libraries have been used only with simple model organisms such as yeast, *Caenorhabditis elegans*, etc. but are not considered for higher organisms, which carry increasing amounts of noncoding genetic material. A limitation of heterologous expression is the lack of proper post-translational modification, which in some instances can be overcome by using genetically engineered host strains and in others may be introduced in vitro.

Recently, global studies were reported that investigated the protein interactomes of whole organisms nonbiased for domain–ligand interactions. These have established systematic proteome interaction maps utilizing large-scale yeast two-hybrid screens for yeast itself, nematode, and *Drosophila* [16–19]; work on the human interactome is underway. Together with the global protein complex analyses mentioned above (Section 20.2), comprehensive catalogs of protein interactions are currently being assembled. These will be mined with bioinformatics tools for known and new module–ligand interactions.

20.3.1
Ligand Repertoires Used with the Yeast Two-Hybrid System

The DNA for a module of interest is cloned into a bait expression plasmid in-frame with the gene for the DNA binding domain of a transcription factor, and a library of cDNA or fragmented genomic DNA is cloned into a prey expression plasmid in-frame with the gene for the transcription activation domain of that transcription factor. Yeast transformants resulting from transfection or mating which carry both plasmids and produce active interacting partners are identified by activation of a reporter gene such as *lacZ* or a nutritional marker under control of the reconstituted transcription factor. Several examples are listed in Table 20.1.

Tong et al. [34] exploited yeast two-hybrid screens with prey libraries in various formats: a genome-wide array of transformants representing all yeast open-reading-frame fusions and conventional yeast genomic and cDNA libraries. The screens were used to systematically confirm module–ligand interactions, which had been derived from consensus binding motifs and computational network analysis based on phage-display screens of random peptide libraries (see Sections 20.4.1 and 20.4.2). This approach was applied to yeast SH3 domains and used to verify 59 interactions for 39 proteins by matching a phage-display network of 394 interactions among 206 proteins and a yeast two-hybrid network of 233 interactions among 145 proteins.

Table 20.1 Screens of ligand repertoires with the yeast two-hybrid system.

Prey library	*Bait*	*References*
HeLa cDNA	SH2SH3 of c-ABL	20
Human fetal brain cDNA	All 3 SH3 of Nck	21
Human brain cDNA	SH3 of αI spectrin	22
Human placenta cDNA	SH3 of $p130^{Cas}$	23, 24
Human mammary gland cDNA	Bovine SF1	25
p53–/– astrocyte cDNA	SH3-1 and SH3-23 of SETA	26
Yeast genomic DNA	SH3 of Abp1	27
Mouse skin cDNA	SH3 of Sh3yl1	28
Mouse keratinocyte cDNA	Fyn	29
Human lung fibroblast cDNA	UD/SH2/SH3 of c-Src	30
T-cell cDNA and 11.5-day mouse embryo	Human p52 Shc	31
Human brain cDNA	WW1 of YAP	32
Human leukocyte cDNA	EVH1 of Mena(S)	33

20.3.2
Module Repertoires Used with the Yeast Two-Hybrid System

In the reverse approach, when cloning a ligand of interest into a bait plasmid, the same prey libraries can be used to select for modules. Several examples are listed in Table 20.2.

Table 20.2 Screens of module repertoires with the yeast two-hybrid system.

Prey library	*Bait*	*Selected domain*	*References*
HeLa cDNA	C-terminal part (aa 545–1149) of c-Abl[a)]	SH3	35
LB-EBV cDNA	$p47^{phox}$	SH3	36
Mouse brain cDNA	Alix	SH3	37
Rat brain cDNA	Proline-rich tail (aa 1015–1308) of the 145-kDa isoform of synaptojanin	SH3	38
Random primed *C. elegans* cDNA	N-terminal part of human huntingtin	SH3	39
Rat brain cDNA	N-terminal part (aa 1–65) of Pγ-rod	SH3	40
Rat adipose cDNA	N-terminally truncated (aa 392–1129) of JAK2	SH2	41
Human brain cDNA	Cytoplasmic region (aa 790–1356) of KDR	SH2	42
8.5–9.5-day mouse embryo cDNA	Cytoplasmic region (aa 790–1356) of KDR	SH2	43
Human fetal brain cDNA	P413 to P713 fragment of athrophin-1	WW	44
Mixture of oligo(dT) and random Primed mouse P19 cell cDNA	Fragment (aa 292–371) of PEBP2αB1	WW	45
Pretransformed human brain cDNA	Human dendrin/KIAA0749	WW	46
Activated human T cell cDNA	Cytoplasmic region (aa 221–327) of CD2	GYF	47
9.5–10.5-day mouse embryo cDNA	N-terminal fragment (1–172 aa) of mouse Grb10δ	GYF	48
Differentiated TA-1 fat-cell cDNA	N-terminal fragment (1–172 aa) of mouse Grb10δ	GYF	48
Lung cDNA library	51 aa of the C terminus of rat CFTR	PDZ	49
Human fetal brain cDNA	Cytoplasmic domain of IGFIR	14-3-3	50

a) aa = amino acids in the tables.

Table 20.3 Screens of phage expression libraries of ligands.

Vector	*Prey library*	*Bait*	*References*
λEXlox	16-day mouse embryo cDNA	GST-SH3 of c-Abl, cCrk, c-Src, n-Src	51
λEXlox	16-day mouse embryo cDNA	GST-Gads	52
λEXlox	16-day mouse embryo cDNA	^{35}S-labeled GST-SH3-SH3 of Crk	53
λEXlox	16-day mouse embryo cDNA	SH3 domains of Src and Abl fused to N terminus of *E. coli* AP[a)]	54
λEXlox	16-day mouse embryo cDNA	^{32}P-labeled GST-WW of YAP	55
λEXlox	16-day mouse embryo cDNA	EH-AP of intersectin	56
λEXlox	12-, 14-, 16-day mouse embryo cDNA	^{32}P-labeled GST-WW of FE65	57
λEXlox	Oligo(dT) and random primed mouse 10.5-day embryonic limb bud cDNA	GST-WW-WW of FBP11	58
λEXlox	Random primed mouse 10.5-day embryonic limb buds cDNA	GST-WW-WW of FBP30	59
λgt11	Mouse pre-B cell line 22D6 cDNA	Biotinylated GST-SH3 of Abl	60, 61
λgt11	Human spleen cDNA	GST-SH3 of CRK	62
λZAP	Macrophage cDNA	GST-SH3 of Nck	63
λZAP	TPA-differentiated HL60 cells cDNA	GST-SH3 of Nck	64

a) AP = alkaline phosphatase.

20.3.3
Phage Expression Libraries of Ligands

The genomic or cDNA library is cloned primarily into a λ expression vector, packaged into phage particles, and plated onto a bacterial lawn. The phage plaques producing the fusion protein are lifted onto nitrocellulose filters, and these are probed with the labeled or tagged protein module of interest. DNA of positive individual phage plaques is then sequenced. Table 20.3 gives an overview of reported screens.

20.3.4
Phage Expression Libraries of Modules

The λ phage-based cDNA libraries mentioned above have also been used to search for proteins containing modules, by probing the filter lifts with labeled ligands such as synthetic peptides or fragments of proteins that carry functional sites. Several examples are listed in Table 20.4.

Table 20.4 Screens of phage expression libraries of modules.

Vector	*Prey library*	*Bait*	*Selected domain*	*References*
λEXlox	16-day mouse embryo cDNA	Biotin–SGSGILAPPVPPRNTR-NH_2 class II Src SH3 ligand from a phage library [65]	SH3[a)]	66
λEXlox	16-day mouse embryo cDNA	Biotin–SGSGSRLTQSKPPLPPKPSWVSR-NH_2 cortactin SH3 ligand from phage library [67]	SH3[a)]	66
λgt11	Human prostate cancer cDNA	Biotin–SGSGILAPPVPPRNTR-NH_2 class II Src SH3 ligand from a phage library [65]	SH3[a)]	66
λEXlox	16-day mouse embryo cDNA	GST-SH2-SH1-(proline-rich region) of cAbl	SH3	68
λZAP II	*Xenopus laevis* oocyte cDNA	Biotin–SGSGILAPPVPPRNTR-NH_2 class II Src SH3 ligand from a phage library [65]	SH3[a)]	56
λEXlox	10.5-day mouse limb bud embryo cDNA	$(His)_6$-proline-rich fragment of formin (518–750 aa)	SH3, WW	69
λEXlox	16-day mouse embryo cDNA	Mixture of biotinylated DHQ*po*Y*po*YNDFPGKEPP, DHQ*po*YYNDFPGKEPP DHQY*po*YNDFPGKEPP DHQYYNDFPGKEPP ELFDDS*po*YVNIQNLD derived from human p52 Shc	SH2[b)]	70
λgt11	Human brain stem cDNA	^{32}P-labeled C-terminal domain of the hEGFR	SH2	71
λTriplEX	Human brain cDNA	Mixture of biotinylated PGTP$_4$YTVGPGY (WBP-1), YVQP$_4$YPGPM (WBP-2A), PGTPYP$_4$EFY (WBP-2B), G$_3$FPPLP$_4$YPPLG (Ras-GAP)	WW[a)]	72
λTriplEX	Human bone marrow cDNA	Mixture of biotinylated PGTP$_4$YTVGPGY (WBP-1), YVQP$_4$YPGPM (WBP-2A), PGTPYP$_4$EFY (WBP-2B), G$_3$FPPLP$_4$YPPLG (Ras-GAP)	WW[a)]	72
λEXlox	10-day mouse embryo cDNA	Biotin–SNEEP$_4$YEDPYWGNG-NH_2 N-terminal prolin-rich motif of LMP2A	WW[a)]	6
λgt11	Chicken embryo cDNA	Proline-rich peptide composed of 3 tandem copies of the 11 aa from the Rous sarcoma virus Gag p2b sequence	WW	73

a) This approach was named cloning of ligand targets (COLT).
b) This approach was named cloning of receptor targets (CORT).

20.3.5
Phage Display of Protein Ligands

DNA libraries are cloned into phage vectors in-frame with one of the phage coat protein genes. The resulting phages display the fusion protein on their surface and are used to select binding ligands by affinity enrichment ('panning') on immobilized modules. Several interesting reports utilizing this approach are described below; however, it is more extensively used with repertoires of small peptide ligands (see Section 20.4). A very informative comparison of the particular properties of several different phage-display systems was given by Castagnoli et al. [74].

Ligands for SH3 domains were isolated from phage-displayed expression libraries of cDNA from the human epithelial breast carcinoma cell line MCF7 and a murine fibroblast NIH 3T3 cDNA using the T7Select1-1b system. Both libraries (10^7 primary recombinants with an average length of 0.7–0.8 kb) were screened with the human Grb2-GST (GST = glutathione S-transferase) fusion protein as bait [75]. A fragmented yeast genomic DNA library expressed and displayed as N-terminal fusion to the pVIII protein of M13 phage (10^8 independent clones representing more than 50 times the whole yeast genome and presenting yeast protein fragments of average length 40 amino acids) was affinity-selected for ligands bound by the full-length yeast Abp1 protein, which was expressed as GST fusion and immobilized to glutathione–*S*-Sepharose [27]. A fragmented human leukocyte cDNA library (10^8 clones, 0.2–0.5 kb) was fused to the bacteriophage M13 pIII gene. This library was first depleted of nonphosphorylated ligands by passage over Sepharose loaded with the SH2 domains of SHP-2, then was phosphorylated in vitro with fyn kinase and panned against the tandem SH2 domains of SHP-2 [76].

Usually, several rounds of panning with intermediate amplification of phage eluates by bacterial infection are performed to enrich over unspecific binders. Kurakin et al. [77] combined phage display and expression library screening formats to overcome this drawback. After the first panning round, eluted phages are plated onto a bacterial lawn, and plaques are transferred onto a nitrocellulose membrane. The protein target tagged with a reporter such as alkaline phosphatase is then used as a one-step detection reagent to screen for interacting plaques on the membrane. They named this procedure TAIS (target-assisted iterative screening). Using this method with a normal human brain T7 phage-displayed cDNA library and several bait modules (PDZ domain of PSD95; SH3 domains of Src, Abl, and Crk; third WW domain of Nedd4), they identified 12 novel putative and two previously described interactions [77].

20.3.6
Phage Display of Protein Domains

Again, in the reverse of the scheme above, protein modules have been searched for by panning phage-displayed cDNA libraries on immobilized ligand baits. SH3 and WW domains were isolated from an expression cDNA library made from human brain displayed on the lambda D capsid protein (9.8×10^7 primary recombinants) by panning on a GST fusion protein of the proline-rich fragment (residues 1058–1119) of synaptojanin 1 [78]. WW domains were isolated from a normal human brain T7 phage-displayed cDNA library screened with the WW-YAP ligand peptide GTP_4YTVG, which was tethered to a spot on a cellulose membrane. This method will be further advanced into an automated process utilizing SPOT arrays of synthetic peptide libraries as multiple baits to systematically scan fragments from hundreds of proteins for interacting modules [79]. The phosphotyrosine-binding domains SH2 and PTB were isolated from MCF7 and NIH 3T3 phage-displayed libraries (for details see Section 20.3.5), which were screened with the EGFR- and Shc-derived synthetic phosphopeptides (EGFR-pY1068 biotin–DDAFLPVPE*po*YVNQSVPKR, EGFR-pY1148 biotin–SHQMSLDNPD*po*YQQDFFPK, and Shc-pY317 biotin–LFDDPS*po*YVNVQNL) as baits [75]. Kringle domains were isolated from a T7 phage-displayed cDNA expression library (10^6 primary recombinants) constructed with mRNA obtained from prion-infected neuroblastoma (ScN2a) cells and panned on the full-length recombinant mouse prion protein expressing a dominant-negative mutation at codon 218 (recMoPrp(Q218K)) [80].

20.3.7
Protein Arrays as Ligand Repertoires

The availability of fully sequenced and annotated genomes of model organisms opens the opportunity to prepare comprehensive sets of individually expressed and purified proteins as bait or prey repertoires. This has been achieved for 5800 open reading frames of *Saccharomyces cerevisiae*, and from these, a yeast proteome microarray containing approximately 80% of the yeast proteins was constructed. The yeast proteins were fused to GST-HisX6 tags at their N termini and spotted onto nickel-coated glass slides. These arrays were probed, e.g., with calmodulin in the presence of calcium. Six of the 12 known calmodulin targets were detected. Of the remaining six, two were not present in this collection, and four were not detectable with anti-GST antibodies. Thirty-three additional potential partners for calmodulin were identified. Sequence comparison revealed that 14 of the 39 calmodulin-binding proteins contain a motif whose consensus is *fn*(I/L)QxxK(K/x)G[+]*fc* [81].

20.3.8
Protein Arrays as Domain Repertoires

A repertoire of 212 GST fusion proteins and protein domains was collected from various sources: 33 WW, 23 SH3, 17 SH2, 23 PH, 23 PDZ, 8 FF, 7 14.3.3, 5 PTB, 4 FHA, 2 KH, and 67 proteins without a canonical domain structure. These were arrayed onto nitrocellulose-coated glass slides to produce a protein domain chip. Five peptides, including one with methylated arginine residues as an example of a post-translational modification, all with known binding specificity to the domains mentioned above were used to confirm that the arrayed proteins retain their binding integrity. Extracts of human HCF7 cells and antibodies against Sam 68 and SmB′, as well as peptides derived from these proteins, were used to demonstrate distinct and reproducible binding to subsets of this array. In this manner, module-specific profiles for cellular proteins were obtained from the binding patterns. These were, at most, similar to those obtained with the respective peptide ligands and only slightly broader, which is to be expected, since proteins may harbor additional interacting residues [82].

In a different approach, sixty-five human WW domains were expressed as GST-fusion proteins and immobilized on the surface of the wells of a microtiter plate [83]. A quantitative ELISA-like binding assay was used to screen these modules sequentially with 1930 putative synthetic peptide ligands that were selected by searching the SwissProt and TrEMBL protein databases. A cross-affinity matrix that represented all the possible combinations of interactions between the human WW domains and their peptide ligands was established and revealed clear ligand preferences for the WW domains: 31 of them were classified into a subgroup that recognize the *fn*PPxY*fc* motif (elm: LIG_WW_1), 11 of them recognize *fn*PPLP*fc*/ *fn*PPR*fc* (elm: LIG_WW_2/ elm:LIG_WW_3), 3 bind *fn*PGM*fc*/*fn*PPR*fc* (elm: LIG_WW_3), and 4 could not be classified in any subgroup.

20.3.9 Mutagenized Domain Libraries

Site-directed mutagenesis with mutagenic oligonucleotide primers or random mutagenesis by error-prone PCR were used to generate repertoires of domain variants. These repertoires were used to investigate the contributions of individual residues to domain activity and/or to search for domains with new properties. Obviously, all above listed methods for DNA library screens are applicable.

20.3.9.1 Site-directed Mutagenesis

Two bacterial expression libraries of mutated YAP WW1 domains with random sequences at the position corresponding either to L190 and H192 together (theoretical degeneracy = td = 400) or to Q195 alone (td = 20) were constructed in the pGEX-2TK. Colony lifts onto nitrocellulose filters were screened with radio-labeled ligands of YAP (PPxY) or FE65 (PPLP). A single mutation, L190W in the YAP WW1 domain, was found to allow binding to both types of ligands, and an additional substitution, H192G, almost completely shifted the selectivity to that of FE65 [84]. A library of Hck SH3 domains (1.3×10^8 recombinant clones) with a random hexapeptide replacement in the RT loop (termed RRT-SH3) was generated by PCR-assisted mutagenesis using a degenerate primer. This library was displayed on the surface of M13 bacteriophage and used for panning with HIV-1 Nef and Nef-F90R mutant. RRT-SH3 domains that can bind to Nef up to 40 times more avidly than Hck-SH3 were identified. Also, RRT-SH3 domains with an opposite specificity for the Arg90 Nef mutant were selected [85]. A repertoire of 10^7 different SH3 domains was created by grafting the residues that are present on the binding surfaces of natural SH3 domains onto the scaffold of the human Abl-SH3 domain (td $\sim 2 \times 10^9$). The library was displayed by fusion to the C terminus of the D protein of bacteriophage λ and screened by affinity selection with APTYP$_3$LPP and

LSSRPLPTLPSP peptides, which are known ligands for human Abl and Src SH3 domains, respectively. As few as two or three amino acid substitutions can lead to dramatic changes in recognition specificity [86]. A GST-fused SH2-scaffolded repertoire was generated by randomly mutagenizing five critical amino acid residues (C715, L726, G727, L746, and Y747) in the specificity-determining region of the PLCγ C-terminal SH2 domain and cloning into pGEX vector (5×10^4 primary clones). Biotinylated phosphopeptides derived from a natural PLCγ-SH2 ligand (EGDND*po*YIIPLPDPK), as well as unrelated Lck- and Grb2-SH2 ligands (EPQ*po*YEEIPIYL, PSAEL*po*YSNALPVGLP, EPPDHQ*po*Y*po*YNDPFPGKE), were used to screen colony lifts of this library. The isolated SH2 domains possessed binding affinity constants for the selected peptides similar to those of cognate SH2 domains, but surprisingly, all isolated SH2 domains retained high affinity for the natural PLCγ-SH2 ligand [87].

20.3.9.2 **Random Mutagenesis**

A yeast reverse two-hybrid system (interactions activate a toxic reporter) was used to select single amino-acid changes that abolish protein–protein interaction. A prey library of the mutated fragment (159–437) of the E2F1 transcription factor was prepared by mutagenic PCR and screened with the full-length DP1 protein as bait in a two-step selection process in which misfolded variants were deselected in the second step. A putative helical region that is conserved among E2F family members was identified as the essential part of the E2F/DP1 heterodimerization domain [88].

In a similar approach, the PDZ domain of mouse AF-6 was mutated by using error-prone PCR. This library was screened by using the yeast two-hybrid system with four different C-terminal expressed peptides (FHAALGAV, FHAALGKYV, FHTQITRYV, and SPANIYYKV) as baits. None of these peptides interact with the native AF-6 PDZ domain. The majority of mutations leading to high-affinity binding to the new peptide ligands were localized in the three secondary structural elements of the PDZ domain that are known to make up the peptide binding groove [89]. Four Pex5p single-point mutants that abrogate binding to the SH3 domain of Pex13p were used as baits in a yeast two-hybrid screen for suppressor mutants from a randomly mutagenized (error-prone PCR) prey library of the Pex13p SH3 domain. Mutations that restore the interaction with Pex5p were found at a new, unique site on the Pex13p-SH3 domain that is distinct from the classical PxxP-binding pocket [90].

Error-prone PCR was also used to assemble a prey library resulting from the PDZ domain of Omi (PDZ_{Omi}) (7.5×10^5 independent yeast clones). PDZ_{Omi} differs from the other PDZ domains in both amino acid sequence and binding specificity: DIELVMI at the C terminus of Mix2 is the only known interactor for PDZ_{Omi}. This library was screened in a yeast two-hybrid system with the C-terminal part of Myc oncoprotein ($Myc_{282–439}$) containing the sequence LRNSCA at its C terminus. Fifteen mutated PDZ domains that could specifically interact with $Myc_{282–439}$ were isolated [91].

20.4 Repertoires of Peptide Ligands Based on Expression Cloning of Oligonucleotide Libraries

A typical feature of the protein modules discussed in this book is their action by binding or post-translational modification (including cleavage) on short target sequences (functional sites) present in other proteins. Such functional sites can be mimicked by appropriate short peptide sequences of completely synthetic origin. This section recapitulates approaches to generating such synthetic repertoires by cloning the corresponding coding double-stranded DNA repertoires made from synthetic oligonucleotides. The common codon scheme NNK or NNS is used for chemical synthesis of degenerate oligonucleotide sequences, where N is an equimolar mixture of A + C + G + T, K of G + T, and S of G + C. In this way, the codons encoding all 20 amino acids are included, and stop codons are restricted to only TAG [92]. A triplet (codon)-based chemical synthesis of random oligonucleotide repertoires is feasible and could avoid the expression of many nonsense peptides [93], but it is used only rarely because the triplet building blocks and their chemical assembly are not trivial. Often, codons for cysteines are included at predefined positions to allow for cyclization of the expressed peptide structures to form constrained loops.

These oligonucleotide repertoires are primarily expressed as fusions to larger proteins, with the phage-display method clearly dominating (see also Section 20.3.5). The majority of libraries are based on the filamentous phage M13, with fusion to the N terminus of either the minor coat protein pIII (five copies at the tip of the phage particle) or the major coat protein pVIII (approximately 2700 copies forming the body of the phage particle). These two variations are distinct in their exploitation or prevention of avidity effects in the selection of low- or high-affinity binders. Other types of phages are rarely used for peptide display, except when constructing fusions to the C termini of proteins, which is a prerequisite, for example, when selecting binders to PDZ domains. The D protein of the λ phage capsid and the 10B coat protein of the T7 phage are two examples. The T7 display system was also reported in the context of a novel target-assisted iterative screening method (TAIS) [77] as described in Section 20.3.5. As a result, a novel unconventional SH3 binding motif, *fn*Px(P/A)xxR*fc*, was discovered and further characterized. The proper display of peptides at the C terminus of M13 phage coat protein VIII can also be achieved with the help of a special 13-amino-acid long peptide linker. Fusion to the C terminus of the *Escherichia coli* Lac repressor was exploited for a plasmid-display system by Stricker et al. [94]. Following bacterial expression, the Lac repressor protein binds to the *lac* operator sequence on the same plasmid, thereby linking each peptide to the plasmid encoding that peptide. Because of the small particle size, much higher diversities of peptides can be screened with such libraries.

20.4.1
Random Peptide Libraries

Random peptide libraries are of a generic nature and applicable to any screening task. Therefore, many random libraries of different structures with respect to the length of the peptide insert and the presence of constraints by cysteine bridges, that were constructed for other studies, were also utilized for selecting ligands for protein modules. One example is the M13 pVIII displayed nonapeptide library X_9 of Felici et al. [95], which was widely used by many researchers for protein domain–peptide ligand interaction studies. Table 20.5 gives an overview of several reports found in the literature. The Felici library has also been modified through an efficient tyrosine residue phosphorylation using PTK $p55^{fyn}$ and has become an important tool for studying *po*Y-binding modules.

Tong et al. [34] took advantage of this library for a systematic compilation of consensus motifs for 24 predicted yeast SH3 domains. This compilation was then used to derive a protein interaction network that links each recognition module to all those proteins found in the database to contain a preferred peptide site (see also Section 20.3.1).

Table 20.5 Screens of random peptide repertoires.

Library	*Domain*	*Protein*	*Ligand*	*References*
M13 display at N terminus of phage coat protein III				
GX_6G	SH3	hSrc	RSLPPIPG (see Table 20.6, entry 3)	96
X_7, CX_7C	Chromo-shadow	HP1	*fn*(P/L)(W/R/Y)VΦΦ*fc*, linear and constr.[a)]	97
CX_9C	SH2	Grb2	*fn*YxN*fc*, constr.	98
$X_{10}C$	EH	frog intersectin	*Fn*NPF*fc*, constr.	56
$X_{10}C$		β_1-syntrophin	*fn*x[+]ETC(L/M)AgxΦC*fc*, constr.	99
X_{12}	WW	hWWP1, hWWP2, hWWP3, hYAP, yRsp5.3, mNedd4	*fn*(L/P)PxY*fc*, class I (see Table 20.6, entry 7)	100
X_{12}	SH3-1, SH3-2	CIN85	*fn*Px(P/A)xxR*fc*	101
X_{15}	SH3	Src	*fn*RPLPxxP*fc*	102
$X_4CX_{10}CX_4$	SH2	Grb7	*fn*YxN*fc*, constr. (see Table 20.6, entry 8)	103
$X_2CX_{14}CX_2$		h14-3-3τ	*fn*WLDL*fc*, constr.	104
X_2CX_{18}		h14-3-3τ	*fn*RSx{1,3}[−]*fc*, linear, constr.	
X_{10}(G/C/R/S) G(V/A/D/G)-X_{10}[b)], $X_{18}PGX_{18}$[b)]	SH3	Src	*fn*RPLPPLP*fc* (see Table 20.6, entry 2)	65

Table 20.5 (continued)

Library	*Domain*	*Protein*	*Ligand*		*References*
M13 display at N terminus of phage coat protein VIII					
X_9	SH3	h amphiphisin	*fn*RPxR*fc*		105
		h endophilin-2	*fn*P[+]RPPxPR*fc*		
X_9	EH1–3	mEps15, mEps15R	*fn*NPF([+]/Φ)*fc* *fn*(N/H)(P/H)F[+]*fc*		106
X_9	EH1	mEps15	*fn*ζNPF([+]/Φ)*fc*	class I	107
	EH2		*fn*ζNPF([+]/Φ)*fc*	class I	
	EH3		*fn*FW*fc*, *fn*NPFW*fc*	class II	
	EH1	mEps15R	*fn*NPFR*fc*	class I	
	EH2		*fn*ζNPF([+]/Φ)*fc*	class I	
	EH3		*fn*(S/T)NPFRQ*fc*	class I	
	EH1	yYBL047C	*fn*Yxxx(F/L)WRP*fc*, *fn*NPF*fc*	class II, class I	
	EH2		*fn*WWxxAD*fc* *fn*NPF*fc*	class II class I	
	EH3		*fn*Rxx*NPFR**fc*	class I	
	EH2	yPan1	*fn*NNPFxD*fc*	class I	
	EH1	yEnd3	*fn*SWGxxxW*fc* *fn*H(T/S)F*fc*	class II	
X_9	WW[c)]	utrophin	*fn*PPxY*fc*		108
X_9	SH3	hEps8, mEps8, hEps8R2, mEps8R2	*fn*PxxDY*fc*		109
X_9	EH	frog intersectin	*fn*NPF*fc*		56
X_9[d)]	PTB	Shc	*fn*F/YxNPT*po*YxxY/W*fc*		110
X_9[d)]	SH2	Shc, Sli, Rai	*fn*N(I/V)*po*Y(E/G)T(I/V/L)(W/F)*fc* *fn*(L/I/V)*po*Y(E/G)T(W/Y/F)*fc* *fn*(V/I)*po*Y(E/G)(T/Y)(I/L)*fc*		111
X_{10}, X_{12}, CX_8C	SH2	hGrb2	*fn*YxN*fc*, linear, constr. (see Table 20.6, entry 9)		112

a) Constr. = sequence constrained via disulfide bridging between cysteines.
b) The non-X positions resulted from the cloning strategy used to generate the oligonucleotide library.
c) Interaction between peptides and WW domain occurred only in the presence of EF hands and ZZ domain.
d) Modified through an efficient tyrosine residue phosphorylation using PTK $p55^{fyn}$.
e) Peptides were fused to the C terminus of the *Escherichia coli* lac repressor.

Table 20.5 (continued)

Library	*Domain*	*Protein*	*Ligand*		*References*
M13 display at C terminus of phage coat protein VIII					
X_7	PDZ2 PDZ3	MAGI3 kinase	*fn*(T/S)WV *fn*FDI	class I class II	113
λ phage display at C terminus of the D protein					
X_9	PDZ1	hINDAL	*fn*(V/R)ΩWΦ	class II	114
	PDZ2		*fn*EΩDF	class II	
	PDZ3		*fn*ΩDΦ, *fn*Ω(D/E)	class II class IV	
	PDZ4		*fn*EΦΩV	class II	
	PDZ5		*fn*(S/T)W(V/L)	class I	
	PDZ6		*fn*ΦSΩV	class I	
	PDZ7		*fn*SxV	class I	
The T7 phage display at C terminus of the 10B phage coat protein					
X_{16}	SH3-1 SH3-2 SH3-3	CIN85	*fn*Px(P/A)xxR*fc*		101
Plasmid display					
X_{15}[e]	PDZ3	PSD95	*fn*E(T/S)xV		94
X_{15}[e]	PDZ	nNOS	*fn*DxV		94

20.4.2
Dedicated Peptide Libraries

The rather short general consensus motifs identified by screening the random libraries above often failed to explain the specificity of proteins pulled down from cell extracts [115]. Specially constructed phage libraries displaying peptide ligand repertoires biased for particular domains have been constructed for more detailed analyses of ligand preferences. Table 20.6 summarizes these experiments, the majority of which were follow-up studies to the experiments listed in Table 20.5. However, the majority of these analyses have been conducted by using chemical peptide synthesis (see Section 20.5). Schmitz et al. [116] prepared a peptide repertoire X_3YX_4 of eight amino acids at the N terminus of M13 pIII to search for tyrosine kinase substrates. After enzymatic phosphorylation of the phage library, this repertoire was then suitable for studying *po*Y-binding modules. A library similar to that of entry 5 (Table 20.6), X_4PxxPX_4, was used by Tong et al. [34] in their systematic yeast SH3 domain study (see also Section 20.3.1.).

Table 20.6 Screens of oriented peptide repertoires.

Entry	*Library*	*Domain*	*Protein*	*Ligand*	*Class*[a]	*References*
M13 display at N terminus of phage coat protein III						
1	X_3YX_4[b]	SH2	Grb2	*fnpo*Y(M/E)NW*fc*	II	118
2	8 (aa)[c]	SH3	Src	*fn*RPLPPLP*fc*	1 R	65
3	X_6PPIPG, RSLRPLX_6, PPPYPPX_6	SH3	hSrc, avian Src, Fyn, Lyn, P13K, mAbl	*fn*RPLPPLPxP*fc* *fn*RPLPP(I/L)P*fc* *fn*RxxRPLPPLPxP*fc* *fn*RxxRPLPPLPPP*fc* *fn*PPPYPPPP(I/V)P*fc*	1 R 1 R 1 R 1 R 1 @	96
4	X_5RPLPPLPPP, RSLRPLPPLPX_5, GAAPPLPPRX_5	SH3	Src, Fyn, Lyn, P13K, Yes	for details see reference	1 R	115
5	X_6PXXPX_6	SH3	Src	*fn*LxxRPLPxΨP*fc*	1 R	67
			Yes	*fn*ΨxxRPLPxLP*fc*	1 R	
			Abl	*fn*PPxΩxPPPΨP*fc*	1 @	
			Cortactin	*fn*[+]PPΨPxKPxWL*fc*	2 K	
			p53bp2	*fn RPX*ΨPΨR[+]SxP*fc*	2 R	
			PLCγ	*fn*PPVPPRPxxTL*fc*	2 R	
			Crk	*fn*ΨPΨLPΨK*fc*	2 K	
			Grb2-N	*fn*[+]ΩxPLPxLP*fc*	I	
6	X_6PPX_6	WW	hYAP, mYAP	*fn*PPPPYP*fc*	I	119
7	X_6PPX_6	WW	hWWP1, hWWP2, hWWP3, hYAP, yRsp5.3, mNedd4	*fn*(L/P)PxY*fc*	I	100
8	X_4CX_4Y (D/A/E/X)-NX_3CX_4	SH2	Grb7	*fn*Y(D/A/E)N*fc*, constr.[d]	II	103
M13 display at N terminus of phage coat protein VIII						
9	X_5YXNX_8	SH2	hGrb2	for details see reference	II	112

a) When available, subclassification of SH3 modules is according to Cesareni et al. [117].
b) Modified through an efficient tyrosine residue phosphorylation using a mix of phosphotyrosine kinases Blk, c-Src and Syk [115].
c) Library containing eight-amino-acid long peptides encoded by the DNA sequence $((C/A)NN)_8$ which restricted the amino acid compositions to the following proportions: 6R:4P:4L:3I:2H:2Q:2K:2N:2S:1M.
d) Constr. = sequence constrained via disulfide bridging between cysteines.

20.5
Synthetic Peptide Repertoires

Combinatorial and simultaneous parallel chemical synthesis techniques have been developed and refined since the early 1980s and have been applied very successfully to the preparation of peptide repertoires for studying molecular recognition events in immunology. This approach concerns the discovery and analysis of immunogenic sites (epitopes) of protein antigens recognized by antibodies and T cells [120]. Principally, the short functional sites recognized by the protein modules discussed in this book can be likewise considered to be linear epitopes. Thus, the sophisticated approaches from immunology for library design and assay formats apply directly here.

Peptides are assembled through stepwise coupling of amino acid derivatives, starting from the C terminus tethered to a solid support material. There is a large variety of support materials and formats available, with PEG–polystyrene (Rapp-Tentagel), polyethylene pins, and cellulose membranes dominating the field. In addition to the 20 genetically coded amino acids, any nonnatural synthetic derivative can be incorporated. Random positions in a library can be freely defined with respect to location and number in the sequence, as can the composition of the amino acid residues, easily yielding theoretical diversities from a few hundred to several billion peptides. These positions are generated either by coupling a mixture of amino acid derivatives or by mixing together portions of support material from appropriate single amino acid coupling reactions (the 'mix and split' approach). Cysteines are often omitted or left protected (Acm) to avoid random disulfide scrambling. After assembly and deprotection, peptides may be cleaved from the support and assayed in solution or are left tethered to the support for in situ solid-phase binding assays. For more information see [121].

These synthetic opportunities have been exploited to generate and screen peptide libraries in many approaches (for a review see [122]), such as (1) selection from the full random pool; (2) iterative or recursive screening by various sublibrary pooling strategies with a combination of one or more defined amino acid positions on a random background, with the best hit of one screening round then being used as the start for the next round; (3) positional scanning of all consecutive positions by single amino acid incorporations or dual positional scanning for two neighboring positions; (4) orthogonal (self-deciphering) pools; (5) epitope mapping (or peptide walking), the systematic scanning of a protein sequence with overlapping peptide fragments; (6) replacement analysis, the systematic single substitution of each epitope residue by one mock amino acid (e.g., an alanine scan); (7) replacement net analysis, the systematic single substitution of each epitope residue by all other amino acid residues; (8) motif scanning, the systematic editing of an epitope sequence on an all-alanine (A) or all-random (X) background to define consensus recognition motives; and (9) finally, large collections of individual peptides derived by searching genomic sequences with epitope-prediction algorithms.

Table 20.7 Screens of soluble peptide libraries.

Library	*Specifications*	*Theoretical degeneracy*	*Screened domain*	*Protein of origin*	*References*
$APXYSP_5$		20	SH3	Abl	123
$APTXSP_5$		20	SH3	Abl	123
$APTYSPXP_3$		20	SH3	Abl	123
$MAX_4YX_4AK_3$	X–C,W	$1.1 \cdot 10^{10}$	PTB	dNumb	124
***po*Y-oriented**					
$G(F_2Pmp)XPXS$-amide	X–C; F_2Pmp[a] as mimic of *po*Y	361	SH2	Pl 3-kinase	125
GDG*po*YX_3SPL_3	X–C,W	5832	SH2	22 proteins	126, 127
GAX_3*po*YX_3K_3	X–C,W	$3.4 \cdot 10^7$	PTB	Shc, Cbl	128, 129
MAX_3NXX*po*$YXAK_2$	X–C,W	$3.4 \cdot 10^7$	PTB	ShcC	130
***po*S/*po*T-oriented**					
MAX_3*po*SX_3AKK	X–C	$4.7 \cdot 10^7$	14-3-3	8 proteins	131
MAX_4*po*SXPXXAKK	X–C	$8.9 \cdot 10^8$	14-3-3	8 proteins	131
MAX_3RXXR*po*SXPAKK	X–C	$4.7 \cdot 10^7$	14-3-3	8 proteins	131
MAXXRXX*po*SXPAKK	X–C	$2.5 \cdot 10^6$	14-3-3	8 proteins	131
MASX*po*SXXAKK	X–C	$6.9 \cdot 10^3$	14-3-3	8 proteins	131
ISRST*po*SX_3NK	X–C	$6.9 \cdot 10^3$	BRCT	5 proteins	132
KAX_3*po*SX_3AK	X–C	$4.7 \cdot 10^7$	BRCT	BRCA1	132
MAX_4*po*TX_4AK_3	X–C,W	$1.1 \cdot 10^{10}$	FHA	6 proteins	133
MAX_4*po*$TXXIXXAK_3$	X–C	$1.7 \cdot 10^{10}$	FHA	3 proteins	133
P-oriented					
$MAX_4PXXPX_3AK_3$	X–C,W	$2 \cdot 10^{11}$	SH3	amphiphysin	134
$MAX_4PXXPX_3AK_3$	X–C,W	$2 \cdot 10^{11}$	WW	4 proteins	59
MAX_4PX_4AKK	X–C	$1.7 \cdot 10^{10}$	WW	5 proteins	59
$MAX_4PPRX_4AK_3$	X–C	$1.7 \cdot 10^{10}$	WW	3 proteins	59
$MAX_4YX_4AK_3$	X–C,Y	$1.1 \cdot 10^{10}$	WW	YAP	59
C-terminal–oriented					
KNX_8-COOH	X–C,W	$1.1 \cdot 10^{10}$	PDZ	6 proteins	135
KNX_6(S/T/Y)XX-COOH	X–C,W	$1.1 \cdot 10^{10}$	PDZ	5 proteins	135

[a] F_2Pmp = (phosphonodifluoromethyl)-phenylalanine.

20.5.1 Soluble Peptide Libraries as Ligand Repertoires

Screening for module binding is achieved by affinity chromatography selection with an immobilized domain or protein of interest. The eluted peptides are sequenced or analyzed by mass spectrometry as a mixture. This readily yields the consensus binding motif as result of the pool analysis. The random library, however, has to be oriented to one or more residues that are decisive for binding, to avoid phase problems if the interacting residues on different peptides are located at different positions. Table 20.7 summarizes some library screens reported in the literature.

20.5.2 Bead-bound Peptide Libraries as Ligand Repertoires

A widely applied format of peptide libraries for bioscreening is the one-bead–one-compound (OBOC) combinatorial library format synthesized on resin beads by the mix-and-split technique (for a review see [136]). The beads are of about 100 μM in size and carry about 100 pmol of a peptide sequence; one gram of resin contains about 3×10^6 beads. They are screened by incubation with a domain or protein of interest, and bound protein is detected by various labels. The positive labeled beads are manually isolated with the help of a microscope and tweezers and are then individually sequenced by Edman degradation. A biased combinatorial library of peptides having the form x_3PPxPxx–resin (x–C; td ~4.7×10^7) was synthesized, and about 2×10^6 beads were screened with the fluorescently labeled SH3 domains of PI3K and Src [137, 138]. A combinatorial library Ax_3*po*Tx_3ABBRM–resin (x–C,M + norleucine as a substitute for M; td ~4.7×10^7) was screened with the biotinylated FHA1 and FHA2 domains of yeast Rad53 to determine their binding requirements [139, 140]. A phosphotyrosine-oriented peptide library DExx*po*Yx_3IBBRM–resin (x–C,M + norleucine; td = 2.5×10^6) was synthesized and about 2.86×10^6 beads were assayed with the N- and C-terminal SH2 domain of SHP-1 [141].

If much smaller beads (~10 μm) are used, rapid FACS (fluorescence-activated cell sorting) technology can be exploited to separate fluorescence-positive beads. However, the peptide material per bead is no longer sufficient for sequencing and, thus, other strategies are required to identify the ligands. Pool sequencing of the sorted beads was used to reveal the binding consensus of the Grb2 and Syk SH2 domain screened with a phosphotyrosine-oriented peptide library EPX^6*po*Y$X^{19}X^7X^{19}X^7X^6$ (the superscript numbers refer to the number of defined amino acids that were coupled at each position (td ~6.4×10^5); for a detailed matrix representation see [142]). A recursive deconvolution strategy was used to investigate the Grb2 SH2 domain with a phosphotyrosine-oriented peptide library AS*po*Yx_4SA (td = 1.6×10^5). First, 20 sublibraries AS*po*Y$O^{+1}x_3$SA with one of the 20 amino acids fixed at position O^{+1} were synthesized and incubated with a mixture of differently labeled target module and negative reference modules. The number of beads that sorted positive for a particular amino acid–module combination was taken as a binding preference scale. In a series of synthesis and screening rounds, amino acid preferences for the consecutive random positions were determined similarly, taking the best hit of each previous screen. To select a high-affinity motif, the library was screened for beads that bind Grb2 but not the Grb2 R67H mutant. To search for the most selective ligand, Grb2-positive beads were sorted that did not bind to the SH2 domains of hAbl, Nck, bPl3Kp85-N, bPl3Kp85-C, mSHP-2N, vCrc, and Grb2 R67H (used as a negative reference mixture) [143].

20.5.3
Peptide Arrays as Ligand Repertoires

Synthetic peptides displayed at high density as an array on a planar support are a potent alternative tool for repertoire screening (for a recent review see [144]). Currently, the major technique for in situ synthesis and screening by this approach is the SPOT method. Individual peptide preparations are easily synthesized at different locations on a cellulose membrane by distribution of small (less than microliter) volumes of the appropriate solutions of activated amino acid derivatives or amino acid mixtures per spot and peptide elongation cycle [145]. Up to 2000 peptides or peptide pools can be synthesized on a single 8×12 cm sheet of cellulose paper (as appropriate for a microtiter plate) and probed simultaneously with a domain or protein of interest. Positive spots in the probed array immediately indicate the sequence of the peptide(s) recognized. Relative intensities of binding signals measured on a SPOT peptide array are, to a good degree, representative of relative affinities (K_d values), as proven by Biacore measurements [146] and competitive ELISA [147]. SPOT synthesis is used to construct various types of libraries.

20.5.3.1 Sublibrary Pools for Iterative A Priori Deconvolution

A library XXO$_a$O$_b$SXV-COO$^-$ (where X = all amino acids combined, except for C, M, and W, and O = all individual natural 20 amino acids) was synthesized as an array of the 400 sublibraries for the O$_a$O$_b$ combinations, each with a td of $17^3 = 4913$. This library was used to determine the amino acid preference of the syntrophin PDZ domain in the −3 and −4 positions with respect to the essential C-terminal Val residue. To display the free carboxy terminus of the peptides, they had to be tethered to the support by the N terminus. This was achieved first by coupling the N terminus of the finished peptide to the support in a cyclization reaction and then cleaving the C-terminal linker. Two descendant libraries xOKESxV-COO$^-$and xxKESOV-COO$^-$ (each sublibrary with td = 289) were synthesized to define requirements at the −1 and −5 positions. Also, heptamer sublibraries of the type x$_4$O$_a$xO$_b$-COO$^-$, x$_5$O$_a$O$_b$-COO$^-$, x$_3$O$_a$xxO$_b$-COO$^-$, x$_3$O$_a$xO$_b$x-COO$^-$, and so forth (each sublibrary with td $\sim 1.4 \times 10^6$) were synthesized, but relevant binding affinities could not be detected [148]. In a different experiment, an oriented library Ax$_3$(*po*S/*po*T)x$_3$A was used to evaluate the binding specificity of four SH2 domains. The importance of different amino acid residues at any given position was estimated by using a positional scanning library format. A total of $19 \times 8 = 152$ sublibraries were spotted for the Ax$_4$(*po*S/*po*T)x$_4$A library; each sublibrary had a td $\sim 1.8 \times 10^9$. This method was designated the oriented peptide array library (OPAL) strategy [149].

20.5.3.2 Protein Scanning Repertoires (Peptide Walking)

With the increasing availability of genomic sequence information and of data from protein–protein interactions, the peptide walking approach is becoming more widely used. Libraries are characterized by peptide length/offset (offset is the number of amino acid residues by which overlapping peptide fragments are shifted along the protein sequence). A 15/6 scan of the *Listeria monocytogenes* ActA protein sequence (amino acids 30–614) was probed with radiolabeled VASP and led to the identification of the *fn*[–]FPPPPx[–]*fc* binding motif of the EVH1 domain [150]. A 15/3 scan of the human zyxin sequence was probed with the EVH1 domain of Mena, and homologous binding sites were found [151]. A 15/3 scan of the 140-kDa isoform of Mena revealed several binding sites for the FE65 WW domain, all containing the PPLP sequence [57]. A 15/1 scan of the C-terminal domain of mGluR5 (amino acids 1138–1169) identified the binding site LTPPSPF for the mVesl EVH1 domain, which recognizes the reverse of the Mena EVH1motif [152]. The binding site PSIDRSTKP of the Gads C-terminal SH3 domain on SLP-76 (amino acids 181–291) was identified with a 15/4 scan [52]. An 11/2 scan of human synaptojanin (amino acids 1060–1312) was probed with the SH3 domains of amphiphysin and endophilin-2; several binding sites were identified, some matching the known consensus motifs and others not [105].

20.5.3.3 Replacement Repertoires

Once a peptide ligand is identified, the contribution of each single residue to the affinity and selectivity of the binding motif can be examined by systematic replacement approaches. Several examples are listed in Table 20.8.

20.5.3.4 Genome/Proteome Scanning

Literature and database searches for functional sites of protein modules are an increasingly important and rich source of peptide ligands that can be synthesized and investigated using the SPOT method. A repertoire of 70 peptides that were described as SH3 or WW ligands was probed with VASP to assess the selectivity of the EVH1 domain [150]. An array of all C-terminal heptapeptide sequences of human proteins found in SwissProt release 34 (3514 sequences) was synthesized with a free carboxyl end by Hoffmüller et al. [157] and probed with the HRP-labeled PDZ domain of syntrophin. Landgraf et al. [158] took the binding motif cores of eight yeast SH3 domains that were obtained by phage-display experiments [34] to define for each one a less-selective, relaxed pattern. The *Saccharomyces* Genome Database was scanned for peptides matching these patterns. Approximately 1500 peptides for each domain were selected for synthesis and probed with the corresponding SH3 domains. This procedure was called WISE (whole interactome scanning experiment). They reproduced this approach also with the SH3 domains of two human proteins, amphiphysin 1 and endophilin 1, for which peptide ligands were selected from the SwissProt/TREMBL database. A total of 3774 peptides were synthesized and tested with these two SH3 domains.

Table 20.8 Screens of replacement repertoires.

Protein–ligand	*Binding site*[a]	*Replacements*	*Domain*	*Protein*	*References*
SLP-76	**PSIDRSTK**P	alanine scan	SH3	Gads	52
APP	**QNGYENPTYKFFEQMQN**	alanine scan	PTB	Dab1	153
Clone 13	**P_7LPAP$_3$QP**	valine scan	WW	FE65	57
Clone 13	P_7**L**PAP$_3$QP	net	WW	FE65	57
APP	QN**GYENPTY**KFFEQMQN	net (except C)	PTB	Dab1	153
ActA	SF**EFPPPPT**D	net	EVH1	VASP	150
mGluR5	LEELVA**LTPPSPF**RD	net	EVH1	mVesl	152
WBP-1	**GTPPPPYTVG**	net	WW	hYAP	154, 155
Artificial	**CPKPPKYPKK**	net	WW	hYAP	119
SLP-76	**PSIDRSTKP**	net	SH3	Gads	52
Synaptojanin	**LPIRPSRAPSR**	net	SH3	amphiphisin, endophilin 2	105
Synaptojanin	**LEPKRP$_4$RP**	net	SH3	amphiphisin, endophilin 2	105
α subunit, the skeletal muscle VGSCs	**GVKESLV**	net	PDZ	syntrophin	148
BACH1	**SRST*po*SPTFNK**	net (except C)	BRCT	BRACA1	132
ActA	**SFEFPPPPTEDEL**	net	EVH1	VASP	156
ActA	**EFPPPPTEDELEII**	net	EVH1	VASP	156

a) aa positions replaced are in bold.

20.5.4
Peptide Arrays as Domain Repertoires

The WW domain, the smallest of all known protein modules, was the subject of a synthetic approach to study structure–activity relations. A replacement net repertoire (cysteine excluded) of 837 individual variants of the 44-residue large hYAP WW domain was synthesized by the SPOT method. This array was probed with peroxidase-labeled EYPPYP$_4$YPSG peptide [159]. Another repertoire of 11 859 variants of the hYAP WW domain was assembled by a combination of the SPOT method with native chemical peptide ligation. One set of $19^3 = 6859$ WW domains (38-mers) having simultaneous substitutions at positions 30, 32, and 35 (making up the ligand binding pocket), by any of the 19 proteinogenic amino acids (excluding C). A second set of 5000 variants was synthesized, bearing combinations of 19 proteinogenic and 20 nonproteinogenic and phosphorylated amino acids in positions 30, 32, and 35. These arrays were probed with 22 different dye-labeled peptide ligands GTP_4xTVG (x–C, but including *po*Y, *po*S, and *po*T) [160].

In conclusion, a variety of techniques, both biologically- and chemically-based, are available for the experimental determination of peptide and protein binding-partners of modular domains.

Acknowledgements

We are thankful to Dr. Anton Dikmans for critically reading the manuscript.

References

1 Weng, Z., Taylor, J. A., Turner, C. E., Brugge, J. S., Seidel-Dugan, C., Detection of Src homology 3-binding proteins, including paxillin, in normal and v-Src–transformed Balb/c 3T3 cells. *J. Biol. Chem.* **1993**, *268*, 14956–14963.

2 Weng, Z., et al., Identification of Src, Fyn and Lyn SH3-binding proteins: implications for a function of SH3 domains. *Mol. Cell Biol.* **1994**, *14*, 4509–4521.

3 Weng, Z., Rickles, R. J., Feng, S., Richard, S., Shaw, A. S., Schreiber, S. L., Brugge, J. S., Structure–function analysis of SH3 domains: SH3 binding specificity altered by single amino acid substitutions. *Mol. Cell Biol.* **1995**, *15*, 5627–5634.

4 Lu, P. J., Zhou, X. Z., Shen, M., Lu, K. P., Function of WW domains as phosphoserine- or phosphothreonine-binding modules. *Science* **1999**, *283*, 1325–1328.

5 Hussain, N. K., Yamabhai, M., Bhakar, A. L., Metzler, M., Ferguson, S. S. G., Hayden, M. R., McPherson, P. S., Kay, B. K., A role for epsin N-terminal homology/AP180 N-terminal homology (ENTH/ANTH) domains in tubulin binding. *J. Biol. Chem.* **2003**, *278*, 28823–28830.

6 Winberg, G., et al., Latent membrane protein 2A of Epstein–Barr virus binds WW domain E3 protein–ubiquitin ligases that ubiquitinate B-cell tyrosine kinases. *Mol. Cell Biol.* **2000**, *20*, 8526–8535.

7 Ikeda, M., Ikeda, A., Longan, L. C., Longnecker, R., The Epstein–Barr virus latent membrane protein 2A PY motif recruits WW domain-containing ubiquitin–protein ligases. *Virology* **2000**, *268*, 178–191.

8 Liu, S. K., Fang, N., Koretzky, G. A., McGlade, C. J., The hematopoietic-specific adaptor protein gads functions in T-cell signaling via interactions with the SLP-76 and LAT adaptors. *Curr. Biol.* **1999**, *9*, 67–75.

9 Gavin, A.-C., et al., Functional organization of the yeast proteome by systematic analysis of protein complexes. *Nature* **2002**, *415*, 141–147.

10 Ho, Y., et al., Systematic identification of protein complexes in *Saccharomyces cerevisiae* by mass spectrometry. *Nature* **2002**, *415*, 180–183.

11 Allen, J. B., Walberg, M. W., Edwards, M. C., Elledge, S. J., Finding prospective partners in the library: the two-hybrid system and phage display find a match. *Trends Biochem. Sci.* **1995**, *20*, 511–516.

12 Johnsson, N., Varshavsky, A., Split ubiquitin as a sensor of protein interactions in vivo. *Proc. Natl. Acad. Sci. USA* **1994**, *91*, 10340–10344.

13 Sambrook, J., Fritsch, E. F., Maniatis, T., *Molecular Cloning: A Laboratory Manual*. Cold Spring Harbor Laboratory Press, Plainview, NY **1989**.

14 Fisch, I., Peptide display in functional genomics. *Comb. Chem. High Throughput Screen* **2001**, *4*, 157–169.

15 MacBeath, G., Protein microarrays and proteomics. *Nat. Genet. Suppl* **2002**, *32*, 526–532.

16 Uetz, P., et al., A comprehensive analysis of protein–protein interactions in *Saccharomyces cerevisiae*. *Nature* **2000**, *403*, 623–627.

17 Ito, T., Chiba, T., Ozawa, R., Yoshida, M., Hattori, M., Sakaki, Y., A comprehensive two-hybrid analysis to explore the yeast protein interactome. *Proc. Natl. Acad. Sci. U S A* **2001**, *98*, 4569–4574.

18 Li, S., et al., A map of the interactome network of the metazoan *C. elegans*. *Science* **2004**, *303*, 540–543.

19 Giot, L., et al., A protein interaction map of *Drosophila melanogaster*. *Science* **2003**, *302*, 1727–1736.

20 Zhu, J., Shore, S. K., c-ABL tyrosine kinase activity is regulated by association with a novel SH3-domain–binding protein. *Mol. Cell Biol.* **1996**, *16*, 7054–7062.

21 Matuoka, K., Miki, H., Takahashi, K., Takenawa, T., A novel ligand for an SH3 domain of the adaptor protein Nck bears an SH2 domain and nuclear signaling motifs. *Biochem. Biophys. Res. Commun.* **1997**, *239*, 488–492.

22 Ziemnicka-Kotula, D., Xu, J., Gu, H., Potempska, A., Kim, K. S., Jenkins, E. C., Trenkner, E., Kotula, L., Identification of a candidate human spectrin Src homology 3 domain-binding protein suggests a general mechanism of association of tyrosine kinases with the spectrin-based membrane skeleton. *J. Biol. Chem.* **1998**, *273*, 13681–13692.

23 Kirsch, K. H., Georgescu, M.-M., Ishimaru, S., Hanafusa, H., CMS: an adapter molecule involved in cytoskeletal rearrangements. *Proc. Natl. Acad. Sci. USA* **1999**, *96*, 6211–6216.

24 Kirsch, K. H., Georgescu, M.-M., Hanafusa, H., Direct binding of $p130^{Cas}$ to the guanine nucleotide exchange factor C3G. *J. Biol. Chem.* **1998**, *273*, 25673–25679.

25 Zhou, D., Quach, K. M., Yang, C., Lee, S. Y., Pohajdak, B., Chen, S., PNRC: a proline-rich nuclear receptor coregulatory protein that modulates transcriptional activation of multiple nuclear receptors including orphan receptors SF1 (steroidogenic factor 1) and ERRα1 (estrogen related receptor α-1). *Mol Endocrinol* **2000**, *14*, 986–998.

26 Borinstein, S. C., Hyatt, M. A., Sykes, V. W., Straub, R. E., Lipkowitz, S., Boulter, J., Bogler, O., SETA is a multifunctional adapter protein with three SH3 domains that binds Grb2, Cbl, and the novel SB1 proteins. *Cell Signal* **2000**, *12*, 769–779.

27 Fazi, B., et al., Unusual binding properties of the SH3 domain of the yeast actin-binding protein Abp1: structural and functional analysis. *J. Biol. Chem.* **2002**, *277*, 5290–5298.

28 Shimomura, Y., Aoki, N., Ito, K., Ito, M., Gene expression of Sh3d19, a novel adaptor protein with five Src homology 3 domains, in anagen mouse hair follicles. *J. Dermatol. Sci.* **2003**, *31*, 43–51.

29 Seykora, J. T., Mei, L., Dotto, G. P., Stein, P. L., Srcasm: a novel Src activating and signaling molecule. *J. Biol. Chem.* **2002**, *277*, 2812–2822.

30 Chang, B. Y., Conroy, K. B., Machleder, E. M., Cartwright, C. A., RACK1, a receptor for activated C kinase and a homolog of the β subunit of G proteins, inhibits activity of Src tyrosine kinases and growth of NIH 3T3 cells. *Mol. Cell Biol.* **1998**, *18*, 3245–3256.

31 Schmandt, R., Liu, S. K., McGlade, C. J., Cloning and characterization of mPAL, a novel Shc SH2 domain-binding protein expressed in proliferating cells. *Oncogene* **1999**, *18*, 1867–1879.

32 Espanel, X., Sudol, M., Yes-associated protein and p53-binding protein-2 interact through their WW and SH3 domains. *J. Biol. Chem.* **2001**, *276*, 14514–14523.

33 Tani, K., Sato, S., Sukezane, T., Kojima, H., Hirose, H., Hanafusa, H.,

Shishido, T., Abl interactor 1 promotes tyrosine 296 phosphorylation of mammalian enabled (Mena) by c-Abl kinase *J. Biol. Chem.* **2003**, *278*, 21685–21692.

34 Tong, A. H. Y., et al., A combined experimental and computational strategy to define protein interaction networks for peptide recognition modules. *Science* **2002**, *295*, 321–324.

35 Ren, R., Ye, Z. S., Baltimore, D., Abl protein-tyrosine kinase selects the Crk adapter as a substrate using SH3-binding sites. *Genes. Dev.* **1994**, *8*, 783–795.

36 Fuchs, A., Dagher, M.-C., Vignais, P. V., Mapping the domains of interaction of p40^{phox} with both p47^{phox} and p67^{phox} of the neutrophil oxidase complex using the two-hybrid system. *J. Biol. Chem.* **1995**, *270*, 5695–5697.

37 Chatellard-Causse, C., Blot, B., Cristina, N., Torch, S., Missotten, M., Sadoul, R., Alix (ALG-2-interacting protein X), a protein involved in apoptosis, binds to endophilins and induces cytoplasmic vacuolization. *J. Biol. Chem.* **2002**, *277*, 29108–29115.

38 Ringstad, N., Nemoto, Y., De Camilli, P., The SH3p4/Sh3p8/SH3p13 protein family: binding partners for synaptojanin and dynamin via a Grb2-like Src homology 3 domain. *Proc. Natl. Acad. Sci. USA* **1997**, *94*, 8569–8574.

39 Holbert, S., Dedeoglu, A., Humbert, S., Saudou, F., Ferrante, R. J., Neri, C., Cdc42-interacting protein 4 binds to huntingtin: neuropathologic and biological evidence for a role in Huntington's disease. *Proc. Natl. Acad. Sci. USA* **2003**, *100*, 2712–2717.

40 Morin, F., Vannier, B., Houdart, F., Regnacq, M., Berges, T., Voisin, P., A proline-rich domain in the gamma subunit of phosphodiesterase 6 mediates interaction with SH3-containing proteins. *Mol Vis* **2003**, *9*, 449–459.

41 Rui, L., Mathews, L. S., Hotta, K., Gustafson, T. A., Carter-Su, C., Identification of SH2-Bβ as a substrate of the tyrosine kinase JAK2 involved in growth hormone signaling. *Mol. Cell Biol.* **1997**, *17*, 6633–6644.

42 Igarashi, K., Shigeta, K., Isohara, T., Yamano, T., Uno, I., Sck interacts with KDR and Flt-1 via its SH2 domain. *Biochem. Biophys. Res. Commun.* **1998**, *251*, 77–82.

43 Warner, A. J., Lopez-Dee, J., Knight, E. L., Feramisco, J. R., Prigent, S. A., The Shc-related adaptor protein, Sck, forms a complex with the vascular-endothelial-growth-factor receptor KDR in transfected cells. *Biochem. J.* **2000**, *347*, 501–509.

44 Wood, J. D., et al., Atrophin-1, the DRPLA gene product, interacts with two families of WW domain-containing proteins. *Mol. Cell Neurosci* **1998**, *11*, 149–160.

45 Yagi, R., Chen, L.-F., Shigesada, K., Murakami, Y., Ito, Y., A WW domain-containing yes-associated protein (YAP) is a novel transcriptional co-activator. *EMBO J.* **1999**, *18*, 2551–2562.

46 Kremerskothen, J., Plaas, C., Büther, K., Finger, I., Veltel, S., Matanis, T., Liedtke, T., Barnekow, A., Characterization of KIBRA, a novel WW domain-containing protein. *Biochem. Biophys. Res. Commun.* **2003**, *300*, 862–867.

47 Nishizawa, K., Freund, C., Li, J., Wagner, G., Reinherz, E. L., Identification of a proline-binding motif regulating CD2-triggered T lymphocyte activation. *Proc. Natl. Acad. Sci. USA* **1998**, *95*, 14897–14902.

48 Giovannone, B., Lee, E., Laviola, L., Giorgino, F., Cleveland, K. A., Smith, R. J., Two novel proteins that are linked to insulin-like growth factor (IGF-I) receptors by the Grb10 adapter and modulate IGF-I signaling. *J. Biol. Chem.* **2003**, *278*, 31564–31573.

49 Cheng, J., et al., A Golgi-associated PDZ domain protein modulates cystic fibrosis transmembrane regulator plasma membrane expression. *J. Biol. Chem.* **2002**, *277*, 3520–3529.

50 Furlanetto, R. W., Dey, B. R., Lopaczynski, W., Nissley, S. P., 14-3-3 proteins interact with the insulin-like growth factor receptor but not the insulin receptor. *Biochem. J.* **1997**, *327*, 765–771.

51 Alexandropoulos, K., Cheng, G., Baltimore, D., Proline-rich sequences that bind to Src homology 3 domains with individual specificities. *Proc. Natl. Acad. Sci. USA* **1995**, *92*, 3110–3114.

52 Berry, D. M., Nash, P., Liu, S. K.-W., Pawson, T., McGlade, C. J., A high-

affinity Arg-X-X-Lys SH3 binding motif confers specificity for the interaction between Gads and SLP-76 in T cell signaling. *Curr. Biol.* **2002**, *12*, 1336–1341.

53 Schumacher, C., Knudsen, B. S., Ohuchi, T., Di Fiore, P. P., Glassman, R. H., Hanafusa, H., The SH3 domain of Crk binds specifically to a conserved proline-rich motif in Eps15 and Eps15R. *J. Biol. Chem.* **1995**, *270*, 15341–15347.

54 Yamabhai, M., Kay, B. K., Examining the specificity of Src homology 3 domain–ligand interactions with alkaline phosphatase fusion proteins. *Anal Biochem* **1997**, *247*, 143–151.

55 Chen, H. I., Sudol, M., The WW domain of yes-associated protein binds a proline-rich ligand that differs from the consensus established for Src homology 3-binding modules. *Proc. Natl. Acad. Sci. USA* **1995**, *92*, 7819–7823.

56 Yamabhai, M., Hoffman, N. G., Hardison, N. L., McPherson, P. S., Castagnoli, L., Cesareni, G., Kay, B. K., Intersectin, a novel adaptor protein with two Eps15 homology and five Src homology 3 domains. *J. Biol. Chem.* **1998**, *273*, 31401–31407.

57 Ermekova, K. S., Zambrano, N., Linn, H., Minopoli, G., Gertler, F., Russo, T., Sudol, M., The WW domain of neural protein FE65 interacts with proline-rich motifs in Mena, the mammalian homolog of *Drosophila* enabled. *J. Biol. Chem.* **1997**, *272*, 32869–32877.

58 Bedford, M. T., Chan, D. C., Leder, P., FBP WW domains and the Abl SH3 domain bind to a specific class of proline-rich ligands. *EMBO J.* **1997**, *16*, 2376–2383.

59 Bedford, M. T., Sarbassova, D., Xu, J., Leder, P., Yaffe, M. B., A novel Pro-Arg motif recognized by WW domains. *J. Biol. Chem.* **2000**, *275*, 10359–10369.

60 Cicchetti, P., Mayer, B. J., Thiel, G., Baltimore, D., Identification of a protein that binds to the SH3 region of Abl and is similar to Bcr and GAP-rho. *Science* **1992**, *257*, 803–806.

61 Ren, R., Mayer, B. J., Cicchetti, P., Baltimore, D., Identification of a ten-amino acid proline-rich SH3 binding site. *Science* **1993**, *259*, 1157–1161.

62 Tanaka, S., et al., C3G, a guanine nucleotide-releasing protein expressed ubiquitously, binds to the Src homology 3 domains of CRK and GRB2/ASH proteins. *Proc. Natl. Acad. Sci. USA* **1994**, *91*, 3443–3447.

63 Rivero-Lezcano, O. M., Sameshima, J. H., Marcilla, A., Robbins, K. C., Physical association between Src homology 3 elements and the protein product of the *c-cbl* proto-oncogene. *J. Biol. Chem.* **1994**, *269*, 17363–17366.

64 Rivero-Lezcano, O. M., Marcilla, A., Sameshima, J. H., Robbins, K. C., Wiskott–Aldrich syndrome protein physically associates with Nck through Src homology 3 domains. *Mol. Cell Biol.* **1995**, *15*, 5725–5731.

65 Sparks, A. B., Quilliam, L. A., Thor, J. M., Der, C. J., Kay, B. K., Identification and characterization of Src SH3 ligands from phage-displayed random peptide libraries. *J. Biol. Chem.* **1994**, *269*, 23853–23856.

66 Sparks, A. B., Hoffman, N. G., McConnell, S. J., Fowlkes, D. M., Kay, B. K., Cloning of ligand targets: systematic isolation of SH3 domain-containing proteins. *Nat. Biotechnol.* **1996**, *14*, 741–744.

67 Sparks, A., Rider, J. E., Hoffman, N. G., Fowlkes, D. M., Quilliam, L. A., Kay, B. K., Distinct ligand preferences of Src homology 3 domains from Src, yes, Abl, cortactin, p53bp2, PLC_γ, Crk, and Grb2. *Proc. Natl. Acad. Sci. USA* **1996**, *93*, 1540–1544.

68 Kadlec, L., Pendergast, A. M., The amphiphysin-like protein 1 (ALP1) interacts functionally with the cABL tyrosine kinase and may play a role in cytoskeletal regulation. *Proc. Natl. Acad. Sci. USA* **1997**, *94*, 12390–12395.

69 Chan, D. C., Bedford, M. T., Leder, P., Formin binding proteins bear WWP/WW domains that bind proline-rich peptides and functionally resemble SH3 domains. *EMBO J.* **1996**, *15*, 1045–1054.

70 Liu, S. K., McGlade, C. J., Gads is a novel SH2 and SH3 domain-containing adaptor protein that binds to tyrosine-phosphorylated Shc. *Oncogene* **1998**, *17*, 3073–3082.

71 Skolnik, E. Y., Margolis, B., Mohammadi, M., Lowenstein, E.,

Fischer, R., Drepps, A., Ullrich, A., Schlessinger, J., Cloning of PI3 kinase-associated p85 utilizing a novel method for expression/cloning of target proteins for receptor tyrosine kinases. *Cell* **1991**, *65*, 83–90.

72 Pirozzi, G., McConnell, S. J., Uveges, A. J., Carter, J. M., Sparks, A. B., Kay, B. K., Fowlkes, D. M., Identification of novel human WW domain-containing proteins by cloning of ligand targets. *J. Biol. Chem.* **1997**, *272*, 14611–14616.

73 Kikonyogo, A., Bouamr, F., Vana, M. L., Xiang, Y., Aiyar, A., Carter, C., Leis, J., Proteins related to the Nedd4 family of ubiquitin protein ligases interact with the L domain of Rous sarcoma virus and are required for gag budding from cells. *Proc. Natl. Acad. Sci. USA* **2001**, *98*, 11199–11204.

74 Castagnoli, L., et al., Alternative bacteriophage display systems. *Comb. Chem. High Throughput Screen* **2001**, *4*, 121–133.

75 Zozulya, S., Lioubin, M., Hill, R. J., Abram, C., Gishizky, M. L., Mapping signal transduction pathways by phage display. *Nat. Biotechnol.* **1999**, *17*, 1193–1198.

76 Cochrane, D., Webster, C., Masih, G., McCafferty, J., Identification of natural ligands for SH2 domains from a phage display cDNA library. *J. Mol. Biol.* **2000**, *297*, 89–97.

77 Kurakin, A., Bredesen, D., Target-assisted iterative screening reveals novel interactors for PSD95, Nedd4, Src, Abl and Crk proteins. *J. Biomol. Struct. Dyn.* **2002**, *19*, 1015–1029.

78 Zucconi, A., Dente, L., Santonico, E., Castagnoli, L., Cesareni, G., Selection of ligands by panning of domain libraries displayed on phage lambda reveals new potential partners of synaptojanin 1. *J. Mol. Biol.* **2001**, *307*, 1329–1339.

79 Bialek, K., Swistowski, A., Frank, R., Epitope-targeted proteome analysis: towards a large-scale automated protein–protein-interaction mapping utilizing synthetic peptide arrays. *Anal Bioanal Chem* **2003**, *376*, 1006–1013.

80 Ryou, C., Prusiner, S. B., Legname, G., Cooperative binding of dominant-negative prion protein to kringle domains. *J. Mol. Biol.* **2003**, *329*, 323–333.

81 Zhu, H., et al., Global analysis of protein activities using proteome chips. *Science* **2001**, *293*, 2101–2105.

82 Espejo, A., Côté, J., Bednarek, A., Richard, S., Bedford, M. T., A protein-domain microarray identifies novel protein–protein interactions. *Biochem. J.* **2002**, *367*, 697–702.

83 Hu, H., et al., A map of WW domain family interactions. *Proteomics* **2004**, *4*, 1–13.

84 Espanel, X., Sudol, M. A single point mutation in a group I WW domain shifts its specificity to that of group II WW domains. *J. Biol. Chem.* **1999**, *274*, 17284–17289.

85 Hiipakka, M., Poikonen, K., Saksela, K., SH3 domains with high affinity and engineered ligand specificities targeted to HIV-1 Nef. *J. Mol. Biol.* **1999**, *293*, 1097–1106.

86 Panni, S., Dente, L., Cesareni, G., *In vitro* evolution of recognition specificity mediated by SH3 domains reveals target recognition rules. *J. Biol. Chem.* **2002**, *277*, 21666–21674.

87 Malabarba, M. G., Milia, E., Faretta, M., Zamponi, R., Pelicci, P. G., Di Fiore, P. P., A repertoire library that allows the selection of synthetic SH2s with altered binding specificities. *Oncogene* **2001**, *20*, 5186–5194.

88 Vidal, M., Braun, P., Chen, E., Boeke, J. D., Harlow, E., Genetic characterization of a mammalian protein–protein interaction domain by using a yeast reverse two-hybrid system. *Proc. Natl. Acad. Sci. USA* **1996**, *93*, 10321–10326.

89 Schneider, S., et al., Mutagenesis and selection of PDZ domains that bind new protein targets. *Nat. Biotechnol.* **1999**, *17*, 170–175.

90 Barnett, P., Bottger, G., Klein, A. T. J., Tabak, H. F., Distel, B., The peroxisomal membrane protein Pex13p shows a novel mode of SH3 interaction. *EMBO J.* **2000**, *19*, 6382–6391.

91 Junqueira, D., Cilenti, L., Musumeci, L., Sedivy, J. M., Zervos, A. S., Random mutagenesis of PDZ_{Omi} domain and selection of mutants that specifically bind the Myc proto-oncogene and induce apoptosis. *Oncogene* **2003**, *22*, 2772–2781.

92 Adey, N. B., Sparks, A. B., Beasley, J., Kay, B. K., Construction of random

peptide libraries in bacteriophage M13. In: *Phage Display of Peptides and Proteins: A Laboratory Manual* (Eds.: Kay, B. K., Winter, J., McCafferty, J.), Academic Press, San Diego **1996**, pp. 67–78.

93 Sondek, J., Shortle, D., A general strategy for random insertions and substitutions mutagenesis: substoichiometric coupling of trinucleotide phosphoramidites. *Proc. Natl. Acad. Sci. USA* **1992**, *89*, 3581–3585.

94 Stricker, N. L., et al., PDZ domain of neuronal nitric oxide synthase recognizes novel C-terminal peptide sequences. *Nat. Biotechnol.* **1997**, *15*, 336–342.

95 Felici, F., Castagnoli, L., Musacchio, A., Jappelli, R., Cesareni, G., Selection of antibody ligands from a large library of oligopeptides expressed on a multivalent exposition vector. *J. Mol. Biol.* **1991**, *222*, 301–310.

96 Rickles, R. J., Botfield, M. C., Weng, Z., Taylor, J. A., Green, O. M., Brugge, J. S., Zoller, M. J., Identification of Src, Fyn, Lyn, PI3K and Abl SH3 domain ligands using phage display libraries. *EMBO J.* **1994**, *13*, 5598–5604.

97 Smothers, J. F., Henikoff, S., The HP1 chromo shadow domain binds a consensus peptide pentamer. *Curr. Biol.* **2000**, *10*, 27–30.

98 Oligino, L., et al., Nonphosphorylated peptide ligands for the Grb2 Src homology 2 domain. *J. Biol. Chem.* **1997**, *272*, 29046–29052.

99 Gee, S. H., Sekely, S. A., Lombardo, C., Kurakin, A., Froehner, S. C., Kay, B. K., Cyclic peptides as non-carboxyl-terminal ligands of syntrophin PDZ domains. *J. Biol. Chem.* **1998**, *273*, 21980–21987.

100 Kasanov, J., Pirozzi, G., Uveges, A. J., Kay, B. K., Characterizing Class I WW domains defines key specificity determinants and generates mutant domains with novel specificities. *Chem Biol* **2001**, *8*, 231–241.

101 Kurakin, A. V., Wu, S., Bredesen, D. E., Atypical recognition consensus of CIN85/SETA/Ruk SH3 domains revealed by target-assisted iterative screening. *J. Biol. Chem.* **2003**, *278*, 34102–34109.

102 Cheadle, C., Ivashchenko, Y., South, V., Searfoss, G. H., French, S., Howk, R., Ricca, G. A., Jaye, M., Identification of a Src SH3 domain binding motif by screening a random phage display library. *J. Biol. Chem.* **1994**, *269*, 24034–24039.

103 Pero, S. C., Oligino, L., Daly, R. J., Soden, A. L., Liu, C., Roller, P. P., Li, P., Krag, D. N., Identification of novel non-phosphorylated ligands, which bind selectively to the SH2 domain of Grb7. *J. Biol. Chem.* **2002**, *277*, 11918–11926.

104 Wang, B., Yang, H., Liu, Y. C., Jelinek, T., Zhang, L., Ruoslahti, E., Fu, H., Isolation of high-affinity peptide antagonists of 14-3-3 proteins by phage display. *Biochemistry* **1999**, *38*, 12499–12504.

105 Cestra, G., et al., The SH3 domains of endophilin and amphiphysin bind to the proline- rich region of synaptojanin 1 at distinct sites that display an unconventional binding specificity. *J. Biol. Chem.* **1999**, *274*, 32001–32007.

106 Salcini, A. E., et al., Binding specificity and in vivo targets of the EH domain, a novel protein–protein interaction module. *Genes. Dev.* **1997**, *11*, 2239–2249.

107 Paoluzi, S., et al., Recognition specificity of individual EH domains of mammals and yeast. *EMBO J.* **1998**, *17*, 6541–6550.

108 Di Vignano, A. T., Di Zenzo, G., Sudol, M., Cesareni, G., Dente, L., Contribution of the different modules in the utrophin carboxy-terminal region to the formation and regulation of the DAP complex. *FEBS Lett.* **2000**, *471*, 229–234.

109 Mongiovi, A. M., Romano, P. R., Panni, S., Mendoza, M., Wong, W. T., Musacchio, A., Cesareni, G., Di Fiore, P. P., A novel peptide–SH3 interaction. *EMBO J.* **1999**, *18*, 5300–5309.

110 Dente, L., Vetriani, C., Zucconi, A., Pelicci, G., Lanfrancone, L., Pelicci, P. G., Cesareni, G., Modified phage peptide libraries as a tool to study specificity of phosphorylation and recognition of tyrosine containing peptides. *J. Mol. Biol.* **1997**, *269*, 694–703.

111 Pelicci, G., et al., A family of Shc related proteins with conserved PTB, CH1 and SH2 regions. *Oncogene* **1996**, *13*, 633–641.

112 Hart, C. P., Martin, J. E., Reed, M. A., Keval, A. A., Pustelnik, M. J., Northrop, J. P., Patel, D. V., Grove, J. R., Potent inhibitory ligands of the GRB2 SH2 domain from recombinant

peptide libraries. *Cell Signal* **1999**, *11*, 453–464.

113 Fuh, G., Pisabarro, M. T., Li, Y., Quan, C., Lasky, L. A., Sidhu, S. S., Analysis of PDZ domain–ligand interactions using carboxyl-terminal phage display. *J. Biol. Chem.* **2000**, *275*, 21486–21491.

114 Vaccaro, P., Brannetti, B., Montecchi-Palazzi, L., Philipp, S., Helmer-Citterich, M., Cesareni, G., Dente, L., Distinct binding specificity of the multiple PDZ domains of INADL, a human protein with homology to INAD from *Drosophila melanogaster*. *J. Biol. Chem.* **2001**, *276*, 42122–42130.

115 Rickles, R. J., Botfield, M. C., Zhou, X. M., Henry, P. A., Brugge, J. S., Zoller, M. J., Phage display selection of ligand residues important for Src homology 3 domain binding specificity. *Proc. Natl. Acad. Sci. USA* **1995**, *92*, 10909–10913.

116 Schmitz, R., Baumann, G., Gram, H., Catalytic specificity of phosphotyrosine kinases Blk, Lyn, c-Src and Syk as assessed by phage display. *J. Mol. Biol.* **1996**, *260*, 664–677.

117 Cesareni, G., Panni, S., Nardelli, G., Castagnoli, L., Can we infer peptide recognition specificity mediated by SH3 domains? *FEBS Lett.* **2002**, *513*, 38–44.

118 Gram, H., Schmitz, R., Zuber, J. F., Baumann, G., Identification of phosphopeptide ligands for the Src-homology 2 (SH2) domain of Grb2 by phage display. *Eur. J. Biochem.* **1997**, *246*, 633–637.

119 Linn, H., Ermekova, K. S., Rentschler, S., Sparks, A. B., Kay, B. K., Sudol, M., Using molecular repertoires to identify high-affinity peptide ligands of the WW domain of human and mouse YAP. *Biol. Chem.* **1997**, *378*, 531–537.

120 Liu, R., Enstrom, A. M., Lam, K. S., Combinatorial peptide library methods for immunobiology research. *Exp. Hematol.* **2003**, *31*, 11–30.

121 Chan, W. C., White, P. D., *Fmoc Solid-phase Peptide Synthesis: A Practical Approach*. Oxford University Press, New York **2000**.

122 Frank, R., The SPOT-synthesis technique. Synthetic peptide arrays on membrane supports: principles and applications. *J. Immunol. Methods* **2002**, *267*, 13–26.

123 Santamaria, F., Wu, Z., Bolègue, C., Pál, G., Lu, W., Reexamination of the recognition preference of the specificity pocket of the Abl SH3 domain. *J. Mol. Recognit.* **2003**, *16*, 131–138.

124 Li, S.-C., et al., High-affinity binding of the *Drosophila* Numb phosphotyrosine-binding domain to peptides containing a Gly-Pro-(p)Tyr motif. *Proc. Natl. Acad. Sci. USA* **1997**, *94*, 7204–7209.

125 Kelly, M. A., Liang, H., Sytwu, I.-I., Vlattas, I., Lyons, N. L., Bowen, B. R., Wennogle, L. P., Characterization of SH2-ligand interactions via library affinity selection with mass spectrometric detection. *Biochemistry* **1996**, *35*, 11747–11755.

126 Songyang, Z., et al., SH2 domains recognize specific phosphopeptide sequences. *Cell* **1993**, *72*, 767–778.

127 Songyang, Z., et al., Specific motifs recognized by the SH2 domains of Csk, 3BP2, fps/fes, GRB-2, HCP, SHC, Syk, Vav. *Mol. Cell Biol.* **1994**, *14*, 2777–2785.

128 Songyang, Z., Margolis, B., Chaudhuri, M., Shoelson, S. E., Cantley, L. C., The phosphotyrosine interaction domain of SHC recognizes tyrosine-phosphorylated NPXY motif. *J. Biol. Chem.* **1995**, *270*, 14863–14866.

129 Lupher, M. L. Jr., Songyang, Z., Shoelson, S. E., Cantley, L. C., Band, H., The Cbl phosphotyrosine-binding domain selects a D(N/D)XpY motif and binds to the Tyr^{292} negative regulatory phosphorylation site of ZAP-70. *J. Biol. Chem.* **1997**, *272*, 33140–33144.

130 O'Bryan, J. P., Martin, C. B., Songyang, Z., Cantley, L. C., Der, C. J., Binding specificity and mutational analysis of the phosphotyrosine binding domain of the brain-specific adaptor protein ShcC. *J. Biol. Chem.* **1996**, *271*, 11787–11791.

131 Yaffe, M. B., et al., The structural basis for 14-3-3:phosphopeptide binding specificity. *Cell* **1997**, *91*, 961–971.

132 Rodriguez, M., Yu, X., Chen, J., Songyang, Z., Phosphopeptide binding specificities of BRCA1 COOH-terminal (BRCT) domains. *J. Biol. Chem.* **2003**, *278*, 52914–52918.

133 Durocher, D., Taylor, I. A., Sarbassova, D., Haire, L. F., Westcott, S. L., Jackson, S. P., Smerdon, S. J., Yaffe,

M. B., The molecular basis of FHA domain: phosphopeptide binding specificity and implications for phospho-dependent signaling mechanisms. *Mol. Cell* **2000**, *6*, 1169–1182.

134 Grabs, D., Slepnev, V. I., Sogyang, Z., David, C., Lynch, M., Cantley, L. C., De Camilli, P., The SH3 domain of amphiphysin binds the proline-rich domain of dynamin at a single site that defines a new SH3 binding consensus sequence. *J. Biol. Chem.* **1997**, *272*, 13419–13425.

135 Songyang, Z., et al., Recognition of unique carboxyl-terminal motifs by distinct PDZ domains. *Science* **1997**, *275*, 73–77.

136 Lam, K. S., Liu, R., Miyamoto, S., Lehman, A. L., Tuscano, J. M., Applications of one-bead one-compound combinatorial libraries and chemical microarrays in signal transduction research. *Acc Chem Res* **2003**, *36*, 370–377.

137 Yu, H., Chen, J. K., Feng, S., Dalgarno, D. C., Brauer, A. W., Schreiber, S. L., Structural basis for the binding of proline-rich peptides to SH3 domains. *Cell* **1994**, *76*, 933–945.

138 Chen, J. K., Lane, W. S., Brauer, A. W., Tanaka, A., Schreiber, S. L., Biased combinatorial libraries: novel ligands for the SH3 domain of phosphatidylinositol 3-kinase. *J. Am. Chem. Soc.* **1993**, *115*, 12591–12592.

139 Liao, H., et al., Structure of the FHA1 domain of yeast Rad53 and identification of binding sites for both FHA1 and its target protein Rad9. *J. Mol. Biol.* **2000**, *304*, 941–951.

140 Byeon, I.-J. L., Yongkiettrakul, S., Tsai, M.-D., Solution structure of the yeast Rad53 FHA2 complexed with a phosphothreonine peptide pTXXL: comparison with the structures of FHA2-pYXL and FHA1-pTXXD complexes. *J. Mol. Biol.* **2001**, *314*, 577–588.

141 Beebe, K. D., Wang, P., Arabaci, G., Pei, D., Determination of the binding specificity of the SH2 domains of protein tyrosine phosphatase SHP-1 through the screening of a combinatorial phosphotyrosyl peptide library. *Biochemistry* **2000**, *39*, 13251–13260.

142 Müller, K., et al., Rapid identification of phosphopeptide ligands for SH2 domains: screening of peptides by fluorescence-activated bead sorting. *J. Biol. Chem.* **1996**, *271*, 16500–16505.

143 Kessels, H. W. H. G., Ward, A. C., Schumacher, T. N. M., Specificity and affinity motifs for Grb2 SH2–ligand interactions. *Proc. Natl. Acad. Sci. USA* **2002**, *99*, 8524–8529.

144 Frank, R., High-density synthetic peptide microarrays: emerging tools for functional genomics and proteomics. *Comb. Chem. High Throughput Screen* **2002**, *5*, 429–440.

145 Frank, R., Overwin, H., SPOT-synthesis: epitope analysis with arrays of synthetic peptides prepared on cellulose membranes. In: Morris, G. E. (Ed.), *Methods in Molecular Biology. Epitope Mapping Protocols*, Humana Press, Totowa, NJ **1996**, *66*, 149–169.

146 Hultschig, C., Two dimensional screening: towards establishing a novel technique to study biomolecular interactions. PhD thesis, http://www.biblio.tu-bs.de/ediss/data/20000207a/20000207a.html **2000**.

147 Kramer, A., Reineke, U., Dong, L., Hoffmann, B., Hoffmüller, U., Winkler, D., Volkmer-Engert, R., Schneider-Mergener, J., SPOT synthesis: observations and optimizations. *J. Pept. Res.* **1999**, *54*, 319–327.

148 Schultz, J., Hoffmüller, U., Krause, G., Ashurst, J., Macias, M. J., Schmieder, P., Schneider-Mergener, J., Oschkinat, H., Specific interactions between the syntrophin PDZ domain and voltage-gated sodium channels. *Nat. Struct. Biol.* **1998**, *5*, 19–24.

149 Rodriguez, M., Li, S. S.-C., Harper, J. W., Songyang, Z., An oriented peptide array library (OPAL) strategy to study protein–protein interactions. *J. Biol. Chem.* **2004**, *279*, 8802–8807.

150 Niebuhr, K., et al., A novel proline-rich motif present in ActA of *Listeria monocytogenes* and cytoskeletal proteins is the ligand for the EVH1 domain, a protein module present in the Ena/VASP family. *EMBO J.* **1997**, *16*, 5433–5444.

151 Drees, B., Friederich, E., Fradelizi, J., Louvard, D., Beckerle, M. C., Golsteyn, R. M., Characterization of the interaction between zyxin and members of the Ena/vasodilator-stimulated

phosphoprotein family of proteins. *J. Biol. Chem.* **2000**, *275*, 22503–22511.

152 Barzik, M., Carl, U. D., Schubert, W.-D., Frank, R., Wehland, J., Heinz, D. W., The N-terminal domain of Homer/Vesl is a new class II EVH1 domain. *J. Mol. Biol.* **2001**, *309*, 155–169.

153 Howell, B. W., Lanier, L. M., Frank, R., Gertler, F. B., Cooper, J. A., The disabled 1 phosphotyrosine-binding domain binds to the internalization signals of transmembrane glycoproteins and to phospholipids. *Mol. Cell Biol.* **1999**, *19*, 5179–5188.

154 Chen, H. I., Einbond, A., Kwak, S.-J., Linn, H., Koepf, E., Peterson, S., Kelly, J. W., Sudol, M., Characterization of the WW domain of human yes-associated protein and its polyproline-containing ligands. *J. Biol. Chem.* **1997**, *272*, 17070–17077.

155 Pires, J. R., et al., Solution structures of the YAP65 WW domain and the variant L30K in complex with the peptides GTPPPPYTVG, N-(n-octyl)-GPPPY and PLPPY and the application of peptide libraries reveal a minimal binding epitope. *J. Mol. Biol.* **2001**, *314*, 1147–1156.

156 Ball, L. J., et al., Dual epitope recognition by the VASP EVH1 domain modulates polyproline ligand specificity and binding affinity. *EMBO J.* **2000**, *19*, 4903–4914.

157 Hoffmüller, U., Russwurm, M., Kleinjung, F., Ashurt, J., Oschkinat, H., Volkmer-Engert, R., Koesling, D., Schneider-Mergener, J., Interaction of a PDZ protein domain with a synthetic library of all human protein C termini. *Angew. Chem. Int. Ed.* **1999**, *38*, 2000–2004.

158 Landgraf, C., Panni, S., Montecchi-Palazzi, L., Castagnoli, L., Schneider-Mergener, J., Volkmer-Engert, R., Cesareni, G., Protein interaction networks by proteome peptide scanning. *PLoS Biology* **2004**, *2*, 94–103.

159 Toepert, F., Pires, J. R., Landgraf, C., Oschkinat, H., Schneider-Mergener, J., Synthesis of an array comprising 837 variants of the hYAP WW protein domain. *Angew. Chem. Int. Ed.* **2001**, *40*, 897–900.

160 Toepert, F., Knaute, T., Guffler, S., Pires, J. R., Matzdorf, T., Oschkinat, H., Schneider-Mergener, J., Combining SPOT synthesis and native peptide ligation to create large arrays of WW protein domains. *Angew. Chem. Int. Ed.* **2003**, *42*, 1136–1140.

21
Computational Analysis of Modular Protein Architectures

Rune Linding, Ivica Letunic, Toby J. Gibson, and Peer Bork

21.1
Introduction

In this chapter we discuss some of the bioinformatics and computational tools that exist for finding modular domains and their cognate ligands. We begin with a general discussion of protein architecture and relate this to the modular model of protein function presented in this book. We pay particular attention to the SMART, ELM, GlobPlot, and DisEMBL resources, because these are the ones we are most familiar with. We apologize to those involved with all the great resources out there that we have not covered in this chapter.

21.2
Protein Architecture: Sequence, Structure, and Function

Nature seems to present protein functions in two states: structured and unstructured. Proteins are heteropolymers of amino acids; the sequence of amino acids determines not only the structure and folding of a protein but also the lack of structure. Molecular functions in proteins are associated with structural units, e.g., modular globular domains. However, an emerging large group of functional sites are found primarily in unstructured parts of proteins.

In this chapter we explore how these sites can be identified in proteins and describe some of the computational tools that can be used to analyze protein sequences.

21.2.1
The Modular Model of Protein Function

Multidomain proteins predominate in eukaryotic proteomes. The basic hypothesis in what one might call the modular model for protein function is that individual functions assigned to different sequence segments (often domains) combine to create a complex function for the whole protein.

Modular Protein Domains. Edited by G. Cesareni, M. Gimona, M. Sudol, M. Yaffe

ISBN: 3-527-30813-X

The term 'modular' refers directly to the autonomous nature of the individual folding units determining these functions, whereas the term 'globular' describes the structural state of a domain whether or not it is modular. A dogma in structural biology is that atomic structure determines function. The modular model has grown out of this paradigm; however, we know that at the fold level this is not always true, one can find structures belonging to the same fold (e.g. in SCOP) that have completely different functions: an example of this is glutathione-S-transferase and S-crystallin. They share 75% sequence similarity and the same fold, but the former is an enzyme and the latter a structural protein [4]. This is mainly a problem when one is trying to infer function from sequence; having the atomic structure solved frequently helps to define the function.

Because single-domain globular proteins are often, although not always, less difficult to crystallize, for a long time they dominated our perceptions of typical protein structure (although fibrous proteins like collagen were of course well known). Gradually, as protein sequences have accumulated, the monodomain view of protein structure has been replaced by the realization that most proteins are multidomain, at least in higher eukaryotes. Multidomain architectures are usual for transmembrane receptors, signaling proteins, cytoskeletal proteins, chromatin proteins, transcription factors, and so forth. Multidomain proteins can be described as consisting of a series of modules or globular domains and a set of short linear functional sites. An archetypal protein of the modular model is Src (Figure 21.1).

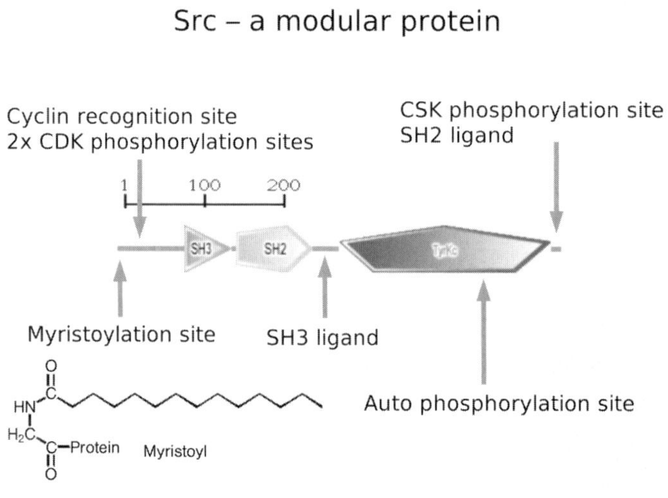

Figure 21.1 Domain and functional site architecture of the well known proto-oncogenic protein kinase Src. About 60%–80% of proteins from higher eukaryotes have analogous modular architectures [6]. Even though 'only' ~30 000 ORFs are predicted to be in the human genome, most of these have several splice variants and, in addition, several functional sites, e.g., post-translational modification sites (PTMs). These sites exist in various states and thereby increase the number of different functional isoforms several fold. The arrows show only the approximate locations of the functional sites.

Src consists of three globular domains: SH3, SH2, and a tyrosine kinase domain (itself consisting of two structural domains), and eight known functional sites including four phosphorylation sites and three ligands of modular domains (SH3, SH2, and cyclin). Please see Chapters 1 and 2 for more information on the SH3 and SH2 domains.

21.2.2 Partitioning of Protein Space

It is becoming increasingly clear that many functionally important protein segments occur outside globular domains [27, 99]. The set (or space) of all observations of protein structure and function is partitioned into two subspaces (Figure 21.2). The first consists of globular units having binding pockets, active sites, and interaction surfaces. The second subspace consists of nonglobular segments such as sorting signals, post-translational modification sites, and protein ligands (e.g., SH3 or WW ligands). Globular units are built of regular secondary structural elements and contribute the majority of the structural data deposited in PDB. The globular function space is described very well by domain databases such as SMART [49] and Pfam [11]. In contrast, the nonglobular subspace encompasses disordered, unstructured, and flexible regions lacking regular secondary structure. Functional sites within the nonglobular space are known as linear motifs catalogued by ELM [71], PROSITE [80], and Scansite [101]. This group of sites includes protein interaction sites, cell compartment targeting signals, post-translational modification sites, and cleavage sites.

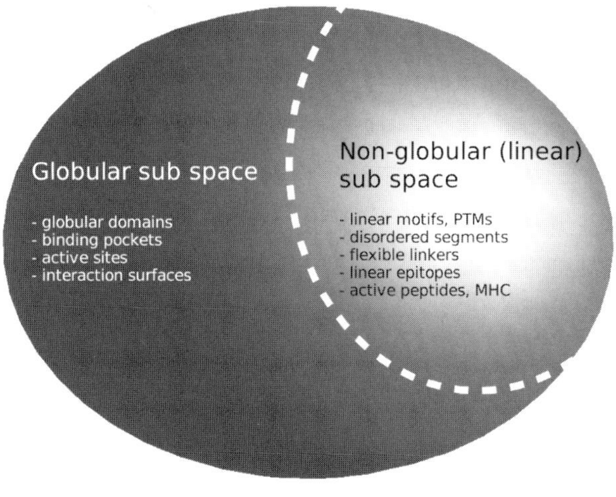

Figure 21.2 A conceptual model of protein structure and function observation space. Many functions and their structures can be assigned to two different subspaces. A linear/nonglobular one and a globular one. It is important to notice that these two subspaces are not detached from each other – a good example is disordered loops that can protrude from a globular domain. Along the borderline (white) we find coiled coils, repeat proteins (often forming rods), and single transmembrane helices (TM1).

Traditionally, protein function and interaction have been studied from a domain-centric view, and in fact, most large datasets that deal with protein interaction have also focused on this type of interaction [2]. This is because methods such as affinity purification tend to isolate 'sticky' and 'nontransient' protein complexes. The methods for isolating transient interactions, such as the binding of a cognate ligand to its modular domain, are different, and much smaller in-vivo datasets exist.

However, the nontransient network is only a part, and perhaps even a subpart, of the interaction networks within the cell. We hope that this book will encourage the scientific community to focus on collecting large amounts of data on domain–ligand interactions, since only by having these can we try to obtain a full picture of the cellular protein networks.

Below, we discuss how to annotate and analyze protein sequences for modular domains and linear motifs. Methods for finding domains and predicting their function are described first, followed by a description of how to identify unstructured regions and their potential functions.

21.3 Analyzing Globular Domains

The past three decades have seen relatively steady levels of domain discovery [18]. It seems likely that most of the more common mobile protein domains have already been described in the literature (see, e.g., this book on signaling domains and the number of their occurrences in the human genome, and Figure 21.3). However,

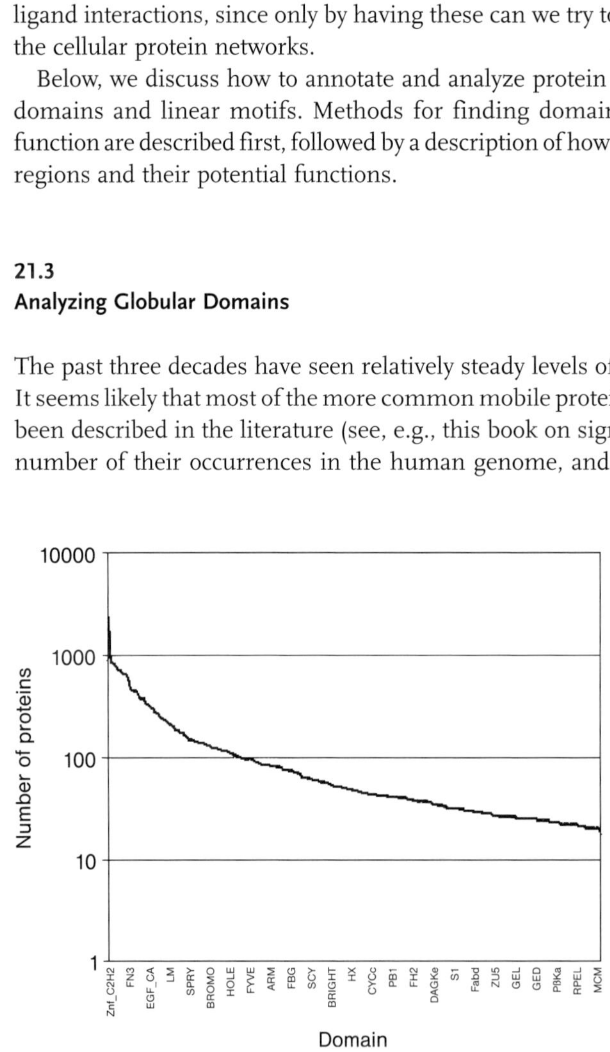

Figure 21.3 Presence of SMART domains in the *Homo sapiens* proteome. Domains present in more than 20 proteins are shown. Domain names are displayed for only a few of the 314 domains.

because there is a large number of domains that are detectable only in a relatively small number of proteins, we predict that various domains are still hiding in many proteins and that each genome might harbor its own repertoire of species-specific or at least lineage-specific domains [26].

Since the late 1970s, homology search has been an extremely powerful computational technique for assigning novel functions to proteins. In addition to database search methods that were introduced in the early 1980s, an awareness of conserved entities such as local motifs was incorporated into software that was able to scan dedicated collections of such motifs in the mid 1980s (see, e.g., PROSITE, 1985, for an early resource). Yet, the statistics of more sophisticated search methods such as BLAST are still struggling with compositional biases due to nonglobularity or multiple occurrences of such structural entities within a search sequence. However, without discrimination of the rationale behind functionality (e.g., short functional motifs that can change quickly in time, such as glycosylation patterns vs. essential catalytic residues that stay conserved over billions of years), homology searches with the aim of function prediction remain limited. Thus, in this chapter we first define globular domains and then describe resources that help to annotate known domains. Finally, we explain the analysis options of one of these resources, SMART.

21.3.1
Globularity of Domains

Both 'domain' and 'globular' are terms defined in structural protein biology. They have since been used in other contexts such as function description and sequence analysis. Domains were first defined in structural terms, since early X-ray structures showed separate entities with defined subfunctions connected by flexible regions. The term globular refers to the globular structural state: a protein globule can be depicted as a soluble sphere having a hydrophobic core. An operational definition of globular proteins can be found in [19]:

> Most natural proteins in solution are much smaller in their dimensions than comparable polypeptides with random or repetitive conformations and have roughly spherical shapes; hence they are generally referred to as globular. Their physical properties do not change gradually as the environment is altered (e.g., by changes in temperature, pH or pressure) as do the properties of random polypeptides. Instead, globular proteins usually exhibit little or no change, until a point is reached at which there is a sudden drastic change and, invariably, a loss of biological function. This phenomenon is known as denaturation.

Structural biologists relate the term globular to domains that are compact and fold independently of the remainder of the host protein. Theoretical definitions of domains exist [85, 86, 88], including several algorithms for classifying domains and folds [43, 87].

Resources for finding globular domains with determined tertiary structure include SCOP (fold level) [54], ASTRAL (domains from SCOP) [16], SUPERFAMILY (hidden

Markov models of SCOP domains) [35], and CATH (architecture level) [37, 65]. However, these resources are confined to the structural knowledge base PDB [96]. Therefore, only a subset of the total fold and structure space is described; the remaining domains are to a certain extent described in Pfam and SMART.

21.3.2
Resources for Analysis of Globular Domains

There are numerous domain databases available that can be useful for detection and analysis of globular domains in your favorite protein sequence. They can be separated into several categories:

- Databases such as SMART (http://smart.embl.de/ [49]) and PFAM (http://www.sanger.ac.uk/software/pfam/ [11]) primarily make use of hand-edited sequence alignments representing single protein domains with well defined borders at the sequence level. PROSITE (http://www.expasy.org/prosite/ [80]) is also a handmade resource, but it contains a much more heterogeneous set of domains and motifs although, for linear motifs, it has been superseded by ELM.
- Other databases rely on various automatic methods to generate their domain signatures. This is so for ProDom (http://www.toulouse.inra.fr/prodom.html [79]) and BLOCKS (http://www.blocks.fhcrc.org/ [38]). Such resources predict domains that do not always correspond to known structural and globular domains and for this purpose may not be as sensitive. However, these resources are of substantial discovery value, since they collect conserved sequence segments that might specify novel functions.
- In addition to the search functionality of the databases themselves, several metaservers allow users to search multiple domain databases. The Interpro database, at the EBI (http://www.ebi.ac.uk/interpro/ [61]) allows searching of the PROSITE, PFAM, PRINTS, ProDom, and SMART model collections, and the Conserved Domain Database (CDD) at the NCBI (http://www.ncbi.nlm.nih.gov/structure/cdd/cdd.shtml [57]) allows searching of profiles derived from SMART and PFAM using a modified version of the BLAST algorithm.

We describe the SMART resource in greater detail, because it focuses on modular signaling domains.

21.3.3
SMART: Simple Modular Architecture Research Tool

The explosion of sequence data increases the need for computational sequence-analysis tools that annotate novel genes with predicted functions. Function prediction, however, is fraught with potential pitfalls, such as variable sequence divergence, nonequivalent functions of homologs, and nonidentical multidomain architectures [25]. Detecting nonenzymatic regulatory domains is essential to predicting a protein's cellular role, binding partners, and subcellular localization.

Such domains can be divergent in sequence and occur in contrasting multidomain contexts. This leads to difficulties in unraveling the evolution and function of multidomain proteins. To help in solving these problems, SMART has been developed to identify and annotate protein domains, particularly those in eukaryotes that are genetically mobile and difficult to detect.

21.3.3.1 The SMART Alignment Set

Domain detection in SMART relies on multiple sequence alignments of representative family members.

Alignment Construction Protocol

The starting point for constructing a multiple sequence alignment that optimally represents a domain family is an alignment of divergent family members based on known tertiary structures, where possible, or from homologs identified in a PSI-BLAST [3] analysis. These alignments are optimized manually and, after construction of a hidden Markov model (HMM), used to search current sequence databases (Figure 21.4). Each sequence of the alignment is also used as a query in a PSI-BLAST search. All sequences that are significantly similar [as detected by HMM ($E < 0.01$) or PSI-BLAST ($E < 0.001$) searches] are added to the alignment using the sequence-versus-HMM alignment method of HMMer. Alignments are checked manually for potential false positives or misassembled protein sequences derived from genomic sources. From this alignment, one of each sequence pair sharing > 67% identity is deleted to reduce redundancy. The resulting alignment is used as a starting point for a subsequent round of searches. This iterative procedure is pursued until no new homologs are detected.

Searching Method

To maximize the sensitivity of domain and repeat detection, SMART uses hidden HMMer models as implemented in the HMMer software package (http://hmmer.wustl.edu/). HMMER provides statistically sound E values, thus giving a robust estimate of the significance of a domain hit. [The E value represents the number of sequences having a score $\geq X$ that would be expected absolutely by chance. The E value connects the score (X) of an alignment between a user-supplied sequence and a database sequence, generated by any algorithm, with how many alignments having similar or greater scores would be expected from a search of a random sequence database of equivalent size.] From a database search with an HMM derived from the SMART alignment, the highest per-protein E value of identified true positives (E_p) and the lowest per-protein E value of predicted true negatives (E_n) are stored within the SMART database. Similarly, for two or more repeats in a protein, the lowest E value of a false positive repeat (E_r) is stored. To ensure that the E value thresholds are independent of database size, the size of the protein database used when deriving the thresholds is also recorded. SMART predicts a domain homolog within any sequence that either has an E value $< E_p$ or else when $E_p <$ E value $< E_n$ and E value < 1.0. If no repeat threshold is defined, all hits in a protein are reported; otherwise only those with E values $< E_r$ are shown.

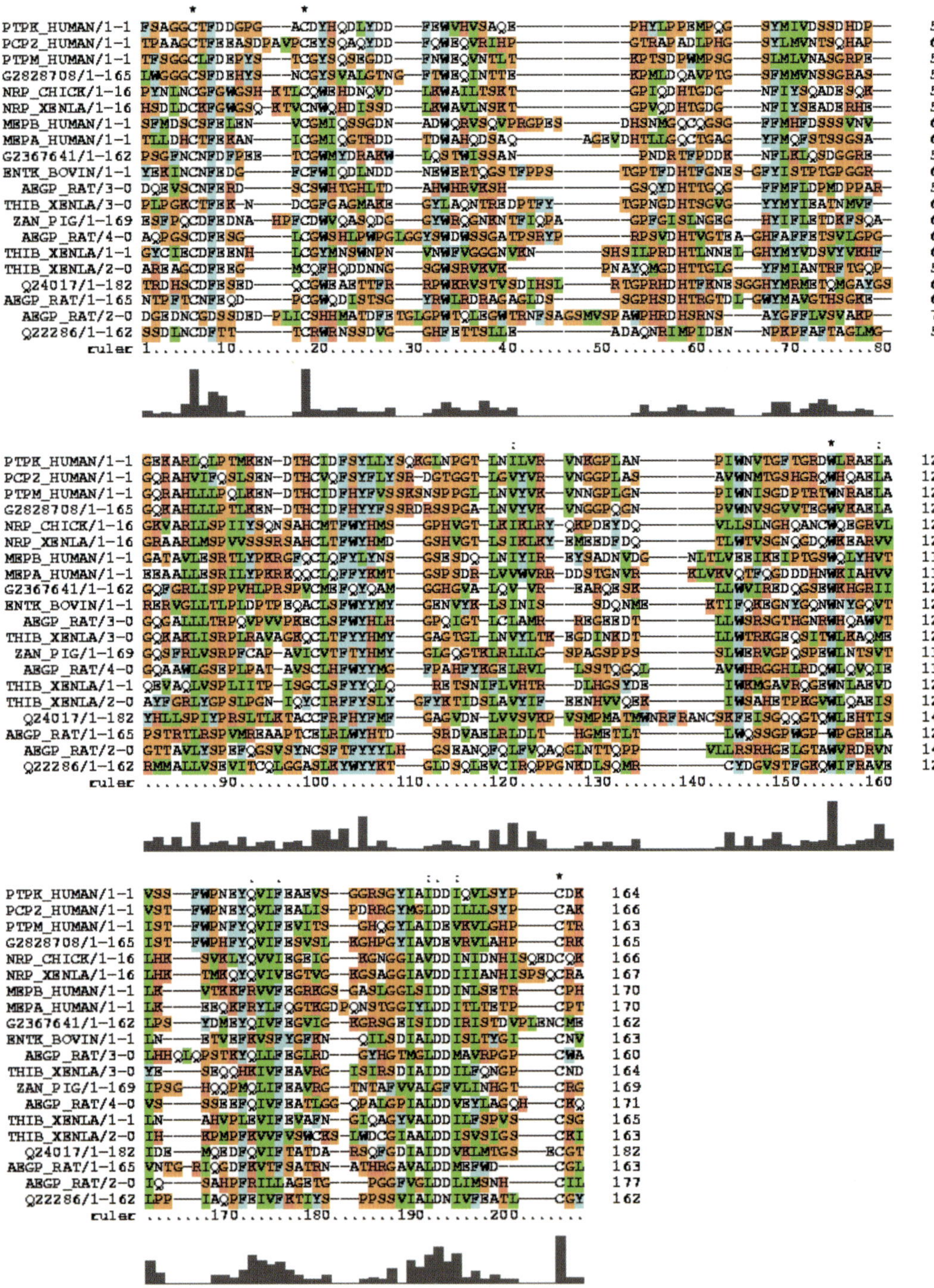

Figure 21.4 SMART alignment representing the MAM domain.

Domain Coverage

Originally, SMART was intended as a tool for the analysis of domains involved in eukaryotic signal transduction [76]; it was later expanded to detect domains of extracellular proteins and bacterial two-component regulatory systems. Gradually, various domains associated with DNA, RNA, chromatin, and cytoskeletal functions have been added. Over the past few years, to augment the SMART domain set, several semi-automatic search methods to identify new and biologically interesting domains were developed. The current release of SMART (version 4.0) includes nearly 700 protein domains.

21.3.3.2 SMART Relational Database System

The core of SMART is a relational database management system (RDBMS) powered by PostgreSQL (http://www.postgresql.org/), which stores information on SMART domains and the underlying nonredundant protein database.

Protein Database

Basic components of SMART's source sequence database are the Swiss-Prot and SP-TrEMBL [10] protein sets, which have been used by SMART since its inception. This set was recently expanded by inclusion of all proteomes available in the Ensembl collection [17]. Sequences from all sources are compared, and a nonredundant set of proteins with multiple identifiers per sequence is generated. Sequences are retrievable and linkable via any of the original identifiers.

Domain Database

The SMART domain database stores information on each domain's presence in all proteins in the relational database. Each domain's hit borders, raw bit score, and Expect (E) value are recorded, together with the protein accession code, description, and species name. In addition to domain information, other intrinsic features of each protein, such as transmembrane regions, coiled coils, signal peptides, and internal repeats are included.

21.3.3.3 Web Interface

SMART provides a web-based interface to its underlying relational database and HMMer-based search engine. There are two principal ways of using SMART: individual sequence analysis and domain architecture analysis. Here we describe major features of the current SMART (version 4.0).

Sequence Analysis

SMART uses the CRC64 algorithm to calculate checksums for all user-supplied sequences. If a matching checksum is found in the SMART database, precalculated results are displayed. If there is no match, HMMer software is used to scan the sequence with all SMART profiles. It is also possible to include Pfam profiles in the search.

Resulting schematic protein representations (Figure 21.5) are easy to interpret: a gray line shows the protein backbone, and different colored shapes represent

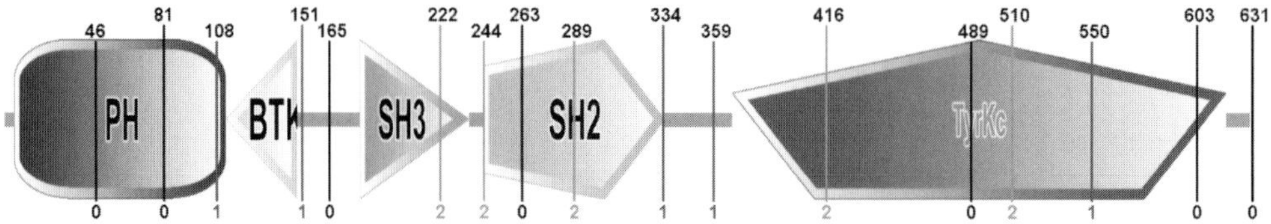

Figure 21.5 SMART representation of mouse tyrosine protein kinase TEC (ENSMUSP00000006349). Gray lines show the protein backbone, with domains represented by different colored shapes. Intron positions are indicated by vertical lines showing the amino acid location and the intron phase. Intron positions are taken from Ensembl gene predictions. See Chapters 13 and 15 for more information on the PH domain and kinase domains, respectively.

domains and features that are confidently predicted. If a user-supplied sequence has a matching checksum identified, several important features become available in the main results page.

Where available, intron positions are shown in schematic protein figures. For proteins that match any of the Ensembl predictions, SMART shows intron positions as vertical colored lines (Figure 21.5). This information is retrieved from a precalculated mapping of Ensembl gene structures to protein sequences.

Extra information may be associated with the sequence. If multiple IDs are associated with the same sequence, users receive a list of all IDs with links to corresponding source databases. Since SMART incorporates Ensembl genomes, users also receive a list of alternative splices of the gene encoding the analyzed protein (if there are any). It is possible to either display SMART protein annotation for any of the alternative splices or obtain a graphical multiple sequence alignment of all of them (Figure 21.6).

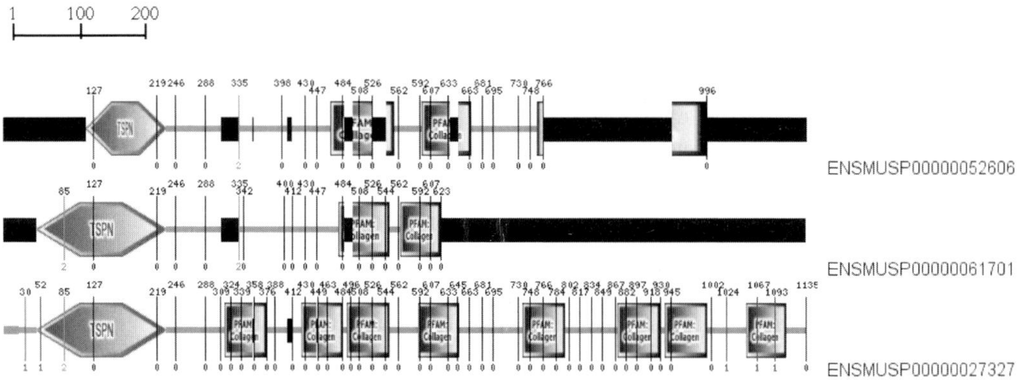

Figure 21.6 SMART graphical alignment of alternative splice variants of *Mus musculus* procollagen gene *ENSMUSG00000026141*. Domain and intron positions are adjusted according to gaps in the alignment (black boxes).

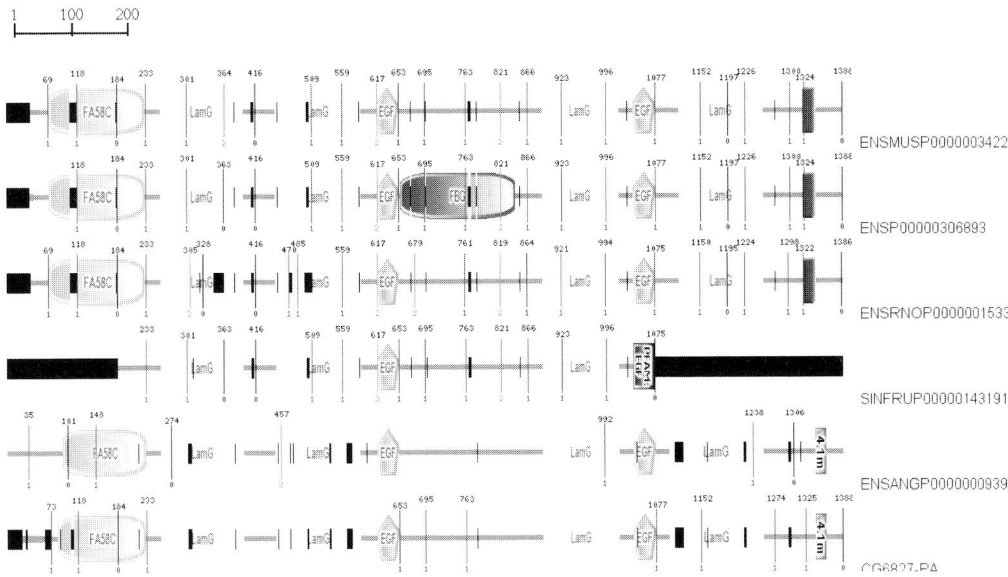

Figure 21.7 SMART representation of an orthologous group alignment. Orthologous proteins from different species are aligned using Clustal W. Domains, intrinsic features, and introns are mapped onto the alignment with their positions adjusted according to gaps (black boxes). This tool allows easy visual comparison of intron positions and their relations to protein features. Proteins displayed: *H. sapiens* ENSP00000306893, *M. musculus* ENSMUSP00000034225, *Rattus norvegicus* ENSRNOP00000015337, *Fugu rubripes* SINFRUP00000143191, *Drosophila melanogaster* CG6827-PA, and *Anopheles gambiae*, ENSANGP00000009390.

Orthology information: SMART provides orthology information for all Ensembl-predicted proteins. These relationships are distinct from those provided by Ensembl. There are two separate sets of orthologs for each protein: 1 : 1 reciprocal best matches in other genomes and orthologous groups with reciprocal best hits from all genomes analyzed (i.e., each of these proteins has exactly one ortholog in all six genomes). Orthologous groups are displayed as graphical multiple sequence alignments (Figure 21.7). All orthology information is extracted from all-against-all Smith–Waterman similarities for combined proteomes, using a previously described method [103].

Domain Architecture Analysis (Architecture SMART and Alert SMART)

Architecture SMART allows users to search for specific domain architectures using AND/NOT logic. Since the SMART database includes intrinsic protein features as well as precalculated results for Pfam [11] domains, these can be used together with SMART domains. For example, it is possible to identify receptor tyrosine kinases by searching for proteins that contain both a tyrosine kinase domain and a predicted transmembrane region (query “TyrKc AND TRANS”, Figure 21.8).

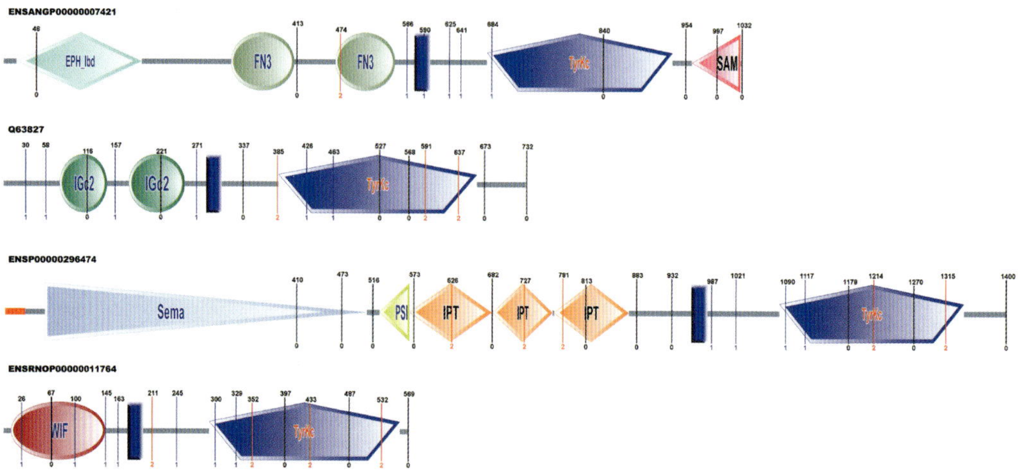

Figure 21.8 Using intrinsic features in domain architecture queries. The SMART database was queried for all proteins containing a tyrosine kinase domain and a transmembrane region ("TyrKc AND TRANS"), and 660 proteins were found, including the four displayed here. The color of domain names correlates with both subcellular localization (blue = extracellular, black = intracellular) and catalytic activity (red = catalytically active).

In addition to standard domain querying, SMART can be used to find proteins based on gene ontology (GO [8]) terms associated with domains. Associations of domains with GO are taken from Interpro [61]. Querying with GO terms is a two-step process. In the first step, the user obtains a list of domains matching the GO terms entered. After selecting the domains of interest from the list, proteins containing those domains are displayed. As with standard domain querying, results can be limited to specific taxonomic ranges.

Finding Proteins with Similar Domain Architecture

SMART can search for all proteins that have the same domain architecture as the query (having all the domains of the query protein in the same co-linear order) or that have an identical set of domains (at least one of all domain types of the query protein, irrespective of order). Identification of proteins having identical or near-identical domain architectures as the query sequence may improve predictions of protein functions. This feature also reveals, by using a taxonomic breakdown, the phyletic distribution of a given architecture.

21.3.3.4 Application of SMART

Apart from its use as a web tool, SMART has been applied to large-scale annotation projects, such as annotation of the human, mouse, and mosquito draft genome sequences [48, 95, 103]. It was also used in the investigation of single domain families in model organisms [40] and for the study of sequence conservation in multiple alignments [67]. In conjunction with genomic data, SMART was used for the study of conservation of gene (i.e., intron/exon) structure [13].

SMART has also been incorporated into other domain and protein family resources that are used for the primary annotation of sequence databases. It is a component database of Interpro [61], which contributes to the annotation of Swiss-Prot sequences [10], and of the Conserved Domain Database (CDD), which contributes to the annotation of RefSeq sequences [70].

21.3.4 Other Features and Resources

21.3.4.1 Globular Repeats

Repeats can be hard to detect in protein sequences: most of them are short, and their sequences are often highly divergent. The numbers of repeats in different proteins are extremely variable. Finally, defining the first and last residues of a repeat is more contentious than for a domain, since repeats are more prone to circular permutation than are domains, particularly within closed structures [74], and to partial truncation, resulting in non-integer repeat numbers.

Repeat detection methods are often incorporated into domain prediction servers (for example, SMART uses the Prospero [60] program from the Ariadne package), but dedicated protein repeat prediction servers also exist, e.g., REPeats (http://www.embl-heidelberg.de/~andrade/papers/rep/search.html [5]). The GlobPlot method described in Section 21.4.1.4 is also fairly capable in detecting repeats.

21.3.4.2 Domain Interaction Prediction

Although homology-based methods are a primary source of globular domain discovery, protein–protein interactions are also being used and becoming more popular for this purpose. One of the servers for exploration of such interactions is STRING. The STRING database (http://string.embl.de/) is dedicated to proteome-wide prediction of protein–protein associations [94]. It is an integrated resource relying on a wide range of experimental and computational datasets to make reliable interaction predictions. It contains genomic context associations (derived from genome comparisons), interactions derived from coexpression analysis, and various types of high-throughput experimental data, all of which are stringently benchmarked by using a common reference.

21.3.4.3 No Domains?

If no domains are found by, e.g., SMART or Pfam, this does not mean that your favorite protein does not contain any higher fold or globular domains. Most often, it simply indicates the presence of nonannotated domains, which of course has potentially higher discovery value. However, several resources exist to help identify potential domain boundaries and give hints as to the structure of what might be hidden in the sequence.

Secondary Structure Prediction: Good

Prediction of secondary structure is the most mature of any structure prediction strategy, and accuracies of up to ~80% can be achieved [20, 21, 68]. An initial

BLAST search to find homologous proteins is important to get a better idea of the function and to build a sequence set for a multiple alignment that can be used on secondary structure prediction servers such as PredictProtein (PROFsec, http://www.predictprotein.org/) and JPRED (http://www.compbio.dundee.ac.uk/~www-jpred/).

Tertiary Structure Prediction: Difficult

Prediction of tertiary structure and folds is still error-prone and difficult; however, having good secondary structure predictions at hand can assist this analysis. Perhaps the best approach is to submit the sequence to one of the homology-based prediction servers, such as the 3D-JURY metaserver (http://bioinfo.pl/Meta/ [34]) or SWISS-MODEL (http://www.expasy.org/swissmod/SWISS-MODEL.html [77]). Other resources can be found on the websites for the evaluation competitions CASP (http://predictioncenter.llnl.gov/casp5/Casp5.html) and CAFASP(http://bioinfo.pl/cafasp/).

Other Sequence Features: Narrowing Down Domain Boundaries

Single transmembrane segments (TM1), coiled coils, and low-complexity regions are all incorporated in the SMART server. Sometimes low-complexity regions are disordered (see Section 21.4.1.3). Coiled coils are also disordered sequences; however, they behave like globular units after the coiled-coil structure is formed, which is a very clear example of disorder–order transition.

At EMBL in Heidelberg we have two additional methods that are useful in the definition of potential domain boundaries: DomCut [84] and GlobPlot [52], see Section 21.4.1.4. Another resource for potential domain boundary prediction is DomPred (http://bioinf.cs.ucl.ac.uk/dompred/ [58]).

Many proteins are entirely and natively unstructured and without globular domains, and the rest of this chapter is dedicated to the analysis of this part of protein space.

21.4
Analyzing Nonglobular Protein Segments

Since most attention in assigning function to proteins has been on globular domains, there are relatively few tools for analyzing the nonglobular protein space. Structural biology has tended to avoid unstructured proteins and regions (e.g., by removing them in recombinants), which has led to a skew toward globular proteins in structural datasets.

However, this neglect is not confined to structural biology – bioinformatics has also tended to keep nonglobular function prediction under the academic carpet. Although resources are readily available for revealing globular domains in sequences, until recently there has not been any comprehensive collection of short functional sites/motifs comparable to the globular domain resources. Yet these are just as important for the function of multidomain proteins. Indeed, it is impossible for a researcher to find a list of currently known motifs – going through the literature to retrieve them is impractical without foreknowledge in more areas than any one person has. This neglect is primarily due to the fact that short sequence motifs are statistically insignificant and difficult to handle compared to domains for which accurate sequence models can be produced.

21.4.1
Unstructured Regions: Protein Disorder

The approach to finding functional sites is fundamentally different from the one described above for globular domains. Since linear motifs are often shorter than 10 amino acids, they overpredict massively even if they are described by using artificial neural networks or other sensitive probabilistic methods. However, linear motifs are context-dependent in the sense that they are functional only if they are exposed for interaction with a modular domain or in the right cell compartment. Structurally they prefer to be in nonglobular or disordered regions of the protein, both of which can be detected fairly accurately. A typical functional site is shown in Figure 21.9; notice the linear unstructured and flexible protein backbone, a requirement for the CSK kinase to be able to modify the tyrosine.

In the following we discuss how to find potentially nonglobular areas, including those that appear structurally disordered, and how to predict functional sites in them.

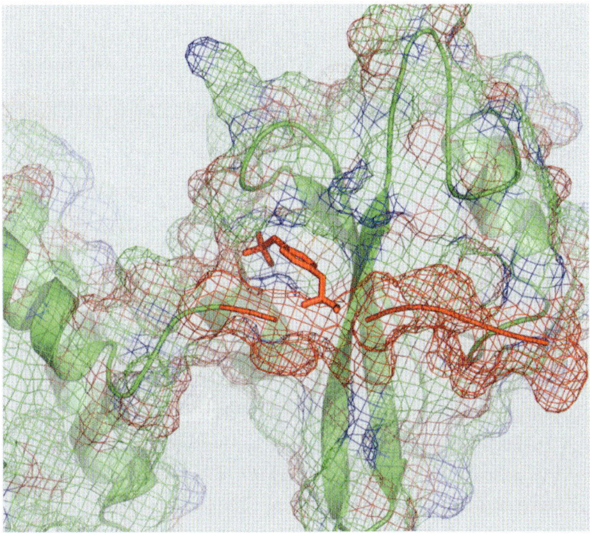

Figure 21.9 The C-terminal CSK Tyr-phosphorylation site in Src (PDB: 1fmk) in the closed conformation bound to the SH2 domain. This linear motif (red) shows the general features of an ELM: it is linear in sequence and structure space. The sequence of the instance of this functional site is TEPQYQPGE. In the ELM resource this is called the MOD_TYR_CSK functional site and is described by the pattern [TAD][EA].Q(Y)[QE].[GQA][PEDLS]. The image was created with PyMOL (Table 21.2).

Recently, it has become possible to analyze natively unstructured proteins by methods such as NMR. Besides their high content of functional sites, disordered and nonglobular regions are exciting for many other reasons.

21.4.1.1 What Role Does Protein Disorder Play in Biology?

Target Selection
In the post-genomic era, discovery of novel domains and functional sites in proteins is of growing importance. One focus of structural genomics initiatives is to solve structures for novel domains and thereby increase the coverage of fold and structure space [14]. During the target selection process in structural genomics/biology, it is important to consider intrinsic protein disorder, because disordered regions (at the N and C termini or even within domains) often lead to difficulties in protein expression, purification, and crystallization. It is therefore essential to be able to predict which regions of a target protein are potentially disordered/unstructured.

IDPs (Intrinsically Disordered Proteins)
Although IDPs (also known as intrinsically unstructured proteins) are under-researched, an increasing number are being found. These are proteins or domains that, in their native state, are either entirely disordered or contain large disordered regions. More than 100 such proteins are known, including Tau, prions, Bcl-2, p53, 4E-BP1, and HMG proteins (see Figure 21.14) [7, 47, 56, 90, 91].

Protein disorder is important for understanding protein function as well as protein folding pathways [67, 92]. Although little is understood about the cellular and structural meaning of IDPs, they are thought to become ordered only when bound to another molecule (e.g., CREB–CBP complex [72]) upon changes in the biochemical environment [27, 29].

Function of Disorder and IDPs
The current view on protein disorder is that it allows for more interaction partners and modification sites [53, 90, 99]. However, we have not been able to confirm this hypothesis by analyzing a large interaction dataset (unpublished results). This might be because such datasets are enriched in nontransient interactions, but interactions carried out by disordered proteins are transient.

Perhaps disordered proteins have evolved to provide a simple solution to having large intermolecular interfaces while keeping smaller protein, genome, and cell sizes [36]. It has been proposed that having several relatively low-affinity linear interaction sites allows for a flexible, subtle regulation as well as accounting for specificity and cooperative binding effects [31]. In light of the modular model described in Section 21.2.1, we can see how these sites can be used in a combinatorial manner to generate a very large set of potential interaction environments.

Protein Disorder and Disease
Structural disorder in proteins is now known to play a central role in diseases mediated by protein misfolding and aggregation [12, 45, 78]. Amylogenic diseases such as Alzheimer's, Type II diabetes, and BSE are thought to be related to the

occurrence of short linear motifs in unfolded regions. These motifs are important for initiation of the formation of the amyloid fibers that cause great harm to the cellular environment, particularly in brain tissue. There are several proposed peptide models for these motifs, and the structural context in which they occur are under investigation [24, 55, 63, 83].

Other diseases such as Parkinson's, Huntington's, and serpinopathies are related to misfolding of proteins. The understanding of protein misfolding is related to analysis of the unstructured ensemble or the unfolded state of a polypeptide. This state can be analyzed in natively disordered proteins [30].

How does one characterize protein disorder and nonglobular regions? The field of protein disorder studies has, so far, failed to reach any agreement on this.

21.4.1.2 **What is Protein Disorder?**

No commonly agreed definition of protein disorder exists. The thermodynamic definition of disorder in a polypeptide chain is the random-coil structural state. The random-coil state can best be understood as the structural ensemble spanned by a given polypeptide in which all degrees of freedom are used within the conformational space. However, even under extremely denaturing solvent conditions, such as 8 M urea, this theoretical state is not observed in solvated proteins [46, 66, 89]. Proteins in solution thus seem to always retain a certain amount of residual structure.

Protein disorder is observed by a variety of experimental methods, such as X-ray crystallography; NMR, Raman, and CD spectroscopy; and hydrodynamic measurements [29, 82]. In vivo studies of disorder are possible with NMR spectroscopy on living cells (e.g., anti-sigma factor FlgM [22]). Each of these methods detects different aspects of disorder, resulting in several operational definitions of protein disorder (see [90] for a review).

Regions without regular secondary structure can be predicted by the NORSp (nonregular structure) server [53]; however, as the authors point out, such regions are not necessarily disordered. Structures such as the Kringle domain (PDB: 1krn) are almost entirely without regular secondary structure in their native state, but they still have tertiary structure in which the basic building block is coils. These loopy proteins are not necessarily IDPs, since they can still form a well defined globular tertiary structure.

In our work we have used four definitions of protein disorder:

- Loops/coils as defined by DSSP [44]. Residues are assigned to one of several secondary structure types. For this definition we consider residues in an α-helix (H), 3_{10} helix (G), or β-strand (E) to be ordered and all other states (T, S, B, I) to be in loops (also known as coils). Loops/coils are not necessarily disordered (e.g., turns); however, protein disorder is found only within loops. It follows that one can use loop assignments as a necessary but not sufficient requirement for disorder; a disorder predictor based entirely on this definition is thus promiscuous.
- Hot loops constitute a refined subset of the above: namely, those loops having a high degree of mobility as determined from Cα temperature factors (B factors).

It follows that highly dynamic loops should be considered disordered. Several attempts have been made to try to use B factors for disorder prediction [15, 28, 32, 93, 104], but there are many pitfalls in doing so, because B factors can vary greatly within a single structure due to the effects of local packing and structural environment. Recent progress in deriving propensity scales for residue mobility based on B factors [81] has encouraged us to use B factors for defining protein disorder. The details for hot loops can be found in the methods part of [51].

- Missing coordinates/remark465 in X-ray structure, as defined by remark465 entries in the PDB. Nonassigned electron densities most often reflect intrinsic disorder and were used early, for disorder prediction [50].
- Russell–Linding propensities are parameters based on the hypothesis that the tendency for disorder can be expressed as P = RC – SS where RC and SS are the propensities for a given amino acid to be in random coil and regular secondary structure, respectively. This scale was defined during the development of the GlobPlot predictor described in Section 21.4.1.4.

Figure 21.10 shows the disorder propensities for each amino acid by our four definitions of disorder. A more detailed discussion of these values can be found in [51], but in general, hydrophobic residues promote order according to all definitions of disorder. Disorder-promoting residues include proline, lysine, serine, threonine, and methionine.

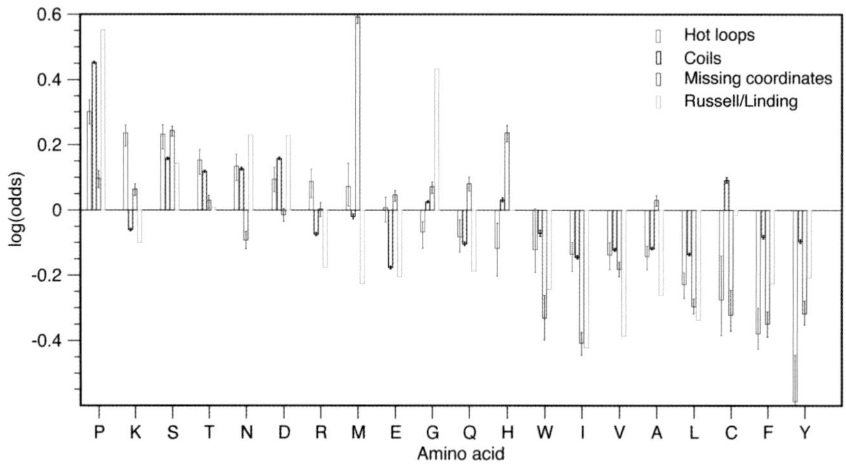

Figure 21.10 Propensities of the amino acids to be disordered, according to the definitions used in DisEMBL and GlobPlot (sorted by hot loop preference). This scale directly reflects the datasets used for training; however, it is only a rough approximation of what the DisEMBL neural networks use in predicting disorder. Error bars correspond to the 25th and 75th percentiles as estimated by stochastic simulation. The Russell–Linding scale is an absolute scale. Methionine suffers a bias in the remark465 dataset for at least two reasons: (1) often the N-terminal methionine is missing; and (2) some structures are solved using selenomethionine derivatives for phasing, which can lead to deletion of the residue in the PDB entry. The same bias is seen in ([29], Figure 10).

21.4.1.3 **Methods for Finding Protein Disorder**

Several other attempts have been made to predict disorder. Perhaps the earliest were methods of finding regions of low complexity. Although many such regions are structurally disordered, the correlation is far from perfect, because regions of low sequence complexity are not always disordered (and vice versa) [27]. Likely the strongest evidence for this correlation comes from the fact that low-complexity regions are rarely seen in protein 3D structures [75]. Methods to predict low complexity, like SEG [98] and CAST [69], are thus often used for this purpose. Methods using hydrophobicity can also give hints about disordered regions, because low-complexity regions are typically exposed and rarely hydrophobic.

The first tool designed specifically for prediction of protein disorder was PONDR (predictor of naturally disordered regions, http://www.pondr.com [32, 33, 73]). It is based on artificial neural networks. PONDR is, however, not freely accessible to academics. Refer to [59] for a recent evaluation of disorder prediction (DisEMBL was published after CASP5).

Prediction of protein tertiary structure may be an alternative route to disorder prediction, although such methods are computationally intensive and error-prone. Moreover, such methods are usually designed to predict the structure of globular domains, and their behavior with other sequences can be unpredictable.

At the EMBL in Heidelberg we have developed methods for finding unstructured regions from sequence data alone. These tools were primarily developed for use in the ELM project to help find regions potentially containing functional sites. However, these tools are now being used by several structural genomics initiatives and laboratories around the world who are either studying IDPs or trying to optimize their recombinant protein expression vectors by cutting out disordered segments.

21.4.1.4 **GlobPlotting**

GlobPlot was invented specifically to aid the ELM project; however, it proved to be of much wider interest [52]. From the beginning we wanted a graphical tool that could generate easy-to-interpret plots of the tendency within a sequence for structured or lack of structure. The basis for GlobPlot was the Russell–Linding scale mentioned earlier in Section 21.4.1.2. The combination of random coil and secondary structure in the Russell–Linding scale enhanced the discrimination of the graphs and was the key factor in the success of this scale at detecting both disorder and globular packing.

GlobPlot is not intended to be a competitor in secondary structure prediction, because it cannot give the same level of detail as can be obtained from secondary structure prediction based on multiple alignment. GlobPlot is an ab initio method, i.e., it requires only one sequence and can therefore be applied to novel sequences having no homologs, i.e., it does not use multiple alignment. The basic algorithm behind GlobPlot is beautifully simple and very fast: each amino acid a_i has a defined propensity $P(a_i)R$ (see Russell–Linding in Figure 21.10). Given a protein sequence of length L, we define a sum function $Dis(a_i)$ as follows:

$$Dis(a_i) = \sum_{j=1}^{L} P(a_i)$$

where $P(a_i)$ is the propensity for the *i*th amino acid. The GlobPlot webserver plots the function, and the graphs are referred to as globplots. Before plotting, the digital-smoothing Savitzky–Golay algorithm is used to reduce noise on the curve.

Analyzing a GlobPlot

Reading globplots is fairly easy, but different from, e.g., hydropathy plots, in that globplots are cumulative-sum curves rather than derivative curves. Because GlobPlot plots this running sum, the graph is analyzed by looking at the slope. The numbers on the ordinate do not matter, they equal the running sum, and we are interested only in whether or not a given segment of the graph is disordered. The latter is seen by the decrease or increase in the slope, because that is how the Russell–Linding scale works: negative values correspond to ordered residues, and positive values indicate disorder-promoting amino acids.

We designed GlobPlot like this because we think it results in profile-like, intuitive plots. In particular, we wanted to avoid a high-variation curve such as the derivative curve. The globplot in Figure 21.11 is a good example of one of these profile-like curves: the GlobPlot plot for mucin predicts that the central part of the protein is almost completely disordered (using the Russell–Linding disorder definition) – this is probably why this protein is so slimy.

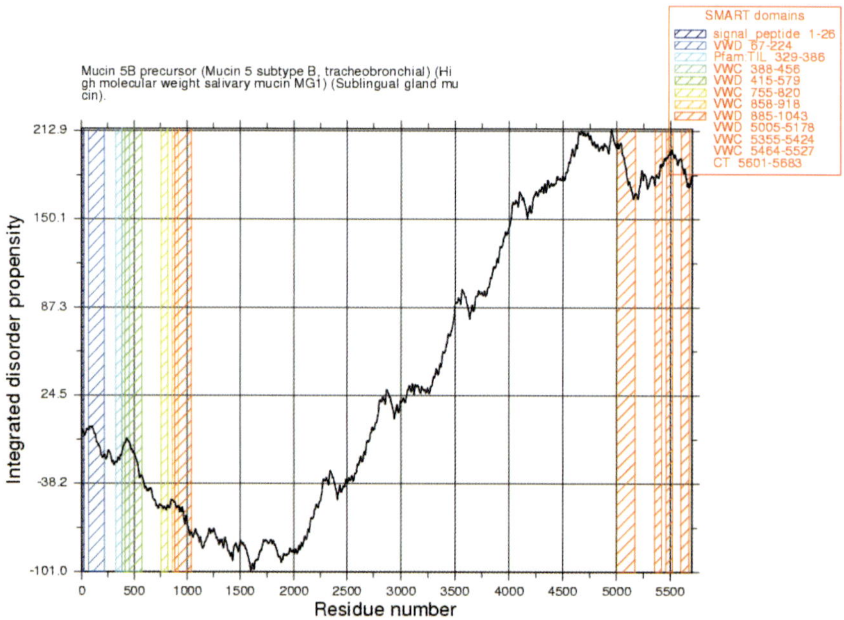

Figure 21.11 Globplot of human mucin 5 protein (Swiss-Prot: MU5B_HUMAN). Most of this slimy protein is highly disordered. Since GlobPlot plots a running sum of the propensity to disorder, the graph is analyzed by looking at its slope. The numbers on the ordinate axis do not really matter, it is the uphill or downhill tendency that should be read. Referring to Figure 21.10 indicates that disorder-promoting propensities are positive, so 'uphill' on the graph is equivalent to disorder.

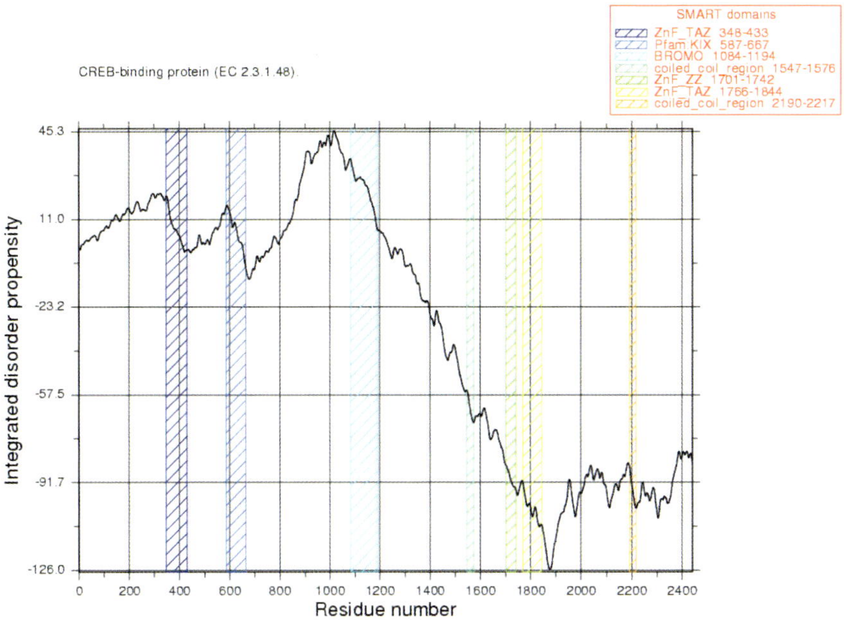

Figure 21.12 Globplot of human CREB-binding protein (CBP_HUMAN). About half of the sequence appears to be disordered, with long flexible regions observed at the N and C termini. The flexible region just after the KIX domain might be important for induced binding of the pKID domain of CREB to CBP [23, 72]. For further discussion of disorder in CBP/CREB see Wright et al. [99]. See Chapter 16 for information on the bromo domain.

Domain detection with GlobPlot is as easy as finding protein disorder, since both features are shown in the plot. To help you to navigate and understand the plots, the webserver overlays the graph with any predicted SMART domains. In domain hunting situations, you would look for downhill regions in the graph. As seen in Figure 21.12, GlobPlot can detect potential domains: notice the downhill slope whenever a domain is found by SMART/Pfam. GlobPlot often detects additional sequence to be ordered, this is because SMART and Pfam use only the most conserved sequence part of a domain to generate their hidden Markov models for the domain. This indicates that GlobPlotting is useful for domain boundary definition.

21.4.1.5 **Prediction of Multiple Types of Disorder with DisEMBL**

The performance of GlobPlot encouraged us to refine our approach and predict disorder in a more traditional biocomputational manner by training artificial neural network predictors for the various definitions of disorder mentioned above. This work led to the DisEMBL disorder predictor ensemble.

DisEMBL is a computational tool for prediction of disordered/unstructured regions within a protein sequence [51]. DisEMBL currently provides three alternative

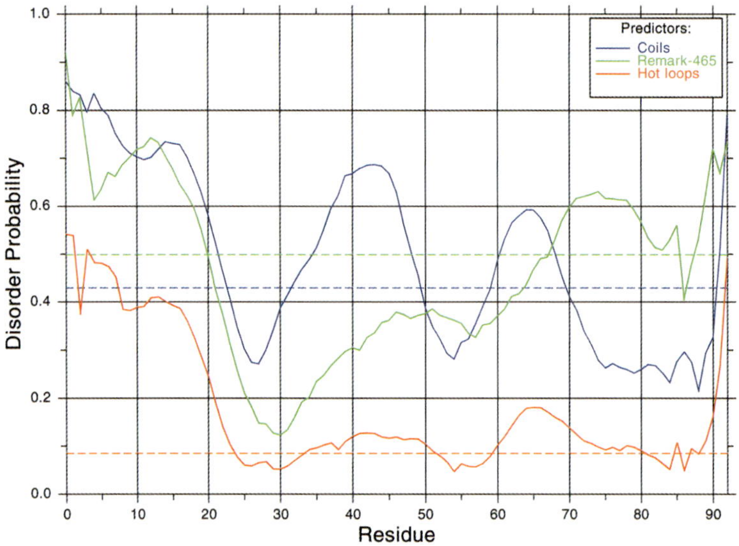

Figure 21.13 Sample output from the DisEMBL web server, showing predictions for yeast nonhistone chromosomal protein 6A (high mobility group protein, Swiss-Prot: NHPA_YEAST). The green curve shows the predictions obtained for missing coordinates, red for the hot loop network, and blue for coils. The horizontal lines correspond to the random expectation level for each predictor: for coils and hot loops the prior probabilities were used, and a neural network score of 0.5 was used for remark465. From this plot it is seen that the N-terminal tail of the protein is especially predicted to be disordered. See Figure 21.14 for a mapping of the hot loop predictions onto the structure of this protein.

disorder definitions: hot loops, coils, and missing coordinates as defined in Section 21.4.1.2. The coils predictor is used primarily as a filter to require disorder to be within coil-predicted regions (see Section 21.4.1.2). DisEMBL is a highly accurate predictor, predicting more than 60% of hot loops with fewer than 2% false positives [51].

Hot Loops

'Hot loops' is a novel definition of disorder based on X-ray data. We think that it will prove difficult to pull out a much more precise definition of disorder based on crystallographic data. An example of hot loop results is shown in Figure 21.14, where we mapped the probabilities shown in Figure 21.13 onto the structure of nonhistone chromosomal protein 6A from yeast. It is remarkable that a definition based on X-ray data can predict so well for NMR structures, arguing that this novel definition of disorder is relevant. We also showed this correlation earlier, as well as a comparison of the correlations between our alternative definitions of protein disorder [51].

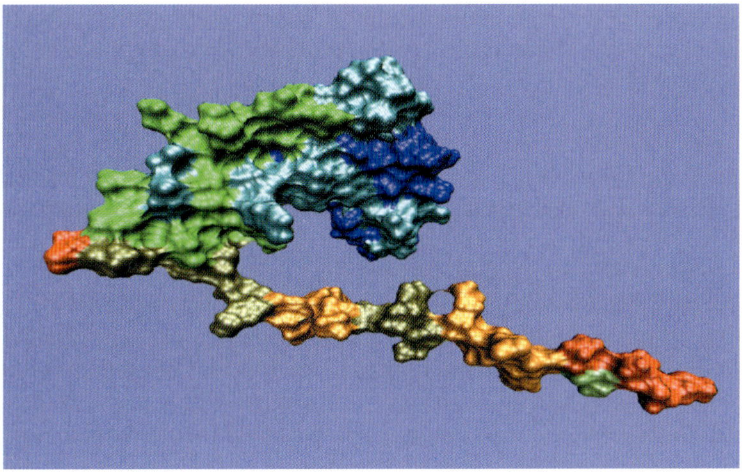

Figure 21.14 DisEMBL hot loop predictions mapped on the NMR structure of nonhistone chromosomal protein 6A (high mobility group protein, PDB: 1cg7; model 1, Swiss-Prot: NHPA_YEAST). The predicted probabilities are indicated with a color scale going from blue to red, where red corresponds to the most likely disordered regions and blue to ordered regions. The unstructured tail clearly shows the highest disorder scores (see Figure 21.13). Surface plot generated with VMD (see Table 21.2).

Using DisEMBL

DisEMBL is freely available via a web interface (http://dis.embl.de/) and can be downloaded for use in large-scale studies. The web interface is fairly straightforward to use, you can submit a sequence or enter the Swiss-Prot/SWALL accession (e.g., P08630) or entry code (e.g., HMG1_HUMAN). The server fetches the sequence and description of the polypeptide from an ExPASy server using Biopython.org software. The probability of disorder is shown graphically, as illustrated at Figure 21.13. The random expectation levels for the different predictors are shown on the graph as horizontal lines, but should merely be considered absolute minima. The default parameters are set for optimal prediction and should be changed only in rare situations. On-line documentation of the various settings is provided at http://dis.embl.de/help.html. If the query protein sequence is very long, > 1000 residues, you can download the predictions and use a local graphing/plotting tool such as Grace or OpenOffice.org to plot and zoom the data. A future version of DisEMBL may include a web applet for interactive plotting and zooming of the graphs.

GlobPlot and DisEMBL

The GlobPlot algorithm is very simple and intuitive, which is appealing. Although it was originally designed for prediction of protein disorder, the Russell–Linding propensity scale functions just as well for detection of domain boundaries, repeats, and other globular features. The Russell–Linding scale and the SMART domain overlay feature are unique to GlobPlot. DisEMBL is more accurate than GlobPlot in coil prediction, which is related to the Russell–Linding scale. It furthermore provides the novel hot-loop definition of disorder. The two methods complement each other, since they approach disorder prediction differently. In general, we urge you to submit your sequences to both tools.

21.4.1.6 **Design of Protein Expression Vectors**

As mentioned earlier, protein disorder is related to problems during protein expression, purification, and crystallization. Other tools such as TANGO (http://tango.embl.de/) deal with protein cross-beta aggregation which is different from the disorder in solvated proteins that our tools predict.

We believe that identification of potential disordered regions should provide a good basis for setting up expression vectors and/or comparing the data with obtained structural data. However, currently we cannot assess which of the definitions of disorder is most appropriate for design of protein expression vectors. We thus strongly encourage feedback on successes and failures in using DisEMBL for expression and structural analysis of proteins.

21.4.2
Function Prediction for Nonglobular Protein Segments

Having identified candidate unstructured regions, one can start searching for function in them. Most functions correlate with short linear peptide motifs that are used for cell compartment targeting, protein–protein interaction, regulation by phosphorylation, acetylation, glycosylation, and a host of other post-translational

Functional Group	Functional Sites (ELMs)
Targeting signals	KDEL, NLS, NES Mitochondrial Import signals PTSI/II, Signal peptides
Linear protein interaction sites	EH, PDZ, SH2, SH3, WW, PCNA, LXXLE Calmodulin/Cyclin binding helix Protease inhibitors Active peptides and hormones
Covalent modification sites	Phosphorylation, methylation, acetylation glycosylation, fucosylation, prenylation myristoylation, ubiquitinylation
RNA interaction motifs	RX, RGY, RGG motifs
Processing sites	Diverse protease cleavage targets in secretion, apoptosis etc.

Figure 21.15 Main classes of functional sites. Functional sites are as varied and numerous as domains are. On a proteome level we expect at least five sites per protein, resulting in about 150 000 instances in the human proteome. This indicates the presence of a gigantic and complex interaction and regulatory system.

modifications. See Figure 21.15 for an overview of the many functions these sites perform. The number of known categories of functional sites has increased dramatically in the past few years, and it is obvious that there are more to be discovered. These sites are usually short and often reveal themselves in multiple sequence alignments as short patches of conservation, leading to their definition as linear motifs. In addition to occurring outside globular domains, some sites, e.g., phosphorylation sites, are often found in exposed flexible loops protruding from globular domains. Considering the abundance of targeting signals and post-translational modification sites, it is reasonable to assume that there are more functional sites than globular domains in a higher eukaryotic proteome.

21.4.2.1 **Available Resources**

ELM is the largest collection of linear motifs, followed by Scansite and PROSITE [64, 80, 101]. Scansite is a very capable resource focusing on cell signaling. It complements ELM in using position-specific scoring matrices (PSSMs) for prediction, which are more sensitive than the regular expressions ELM uses. However, Scansite does not provide an annotated database similar to ELM.

A series of individual predictors of functional sites can be found at http://www.cbs.dtu.dk/services/ which is hosted by the Center for Biological Sequence Analysis in Denmark. The CBS focuses on providing high-performance neural network predictors but without any annotated knowledgebase interface, taking a complementary approach to the other resources.

The PROSITE database has collected a number of linear protein motifs, representing them as regular expressions. PROSITE patterns have been very useful but suffered from severe overprediction; more recently the database has emphasized globular domain annotation at the expense of linear motifs.

Also of interest are protein interaction databases such as BIND and DIP [9, 100]. More informative protein interaction databases that store known instances of linear motifs include MINT [102] and Phospho.ELM at http://phospho.elm.eu.org/. Databases of instances are not directly useful for prediction but provide valuable data-mining resources.

It was recently demonstrated that short functional sites or protein features are crucial for the classification of protein function [41]. The Protfun method is an ab initio method for prediction of higher functional classes based on sequence features alone [42].

21.4.3
The Eukaryotic Linear Motif Resource: ELM

In this section we describe the ELM resource in detail, since it is the largest resource for linear motifs.

The Eukaryotic Linear Motif server (http://elm.eu.org/ [71]) is a new bioinformatics resource for investigating candidate short functional motifs in eukaryotic proteins. Some of the concepts used within ELM are defined in Table 21.1. An example of the concepts used in practice can be seen in Figures 21.16 and 21.17.

Table 21.1 Definitions of concepts used in the ELM resource. Functional sites are, as opposed to, e.g., active sites, short and linear in sequence and structure space. In the ELM resource we describe functional sites as linear motifs. Here, the linear motif is shown as a regular expression or pattern, but it could as well have been another type of sequence model, e.g., a hidden Markov model.

Concept	*Definition*	*Example*
A functional site	A set of short linear (sub)sequences that can be related to a molecular function	LIG_RBBD: Rb pocket-interacting sequence
An ELM	The common pattern of a set of linear (sub)sequences that can be related to a molecular function	[LI].C.[DE]
An ELM instance	An instance of an ELM in a particular polypeptide	RBB1_HUMAN: LVCHE

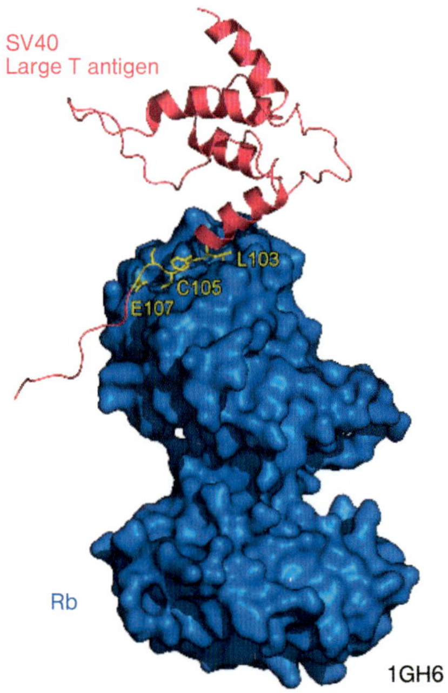

Figure 21.16 LIG_RBBD is a functional site responsible for interaction with retinoblastoma (Rb) family proteins. Rb proteins are known for repression of E2F proteins, which are required for transcription of proteins important in the cell cycle. The figure shows the SV40 large T antigen interacting with the Rb pocket (PDB: 1gh6).

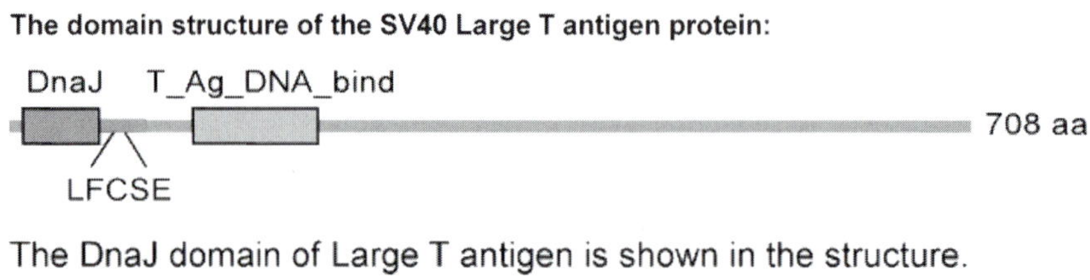

Figure 21.17 The location of the instance (LFCSE) of LIG_RBBD in the SV40 large T antigen protein is C-terminal to the DnaJ globular domain.

Linear motifs are short (usually < 10 amino acids) and therefore difficult to evaluate, since the usual significance assessments are inappropriate. Therefore, the ELM resource deploys logical context filters to eliminate false positives. The prediction strategy ELM uses is what we call knowledge-based decision support (KBDS). The basic idea is that, since we cannot discriminate ELMs based on sequence matching, we can use a knowledge base of contextual information regarding functional sites and ELMs to filter out false positives. This knowledge base is created/curated manually from the scientific literature. Currently, KBDS filters are in place for cell compartment, globular domain clash, and taxonomic range. In favorable instances, the filters can reduce the number of retained matches by an order of magnitude or more.

21.4.3.1 **ELM Annotation – 'Site seeing'**

All data input is by hand curation, performed by trained molecular biologists. Annotating each ELM is called 'Site seeing' and includes the processes shown in Figure 21.18. To promote interoperability with other bioinformatics resources, ELM uses three public annotation standards. Gene ontology (GO) identifiers are used for cell compartment, molecular function, and biological process [8, 39], and the NCBI taxonomy database identifiers [97] are used for taxonomic nodes at the apex of phylogenetic groupings in which an ELM occurs. Annotations of ELM instances are assigned ontology terms from the Proteomics Standards Initiative Molecular Interaction ontologies for evidence methods (HUPO.org). In the future the ELM resource will be able to report known instances of ELMs with details about what kind of experiments were performed to show the instance, with links to the relevant literature.

The motif patterns are currently represented as POSIX regular expressions (usable in the Python and PERL languages), analogous to PROSITE patterns, but with a different syntax. For example, the FxDxF motif, which is responsible for the binding of accessory endocytic proteins to the alpha subunit of adaptor protein complex AP-2, has a consensus sequence of F-x-D-x-F and is written F. D. F. Linear motifs in ELM will in the future include motif descriptions according to the Seefeld convention nomenclature for linear motifs (see Chapter 22 in this book and [1]).

In the future, ELM might incorporate HMMs or other sensitive search methods; nevertheless, linear motifs will continue to overpredict and require alternative approaches for reducing the levels of false positives.

21.4.3.2 **ELM Resource Architecture**

The core of the ELM resource is a relational database, powered by PostgreSQL, storing data about linear motifs. Figure 21.19 outlines how the ELM server is implemented. The user submits a protein sequence to the server and receives a list of matching ELMs that have been filtered to remove false positives (it may naturally include false negatives and residual false positives). Matched motifs are usually not statistically significant, and overprediction occurs despite filtering; hence matches should be considered to represent potential true instances of functional sites and should be used as guides to experimental determination.

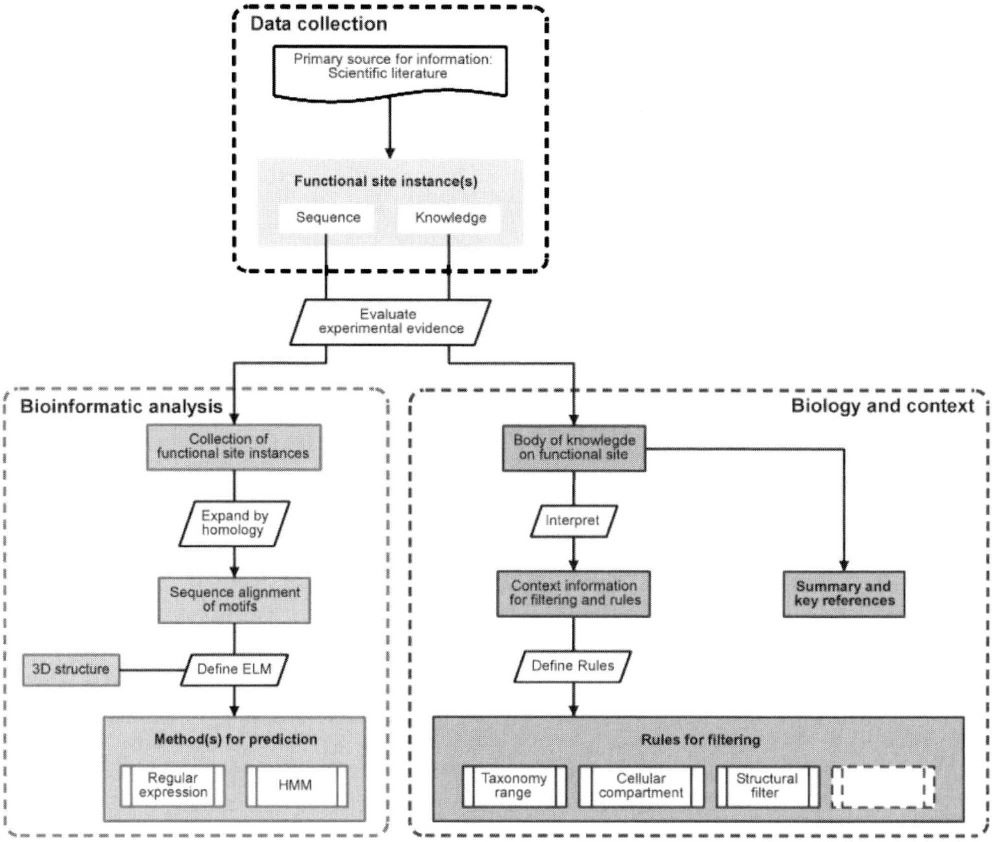

Figure 21.18 The flow of the 'siteseeing' process typically involves extensive literature searches, BLAST runs, multiple alignment of relevant protein families, perusal of Swiss-Prot and other online databases, and, where practical, discussion with experimentalists from the field. The empty box symbolizes additional future strategies.

21.4.3.3 **Knowledge-based Decision Support (KBDS): ELM Filtering**

Sequence-matching methods find many false – but apparently plausible – instances of ELMs that somehow are not recognized by their cognate binding/modification domains. There are two explanations for this:

- One obvious reason why a sequence that matches a motif is not a true functional site is that the motif does not fully and accurately represent the functional site. This can partly be solved by deploying more sophisticated sequence models such as PSSMs or artificial neural networks, an approach used by Scansite and CBS.

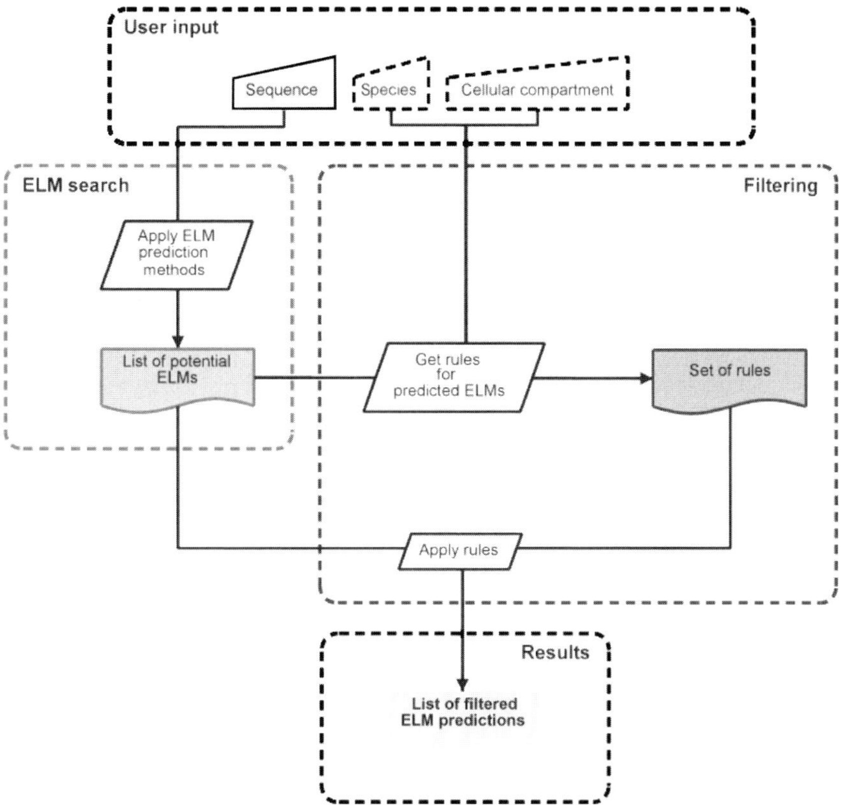

Figure 21.19 Flowchart of the ELM server. Dashed boxes indicate the four stages from input to results. As the server is further developed, more filters will be added to allow more query-dependent data to be retrievable.

- Another reason is that the sequence matches (potential ELM instances) occur in an irrelevant context. They may match a sequence from a wrong cellular compartment or from a species that does not use this functional site. As we have seen, the structural context is also of great importance for linear motifs to be reachable so as to be functional.

It is possible to develop context filters that remove such false positives. In ELM we do most of this inside the knowledge database that the 'siteseeing' process is building. The ELM database is designed to accommodate these types of filters or KBDS modules.

Currently, three filters are installed on the ELM server. These filters are not completely accurate and introduce false negatives occasionally, although we try to avoid this as much as possible. In general, the approach in ELM is to predict as few false positives as possible, but it is even more important to avoid false negatives.

Cell Compartment Filter

In ELM every linear motif is annotated with GO terms for the set of cell compartments in which it is known to function. For example, KDEL is a signal for retention of the host protein by the endoplasmic reticulum, whereas the SUMO site applies to proteins in the nucleus and the PML body. The user specifies the compartments in which the query protein functions, and all matches for ELMs not found in these compartments are filtered out. In the future ELM may support prediction of compartments using LOC3D [62].

Globular Domain Filter: A Two-track Filtering Strategy

Globular domains identified with the SMART and Pfam (domain subset) resources are used for filtering out ELMs. This filter has two tracks:

- a domain filter,
- an ELM rescue or reinstate module.

The domain filter works simply by removing all ELMs within the boundaries of the SMART/Pfam domains matching the same sequence, since they are false positives. The primitive assumption here is that sites within globular units are not accessible and therefore not functional, clearly an oversimplification.

ELMs can occur inside certain domains, e.g., the internal tyrosine phosphorylation sites in the active loops of tyrosine kinase domains, as is described in Section 21.2.1. This later group of ELMs are to a certain extent being 'rescued' by the ELM rescue module, i.e., for some ELMs certain SMART/Pfam domains are simply not used for filtering in the domain filter.

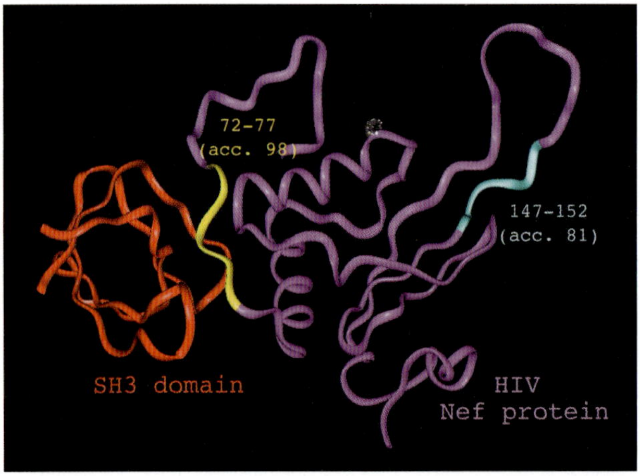

Figure 21.20 The V-1 Nef protein (magenta) in complex with wild-type Fyn SH3 (red) domain (PDB: 1avz) contains two potential SH3 ELMs. Residue numbers are given as well as the accessibility (acc) of the highlighted fragments. The yellow sequence is the only one that binds to the Fyn partner. The cyan putative ELM is covered by a loop and has a accessibility of 81% (compared to the 98% of the true binding domain).

Given the limited accuracy of the domain filter, the unfiltered results are provided on the results front page. In many situations, users can investigate surface accessibility by examining an available 3D structure, by using a good-quality 2D structure prediction [20, 21, 68], or perhaps by using a homology modeling server such as SWISS-MODEL or the 3D-JURY metaserver [34, 77]. We are currently developing better domain filters, e.g., using surface accessibility from known structures to discriminate false from true positives. A good example of how this might work is shown in Figure 21.20.

Taxonomic Filtering

Some types of functional sites are found in all eukaryotes, e.g., the ER retention signal KDEL is universal. But others are restricted to specific eukaryotic taxa. Perhaps most strikingly, the large receptor tyrosine kinase multigene family is found only in metazoa. Each ELM is annotated with one or more NCBI taxonomy nodes to indicate its known phylogenetic distribution. The user provides the query species, and all ELMs that are not assigned to its lineage are filtered out.

21.4.3.4 **Using ELM**

The public ELM webserver allows you to retrieve filtered as well as unfiltered raw results. This approach should encourage you to think critically about ELM server results. Figure 21.21 shows the ELM server output using the human Src sequence as a query. This example indicates the potential of the KBDS approach for improving motif searches. A pipeline interface to ELM prediction for use in proteome analysis is currently being developed and implemented; this pipeline and the results will be made available as soon as possible.

The predictive power of the ELM resource can be enhanced by harnessing it to other data, including experimental results. For example, many protein kinase recognition sites are among those which severely overpredict. If a protein is known not to be phosphorylated, kinase sites can all be ignored; but if it is known to be phosphorylated, then the kinase-site matches can be targeted for experimental testing. Mass spectrometry can be a useful tool for revealing post-translational modifications. ELM can provide synergism with appropriate experiments and can help in mapping out a research program. In this way, the ELM resource should become increasingly useful to the research community

21.5 URLs

In Table 21.2 we have listed some URLs we thought might be useful for you to explore. Many more links can be found in the annual database (http://nar.oupjournals.org/content/vol31/issue1/) and webserver (http://nar.oupjournals.org/content/vol31/issue13/) open access issues of *Nucleic Acids Res.earch.*

Table 21.2 Some resources referred to in this chapter. For more information please see the individual websites.

Resource	*Function classes*	*http://*
SMART	globular modular domains	smart.embl.de
Pfam	globular modular domains	www.sanger.ac.uk/Software/Pfam/
Interpro	Globular domain meta server	www.ebi.ac.uk/interpro/
CDD	globular modular domains	web.ncbi.nlm.nih.gov/Structure/odd/odd.shtml
PROSITE	Domain signatures and a few linear motifs	www.expasy.ch/sprot/prosite.html
PredictProtein	Secondary structure prediction	www.predictprotein.org
ELM	Functional sites, ELMs, linear motifs	elm.eu.org
PyMOL	Very nice and easy to use molecule viewer and renderer	pymol.sourceforge.net
VMD	Feature rich molecule viewer	www.ks.uiuc.edu/Research/vmd/
Scansite	Phosphorylation and signaling motifs	scansite.mit.edu
Protfun	Enzyme categories and higher functional classes	www.cbs.dtu.dk/services/ProtFun
NetNglyc	N-glycosylation motifs	www.cbs.dtu.dk/services/NetNGlyc
PredictNLS	Nuclear localization signals	cubic.bioc.columbia.edu/predictNLS
SignalP	Cleavage sites & signal/non-signal peptide prediction	www.cbs.dtu.dk/services/SignalP
PSORT	Protein sorting signals	psort.nibb.ac.jp
Sulfinator	Tyrosine sulfation motifs	us.expasy.org/tools/sulfinator
GlobPlot	Protein disorder and globularity	globplot.embl.de
DisEMBL	Protein disorder	dis.embl.de
GO	biological function, component and process	www.geneontology.org
Ensemble	Genome browsing	www.ensembl.org
Phospho.ELM	Instances of Ser/Thr/Tyr phosphorylation	phospho.elm.eu.org
Perl	Script oriented language widely used in bioinformatics	www.perl.com
Python	Highly object oriented language designed for large projects	www.python.org
HMMER	Hidden Markov Model software suite	hmmer.wustl.edu
Biopython	Bioinformatics modules for perl	www.biopython.org
Bioperl	Bioinformatics modules for Python	www.bioperl.org

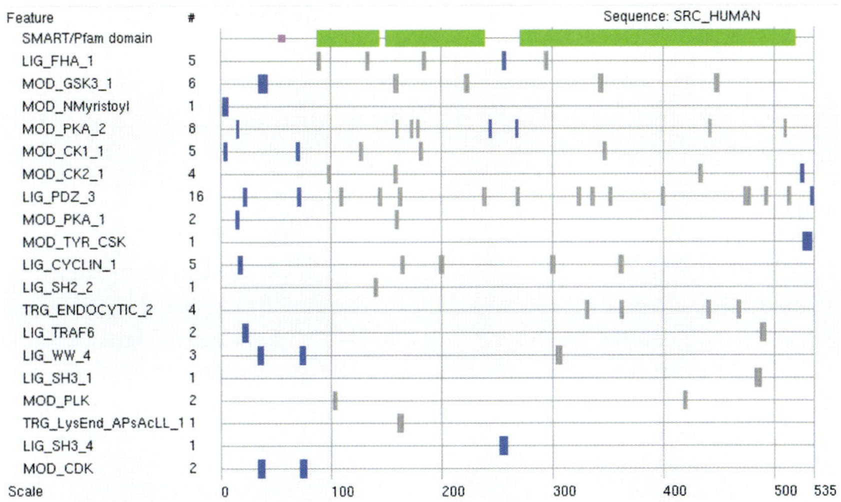

Figure 21.21 Sample output from the ELM server. The query sequence was Src (Swiss-Prot: SRC_HUMAN). The surviving ELMs are shown in blue, and the motifs that have been filtered out are shown in grey. This figure illustrates only how the globular domain (green) filter works: of the 103 ELMs in the resource at the time of writing, 27 match the sequence, but two are removed by the species filter, seven by the compartment filter, and five by the SMART/Pfam domain filter. Of the remaining 14 ELMs that survive the filtering, six are known to be true, and two are false negatives, i.e., not predicted by ELM (the C-terminal SH2 ligand and the autophosphorylation site within the tyrosine kinase domain; compare with Figure 21.1). The functionality of the ELM rescue module is not shown in this figure.

21.6 Concluding Remarks

We hope that you now have a pretty clear idea of how to approach the analysis of your favorite modular protein. In this chapter we have not discussed all available resources for analysis of proteins, we apologize to the authors of these resources.

The mapping of globular domains should be considered mature – methods such as Pfam and SMART are highly reliable for determining potential domains in a sequence. The prediction of functional sites is a much younger field, although advancements have been made with nonsequence approaches such as the KBDS system in ELM.

The paradigm behind this chapter and the modular model of protein function is that sequence determines structure, which again determines function. This is clearly true in many instances; however, like any dogma, it is ultimately wrong and misleading. Our view of protein function is still very primitive. We expect the modular model to be enveloped by a more holistic model.

It does indeed seem as if nature is presenting molecular functions in two modes: structured domains that are folded and in which the fold/structure determines the function of the domain/protein, and an unstructured mode like the one we see for

ELMs. These are modular units which seem to behave like autonomous bit/information strings carried within the host protein to accommodate certain functions or tuning of the host structure – they are themselves unstructured and only their sequence determines their function.

Acknowledgements

This work was partly supported by EU grant QLRI-CT-2000-00127. Thanks to Kresten Lindorff-Larsen, Sophie Chabanis-Davidson, Sara Quirk, and Francesca Diella for commenting on this chapter. Thanks to Pål Puntervoll and Manuela Helmer Citterich for figures. Finally, we are deeply grateful to FreeBSD.org, (bio)Python.org, PostgreSQL.org, Debian.org, Gentoo.org, and Apache.org for fantastic open free software.

References

1 Aasland, R., et al., Normalization of nomenclature for peptide motifs as ligands of modular protein domains. *FEBS Lett.* **2002**, *513*, 141–144.

2 Aloy, P., Russell, R., The third dimension for protein interactions and complexes. *Trends Biochem. Sci.* **2002**, *27*, 633–638.

3 Altschul, S., Madden, T., Schaffer, A., Zhang, J., Zhang, Z., Miller, W., Lipman, D., Gapped BLAST and PSI-BLAST: a new generation of protein database search programs. *Nucleic Acids Res.* **1997**, *25*, 3389–3402.

4 Andrade, M., Bioinformatics and Genomes Current Perspectives. Horizon Scientific Press **2003**.

5 Andrade, M., Ponting, C., Gibson, T., Bork, P., Homology-based method for identification of protein repeats using statistical significance estimates. *J. Mol. Biol.* **2000**, *298*, 521–537.

6 Apic, G., Gough, J., Teichmann, S., An insight into domain combinations. *Bioinformatics* **2001**, 17 Suppl 1, S83–89.

7 Aritomi, M., Kunishima, N., Inohara, N., Ishibashi, Y., Ohta, S., Morikawa, K., Crystal structure of rat Bcl-xL: implications for the function of the Bcl-2 protein family. *J. Biol. Chem.* **1997**, *272*, 27886–27892.

8 Ashburner, M., Gene ontology: tool for the unification of biology. The Gene Ontology Consortium. *Nat. Genet.* **2000**, *25*, 25–29.

9 Bader, G., Betel, D., Hogue, C., BIND: the Biomolecular Interaction Network Database. *Nucleic Acids Res.* **2003**, *31*, 248–250.

10 Bairoch, A., Apweiler, R., The SWISS-PROT protein sequence database and its supplement TrEMBL in 2000. *Nucleic Acids Res.* **2000**, *28*, 45–48.

11 Bateman, A., et al., The Pfam protein families database. *Nucleic Acids Res.* **2002**, *30*, 276–280.

12 Bates, G., Huntingtin aggregation and toxicity in Huntington's disease. *Lancet* **2003**, *361*, 1642–1644.

13 Betts, M., Guigo, R., Agarwal, P., Russell, R., Exon structure conservation despite low sequence similarity: a relic of dramatic events in evolution? *EMBO J.* **2001**, *20*, 5354–5360.

14 Brenner, S., Target selection for structural genomics. *Nat. Struct. Biol.* **2000**, 7 Suppl, 967–969.

15 Brooks, B., Karplus, M., Normal modes for specific motions of macromolecules: application to the hinge-bending mode of lysozyme. *Proc. Natl. Acad. Sci. USA* **1985**, *82*,4995–4999.

16 Chandonia, J., Walker, N., Lo Conte, L., Koehl, P., Levitt, M., Brenner, S., ASTRAL compendium enhancements. *Nucleic Acids Res.* **2002**, *30*, 260–263.

17 Clamp, M., et al., Ensembl 2002: accommodating comparative genomics. *Nucleic Acids Res.* **2003**, *31*, 38–42.

18 Copley, R., Doerks, T., Letunic, I., Bork, P., Protein domain analysis in the era of complete genomes. *FEBS Lett.* **2002**, *513*, 129–134.

19 Creighton, T., Proteins Structures and Molecular Properties, 2nd edit. Freeman, New York **1993**.

20 Cuff, J., Barton, G., Application of multiple sequence alignment profiles to improve protein secondary structure prediction. *Proteins* **2000**, *40*, 502–511.

21 Cuff, J., Clamp, M., Siddiqui, A., Finlay, M., Barton, G., JPred: a consensus secondary structure prediction server. *Bioinformatics* **1998**, *14*, 892–893.

22 Dedmon, M., Patel, C., Young, G., Pielak, G., FlgM gains structure in living cells. *Proc. Natl. Acad. Sci. USA* **2002**, *99*, 12681–12684.

23 Demarest, S., Martinez-Yamout, M., Chung, J., Chen, H., Xu, W., Dyson, H., Evans, R., Wright, P., Mutual synergistic folding in recruitment of CBP/p300 by p160 nuclear receptor coactivators. *Nature* **2002**, *415*, 549–553.

24 Dobson, C., Protein misfolding and human disease. *ScientificWorldJournal* **2002**, 2 (1 Suppl 2), 132.

25 Doerks, T., Bairoch, A., Bork, P., Protein annotation: detective work for function prediction. *Trends Genet* **1998**, *14*, 248–250.

26 Doerks, T., Copley, R., Schultz, J., Ponting, C., Bork, P., Systematic identification of novel protein domain families associated with nuclear functions. *Genome Res.* **2002**, *12*, 47–56.

27 Dunker, A., Brown, C., Lawson, J., Iakoucheva, L., Obradovic, Z., Intrinsic disorder and protein function. *Biochemistry* **2002**, *41*, 6573–6582.

28 Dunker, A., et al., Protein disorder and the evolution of molecular recognition: theory, predictions and observations. *Pac. Symp. Biocomput.* **1998**, 473–484.

29 Dunker, A., et al., Intrinsically disordered protein. *J. Mol. Graph. Model* **2001**, *19*, 26–59.

30 Dyson, H., Wright, P., Equilibrium NMR studies of unfolded and partially folded proteins. *Nat. Struct. Biol.* **1998**, 5 Suppl, 499–503.

31 Evans, P., Owen, D., Endocytosis and vesicle trafficking. *Curr. Opin. Struct. Biol.* **2002**,12, 814–821.

32 Garner, E., Cannon, P., Romero, P., Obradovic, Z., Dunker, A., Predicting disordered regions from amino acid sequence: common themes despite differing structural characterization. *Genome Inform Ser Workshop Genome Inform* **1998**, *9*, 201–213.

33 Garner, E., Romero, P., Dunker, A., Brown, C., Obradovic, Z., Predicting binding regions within disordered proteins. *Genome Inform. Ser. Workshop Genome Inform.* **1999**, *10*, 41–50.

34 Ginalski, K., Elofsson, A., Fischer, D., Rychlewski, L., 3D-Jury: a simple approach to improve protein structure predictions. *Bioinformatics* **2003**, *19*, 1015–1018.

35 Gough, J., The SUPERFAMILY database in structural genomics. *Acta Crystallogr. D Biol. Crystallogr.* **2002**, *58*, 1897–1900.

36 Gunasekaran, K., Tsai, C., Kumar, S., Zanuy, D., Nussinov, R., Extended disordered proteins: targeting function with less scaffold. *Trends Biochem. Sci.* **2003**, *28*, 81–85.

37 Harrison, A., Pearl, F., Sillitoe, I., Slidel, T., Mott, R., Thornton, J., Orengo, C., Recognizing the fold of a protein structure. *Bioinformatics* **2003**, *19*, 1748–1759.

38 Henikoff, J., Greene, E., Pietrokovski, S., Henikoff, S., Increased coverage of protein families with the blocks database servers. *Nucleic Acids Res.* **2000**, *28*, 228–230.

39 Hill, D., Blake, J., Richardson, J., Ringwald, M., Extension and integration of the gene ontology (GO): combining GO vocabularies with external vocabularies. *Genome Res.* **2002**, *12*, 1982–1991.

40 Hill, E., Broadbent, I., Chothia, C., Pettitt, J., Cadherin superfamily proteins in *Caenorhabditis elegans* and

Drosophila melanogaster. *J. Mol. Biol.* **2001**, *305*, 1011–1024.

41 Jensen, L., Ussery, D., Brunak, S., Functionality of system components: conservation of protein function in protein feature space. *Genome Res.* **2003**, *13*, 2444–2449.

42 Jensen, L. J., et al., Prediction of human protein function from post-translational modifications and localization features. *J. Mol. Biol.* **2002**, *319*, 1257–1265.

43 Jonassen, I., Eidhammer, I., Taylor, W., Discovery of local packing motifs in protein structures. *Proteins* **1999**, *34*, 206–219.

44 Kabsch, W., Sander, C., Dictionary of protein secondary structure: pattern recognition of hydrogen-bonded and geometrical features. *Biopolymers* **1983**, *22*, 2577–2637.

45 Kaplan, B., Ratner, V., Haas, E., Alpha-synuclein: its biological function and role in neurodegenerative diseases. *J. Mol. Neurosci.* **2003**, *20*, 83–92.

46 Klein-Seetharaman, J., et al., Long-range interactions within a nonnative protein. *Science* **2002**, *295*, 1719–1722.

47 Kussie, P., Gorina, S., Marechal, V., Elenbaas, B., Moreau, J., Levine, A., Pavletich, N., Structure of the MDM2 oncoprotein bound to the p53 tumor suppressor transactivation domain. *Science* **1996**, *274*, 948–953.

48 Lander, E., et al., Initial sequencing and analysis of the human genome. *Nature* **2001**, *409*, 860–921.

49 Letunic, I., et al., Recent improvements to the SMART domain-based sequence annotation resource. *Nucleic Acids Res.* **2002**, *30*, 242–244.

50 Li, X., Obradovic, Z., Brown, C., Garner, E., Dunker, A., Comparing predictors of disordered protein. *Genome Inform Ser Workshop Genome Inform* **2000**, *11*, 172–184.

51 Linding, R., Jensen, L., Diella, F., Bork, P., Gibson, T., Russell, R., Protein disorder prediction: implications for structural proteomics. *Structure* **2003**, *11*, 1453–1459.

52 Linding, R., Russell, R., Neduva, V., Gibson, T., GlobPlot: exploring protein sequences for globularity and disorder. *Nucleic Acids Res.* **2003**, *31*, 3701–3708.

53 Liu, J., Tan, H., Rost, B., Loopy proteins appear conserved in evolution. *J. Mol. Biol.* **2002**, *322*, 53–64.

54 Lo Conte, L., Brenner, S., Hubbard, T., Chothia, C., Murzin, A., SCOP database in 2002: refinements accommodate structural genomics. *Nucleic Acids Res.* **2002**, *30*, 264–267.

55 Lopez De La Paz, M., Goldie, K., Zurdo, J., Lacroix, E., Dobson, C., Hoenger, A., Serrano, L., De novo designed peptide-based amyloid fibrils. *Proc. Natl. Acad. Sci. USA* **2002**, 99, 16052–16057.

56 Lopez Garcia, F., Zahn, R., Riek, R., Wuthrich, K., NMR structure of the bovine prion protein. *Proc. Natl. Acad. Sci. USA* **2000**, *97*, 8334–8339.

57 Marchler-Bauer, A., et al., CDD: a curated Entrez database of conserved domain alignments. *Nucleic Acids Res.* **2003**, *31*, 383–387.

58 Marsden, R., McGuffin, L., Jones, D., Rapid protein domain assignment from amino acid sequence using predicted secondary structure. *Protein Sci.* **2002**, *11*, 2814–2824.

59 Melamud, E., Moult, J., Evaluation of disorder predictions in CASP5. *Proteins* **2003**, 53 Suppl 6, 561–565.

60 Mott, R., Accurate formula for P-values of gapped local sequence and profile alignments. *J. Mol. Biol.* **2000**, *300*, 649–659.

61 Mulder, N., et al., The InterPro Database, 2003, brings increased coverage and new features. *Nucleic Acids Res.* **2003**, *31*, 315–318.

62 Nair, R., Rost, B., LOC3D: annotate subcellular localization for protein structures. *Nucleic Acids Res.* **2003**, *31*, 3337–3340.

63 Nilsson, M., Dobson, C., In vitro characterization of lactoferrin aggregation and amyloid formation. *Biochemistry* **2003**, *42*, 375–382.

64 Obenauer, J., Cantley, L., Yaffe, M., Scansite 2.0: Proteome-wide prediction of cell signaling interactions using short sequence motifs. *Nucleic Acids Res.* **2003**, *31*, 3635–3641.

65 Orengo, C., Pearl, F., Thornton, J., The CATH domain structure database. *Methods Biochem Anal* **2003**, *44*, 249–271.

66 Pappu, R., Srinivasan, R., Rose, G., The Flory isolated-pair hypothesis is not valid

for polypeptide chains: implications for protein folding. *Proc. Natl. Acad. Sci. USA* **2000**, *97*, 12565–12570.

67 Pei, J., Grishin, N., AL2CO: calculation of positional conservation in a protein sequence alignment. *Bioinformatics* **2001**, *17*, 700–712.

68 Pollastri, G., Przybylski, D., Rost, B., Baldi, P., Improving the prediction of protein secondary structure in three and eight classes using recurrent neural networks and profiles. *Proteins* **2002**, *47*, 228–235.

69 Promponas, V., Enright, A., Tsoka, S., Kreil, D., Leroy, C., Hamodrakas, S., Sander, C., Ouzounis, C., CAST: an iterative algorithm for the complexity analysis of sequence tracts: complexity analysis of sequence tracts. *Bioinformatics* **2000**, *16*, 915–922.

70 Pruitt, K., Maglott, D., RefSeq and LocusLink: NCBI gene-centered resources. *Nucleic Acids Res.* **2001**, *29*, 137–140.

71 Puntervoll, P., et al., ELM server: a new resource for investigating short functional sites in modular eukaryotic proteins. *Nucleic Acids Res.* **2003**, *31*, 3625–3630.

72 Radhakrishnan, I., Perez-Alvarado, G., Parker, D., Dyson, H., Montminy, M., Wright, P., Solution structure of the KIX domain of CBP bound to the transactivation domain of CREB: a model for activator:coactivator interactions. *Cell* **1997**, *91*, 741–752.

73 Romero, P., Obradovic, Z., Kissinger, C. R., Villafranca, J., Dunker, A., Identifying disordered proteins from amino acid sequences. *Proc. IEEE Int. Conf. Neural Networks* **1997**, *1*, 90–95.

74 Russell, R., Ponting, C., Protein fold irregularities that hinder sequence analysis. *Curr. Opin. Struct. Biol.* **1998**, *8*, 364–371.

75 Saqi, M., Sternberg, M., Identification of sequence motifs from a set of proteins with related function. *Protein Eng.* **1994**, *7*, 165–171.

76 Schultz, J., Milpetz, F., Bork, P., Ponting, C., SMART, a simple modular architecture research tool: identification of signaling domains. *Proc. Natl. Acad. Sci. USA* **1998**, *95*, 5857–5864.

77 Schwede, T., Kopp, J., Guex, N., Peitsch, M., SWISS-MODEL: An automated protein homology-modeling server. *Nucleic Acids Res.* **2003**, *31*, 3381–3385.

78 Schweers, O., Schonbrunn-Hanebeck, E., Marx, A., Mandelkow, E., Structural studies of tau protein and Alzheimer paired helical filaments show no evidence for beta structure. *J. Biol. Chem.* **1994**, *269*, 24290–24297.

79 Servant, F., Bru, C., Carrere, S., Courcelle, E., Gouzy, J., Peyruc, D., Kahn, D., ProDom: automated clustering of homologous domains. *Brief Bioinform* **2002**, *3*, 246–251.

80 Sigrist, C., Cerutti, L., Hulo, N., Gattiker, A., Falquet, L., Pagni, M., Bairoch, A., Bucher, P., PROSITE: a documented database using patterns and profiles as motif descriptors. *Brief Bioinform.* **2002**, *3*, 265–274.

81 Smith, D., Radivojac, P., Obradovic, Z., Dunker, A., Zhu, G., Improved amino acid flexibility parameters. *Protein Sci.* **2003**, *12*, 1060–1072.

82 Smyth, E., Syme, C., Blanch, E., Hecht, L., Vasak, M., Barron, L., Solution structure of native proteins with irregular folds from Raman optical activity. *Biopolymers* **2001**, *58*,138–151.

83 Stefani, M., Dobson, C., Protein aggregation and aggregate toxicity: new insights into protein folding, misfolding diseases and biological evolution. *J. Mol. Med.* **2003**, *81*, 678–699.

84 Suyama, M., Ohara, O., DomCut: prediction of inter-domain linker regions in amino acid sequences. *Bioinformatics* **2003**, *19*, 673–674.

85 Taylor, W., Protein structural domain identification. *Protein Eng.* **1999**, *12*, 203–216.

86 Taylor, W., A deeply knotted protein structure and how it might fold. *Nature* **2000**, *406*, 916–919.

87 Taylor, W., Defining linear segments in protein structure. *J. Mol. Biol.* **2001**, *310*, 1135–1150.

88 Taylor, W., Lin, K., Protein knots: a tangled problem. *Nature* **2003**, *421*, 25.

89 Teilum, K., Kragelund, B., Poulsen, F., Transient structure formation in unfolded acyl-coenzyme A–binding

protein observed by site-directed spin labeling. *J. Mol. Biol.* **2002**, *324*, 349–357.

90 Tompa, P., Intrinsically unstructured proteins. *Trends Biochem. Sci.* **2002**, *27*, 527–533.

91 Uversky, V., Natively unfolded proteins: a point where biology waits for physics. *Protein Sci.* **2002**, *11*, 739–756.

92 Verkhivker, G., Bouzida, D., Gehlhaar, D., Rejto, P., Freer, S., Rose, P., Simulating disorder–order transitions in molecular recognition of unstructured proteins: where folding meets binding. *Proc. Natl. Acad. Sci. USA* **2003**, *100*, 5148–5153.

93 Vihinen, M., Torkkila, E., Riikonen, P., Accuracy of protein flexibility predictions. *Proteins* **1994**, *19*, 141–149.

94 von Mering, C., Huynen, M., Jaeggi, D., Schmidt, S., Bork, P., Snel, B., STRING: a database of predicted functional associations between proteins. *Nucleic Acids Res.* **2003**, *31*, 258–261.

95 Waterston, R., et al., Initial sequencing and comparative analysis of the mouse genome. *Nature* **2002**, *420*, 520–562.

96 Westbrook, J., Feng, Z., Chen, L., Yang, H., Berman, H., The Protein Data Bank and structural genomics. *Nucleic Acids Res.* **2003**, *31*, 489–491.

97 Wheeler, D., et al., Database resources of the National Center for Biotechnology Information: 2002 update. *Nucleic Acids Res.* **2002**, *30*, 13–16.

98 Wootton, J., Nonglobular domains in protein sequences: automated segmentation using complexity measures. *Comput Chem* **1994**, *18*, 269–285.

99 Wright, P., Dyson, H., Intrinsically unstructured proteins: reassessing the protein structure–function paradigm. *J. Mol. Biol.* **1999**, *293*, 321–331.

100 Xenarios, I., Salwinski, L., Duan, X., Higney, P., Kim, S., Eisenberg, D., DIP, the Database of Interacting Proteins: a research tool for studying cellular networks of protein interactions. *Nucleic Acids Res.* **2002**, *30*, 303–305.

101 Yaffe, M., Leparc, G., Lai, J., Obata, T., Volinia, S., Cantley, L., A motif-based profile scanning approach for genome-wide prediction of signaling pathways. *Nat. Biotechnol.* **2001**, *19*, 348–353.

102 Zanzoni, A., Montecchi-Palazzi, L., Quondam, M., Ausiello, G., Helmer-Citterich, M., Cesareni, G., MINT: a molecular interaction database. *FEBS Lett.* **2002**, *513*, 135–140.

103 Zdobnov, E., et al., Comparative genome and proteome analysis of *Anopheles gambiae* and *Drosophila melanogaster*. *Science* **2002**, *298*, 149–159.

104 Zoete, V., Michielin, O., Karplus, M., Relation between sequence and structure of HIV-1 protease inhibitor complexes: a model system for the analysis of protein flexibility. *J. Mol. Biol.* **2002**, *315*, 21–52.

22 Nomenclature for Protein Modules and Their Cognate Motifs

Pål Puntervoll and Rein Aasland

22.1 Introduction

More than 80% of eukaryotic proteins have a modular architecture, in the sense that they contain two or more globular protein domains [1]. In general, globular domains are considered structurally independent, and they can often be associated with one or more molecular functions. For this reason, molecular biological research often takes a protein domain-oriented approach. Similarly, bioinformatics analysis of protein sequence and structure is also, to a large extent, focused on protein domains. This is illustrated by the fact that the most important part of the annotation of newly sequenced genomes and their cognate proteomes is based on protein domain-based sequence comparisons.

Many globular protein domains act on short sequences present in other proteins, by binding or post-translational modification (including cleavage). We refer to such short sequence motifs as functional sites [2]. Most chapters in this book deal with globular domains and their interactions with functional sites. Following a recent meeting on protein modules, a nomenclature for description of such short sequence motifs was developed and has become known as the Seefeld convention [3]. This nomenclature was developed primarily as a tool for standardized description of short motifs in scientific papers and other communications. A computer-readable version of the Seefeld convention was also developed. Here we discuss the properties of the Seefeld convention and other methods for representation of functional sites and their cognate recognition modules.

22.2 Protein Modules

Let us first briefly consider the terminology for parts of proteins. Some of the terms used to describe proteins have been used for a very long time, and their meanings have become less precise. The term 'protein' is currently used to denote both single

Modular Protein Domains. Edited by G. Cesareni, M. Gimona, M. Sudol, M. Yaffe

ISBN: 3-527-30813-X

polypeptides as well as protein complexes and even collections of different proteins. Similarly, the term 'domain' is used with quite different meanings. Many molecular biologists often use 'domain' in a loose sense to denote any subpart of a protein that has a particular property. Bioinformatics people are more likely to refer to a domain as "an independent structural unit found alone or in combination with other domains or repeats" [4]. Structural biologists might define a domain even more precisely as "a relatively rigid region, moving as a rigid body that is separated from the other domains by more flexible inter-domain regions" [5]. Similarly, the terms 'sequence motif' and 'pattern' are used with different and overlapping meanings.

For the purpose of our discussion, we first explain how we use the various terms for the parts of proteins. All parts of a protein are considered to be regions or features. We use the term 'region' to denote any subsequence (or set of subsequences) of a protein and 'feature' to denote structural or functional units. We then distinguish between structural and functional features, so that terms of both classes can be meaningfully used for the same protein part.

The description of structural parts of proteins has long traditions, with terms for description of primary, secondary, and tertiary structure. Furthermore, different levels of description of protein structure, such as architectures, folds, and domains, are hierarchically classified in resources such as CATH [6] and SCOP [7]. The term 'protein domain' is, however, used to denote both structural and functional units. However, the increasing prominence of data resources such as Pfam [8], SMART [9], and InterPro [10], for detection of globular protein domains based on multiple sequence alignments, is causing this term to become more strongly associated with functional units (see also [11]). This is underscored by the fact that Pfam entries for which no function is yet known are denoted 'domain of unknown function' (DUF). Globular protein domains are therefore usually denoted by names from these databases. For explicit reference to such protein domains, one should provide an appropriate database reference (e.g., 'chromo domain', InterPro: IPR000953 or Pfam: PF00385). Similarly, a domain instance can be specified by giving both sequence reference and domain reference, e.g., sw: HP1_DROME/Pfam: chromo/ or sw: HP1_DROME/InterPro: IPR000953/.

The term 'protein module' is often used synonymously with 'protein domain'. The term module, however, strongly implies that modules are parts that can be connected to other modules in various combinations. As such, the term module relates to the evolutionary process that results in new module combinations. It is also interesting to note that 'module' is most often used in conjunction with a qualifying term, as in 'structural module', 'functional module', and 'recognition module'. As an extension to the concept of modular architecture of proteins, we also consider other parts of proteins to be modules, such as transmembrane helices, coiled-coil regions, and functional sites.

To refer precisely and unambiguously to protein parts, it is necessary to have a controlled vocabulary of terms, also known as an ontology. An ontology for protein parts is currently being developed by the Sequence Ontology (SO) consortium [12] (S. Lewis and M. Ashburner, personal communication).

22.3
Functional Sites and Their Recognition Modules

We define functional sites in proteins as short linear sequence motifs that serve as ligands or post-translational modification sites. Ligands represent those motifs that undergo noncovalent interactions with other proteins, and post-translational modification sites are motifs that are subject to covalent modifications (Figure 22.1). We illustrate this with four examples: (1) the LxxLL motif (elm: LIG_NRBOX, where x is any amino acid), which mediates interaction between nuclear receptors and their cofactors; (2) the KDEL motif (elm: TRG_ER_KDEL), which serves as an ER-retention signal; (3) post-translational modification sites, which include sites for phosphorylation, acetylation, etc.; and (4) protease cleavage sites [13].

A particular sequence can serve multiple functions, depending on its molecular and cellular context, e.g., Lys9 on histone H3 is a target for histone acetyl transferases. Once acetylated, it can serve as binding site for a bromo domain as well as being a substrate for a histone deacetylase (see Figure 22.4). The motif recognized by each of the three proteins may, however, be different.

Functional sites are different from binding sites and active sites in enzymes (see Figure 22.1). Active sites usually reside in globular domains and are composed of several amino acid residues located in different regions of the protein. Binding sites are similar to active sites, except that they do not contain catalytic residues. We use the term 'recognition module' to denote protein domains (or other protein modules) that recognize functional sites. Residues in binding sites and active sites

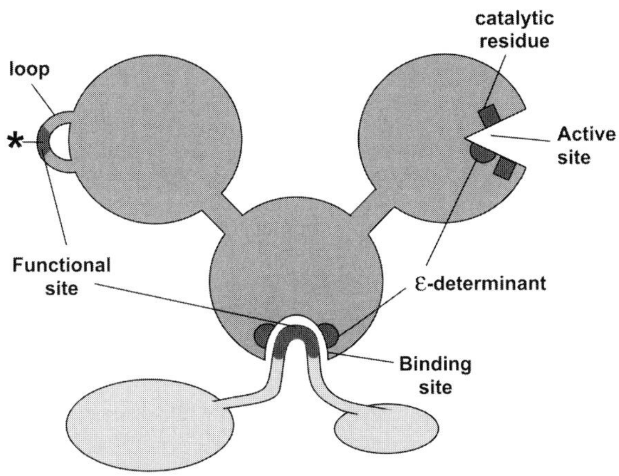

Figure 22.1 Functional sites and recognition modules. Functional sites are defined here as short linear motifs and are found in two molecular contexts – in loops of globular domains (as shown in the dark-grey recognition module protein) or in nonglobular regions (as shown in the light-grey protein). The two main classes of functional sites are shown: the covalently modifiable type (marked with an asterisk) is modified by the active site of a recognition module, and the noncovalent ligand type is recognized by the binding site. ε determinants are residues that are important in the recognition of functional sites.

that are important for recognition of functional sites have been termed ε determinants (see Figure 22.1) [14]. This terminology is implemented in the Seefeld convention.

It is possible for one protein domain to serve as a recognition module for more than one functional site. The ligand-binding domain of nuclear receptors (Pfam: HOLI) is an interesting example. When the peroxisome proliferator-activated receptor-α (PPARα) is *not* bound to its agonist ligand, its ligand-binding domain has a second binding site for transcriptional corepressors such as SMRT, which use the functional site called the corepressor NR box (elm: LIG_CORNRBOX). When the agonist docks in its binding site, the surface of the ligand binding domain changes, the binding site for SMRT disappears, and a new binding site appears in the same region, namely a binding site for interaction with transcriptional coactivators such as SRC-1, which uses the functional site known as the NR box (elm: LIG_NRBOX) [15].

As illustrated in Figure 22.1, functional sites are often found in otherwise disordered regions in proteins. When functional sites reside in globular domains, they appear in surface-exposed loops. This bias toward disordered regions is employed in a simple context filter in the ELM resource by which functional sites predicted in globular domains are removed [2, 11].

22.4
Representation of Motifs and Functional Sites

Functional-site motifs can be represented in several ways for different purposes (Figure 22.2). The first part of this section discusses the presentation of a motif in a format suitable for visual inspection, and the second part deals with computer-readable representations.

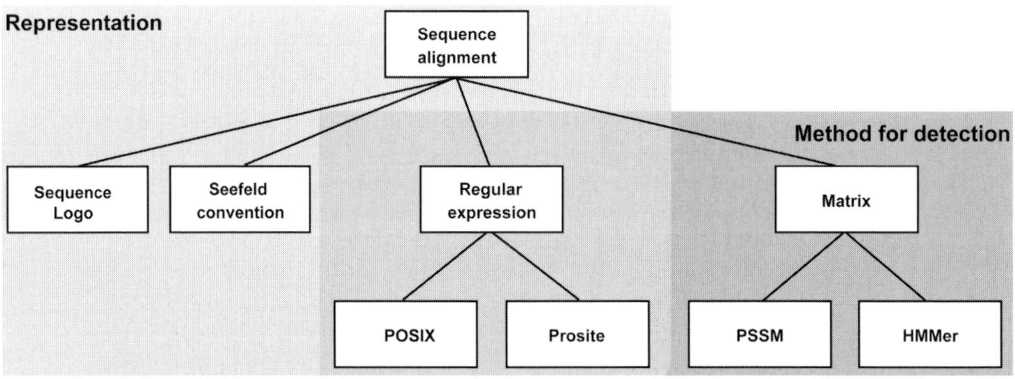

Figure 22.2 Types of representations of motifs and functional sites. The relationships between different types of methods for representation and detection of motifs and functional sites are shown. See text for further details.

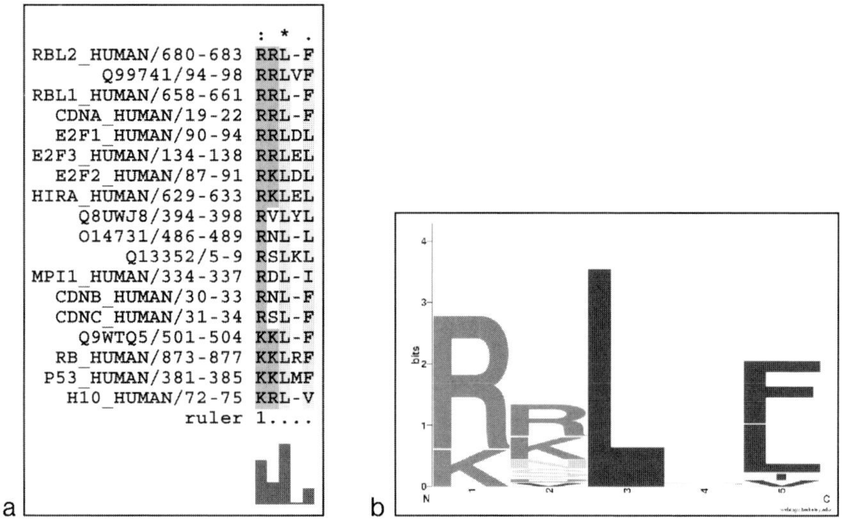

c

Seefeld Convention:

(i) *fn* [+] x L (F/L/I/V) *fc* (ii) *fn* [+] x L Φ *fc*

(iii) *fn* [+] x L x (F/L/I) *fc* (iv) *fn* [+] x L x Φ *fc*

Prosite regular expression:

[RK]-x-L-x(0,1)-[FLIV]

Posix regular expression:

[RK].L.{0,1}[FLIV]

Figure 22.3 Various ways of representing the cyclin ligand motif (elm: LIG_CYCLIN). (a) The cyclin ligand motif in 18 human proteins is shown as represented by a ClustalX alignment of the core-motif residues only. (b) The alignment was submitted to the WebLogo application (http://weblogo.berkeley.edu/logo.cgi), and the resulting sequence logo representation is shown. (c) The Seefeld convention representation of the cyclin ligand motif: (i) and (iii) together describe the motif accurately; (ii) and (iv) are more generalized representations, in which the symbol for any hydrophobic residue (Φ) is used to denote the last position. Two Seefeld representations are needed because of the inability of the current Seefeld convention to express gaps (see text).

It is traditional in molecular biology to present motifs as consensus sequences. The nuclear receptor binding motif found in transcriptional coactivators is an example of a motif that can be described by its consensus sequence, namely LxxLL. Although this is often sufficiently precise for simple motifs, representations of more complex motifs often become imprecise or unreadable. The example shown in Figure 22.3 illustrates this. The cyclin ligand peptide (elm: LIG_CYCLIN), found in a number of proteins that interact with cyclin/cyclin-dependent kinase (cyclin/CDK), is frequently referred to as the RxL motif in the literature (e.g., [16]), referring to the consensus sequence. However, the RxL consensus represents only a subset of the experimentally verified cyclin ligand peptides (Figure 22.3a), since many have K instead of R (i.e., KxL). Furthermore, the motif is 4–5 residues long; thus, the commonly used consensus sequence describes the motif only partially. In fact, a multiple sequence alignment is the representation of choice if full information on all observed motif instances is needed. The sequence logo representation (Figure 22.3b) [17] can often be more intuitive. In this graphical representation the one-letter amino acid codes for each position are stacked on top of each other, with letter heights corresponding to the relative frequency of each amino acid at that position.

Both multiple sequence alignments and sequence logos must be presented as graphics. The Seefeld convention is designed to represent a motif embedded in normal text. The rules (syntax) for the Seefeld convention are given in Table 22.1.

Table 22.1 Summary of the Seefeld convention rules and symbols and equivalent POSIX regular expressions for representing consensus sequences of functional sites.

Summary of Seefeld rule	*Amino acids*[a]	*Seefeld symbol*	*ASCII version of symbol*	*POSIX regular expression*[b]
Hydrophobic	VILFWYM	Φ	%	[VILFWYM]
Aromatic	FWY	Ω	@	[FWY]
Hydrophilic	NQSTEDKRH	ζ	&	[NQSTEDKRH]
• Positively charged	KR	[+]	[+]	[KR]
• Negatively charged	DE	[−]	[−]	[DE]
Aliphatic	VILM	Ψ	#	[VILM]
Small	PGAS	π		[PGAS]
Phosphorylated	Y	*po*Y	[Y:po]	n.p.
Sulfated	Y	*su*Y	[Y:su]	n.p.
O-glycosylated	S	*gl*S	[S:gl]	n.p.
N-glycosylated	N	*gl*N	[N:gl]	n.p.
Methylated	R	*me*R	[R:me]	n.p.
• Symmetrical	R	*sme*R	[R:sme]	n.p.
• Asymmetrical	R	*ame*R	[R:ame]	n.p.
Acetylated	K	*ac*K	[K:ac]	n.p.
Hydroxylated	P	*hy*P	[P:hy]	n.p.
Excluded	n.p.			
Unknown/any	n.p.	x	x	.
C-terminal ligand core flanking sequence	n.p.	*fc*	fc	n.p.
N-terminal ligand core flanking sequence	n.p.	*fn*	fn	n.p.

a) Amino acids are grouped into subsets based on their physiochemical properties, and the amino acids belonging to each subset are listed. For post-translational modifications, only one representative amino acid is shown.

b) n.p.: not possible.

In Figure 22.3c we illustrate the use of the Seefeld convention with the cyclin ligand motif as an example. Note that the Seefeld convention in its present form does not handle gaps. Gapped motifs must therefore be described by two or more representations. Hence, we suggest that ranges could be added to the Seefeld convention rules, using the same representation as in POSIX regular expressions: {*min, max*} (see [18]). The cyclin ligand motif can then be reformulated as *fn* [+] x L x{0,1} Φ *fc*. Although the Seefeld convention is well suited for describing such unmodified functional site motifs, its true strength lies in its ability to describe covalently modified residues (see below).

For computational purposes, the Seefeld convention has an ASCII version, in which Greek symbols are replaced by standard ASCII symbols and text formatting (such as italics) is not allowed (Table 22.1). Currently, there is no tool that can search for Seefeld motif matches within a protein sequence, but in principle, this should be possible. The Seefeld convention is in many respects similar to regular expressions (Table 22.1), which are used for detecting short functional site motifs, e.g., in the web resources ELM [2] and Prosite [19]. ELM uses the POSIX (UNIX-style) regular expression, whereas Prosite has its own syntax [20]. The cyclin ligand motif represented by both Prosite and POSIX regular expressions is also shown in Figure 22.3c. This example illustrates that regular expressions are also suitable for easy presentation of motifs. Whereas use of regular expressions seems to be the most efficient method for detecting short functional site motifs, traditional matrix-based methods, such as position-specific scoring matrices (PSSM) and hidden Markov models (HMM), can be applied to longer motifs [21–23].

22.5 Application of the Seefeld Convention to a Complex Example

As an interesting test case for application of the Seefeld convention, we apply it to one of the regulatory switches in the N-terminal tail of histone H3 (Figure 22.4 [24]). This example nicely illustrates the interconnections that may occur between different functional sites. For example, when the (unmodified) Lys9 in histone H3 is recognized by a histone methyltransferase, it becomes methylated (i.e., H3: *dme*K9), and a new functional site is formed, which can be recognized by the HP1 chromo domain [25, 26]. This example can also be used to illustrate the use of ε determinants [3, 14] with the Seefeld convention. The NMR structure of H3: *dme*K9 bound to the *Drosophila* sw: HP1_DROME/chromo shows that the methylated lysine is recognized by the three aromatic sidechains sw: HP1_DROME/chromo: Y24,W45,Y48, which in Seefeld nomenclature would be represented as β0–2 = Y, β2–7 = W, and β2:β3–3 = Y (see pdb: 1kna [26]). Thus, to denote a particular instance of an ε determinant, we propose the notation: sw: HP1_DROME/chromo: β0–2 = Y. Here we have shown a simple and unambiguous way to denote functionally important residues both with respect to absolute position in the polypeptide chain and with reference to the secondary structural elements in the chain. The nomenclature for the ε determinants does not, however, take into account the fact that

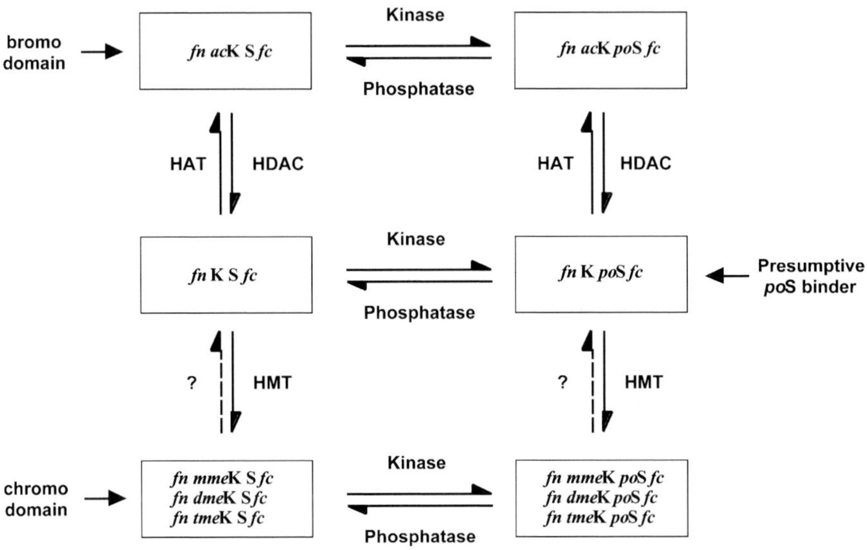

Figure 22.4 A complex example: application of the Seefeld convention to a regulatory switch on the N-terminal tail of histone H3. The figure illustrates both experimentally verified events taking place at Lys9 and Ser10 on H3, as well as postulated events. The enzymes responsible for each modification are HAT: histone acetyltransferase; HDAC: histone deacetylase; kinase: Ser/Thr protein kinase; phosphatase; HMT: histone methyltransferase. The '?' indicates the possible existence of a histone lysine demethylase. To include the three different methylated lysines, we propose an extension to the Seefeld convention in that *mme*, *dme*, and *tme* would denote mono-, di-, and trimethylated lysines, respectively. The bromo and chromo domains are protein modules known to recognize acetylated and methylated lysines, respectively. It is not known how the phosphorylated Ser10 is recognized.

different criteria are used for the assignment of secondary structural elements. Therefore, this notation must be used with reference to the source of the secondary structure definition, such as a pdb file.

22.6 New Directions

The increasing amount and variety of proteomics data being generated challenges both our understanding of proteins and the computational handling of this type of information. The Proteomics Standards Initiative (PSI) of the Human Proteome Organisation (HUPO) aims to develop ontologies for precise reference to proteomics data [27]. As discussed above, post-translational modifications require particular attention.

The SwissProt database contains information on modifiable residues in its feature tables, and PDB contains information on modified residues observed in solved structures. Beyond this, only specialized databases contain information on modified sites, for example, PhosphoBase and O-GLYCBASE [28, 29]. Furthermore, we know from several examples such as the histone tails that post-translational modifications can modulate the function of a protein in a temporal fashion. It is therefore necessary to develop a system for precise representation of modifications both in time and space. One aspect of protein function that is affected by post-translational modifications is the ability of proteins to interact with other proteins. Protein–protein interaction databases, such as MINT [30], BIND [31], and DIP [32], currently do not contain information on such conditional interactions. We hope that the Proteins Standard Initiative [27] will address these issues.

Handling different combinations of post-translational modifications also represents a challenge when predicting functional sites. Today, the only option when using resources such as ELM, Prosite, and Scansite [33] is to submit unmodified amino acid sequences. This implies that functional sites that actually contain modifications are represented in the nonfunctional, unmodified state. To take advantage of information on post-translational modifications in the prediction of functional sites in the ELM resource as well as in other bioinformatics tools, a format for representing such modifications is needed. Although the Seefeld convention (and its ASCII version) has the flexibility to achieve this, it does not currently allow for description of all known modifications (note that we have proposed a notation for mono-, di-, and trimethylation of lysine in Figure 22.4). Furthermore, the Seefeld convention does not allow for representation of modifiable residues. It is, however, beyond the scope of this chapter to discuss all implications of this urgent problem.

Acknowledgements

Thanks to Morten Mattingsdal, Nathalie Reuter, Rune Linding, and Toby J. Gibson for helpful comments and to Michael Ashburner and Suzanna Lewis for information on Sequence Ontology. P. P. and R. A. are supported by a grant from the EU: QLRI-CT-2000-0127.

References

1 Apic, G., Gough, J., Teichmann, S. A., An insight into domain combinations. *Bioinformatics* **2001**, *17 Suppl 1*, S83–89.

2 Puntervoll, P., et al., ELM server: a new resource for investigating short functional sites in modular eukaryotic proteins. *Nucleic Acids Res.* **2003**, *31*, 3625–3630.

3 Aasland, R., et al., Normalization of nomenclature for peptide motifs as ligands of modular protein domains. *FEBS Lett.* **2002**, *513*, 141–144.

4 Apweiler, R., et al., (**2003**) InterPro User Manual.

5 Reuter, N., Hinsen, K., Lacapere, J. J. Transconformations of the SERCA1 Ca-ATPase: a normal mode study. *Biophys. J.* **2003**, *85*, 2186–2197.

6 Orengo, C. A., Michie, A. D., Jones, S., Jones, D. T., Swindells, M. B., Thornton, J. M., CATH: a hierarchic classification of protein domain structures. *Structure* **1997**, *5*, 1093–1108.

7 Murzin, A. G., Brenner, S. E., Hubbard, T., Chothia, C., SCOP: a structural classification of proteins database for the investigation of sequences and structures. *J. Mol. Biol.* **1995**, *247*, 536–540.

8 Bateman, A., et al., The Pfam protein families database. *Nucleic Acids Res.* **2002**, *30*, 276–280.

9 Letunic, I., et al., Recent improvements to the SMART domain-based sequence annotation resource. *Nucleic Acids Res.* **2002**, *30*, 242–244.

10 Mulder, N. J., et al., The InterPro database 2003, brings increased coverage and new features. *Nucleic Acids Res.* **2003**, *31*, 315–318.

11 See Chapter 21 of this volume by Linding et al.

12 http://song.sourceforge.net/.

13 In the ELM resource, four types of functional sites are considered: ligands (LIG), subcellular targeting motifs (TRG), post-translational modification sites (MOD), and protease cleavage sites (CLV). We use ELM identifiers in this text to refer to different functional sites.

14 Sudol, M., From Src homology domains to other signaling modules: proposal of the 'protein recognition code'. *Oncogene* **1998**, *17*, 1469–1474.

15 Xu, H. E., et al., Structural basis for antagonist-mediated recruitment of nuclear co-repressors by PPARalpha. *Nature* **2002**, *415*, 813–817.

16 Takeda, D. Y., Wohlschlegel, J. A., Dutta, A., A bipartite substrate recognition motif for cyclin-dependent kinases. *J. Biol. Chem.* **2001**, *276*, 1993–1997.

17 Schneider, T. D., Stephens, R. M., Sequence logos: a new way to display consensus sequences. *Nucleic Acids Res.* **1990**, *18*, 6097–6100.

18 http://elm.eu.org/help.html.

19 Falquet, L., Pagni, M., Bucher, P., Hulo, N., Sigrist, C. J., Hofmann, K., Bairoch, A., The PROSITE database, its status in **2002**. *Nucleic Acids Res.* **2002**, *30*, 235–238.

20 http://us.expasy.org/tools/scanprosite/scanprosite-doc.html.

21 Krogh, A., Brown, M., Mian, I. S., Sjolander, K., Haussler, D., Hidden Markov models in computational biology: applications to protein modeling. *J. Mol. Biol.* **1994**, *235*, 1501–1531.

22 Eddy, S. R., Profile hidden Markov models. *Bioinformatics* **1998**, *14*, 755–763.

23 Bailey, T. L., Gribskov, M., Methods and statistics for combining motif match scores. *J. Comput. Biol.* **1998**, *5*, 211–221.

24 Fischle, W., Wang, Y., Allis, C. D., Binary switches and modification cassettes in histone biology and beyond. *Nature* **2003**, *425*, 475–479.

25 Nielsen, P. R., et al., Structure of the HP1 chromodomain bound to histone H3 methylated at lysine 9. *Nature* **2002**, *416*, 103–107.

26 Jacobs, S. A., Khorasanizadeh, S., Structure of HP1 chromodomain bound to a lysine 9-methylated histone H3 tail. *Science* **2002**, *295*, 2080–2083.

27 Orchard, S., Hermjakob, H., Apweiler, R., The proteomics standards initiative. *Proteomics* **2003**, *3*, 1374–1376.

28 Gupta, R., Birch, H., Rapacki, K., Brunak, S., Hansen, J. E., O-GLYCBASE version 4.0: a revised database of O-glycosylated proteins. *Nucleic Acids Res.* **1999**, *27*, 370–372.

29 Kreegipuu, A., Blom, N., Brunak, S., PhosphoBase, a database of phosphorylation sites: release 2.0. *Nucleic Acids Res.* **1999**, *27*, 237–239.

30 Zanzoni, A., Montecchi-Palazzi, L., Quondam, M., Ausiello, G., Helmer-Citterich, M., Cesareni, G., MINT: a molecular interaction database. *FEBS Lett.* **2002**, *513*, 135–140.

31 Bader, G. D., Betel, D., Hogue, C. W., BIND: the biomolecular interaction network database. *Nucleic Acids Res.* **2003**, *31*, 248–250.

32 Xenarios, I., Salwinski, L., Duan, X. J., Higney, P., Kim, S. M., Eisenberg, D., DIP, the database of interacting proteins: a research tool for studying cellular networks of protein interactions. *Nucleic Acids Res.* **2002**, *30*, 303–305.

33 Yaffe, M. B., Leparc, G. G., Lai, J., Obata, T., Volinia, S., Cantley, L. C., A motif-based profile scanning approach for genome-wide prediction of signaling pathways. *Nat. Biotechnol.* **2001**, *19*, 348–353.

Epilogue:
New Levels of Complexity in the Functional Roles of Modular Protein Interaction Domains: Switches and Sockets in the Circuit Diagrams of Cellular Systems Biology

Harel Weinstein

It has not been that long since the modular interaction domains presented in this book appeared merely as multicolored, oddly shaped beads on linear strings representing multidomain proteins. As the chapters of the book document, however, sustained efforts during the past decade have imparted to each of these beads a specific 3D structure at atomic resolution and a specific character reflecting their individual biochemical properties and cellular functions. And yet, as noted by Elia and Yaffe in Chapter 8 on phosphoserine/threonine binding domains, these modules can also exhibit "a remarkable amount of functional and structural diversity" in the various genomes, so that the study of such domains provides insights into the manner in which their function "has been repeatedly rediscovered during evolution", embedded in different structural contexts. Conversely, it is also clear from the description of the modules throughout this book that most of them cannot be considered to be monofunctional. Even after being educated by this compendium about the remarkable functional plasticity of the individual domains, it is still striking to witness in the rapidly growing literature evidence for additional varieties of unexpected functions that can be achieved by *combinations* of the modules described in this book. This remarkable new development in the functional understanding of modular domains comes both from artificial constructs produced by protein engineering of module combinations and from the discovery and analysis of combinations brought together by evolution. These findings generate a new image of the modular interaction domains – already discernable in the wealth of structural and functional information contained in every chapter of this book – that suggests that they function as organizing units for protein–protein interactions in the cell. Unlike other protein families involved in processes requiring regulated protein–protein interactions, these modular domains may not exhibit intramolecular allosteric behavior, since most of them are likely to lack the flexibility and dynamic properties exhibited by other types of conserved modules, such as the nucleotide binding subunit (G_α) of the heterotrimeric G proteins [1] (see, however [2]). Together, the interactive selectivity and functional versatility of these domains, their compact and parsimonious structures, and the fact that the sequences of multidomain proteins contain repetitive (albeit not always identical) copies of such modules,

Modular Protein Domains. Edited by G. Cesareni, M. Gimona, M. Sudol, M. Yaffe

ISBN: 3-527-30813-X

inevitably lead to the suggestion that the repeated positioning of such interaction domains is most likely to foster and/or sustain the organization of functional protein assemblies in the cell.

The systematic evolution of this realization about the possible role of combinations of domains is elegantly illustrated in Chapter 1 by Tony Pawson et al. on the SH2 domain – one of the first examples of such structurally simple domains with a biologically complex functionality. The accumulating evidence indicates that the special attributes of such domains may allow them to elicit even more complex functions than those related to recognition of single protein motifs. This could be achieved by engaging the multiple copies of the domains in a variety of spatial contexts into which they would attract and organize their binding partners. In fact, the combinations of the PDZ domains discussed by Lasky and colleagues (Chapter 13) offer clear evidence gathered from the literature that scaffolding functions can indeed be implicated in such specialized cellular processes, specifically phototransduction, organization of postsynaptic density, and membrane protein activity. The discovery in signaling pathways of a variety of such scaffolding and anchoring components, consisting of an assortment of adaptor modules, points to their importance in the colocalization of proteins involved in the pathway interactions. Recent quantitative analyses [3] suggest that such colocalization is a necessity in signaling, to facilitate protein interactions and prevent cross talk. This reasoning, and the valuable information gathered in this book about the properties of the modules, suggest that the time has now come to discover and systematically characterize the biological role of the various combinations of modules, perhaps of the strings of domains associated with the modular proteins as a whole, or even of combinations of such strings. To achieve this new level of understanding, it seems necessary to reconsider the deep insights offered by the structural properties of the individual modules and the details of the pairwise interaction mechanisms in which they have been shown to participate, but now in the broader context of assemblies. This is likely to require an integrative systems outlook and comprehensive descriptions (or maps) of cell system mechanisms, such as signaling pathways and transcription networks.

Such map-like descriptions of the protein interactions in signaling pathways and other cell system networks are now commonly referred to as a 'wiring diagrams', by analogy to representations of engineering systems (e.g., electrical circuits). A wiring diagram is expected to convey detailed information about the communications among the components, in this case, the proteins in the cell. For protein interaction systems, these communications entail complex time- and location-dependent assembly and disassembly of intracellular protein aggregates. It is precisely this complexity that makes representations in terms of wiring diagrams both necessary and desirable. The added advantage is that such a representation of mechanisms in cell systems physiology enables the application of (1) established engineering principles and (2) quantitative analyses with the powerful algorithms developed in systems engineering to the study of properties and attributes of protein interaction schemes. Such wiring diagram representations are not mere representations of pairwise protein interactions discovered with screening methods, but functional connectivity maps in which the quantitative kinetic and spatiotemporal parameters of protein interactions in the cell signaling pathways become analogous to the strength and timing parameters of an electrical circuit. Consequently, the information processing system of the cell (like the electrical circuit) represented by these mechanistic maps can be subjected to mathematical modeling and simulation approaches that have been developed in engineering. Given the growing perception of (and evidence for) the key involvement of modular domains in the organization of the protein interactions that determine cell physiology and function at an integrated system level, we gather even more from the analogy to these circuits. Thus, the engineering principles underlying their mode of operation and the expectation that such circuits must contain functional modules such as gating switches, for example, point to new and unexpected directions for research on the properties of modular protein interaction domains.

The practical importance of these new directions of inquiry is suggested by evidence from novel 'designer' modular protein systems, which exhibit new types of functional properties [4]. In these designed systems, combinations of modular protein interaction domains were shown to act as facilitators and putative signaling switches that modulate the output of various functional domains (e.g., catalytic) in the protein [5]. The novel emergent properties elicited from such combinatorial constructs of various domains parallel those demonstrated by component modules of electrical and mechanical systems, thus strengthening the analogy of the cellular systems with the engineering systems represented by the wiring diagrams. Moreover, a 'modular logic' for the signaling that can be regulated by combinations of modules has been described in compelling schemes by Lim [6]. These schemes emphasize the tantalizing possibilities of signal integration obtainable by linking the modular domains in various combinations. The rich variety of switching combinations made possible by such constructs is illustrated in Figure 1, compiled by W. Lim (personal communication) from recently described experiments in his laboratory [4]. An astounding diversity of functional behaviors is evident even in this short list of multidomain constructs, testifying to the richness of switching phenotypes that could be achieved in the cell by the various modular domains acting in combination. It is noteworthy, moreover, that, whether they act as adaptors of various proteins or scaffolds for protein assemblies, the functions of the modular domains and their combinations can in turn be modulated by selective ligand binding. This is clearly demonstrated by the switches designed in the Lim laboratory [4]. This ligand dependence can be exploited in the sorting of functional units and feedback mechanisms that can be utilized both for the further study of signaling pathways and for the engineering of systems with switching properties that can modulate the output in a desired fashion as a function of ligand specificity and concentration.

The properties of the modular domains and their combinations have evolved as described in the chapters of this book to constitute key organizing elements in high-level physiological mechanisms in the cell. Before a complete map of the cell physiology can be obtained, it will therefore be necessary to achieve an understanding of the properties, processes, and outputs of the integrated functional units

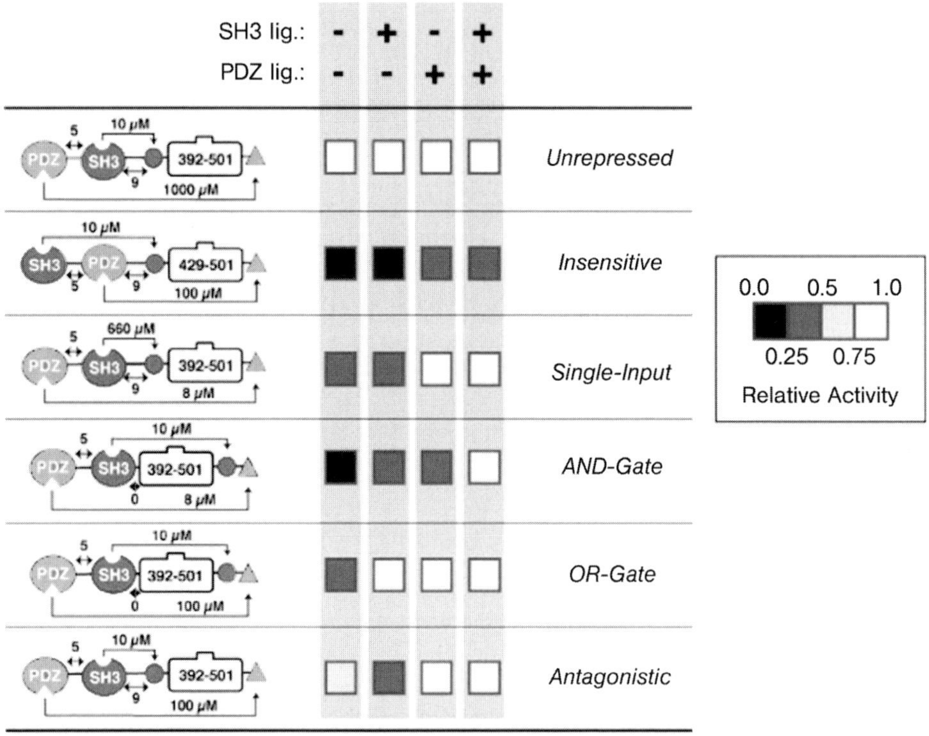

Figure 1 Switch types obtainable from domain recombinations as described in [3]. The switch architecture is shown on the left, including the two input domains (PDZ, SH3), their respective ligands (triangle and circle, respectively), and the output components that differ in sequence length, as shown. The ligand concentrations are indicated for each example, and the relative activity for each parameter set is indicated by different shading (from low to high activity: black, dark gray, light gray, white).

produced by combinations of the modular domains discussed in this book. The quantitative analysis and simulation of protein interaction pathways, represented as wiring diagrams of the signaling networks in the cell, is likely to be a major tool in the investigation of the functional roles of modular protein interaction domain combinations in cell physiological mechanisms. The results should offer surprising new and nonintuitive insights into cell system function and should open completely new directions in the quest for novel types of therapeutic interventions.

References

1 Ceruso, M. A., Periole, X., Weinstein, H., Molecular dynamics simulations of transducin: interdomain and front to back communication in activation and nucleotide exchange. *J. Mol. Biol.* **2004**, *338*, 469–481.

2 Fuentes, E. J., Der, C. J., Lee, A. L., Ligand-dependent dynamics and intramolecular signaling in a PDZ domain. *J. Mol. Biol.* **2004**, *335*, 1105–1115.

3 Batada, N. N., Shepp, L. A., Siegmund, D. O., Stochastic model of protein–protein interaction: why signaling proteins need to be colocalized. *Proc. Natl. Acad. Sci. USA* **2004**, *101*, 6445–6449.

4 Dueber, J. E., et al., Reprogramming control of an allosteric signaling switch through modular recombination. *Science* **2003**, *301*, 1904–1908.

5 Park, S. H., Zarrinpar, A., Lim, W. A., Rewiring MAP kinase pathways using alternative scaffold assembly mechanisms. *Science* **2003**, *299*, 1061–1064.

6 Lim, W. A., The modular logic of signaling proteins: building allosteric switches from simple binding domains. *Curr. Opin. Struct. Biol.* **2002**, *12*, 61–68.

Subject Index

Modular Protein Domains. Edited by G. Cesareni, M. Gimona, M. Sudol, M. Yaffe

ISBN: 3-527-30813-X